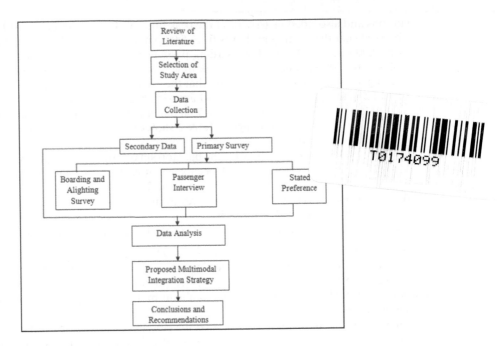

Figure 2. Methodology adopted for the study.

3.1 Data collection

The surveys conducted to collect relevant data regarding traffic volume and passenger demand existing in the study areas were:

Bus Stop Boarding and Alighting Survey: This survey was conducted for 14 hours at the major bus stops of the study areas covering morning and evening peak periods. The survey format captured the time of arrival at the bus stop, route number/name, origin of bus, destination of bus, number of passengers boarding and alighting the bus and occupancy details. Trained enumerators were stationed at identified bus stops capturing the requisite data.

Bus Passenger Interview Survey: The passengers waiting to board the buses and alighting from the buses were interviewed to capture the passenger travel characteristics throughout the study area. The surveys were carried out at the identified locations for one working day devoid of any rainfall/festivities and so on, for a period of fourteen hours (06:00 am to 8:00 pm) capturing the bus passengers on either side of the road, best representing the regular commuting pattern of the passengers. Data corresponding to socio economic as well as travel characteristics of the commuters were collected. The major socio economic characteristics collected included age, gender, monthly income, education, occupation and monthly expenses on transport.

Stated Preference Survey (SP Survey): The SP survey questionnaire was designed to identify the travel patterns and preferences of the interviewed respondents. The first section of the questionnaire was designed to ask questions covering respondents'| existing travel patterns and socio-economic characteristics. The second section of the questionnaire was related to the perception of the existing public transport system in the study area. The last section on both the questionnaires was supported by five SP survey travel attribute scenarios. It was intended to provide the respondents with varying scenarios with existing and proposed system attributes such as travel time, waiting time, fare/cost and comfort level.

3.2 Data analysis

a. **Bus Stop Boarding and Alighting:** The survey helped in evaluating bus demand in each time interval. It gave the peak number of commuters boarding and alighting at both of the stations and also aided in arriving at the peak hour of demand for buses. At Kaloor, it was observed that the evening peak was from 5:00 pm to 6:00 pm and morning peak time was from 9:00 am to 10:00 am for boarding passengers, whereas the evening peak was from 4:00 pm to 5:00 pm and morning peak time was from 9:00 am to 10:00 am for alighting passengers. At Edapally, it was found that the evening peak was from 5:00 pm to 6:00 pm and morning peak time was from 8:00 am to 9:00 am for boarding passengers, whereas the evening peak was from 6:00 pm to 7:00 pm and morning peak time was from 8:00 am to 9:00 am for alighting passengers. Table 1 shows the peak traffic volume from each bus stop of the study area.

b. **Bus Occupancy Survey:** Along with the bus stop survey, the bus occupancy survey was also carried out. Higher proportions of occupancy at each of the stations during both peak hours and off-peak hours are as shown in Table 2.

c. **Bus Passenger Interview Survey:** Data regarding socio-economic characteristics of the travelers were collected. The following details of the passengers traveling by bus regarding their access as well as egress trips were also obtained as part of the survey:

Access and Egress Modes: Share of access mode from origin to boarding point and dispersal mode from alighting point to destination of the boarding bus passengers are shown in Figures 3 and 4. It was found that bus is the predominant mode of travel from origin to boarding point while walking is the predominant mode from alighting point to destination of the bus users.

Proximity to the Bus Stops: This showed the closeness of bus stops either to the point of origin or destination. To analyze the proximity of bus stops, the distance from the origin to the bus stop and from the bus stop to the destination were considered. Figure 5 shows the distribution of access and egress trip distances.

Table 1. Peak hour volume of boarding and alighting passengers.

Location	Peak volume of boarding passengers	Peak volume of alighting passengers
Edappally	1412	821
Kaloor	577	1116

Table 2. Occupancy level.

Location	Peak hour occupancy level	Off-peak hour occupancy level
Edappally	45% of the buses have an occupancy level of full sitting plus full standing	30% of the buses are full sitting
Kaloor	40% of the buses are full sitting plus half standing	80% of the buses are half sitting

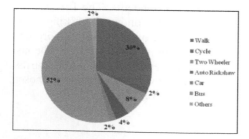

Figure 3. Access mode from origin to boarding point.

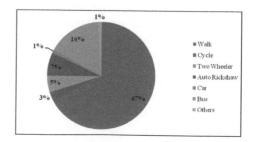

Figure 4. Egress mode from alighting point to destination.

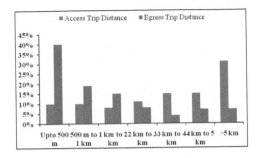
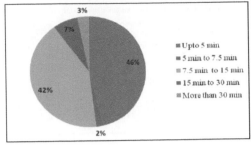

Figure 5.   Proximity to the bus stops for bus passengers.

Figure 6.   Waiting time at the bus stop.

It was found that approximately 31% of the access trips to the bus stops were more than 5 km and 40% of the egress trips from the bus stops were within 500 m. Another key observation is that more than 30% of the access and egress trips have a trip distance greater than 1 km.

Waiting Time at the Bus Stop: Figure 6 represents the distribution of commuters according to the waiting time at the bus stop. It was found that about 46% of the boarding passengers wait up to five minutes to board the bus followed by 42% of commuters having a waiting time of 7.5 minutes to 15 minutes.

S P Survey: The major observations from the survey are as follows:

Mode and Trip Purpose: Figure 7 gives the distribution of trip purpose among commuters considering different types of mode. The majority of the commuters travel with a trip purpose of work followed by business and education.

Mode and Travel Time: Figure 8 gives the distribution of travel time among respondents considering different types of mode. It was found that the majority of the commuters have a travel time of 20–40 minutes. It was also found that the majority of the commuters traveling by bus have a travel time greater than 40 minutes.

Mode and Travel Cost: Figure 9 gives the distribution of travel cost among respondents considering different types of mode. It was found that the majority of trips involving a car have travel costs greater than 40 Rs and trips using autorickshaws have travel costs less than 20 Rs.

Mode and Travel Distance: Figure 10 represents the distribution of trip length by different modes of transport. It was found that the trips having a travel distance greater than 20 km are mostly performed by bus. It was also found that the commuters having a travel distance less than 20 km prefer cars and two-wheelers as their modes.

Monthly Income and Mode of Travel: The main purpose of comparing monthly income with mode of travel is to observe the category of commuters with different income levels using different types of mode. It was found that the higher income commuter groups mainly prefer cars and two-wheelers. Figure 11 represents a comparison of different modes of travel with income. It was found that the higher income groups did not prefer public transport, whereas low income groups prefer public transport for their travel.

Daily Travel Cost and Monthly Income: Figure 12 represents the comparison of monthly income with travel cost.

Stated Preference: A S P survey was also conducted by giving five different scenarios to the passengers and their opinions were sought regarding their willingness to shift to a public transport system varying in terms of speed, travel time, as well as comfort and convenience. The details of each scenario with its characteristics are shown in Table 3. It was found that about 49% of the users were satisfied with the existing scenario. About 27% of the commuters were willing to shift to public transport if travel time and waiting time are reduced by 25%.

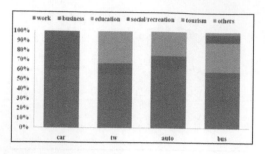

Figure 7. Comparison of mode with trip purpose.

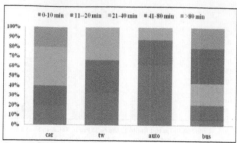

Figure 8. Comparison of mode with travel time.

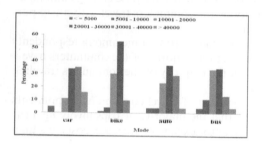

Figure 9. Distribution of travel costs by mode.

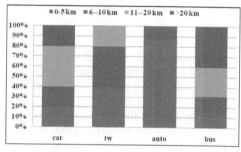

Figure 10. Comparison of mode with travel distance.

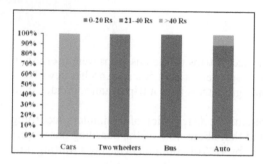

Figure 11. Comparison of mode with income.

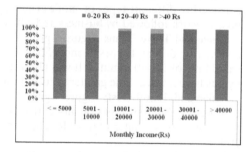

Figure 12. Comparison of monthly income with daily travel cost.

Table 3. Choice set for public transport users.

Parameters	Scenarios				
	1	2	3	4	5
Waiting time (min)	Existing	Existing	Reduced by 25%	Reduced by 25%	Reduced by 50%
Travel time (min)	Existing	Existing	Same as existing	Reduced by 25%	Reduced by 25%
Travel cost (Rs.)	Existing	Existing	Increased by 25%	Increased by 25%	Increased by 50%
Transfers	Existing	Yes	Yes	No	No
Comfort level	Existing	Crowded + Non-AC	Sitting + Standing + Non-AC	Sitting + Non-AC	Sitting + AC
Percentage of users opting	49%	11%	10%	27%	3%

4   INTEGRATION STRATEGIES

The major integration strategies adopted for the study are the following:

a. Physical Integration

1. **Walkability:** Walking was found to be the predominant mode from alighting point to destination of the bus users. Pedestrian facilities need to be enhanced and the key factors to making walking appealing are safety, activity and comfort. For comfortable and safe walking, footpaths of a minimum width of 1.8 m should be provided.
2. **Cyclability:** Cycling is an emission free, healthy and affordable transport option that is highly efficient and consumes little space and few resources. It combines the convenience of door to door travel, the route and schedule flexibility of walking and the range and speed of many local transit services.
3. **Connectivity:** Short and direct pedestrian and cycling routes require highly connected networks of paths and streets around small, permeable blocks. A tight network of paths and streets offering multiple routes to many destinations can also make walking and cycling trips varied and enjoyable. Frequent street corners and narrower rights of way with slow vehicular speed and many pedestrians encourage street activity and local commerce.
4. **Multimodal Integration:** Integration of mass transit with the city bus system and extended to the Non Motorized Modes such as bicycles as well as Intermediate Para transit system not only ensure first and last mile connectivity but also the seamless transfer of network users. Multimodal integration exists at various stages such as: institutional, physical, fare, operational. The level at which multiple modes need to be integrated shall depend on the local city authorities.
5. **Information Integration:** This is the information system placed on-board in the bus. It is a real time information display unit which helps in providing passengers with the necessary information related to their commute. The information displayed varies from the next/current stop information, current location, expected time to reach the destination or the nearest metro station and so on. Route data may be presented as a linear map, highlighting the current position of the bus and the next stop that it is approaching. These maps are held as image files on the vehicle computer and displayed when the bus reaches a pre-specified geolocation. Likewise, audio files can be delivered in exactly the same way, notifying passengers of a particular stop as the vehicle approaches.

b. Fare Integration

In the current scenario, it is imperative that the fare collection system for the city bus transport, feeder and metro be integrated for the ease of the passengers. The data from ticket issuing machines should be downloaded into the central room server for further analysis. These data which include denomination wise ticket sales, number of passengers, load factor at each stage of the route and so on, are stored in the computer for further analysis. Based on these data, the operations of the bus system can be monitored and documented for future planning of operations. These data will be of great help for route restructuring, route analysis, introduction/curtailment of routes and so on. Two important components of fare integration are:

1. Hand held ticketing devices are small ticket vending units which are capable of producing tickets as per the specified distance or stoppages.
2. Smart cards are small plastic cards which hold a certain value of money and can be used as a fare paying system at metro stations.

5   CONCLUSIONS

The existing transport system of Kochi city was analyzed as part of the study. At Edappally and Kaloor, the majority of the public transport buses had an occupancy level exceeding their capacity. During peak hours, about 45% of the buses had an occupancy level of full sitting plus full standing at Edappally, whereas the occupancy level was full sitting plus half standing at Kaloor. The results of the S P survey indicate that the majority of the commuters, which constituted about 49% of the total, were satisfied with the existing transport conditions. About 27% of the commuters were willing to shift to a public transportation system with travel attributes such as travel time and waiting time being reduced by 25%, inspite of the increase in travel costs. Moreover, walking was found to be the predominant mode from alighting point to destination of the bus users. Thus, pedestrian facilities need to be enhanced and the key factors to making walking appealing are safety, activity and comfort. Significant integration strategies were also recommended as a part of the study involving both fare integration using hand held ticketing devices and smart cards, as well as physical integration strategies involving walking, cycling and connectivity.

REFERENCES

Arentze, T.A. & E.J.E. Molin (2013), Travelers' preference in multimodal networks: Design and results of a comprehensive series of choice experiments, Transportation Research Part A 58 (2013), 15–28.

Cheryan, C.A. & S. Sinha (2015). Assessment of transit transfer experience: Case of Bangalore, 8th Urban Mobility India Conference & Expo.

Li, L., J. Xiong, A. Chen, S. Zhao, & Z. Dong (2014). Key strategies for improving public transportation based on planned behaviour theory: Case study in Shanghai, China, Journal of Urban Planning and Development 141(2): 04014019.

Muley, B.R. & C.S.R.K. Prasad (2014). Integration of public transportation systems. Urban Mobility India Conference & Expo, Transportation Division NIT Warangal.

Najeeb, P.M.M. (2008). Study of cognitive behavior therapy for drivers improvement, Australasian Road Safety Research, Policing and Education Conference, Adelaide, South Australia.

Patni, S. & Ghuge, V.V. (2015). Towards achieving multimodal integration of transportation systems for seamless movement of passengers: Case study of Hyderabad City. *Urban Mobility India Conference & Expo.*

Emerging Trends in Engineering, Science and Technology for Society,
Energy and Environment – Vanchipura & Jiji (Eds)
© *2018 Taylor & Francis Group, London, ISBN 978-0-8153-5760-5*

Quantitative evaluation of bus terminal using time-space analysis

V. Hridya & George Geeva
Department of Civil Engineering, Rajiv Gandhi Institute of Technology, Kottayam, Kerala, India

ABSTRACT: Bus terminals are enclosures where the interactions among passengers are very high. Activities like walking, waiting, route choice, boarding and alighting of buses occur. Pedestrian level of service is an overall measure of walking conditions on a path, facility or route. In this study, the quantitative evaluation of a terminal is conducted using time space analysis. The selected site for the study is Aluva Kerala state road transport corporation bus terminal. The main aim of the study is to determine the level of service of the terminal. Queuing and walking are considered as the major passenger activities inside a terminal. Different activities require a different amount of space and time. The time and space utilized by passengers for standing and circulation is considered.

1 INTRODUCTION

Transportation is defined as a system of interacting components or modules that are specified as passenger processing elements which aid pedestrian movement, and the environment which encompasses those dimensions with which the pedestrian associates his or her personal comfort, convenience, safety and security [Demestsky, M.J., Hoel, L.A. & Virkler, R.M. 1977] The combined performance of subsystems for passenger processing and environment conditions accounts for the overall effectiveness of station design [Demestsky, M.J., Hoel, L.A. & Virkler, R.M. 1977].

A bus terminal is the point of entry or exit to the public transportation system. It may also act like a point for concentration, dispersion, and interchange of modes, storage space for passengers and vehicles, as well as a maintenance area. A terminal that accommodates a high passenger volume can be looked upon as two interacting components—passenger processing elements and environmental conditions. The effective interactions of these components control the performance of the bus terminal.

The Highway Capacity Manual's (HCM) methods for analyzing pedestrian Level of Service (LOS) are based on the measurement of pedestrian flow rate and sidewalk space. The pedestrian flow rate, which incorporates pedestrian speed, density and volume, is equivalent to vehicular flow. Pedestrian speed as observed in HCM 2000 as follows: "As volume and density increase, pedestrian speed declines. As density increases and pedestrian space decreases, the degree of mobility afforded to the individual pedestrian declines, as does the average speed of the pedestrian stream."

1.1 *Need for study*

The cities and nations who had invested more in efficient transportation systems are now very weak in their environmental sustainability. A terminal is a point where the bus starts or ends its schedule and hence it makes a clear contribution to the overall journey experience of the passenger. The passenger may have some expectations about the terminal, about the access of the terminal, frequency of the buses, the services and the facilities provided and so on. If these criteria do not meet the expectations of the passenger, this may result in a shifting of travel mode from public to private.

The relevance of shifting from private to public vehicles is increasing nowadays. A study on the quality of facilities provided at the terminal will help to gain ideas regarding the improvements to be done by the authority.

1.2 Literature review

1.2.1 Different types of users

The different types of passengers using a transit platform are identified as commuters/business travelers, leisure travelers, passengers in wheelchairs, those with a physical or cognitive mobility impairment, people with medium sized luggage, people with large sized luggage and parents with small children [Department of New York Rails, 2011]. Passengers can also be categorized as general users and special users. The physically handicapped and elderly are considered as special users [Demestsky, M.J., Hoel, L.A. & Virkler, R.M. 1977].

1.2.2 Factors considered for evaluation

As a terminal is a place of integration of passengers, buses, other services and so on, the facilities provided are of greater importance. Various factors identified are physical characteristics, location facilities and user factors as per the Guidelines for Assessing Pedestrian Level of Service-(Main roads Western Australia, 2006). Suitability for use, path width, surface quality, crossing opportunities and support facilities are the physical characteristics. Location facilities include connectivity, path environment and potential for vehicular conflict. User factors selected are pedestrian volume, mix of path users and personal security. The physical environment includes lighting, air quality, temperature, aesthetics advertising, cleanliness, music and so on. Non-transport businesses and services, private concessions such as newspaper stands, coffee shops, barber shops and other small businesses and services, restrooms and lounges, first aid stations, public telephones, weather protection facilities can also be considered as performance evaluators in a bus terminal [Demestsky, M.J., Hoel, L.A. & Virkler, R.M. 1977]. Safety, amenities, efficient movement are the other factors described in the Public Transport Infrastructure Manual, 2016.

1.2.3 Time space analysis

In qualitative analysis the time and space utilized by passengers is considered. Time space analysis is a method of analysis used to determine the pedestrian traffic in urban streets which exhibits the need for holding space and circulation space. Pedestrian LOS is an overall measure of walking conditions on a path, facility or route. This is directly linked to factors that affect pedestrian mobility, comfort and safety. In a study by Grigoriadou and Braaksma (1986), the pedestrian activities like queuing and circulation were taken into consideration. In platforms there would be holding space for standing and circulation space for walking. The time and space occupied by the passenger activities are incorporated to determine the space requirement of pedestrians. The passenger LOS of terminal areas are developed based on average pedestrian space, personal comfort and degrees of internal mobility. The standards are presented in terms of average area per person and average interpersonal space.

2 STUDY AREA

Aluva formerly called as Alwaye, is the second biggest town of Greater Cochin City in the Ernakulam district of Kerala. It is also considered as the industrial and commercial city of Kochi and it is the industrial epicenter of the state. Aluva is located on the banks of River Periyar. Aluva is also a major transportation hub, with easy access to all major forms of transportation. Aluva acts as a corridor which links the highland districts to the rest of the state. Aluva is more famous for its accessibility through rail, air, metro and by bus.

Aluva KSRTC bus terminal is one of the major bus depos in Kerala. Buses from all other parts of Kerala have services to this station. Bus services from places like Mysore, Mangalore, Bangalore, Trichy, Coimbatore, Salem, Palani, Kodaikanal and so on are also available. At present, the depo has 107 buses of its own and seven interstate services.

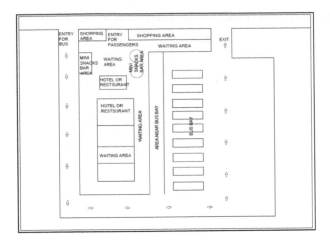

Figure 1.   Layout of Aluva KSRTC bus terminal.

After the pilot survey of different bus stations, it was noticed that Aluva KSRTC bus terminal is the one with less facilities compared to the other stations nearby. The study of this bus terminal was very timely in order to determine the pedestrian activities within it and thereby, to determine the LOS. Aluva KSRTC bus terminal acts as a center that connects different parts of the state. Because of that, many categories of passengers use the bus service from this station for different purposes such as work trips, business trips, shopping and recreational trips. The major share of the passengers includes students and private sector officials.

Figure 1 shows the layout of Aluva KSRTC bus terminal. The entry and exit points for buses and passengers are shown and also the different zones selected for study are depicted in the layout.

3   METHODOLOGY

3.1   *Quantitative evaluation using time-space analysis*

Bus terminals are critical nodes in a bus transit network. "Passengers" safety, comfort, convenience and so on will be affected by the platform activities. Pedestrian activities like queuing and circulation that is standing and walking, are mainly occur on a terminal platform. The walking and circulation presents different time as well as space characteristics. The time-space method works on the principle that the walking pedestrians and the standing pedestrians will use a different amount of space for different time periods. This method of evaluation is a type of quantitative evaluation since the quantity of time and space available in the bus terminal are considered. Table 1 shows the LOS for different pedestrian modules for a queuing area [Transport Research Board, 2000].

The selection of the study site was very important since the waiting area was very congested and limited without any proper planning. Therefore, it was necessary to determine where all the passengers would wait and spend their time in between journeys. This would help to identify how people utilize the available space.

3.2   *Calculation steps*

The calculations were done using the following equations [Grigoriadou, M. & Braaksma, J.P. 1986].

i.   Calculate the total available time space (TS)

$$TS = A \times AP \qquad (1)$$

181

Table 1. LOS, pedestrian module and interpersonal space as per HCM.

LOS	Module (m²/ped)	Interpersonal space (m)	Comments
A	1.2 or more	1.2 or more	free circulation much interpersonal space no crowding
B	0.9–1.2	1.1–1.2	very little crowding sufficient interpersonal space
C	0.7–0.9	0.9–1.1	restricted circulation small interpersonal space walking speed controlled
D	0.3–0.7	0.6–0.9	severely restricted circulation limited interpersonal space no personal contact
E	0.2–0.3	0.6 or less	standing room only little interpersonal space no circulation possible
F	0.2 or less	Touching	standing only no interpersonal space jammed conditions

where TS = total available time-space, in m².min; A = visible platform area, in m²; and AP = analysis period, in min.

ii. Calculate holding area time space requirements (TS_h)

$$TS_h = SP \times AST \times AASP \qquad (2)$$

where TS_h = holding area time space requirements, in m².sec; SP = number of standing pedestrians; AST = average standing time, in sec; AASP = standing pedestrian module, in m²/ped.

iii. Calculate the net circulation area time-space (TS_c)

$$TS_c = TS - (TS_h/60) \qquad (3)$$

where TS_c = net circulation area time-space, in m².min; TS = total available time-space, in m².min; and TS_h = holding area time space requirements, in m².sec.

iv. Calculate the total circulation time (TC)

$$TC = CP \times ACT \qquad (4)$$

where TC = total circulation time, in ped-min; CP = total number of circulating pedestrian; and ACT = average circulation time, in min.

v. Calculate the pedestrian module for walking pedestrian (MOD)

$$MOD = (TS_c/TC) \qquad (5)$$

where MOD = pedestrian module for walking pedestrian, in m²/ped; TS_c = net circulation area time-space, in m².min; and TC = total circulation time, in ped-min.

vi. Compare the calculated pedestrian module with the standard values shown in Table 1 to determine LOS.

4 DATA USED

The study site was Aluva KSRTC bus terminal. The quantitative analysis of the terminal would help us to identify the real conditions of the terminal. After conducting a pilot survey, the different zones inside the terminal were identified. From the survey, five different zones where major pedestrian activities occur were selected. The zones are the waiting area, area near the bus bay, hotel or restaurant, mini snacks bar area and shopping area. Areas of all the zones were measured manually. A pedestrian volume survey was conducted to obtain the

Table 2. Peak hour passenger count and average time spent by passengers at each zone.

Name of the zone	No. of pedestrians	Thursday		Friday		Saturday		Average time (min)
		Morning	Evening	Morning	Evening	Morning	Evening	
Area near bus bay	Standing pedestrians	192	228	296	272	116	200	12.25
	Walking pedestrians	224	164	272	312	168	188	
Waiting area	Standing pedestrians	228	256	372	300	240	184	10.8
	Walking pedestrians	220	180	180	272	140	176	
Shopping area	Standing pedestrians	108	164	104	228	44	80	8.2
	Walking pedestrians	76	80	64	140	88	88	
Hotel or restaurant	Standing pedestrians	32	40	60	76	48	76	15.58
	Walking pedestrians	24	24	8	60	60	72	
Mini snacks bar area	Standing pedestrians	40	56	80	108	64	80	6.8
	Walking pedestrians	68	60	68	100	96	92	

number of passengers occupying each zone and the average time spend by passengers for each activity. Standing and circulating activities of passengers are included in this study.

The survey was conducted manually for three different days, including a typical working day, a weekend and a holiday. The survey was conducted during the morning peak hours and evening peak hours in a time interval of 15 minutes. Table 2 shows the number of passengers present in each zone during the time of the survey. The morning peak time was identified from a pilot study from 8.00 am to 10.00 am and evening peak time was from 3.30 pm to 6.00 pm. Data in the table depicts the passengers in every hour of survey. The data collected include the area of all the zones, number of standing pedestrians, number of circulating pedestrians, average standing time of pedestrian, average circulation time of pedestrians and standing pedestrian module. After the data collection, the calculations were done using the equations given in section 3.2.

The calculation steps were automated using a tool called the Structured Query Language developer. The SQL developer uses computer programming language called SQL. SQL is a set of statements with which all programs and users access data in an Oracle database. The SQL provides benefits to managers, end users, application programmers and database administrators. Technically speaking, SQL is a data sublanguage. It works with a set of data rather than individual units. For the data input we have to create a survey table which includes all the data collected from the survey. Then, we have to prepare a query with equations used for calculation for reading the data from the table and applying it to the calculation steps. The final result is that the pedestrian module will compare with Table 1 and then the LOS is determined.

5 RESULTS AND DISCUSSIONS

The MOD calculated for each zone is in the range of 0.2 or less which indicates the LOS grade is 'F'. This indicates that the bus terminal has a poor performance with congested conditions. There is interpersonal space and a stand only condition exists at peak hours.

6 CONCLUSION

- Walking and queuing were the major activities occurring in the terminal.
- The evaluation was done using time-space analysis.
- The time and space utilized by passengers were considered.
- Five zones of the terminal were identified where the major passenger activities occur.
- The area of all the zones, number of standing pedestrians, number of circulating pedestrians, average standing time of pedestrians, average circulation time of pedestrians and standing pedestrian module were determined.
- In a quantitative aspect or by means of time-space analysis, the LOS was found as LOS grade F, which indicates a congested and a standing only possible condition.
- The results indicate low quality of terminal and lack of space availability during peak hours.

7 LIMITATIONS AND SCOPE

The data collection was done manually due to the complex structure of the bus terminal and due to some administrative implications. This could be a videographic survey to gain more precise values.

The work could be extended to greater number of terminals.

The comparison of the LOS values of different terminals could be done.

Only a quantitative evaluation was carried out in this study. Quantitative as well as qualitative analysis could be incorporated in the same study.

Classification on the basis of long distance travelers, short distance travelers and so on could be done.

ACKNOWLEDGMENTS

The authors gratefully acknowledge the effort of classmates of the final semester transportation engineering programme at RIT, Kottayam, who played a great part in the data collection.

REFERENCES

Chang, C.C. (2009). A model for the evaluation of airport service quality. *Proceedings of the Institution of Civil Engineers transport 162, 4,* 207–213.

Correia, A.R. & Wirasinghe, S.C. (2008). Analysis of level of service at airport departure lounges: user perception approach. *Journal of Transportation Engineering,* 134, 105–109.

Demestsky, M.J., Hoel, L.A. & Virkler, R.M. (1977). *A procedural guide for the design of transit stations and terminals.* Department of civil engineering, University of Virginia, Charlottesvi, le, Virginia.

Department of NewYork Rails. (2011). *Station capacity assessment guidance.*

Eboli, L. & Mazzulla, G. (2007). Service quality attributes affecting customer satisfaction for bus transit. *Journal of Public Transportation, 10 (3), 21–34.*

Geeva, G. & Anjaneyulu, M.V.L.R. (2015). Development of quality of service index for bus terminals, *Proc. 2nd Conference on Transportation Systems Engineering and Management. NIT Tiruchirappalli, India,* May 1–2.

Grigoriadou, M. & Braaksma, J.P. (1986). Application of the time-space concept in analyzing metro station platforms. *Journal of Institute of Transportation Engineers, 33–37.*

Litman, T. (2008). Valuing transit service quality improvements. *Journal of Public Transportation,* 11, 2.

Main Roads Western Australia. (2006). *Guidelines for assessing pedestrian level of service.* 1–9.

Transit Translink Authority. (2012). *Public Transport Infrastructure Manual.*

Transport Research Board. 2000. *Highway capacity manual,* National Research Council, Washington D.C.

Emerging Trends in Engineering, Science and Technology for Society,
Energy and Environment – Vanchipura & Jiji (Eds)
© 2018 Taylor & Francis Group, London, ISBN 978-0-8153-5760-5

Modeling of the transport of leachate contaminant in a landfill site: A case study in Mangaluru

Anand Divya & S. Shrihari

Department of Civil Engineering, National Institute of Technology, Surathkal, Mangaluru, Karnataka

H. Ramesh

Department of Applied Mechanics, National Institute of Technology, Surathkal, Karnataka, India

ABSTRACT: Ground water flow and the solute transport model MODFLOW and MT3DMS were established to determine the spread of contamination from a landfill maintained by Mangaluru City Corporation at Vamanjoor, located nearly 8.5 km from the center of the city. As Vamanjoor is home for many educational institutes and also a residential area, the spread of the contamination has to be analyzed. For this study, the aquifer considered is a subbasin of the Gurupur basin. This study has focused on handling the data available in the most efficient way to develop a consistent simulation model. The model was calibrated successfully with RMSE value of observed versus simulated head as 0.32 m. The evaluation of model was also done by comparing with the measured water head and chloride level from the field on a seasonal basis. After validating successfully, the model was run to determine the extent of contamination and also to forecast a scenario for maximum rainfall. The results show that the contamination has spread to a distance of 1 km from the landfill and with maximum rainfall the spread will be around 1.8 km from the landfill.

1 INTRODUCTION

Solid waste, when dumped into an uncontrolled landfill, can cause a serious threat to an underlying aquifer due to the migration of leachate generated from the landfill into the groundwater. The amount and nature of waste dumped in a landfill depends upon various factors such as inhabitants of the city, people's lifestyle, food habits, standard of living, the degree of industrialization and commercialization of that area, culture and tradition of inhabitants, and also the climate of the area. Due to unscientific collection, transportation and disposal of solid waste without environmentally friendly methods such as composting, incineration and so on, dumping of waste in India has become more chaotic. The leachate characteristics also depend upon the pre-treatment of the solid waste such as separation of recyclable material like plastics, paper, metals, glass and so on, as well as grinding or bailing of the waste (Kumar & Alappat, 2005). As time progresses, organic components tend to degrade and become stable, whereas conservative elements such as various heavy metals, chloride, sulfide and so on, will remain long after waste stabilization occurs. Metals are found in high concentration in leachate as they are usually precipitated within the landfill. The disposal of waste generated from domestic and industrial areas makes landfill sites a necessary component of a metropolitan life cycle. However, low-lying disposal sites which are lacking in proper leachate collection systems, observation of landfill gas and collection tools, are a potential threat to underlying groundwater resources.

Landfill sites are complex environments characterized by many interacting physical, chemical and biological processes. Leachate from landfills exhibits a major potential environmental impact for groundwater and surface water pollution and represents a potential health risk to both surrounding ecosystems and human populations. Mangaluru generates around 250 tons of solid waste every day, of which 200 tones is collected and disposed in the landfill located at Vamanjoor. Vamanjoor, which is located 15 km from the city, is along a national

Figure 1. Google Earth image of the study area.

highway (NH13) and is home to many educational institutes. The dumping yard has an area of 28.32 hectares which is poorly managed. In the vicinity, the ground water is getting polluted because of contamination by leachate.

The groundwater flow model is simulated with the help of Visual **MODFLOW**, and the movement of the contaminant subjected to a variety of boundary conditions has been simulated by using **MODFLOW**, **MT3DMS** and **MODPATH** which are widely used (Rejani et al., 2008; Da An et al., 2013). All the complex processes such as advection, dispersion and chemical reaction are well addressed with the software.

2 STUDY AREA

Vamanjoor is located in the Gurupur basin which covers an area of 841 km^2 (Figure 1). The river Gurupur is a major river flowing in a westerly direction in the Dakshina Kannada district in Karnataka State. The basin covers the foothills of the Western Ghats; in the middle portion lies the lateritic plateaus and a flat coastal alluvium at its mouth. The area lies between 12°50' to 13°10' north latitude and 74°0' to 75°5' east longitude. The study area is tropical, with a humid type of climate and gets an annual average precipitation of 3,810 mm. From the previous studies conducted in the area, the aquifer is categorized as unconfined with rich lateritic formation and having good groundwater potential (Harshendra, 1991). The transmissivity of the aquifer and its specific yield was determined to be 10 and 213 m^2/day and 7.85%, respectively.

3 METHODOLOGY

3.1 Conceptual model development

To make the simulation model, a finite difference codes MODFLOW 2000 (Harbaugh et al., 2000) a constant density flow model and MT3DMS 5.2 (Zheng, 2006; Zheng & Wang, 1999) were used to solve the space and time-based heads of groundwater and concentration of

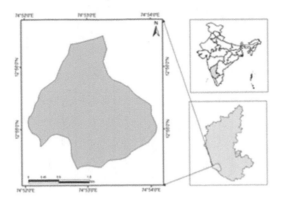

Figure 2.　Location of the subbasin.

chloride of the subbasin. The model domain is enclosed by 1,622 active cells which includes the total area with cell dimensions of 100 m × 100 m. The domain extends 30 m in the vertical direction which is representing a mono layered unconfined aquifer. In order to make a digital elevation model (DEM), the contour lines of the toposheets numbered 48/L/13/NE and 48/L/13/NW (scale 1:25000) were considered. The spatial discretization details of the model are given by X or longitudinal direction 485640 E and Y or latitudinal direction 1427956 N (with respect to origin of UTM WGS 1984 Zone 43), the number of cells in X and Y direction being 41 (Figure 2). For modeling purposes, everyday time step was used. The discretization based on time as well as space was arrived at after the first stages which were based on the precision of the results.

3.2　*Input parameters and boundary conditions*

As there was some difficulty in obtaining data, whatever data were available were collected and the model was built in the most appropriate way. The rivers were assigned with an arc feature and assigned a river boundary condition using the RIV package which is available in MODFLOW. The northern boundary was assigned with the river boundary for which data such as river bed conductance, its thickness and stage data were assigned as per available data from previous research and field visits. The Drain (DRN) package of MODFLOW was used to simulate the effect of a drainage network in the model. All boundaries, other than the northern boundary, were assigned as no flow boundaries or impermeable boundaries. They were simulated by specifying cells for which a flow equation is not solved. No flow cells were used to delete the portion of the array of cells beyond the aquifer boundary.

The annual average rainfall in the region is about 3,810 mm (Figure 3), which is from the nearest rainfall station located at Bajpe airport at a radial distance of 5 km from the landfill. Rainfall recharge is considered as the main source of replenishment for the aquifer. From earlier studies, it was found that the recharge co-efficient appropriate to this region varies from 10 to 20%. Hence, for the present study, a groundwater recharge co-efficient of 20% was taken. A total number of 68 wells were considered in the model domain which was based on available data. The water drawn per well was calculated based on the irrigation requirements and also the domestic needs.

The chloride concentration was taken as the indicator of pollution due to leachate because it is considered as the most consistent parameter in order to find leachate contamination (Papadopoulou et al., 2007). The concentration of chloride measured from landfill leachate during October 2016 was analyzed and the value assigned to each cell of the landfill and taken as an initial concentration for the transport model. The concentration of chloride in the groundwater samples collected from observation wells was analyzed.

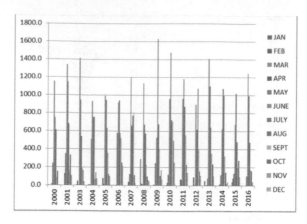

Figure 3. Annual rainfall data obtained from Bajpe airport (Obtained from IMD, Bangalore).

3.3 *Model simulation*

The starting values assigned for the dependent variable, such as freshwater head for groundwater flow and concentration for the solute transport were represented by the initial condition. In the present study, steady state calibration was carried out for October 2016 and simulated water levels were assigned as initial conditions. First, the aquifer parameters were assigned in a random manner and calibration was achieved by a trial and error method until a satisfactory match was reached between the observed as well as the simulated head. The head of groundwater head got as a result of steady state simulation is compared with the one third of the data of observation well got from the subbasin and also the record of previous research of Honnanagoudar (2015). The obtained head is taken as the initial head condition for transient groundwater flow simulation done as a basis of daily time.

4 RESULTS

The scatter plot of observed and model simulated values were plotted with the x and y axes having the same intervals and a 1:1 trend line (or 45° line) was fitted diagonally at point (0,0) across the plot area (Figure 4). The figure reveals that the model fits the observed groundwater heads as all points are lying close to the diagonal line. The RMSE value for the observed and simulated values of head was obtained as 0.325 m, which also reveals that it is a perfect fit and can be used for further applications. The groundwater head thus obtained was compared with a one third value of observation wells of the subbasin and also with the data from previous research (Honnanagoudar, 2015). The graph showed a convincingly good agreement between the observed and simulated head with a RMSE value of 0.625 m. The process of validation was then carried out by taking the water head obtained from the three observation wells in the area maintained by the Central Ground Water Board and the Department of Mines and Geology, Government of Karnataka. An RMSE value of 1.15 m was obtained after analyzing the observed and calibrated groundwater head. The results were found to be consistent with that of the calibrated results and therefore, the model was considered to be reliable for future prediction. After successful calibration and validation, parameters such as recharge co-efficient 10%, porosity 30%, bed conductance of 15 m/day, horizontal conductivity 7 m/day, specific yield 7.85% and transmissivity of 213 m²/day were taken for future application of the model. The hydrodynamic dispersivity was given initially and adjusted by a trial and error method during calibration. Horizontal transverse dispersivity of one tenth was suggested by Cobaner et al. (2012). As per Bhosale and Kumar (2001) for related coastal aquifer conditions, the value of longitudinal dispersivity can be taken between 15 to 150 m. Similar to the calibration of the

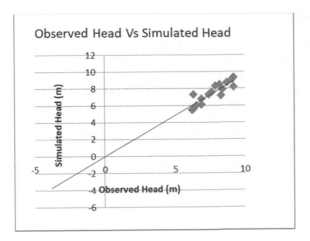

Figure 4.　Scatter plot of steady state calibration.

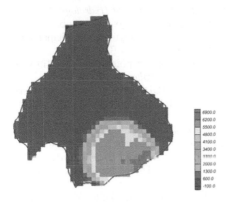

Figure 5.　Simulated chloride levels which shows the spread of the contaminant around the landfill.

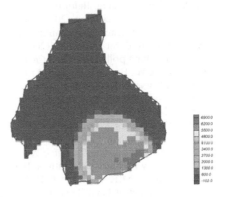

Figure 6.　Simulated chloride levels for maximum recharge level.

flow parameters, calibration of the transport parameter was also completed. Only four wells were located in the contaminated region which were chosen as observation wells. The chloride level of the landfill leachate was identified as 6,000 mg/l during October 2016. The chloride level of water in the observation wells was analyzed and compared with the simulated values. After successful calibration, the transport parameter such as longitudinal and transverse dispersivity was taken as 25 m and 2.5 m, respectively. The model was validated by comparing the values of that maintained by the Karnataka State Pollution Control Board during 2016. Since the results were reliable, the model was taken for forecasting the future scenario.

4.1　*Predictive simulation*

An effort was made to determine the pollution due to landfill as shown in Figure 5, which indicates that the pollutant has spread for a distance of 1 km around the landfill. An attempt was also made to forecast a scenario for maximum rainfall. A maximum annual rainfall of 4.75 m had occurred in 2001 in the study area in the last 15 years (Figure 6). The resulting simulation shows that the spread of the contaminant will reach 1.8 km around the landfill during such heavy rainfall.

5 CONCLUSION

Groundwater, which is one of the world's most important natural resources, is under constant threat due to various human activities. The purpose of the present study is to understand the extent of contamination due to landfill by using the software MODFLOW and MT3DMS. After successfully calibrating the model with a RMSE value for observed and simulation head equal to 0.32 m, the model was applied for predicting various scenarios. The results of the study show that the spread of the contaminant has reached almost a 1 km radius around the landfill. Additionally, the model predicted that contamination could reach a distance of 1.8 km when the value of maximum annual rainfall of 4.75 m is included. Since the area of the current study is home to many educational institutes and also a residential area, urgent attention must be given to prevent the spread of the contaminants.

REFERENCES

Bhosale, D.D. & Kumar, C.P. Simulation of seawater intrusion in Ernakulam coast. Retrieved from: http://www.angelfire.com/nh/cpkumar/publication/ernac.pdf. Accessed: 10/10/2017.

Cobaner, M., Yurtal, R., Dogan, A. & Motz, L.H. (2012). Three dimensional simulation of seawater intrusion in coastal aquifers: A case study in the Goksu Deltaic Plain. *Journal of Hydrology*, 262–280.

Da An, Yonghai Jiang, Beidou Xi, Zhifei Ma, Yu Yang Queping Yang Mingxiao L, Jinbao Zhang Shunguo Ba & Lei Jiang (2013). Analysis for remedial alternatives of unregulated municipal solid waste landfills leachate-contaminated groundwater front. *Earth Sci. 7*(3), 310–319.

Dinesh Kumar Babu J. Alappat (2005). Evaluating leachate contamination potential of landfill sites using leachate pollution index. *Clean Technologies and Environmental Policy*, 7, 190–197.

Harbaugh A.W., Banta, E.R., Hill, M.C. & Mc Donald, M.G. (2000). MODFLOW – 2000, the US Geological Survey Modular Groundwater Model—User guide to modularization concepts and ground water flow process. *US Geological Survey, Open File Report*, 1–92.

Harshendra K (1991). *Studies on water quality and soil fertility in relation to crop yield in selected river basins of D.K District of Karnataka State* (Ph.D. Thesis, Mangalore University, Karnataka, India, 146–147).

Honnanagoudar, S.A. (2015). *Studies on aquifer characterization and seawater intrusion vulnerability assessment of coastal Dakshina Kannada district, Karnataka* (Ph.D. Thesis, National Institute of Technology Karnataka, Surathkal India, 184–185).

Papadopoulou, M.P., Karatzas, G.P. & Bougioukou, G.G. (2007). Numerical modeling of the environmental impact of landfill leachate leakage on groundwater quality – a field application. *Environment Modeling and Assessment*, 12, 43–54.

R. Rejani, Madan K. Jha S.N. Panda R. Mull (2008). Simulation modeling for efficient groundwater management in Balasore Coastal Basin, India. *Water Resources Management*, 22(1), 23–50.

Zheng, C. (2006). *MT3DMS v 5.2. Supplemental user's guide*. Department of Geological Sciences, University of Alabama.

Zheng, C. & Wang, K. (1999). *A modular three dimensional multispecies transport model for simulation of advection, dispersion and chemical reactions of contaminants in groundwater systems.* Contract Report SERD 99–1, U.S. Army Corps of Engineers, United States.

Emerging Trends in Engineering, Science and Technology for Society,
Energy and Environment – Vanchipura & Jiji (Eds)
© *2018 Taylor & Francis Group, London, ISBN 978-0-8153-5760-5*

Critical review of water quality modeling and farm scale nutrient transport models widely used

Nishchhal Nihal Pandey, Vishnu Sharma & H.N. Udayashankar
Manipal Institute of Technology, Manipal, India

ABSTRACT: Use of software engineering concepts in environmental modeling initiated many researchers to develop new models to predict the future scenario in water quality management. To understand the basics in numerical water quality modeling, a general differential equation involving advection, diffusion and source term is explained along with a modeling framework. We reviewed the present framework of farm-scale nutrient transport models that are used to simulate and predict the nutrient concentration in each water body. Each model addresses the loss of nitrogen and phosphorus from rural population and nutrient load and concentration in waterways. Some of the key farm scale nutrient transport models that are widely used at the present time are discussed.

Keywords: Water quality modeling, Nutrient load, Nutrient transport

1 INTRODUCTION

At present, the environmental management has become more complex (Purandara et al. 2012). This is mainly due to anthropogenic activities, rapid urbanization and fleeting growth in agricultural fields, environmental mandates, recreational interests, hydropower generation, over-allocation and land use patterns which results in climate change and fragmented nature of available information (Purandara et al. 2012) (Welsh et al. 2013). Water quality modeling of freshwaters is a trending research area in the present scenario which focuses on the evaluation of biological and chemical status of the water bodies (Altenburger et al. 2015). Point source pollution is found to be managed well compared to non-point source pollution. Non-point source pollutants do not originate from a statutory point source, but dispersed into the receiving water by various means. NPS pollutants include components from evapotranspiration, percolation, interception, absorption, and vegetative covers which can bring about changes in the hydrologic cycle, water balance, land surface characteristics and surface water characteristics (Tong & Chen 2002) (LeBlanc et al. 1997) (Lai et al. 2011).

Water quality models are effective tools for understanding the fate and transport of contaminants in a river system (Wang et al. 2013). Many site-specific river basin models were developed and used by the engineers in the early days for decision making purpose. Numerous general water quality models were developed in the recent years for understanding various hydrological processes. Some of them include, QUAL 2 K (Esterby 1996), WASP7, CE-QUAL-ICM (Chuco 2004) (Bahadur et al. 2013) (D., McKinnon, A., Brinkman, R., Trott, L., Undu, M., Muawanah and Rachmansyah (2009)) (Cameira, M., Fernando, R., Ahuja, L. and Ma, L. (2007)), HEC-RAS, MIKE11, DUFLOW, AQUASIM, DESERT, EFDC model (U.S. Environmental Protection Agency 1999) (The U.S. Environmental Protection Agency 1997), GSTAR-1D, CASC2D.

Computer models used in integrated water quality modeling are capable of combining various spatial and environmental data for complex studies. To understand these water quality models, a basic knowledge on mathematical modeling is required. For example, models have been developed to know the fate of organic or inorganic contaminants transport or to know the transport of nutrients, pesticide, sediment loss, erosion for informing land management practices including supplying rates and the application of fertilizer and irrigation; and for both urban and rural outlines. In the present study, the scope is limited to the models currently used for modeling nitrogen and phosphorus (these are the parameters of concern in most of the circumstances) loss from rural land and the concentrations and loads in freshwater.

Process-based models are extensively used to inspect nutrient dynamics for water management purposes. Simulating nutrient transport and transformation processes from agricultural land into water bodies at the catchment scale are particularly relevant and challenging tasks for water authorities (Tuo et al. 2015) (Hashemi, F., Olesen, J., Dalgaard, T. and Børgesen, C. (2016)).

2 MATERIALS AND METHODS

2.1 *Basic model development*

All the models are primarily based on the principle of conservation of mass. For a variable within a control volume, a one-dimensional advection-dispersion equation with source term can be derived.

$$\frac{\partial C}{\partial t} = -U\frac{\partial C}{\partial x} + E\frac{\partial^2 C}{\partial x^2} + S_c \tag{1}$$

where, t is time, U is the steady-state average velocity of the water in the flow direction (x) and E is the dispersion coefficient. U and E are assumed to be in the direction of flow (x) of the water body. In a great number of water bodies, eq. (1) can be applied where flow can be approximated as one-dimensional. First term represents the rate of change of concentration (C) of a component, second term represents advection (first order), third term represents dispersion (second order) and the fourth term represents the source term which incorporate inflows, outflows, and reactions due to physical, chemical, and/or biological processes (Stamou & Rutschmann 2011).

Eq. (1) can be solved numerically using finite difference method. The source term may vary according to the type of water body under consideration (Stamou & Rutschmann 2011) (Gelda & Effler 2000).

2.2 *Assumptions and inputs for the model*

All mathematical models for environment studies are generated based on certain practical systems or equations which means that it can only accept a particular set of data as inputs. Moreover most of the mathematical models have to be developed through computers, a valid coding program is needed to be generated to process the input for the model. Before developing any mathematical framework there are a set of assumptions which are made and the mathematical model is valid only when these sets of assumptions are strictly followed. The assumptions made can be of Numerical in nature which are made during the implementation or problem solving part of the mathematical model or logical/qualitative type which is made during the generation of the equation or the model itself. As we make all our calculations based on the assumptions made, it is of utmost importance that the assumptions are followed during the calculations part or the model becomes insignificant. (Shukla, j., hallam, t. g. and capasso, v) (holzbecher, e. o) (wang, q., li, s., jia, p., qi, c. and ding, f) (A. Mudgal, C. Baffaut, S. H. Anderson, E. J. Sadler and A. L. Thompson (2010)), (Cherry, K., Shepherd, M., Withers, P. and Mooney, S. (2008)).

2.3 *Farm scale nutrient models*

Farm scale models consider nutrient loss in reaction to land management decisions on a farm, paddock, or plot scale. Sometimes called "root zone" models, these models calculate

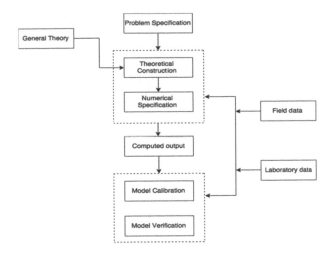

Figure 1.   Principal components of modeling framework.

the quantity of nutrients that are lost from the top-most soil layers instead of being absorbed by plant roots.

2.3.1   *FARMSIM*

The model is developed especially for the complex, dynamic systems with many interacting biophysical components along a wide range of soil, climatic and socioeconomic conditions (Wijk et al. 2004).

2.3.2   *NUANCES-FARMSIM*

Nutrient Use in ANimal and Cropping systems: Efficiencies and Scales-FARM SIMulator is an integrated crop—livestock model developed to analyse African smallholder farm systems.

2.3.3   *SWATRE*

A transient one-dimensional finite-difference model for vertical unsaturated flow with water uptake by roots is presented. In the model a number of boundary conditions are given for both top and bottom of the system. At the top, 24-hr. data on rainfall, potential soil evaporation and potential transpiration are needed. When the soil system remains unsaturated, one of three bottom boundary conditions can be used: pressure head, zero flux or free drainage. When the lower part of the system remains saturated, one can either give the groundwater level or the flux through the bottom of the system as input. In the latter case the groundwater level is computed (Belmans et al, 1983).

2.3.4   *ANIMO*

Simulation of the nitrogen behaviour in the soil and the nitrogen uptake by winter wheat was performed using the model ANIMO (Rijtema & Kroes, 1991). It is a detailed process oriented simulation model for evaluation of nitrate leaching to groundwater, N- and P-loads on surface waters and Green House Gas emission. The model is primarily used for the ex-ante evaluation of fertilization policy and legislation at regional and national scale. The output of SWATRE model is taken as an input for the ANIMO mode.

2.3.5   *OVERSEER*

The OVERSEER nutrient budget model is a decision support tool (Wheeler et al. 2003) to assist farmers and consultants develop nutrient plans (Wheeler et al. 2006). It produces estimates of long-term average nutrient losses via drainage and runoff at a farm and farm block level. Overseer also estimates greenhouse gas emissions and aids in planning fertiliser applications.

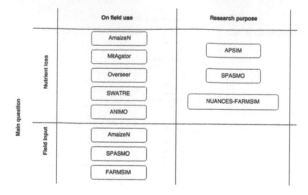

Figure 2.   Classification of farm scale models based on intended use and the question of focus.

2.3.6   *AMAIZEN*

The AmaizeN is a maize growth simulation software. The core of the system is the daytime step simulation model of maize growth and development, driven by solar radiation and interacting with soils. This model is an extension of the maize growth simulation model which is modified for cooler situations.

2.3.7   *APSIM*

Agricultural Production System Simulator is a software system which allows models of crop and pasture production, residue decomposition, soil water and nutrient flow, and erosion to be readily re-configured to simulate various production systems and soil and crop management to be dynamically simulated using conditional rules (Mccown et al. 1996). The main objective behind the development of the model was to simulate biophysical process in farming systems, in particular where there is interest in the economic and ecological outcomes of management practice in the face of climatic risk (Keating et al. 2003). The model simulates the dynamics of soil-/plant-management interactions within a single crop or a cropping system (Wang et al. 2002).

2.3.8   *MitAgator*

MitAgator is a farm scale GIS (Geographic Information System) based DST (Decision Support Tool) that has been developed to identify and estimate nitrogen, phosphorus, sediment, and E. coli loss spatially across a farm landscape. This model is a spatially explicit model that extends the results produced by Overseer to identify where on the farm property nutrient loss is occurring (Anastasiadis et al. 2013).

2.3.9   *SPASMO*

Soil Plant Atmosphere System Model is a physics based model of plant growth and nitrogen leaching, with some estimation of phosphorus. It focuses on arable and 15 horticultural activities. SPASMO produces estimates of crop production, drainage, and nutrient leaching from soil on a daily time step. Its inputs include spatial data on climate, irrigation, soil properties, land use, and crop and stock management (Anastasiadis et al. 2013).

3   CONCLUSIONS

A brief description of numerical water quality models has been explained. Farm scale models and the use of models for understanding nutrient loss to waterways have been discussed. The practice of modelling is well established among scientists as it provides an effective way to think about and understand complex phenomena. Models provide structure to guide new research by combining existing knowledge and identifying areas where new knowledge is needed. In this way, models are a summary of the science and a guide for further science.

Models are established, used and sophisticated in the context of a wider scientific community. This community both updates the design of models and tests the validity of their results. This is an integral part of the accepted processes by which scientific understanding improves.

REFERENCES

Argent, R., Sojda, R., Giupponi, C., McIntosh, B., Voinov, A., & Maier, H. (2016). Best practices for conceptual modelling in environmental planning and management. *Environmental Modelling & Software*, *80*, 113–121. http://dx.doi.org/10.1016/j.envsoft.2016.02.023.

Bahadur, R., Ziemniak, C., Amstutz, D., & Samuels, W. (2013). Global River Basin Modeling and Contaminant Transport. *American Journal of Climate Change*, *02*(02), 138–146. http://dx.doi.org/10.4236/ajcc.2013.22014.

Belmans, C., Wesseling, J., & Feddes, R. (1983). Simulation model of the water balance of a cropped soil: SWATRE. *Journal of Hydrology*, *63*(3–4), 271–286. http://dx.doi.org/10.1016/0022-1694(83)90045-8.

Cameira, M., Fernando, R., Ahuja, L. and Ma, L. (2007). Using RZWQM to simulate the fate of nitrogen in field soil–crop environment in the Mediterranean region. *Agricultural Water Management*, *90*(1–2), pp.121–136.

Cherry, K., Shepherd, M., Withers, P. and Mooney, S. (2008). Assessing the effectiveness of actions to mitigate nutrient loss from agriculture: A review of methods. *Science of the Total Environment*, *406*(1–2), pp.1–23.

Cunderlik, J. (2007). River Bank Erosion Assessment using 3D Hydrodynamic and Sediment Transport Modeling. *Journal of Water Management Modeling*. http://dx.doi.org/10.14796/jwmm.r227-02.

Esterby, S. (1996). Review of methods for the detection and estimation of trends with emphasis on water quality applications. *Hydrological Processes*, *10*(2), 127–149. http://dx.doi.org/10.1002/(sici)1099-1085(199602)10:2<127::aid-hyp354>3.0.co;2–8.

Gelda, R., & Effler, S. (2003). Application of a Probabilistic Ammonia Model: Identification of Important Model Inputs and Critique of a TMDL Analysis for an Urban Lake. *Lake and Reservoir Management*, *19*(3), 187–199. http://dx.doi.org/10.1080/07438140309354084.

Guagnano, D., Rusconi, E., & Umiltà, C. (2013). Joint (Mis-)Representations: A Reply to Welsh et al. (2013). *Journal of Motor Behavior*, *45*(1), 7–8. http://dx.doi.org/10.1080/00222895.2012.752688.

Gyawali, S., Techato, K., Monprapussorn, S., & Yuangyai, C. (2013). Integrating Land Use and Water Quality for Environmental based Land Use Planning for U-tapao River Basin, Thailand. *Procedia—Social and Behavioral Sciences*, *91*, 556–563. http://dx.doi.org/10.1016/j.sbspro.2013.08.454.

Hashemi, F., Olesen, J., Dalgaard, T. and Børgesen, C. (2016). Review of scenario analyses to reduce agricultural nitrogen and phosphorus loading to the aquatic environment. *Science of the Total Environment*, 573, pp. 608–626.

Holzbecher, E. (2012). *Environmental modeling*. Heidelberg: Springer.

Keating, B., Carberry, P., Hammer, G., Probert, M., Robertson, M., & Holzworth, D. et al. (2003). An overview of APSIM, a model designed for farming systems simulation. *European Journal of Agronomy*, *18*(3–4), 267–288. http://dx.doi.org/10.1016/s1161-0301(02)00108-9.

McKinnon, D., A. Brinkman, R. Trott, L. Undu, M. Muawanah and Rachmansyah (2009). The fate of organic matter derived from small-scale fish cage aquaculture in coastal waters of Sulawesi and Sumatra, Indonesia. *Aquaculture*, 295(1–2), pp. 60–75.

Mudgal, A., C. Baffaut, S.H. Anderson, E.J. Sadler and A.L. Thompson (2010). APEX Model Assessment of Variable Landscapes on Runoff and Dissolved Herbicides. *Transactions of the ASABE*, 53(4), pp.1047–1058.

Purandara, B., Varadarajan, N., Venkatesh, B., & Choubey, V. (2011). Surface water quality evaluation and modeling of Ghataprabha River, Karnataka, India. *Environmental Monitoring and Assessment*, *184*(3), 1371–1378. http://dx.doi.org/10.1007/s10661-011-2047-1.

Rabi, A., Hadzima-Nyarko, M., & Šperac, M. (2015). Modelling river temperature from air temperature: case of the River Drava (Croatia). *Hydrological Sciences Journal*, *60*(9), 1490–1507. http://dx.doi.org/10.1080/02626667.2014.914215.

Rijtema, P., & Kroes, J. (1991). Some results of nitrogen simulations with the model ANIMO. *Fertilizer Research*, *27*(2–3), 189–198. http://dx.doi.org/10.1007/bf01051127.

Shukla, J., Hallam, T., & Capasso, V. *Mathematical modelling of environmental and ecological systems*.

Stamou, A., & Rutschmann, P. (2011). Teaching simple water quality models. *Education for Chemical Engineers*, *6*(4), e132–e141. http://dx.doi.org/10.1016/j.ece.2011.08.005.

Wang, Q., Li, S., Jia, P., Qi, C., & Ding, F. (2013). A Review of Surface Water Quality Models. *The Scientific World Journal*, *2013*, 1–7. http://dx.doi.org/10.1155/2013/231768.

WANG Tian-Xiao (2013) Chemical constituents from Psoraleacorylifolia and their antioxidant α-glucosidase inhibitory and antimicrobial activities. (2013). *China Journal of Chinese Materia Medica*. http://dx.doi.org/10.4268/cjcmm20131423.

Emerging Trends in Engineering, Science and Technology for Society,
Energy and Environment – Vanchipura & Jiji (Eds)
© 2018 Taylor & Francis Group, London, ISBN 978-0-8153-5760-5

A novel hybrid material for the trace removal of hexavalent chromium [Cr(VI)] from contaminated water

A.R. Laiju
National Institute of Technology, Srinagar, Uttarakhand, India

S. Sarkar
Indian Institute of Technology, Roorkee, Uttarakhand, India

ABSTRACT: This study investigated the performance of two anion exchange resins and also synthesized and evaluated the performance of a hybrid material consisting of a strong base anion exchange resin dispersed with FeS (iron sulphide) nanoparticles for the removal of hexavalent chromium, Cr(VI). The synthesis of the hybrid was carried out by using an *in situ* process, where the strong base anion exchange resin serves as a nanoreactor and provides a confined medium for synthesis. They stabilize and isolate the synthesized nanoparticles preventing their aggregation. Equilibrium batch studies, adsorption isotherm and column studies were performed to determine the maximum uptake capacity for Cr(VI). Comparison of a fixed bed column run between the hybrid material and parent resin confirmed that the Cr(VI) was selectively removed and the hybrid showed higher capacity. The wide availability of resin and low-cost chemicals for synthesis and regeneration will make hybrid material an attractive option for the removal of Cr(VI) from contaminated water.

1 INTRODUCTION

Chromium is unique among toxic heavy metals in the environment and toxicity is regulated on the basis of its oxidation state and total concentration. In water distribution systems and in water treatment processes, chromium exists mainly as trivalent chromium, Cr(III) and hexavalent chromium, Cr(VI) (Kimbrough et al., 1999). Cr(VI), an inorganic contaminant which received public attention recently, can be considered as a potential human carcinogen (International Agency for Research on Cancer, 2012). Public concern and potential adverse health effects prompt the investigation of Cr(VI) in drinking water supplies below Maximum Contaminant Level (MCL). In India, the maximum permissible limit for Cr(VI) in drinking water is 50 μg/L. Most of the regulations by different agencies are based on total chromium not just Cr(VI), due to the absence of sufficient risk analysis for the oral intake of Cr(VI) (Anderson, 1997). To reduce the contamination of Cr(VI) from water and wastewater, various methods have been used mainly involving reduction followed by precipitation, adsorption, ion exchange and membrane process (Sharma et al., 2008).

The objectives of the present study are to evaluate the performance of a strong base anion exchange resin, Amberlite IRA 400 (IRA 400), a weak base anion exchange resin, Lewatit MP 64 (LMP 64) and to synthesize a novel Hybrid Ion exchange Material (HIM) and to validate the performance of trace concentration of Cr(VI) from contaminated water. The physical properties of the resin are shown in Table 1.

Table 1. Physical properties of resins.

Resin	Amberlite IRA-400	Lewatit MP 64
Type	Strong base anion exchange	Weak base anion exchange
Functional group	Quaternary ammonium	Tertiary amine
Matrix	Styrene–divinyl benzene (gel)	Styrene–divinyl benzene (macro porous)
Ionic form	Chloride	Chloride
Temperature limit (°C)	77	70
Moisture holding capacity (%)	40–47	50–60
Particle size (µm)	600–750	300–1250
Total exchange Capacity (eq//L)	1.4	1.3

2 EXPERIMENTAL METHODOLOGY

2.1 Synthesis of HIM

For the synthesis of HIM, iron sulphide (FeS) nanoparticles were immobilized in IRA 400 using an *in situ* synthesis process (Sarkar et al., 2012) via two steps as shown in Equations 1 and 2. The over bar denotes theres in phase. In total, 10 gm of IRA 400 was added to 100 ml of 0.3 M Na_2S solution and shaken in an orbital shaking incubator for one hour at a temperature of $25 \pm 0.5°C$ and the resultant solution was filtered and washed thoroughly with distilled water. The resin was then transferred into a solution containing 30 g/L of $FeCl_2$ solution of 100 ml and shaken for one hour at a temperature of $25 \pm 0.5°C$. The resultant material, now expected to be impregnated with FeS nanoparticles, was drained and washed with distilled water and stored in an ethanol solution to prevent further oxidation. Figure 1 shows a diagram of HIM containing FeS nanoparticles.

$$\overline{2R(CH_3)_3Cl} + S^{2-} \leftrightarrow \overline{\left[R(CH_3)_3\right]_2 S^{2-}} + 2Cl^- \tag{1}$$

$$\overline{\left[R(CH_3)_3\right]_2 S^{2-}} + Fe^{2+} + 2Cl^- \leftrightarrow \overline{2R(CH_3)_3 Cl^-} + FeS(s) \tag{2}$$

2.2 Batch experiments

In order to evaluate the performance of resin and hybrid material for Cr(VI) removal, the effect of pH, initial concentration, resin dosage and competing anions (SO_4^{2-}, Cl^-, HCO_3^-) were investigated. The removal efficiency, η (%) and the amount of Cr(VI) ions adsorbed at equilibrium q_e, (mg of Cr(VI)/gm of resin) can be calculated using Equations 3 and 4, respectively, where C_0 and C_e are initial and equilibrium concentrations of Cr(VI), m is mass of material added, V is the volume of solution.

$$\eta(\%) = \frac{C_0 - C_e}{C_0} \times 100 \tag{3}$$

$$q_e = \frac{V(C_0 - C_e)}{m} \tag{4}$$

2.3 Fixed bed column run

Fixed bed column studies were performed in a plexiglass column of 11 mm internal diameter and 300 mm height. With the help of a microprocessor based peristaltic pump (Ravel RH P100 L), feed water was passed through the column with IRA 400, LMP 64 and HIM at a certain pH until breakthrough. Figure 2 shows a schematic diagram of a fixed bed column run. Effluent samples were collected at definite intervals during the column run and were

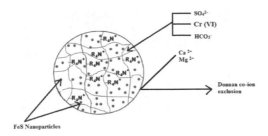

Figure 1. Diagram of HIM containing FeS nanoparticles.

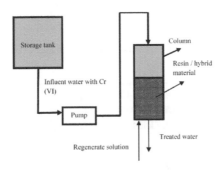

Figure 2. Schematic diagram for fixed bed column run.

analyzed for Cr(VI) concentration by the 1,5 diphenylcarbazide method by using a split beam UV visible spectrophotometer (T60 UV, PG Instruments, UK) at 540 nm.

3 RESULTS AND DISCUSSION

3.1 *Characteristics of HIM*

The characteristics of IRA 400 and HIM were investigated by SEM (Scanning Electron Microscopy) with EDX (Energy-Dispersive X-ray analysis) as presented in Figure 3.

SEM and EDX results of HIM shows that the FeS particles were uniformly precipitated in side the HIM and the Fe content was approximately 6% by weight, as revealed by the EDX study.

3.2 *Effect of resin dosage*

Increasing the dosage of resin and HIM, increased the efficiency of Cr(VI) removal. After 0.8 g/L, the increment in Cr(VI) removal was not increased as significantly. The increase in removal efficiency is due to an increase in number of ion exchange sites for exchange between the liquid phase and the exchanger phase (Babu & Gupta, 2008). If the dosage is increased by keeping the initial concentration constant, the amount of Cr(VI) absorbed per unit mass showed a decrease due to the unavailability of Cr(VI) ions for the exchange sites. The uptake capacity dropped from 92.6 to 4, 45.27 to 2.38 and 107.5 to 5.32 mg/gm for IRA 400, LMP 64 and HIM, respectively, by increasing the dosage from 0.2 to 1.2 g/L (Figure 4). The reason for the drop in the uptake capacity while increasing the dosage is due to ion exchange sites remaining unsaturated during the exchange process.

3.3 *Effect of pH*

From Figure 5, the maximum uptakes of 11.08, 7.01 and 14.41 mg/gm were obtained for IRA 400, LMP 64 and HIM at pH 4. For resin and hybrid, the uptake, as well as efficiency, increases as the pH decreases.

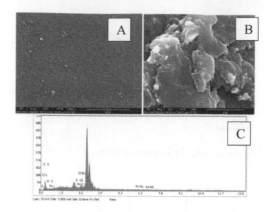

Figure 3.   SEM image of IRA 400 (A), HIM (B) and EDX of HIM (C).

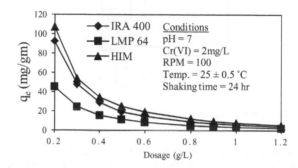

Figure 4.   Effect of resin dosage.

From the predominance diagram, H_2CrO_4 is predominant for a pH less than 1, $HCrO_4^-$ for a pH between 1 and 6.5 and divalent CrO_4^{2-} for a pH above 6.5. Dimerization of $HCrO_4^-$ ions ($Cr_2O_7^{2-}$) is possible if the concentration is higher than 1 g/L (Saleh et al., 1989). The increased Cr(VI) removal efficiency at an acidic pH is mainly due to the fact that $HCrO_4^-$, being mono-valent, can attach to a single ion exchange functional group and CrO_4^{2-}, a divalent, needs to bind to two ion exchange functional groups (Sengupta & Clifford, 1986). At an alkaline pH, the sorption trend is likely decreased due to the competition between CrO_4^{2-} and OH^- for the binding sites on the exchanger which results in lower uptake. The type of functional group, quaternary ammonium moiety of IRA 400, has a significant effect on the uptake of Cr(VI) rather than the tertiary amine of LMP 64 (Pehlivan & Cetin, 2009; McGuire et al., 2007).

3.4   *Effect of initial concentration*

Initial Cr(VI) concentration varied from 1 mg/L to 15 mg/L at pH 7 with a dosage of 0.8 g/L and the results are shown in Figure 6. Maximum uptakes of 4.71, 2.8 and 6.07 mg/gm were obtained for IRA 400, LMP 64 and HIM, respectively, for an initial concentration of 1 mg/L and as the initial concentration of Cr(VI) increases, the Cr(VI) uptake decreases gradually. At lower concentrations, all Cr(VI) molecules present in the solution interact with the binding sites of the resin and as the concentration increases, the binding sites become saturated lead-ing to decreased Cr(VI) removal efficiency (Balan et al., 2013; Neagu et al., 2003).

3.5   *Effect of competing anions*

The main competing anions for Cr(VI) in ground water, SO_4^{2-}, Cl^- and HCO_3, were investi-gated under different concentrations from 0 to 200 mg/L. Chloride and bicarbonate did not interfere in Cr(VI) removal, so the result is not included but as the concentration of sulfate increased the uptake capacity decreased, showing that the affinity for sulfate increases as the concentration of sulfate increases for IRA 400, LMP 64 and HIM (Figure 7).

3.6   Adsorption isotherm

Adsorption capacities can be described by using Langmuir and Freundlich models indicated by Equations 5 and 6, and the results are shown in Figure 8.

$$\frac{1}{q_e} = \frac{1}{Q_{max}bC_e} + \frac{1}{Q_{max}} \tag{5}$$

$$loq\,q_e = \frac{1}{n}\log C_e + loqk_f \tag{6}$$

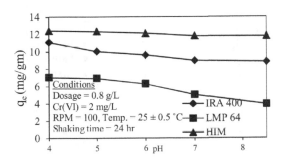

Figure 5.   Effect of pH.

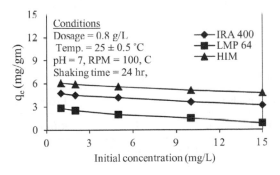

Figure 6.   Effect of initial concentration.

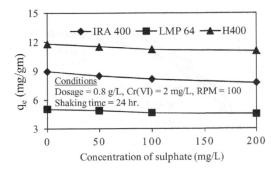

Figure 7.   Effect of sulfate.

where Q_{max} is the maximum uptake capacity of Cr(VI) (mg/gm), b is the Langmuir constant (g/L), k_f is the Freundlich constant (mg/g), and 1/n is the adsorption intensity. The correlation coefficients (R^2 value) of both Langmuir and Freundlich models are close to 1, but the Freundlich model shows better correlation for the whole initial high concentration of Cr(VI) than the Langmuir model. Thus, it indicates that the adsorption is not a monolayer layer adsorption, rather it is a multilayer adsorption. The parameters obtained for the Langmuir and Freundlich models are shown in Table 2.

3.7 *Kinetic study*

From Figure 9, observed that during the initial stage the rate removal is faster and further it took longer time to reach equilibrium after 360 minutes from the start of experiment. Interdiffusion of ions take place in two steps, (1) liquid film diffusion and (2) intraparticle

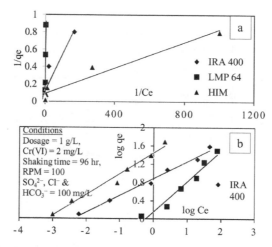

Figure 8. (a) Langmuir isotherm (b) Freundlich isotherm.

Table 2. Parameters for Langmuir and Freundlich isotherm.

	Langmuir isotherm			Freundlich isotherm		
	Q_{max} (mg/g)	b	R^2	k_f (mg/g)	n	R^2
IRA 400	8.143	29.24	0.8742	7.72	2.68	0.9869
LMP 64	7.183	0.4	0.8384	1.42	1.50	0.956
HIM	10.93	130.7	0.9466	27.97	2.19	0.978

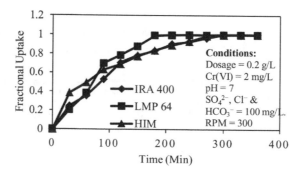

Figure 9. Fractional uptake of aqueous Cr(VI).

202

diffusion and slowest among the two process will control the overall exchange process. (Li & Sengupta, 2000; DeMarco et al., 2003; Greenleaf et al., 2006; Chanthapon et al., 2017).

Normalized fractional loading or uptake (F) by the exchanger during the batch kinetic study changes with time (Figure 9) and can be defined by Equation 7, where $C_{cr,o}$, $C_{cr,t}$, and $C_{cr,e}$ represent concentration of Cr(VI) at initial, timet and equilibrium respectively and $q_{cr,t}$, $q_{cr,e}$ are Cr(VI) uptake at time t and at equilibrium respectively.

$$F = \frac{q_{Cr,t}}{q_{cr,e}} = \frac{\left(C_{Cr,0} - C_{Cr,r}\right)}{\left(C_{Cr,0} - C_{Cr,e}\right)} \tag{7}$$

3.8 *Fixed bed column run and regeneration*

Figure 10 shows the effluent concentration of Cr(VI) during separate column runs with IRA 400, LMP 64 and HIM, with identical influent characteristics. Effluent Cr (VI) concentration of 50 µg/L occurred after approximately 2100, 650 and 5500 bed volumes (BVs, volume of water treated per volume of resin) and exhausted at 4275, 1532 and 10268 BVs for IRA 400, LMP 64 and HIM respectively. Interruption test was carried out during the fixed column run for HIM, marked with circles in Figure 9. When the flow resumed, the effluent concentration of Cr(VI) dropped significantly to 48.2 µg/L and it took approximately 1,250 BVs for the effluent concentration to reach the concentration present prior to the interruption. The interruption allows the sorbed C(VI) to evenly spread out inside the host ion exchanger. Faster uptakes occurred after the interruption, providing evidence in support of intraparticle diffusion being the rate limiting step for the hybrid material.

The mechanism behind the superior removal of Cr(VI) by HIM may be due to the high positive charge of the functional group of the parent anion exchanger, Cr(VI) species either $HCrO_4^-$ or CrO_4^{2-}, will be attracted inside the host material due to the Donnan membrane effect. The Cr (VI) species inside the HIM undergo redox reaction with the impregnated iron sulphide nanoparticles and this reaction frees up the active ion exchange site were Cr (VI) initially occupied and can sorb new set of Cr (VI). Equations 8 and 9 show the redox reactions between FeS nanoparticles and Cr (VI) inside the host material.

$$3FeS + 8HCrO_4^- + 32H^+ \rightarrow 3Fe^{3+} + 8Cr^{3+} + 3SO_4^{2-} + 20H_2O \tag{8}$$

$$Cr^{3+} + 3O^{H-} \rightarrow Cr(OH)_3 \tag{9}$$

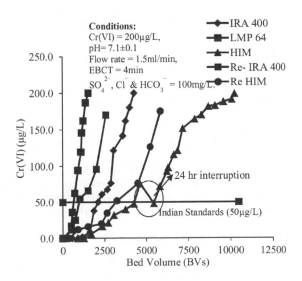

Figure 10. Fixed bed column run.

Regeneration of exhausted resin and HIM were carried out by using 5% NaCl. Results show that regenerated hybrid material (Re-HIM) was able to remove 5,869 BVs up to 175 µg/L effluent Cr(VI) concentration.

4 CONCLUSION

Iron sulphide impregnated hybrid material were synthesized by *in-situ* synthesis process on functional polymer supported host strong base anion exchangeres in. The removal process was found to be strongly dependent on the pH, dosage, initial concentration and sulfate concentration. HIM shows high selectivity toward Cr(VI), with a high adsorption capacity found by fitting the Freundlich model. A fixed bed column containing HIM with feed water composition of 200 µg/L of Cr(VI), sulphate, chloride and bicarbonate of 100 mg/L can treat up to 5,500 BVs before reaching Cr (VI) concentration corresponding to Indian standards (50 µg/L). The column study identified that the rate limiting step is the intraparticle solute transport mechanism. A fixed bed column study and the possibility of regeneration shows that HIM can be considered as a promising hybrid material for the trace removal Cr(VI) from contaminated water.

REFERENCES

Anderson, R.A. (1997). Chromium as an essential nutrient for humans. *Regulatory Toxicology and Pharmacology*, *26*(26), s35–s41.

Babu, B.V. & Gupta, S. (2008). Adsorption of Cr(VI) using activated neem leaves: Kinetic studies. *Adsorption,14*(1), 85–92.

Balan, C., Volf, I. & Bilba, D. (2013). Chromium (VI) removal from aqueous solutions by purolite base anion-exchange resins with gel structure. *Chemical Industry and Chemical Engineering Quarterly,19*(4), 615–628.

Chanthapon, N., Sarkar, S., Kidkhunthod, P. & Padungthon, S. (2017). Lead removal by a reusable gel cation exchange resin containing nano-scale zero valent iron. *Chemical Engineering Journal*, *331*, 545–555.

DeMarco, M.J., Sengupta, A.K. & Greenleaf, J.E. (2003). Arsenic removal using a polymeric/inorganic hybrid sorbent. *Water Research*, *37*(1), 164–176.

Greenleaf, J.E., Lin, J.C. & Sengupta, A.K. (2006). Two novel applications of ion exchange fibers: Arsenic removal and chemical-free softening of hard water. *Environmental Progress*, *25*(4), 300–311.

International Agency for Research on Cancer. (2012). *IARC Monograph: Chromium (VI) Compounds*. 147–168.

Kimbrough, D.E., Cohen, Y., Winer, A.M., Creelman, L. & Mabuni, C. (1999). A critical assessment of chromium in the environment. *Critical Reviews in Environmental Science and Technology*, *29*(1), 1–46.

Li, P. & Sengupta, A.K. (2000). Intraparticle diffusion during selective ion exchange with a macroporous exchanger. *Reactive and Functional Polymers*, *44*(3), 273–287.

McGuire, M.J., Blute, N.K., Qin, G., Kavounas, P., Froelich, D. & Fong, L. (2007). Hexavalent chromium removal using anion exchange and reduction with coagulation and filtration. *Water Research Foundation,* Project #3167, 140.

Neagu, V., Untea, I., Tudorache, E. & Luca, C. (2003). Retention of chromate ion by conventional and N-ethylpyridinium strongly basic anion exchange resins. *Reactive and Functional Polymers*, *57*(2–3), 119–124.

Pehlivan, E. & Cetin, S. 2009. Sorption of Cr(VI) ions on two Lewatit-anion exchange resins and their quantitative determination using UV-visible spectrophotometer. *Journal of Hazardous Materials*, *163*(1), 448–453.

Saleh, F.Y., Parkerton, T.F., Lewis, R.V., Huang, J.H. & Dickson, K.L. (1989). Kinetics of chromium transformation in the environment. *Science of the Total Environment*, *86*, 25–41.

Sarkar, S., Guibal, E., Quignard, F. & Sengupta, A.K. (2012). Polymer-supported metals and metal oxide nanoparticles: Synthesis, characterization, and applications. *Journal of Nanoparticle Research*, *14*(2), 1–24.

Sengupta, A.K. & Clifford, D. (1986). Important process variables in chromate ion exchange. *Environmental Science & Technology*, *20*(2), 149–55.

Sharma, S.K., Petrusevski, B. & Amy, G. (2008). Chromium removal from water: A review. *Journal of Water Supply: Research and Technology—AQUA*, *57*(8), 541–553.

Emerging Trends in Engineering, Science and Technology for Society,
Energy and Environment – Vanchipura & Jiji (Eds)
© 2018 Taylor & Francis Group, London, ISBN 978-0-8153-5760-5

Solid waste management practices and decision-making in India

R. Rajesh & B.K. Bindhu
Rajiv Gandhi Institute of Technology, Kottayam, Kerala, India

ABSTRACT: One of the most significant problems in urban India is poor Solid Waste Management (SWM) systems that are leading to environmental, health and economical issues. The aim of this paper is to provide an insight into the SWMs and practices in India, and the challenges faced. The review focuses on three major areas: solid waste characteristics; SWM systems; and decision-making. The two major issues at source are the increasing amount of waste, and the changing composition of the waste streams. The issues in waste logistics are due to poor waste segregation, unregulated waste transportation, and incompatible and outdated transportation technologies. Sustainable waste processing and disposal practices are yet to be adopted and are facing numerous implementation and operational challenges. There is a need to understand the operational issues in waste processing and disposal so that better technological development and adoption at a macro level is made possible. The decision-making practices and solutions generated in India are unscientific or based on piece meal approaches. Inefficiencies in waste management systems are primarily due to poor alignment between requirements, constraints and operations in the SWM system. A scientific SWM design that meets the fundamental requirements to handle a variety of waste, to be environmentally friendly, and is economically and socially acceptable, is needed. A robust SWM system design places emphasis on active public participation in decision-making, and operational integration of its subcomponents.

1   INTRODUCTION

Solid waste generation, and its management, is a worldwide problem. In India, it is anticipated that about 260–300 million tons of solid waste per annum will be generated by 2050 (Joshi & Ahmed, 2016). This is primarily due to population growth, increasing urbanization and socio-economic development. Solid waste leads to different environmental problems and ecosystem changes, human health issues, and socio-economic issues. There is an increasing focus on using Solid Waste Management (SWM) systems in the movement toward an environmentally sustainable society. Over the last decades, there have been continuous changes in the solid waste management practices evolving from the simple form of collection and dumping, to integrated solid waste management arising from learned experiences. However, there is a need to understand the operational issues too.

The aim of this paper is to provide an insight into the SWM systems and practices in India.

2   LITERATURE REVIEW

Peer-reviewed articles published since 2000 from Scopus and Google Scholar databases were selected using the keywords 'e-waste', 'municipal solid waste', 'solid waste management', 'waste recycling', 'life cycle assessment', 'waste disposal', 'environment assessment', and 'multi-criteria decision-making'. The articles were included after careful review of the abstracts.

2.1 Solid waste characteristics

Solid waste is the waste generated from domestic, commercial, and construction activities. The source of any waste is from the process or activity undertaken by an individual as in their home or from an enterprise (e.g. hospital, hotels, and industries). Solid waste can be classified based on origin, source, risk, characteristics or treatability. Solid waste consists of organic matter, fine earth, paper, plastic, glass, metals, rubber, textiles, leather, chemicals and inert materials. Types of waste include biodegradable waste, recyclable material, inert waste matter, domestic, and hazardous waste (Joshi & Ahmed, 2016; Nandy et al., 2015; Sharholy et al., 2008).

During the last decade, solid waste generation has doubled (CPCB, 2013). Per capita waste generation in cities varies from 0.2 kg to 0.6 kg per day depending upon the size of population, and is estimated to increase by 1.33% annually (Joshi & Ahmed, 2016). The Indian industrial sector generates an estimated 100 million tons/year of non-hazardous solid waste (EBTC, 2014). Annually, about 12 million tons of inert waste is generated in India from street sweepings and construction and demolition waste, and it occupies about one-third of the total municipal solid waste in landfill sites (Joshi & Ahmed, 2016).

In India, solid waste differs greatly with regard to its composition and hazardous nature, when compared to MSW in western countries (Sharholy et al., 2008). Solid waste consists of 30–60% organic matter, 3–10% recyclables, and 30–40% inert matter (Kumar et al. 2009; Sharholy et al., 2008; Singh et al., 2011; Zhu et al., 2008). There is also a marked distinction between the solid waste from urban and rural areas. Rural waste is largely agricultural in nature, but rural areas suffer from 'pollution sinks' from the encroaching urban sprawl (EBTC, 2014). Sharholy et al. (2008) pointed out that the physical and chemical characteristics of solid waste changes with population density, and indicates the effect of urbanization and development. The relative percentage of organic waste in solid waste generally increases with decreasing socio-economic status; so rural households generate more organic waste than urban households (Sharholy et al., 2008). The per capita generation rate is high in some states, such as Gujrat, Delhi and Tamil Nadu, and cities, such as Chennai, Kanpur, Lucknow and Ahmedabad. On the contrary, it is low in states, such as Meghalaya, Assam, Manipur and Tripura, and cities, such as Nagpur, Pune and Indore (Sharholy et al., 2008). Population explosion, coupled with changing lifestyles and improved living standards, has resulted in an increase in the generation of solid waste. The two major current waste management challenges are an increasing amount of waste, and the changing composition of the waste stream.

2.2 Solid waste management systems in india

SWM systems comprise of the following components: storage; collection-transportation; processing; and disposal. A SWM system focuses on strategies to reduce environmental impacts and generate value from the waste flow. Figure 1 shows the SWM practices in India. Wastes generated from process-activities at any house are accounted for through practices, such as animal feeding, burning, on-site dumping and sales to recyclers. The waste not accounted for through the above practices are dumped in streets or ditches, or moved to municipal bins. Commercial waste also flows in a similar manner except for the burying of waste. The waste collected in municipal bins is transported to waste treatment plants or landfill sites managed by civic bodies. Ragpickers affect the material flow of waste streams in terms of segregation and recovery of valuable materials from open dumping sites, community storage bins or from municipal landfill sites. The material recovered from the rag pickers or from the waste treatment plants then reaches the recyclers (Section 2.2.1). The waste processing and treatment stabilizes the waste materials, so that the final residues produced are not harmful, and incapable of further change or able to find ready entry into the various natural biogeochemical cycles (Section 2.2.2). Waste management is undergoing drastic change to offer more options that are sustainable. A short review of recent literature focusing on the operational dimension of SWM in India is presented in this study.

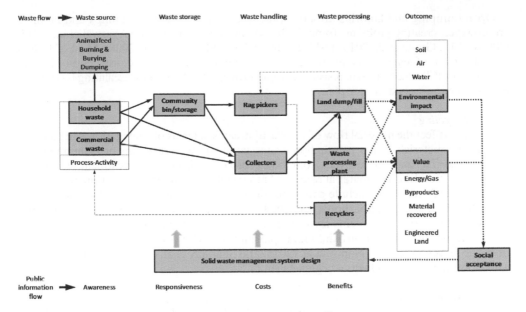

Figure 1.   Schematic of solid waste management practice in India.

2.2.1 *Waste logistics*

First level collection, initial segregation and recycling is undertaken by domestic servants, garbage collectors, rag pickers and waste merchants. Second level collection is completed by non-governmental organizations (NGO), private contractors and welfare associations (Joshi & Ahmed, 2016; Nandy et al., 2015; Sharholy et al., 2008). Modes of storage include storage bins, disposal centers and open collection points. The modes of transport to the transfer station include bullock drawn carts, tractor-trailers, tricycles, wheelbarrows, hand rickshaws, compactors, trucks and dumpers/hydraulic vehicles (Joshi & Ahmed, 2016). Collection and transportation activities constitute approximately 80–95% of the total budget of the municipal SWM system (Sharholy et al., 2008). Interestingly, collection efficiency is higher in the cities and states where private contractors and NGOs are employed for this role (Sharholy et al., 2008). The solid waste flow complexity arises from different sources, waste composition and flow networks which makes waste management very difficult. Waste segregation in the waste storage phase, unregulated waste flow in the waste handling phase (Figure 1) and incompatible and outdated transportation technologies, result in poor waste logistics systems.

2.2.2 *Waste processing and disposal*

MSW processing and disposal practices now include a sophisticated range of options, such as reuse, recycling, incineration with energy recovery, advanced landfill design and a range of alternative technologies. In India, the waste treatment practices include aerobic composting in cities, such as Mumbai, Delhi, Bangalore, Ahmedabad, Hyderabad, Bhopal, Luknow, Vijaywada and Gwalior; vermicomposting in cities, such as Hyderabad, Bangalore, Mumbai and Faridabad; anaerobic digestion systems in Delhi, Bangalore and Lucknow; small incinerators for the burning of hospital waste in various cities; gasification system in Thiruvananthapuram, Nohar and New Delhi; and refuse-derived fuels in Hyderabad, Guntur, and Vijayawada (Agarwal et al., 2015; Joshi & Ahmed, 2016; Kalyani & Pandey, 2014; Sharholy et al., 2008). These have all produced varying degrees of success in terms of economic viability and environmental impact. The nature of the technology for processing is region specific and depends on the waste characteristics. The energy recovery method appears to be the most preferred and valued process, but there is a need to understand and use the learning experiences for generating drivers for technology development and adoption at a macro level.

Open dumping and landfills are both practiced. Unscientific open dumps and landfills are however, creating problems to public health and the environment (Agarwal et al., 2015; CPCB, 2013; Gupta et al., 2015; Joshi & Ahmed, 2016; Sharholy et al., 2008). Most of India's cities/towns are facing problems in the identification of sites for the construction of more sanitary landfills due to public resistance, rapid growth of urban areas, escalating land prices and not having a good master plan.

2.2.3 *Recycling*

Ragpickers affect the material flow of the waste streams in terms of segregation and recovery of valuable materials from open dumping sites, community storage bins or from municipal landfill sites. The material recovery from the rag pickers or from waste treatment plants then reaches the recyclers. The material recovered moves into subsequent production cycle. Recyclable materials recovered include paper, plastic, glass, metals, and e-waste (Joshi & Ahmed, 2016; Nandy et al., 2015; Sharholy et al., 2008). Rag pickers, scrap dealers, waste traders and recycling plants are the elements involved in recycling, and recycling points include houses, open dumps, bus/train stations and municipal landfills. With respect to the value generated from the recycling process, Nandy et al. (2015) and Sharholy et al. (2008) take a positive view while Gupta et al. (2015) takes a negative view.

2.2.4 *Organizational framework*

Solid waste management in India is a part of public health and sanitation, and is an obligatory function of the municipal bodies. At the national policy level, a number of ministries, such as the Ministry of Environment and Forests, Ministry of Urban Development, Ministry of Health and Family Welfare, Ministry of Non-Conventional Energy Sources and Ministry of Agriculture, are responsible for different aspects of solid waste management (Joseph et al., 2012). The Central Pollution Control Board and the State Pollution Control Boards form the administrative core of the SWM sector. Municipal, industrial, biomedical and electronic waste are governed by different laws and policies: the Municipal Solid Waste Rules 2000; the Plastics Manufacture and Usage Rules 2003; the Biomedical Waste Rules 1998; and the Hazardous Wastes Amendment Rules 2000 (Agarwal et al., 2015), for example.

Within the above organizational framework there are four major sectors that operate the SWM system: the public sector; the private-formal sector; the private-informal sector; and the community representatives. The public sector comprises of local authorities and local public departments at city level. The private-formal sector constitutes are large and small registered enterprises doing collection, transport, treatment, and disposal and recycling. The private-informal sector consists of small-scale, non-recognized private sectors which comprise of waste pickers, itinerant waste buyers, traders and non-registered small-scale enterprises. The community representatives are in the form of NGOs.

2.3 *Decision-making*

SWM systems aim for overall waste management which is the best environmentally, economically sustainable for a particular region, and socially acceptable (Agarwal et al., 2015; Guerrero et al., 2013). A balanced coordination among different factors (social, institutional, environmental, financial and technical) is needed to achieve an optimal waste management plan (Guerrero et al., 2013; Srivastava et al., 2014). Decision makers need to be well informed when developing integrated waste management strategies that are adapted to the needs of a city, and consider the ability of citizens to pay for the services (Guerrero et al., 2013). The concept of sustainable waste management is gaining more focus and there is emphasis on multi-criteria decision analysis.

Decisions pertaining to SWM systems are complex as multiple criteria and multi-actors are involved. Facility location problems for transfer stations or treatment plant locations are solved using geographical information systems, mixed integer linear programs, analytic hierarchy process (ANP) and Technique for Order of Preference by Similarity to Ideal Solution (TOPSIS) techniques (Choudhary & Shankar, 2012; Khan & Samadder, 2015; Sumathi et al., 2008; Yadav et al., 2016). Multi-criteria decision-making techniques, such as analytic network process (ANP) and TOPSIS, are used for the selection of alternatives pertaining to disposal methods, power generation technologies and landfill sites (Khan & Faisal, 2008; Kharat et al., 2016; Nixon et al., 2013; Thampi & Rao, 2015). Operation research techniques for a waste management strategy include linear programming, advanced locality models, goal programming, integer programming, neuro-fuzzy inference systems, artificial neural networks models, and mathematical models (Phillips & Mondal, 2014; Rathi, 2007; Singh & Satija, 2015). Operation research techniques can help decision makers to achieve cost savings as well as to improve waste recovery.

Environmental impact assessments are a widely applied tool used in the selection of a disposal site or processing strategies that account for ecological, social, cultural and economical factors. Life cycle assessments have also been widely used to evaluate the environmental benefits and drawbacks of waste management (Kiddee et al., 2013). Economic evaluations (Yadhav et al., 2010), full cost accounting (Debnath & Bose, 2014), regression (Parthan et al., 2012), Strengths-Weaknesses-Opportunities-Threats analysis (Srivastava, 2005), and cost-benefit analysis (Pandyaswargo & Premkumara, 2014) are a few other generic techniques that are used for analysis, selection of plans or assessment of services.

Initiatives by the government, NGOs, private companies and the local public toward sustainable management practices has increased significantly in recent years, and there is a growing emphasis on community participation in decision-making and operational aspects of a SWM system (Joseph et al., 2012; Srivastava et al., 2005; Srivastava et al., 2015). Notwithstanding the recent initiatives, there are numerous operational and sustainability issues in the working of SWM systems. Primary causes of inefficient systems include lack of planning, coordinated efforts, leadership, financial constraints, technological issues, institutional problems within the departments, fragile links with networking agencies, poor public awareness, lack of trained staff and lack of community participation. Such issues arise due to poor alignment between requirements, constraints and operations in the SWM system.

An integrated and sustainable SWM system design needs to account for public awareness on waste, environmental impact, value generated from waste processing and disposal, cost-benefits, social acceptance, and responsiveness of the operational component (Figure 1). The basic design requirements should be: able to handle a variety of waste; environmentally friendly; and economically and socially acceptable. This calls for research and developments in: (i) logistic systems that can handle a variety of waste and are responsive to the public's operational demands; (ii) waste processing and disposal technologies that are economical and environmentally friendly; and (iii) robust decision-making frameworks that enable public participation.

3 CONCLUSION

This paper provides an insight into the SWM systems and practices in India, and the challenges faced. The review focused on three major areas: solid waste characteristics; SWM systems; and decision-making. Based on the literature review, it is evident that SWM is a complex and challenging problem. The two major waste management challenges are the increasing amount of waste and the changing composition of the waste stream. Waste segregation, unregulated waste transportation, and incompatible and outdated transportation technologies result in a poor waste logistics system. There is a need to understand the operational issues in waste processing and disposal so that better technological development and adoption, at a macro level, is made possible.

Decision-making processes pertaining to SWM systems are complex because multiple criteria are involved. Invariably, the decision-making practices and solutions generated in India are unscientific or based on piecemeal approaches. But recent literature provides evidence in the use of multi-criteria decision-making techniques. The poor performance of SWM systems are primarily due to poorly aligned requirements, and constraints and operations. A scientific SWM design is needed which meets the fundamental requirements of being able to handle a variety of waste, be environmentally friendly, and be economically and socially acceptable. A robust integrated SWM system design places emphasis on active public participation in decision-making, and operational integration of subcomponents of a SWM system.

REFERENCES

Agarwal, R., Chaudhary, M. & Singh, J. (2015). Waste management initiatives in India for human well being. *European Scientific Journal, 11*(10).

Central Pollution Control Board (CPCB, 2013). *Status report on municipal solid waste management.* Retrieved from http://www.cpcb.nic.in/divisionsofheadoffice/pcp/MSW_Report.pdfhttp://pratham.org/images/paper_on_ragpickers.pdf.

Choudhary, D. & Shankar, R. (2012). An STEEP-fuzzy AHP-TOPSIS framework for evaluation and selection of thermal power plant location: A case study from India. *Energy, 42*(1), 510–521.

Debnath, S. & Bose, S.K. (2014). Exploring full cost accounting approach to evaluate cost of MSW services in India. *Resources, Conservation and Recycling, 83*, 87–95.

European Business and Technology Centre (EBTC, 2014). *Snapshot: Waste management in India.* Retrieved from http://ebtc.eu/index.php/knowledge-centre/publications/environment-publications/174-sector-snapshots-environment/255-waste-management-in-india-a-snapshot.

Guerrero, L.A., Maas, G. & Hogland, W. (2013). Solid waste management challenges for cities in developing countries. *Waste Management, 33*(1), 220–232.

Gupta, N., Yadav, K.K. & Kumar, V. (2015). A review on current status of municipal solid waste management in India. *Journal of Environmental Sciences, 37*, 206–217.

Joseph, K., Rajendiran, S., Senthilnathan, R. & Rakesh, M. (2012). Integrated approach to solid waste management in Chennai: An Indian metro city. *Journal of Material Cycles and Waste Management, 14*(2), 75–84.

Joshi, R. & Ahmed, S. (2016). Status and challenges of municipal solid waste management in India: A review. *Cogent Environmental Science, 2*(1), 1139434.

Kalyani, K.A. & Pandey, K.K. (2014). Waste to energy status in India: A short review. *Renewable and Sustainable Energy Reviews, 31*, 113–120.

Khan, S. & Faisal, M.N. (2008). An analytic network process model for municipal solid waste disposal options. *Waste Management, 28*(9), 1500–1508.

Khan, D. & Samadder, S.R. (2015). A simplified multi-criteria evaluation model for landfill site ranking and selection based on AHP and GIS. *Journal of Environmental Engineering and Landscape Management, 23*(4), 267–278.

Kharat, M.G., Raut, R.D., Kamble, S.S. & Kamble, S.J. (2016). The application of Delphi and AHP method in environmentally conscious solid waste treatment and disposal technology selection. *Management of Environmental Quality: An International Journal, 27*(4), 427–440.

Kiddee, P., Naidu, R. & Wong, M.H. (2013). Electronic waste management approaches: An overview. Waste Management, *33*(5), 1237–1250.

Kumar, S., Bhattacharyya, J.K., Vaidya, A.N., Chakrabarti, T., Devotta, S. & Akolkar, A.B. (2009). Assessment of the status of municipal solid waste management in metro cities, state capitals, class I cities, and class II towns in India: An insight. *Waste Management, 29*(2), 883–895.

Nandy, B., Sharma, G., Garg, S., Kumari, S., George, T., Sunanda, Y. & Sinha, B. (2015). Recovery of consumer waste in India—A mass flow analysis for paper, plastic and glass and the contribution of households and the informal sector. *Resources, Conservation and Recycling, 101*, 167–181.

Nixon, J.D., Dey, P.K., Ghosh, S.K. & Davies, P.A. (2013). Evaluation of options for energy recovery from municipal solid waste in India using the hierarchical analytical network process. *Energy, 59*, 215–223.

Pandyaswargo, A.H. & Premakumara, D.G.J. (2014). Financial sustainability of modern composting: The economically optimal scale for municipal waste composting plant in developing Asia. *International Journal of Recycling of Organic Waste in Agriculture, 3*(3), 1–14.

Parthan, S.R., Milke, M.W., Wilson, D.C. & Cocks, J.H. (2012). Cost function analysis for solid waste management: A developing country experience. *Waste Management & Research, 30*(5), 485–491.

Phillips, J., & Mondal, M. K. (2014). Determining the sustainability of options for municipal solid waste disposal in Varanasi, India. *Sustainable Cities and Society*, 10, 11–21.

Rathi, S. (2007). Optimization model for integrated municipal solid waste management in Mumbai, India. *Environment and Development Economics*, 12(1), 105–121.

Sharholy, M., Ahmad, K., Mahmood, G. & Trivedi, R.C. (2008). Municipal solid waste management in Indian cities—A review. *Waste Management*, 28(2), 459–467.

Singh, D. & Satija, A. (2015). Optimization models for solid waste management in Indian cities: A study. In K.N. Das, K. Deep, M. Pant, J.C. Bansal, A. Nagar (Eds.), *Proceedings of Fourth International Conference on Soft Computing for Problem Solving* (pp. 361–371). New Dehli, India: Springer.

Singh, R.P., Tyagi, V.V., Allen, T., Ibrahim, M.H. & Kothari, R. (2011). An overview for exploring the possibilities of energy generation from municipal solid waste (MSW) in Indian scenario. *Renewable and Sustainable Energy Reviews*, 15(9), 4797–4808.

Srivastava, V., Ismail, S.A., Singh, P. & Singh, R.P. (2014). Urban solid waste management in the developing world with emphasis on India: Challenges and opportunities. *Reviews in Environmental Science and Bio/Technology*, *14*(2), 317–337.

Srivastava, V., Ismail, S.A., Singh, P. & Singh, R.P. (2015). Urban solid waste management in the developing world with emphasis on India: Challenges and opportunities. *Reviews in Environmental Science and Bio/Technology*, *14*(2), 317–337.

Srivastava, P.K., Kulshreshtha, K., Mohanty, C.S., Pushpangadan, P. & Singh, A. (2005). Stakeholder-based SWOT analysis for successful municipal solid waste management in Lucknow, India. *Waste Management*, *25*(5), 531–537.

Sumathi, V.R., Natesan, U. & Sarkar, C. (2008). GIS-based approach for optimized siting of municipal solid waste landfill. *Waste Management*, *28*(11), 2146–2160.

Thampi, A. & Rao, B. (2015). Application of multi-criteria decision making tools for technology choice in treatment and disposal of municipal solid waste for local self government bodies—A case study of Kerala, India. *The Journal of Solid Waste Technology and Management*, *41*(1), 84–95.

Yadav, I.C., Devi, N.L., Singh, S. & Devi Prasad, A.G. (2010). Evaluating financial aspects of municipal solid waste management in Mysore City, India. *International Journal of Environmental Technology and Management*, *13*(3–4), 302–310.

Yadav, V., Karmakar, S., Dikshit, A.K. & Vanjari, S. (2016). A feasibility study for the locations of waste transfer stations in urban centers: A case study on the city of Nashik, India. *Journal of Cleaner Production*, *126*, 191–205.

Zhu, D., Asnani, P.U., Zurbrügg, C., Anapolsky, S. & Mani, S. (2008). *Improving municipal solid waste management in India: A sourcebook for policymakers and practitioners*. Washington, DC: World Bank Publications.

Emerging Trends in Engineering, Science and Technology for Society,
Energy and Environment – Vanchipura & Jiji (Eds)
© 2018 Taylor & Francis Group, London, ISBN 978-0-8153-5760-5

Powering India's villages sustainably: A case study of Bihar

D. Kamath & A. Anctil
Department of Civil and Environmental Engineering, Michigan State University, East Lansing, Michigan, USA

ABSTRACT: Since independence, the Indian government has been trying to electrify all rural areas—a daunting task. Bihar, with less than 50% of households electrified, has ambitious plans for increased solar power use. This study compared the environmental and economic benefits of centralized and decentralized solar power options to electrify Bihar's rural households. A centralized scenario with utility-scale, photovoltaic plants was compared with decentralized residential rooftop photovoltaic systems. A comparative environmental and cost life cycle assessment was conducted with a functional unit of 1 kWh electricity to a rural household in Bihar. The centralized scenario had lower environmental impacts and costs. However, Bihar's electricity consumption is mainly residential, which could lead to unutilized electricity. Considering this made the centralized scenario the worse option. This study tried to understand the effect of electricity consumption profiles on a system's environmental impacts and costs and the role it plays in policy decisions regarding generation capacity increases.

Keywords: Rural electrification, environmental benefits, Life Cycle Assessment, centralized and decentralized power systems, photovoltaics, rooftop solar, policy, Sustainable Development Goals

1 INTRODUCTION

One of the main goals of the Indian Government, since independence, has been to provide electricity to all its households, especially rural ones. At the village level, six of the 31 Indian states have 100% electrification. However, some other states are far behind. One such state is Bihar. With around 47% of households electrified, the state is planning to increase its generation capacity and transmission and distribution infrastructure (Open Government Data Platform of India, 2017). The electrification process has been slow, leading to gaps in supply, which are being met by mushrooming generator businesses (Oda, 2012).

At present, India is trying to increase the renewable fraction of its energy mix and is aggressively pursuing solar power. The National Solar Mission has a revised aim of deploying 100,000 MW of grid-connected solar power by 2022 (Press Trust of India, 2015). Bihar has also aimed at increasing its installed capacity and is looking forward to solar as its main option (Verma, 2017). The Bihar Government is looking at centralized, utility-scale solar power plants based on photovoltaic (PV) technology as an answer. Two locations, Kajra and Pirpainti, which had been chosen for thermal power plants are now being considered for PV power plants (Verma, 2017). At the same time, India is also trying to increase rooftop PV installations by providing incentives (Ministry of New and Renewable Energy, 2017c). Many studies have compared decentralized PV systems to other methods of power generation, finding them superior economically and environmentally (Gmünder et al., 2010; Molyneaux et al, 2016). A comparison between centralized and decentralized PV systems has not been conducted.

The present study compared solar PV installations in centralized, utility-scale and decentralized, rooftop scenarios to provide electricity for Bihar's rural households. The total capacity was assumed to be 400 MWp, similar to what might be installed at Kajra and Pirpainti. The two scenarios were compared using Life Cycle Assessment (LCA) to understand the environmental impacts and the Levelized Cost of Electricity (LCOE) to understand the economics.

2 METHODS

2.1 Rural electricity use

Two cases were considered for this study. Case 1 provides electricity to the rural households for a restricted time frame (18:00–06:00) and Case 2 for 24 hours. Two synthetic electricity demand curves were developed with a bottom-up approach, with appliance use as shown in Table 1, based on a methodology in a previous study (Blum et al., 2013). Assuming a four-member house hold, this would be on par with Bihar's per capita consumption of 203 kWh in 2014–2015 and 258 kWh in 2015–2016 (Tripathi, 2017).

2.2 System design

For rural electrification using solar power, this study considered centralized and decentralized scenarios, modeled using Homer Pro software (Homer Energy, 2017). A deployment of a total of 400 MWp was modeled. The solar Global Horizontal Irradiance data for the locations were obtained from the US National Renewable Energy Laboratory's National Solar Radiation Database (National Renewable Energy Laboratory, n.d.). The temperature data were obtained from the NASA Surface Meteorology and Solar Energy Database (NASA, n.d.). Subsequently, 17.5% efficiency multi-Si PV panels with a lifetime of 30 years, along with a 98% efficiency inverter, which has a lifetime of 15 years (Fu et al., 2017), were used as inputs. The PV panels were assumed to be latitude tilted. Since the roof angle in India usually lies between 10–30° (B.I.S. IS 875–1987), latitude tilting is easily obtained from the residential rooftops.

The centralized scenario had two PV power plants located in Kajra and Pirpainti, Bihar, selected because of the plans to set up thermal power plants here, which fell through. The land procured for these projects was approximately 1000 acres each (Verma, 2017), providing enough area for a 200 MWp capacity solar power plant at both locations (International Finance Corporation, 2015). Both PV plants were assumed to be connected to the existing

Table 1. Appliance use considered for Bihar's rural households (IIII).

Case		Case 1: 0.92 kWh/day			Case 2: 1.88 kWh/day		
	Power consumption	Quantity per HH	Usage duration per day	Season	Quantity per HH	Usage duration per day	Season
Appliances	W	No.	NA	NA	No.	NA	NA
Light bulb indoor	16	2	18:00–00:00	All year	2	18:00–00:00	All year
Light bulb outdoor	16	1	18:00–06:00	All year	1	18:00–06:00	All year
TV 19″	80	0.2 (1 every 5 HH)	18:00–23:00	All year	1	18:00–23:00	All year
Heater	400	0.2 (1 every 5 HH)	23:00–0:00, 4:00–05:00	Winter	0.6 (3 every 5 HH)	23:00–0:00, 4:00–05:00	Winter
Cooler	300			Summer	0.2 (1 every 2 HH)	22:00–23:00	Summer
Fan	25	0.5 (1 every 2 HH)	22:00–06:00	Summer	1	20:00–21:00, 00:00–06:00	Summer
Refrigerator	100	NA	NA	NA	4 per 30 HH	17:00–09:00	All year
Street light	80	0.2	18:00–06:00	All year	1	18:00–06:00	All year

electricity grid and included a transmission and distribution system, based on the planned transmission and grid system extension in preparation for the thirteenth development plan (Central Electricity Authority, 2016). The energy losses were assumed to be 20% based on various reports (Central Electricity Authority, 2017; The World Bank, n.d.).

The decentralized scenario hadhousehold-level rooftop solar systems with PV panels and lead-acid batteries as storage, because of the reasonable costs (Ministry of New and Renewable Energy, 2017). Loma (Bhijrauli) was taken to be a proxy location to obtain the solar insolation. This system was designed by optimizing for net present cost. The costs of the PV panels, inverter and batteries were assumed based on US benchmark data as proxy and market rates (Fu et al., 2017; Vijayakumar, 2015).

Based on the design requirement for one household, the number of households equaling 400 MWp PV installations was calculated. This demand was assumed to be the first scenario's demand. The centralized and decentralized scenarios were compared with the assumption that both had a total PV installed capacity of 400 MWp and took care of the demand of the same number of households.

2.3 *Life cycle assessment*

LCA was used to compare the greenhouse gas emissions and the embodied energy of the PV systems using the ReCipe Midpoint (Heirarchist) and the Cumulative Energy Demand (CED) method. The life cycle inventories for the PV system and the electricity grid were based on the Ecoinvent 3 database (Wernet et al., 2016). The transmission and distribution system inventory and the lead-acid battery inventory were based on previous publications (Jorge et al., 2012a, 2012b; Rydh, 1999).

For this study, the functional unit, which is the reference unit used to compare systems or products in LCA, is 1 kWh of electricity to a rural household in Bihar. The assessment excluded transportation and the end-of-life of the system.

2.4 *Levelized cost of electricity*

The LCOE was used to calculate the average cost of useful electricity produced. It was calculated by dividing the annual costs incurred during the lifetime of the system (annualized costs) by the electricity produced per year. Its unit was Rs/kWh, which allowed for an easy comparison with conventional electricity costs or tariffs. The annualized costs were based on the net present costs which were the discounted costs incurred in the lifetime of the system, including the capital, operation and maintenance, and replacement costs (Homer Energy, 2017).

The nominal discount rate considered was 11.75% (Clean Development Mechanism Executive Board, 2011) and the expected inflation rate was 4.6% (Organisation for Economic Cooperation and Development, n.d.). Data for the costs incurred in the lifetime of the system were obtained from governmental benchmarks ((Deo et al., 2010a, 2010b; Gakkhar, 2017; Pradhan et al., 2016).

3 RESULTS AND DISCUSSION

3.1 *System design*

Homer Pro was used to find the PV generation from the centralized scenario and the amount of demand that was met by the PV system and the electricity grid. For the decentralized scenario, Homer Pro was used to optimize the system's net present cost. Tables 2 and 3 show the results. The number of households accounting for 400 MWp PV installation was also determined.

For Case 1, 800,000 households were considered as the rooftop PV size was 0.5 kWp, whereas, for Case 2, the number decreased to 266,667. This caused the total consumption for Case 1 with 800,000 households to be higher, even though the household consumption was much lower than that in Case 2.

Table 2. Centralized scenario generation for both locations.

Case	Case 1: 0.92 kWh/day			Case 2: 1.88 kWh/day		
Location	Kajra	Pirpainti	Total	Kajra	Pirpainti	Total
PV (MWp)	200	200	400	200	200	400
Converter (MW)	200	200	400	200	200	400
Total use (GWh/year)	134.3	134.3	268.6	91.5	91.5	183
Electricity needed from grid (GWh/year)	128.6	128	256.5	83.9	83.7	167.6
Electricity sent to grid (GWh/year)	331.7	336.5	668.2	340.4	324.7	665.1

Table 3. Design for decentralized scenario for one household.

Case	Case 1: 0.92 kWh/day	Case 2: 1.88 kWh/day
PV (kWp)	0.5	1.5
Converter (kW)	0.25	0.75
Battery capacity (kWh)	5	10
Electricity sold yearly when grid connected (kWh)	380	1429
No. of HH considered (total installation of 400 MWp PV)	800000	266667

3.2 Life cycle assessment

A comparative LCA was conducted with the existing electricity grid as the baseline. For each case, five scenarios were compared for sensitivity analysis. The five scenarios can be described as follows:

1. Existing electricity grid utilized to meet demand (centralized conventional).
2. Centralized system with all electricity utilized either by households or sent to grid (centralized PV).
3. Centralized system with excess electricity left unutilized (centralized PV excess).
4. Decentralized system, off-grid, causing excess electricity to go unutilized (decentralized off-grid).
5. Decentralized system, on-grid, causing excess electricity to be sent to grid (decentralized on-grid).

A mismatch between the load and the generation is seen. In the centralized scenario, the load is met by the existing electricity grid and the excess electricity from the PV generation is utilized elsewhere (Scenario 2). Scenario 3 is modeled to consider the case when there is no utilization of the excess electricity. In the decentralized system, the lack of match between the load and generation is compensated using batteries (Scenario 4). Scenario 5 considers the utilization of the excess electricity still produced.

The Global Warming Potential (GWP) and CED for all scenarios are shown in Figures 1(a) and 1(b). When the household load was 0.92 kWh/day, as in Case 1, the GWP and CED were maximum for Scenario 3 (centralized PV excess), higher than Scenario 1 (centralized conventional). When the excess electricity was utilized, as in Scenario 2, it was one of the best scenarios, with negative impacts compared to Scenario 1. Scenario 5 offered the most environmental benefits in this case. However, when the load is 1.88 kWh/day, design changes led to Scenario 3 having impacts similar to Scenario 1, and utilizing the excess electricity led to Scenario 2 having the lowest impacts, followed by Scenario 5.

Table 4 describes the scenarios diagrammatically along with their system boundaries, typical demand and PV generation, showing the mismatch between the load and the generation.

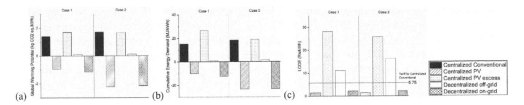

(a) (b) (c)

Figure 1. Figure shows (a) GWP, (b) CED and (c) comparison of LCOE with the present tariff.

Table 4. System boundaries for each scenario with a comparison of demand and PV generation.

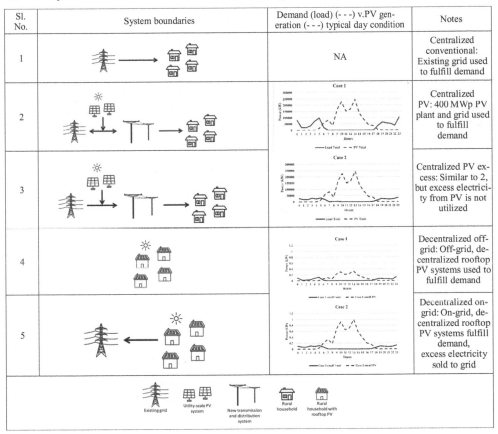

Sl. No.	System boundaries	Demand (load) (- - -) v.PV generation (- - -) typical day condition	Notes
1		NA	Centralized conventional: Existing grid used to fulfill demand
2		Case 1	Centralized PV: 400 MWp PV plant and grid used to fulfill demand
3		Case 2	Centralized PV excess: Similar to 2, but excess electricity from PV is not utilized
4		Case 1	Decentralized off-grid: Off-grid, decentralized rooftop PV systems used to fulfill demand
5		Case 2	Decentralized on-grid: On-grid, decentralized rooftop PV systems fulfill demand, excess electricity sold to grid

3.3 Levelized cost of electricity

Scenarios in which the excess electricity was not sent to the grid, such as with high electricity wastage, had higher LCOE, as seen in Figure 1(c). Similar to the LCA results, Scenario 2 had the lowest LCOE. However, when the excess electricity was not utilized, the costs were much higher in all the scenarios. Comparing both the LCA and LCOE results, the best scenario for both cases was the centralized PV system. However, it is important to consider the unutilized excess electricity, because 66% of Bihar's total electricity generated is consumed by the residential sector (Molyneaux et al., 2016). Hence, the peak demand occurs during the evening when there is no PV electricity generation as shown in Table 4. Bihar is already in a situation where the power generation follows the load profile closely. PV electricity generation without any storage could aggravate this situation, leading to a large amount of unutilized excess electricity, as in Scenario 3 in this study. This situation could be averted with more industrial and commercial projects or grid storage.

Grid storage is picking up in popularity in India recently with a 10 MW grid storage array planned for Delhi (Maloney, 2017). However, it is still not cost competitive when compared to PV projects with no storage, leading to the cancellation of tenders for solar farms with storage (Chandrasekaran, 2017). Until then, decentralized systems, either on- or off-grid, might be a good option for rural electrification, depending on the completion of transmission and distribution projects and access to interconnection points, as well as the level of regional development.

4 CONCLUSION

This study compared rural electrification options for Bihar using PV technology, with the existing grid as the baseline. Overall, the best scenario in terms of GWP, CED and LCOE was the centralized PV system. However, in Bihar, such a system could lead to significant wastage of electricity, increasing the GWP, CED and LCOE compared to the off-grid decentralized PV system. This must be considered when comparing rural electrification options as most of Bihar's consumption is from the residential sector, with a peak demand at times of low or no solar insolation.

The electricity consumption profile could impact the system's environmental impacts and costs, turning a seemingly best scenario to the worst, because of unutilized generated electricity. It plays a crucial role in policy decisions and should be considered when looking at capacity increases for energy generation.

Previous studies have also cited high connection costs leading to low connection rates (Cook, 2011), which is the case when there is no correlation between the consumption and generation profiles. Electricity would be unaffordable and hence, inaccessible to the poor. However, the cost might be reduced by increasing industrial development in the region. Future studies should be aimed to understand the effect of such a development in the area and if the benefits of a rural electrification program would reach the intended population, and what policies would aid this endeavor.

REFERENCES

B.I.S. IS 875–1987, *"Code of Practice for design loads (other than earthquake) for building and structure"*, *Part, 3*. New Delhi.

Blum, N.U., Sryantoro Wakeling, R., & Schmidt, T.S. (2013). Rural electrification through village grids—Assessing the cost competitiveness of isolated renewable energy technologies in Indonesia. *Renewable and Sustainable Energy Reviews, 22*, 482–496.

Central Electricity Authority. (2016). *Bihar: Transmission System Study for 13th Plan*. New Delhi.

Central Electricity Authority. (2017). *Power Sector: Executive Summary for the Month of Jan, 2017*. New Delhi.

Chandrasekaran, K. (2017, July 20). Crashing solar tariffs crush storage plans. *The Economic Times*. New Delhi.

Clean Development Mechanism Executive Board. (2011). *Guidelines on the Assessment of Investment Analysis*. Bonn.

Cook, P. (2011). Infrastructure, rural electrification & development. *Energy for Sustainable Development, 15*(3), 304–313.

Deo, P., Jayaraman, S., Verma, V.S., & Dayalan, M.D. (2010a). *Benchmark Capital Cost for 400/765 kV Transmission Lines*. New Delhi.

Deo, P., Jayaraman, S., Verma, V.S., & Dayalan, M.D. (2010b). *Benchmark Capital Cost for Substation associated with 400/765 kV Transmission System*. New Delhi.

Fu, R., Feldman, D., Margolis, R., Woodhouse, M., & Ardani, K. (2017). *U.S. Solar Photovoltaic System Cost Benchmark : Q1 2017. National Renewable Energy Laboratory.* Golden, CO.

Gakkhar, N. (2017). *Benchmark Cost for "Grid Connected Rooftop and Small Solar Power Plants Programme" for the year 2017–18*. New Delhi.

Gmünder, S.M., Zah, R., Bhatacharjee, S., Classen, M., Mukherjee, P., & Widmer, R. (2010). Life cycle assessment of village electrification based on straight jatropha oil in Chhattisgarh, India. *Biomass and Bioenergy, 34*(3), 347–355.

Homer Energy. (2017). Levelized cost of electricity. *Homer Pro Support Documentation.* Boulder, Colorado: Homer Energy.

International Finance Corporation. (2015). *Utility-Scale Solar Photovoltaic Power Plants: A Project Developer's Guide.* Washington, D.C.

Jorge, R.S., Hawkins, T.R., & Hertwich, E.G. (2012a). Life cycle assessment of electricity transmission and distribution—part 1: power lines and cables. *The International Journal of Life Cycle Assessment, 17*(1), 9–15.

Jorge, R.S., Hawkins, T.R., & Hertwich, E.G. (2012b). Life cycle assessment of electricity transmission and distribution—part 2: transformers and substation equipment. *The International Journal of Life Cycle Assessment, 17*(2), 184–191.

Maloney, P. (2017). AES and Mitsubishi pair up for India's first grid-scale storage project. *Utility Drive.* New York.

Ministry of New and Renewable Energy. (2017). National Solar Mission—An Appraisal. New Delhi: Lok Sabha Secretariat.

Molyneaux, L., Wagner, L., & Foster, J. (2016). Rural electrification in India: Galilee Basin coal versus decentralised renewable energy micro grids. *Renewable Energy, 89*, 422–436.

NASA. (n.d.). Surface Meteorology and Solar Energy Database. Hampton, VA.

National Renewable Energy Laboratory. (n.d.). National Solar Radiation Data Base 1991–2005 Update: Typical Meteorological Year 3. Lakewood, CO.

Oda, H. (2012). Progress and Issues in Rural Electrification in Bihar, India: A Preliminary Analysis. *IDE Discussion Papers.* Chiba, Japan.

Open Government Data Platform of India. (2017). *State-wise Village Electrification state as on date.* New Delhi.

Organisation for Economic Co-operation and Development. (2017). Inflation Forecast (indicator). *OECD Data.*

Pradhan, G.B., Singhal, A.K., Bakshi, A.S., & Iyer, M.K. (2016). *Determination of Benchmark Capital Cost Norm for Solar PV power projects and Solar Thermal power projects applicable during FY 2016–17.* New Delhi.

Press Trust of India. (2015, June 17). Solar Power Target Reset to One Lakh MW. *Business Standard.* New Delhi.

Rydh, C.J. (1999). Environmental assessment of vanadium redox and lead-acid batteries for stationary energy storage. *Journal of Power Sources, 80*(1–2), 21–29.

The World Bank. (2017). Electric Power Transmission and Distribution Losses (% of output).

Tripathi, P. (2017, February 25). No Additional power capacity to Bihar by 2018–19. *The Economic Times.* Patna, Bihar.

Verma, S.K. (2017, August 6). Solar power plants on Nitish radar. *The Telegraph.* Patna, Bihar.

Vijayakumar, S. (2015, May 10). Scaling the Powerwall. *The Hindu.* Chennai, Tamil Nadu.

Wernet, G., Bauer, C., Steubing, B., Reinhard, J., Moreno-Ruiz, E., & Weidema, B. (2016). The ecoinvent database version 3 (part I): overview and methodology. *International Journal of Life Cycle Assessment, 21*(9), 1218–1230.

Geomechanics and foundation engineering

Emerging Trends in Engineering, Science and Technology for Society,
Energy and Environment – Vanchipura & Jiji (Eds)
© *2018 Taylor & Francis Group, London, ISBN 978-0-8153-5760-5*

Role of sodium silicate in strength development of cement treated clayey soil admixed with composite promoter

K. Raj Keerthi & J. Bindu
Department of Civil Engineering, College of Engineering, Thiruvananthapuram, Kerala, India

ABSTRACT: Friendly sodium silicate and promoters, which are compatible with cement are used to obtain improved soil properties. The possibility of using cement and sodium silicate admixed with composite promoters to improve the strength of soft clay was analysed in the present study. The influential factors involved in this study are the proportion of sodium silicate binding agent and the curing time. The unconfined compressive strength of stabilized clay at different ages is tested. Based on literature study, the selected composite promoters for the present study comprises of $CaCl_2$ & NaOH. For the ordinary Portland cement (OPC) and sodium silicate admixed with composite promoters system, the permeation of the $CaCl_2$ and NaOH solutions is expected to facilitate the precipitation of $Ca(OH)_2$ at a molar ratio 1:1 and found significantly improves strength of soft clay. More importantly, it is found that the selected clay stabilizer in much less dosage is needed to achieve the equivalent improvement in strength compared with cement, hence can be a more effective and ecofriendly clay stabilizer.

1 INTRODUCTION

Construction of buildings and other civil engineering structures on weak or soft soil is highly risky because such soil is susceptible to differential settlements due to its poor shear strength and high compressibility. Soil stabilization is the process of improving the physical and engineering properties of problematic soils to some predetermined targets. Sodium silicates have been widely used as supplementary cementing materials substituting ordinary Portland cement to improve the soil properties. OPC is the most commonly used stabilizer since it is readily available at reasonable cost. Nevertheless, a major issue with using OPC is that its production processes are energy intensive and emit a large quantity of CO_2. To improve the environmental acceptability and to reduce the construction cost of the deep mixing method, the partial replacement of the cement by supplementary cementing materials such as sodium silicate is one of the best alternative ways.

Application of sodium silicate for geotechnical works has been reported by many researchers. Used as a component of soil stabilizer, sodium silicate has unique advantages:

(i) reliable and proven performance, (ii) safety and convenient for construction, and (iii) environmental acceptability and compatibility (Rowles and O'Connor, 2003; Ma et al., 2014). In order to investigate the possibility of using cement and sodium silicate admixed with composite promoters to improve the strength of soft clayey soil, Thonnakkal soft clay is considered. These deposits are composed of silty clay, having extremely low shear strength and high compressibility.

The aim of this present study is to achieve an OPC-based clay stabilizer which has the equivalent enhancement of the mechanical properties as a higher content of OPC. The effect of sodium silicate on the strength development of samples stabilized with OPC and composite promoter was investigated. The unconfined compressive strength was used as a practical indicator to investigate the strength development. The binders consisting of OPC, sodium silicate, and composite promoters. The present study aimed to obtain an optimum dosage of sodium silicate and to study its effect on strength development upon curing, since there are conflicts exist on optimum dosage of sodium silicate additive. The use of single promoters used as the component of clay stabilizers have several disadvantages, restrict the application of single promoters in the OPC and sodium silicate system. The composite promoters, comprised of $CaCl_2/Ca(OH)_2$ or $Ca(OH)_2/NaOH$, have no advantage in terms of strength development in comparison with the same addition of single $Ca(OH)_2$ or NaOH. The selected composite promoter (CN) for the present study consists of $CaCl_2$ and NaOH at the mass ratio of 1:1, gives better strength improvement (Cong Ma et al., 2015).

2 MATERIALS

2.1 Clay soil

The samples were collected from Thonnakkal region in the Trivandrum district of Kerala. Index properties of the soils are summarized in Table 1.

2.2 Cement

A 43 grade ordinary Portland cement (OPC) was used in this study. Properties are listed in Table 2.

2.3 Sodium silicate

Sodium silicate (SS), in powdered form is used. It consists of SiO_2 (29.48%) and Na_2O (9.52%).

2.4 Composite promoter

Composite promoter of $CaCl_2$ & NaOH in the molar ratio 1:1 is used for the present study. Sodium hydroxide (NaOH), a flaked solid at room temperature. Calcium chloride ($CaCl_2$), an anhydrous powder, was used as an accelerator. It can serve as a source of calcium ions in the solution. Unlike many other calcium compounds, calcium chloride is soluble. These powdered promoters are all chemically pure and are obtained from Laboratory suppliers Trivandrum.

3 METHODOLOGY

3.1 Sample preparation

The samples were prepared by varying dosages of sodium silicate (0.5%, 1%, 1.5%, 2%, 2.5% and 3%), cement content and composite promoter ($CaCl_2$ & NaOH) dosage were kept constant throughout the test. Cement dosage is fixed as 10% and composite promoter dosage is fixed

Table 1. Basic properties of clay soil.

Soil property	Values
Liquid limit (w_l) (%)	45.2
Plastic limit (w_p) (%)	25.6
Plasticity index (Ip) (%)	19.6
Shrinkage limit (w_s) (%)	20.8
Specific gravity G	2.5
Undrained shear strength, Su (kPa)	124.8
Clay content (%)	55
Silt content (%)	45
IS classification	CI

Table 2. Properties of OPC.

Properties	Values
Specific surface (m^2/kg)	280
Initial setting time (min)	180
Final setting time (min)	260
Compressive strength (MPa)	
3 days	37
7 days	45
28 days	55

as 1% at 1:1 molar ratio for the present study. Samples were prepared at optimum moisture content and maximum dry density obtained from standard proctor test. For each variations of sodium silicate dosage, proctor test were conducted in order to obtain the moulding water content and dry density of samples. Samples were prepared on PVC mould of 4 cm diameter and length of 8 cm (aspect ratio 1:2). The samples was allowed to cure for a required curing period in an air tight moisture proof desiccators maintained at a relative humidity of more than 95%.

3.2 *Experimental programs*

Unconfined compressive strength test (UCS Studies) was conducted to estimate the effect of sodium silicate on the strength development of samples stabilized with OPC and composite promoter (OPC-10% and CN-1%). The UCS tests were carried out according to IS 2720 (part 10):1991. The samples were tested using a compression device fitted with a proving ring for measuring the applied load. The axial displacements were measured using a dial gauge. A constant strain rate was adopted in testing the samples. UCS test were conducted on samples with varying sodium silicate dosage and on varying curing periods.

4 RESULT AND DISCUSSION

4.1 *Compaction results*

The moisture-density (compaction) curves for the cement treated soils were determined using the method described in ASTM D558-04. Figure 2 shows the variation of the maximum dry density with the moisture content for the tested samples. Upon addition of sodium silicate maximum dry density is found to be higher compared with samples treated with cement alone. Also there is a marked reduction in optimum moisture content, thus showing improvement in soil strength. Best compaction characteristics is shown by samples treated with 1%

Figure 1.   UCS experimental set up.

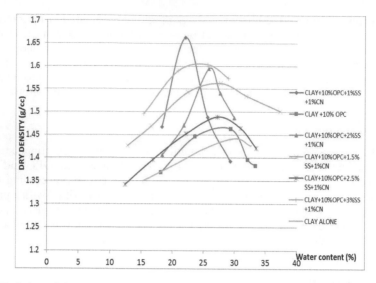

Figure 2. Variation of the maximum dry density with the moisture content for the tested samples.

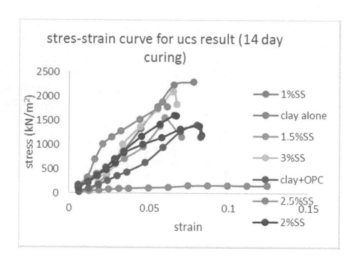

Figure 3. Stress-strain curve for samples treated with different sodium silicate dosage for a curing period of 14 days.

sodium silicate. A percentage increase of 15.35% in dry density and a percentage decrease of 26.7% in moisture content was observed in samples treated with 1% sodium silicate additive.

4.2 UCS test results

4.2.1 Effect of sodium silicate dosage

The UCS tests were conducted with varying percentages of sodium silicate additive 0.5%, 1%, 1.5%, 2%, 2.5% and 3% and samples were prepared for a cement content of 10% and CN of 1%. Figure 3 represent UCS test result for varying sodium silicate additive dosages for a curing period of 14 day. It can be seen that when sodium silicate is used along with cement, the unconfined compressive strength is higher compared with samples treated with cement alone. It observed that the stiffness of the cement treated soil increases after sodium silicate addition and the stiffness continues to increase with higher dosages of sodium silicate. For a particular cement content of 10% and composite promoter (CaCl$_2$ & NaOH) dosage of 1% at 1:1 molar ratio, the optimum sodium silicate dosage is found to be 1%, where strength improvement is maximum. Further increase in sodium silicate dosage from 1%, there is reduction in strength development. Strength development is found to be higher on whole number fraction dosage of sodium silicate when compared to decimal fraction dosage of sodium silicate for a given composite promoter (CaCl$_2$ & NaOH) dosage of 1%.

4.2.2 *Effect of curing on strength development*

Table 3, shows data representing strength development of OPC stabilized clay stabilizer upon curing periods. Strength development of samples increases up on increasing the curing

Table 3. Variation of unconfined compressive strength for different dosages of sodium silicate for a curing period of 14 days.

Sample proportion	14 day unconfined compressive strength (kN/m^2)
Clay alone	163.3
Clay+10%OPC	1407.2
Clay+10%OPC+0.5%SS+1%CN	1327.6
Clay+10%OPC+1%SS+1%CN	2316.2
Clay+10%OPC+1.5%SS+1%CN	1518.0
Clay+10%OPC+2%SS+1%CN	1873.3
Clay+10%OPC+2.5%SS+1%CN	1377.1
Clay+10%OPC+3%SS+1%CN	2073.7

Table 4. Unconfined compressive strength of different samples for varying curing periods.

Sample proportion	Unconfined strength for varying curing days (qu) (kN/m^2)			
	0th day	3rd day	7th day	14th day
Clay alone	124.1	136.2	155.3	163.3
Clay+10%OPC	262.1	882.5	1155.6	1407.2
Clay+10%OPC+0.5%SS+1%CN	136.9	765.9	1327.6	1507.3
Clay+10%OPC+1%SS+1%CN	410.5	1410.2	2316.2	2555.8
Clay+10%OPC+1.5%SS+1%CN	148.7	1176.8	1518.0	1700.0
Clay+10%OPC+2%SS+1%CN	561.6	1378.5	1873.3	2131.2
Clay+10%OPC+2.5%SS+1%CN	171.1	1082.1	1377.1	1650.6
Clay+10%OPC+3%SS+1%CN	142.2	1260.1	2073.7	2290.8

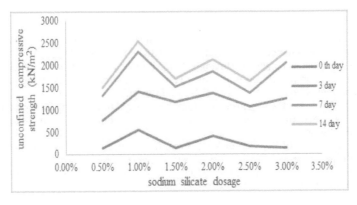

Figure 4. Effect of curing on strength development for samples treated with varying sodium silicate dosages.

Figure 5b.   Failure pattern of sample after curing (at 1% SS content).

Figure 5a.   Failure pattern of sample without curing (at 1% SS content).

period and a significant increase in strength is found after 7 days of curing. More brittle and a sudden failure was observed with curing.

5   CONCLUSION

The conclusions made based on the present study on clayey soil are:

This paper analysed the strength development in OPC and sodium silicate-stabilized clay with composite promoter (NaOH & CaCl₂). NaOH and CaCl₂ as a composite promoter at mass ratio 1:1 along with addition of sodium silicate on cement stabilized clay significantly improves the strength

- Strength development is found to be higher on whole number fraction dosage of sodium silicate when compared to decimal fraction dosage of sodium silicate for a given composite promoter (CaCl₂ & NaOH) dosage of 1%.
- It can be observed that for a particular cement content of 10% and a composite promoter (CaCl₂ & NaOH) at combination of 1:1 molar ratio, 1% sodium silicate was found to be most effective dosage for strength development and observed that upon increasing sodium silicate dosage bonding gel related to sodium silicate become weaker.
- Strength development of samples increases upon increasing the curing period and a significant increase in strength is found after 7 days of curing.

REFERENCES

Bindu, J., and Ramabhadran, A. (2011). "Study on cement stabilized Kuttanad clay." *Proc., Indian Geotechnical Conf.*, Indian Geotechnical Society (IGS), Kochi, India, 465–68.

Cong Maa, Zhaohui Qina, Yingchun Zhuangb, Longzhu Chena, Bing Chena. (2015). "Influence of sodium silicate and promoters on unconfined compressive strength of Portland cement-stabilized clay." *Soils and Foundations* 2015; 55(5): 1222–232, Elsevier.

Mohammad Vali Vakili, Amin Chegenizadeh, Hamid Nikraz, Mahdi Keramatikerman. (2016). "Investigation on shear strength of stabilised clay using cement, sodium silicate and slag." *Applied Clay Science* 03768; Elsevier.

Saroglou I. Haralambos. (2013). "Compressive Strength of Soil Improved with Cement." ascelibrary. org on 06/06/13.

Sina Kazemian, Arun Prasad, Bujang B.K. Huat, Vahed Ghiasi, Soheil Ghareh (2012). "Effects of Cement–Sodium Silicate System Grout on Tropical Organic Soils." Arab J Sci Eng (2012) 37: 2137–2148.

Suganya, K., P.V. Sivapullaiah (2016). "Role of Sodium Silicate Additive in Cement-Treated Kuttanad Soil." *Journal of Materials in Civil Engineering,* ASCE, ISSN 0899-1561.

*Emerging Trends in Engineering, Science and Technology for Society,
Energy and Environment – Vanchipura & Jiji (Eds)*
© 2018 Taylor & Francis Group, London, ISBN 978-0-8153-5760-5

Stabilisation of Kuttanad clay by environment friendly methods—a review

K. Kannan
Department of Civil Engineering, Marian Engineering College, Trivandrum, India

S. Gayathri
Department of Civil Engineering, John Cox C.S.I Memorial Institute of Technology, Trivandrum, India

Abiya K. Jose, Lakshmi P. Nair & Greeshma Das
Department of Civil Engineering, Marian Engineering College, Trivandrum, India

P. Vinod
Department of Civil Engineering, Government Engineering College, Trichur, India

ABSTRACT: Kuttanad clay is a well known soil group, known for its low shear strength and high compressibility, making it almost always unusable and inconstructible on in its natural state. The traditionally used stabilization techniques such as preloading and chemical grouting are unsuitable in the present scenario, as they are either outpaced or affect the environment aggressively. This paper reviews some of the sustainable methods of ground improvement which have been used and reported in highly plastic clays. These techniques can potentially be used in Kuttanad clay also, provided further studies are conducted. The paper also reports the effect of one technique, Microbial Induced Calcite Precipitation (MICP), on the liquid limit of Kuttanad clay.

1   INTRODUCTION

When only poor quality soil is available at the construction site, the best option is to modify the properties of the soil so that it meets the design requirements. The process of improving the strength and durability of soil is known as soil stabilization. It is the alteration of soils to enhance their physical properties. Stabilization can increase the shear strength and control the shrink-swell properties, thus improving the load bearing capacity of the soil.

Kuttanad is situated in the central half of Kerala covering an area of approximately 1100 sq.km and lies 0.6 m to 2.2 m below the mean sea level. Kuttanad clay is an important soil group, well known for its low shear strength and high compressibility. Soil in this region is soft black or grey marine clay composed of minerals such as montmorillonite, kaolinite, iron oxide and aluminum oxide (Vinod and Bindu, 2010). The natural water contents of this soil are very high and close to liquid limit, sometimes even exceeding it. The typical kuttanad soil consists primarily of silt and clay fraction. It is a weak foundation material, with a number of failures to structures and embankment reported. Since Kuttanad is the rice bowl of Kerala, any ground improvement technique adopted in this region should be eco-friendly and should never cause any harm to the environment, especially to the soil and water. Thus, the presently used methods of physical improvement such as preloading and chemical improvement by addition of adulterants might prove to be inefficient in the present scenario, wherein the focus is on fast sustainable technologies. All cement based techniques may seem harmless to the public eye, but add to carbon footprint heavily during manufacture. The improvement of subsoil using alternative biological or ecofriendly chemical methods is thus a growing concern, and the focus of the present paper. In particular, the paper focuses on some alternative sustainable techniques which can be potentially utilized to improve Kuttanad clay.

Table 1. Constituents of Bagasse Ash (Kharade et al., 2014).

Constituent	Percentage
Silica (SiO$_2$)	64.38
Magnesium Oxide (MgO)	0.85
Calcium Oxide (CaO)	10.26
Iron Oxide (Fe$_2$O$_3$)	4.56
Sodium Oxide (Na$_2$O)	1.05
Pottassium Oxide (K$_2$O)	3.57
Alumina (Al$_2$O$_3$)	11.67

2 BAGASSE ASH

Bagasse is a residue obtained from the burning of bagasse in sugar producing factories. Bagasse is the cellular fibrous waste product after the extraction of the sugar juice from cane mills. For each 10 tons of sugarcane crushed, a sugar factory produces nearly 3 tons of wet bagasse which is a byproduct of the sugar cane industry (Kharade et al., 2014). When this bagasse is burned the resultant ash is bagasse ash. Bagasse shows the presence of amorphous silica, which is an indication of pozzolonic properties, responsible in holding the soil grains together for better shear strength. Pozzolanic material is very rich in the oxides of silica and alumina and sometimes calcium. Pozzolans usually require the presence of water in order for silica to combine with calcium hydroxide to form stable calcium silicate which has cementitious properties, which can then develop good bonding between soil grains in case of weak soil. Table 1 shows the constituents of bagasse ash (Kharade et al., 2014).

2.1 *Effects of bagasse ash on clays*

The stabilization of soil using bagasse ash in different percentage has been reported. With the increase of bagasse ash the MDD of mix is reported to be decreasing and CBR value of the mix is increasing up to an optimum percentage of bagasse ash in mix. As CBR value increased, the strength characteristics of clayey soil also increased (Kanchan & Jaiwid, 2013). The results of different tests which are conducted with varying percentage of bagasse ash to check the effect on swelling pressure and on basic properties shows that the bagasse ash effectively dries wet soils, by producing a significant reduction in water content, and provides an initial rapid strength gain. It also decreases the swell potential of expansive soils by replacing some of the volume and by cementing the soil particles together (K.S. Gandhi, 2012). From the reported tests, it was obtained that the index properties like Atterberg limits and Plasticity Index, and free swell index decreases to a lowest value at 20% bagasse ash for black cotton. The bagasse ash is added in lower percentage and it gives good results, showing that bagasse ash is an effective adulterant (K.S. Gandhi, 2014).

3 BIOENZYME

Bioenzyme in soil stabilization have emerged as a revolutionary technique which is becoming popular worldwide. Bioenzymes are nontoxic, nonflammable substances that are obtained from plant extracts (Saini and Priyanka, 2015). Bioenzyme reduces the voids between the particles of soil and minimizes the amount of absorbed water in the soil so that compaction can be maximum. These enzymes have been proven to be very effective and economical. The use of bioenzyme in soil stabilization is not very popular, mostly due to a lack of awareness. Reported literature shows soil types stabilized by the Bioenzymes include sandy clay, silty clay, sandy silt, plastic and non-plastic clay, sandy loam, fine loam, and loam mixed with clay. The dosage levels of Bio-enzymes vary from 1 to 5 litres for 5 m^3 of soil depending on the soil type and characteristics. The amount of

Table 2. Properties of Terrazyme (Shirsath et al., 2017).

Property	Value/specification
Boiling point	212°F
Specific gravity	1 to 1.09
pH value	4.3 to 4.6
Hazardous content	None
Appearance	Brown colored liquid
Reactivity Data	Stable

dilution water depends on in-situ moisture content of soil. There are many bioenzymes available namely renolith, permazyme, terrazyme, fujibeton etc. Most commonly used one is terrazyme.

3.1 *Terrazyme*

Terrazyme modified from vegetable extracts is specially formulated to modify the engineering properties of soil, usually applied as a mixture with water. Terrzayme acts on the soil by reducing the voids between the particles minimizing the adsorbed layer of water and maximizing compaction. It reacts with the organic materials to form cementatious material, bringing about a decrease in the permeability and increase in chemical bond, creating a permanent structure resistant to weather, wear and tear (Gupta etal., 2017). Table 2 shows the properties of terrazyme (Shirsath et al., 2017).

3.2 *Mechanism of terrazyme*

Soil (clay) has surface negative charge particles making the soil attracted to the positive charges and get neutralized. It is found that the adsorbed layer of water contains metals like Na, K, Al, Mg which is the reason for the bond between negative clay particles and water molecules. Thus a significant layer of water is created around soil particles and to attain compaction it is necessary to eliminate the water layer around the particle.

Terrazyme reduces the dielectric charge in water molecule thus creating pressure on the positive metal ions to release the free water. This breaks the electrostatic potential barrier thereby reducing the thickness of the diffused double layer so that the soil particles come closer and attain greater compaction with less compactive effort (Eujine et al, 2014).

3.3 *Effects of terrazyme on clays*

Effect of terrazyme is different for different types of soil and mainly depends on dosage and curing time. With the addition of terrazyme a significant increase in the values of both soaked and unsoaked CBR has been observed, by as much as 4 times for clayey silts (Saini and Vaishnava, 2015). This is because of increased compaction creating stronger bond which resist penetration. The OMC and consistency limits have also been found to decrease, indicating the denseness of the soil. Permeability of the soil also has decreased with the increase in curing time compared to that of sample untreated with terrazyme. It is due to the decrease in voids after enzyme action thereby not allowing the water to flow through the soil (Gupta etal., 2016). Terrazyme decreased liquid limit by as much as 28% and shrinkage limit by 30% in two weeks on high liquid limit clays, but had little effect on plastic limit (Eujine et al., 2014). Effect of terrazyme for different types of soil varies and it mainly depends on dosage. Studies have been reporting the variations on the geotechnical properties for different dosages of terraazyme for (type of soil) (Saini and Vaishnava, 2015).

231

4 BIOCEMENTATION

An alternative to the addition of chemicals such as bagasse ash and various bioenzymes is by the use of some biomediated stabilization technique. Microbial Induced Calcite Precipitation (MICP) is one such technique wherein some bacteria are used to catalyse chemical reactions, resulting in precipitation of calcite within the soil pores. MICP can be induced in soils by many methods, but are usually simulated by one of the four methods—urea hydrolysis, denitrification, iron reduction, and sulphate reduction. Predominance of these various mechanisms depends on their associated reaction's propensity to occur in the environment. Ureolysis is predominant in manipulated soil amongst others because the reaction changes the environmental conditions of a system (i.e. increase in pH), which inhibits other competitive processes. The basic concept of urea hydrolysis involves the hydrolysis of urea to produce carbonate, which combine with the supplied calcium substrate to precipitate calcite. A bacterium with an active urease enzyme is chosen for this purpose, which when introduced along with urea, hydrolyses it into ammonia and carbonate.

$$CO(NH_2)_2 + 2H_2O \rightarrow 2NH_4^+ + CO_3^{2-} \tag{1}$$

The produced carbonate ions precipitate in the presence of calcium ions as calcite crystals, which form cementing bridges between the existing soil grains.

$$Ca_2^+ + CO_3^{2-} \rightarrow CaCO_{3(s)} \downarrow \tag{2}$$

Calcium carbonate thus formed is precipitated at particle-particle interfaces and thus bring improvement in engineering properties of the treated soil into picture. Further, once precipitated, the calcium carbonate will only dissolve very slowly (at a geological time scale), either when continuously flushed by buffered acidic groundwater or as a result of acidifying processes in the pores (e.g. degradation of biomass). When sufficient calcium carbonate is precipitated, durable soil stabilization can be achieved. (Van Paassen et al., 2009).

4.1 *Effect of MICP on marine clays*

Very few literature is available on the effect of MICP on clays, much less in marine clays. Rebata-Landa (2007) showed a relation between grain size and $CaCO_3$ content, and maximum carbonate deposition observed on grains was approximately 100 μm in size. This precisely puts Kuttanad clay at a potentially viable range, as Kuttanad clay although clay contains large quantities of silt. Vinod et al. (2017) have identified a suitable laboratory application procedure for MICP in clays. Kannan et al. (2017) have identified optimums for reagent concentrations and treatment durations in a laboratory scale for marine clays on the basis of variation of liquid limit.

Tests were conducted to identify the variation of liquid limit upon treatment in Kuttanad clays. Liquid limit tests were conducted as per IS 2720 part V – 1985 on 4 clay samples collected from across Kuttanad region both before treatment and 15 days after treatment. Results are tabulated in Table 3.

The clear reduction in liquid limit brought about by the treatment is indication of the effectiveness of biomediated techniques, in particular MICP, as stabilization techniques for Kuttanad clays.

Table 3.   Effect of MICP on liquid limit of Kuttanad clay.

Sample	Initial w_L (%)	Final w_L (%)
1	118.8	88.3
2	126	95.9
3	106.4	80
4	174.2	132.3

5 BIOENCAPSULATION

The dredged or excavated marine clays can be strengthened by an innovative method known as bioencapsulation, which can convert the dredged clay wastes into value added construction materials. It is a process to increase strength of soft clayey soil through the formation of strong shell around a piece of soft material by the action of urease producing bacteria (UPB). Thus, bioencapsulation can be a more effective alternative to MICP for improving Kuttanad clay for applications as a fill material, especially for pavement purposes.

5.1 Mechanism of bioencapsulation

As reported in the literature of Ivanov et al., (2015), marine clay samples were subjected to various tests such as unconfined compression tests, slaking tests etc. before and after treatment. Primarily, the clay was mixed with dry biocement and then fed into a granulator to produce 5 mm clay balls. The clay balls were immersed into a solution of calcium salt and urea for bioencapsulation. The clay aggregates were then removed from the solution and incubated for one day before testing. The calcium concentrations were reportedly measured by optical emission spectrometry.

5.2 Effect of bioencapsulation on clays

Reported study shows that the unconfined compressive strength increased from zero to a considerable value of about 40 kPa, which resembles the strength of a sandstone. The strength of aggregates mainly came from the shell, as the change in strength of nucleus is not much. The water content of clay aggregates decreased from 55% to 41% after bioencapsulation, which was partially due to the use of water for urea hydrolysis. This decrease in the water content led to an increase in the compressive strength. From the slaking test, it was reported that the clay aggregates with UPB showed much higher resistance to disintegration than untreated clay balls. But the strength of bioencapsulated aggregates decreased with an increase in the size more than 5 mm (Ivanov et al., 2015).

6 CONCLUSIONS

This paper reviews some of the environmental friendly methods that can be potentially used to stabilise Kuttanad clays. Consideration of soil as a living ecosystem offers the potential for innovative and sustainable solutions to geotechnical problems. This is a new paradigm for many in geotechnical engineering. Realising the potential of this paradigm requires a multidisciplinary approach that embraces biology and geochemistry to develop techniques for beneficial ground modification. The suggested biological methods possess the potential as mentioned, but needs fine tuning to potentially apply it in a field scale. The laboratory study showed a significant reduction in liquid limit, which underlines the potential of the said techniques. The implementation of the same will thus take time. But when compared with the presently used cement-based techniques, thought to be harmless in spite of the energy inducive, carbon producing manufacturing process, these techniques go a long way into building a sustainable and energy efficient future. Bioencapsulation is a modification further to the biological method which could be an even better alternative. There are undoubtedly many such processes yet to be discovered and further research is required to delineate them. Clear however, that the biological processes influence engineering soil properties and is the future of ground improvement, especially improvement of Kuttanad Clay.

REFERENCES

Eujine, G.N., Somervell, L.T., Chandrakaran, S., & Sankar, N. 2014. Enzyme Stabilization of High Liquid Limit Clay. *European Journal of Geotechnical Engineering* 19: 6989–6995.

Gandhi, K.S. 2012. Expansive Soil Stabilization Using Bagasse Ash, *International Journal of Engineering Research and Technology* 1(5): 1–3.

Ivanov, V., Chu, J., Stabnikov, V., & Li, B. 2015. Strengthening of soft marine clay using biocementation. *Marine Georesources and Geotechnology* 33(4): 320–324.

Kanchan, L.S., & Jawaid, S.M.A. 2013. Geotechnical Properties of Soil Stabilized with Bagasse Ash, *International Journal of Biological Research and Technological Research* 1(9): 5–8.

Kannan, K., Bindu, J. & Sajna, S. 2017. Stabilisation of Marine Clays by MICP: Optimum Reagent Concentration and Treatment Duration, *Proceedings of 3rd International Conference on Materials Mechanics and Management (IMMM 2017), Thiruvananthapuram.*

Kharade, A.S., Salunkhe, S.K., & Dadage, M.D. 2014. Effective Utilization of Sugar Industry Waste Bagasse Ash in Improving Properties of Black Cotton Soil. *International Journal of Engineering Research and Technology* 3(4): 265–270.

Rebata-Landa, V. 2007. Microbial activity in sediments: Effects on soil behavior, Ph.D. dissertation, Georgia Institute of Technology.

Saini, V. & Vaishnava, P. 2015. Soil Stabilization by Using Terrazyme. *International Journal of Advances in Engineering & Technology* 8(4): 566–573.

Shirsath, H.A., Joshi, S.R., & Sharma, V. 2017. Effect of Bio-Enzyme (Terrazyme) on the Properties of Subgrade Soil of Road. *Proceedings of the International Conference on Recent Trends in Civil Engineering, Science and Management*: 53–58.

Van Paassen, L.A., Harkes, M.P., van Zwieten, G.A., Van der Zon, W.H., Van der Star, W.R.L., & Van Loosdrecht, M.C.M. 2009. Scale up of BioGrout: a biological ground reinforcement method. *Proceedings of the 17th International Conference on Soil Mechanics and Geotechnical Engineering*: 2328–2333.

Vinod, P., & Bindu, J. 2010. Compression Index of Highly Plastic Clay—an Emperical Correlation. *Indian Geotechnical Journal* 40(3): 174–180.

Vinod, P., Bindu, J. and Kannan, K. 2017. Potential Utilisation of MICP in clays; Development of a Suitable Application Method, *Proceedings of International Conference on Geotechniques for Infrastructure Projects (GIP 2017), Thiruvananthapuram.*

Emerging Trends in Engineering, Science and Technology for Society,
Energy and Environment – Vanchipura & Jiji (Eds)
© 2018 Taylor & Francis Group, London, ISBN 978-0-8153-5760-5

Development of critical state line concept from hypoplastic model simulations on triaxial strength of silty sand

M. Akhila, Kodi Ranga Swamy & N. Sankar
Department of Civil Engineering, NIT Calicut, Kerala, India

ABSTRACT: Earthquakes can cause liquefaction-induced damages of civil infrastructures supported on the foundation soil. Hence it is required to analyze the susceptibility of lique-faction in foundation soils including sands, silts, and low plastic clays. Based on Critical State Line (CSL) concept, it is possible to identify whether the soil can be susceptible to liquefac-tion or not. The critical state line can be developed based on drained triaxial compression testing data. In this study, CSL of silty sand is developed from drained triaxial test simula-tions based on the hypoplastic model.

1 INTRODUCTION

The liquefaction may occur in fully saturated sands, silts, and low plastic clays. When the saturated soil mass is subjected to seismic or dynamic loads, there is a sudden build-up of pore water within a short duration. If the soil could not dissipate the excess pore pressure, it will result in a reduction of the effective shear strength of soil mass. In this state, the soil mass behaves like a liquid and causes large deformations, settlements, flow failures, etc. This phenomenon is called soil liquefaction. As a result, the ability of soil deposit to support the foundations of buildings, bridges, dams, etc. are reduced. Liquefiable soil also exerts a higher pressure on retaining walls, which can cause them to tilt or slide. The lateral movement could prompt settlement of the retained soil and distraction of structures constructed on various soil deposits. A sudden build-up of pore water pressure during earthquake also triggers land-slides and cause the collapse of dams. Liquefaction effects on damages of structures are com-monly observed in low-lying areas near the water bodies such as rivers, lakes, and oceans.

The CSL is the boundary line which separates the liquefiable and non-liquefiable soil states (Kramer, 1996). Vu To-Anh Phan et al. (2016) have studied the critical state line (CSL) of sand-fine mixtures experimentally. Their result indicates that there is a unique line CSL obtained with specific fines content for various confining pressures and different initial glo-bal void ratios. Aminaton Marto et al. (2014) suggested that neither the fines percentages nor other corresponding compositional characteristics are adequate to be correlated with the critical state parameters of sand matrix soils. Sadrekarimi A and Olson S M (2009) used ring shear tests to find the CSL of soils as the limited displacement that the triaxial device is capable of imposing on a specimen is insufficient to reach a critical state where particle rearrangement and potential crushing are complete. The present paper attempts to use CSL developed from the hypoplastic model simulations to analyze the liquefaction susceptibility of silty sand under static triaxial loading.

2 HYPOPLASTIC MODEL

It is found that hypoplastic models are more advanced to elastic-plastic models for con-tinuum modelling of granular materials. In contrast to elastoplastic models, a decomposition of deformation components into elastic and plastic parts, yield surface, plastic potential, flow rule and hardening rule are not needed in the hypoplastic model. The hypoplastic constitutive law represents the deformation behaviour of cohesionless soils up to certain fines, including the nonlinearity and inelasticity. The first version of the hypoplastic constitutive model was proposed by Kolymbas (1985). It used a single state variable, the current Cauchy stress T_s. Later another state variable the void ratio e was added. The hypoplastic constitutive equation in general form is given by:

$$\dot{T}_s = F\left(T_s, e, D\right) \tag{1}$$

Herein $\dot{T}_s$ represents the objective stress rate tensor as a function of the current void ratio e, the Cauchy granulate stress tensor T_s and the stretching tensor of the granular skeleton D. $D = (L + L^T)/2$ is the symmetric part of the deformation gradient $L = \frac{\partial v(x,t)}{x}$, v being the velocity vector of the continuum representing the grain skeleton at a point x.

3   MODEL PARAMETERS

The silty sand used in this study is processed by mixing 40% quarry dust into the fine sand. The fine sand is collected from Cherthala, Kerala and quarry dust is procured from Blue Diamond M-sand manufacturers, Kattangal, Kerala. All the basic properties tests were performed on the soil combinations, and the properties are listed in Table 1. A combined dry sieve and hydrometer analysis were carried out to obtain the particle size distribution (Figure 1).

The standard routine laboratory tests were conducted on the non-plastic silty sand to determine the model parameters. The limited eight numbers of hypoplastic model parameters (φ_c, h_s, n, e_{i0}, e_{d0}, e_{c0}, β, and α) are determined based on the detailed procedure for the determination of model parameters explained by Herle and Gudehus (1999).

3.1   Critical state friction angle

The critical friction angle φ_c can be obtained from the angle of repose of dry granular material from cone pluviation tests. If the portion of grains with diameter <0.1 mm is too large, conventional shear tests are recommended for determination of φ_c. Hence, for the present study, direct shear box tests were conducted on loose soil at different consolidation pressures to determine the critical friction angle φ_c. Direct shear box tests were conducted on silty sand

Table 1.   Basic properties of silty sand.

Index property	Value
Specific gravity	2.69
D_{10} (mm)	0.03
D_{50} (mm)	0.15
C_u	7.33
C_c	0.74
γ_{max} (kN/m³)	17.56
γ_{min} (kN/m³)	13.93

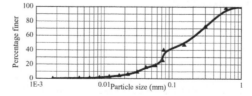

Figure 1.   Particle size distribution curve of silty sand.

samples of loosest and densest possible conditions at different consolidation pressures. The critical friction angle of a sample is determined from the slope of linear failure envelope in loosest possible state.

3.2 Minimum, maximum and critical void ratios at zero pressure state

Based on limited densities, the maximum and minimum void ratios were estimated by using the empirical equations. Three limit void ratios at zero pressures, i.e., e_{i0} (during isotropic compression at the minimum density), e_{c0} (critical void ratio) and e_{d0} (maximum density) were estimated.

3.3 Peak state coefficient

The exponential coefficient of peak state was determined after conducting a series of direct shear box tests on the silty sand in densest possible condition at different consolidation pressures. The peak friction angles of each sample are determined from the slope of linear failure envelope in the densest possible state, and they used to estimate the peak state exponent coefficient of model parameter (α).

3.4 Hardness coefficient and exponent "n."

Oedometer compression tests were carried out on the silty sand in loosest possible condition to arrive the hardness parameters. From this experimental e-log p curve and by using mathematical formulations the hardness parameters of silty sand was determined.

3.5 Stiffness parameter

Oedometric compression tests were carried out on the silty sand in loosest and densest possible conditions (see Figure 2) to arrive the stiffness parameter. From the experimental e-log σ curves and by using mathematical formulations the stiffness parameter is found. All the model parameters are listed in Table 2.

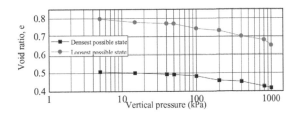

Figure 2. Particle experimental e-log p curves of silty sand in both loosest and densest states.

Table 2. Model parameters.

Parameter	Value
ϕ_c	33°
e_{do}	0.413
e_{co}	0.890
e_{io}	1.068
A	0.035
h_s	43 MPa
N	0.509
B	0.5

4   VALIDATION OF THE MODEL

The element test program has been prepared by Herle using mathematical formulations involved in Hypoplastic soil model. The test program requires three input files namely material parameters, initial state parameters and test conditions. Initially, the hypoplastic model simulations were performed on oedometric compression of both the loose and dense silt soil samples.

The overlapped curves of both the experimental as well as model simulated tests are presented to check the validity of the model. Figures 3 shows the combined overlapped e-log p curves under oedometeric loading. It can be seen that the model simulation results well coincide with experimental curves.

Before performing the numerical simulations on consolidated triaxial compression, the numerical model is validated with the experimental data of CD triaxial test conducted on the silty sand at a void ratio of 0.5 corresponding to the relative density of 85%. The overlapped stress-strain relationships on silty sand from both the experimental and numerical model are presented in Figure 4. It can be seen that the model simulation results well coincide with experimental curves. Therefore, the present model study is extended to perform the triaxial loading simulations under the drained conditions to examine the liquefaction susceptibility of silty sand based on CSL concept.

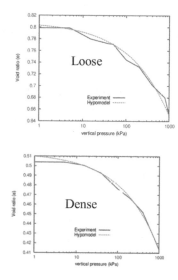

Figure 3.   e-log p curves on loose and dense silty sand under oedometeric compression loading.

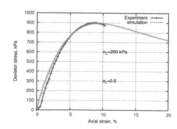

Figure 4.   Comparison of experimental and numerically simulated stress-strain relationship on the silty sand at e = 0.5.

5 RESULTS AND DISCUSSIONS

5.1 *Effect of density*

To study the effect of void ratio on the drained response of silty sands, a series of static tri-axial test simulations were performed on silty sand subjected to consolidation pressures of 100, 200 and 400 kPa by varying the void ratios in the range of 0.5 to 0.98. (Due to space limitations, only results of 200 kPa are presented as graphs in this paper). This range of void ratios was chosen in such a way that the soil state has to change from very loose to the dense condition. It is observed that the dense silty sands experience the dilation behaviour by indicating the sharp peak deviator stress and increase in the volume of soil, i.e., undergo a sudden expansion.

However, loose silty sands experience the contraction behaviour that shows the continuous decrease in volume, i.e., compression. At all tested consolidation pressures, the soil behaviour changes from dilation to contraction (liquefaction) state with an increase in the void ratio. In dense silty sands, it exhibits initial contraction up to a certain axial strain limit and then causes to volumetric expansion leads to dilation. The reason may be due to the re-adjustment of solid particles.

The deviator stresses are increasing with decrease in void ratios at the same axial strain level. A sharp peak stress was observed for dense silty sands at the low strain levels of 7–12% and strain level at the peak stress is increasing with increase the consolidation pressure. For loose silty sands ($e_c = 0.98$~0.804), the constant ultimate stress is reached at the limited failure strain of 20% level. Herein, 20% failure strain is assumed as the critical state level that indicates where the deviator stress almost becomes constant. For the medium dense silty sands, ($e_c = 0.65$), a slight peak deviator stress is observed at low strain levels and further decreased to residual values.

From the volume change responses of silty sand at different void ratios consolidated under each pressure application of 100, 200 and 400 kPa respectively. It demonstrates that the loose silty sands are exhibiting fully contraction behaviour, i.e., volume reduction. The dilative behaviour, i.e., volume expansion is increasing with decrease in the void ratios from 0.65 to 0.5 representing the state of soil changes from medium dense to very dense state. In loose silty sands, the quantitative values of shear strength and volume changes are increasing with increasing the applied consolidation pressures. However, in medium and dense silty sands,

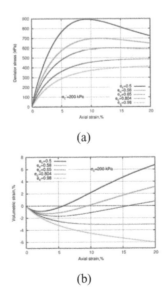

(a)

(b)

Figure 5. (a) Stress-strain characteristics and (b) volume change response of silty sand (at different void ratios and $\sigma_3 = 200$ kPa).

the amount of expansion volume change is decreases with increasing the applied pressures. In summary, the result concludes that the drained response, i.e., either contraction or dilation (liquefaction susceptibility) is dependent on density/void ratio of the soils consolidated at particular low, medium, and high pressures.

5.2   *Effect of consolidation pressure*

To study the effect of consolidation pressure on the drained triaxial response of loose, and dense silty sands, the numerical simulations were performed on silty sands consolidated at different pressures in the range of 50 to 400 kPa. The response of loose silty sand is presented in Figure 6 and discussed here.

Figure 6 shows the drained response in loose ($e_c = 0.98$) silty sands at different consolidation pressures. Figure 6(a) shows the stress-strain characteristics of silty sand consolidated at different pressures. It can be inferred from the figure that the deviator stresses are increases with increase in applied consolidation pressures. It is observed that the ultimate residual constant stress state was reached at about 20% strain level. In the previous section, the effect of density on the drained response is explained in terms of stresses because the tested samples are subjected to unique confining pressures. The effect of density on trends of stresses (stress ratios) is similar due to the unique applied pressure. However, the effect of confining pressure is to be expressed in terms of normalized stress ratios due to the test samples are subjected to different pressures.

For example, the effect of confining pressure on the drained response in terms of stress-strain relation shown in Figure 6(a) is compared with drained response in terms of normalized stress ratios varies with axial strain shown in Figure 6(b). It can be seen from Figure 6(b)

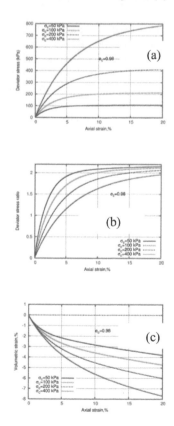

Figure 6.   (a) Stress-strain characteristics, (b) Stress ratio characteristics and (c) Volume change response (loose silty sand consolidated at different pressures).

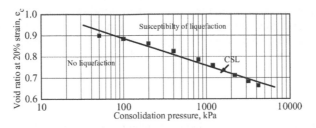

Figure 7.   Identification of liquefaction susceptibility from CSL.

that the normalized stress ratios are decreasing with an increase in consolidation pressures indicate the reverse trends compared to Figure 6(a). In loose silty sands, the ultimate constant deviator stress ratios are obtained at a large strain level of 20% that indicates the silty sand behaves as contraction and stress ratio is more at low consolidation pressures.

Figure 6(c) presents the volume change response of silty sand consolidated at different pressures. It demonstrates that the low consolidated silty sands ($\sigma_c = 50$ kPa) exhibiting less contraction behaviour i.e., less volume reduction. The stress ratios are high at low consolidation pressures due to less contraction behaviour. However, the more contraction behaviour is observed at a high consolidated pressure of 400 kPa. It indicates the volume reduction is increasing with increase in applied pressures from 50 to 400 kPa. The result concludes that the soil behaviour changes from less contraction to highly contraction state in loose silty sands while increasing the applied pressures. The high contraction may take place due to the crushing of particles under the application of high pressures. The contraction soils are not stable and susceptible to liquefaction.

5.3   Critical state line

From the hypoplastic simulations, critical state void ratios are determined at constant volume change corresponding to 20% failure strain level under each application of consolidation pressure on the silty sand in loosest possible condition ($e_c = 0.98$). CSL is the line connecting the critical state void ratios at consolidation pressures in the range of 50 to 4200 kPa. As the range of the consolidation pressures is very large, for better representation, a log scale is required to make the linear relation between pressures and void ratios as shown in Figure 7. It can be seen that the initial critical void ratio at loosest state decreases with increasing the applied pressures consolidated by the soil. The trends are well matched with published research works (Kramer, 1996).

6   SUMMARY AND CONCLUSIONS

In this study, CSL of silty sand is developed from drained triaxial test simulations based on the hypoplastic model. The major findings from the study are given below:

- The effect of density on the drained response of contraction and dilation state of silty sand depends on the applied range of consolidation pressures; similarly, the effect of consolidation pressure on the response of contraction and dilation state of silty sand again depends on denseness of soil.
- The dense silty sands experience the dilation behaviour by indicating a continuous increase in deviator stress to higher values and decrease in pore water pressure towards negative values. However loose silty sands experience the contraction behaviour that shows the continuous increase in pore water pressures and reduction in deviator stress.
- A sharp peak stress was observed for loose silty sands at the low strain levels of 2–4% and then further decreases towards residual stress levels. For dense and medium dense silty sands the continuous increase in deviator stress was observed up to the failure strain level of 25%.

- The CSL was developed from the triaxial drained response of silty sands. CSL act as the boundary to separate the state of no liquefaction and liquefaction regions. The soil behaviour change from the dilative state (No liquefaction) to contraction state (Susceptibility of liquefaction) depends on both the void ratio and applied consolidation pressure.

REFERENCES

Aminaton Marto, Choy Soon Tan, Ahmad Mahir Makhtar and Tiong Kung Leong. 2014. Critical State of Sand Matrix Soils. *The Scientific World Journal*, Hindawi Publishing Corporation.

Atkinson, John, H., and Bransby, P.L. 1978. *The Mechanics of Soils: An Introduction to Critical State Soil Mechanics*, McGraw-Hill.

Gudehus, G. 1996. A comprehensive constitutive equation for granular materials. *Soils and Foundations*, 36(1), 1–12.

Herle, I., and G. Gudehus. 1999. Determination of parameters of a hypoplastic constitutive model from properties of grain assemblies. *Mechanics of Cohesive-Frictional Materials*, 4, 461–486.

Kolymbas, D. 1985. A generalized hypoelastic constitutive law. In Proceedings of the eleventh *International Conference on Soil Mechanics and Foundation Engineering*.

Sadrekarimi, A., and Olson, S.M. 2009. Defining the critical state line from triaxial compression and ring shear tests. In *Proceedings of the 17th International Conference on Soil Mechanics and Geotechnical Engineering: The Academia and Practice of Geotechnical Engineering, 1, 36–39*.

Steven, L., Kramer. 1996. *Geotechnical Earthquake Engineering*. Prentice Hall, New Jercy.

Vu To-Anh Phana, Darn-Horng Hsiaob and Phuong ThucLan Nguyenc. 2016. Critical State Line and State Parameter of Sand-Fines Mixtures, Sustainable Development of Civil, *Procedia Engineering—Urban and Transportation Engineering Conference*, 142, 299–306.

Emerging Trends in Engineering, Science and Technology for Society,
Energy and Environment – Vanchipura & Jiji (Eds)
© 2018 Taylor & Francis Group, London, ISBN 978-0-8153-5760-5

Design of an embankment laterally supported with secant piles at Kuttanad

J. Jayamohan & N.R. Arun
LBS Institute of Technology for Women, Thiruvananthapuram, India

K. Balan
Rajadhani Institute of Engineering and Technology, Thiruvananthapuram, India

S. Aswathy Nair, L.K. Vaishnavi, Megha S. Thampi, D.R. Renju & Chithra Lekshmi
LBS Institute of Technology for Women, Thiruvananthapuram, India

ABSTRACT: This paper presents a case study of the investigation of recurring breaches of the embankment at Puthenarayiram Padasekharam, Kuttanad and the design of its reconstruction. At a location called Kundarikund in D Block of Kuttanad, a section of embankment frequently collapses, inundating the cultivable land. A hydrographic survey was conducted to determine the bed profile and the velocity of water was measured using a current meter. The presence of two depressions on the paddy field side of the embankment where the breach occurs is detected. Laboratory tests on soil samples obtained from boreholes indicated a very high void ratio and the water content was much higher than the liquid limit. Hence, the *insitu* undrained shear strength of the soil was determined by conducting a field vane shear test. The global and internal stability were checked as per standard geotechnical practices. Only very less passive resistance could be generated in the weak clay. To enhance the passive resistance, a berm made of Geobags filled with soil was provided on the paddy field side of the embankment. Contiguous/ secant piles were provided on the right face of the embankment throughout the breached portion. At the canal side, soldier piles are provided at a spacing of two meters connected together by RCC (Reinforced Cement Concrete) precast slabs. The secant piles on the right side are connected to the soldier piles on the left side by transverse RCC beams at a spacing of two meters.

Keywords: Kuttanad, secant piles, field vane shear test, soldier piles, Geobags, berm

1 INTRODUCTION

Kuttanad is a region covering the Alappuzha and Kottayam Districts, in the state of Kerala, well known for its vast paddy fields and geographical peculiarities. The region has the lowest altitude in India, and is one of the few places in the world where farming is carried around 1.2 to 3.0 meters below sea level. Kuttanad is historically important and is a major rice producer in the state. Farmers of Kuttanad are famous for bio saline farming. Kuttanad clay is a soft soil with associated problems of low shear strength and compressibility (Vinayachandran et al., 2013). The soil has a unique combination of minerals such as metahalloysite, kaolinite, iron oxides and aluminum oxides. The diatom frustules present in the soil indicate biological activity during the sediment formation and this also accounts for the nature of organic matter predominantly present in the soil, which is mostly derived from planktonic organisms. A considerable amount of organic matter is present in the soil and the magnitude measured accounts for about 14% by mass (Suganya & Sivapullaiah, 2015, 2017). Kuttanad soil is expansive clay having a high void ratio and low density.

The cultivable land of Kuttanad is 1.2 to 3 meters below mean sea level. These lands are kept submerged for about six to eight months of the year; during which a lot of organic matter settles on the soil making it fertile. During the farming season, the water in the paddy fields is pumped out into the canal which joins a lake. The water level in the canal will be much higher than the adjoining paddy fields. The paddy fields and canal are separated by an embankment constructed on this weak soil. The embankments are usually constructed by driving precast Reinforced Cement Concrete soldier piles at a spacing of two meters connected to each other by RCC precast slabs. These embankments are usually four meters wide and three meters high. The construction and maintenance of these embankments are carried out by the Irrigation Department, Government of Kerala. At a location called Kundarikund in D Block of Kuttanad, a section of embankment collapses quite often, inundating the cultivable land. The government spends crores of Rupees every year on restoration works.

The Irrigation Department requested the Department of Civil Engineering, LBS Institute of Technology for Women, to study the matter and to suggest remedial measures.

2 SITE INSPECTION

The site was inspected on 31-05-2017 by the authors along with the Engineers from Irrigation Department. From the site inspection, it was observed that the Pile-Slab system and the coconut piles are tilted towards the paddy area. Water was flowing from the paddy area to Kochar river. The surface of the water body in the paddy area is very calm and hence possibility of under currents cannot be ruled out. From the observation it is presumed that surface of soil profile in the paddy area is much deeper than that shown in the drawings provided.

One of the possible solutions which can be practically implemented cost effectively is the pile-slab system supported by inclined piles. The inside portion of the pile-slab system should be filled with clay reinforced with woven coir geotextiles (700 gsm).

To assess whether this solution is practically possible or not, the depth of soil in the paddy area from water surface needs to be determined; for which a hydrographic survey was carried out.

It is ideal to also know the water velocity as turbulent flow may occur at the site. In order to design the inclined pile, the undrained shear strength of the soil is also required. This was obtained by conducting a field vane shear test on the natural soil below the existing bund.

3 HYDROGRAPHIC SURVEY

A hydrographic survey was conducted to determine the bed profile and the velocity of water was measured using a current meter. The contour map of the bed obtained from the hydrographic survey is presented in Figure 1. The results of the hydrographic survey revealed the presence of

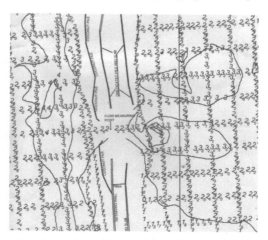

Figure 1.   Contour map of bed surface obtained from the hydrographic survey.

two depressions on the right side of the embankment where breaches frequently occurs. The maximum depth of the depression was up to five meters. The velocity of water flow was measured with a current meter. The maximum velocity of flow was observed to be 0.824 m/sec.

4 SOIL INVESTIGATION

4.1 *Soil properties*

The results of laboratory tests on soil samples are presented in Table 1.

4.2 *Field vane shear test*

Laboratory tests on the soil samples obtained from the boreholes indicated a very high void ratio (2.36) and the water content was much higher than the liquid limit. Hence, the *insitu* undrained shear strength of the soil was determined by conducting a field vane shear test as shown in Figure 2. A bore hole of 30 cm diameter was drilled up to the base level of the embankment using a helical auger.

The blade of the vane shear apparatus had a width of 50 mm and depth of 100 mm. The vane was inserted at the bottom of the borehole and the test was carried out as per IS 4434 - 1997. The undrained shear strength obtained was 9 kPa, which is a very low strength.

5 DESIGN OF EMBANKMENT

Based on the observations from the hydrographic survey, field vane shear test and the soil properties, the following recommendations are made.

Table 1. Soil properties.

Properties	Clay
Class	MH-OH
Specific gravity	2.4
Natural moisture content (%)	116
Bulk density (g/cc)	1.35
Void ratio	2.36
Porosity (%)	70
Liquid limit (%)	71
Plastic limit (%)	38
Plasticity index (%)	32
Cohesion (kPa)	0.88
% Clay & silt	94.3
% Sand	5.6

Figure 2.   Field vane shear test.

5.1 Reconstruction of the collapsed portion of the bund (length 50 m)

River side

Provide RCC precast soldier piles of cross section 35 × 30 cm and length of 10 m at a spacing of 2 m, connected by RCC precast slabs. The top of the soldier piles may be connected together by an RCC beam with a cross section of 40 × 40 cm.

A small berm (Pilla Bund) with a top width of 2 m may be constructed by driving a row of coconut piles and filling with soil. At the external side of the Pilla Bund, woven multifilament polypropylene geotextile must be provided at the external surface as shown in the drawing. The specifications of this geotextile are as follows:

Tensile strength—Machine direction – 55 kN/m, Cross machine direction – 40 kN/m,
Trapezoidal tear strength—Machine direction – 0.73 kN, Cross machine direction – 0.52 kN,
Puncture Strength – 0.62 kN, Mass per unit area – 240 g/m^2,
Average opening size – 0.15 mm.

Padasekharam side

The major stabilizing force is the passive pressure from the soil on the Padasekharam side. To utilize the passive resistance of the soil, there must be a continuous wall underneath the bed surface for a suitable depth. Hence, it is recommended to install driven precast RCC contiguous piles with a cross section of 40 cm × 40 cm and length of 12 m. These contiguous piles must be installed for a length of 50 m including the breached portion of the bund. The top of the contiguous piles must be joined together by providing an RCC beam of 60 × 40 cm. The beam must be connected to the RCC top beam at the river side by transverse beams at 2 m intervals. To increase the passive resistance from the paddy side, a berm, with geobags filled with locally available clayey soil, must be constructed.

RCC precast contiguous/secant piles

The RCC precast contiguous piles of length 12 m may be driven in two stages. At first, a precast pile of length 8 m may be driven to the correct alignment, with the top 1.2 m projecting above water level. Then, the top 1 m portion must be mechanically chipped off. Care must be taken that the reinforcement in the pile does not get damaged due to this chipping of concrete. In the second stage, a 4 m pile should be cast with 1 m development length into the first stage (total length of cast *in situ* portion is 5 m including the development length portion). The reinforcements of the first stage and second stage must be welded together throughout the lap length (1 m) with a 6 mm weld so as to have a better resistance to the impact which occurs during driving. The finished pile should be driven only after 28 days from the concreting of the second stage. The concrete of second stage must be cured as per relevant IS codes of practice.

Geobags

The geobags must have a width of 1.2 m, length of 1.2 m and height of 1 m. There should be four lifting points with two straps on each lifting point. The tensile strength of each strap should at least be 15 kN. The seam strength should be 25 kN/m. The fabric of the geobag should be woven polypropylene with a density of 325 gsm. The fabric should have a wide width tensile strength of 55 kN/m in both machine and cross machine directions. The CBR puncture strength should be 6 kN.

Calculations

The *in situ* undrained shear strength of the soil is very low. If the passive resistance of clay alone is considered, the length of the piles will be very large and will not be feasible. Hence, it is required to enhance the passive resistance by providing a berm. The thickness of the berm proposed is 3 m. The top width of the berm was determined based on the width required for the formation of a full passive wedge. The depth of embedment 'D' below the bottom level of berm is calculated by taking moments of all the forces about top. The details of the various forces acting on the wall are shown in the pressure diagram (Figure 3).

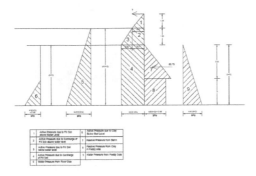

Figure 3. Pressure diagram.

$\phi = 1°$, c = 8.83 kPa
$k_a = 0.966$
$k_p = 1.04$
Unit weight of soil, $\gamma = 13$ kN/m³
Saturated unit weight, $\gamma_{sat} = 14.3$ kN/m³
Submerged unit weight, $\gamma' = 4.5$ kN/m³
Taking moments about top of the pile and simplifying we get

$0.75D^3 + 11.06D^2 - 16.04D - 286.12 = 0$

Solving, D = 5 m

The total depth works out to be 10.75 m. However a total length of 12 m from the top of the bund may be provided for the pile including factor of safety.

The top width of the berm was determined based on the width required for the formation of a full passive wedge. The minimum top width comes to 5.20 m. However the top width provided is 7 m, considering the factor of safety. The bottom width was determined based on the angle of repose from the top outer edge.

The time schedule to be followed for filling of soil inside the bund is

Stage-1: First one meter filling inside the bund
Stage-2: Second one meter after two months from the completion of first layer
Stage-3: Third one meter after two months from the completion of second layer
Stage-4: Top (fourth) layer after six months from the completion of third layer

The berm must also be constructed simultaneously during the filling of the bottom layers inside the bund.

The soil for filling the bund/berm must be taken from sites at least 150 m away from the bund.

The soil fill inside the bund must be reinforced with woven coir geotextile of 700 gsm with 12.5 mm opening size. The vertical spacing of reinforcement must be 1 m. The cross section of the embankment is presented in Figure 4.

5.2 *Reconstruction of the bund for a length of 35 m on either sides of collapsed portion (length 70 m)*

Provide RCC precast soldier piles with a cross section of 30 × 30 cm and length 7.8 m at a spacing of 2 m, connected by RCC precast slabs. The top of the soldier piles must be connected together by an RCC beam with a cross section of 40 × 40 cm.

Berms (Pilla Bund) with a top width of 2 m (similar to the collapsed portion) must be constructed by driving a row of coconut piles and filling soil on both sides of the bund as detailed in Figure 5.

Woven multifilament polypropylene geotextile must be provided at the external surface as shown in Figure 4. The specifications of this geotextile are as follows:

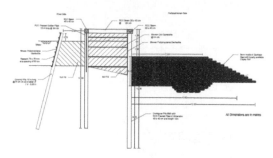

Figure 4.   Cross section of proposed embankment at the breached portion.

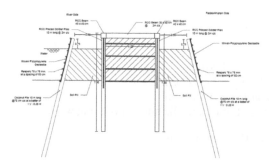

Figure 5.   Cross section of proposed embankment on either sides of breached portion for a length of 35 m.

Tensile strength—Machine direction – 55 kN/m, Cross machine direction – 40 kN/m,
Trapezoidal tear strength—Machine direction – 0.73 kN, Cross machine direction – 0.52 kN,
Puncture Strength – 0.62 kN, Mass per unit area – 240 g/m^2,
Average opening size – 0.15 mm.

6   CONCLUSIONS

The reasons for the recurring breaches of the embankment at Puthenarayiram Padasekharam, Kuttanad have been investigated. The findings are:

- The presence of two depressions of about 5 m in depth on the paddy field side of the embankment were hampering the global stability
- The pile-slab construction method practiced by the Irrigation Department could not harness the passive resistance of the soil below the embankment
- A revised design of embankment has been proposed with secant piles and a berm on the paddy field side.

REFERENCES

Suganya, K. & Sivapullaiah. P.V. Effect of changing water content on the properties of Kuttanad soil, *Geotechnical and Geological Engineering* 33, 913–921.
Suganya, K., & Sivapullaiah. P.V. Role of composition and fabric of Kuttanad clay: a geotechnical perspective, *Bulletin of Engineering Geology and the Environment* 76, 371–381.
Vinayachandran. N., Narayana. A.C., Najeeb. K.M., & Narendra. P. (2013). Disposition of aquifer system, geo-electric characteristics and gamma-log anomaly in the Kuttanad alluvium of Kerala, *Journal of the Geological Society of India* 81, 183–191.

Emerging Trends in Engineering, Science and Technology for Society,
Energy and Environment – Vanchipura & Jiji (Eds)
© 2018 Taylor & Francis Group, London, ISBN 978-0-8153-5760-5

Effect of anchorage of geosynthetic reinforcement on the behaviour of reinforced foundation bed

Aleena Mariam Saji, Alen Ann Thomas, Greema Sunny, J. Jayamohan & V.R. Suresh
LBS Institute of Technology for Women, Thiruvananthapuram, India

ABSTRACT: Geosynthetics demonstrate their beneficial effects only after considerable settlement, since the strains occurring during initial settlements are insufficient to mobilize significant tensile load in the geosynthetic. This is not a desirable feature since for foundations of certain structures, the values of permissible settlements are low. Anchoring the reinforcement is a promising technique yet to be comprehensively studied. This paper presents the results of a series of finite element analyses carried out to investigate the improvement in load-settlement behaviour of a strip footing resting on a Reinforced Foundation Bed due to anchoring the Geosynthetic Reinforcement. It is observed that the bearing capacity can be considerably increased without the occurrence of excessive settlement by anchoring the geosynthetic reinforcement with micropiles.

Keywords: Geosynthetics, Finite Element Analyses, Anchoring, Load-Settlement Behaviour, Reinforced Foundation Bed

1 INTRODUCTION

The decreasing availability of proper construction sites has led to the increased use of marginal ones, where the bearing capacity of the underlying deposits is very low. By the application of geosynthetics it is possible to use shallow foundations even in marginal soils instead of expensive deep foundations. This is done by either reinforcing cohesive soil directly or replacing the poor soils with stronger granular fill in combination with geosynthetic reinforcement. In low-lying areas with poor foundation soils, the geosynthetic reinforced foundation bed can be placed over the weak soil. The resulting composite ground (reinforced foundation bed) will improve the load carrying capacity of the footing and will distribute the stresses on a wider area on the underlying weak soils, hence reducing settlements. During the past 30 years, the use of reinforced soils to support shallow foundations has received considerable attention. Many experimental and analytical studies have been performed to investigate the behaviour of reinforced foundation beds for different soil types (eg. Binqet and Lee (1975), Shivashankar et al. (1993)). Several experimental and analytical studies were conducted to evaluate the bearing capacity of footings on reinforced soil (eg. Shivashankar and Setty (2000); Shivashankar and Reddy (1998); Madhavilatha and Somwanshi (2009); Alamshahi and Hataf (2009); Vinod et al. (2009) Arun et al. (2008) etc).

It is now known that geosynthetics demonstrate their beneficial effects only after considerable settlements, since the strains occurring during initial settlements are insufficient to mobilize significant tensile load in the geosynthetic. This is not a desirable feature since for foundations of certain structures; the values of permissible settlements are low. Thus there is a need for a technique which will allow the geosynthetic to increase the load bearing capacity of soil without the occurrence of large settlements. Lovisa et al 2010 conducted laboratory model studies and finite element analyses on a circular footing resting on sand reinforced with prestressed geotextile. It was found that the addition of prestress to reinforcement resulted in a significant improvement in the load bearing capacity and reduction in settlement of foundation. Lackner et al. (2013) conducted about 60 path controlled static load displacement tests and 80 cyclic load displacement tests to determine the load-displacement behaviour of prestressed reinforced soil structures. Unnikrishnan and Aparna Sai (2014) carried out extensive studies on footings in clay supported on encapsulated granular trench.

In this research, the effects of anchoring the geosynthetic reinforcement, on the load-bearing capacity and settlement response of a reinforced foundation bed overlying weak soil are investigated by carrying out nonlinear finite element analysis using the FE software *PLAXIS 2D*. The effects of anchoring the reinforcement on the stress distribution at the interface between foundation bed and weak soil, axial force distribution in the reinforcement, stress distribution at the interface between reinforcement and surrounding granular soil etc. are particularly studied.

2 FINITE ELEMENT ANALYSES

Finite element analyses are carried out using the commercially available finite element software *PLAXIS* 2D. For simulating the behaviour of soil, different constitutive models are available in the FE software. In the present study Mohr-Coulomb model is used to simulate soil behaviour. This non-linear model is based on the basic soil parameters that can be obtained from direct shear tests; internal friction angle and cohesion intercept. Since strip footing is used, a plain strain model is adopted in the analysis. The settlement of the rigid footing is simulated using non zero prescribed displacements.

The displacement of the bottom boundary is restricted in all directions, while at the vertical sides; displacement is restricted only in the horizontal direction. The initial geostatic stress

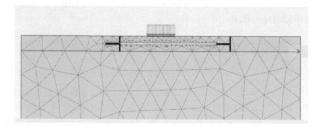

Figure 1.   Geometric model.

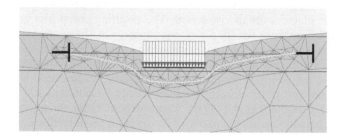

Figure 2.   Typical deformed shape.

Table 1.   Properties of soil.

Property	Sand	Clay
Dry Unit Weight (kN/m³)	17	16
Specific Gravity	2.3	2.1
Cohesion (kPa)	2	20
Angle of Shearing Resistance (°)	30	0
Modulus of Elasticity (kPa)	15000	7000
Poisson's Ratio	0.25	0.3

states for the analyses are set according to the unit weight of soil. The soil is modelled using 15 noded triangular elements. Mesh generation can be done automatically. Medium mesh size is adopted in all the simulations. The size of the strip footing (B) is taken as one metre and the width and depth of soil mass are taken as 10B in all analyses.

The reinforcement is modelled using the 5-noded tension element. To simulate the interaction between the reinforcement and surrounding soil, an interface element is provided on both upper and lower surface of reinforcement. The interaction between soil and reinforcement is simulated by choosing an appropriate value for strength reduction factor R_{inter} at the interface. The geometric model is shown in Fig. 1 and the typical deformed shape in Fig. 2. The soil is modeled using 15-node triangular elements. Poisson's ratio of the soil is assumed to be 0.25 for all cases. Properties of locally available sand and clay are adopted in the analyses. The material properties adopted are outlined in Table 1. The properties required for the geosynthetic is Elastic Axial Stiffness (EA) and the value adopted is 40 kN/m. Footing is modelled as a plate.

The anchorage provided by micropiles is modelled with fixed end anchors. The effect of change in spacing and length of micropiles is simulated by varying the stiffness of fixed end anchors. Various values of stiffness (EA) of the anchors adopted in the analyses are 1000, 1500, 2000 and 3500 MN/m respectively.

3   RESULTS AND DISCUSSIONS

3.1   *Load-settlement behaviour*

Vertical Stress vs Settlement curves for strip footing resting on Reinforced Foundation Bed with anchored geosynthetic reinforcement; obtained from FEA, for various values of stiffness of anchor are presented in Figure 3. It is seen that the load-settlement behaviour considerably improves due to anchoring the reinforcement.

3.2   *Improvement factor*

To quantify the improvement in load-settlement behaviour attained due to anchoring the reinforcement, an Improvement Factor (If) is defined as the ratio of stress with anchored reinforcement to that with unanchored reinforcement; at 2 mm settlement.

Figure 4 presents the variation of Improvement Factor with Stiffness of Anchor, obtained from Finite Element Analyses. It is seen that as the stiffness of anchor increases, the improvement factor increases.

When the footing settles the mid portion of the reinforcement moves down and its edges moves inwards and takes a deformed shape as shown in Figure 2. When the reinforcement is anchored at its edges, there will be an additional resistance to this deformation and more tensile stress gets mobilized in it. This additional mobilized tensile stress improves the load-settlement

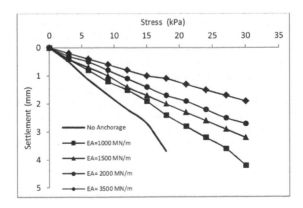

Figure 3.   Vertical stress vs settlement curves.

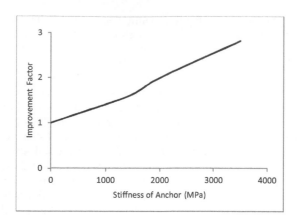

Figure 4.   Improvement factor vs stiffness of anchor.

behaviour. It is seen from Figures 3 and 4 that as the stiffness of anchor increases, the improvement in load-settlement behaviour increases.

4   CONCLUSIONS

Based on the results of Finite Element Analyses carried out the following conclusions are drawn.

- Anchoring the Geosynthetic Reinforcement considerably improves the load-settlement behaviour
- The improvement factor increases with stiffness of the anchor

REFERENCES

Alamshahi, S. and Hataf, N. (2009). Bearing capacity of strip footings on sand slopes reinforced with geogrid and grid-anchor, *Geotextiles and Geomembranes*, 27 (2009) 217–226.
Arun Kumar Bhat, K., Shivashankar, R. and Yaji, R.K. (2008), "Case study of land slide in NH 13 at Kethikal near Man galore, India", 6th *International Conference on Case histories in Geotechnical Engineering*, Arlington, VA, USA, paper no. 2.69.
Binquet, J. and Lee, K.L. (1975). Bearing capacity tests on reinforced earth slabs. *Journal of Geotechnical Engineering Division*, ASCE 101 (12), 1241–1255.
Lackner, C., Bergado, D.T. and Semprich, S. (2013). Prestressed reinforced soil by geosynthetics—concept and experimental investigations, *Geotextiles and Geomembranes*, 37, 109–123.
Lovisa, J., Shukla, S.K. and Sivakugan, N. (2010). Behaviour of prestressed geotextile-reinforced sand bed supporting a loaded circular footing, *Geotextiles and Geomembranes*, 28 (2010) 23–32.
Madhavilatha, G. and Somwanshi, A. (2009). Bearing capacity of square footings on geosynthetic reinforced sand, *Geotex* tiles and Geomembranes, 27 (2009) 281–294.
Shivashankar, R., Madhav, M.R. and Miura, N. (1993). Rein forced granular beds overlying soft clay, *Proceedings of 11th* South East Asian Geotechnical Conference, Singapore, 409–414.
Shivashankar, R. and Reddy, A.C.S. (1998). Reinforced granular bed on poor filled up shedi ground, *Proceedings of the Indian Geotechnical Conference* – 1998, Vol. 1, 301–304.
Shivashankar, R. and Setty, K.R.N.S. (2000). "Foundation problems for ground level storage tanks in and around Mangalore", *Proceedings of Indian Geotechnical Conference 2000*, IIT Bombay, Mumbai.
Unnikrishnan and Aparna Sai. (2014), "Footings in Clay soil supported on Encapsulated Granular Trench", *Proceedings of Indian Geotechnical Conference 2014*, Kakinada.
Vinod, P., Bhaskar, A.B. and Sreehari, S. (2009). Behaviour of a square model footing on loose sand reinforced with braided coir rope, *Geotextiles and Geomembranes*, 27 (2009) 464–474.

Emerging Trends in Engineering, Science and Technology for Society,
Energy and Environment – Vanchipura & Jiji (Eds)
© 2018 Taylor & Francis Group, London, ISBN 978-0-8153-5760-5

An artificial neural network-based model for predicting the bearing capacity of square footing on coir geotextile reinforced soil

Dharmesh Lal, N. Sankar & S. Chandrakaran
Department of Civil Engineering, NIT Calicut, Calicut, India

ABSTRACT: Geosynthetic materials like geotextiles, geogrids and geocells have gained widespread acceptance over recent years due to their superior engineering characteristics and quality control. The rising cost and the environmental concerns created by these synthetic reinforcement materials makes it necessary to explore alternate resources for soil reinforcement. Coir is an eco-friendly, biodegradable, organic material which has high strength, stiffness and durability characteristics compared to other natural reinforcement materials. This paper deals with a systematic series of plate load tests on unreinforced sand and sand reinforced with coir geotextiles. A significant enhancement in strength and stiffness characteristics was obtained with the provision of coir reinforcement. Based on the test results, an Artificial Neural Network (ANN) model has also been established for predicting the strength of sand beds when these reinforcement elements are applied practically. The predicted values from the model and those obtained from the experimental study are found to have a good correlation.

Keywords: Coir geotextile, artificial neural network, bearing pressure, subgrade stabilization

1 INTRODUCTION

In recent years, geosynthetics have become increasingly popular for their use as a reinforcement in earth structures. The use of geosynthetics in reinforcing sand beds has been studied by various researchers (Abu-Farsakh et al., 2013; Guido et al., 1986; Akinmusuru & Akinbolade, 1981; Omar et al., 1993; Ghosh et al., 2005; Latha & Somwanshi, 2009; Sharma et al., 2009). Synthetic fibers have a longer life and do not generally undergo biological degradation, thus minimizing environmental concern. Coir geotextiles can be considered as an efficient replacement for their synthetic counterparts due to their economy and excellent engineering properties. The use of coir as a reinforcement material has been studied by various researchers (Lekha, 1997; SivakumarBabu et al., 2008; Vinod et al., 2009; Subaida et al., 2008). India is one of the leading coir producing countries. While the world focus is shifting to natural geotextiles, India as a producer of coir geotextiles, has much to gain by using it for meeting domestic as well as global demands. Natural geotextiles are becoming increasingly popular in various geotechnical applications like construction of embankments, subgrade stabilization, slope protection work, weak soil improvement and so on. From the studies reported so far, it is perceived that the potential of coir products as a reinforcement material is under-utilized.

Empirical models based on Artificial Neural Networks (ANN) have been widely used for numerous applications in geotechnical engineering. Neural networks have proved to be an efficient tool to predict the behavior of soils under different test conditions, especially, since the relationship between the input and output variables is complex. Although models with different input and output variables are required to be created for the different test conditions, the process is comparatively easy and more realistic than numerical models. The present study deals with the formulation of empirical models based on artificial neural networks for predicting the behavior of sands reinforced with coir geotextiles, under plate load testing conditions.

2 ARTIFICIAL NEURAL NETWORK (ANN)

An Artificial Neural Network (ANN) is a form of artificial intelligence, which tries to simulate the behavior of the human brain and nervous system. In recent years, many researchers have investigated the use of artificial neural networks in geotechnical engineering applications and have obtained reassuring results (e.g. Kung et al., 2007; Kuo et al., 2009; Ornek et al., 2012; Harikumar et al., 2016). MATLAB software was used for formulating the ANN model. The parameters used for creating the model are listed in Table 1.

Table 1. Parameters used for creating model.

Settlement level (s/B)	15%
Reinforcement parameters used for modeling	a) Depth to the first layer of reinforcement (u/B) b) Width of reinforcement c) Number of reinforcement layers
Training function	Gradient descent with momentum and adaptive learning rate back propagation technique
Number of nodes in the input, hidden and output layer	3, 10 and 1, respectively
Learning rate	0.01
Momentum constant	0.9

Figure 1.   Photograph of coir geotextile used for the study.

Figure 2. Photograph of test set up.

3 MATERIALS AND METHODOLOGY

The sand used for the tests had a specific gravity of 2.65, effective size 0.32, coefficient of uniformity 2.56, coefficient of curvature 0.88 and angle of friction of 38.5° (at 60% relative density). Classification according to the Unified Soil Classification System (USCS) is SP (poorly graded sand). All tests were done at a relative density of 60% to simulate medium dense condition. Figure 1 shows a photograph of the coir geotextiles used for the study. The properties of woven coir geotextiles were determined as per Indian standards (IS: 13162, 1992; IS: 14716, 1999).

The model test was conducted on a steel tank 750 mm × 750 mm × 750 mm. The model footing was a square steel plate 150 mm × 150 mm, with a thickness of 25 mm. A hand operated hydraulic jack was used for loading the footing and a pressure gage of 100 kN was fitted to measure the load applied. Figure 2 shows a photograph of the test set up. The objectives of the tests were to study the influence of coir geotextiles on improving the overall performance of the sand foundations. The test series included varying the depth of the reinforcement layer from the top of footing (u). An artificial neural network model has also been established based on the test results. MATLAB software was used to formulate the ANN model. The technique of formulating empirical models is reliable and relatively easy compared with a numerical study.

4 RESULTS AND DISCUSSION

Figure 3 shows the variation of measured values with the values predicted by the model. From the figure, it can be seen that the bearing capacity predicted by the model and those obtained from the experimental study are in good agreement.

It can be further seen that the provision of coir reinforcement enhances the strength characteristics of reinforced soil (an almost 95% increase in strength can be observed, even with a single layer of reinforcement). Additionally, maximum improvement was observed when the reinforcement was provided at a depth of 0.25 times the width of the foundation (0.25 B). Furthermore, the predicted and observed values were found to have a good correlation (see Figure 4), thus establishing the validity of the proposed model.

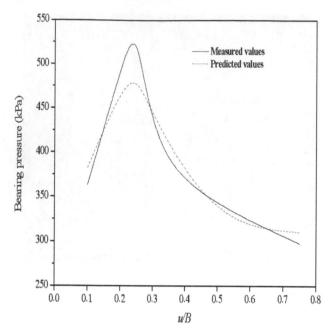

Figure 3. Bearing pressure vs. placement depth (u/B).

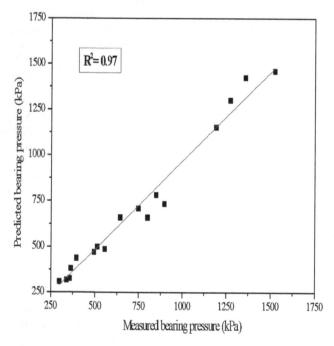

Figure 4. Measured vs. predicted bearing capacity.

5 CONCLUSIONS

A detailed, comprehensive study on the performance of coir geotextiles was conducted using a plate load testing apparatus. The results demonstrated that the coir reinforcement inclusions increase the strength and deformation characteristics of sand. An Artificial Neural Network (ANN) model has also been established for predicting the strength of sand beds when these reinforcement elements are applied practically. The predicted values from the model and those obtained from the experimental study were found to have a good correlation.

REFERENCES

Abu-Farsakh, M., Chen, Q. & Sharma, R. (2013). An experimental evaluation of the behavior of footings on geosynthetic-reinforced sand. *Soils and Foundations, 53*(2), 335–348.

Akinmusuru, J.O. & Akinbolade, J.A. (1981). Stability of loaded footings on reinforced soil. *Journal of Geotechnical Engineering Division, ASCE, 107*, 819–827.

Ghosh, A., Ghosh, A. & Bera, A.K. (2005). Bearing capacity of square footing on pond ash reinforced with jute-geotextile. *Geotextiles and Geomembranes, 23*(2), 144–173.

Guido, V.A., Chang, D.K. & Sweeney, M.A. (1986). Comparison of geogrid and geotextile reinforced earth slabs. *Canadian Geotechnical Journal, 23*, 435–440.

Harikumar, M., Sankar, N. & Chandrakaran, S. (2016). Behavior of model footing on sand bed reinforced with multi-directional reinforcing elements. *Geotextiles and Geomembranes, 44*, 568–578.

IS: 13162. 1992. Geotextiles—methods of test for determination of thickness at specified pressure. New Delhi: Bureau of Indian Standards.

IS: 14716. 1999. *Geotextiles—determination of mass per unit area*. New Delhi: Bureau of Indian Standards.

Kung, G.T.C., Hsiao, E.C.L., Schuster, M. & Juang, C.H. (2007). A neural network approach to estimating deflection of diaphragm walls caused by excavation in clays. *Computers and Geotechnics, 34*(5), 385–396.

Kuo, Y.L., Jaksa, M.B., Lyamin, A.V. & Kaggwa, W.S. (2009). ANN-based model for predicting the bearing capacity of strip footing on multi-layered cohesive soil. *Computers and Geotechnics, 36*(3), 503–516.

Latha, G.M. & Somwanshi, A. (2009). Bearing capacity of square footings on geosynthetic reinforced sand. *Geotextiles and Geomembranes, 27*(4), 281–294.

Lekha, K.R. (1997). Coir geotextiles for erosion control along degraded hill slopes. *Proceedings of Seminar on Coir Geotextiles, Coimbatore, India.*

Omar, M.T., Das, B.M., Puri, V.K. & Yen, S.C. (1993). Ultimate bearing capacity of shallow foundations on sand with geogrid reinforcement. *Canadian Geotechnical Journal, 30*, 545–549.

Ornek, M., Laman, M., Demir, A. & Yildiz, A. (2012). Prediction of bearing capacity of circular footings on soft clay stabilized with granular soil. *Soils and Foundations, 52*(1), 69–80.

Sharma, R., Chen, Q., Abu-Farsakh, M. & Yoon, S. (2009). Analytical modelling of geogrid reinforced soil foundation. *Geotextiles and Geomembranes, 27*, 63–72.

SivakumarBabu, G.L., Vasudevan, A.K. & Sayida, M.K. (2008). Use of coir fibres for improving the engineering properties of expansive soils. *Journal of Natural Fibers, 5*(1), 1–15.

Subaida, E.A., Chandrakaran, S. & Sankar, N. (2008). Experimental investigations on tensile and pull-out behaviour of woven coir geotextiles. *Geotextiles and Geomembranes, 26*(5), 384–392.

Vinod, P., Ajitha, B. & Sreehari, S. (2009). Behavior of a square model footing on loose sand reinforced with braided coir rope. *Geotextiles and Geomembranes, 27*, 464–474.

Emerging Trends in Engineering, Science and Technology for Society,
Energy and Environment – Vanchipura & Jiji (Eds)
© *2018 Taylor & Francis Group, London, ISBN 978-0-8153-5760-5*

Interference effect of adjacent footings resting on granular beds overlying weak soils

Aqila Abdul Khader
Department of Civil Engineering, Marian Engineering College, Trivandrum, Kerala, India

Jayamohan Jayaraj
LBS Institute of Technology for Women, Trivandrum, Kerala, India

S.R. Soorya
Department of Civil Engineering, Marian Engineering College, Trivandrum, Kerala, India

ABSTRACT: Due to limited space available for the construction of structures and support of heavy loads, foundations are often placed close to each other; the footings interact with each other and their behavior is thus not dissimilar to that of a single isolated footing. This study aims to determine experimentally the effect of interference of closely spaced shallow footings (strip footings), resting on granular beds overlying a 'weak soil'. The laboratory model tests were carried out at different 'center to center' spacing between the footings. The ultimate bearing capacity of footings increased up to a certain critical spacing and thereafter decreased. The bearing capacity of interfering footings improved due to the provision of a granular bed.

Keywords: laboratory model tests, interference effect, critical spacing, bearing capacity, granular bed

1 INTRODUCTION

In urban areas, due to the limited space available, foundations often are placed close to each other resulting in an interference with each other. The interference of the failure zones of the footings alters the bearing capacity and load-settlement characteristics of the closely spaced footings.

Stuart (1962) was the first pioneer who investigated exclusively the effect of interference of closely spaced strip footings on ultimate bearing capacity. Using the limit equilibrium technique, he indicated that the interference of two footings on sand leads to an increase in their ultimate bearing capacity. Also, he demonstrated that there existed a certain critical spacing between two footings for which the ultimate bearing capacity becomes maximum. The behavior of the interference effect is attributed to the phenomenon called the 'blocking effect' or 'arching effect'. According to this phenomenon, the soil between the two footings forms an inverted arch, and the combined system of soil and two footings moves down upon loading as a single unit. Since the area of this single unit is greater than that of the sum of the areas of two footings, it results in greater bearing capacity. On the other hand, Stuart stated that the interference effect of adjacent footings on clay would act differently to that on sand, and concluded that the interference effect would not exhibit any change in bearing capacity as the spacing between the footings decreased.

Selvadurai and Rabbaa (1983) studied the contact stress distribution beneath two interfering rigid strip footings of equal width, resting in frictionless contact with a layer of dense sand underlaid by a smooth, rigid base. The study showed that the contact stress distribution for a single isolated foundation has a symmetrical shape and as the spacing between the adjacent footings decreases, the contact stress distribution exhibits an asymmetrical shape. Das and Larbi-Cherif (1984) conducted laboratory model tests on two closely spaced strip foundations on sand and concluded that the ultimate bearing capacity increased up to a critical spacing, but settlement increased due to the interference effect. Graham et al. (1984) used the method of stress characteristics to calculate the bearing capacity of a series of parallel footings and thesewere compared with the laboratory tests conducted on three parallel surface footings resting on sand. Efficiencies were highest at the closer spacing, and it decreased until the footings behaved independently. Kouzer and Kumar (2008) examined the interference effect due to equally spaced multiple strip footings on cohesionless soils, on the ultimate bearing capacity with the help of upper bound limit analysis in conjunction with finite elements and linear programming, and concluded that the efficiencies were higher at closer spacings between the footings. Kumar and Bhoi (2009) experimentally studied the interference effect of two closely spaced strip footings on sand without having any provision of tilt, and found that the bearing capacity of the footingreached its maximum at a certain critical spacing between the footings. The interference effect is even moreextensivefor higher relative densities of sand. Ghosh and Sharma (2010) used the theory of elasticity to model the settlement behavior on layered cohesionless soil beds and concluded that the settlement of closely spaced footings was higher than that of a single isolated footing, which further decreased with the increase in the spacing between the footings. Reddy et al. (2012) conducted a series of model tests using square and circular footings resting on sand to study the interference effect, and concluded that bearing capacity and settlement increased with a decrease in spacing between the footings. Pusadkar et al. (2013) evaluated the influence of interference of symmetrical footings (square, circular and rectangular footings) on the bearing capacity of sandy soil and concluded that the bearing capacity improved with a decrease in the spacing between the footings, whereas, the settlement of the footing increased. Desai and Moogi (2016) investigated the interference of two strip footings on three types of soil (soft clay, sandy clay and medium sand) using PLAXIS-2D and found similar results.

This paper depicts the results of a series of laboratory scale modeling carried out to determine the interference effect of closely spaced strip footings on granular bed overlying clayey soil.

2 LABORATORY MODEL TESTS

2.1 *Materials*

For the laboratory tests, well-graded medium sand was used for the granular bed, and locally available clay was used for a 'weak soil'. The properties of soils are presented in Table 1 and Table 2.

Table 1. Properties of the clay used asthe weak soil.

SI No.	Property	Value
1	Specific gravity	2.5
2	Liquid limit (%)	58
3	Plastic limit (%)	22
4	Shrinkage limit (%)	16.16
5	Percentage of clay (%)	67.2
6	Percentage of silt (%)	30.095
7	Plasticity index (%)	36
8	Permeability (m/s)	3.03×10^{-6}
9	UCC strength (kg/cm²)	1.428
10	Friction angle (Φ)	5°
11	Cohesion (kPa)	25
12	Soil classification	CH
13	Water content during model test (%)	25
14	Average dry unit weight during model test (kN/m³)	14.6

Table 2. Properties of the sand used as the granular bed.

SI No.	Property	Value
1	Specific gravity	2.66
2	Effective grain size for D_{10} (mm)	0.208
3	Effective grain size for D_{60} (mm)	1.074
4	Effective grain size for D_{30} (mm)	0.487
5	Coefficient of uniformity (C_u)	5.16
6	Coefficient of curvature (C_c)	1.062
7	Permeability (m/s)	1.07×10^{-4}
8	Angle of internal friction (Φ)	$31.2°$
9	Cohesion (c in kPa)	0
10	Soil classification	well-graded medium sand
11	Water content during model test (%)	0
12	Average dry unit weight during model test (kN/m³)	19.59
13	Void ratio during model test	0.33

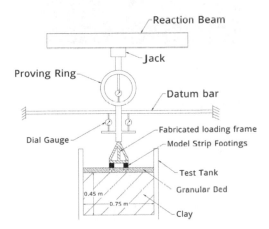

Figure 1. Schematic representation of the test setup.

2.2 Test setup

The test setup consisted of a masonry tank of size 750 mm × 750 mm × 1000 mm assembled on a loading frame. The model strip footings were of size 50 mm × 50 mm × 670 mm. The bottom of the footings were made smooth. Also, a loading frame was fabricated to load the two footings simultaneously. The schematic representation of the test setup is as shown in Figure 1. The load was applied by a hand operated mechanical jack of 50 kN capacity supported against a reaction frame. The load was measured using a proving ring and deformation using two dial gauges placed diametrically opposite to each other. Figure 2 shows the photographic view of the test setup.

2.3 Preparation of the test bed and testing procedure

First, the tank was filled to the required level with the clay representing weak soil. To achieve the desired soil density, a layered filling technique was used. The predetermined density of the clay was used to calculate the weight of the soil required to fill the tank in layers of 50 mm in height. The sand was added similarly and up to a thickness equal to the width of the footing (50 mm). Adjacent model footings were placed at varying spacings. The load was applied in equal increments, and each load increment was maintained until the settlement stopped. The loading was continued until failure. The Center to Center Spacing (S) between the footings

Figure 2.   Photographic view of the test setup.

Table 3.   Experimental program.

Phases	Type	Spacing to footing width ratio (S/B)
A	Clay (weak soil)	
B	Granular bed (GB)	1, 1.5, 2, 2.5, 3, 3.5, 4

of Width (B) was expressed as a Spacing to Footing Width Ratio (S/B). The experimental program was as shown in Table 3.

3   RESULTS AND DISCUSSION

3.1   *Load-settlement behavior*

Laboratory scaled model tests were conducted on two symmetrical strip footings of width 50 mm resting on a granular bed overlying a weak soil. Vertical Stress Versus Normalized Settlement (δ/B), and settlement to footing width ratio curves for two model strip footings resting on clay at various spacings are presented in Figure 3. The behavior of a single footing is also presented for comparison. It is seen from Figure 3? that the closer spacing of footings adversely affected the load-settlement behavior. The load carrying capacity improved when the spacing was increased up to three times the width of the footing, indicating that the blocking effect was felt up to S/B = 3. The plot in Figure 3 highlights the point at which critical spacing occurs. The critical spacing is defined as the spacing at which maximum bearing capacity occurs. As the spacing between the footings was increased past the critical spacing, the bearing capacity decreased.

In Figure 4, which presents the vertical stress versus normalized settlement curves for footings resting on a granular bed overlying clay, a similar behavior was observed. The load-settlement behavior improved until the spacing was increased up to S/B = 2.5 and until which the blocking effect had been felt.

Test results showed that by the provision of a granular bed, the interference effect between the adjacent strip footings was enhanced. Figure 5 shows the improvement in bearing capacity of interfering footings at a spacing of S/B = 2.5.

262

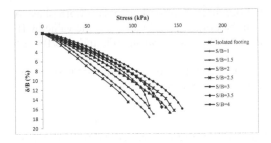

Figure 3. Vertical stress versus normalized settlement curves for footings resting on clay.

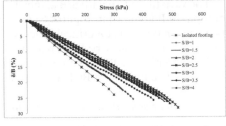

Figure 4. Vertical stress versus normalized settlement curves for footings resting on a granular bed overlying clay.

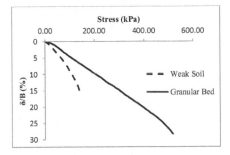

Figure 5. Vertical stress versus normalized settlement curves of phase A and phase B at S/B = 2.5.

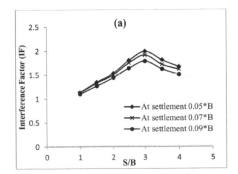

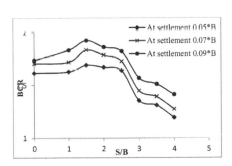

Figure 7. Variation of bearing capacity ratio with S/B at different footing settlements.

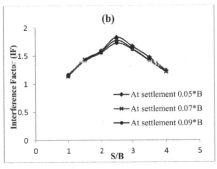

Figure 6. Variation of interference factor with S/B at different footing settlements: (a) interference effect on weak soil; and (b) interference effect on a granular bed.

3.2 Interference Factor (IF) and Bearing Capacity Ratio (BCR)

To evaluate the influence of the interference effect of adjacent strip footings and beneficial effects of the provision of a granular bed, two non-dimensional parameters were used namely: the Interference Factor (IF); and the Bearing Capacity Ratio (BCR).

IF is defined as: the ratio of average pressure on interfering footings of a given size, at a given magnitude of settlement, to the average pressure on an isolated footing of the same size and at the same magnitude of settlement.

BCR is defined as: the ratio of average pressure on an isolated or interfering footing of a given size at a given magnitude of settlement on the granular bed overlying weak soil, to the average pressure on an isolated or interfering footing of the same size and at the same magnitude of settlement on the weak soil. The IF and BCR were defined for different footing settlements at s/B = 5%, 7% and 9% (Kumar & Bhoi, 2009).

The variation of IF with S/B associated with different footing settlements for both phase A and phase B is shown in Figures 6a and 6b. It can be noticed that in all cases the value of IF becomes maximum at a critical spacing of S/B = 3 in the case of weak soil, and at S/B = 2.5 in the case of granular bed overlying weak soil.

The variation of BCR with S/B associated with different footing settlements is shown in Figure 7. The maximum BCR is obtained at a spacing of S/B = 1.5.

3.3 *Comparison with available literature*

A large amount of research has been conducted on the interference effect and bearing capacity of closely spaced footings on sand. Across the majority of the research, it was concluded that when two footings have no clear spacing between them, the footings will act as a single footing with twice the width.

As the spacing between footings increased, the bearing capacity of the footings increased, up to a certain spacing where the interference effect no longer played a part and the bearing capacity decreased. The trend displayed in Figure 3 and Figure 4 is very similar to the trend discussed in available literature on the interference effect of two footings on sand.

4 CONCLUSIONS

In the present study, a series of laboratory scaled model tests were performed to determine the load-settlement behavior of interfering strip footings resting on a granular bed overlying a weak soil. The following conclusions are drawn:

- Interference of adjacent footings significantly affects the load-settlement behavior. There is an optimum distance between the footings at which the bearing capacity has the maximum value due to a blocking effect. In footings resting on clay, the blocking effect appears at S/B = 3, and for footings resting on the granular bed, the blocking effect appears at S/B = 2.5. However, the critical value of S/B at which IF becomes the maximum is not a fixed.
- The trend obtained from the laboratory study is similar to the trend discussed in the literature review on the interference effect of two footings on sand.
- The bearing capacity of closely spaced strip footings improved with the provision of granular bed over a weak soil.

REFERENCES

Das, B.M. & Larbi-Cherif, S. (1984). Bearing capacity of two closely-spaced shallow foundations on sand.

Desai, M.V.G. & Moogi, V.V. (2016). Study of interference of strip footing using PLAXIS-2D. *International Advanced Research Journal in Science, Engineering and Technology*, 3(9), 13–17.

Ghosh, P. & Sharma, A. (2010). Interference effect of two nearby strip footings on layered soil: Theory of elasticity approach. *Acta Geotechnica*, 5, 189–198.

Graham, J., Raymond, C.P. & Suppiah, A. (1984). Bearing capacity of three closely-spaced footings on sand. *Geotechnique*, 34(2), 173–182.

Kouzer, K.M. & Kumar, Jyant. (2008). Ultimate bearing capacity of equally spaced multiple strip footings on cohesionless soils without surcharge. *International Journal for Numerical and Analytical Methods in Geomechanics*, 32, 1417–1426.

Kumar, J. & Bhoi, M.K. (2009). Interference of two closely spaced strip footings on sand. *Journal of Geotechnical and Geoenvironmental Engineering*, 135(4), 595–604.

Pusadkar, S.S., Gupta, R.R. & Soni, K.K. (2013). Influence of interference of symmetrical footings on bearing capacity of soil. *International Journal of Engineering Inventions*, 2(3), 63–67.

Reddy, S.E., Borzooei, S. & Reddy, N.G.V. (2012). Interference between adjacent footings on sand. *International Journal of Advanced Engineering Research and Studies*, 1(4), 95–98.

Selvadurai, A.P.S. & Rabbaa, S.A.A. (1983). Some experimental studies concerning the contact stresses beneath interfering rigid strip foundations resting on a granular stratum. *Canadian Geotechenical-Journal*, 20, 406–415.

Stuart, J.G. (1962). Interference between foundations, with special reference to surface footings in sand. *Geotechnique*, 2(1), 15–22.

Emerging Trends in Engineering, Science and Technology for Society,
Energy and Environment – Vanchipura & Jiji (Eds)
© 2018 Taylor & Francis Group, London, ISBN 978-0-8153-5760-5

Influence of shape of cross section of footing on load-settlement behaviour

B. Anusha Nair, Akhila Vijayan, S. Chandni, Shilpa Vijayan, J. Jayamohan & P. Sajith
LBS Institute of Technology for Women, Thiruvananthapuram, India

ABSTRACT: In general the shape of cross section of footings provided for structures is rectangular. There are well accepted theories to determine the bearing capacity and settlements of footing with a flat base. By altering the cross sectional shape of the footing, better confinement of underlying soil can be attained thereby improving bearing capacity and reducing settlements. In this investigation a series of finite element analyses are carried out to determine the influence of shape of cross section of the footing on the load-settlement behaviour of strip footings. It is observed that by altering the shape of cross section of footings, better confinement of underlying soil can be achieved, thereby improving the load-settlement behaviour.

Keywords: Shape, Cross section, Footing, Load-Settlement behaviour

1 INTRODUCTION

The foundation transmits the load of the structure safely to the ground, without undergoing any shear failure or excessive settlement. The bearing capacity of footings has been extensively studied, both theoretically and experimentally, over the past many decades. The theoretical approach was initiated by Prandtl (1921) and Reissner (1924), and the design-oriented bearing capacity equation (fully considering the soil unit weight, cohesion, and friction angle) was proposed by Terzaghi (1943). After the early development of the bearing capacity solution, most efforts have focused mainly on a more realistic derivation of

Table 1. Cross sectional shapes of footings.

Shape of footing		d/B
	Footing with rectangular cross section	–
	Rectangular cross section with flanges	0.2, 0.4, 0.6, 0.8
	Sloped cross section	0.2, 0.4, 0.6, 0.8

bearing capacity and various correction factors [Meyerhof G. (1965), Taiebat and Carter (2000), Griffiths et al. (2002), Ericson and Drescher (2002), Michalowski and Dawson (2002), Salgado et al. (2004), Lee and Salgado (2005)].

The vertical bearing capacity of a shallow foundation is a classical geotechnical problem. The mechanism of soil failure transforms from a general shear failure for a surface footing (Craig 2004) to a localized failure for a buried footing (Hu et al. 1999; Wang and Carter 2002; Hossain and Randolph 2009, 2010; Zhang et al. 2012, 2014), with the ultimate bearing capacity for a deeply buried foundation demonstrated to be considerably larger than that of a shallow foundation (Merifield et al. 2001; Zhang et al. 2012). Ming and Radoslaw (2005) proposed new shape factors for square and rectangular footings based on elasto-plastic model of the soil and finite element analyses.

Singh and Monika (2016) carried out studies on Shell foundations of various shapes of cross section. In all the types of footings; except shell foundations, the shape of cross section is rectangular and the base has been considered as a plane surface. The pattern of soil movement beneath the footing during loading is a significant factor contributing to load-settlement behaviour. By altering the shape of cross section of footing, it would be possible to provide better confinement of underlying soil thus improving bearing capacity.

The purpose of this paper is to investigate numerically the influence of shape of cross section on the load-settlement behaviour. A series of non-linear finite element analyses are carried out with the FE software PLAXIS 2D. Three different shapes of cross section of footing, shown in Table 1, are considered for the analyses.

2   FINITE ELEMENT ANALYSES

Finite element analyses are carried out using the commercially available finite element software *PLAXIS* 2D. For simulating the behaviour of soil, different constitutive models are available in the FE software. In the present study Mohr-Coulomb model is used to simulate soil behaviour. This non-linear model is based on the basic soil parameters that can be

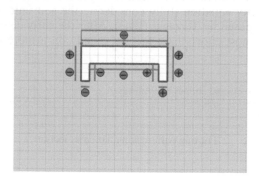

Figure 1.   Geometric model.

Table 2.   Properties of soil.

Property	Clay
Dry Unit Weight (kN/m³)	16
Specific Gravity	2.1
Cohesion (kPa)	20
Angle of Shearing Resistance (°)	0
Modulus of Elasticity (kPa)	7000
Poisson's Ratio	0.3

obtained from direct shear tests; internal friction angle and cohesion intercept. Since strip footing is used, plane strain model is adopted in the analyses. The settlement of the rigid footing is simulated using non zero prescribed displacements.

The displacement of the bottom boundary is restricted in all directions, while at the vertical sides; displacement is restricted only in the horizontal direction. The initial geostatic stress states for the analyses are set according to the unit weight of soil. The soil is modelled using 15 noded triangular elements. Mesh generation can be done automatically. Medium mesh size is adopted in all the simulations.The size of the strip footing (B) is taken as one metre and the width and depth of soil mass are taken as 10B in all analyses The footing and the confining walls are modelled using plate elements. To simulate the interaction between the footing and underlying soil, an interface element is provided at the bottom surface of footing. The geometric model is shown in Fig. 1 and the typial stress distribution after loading is shown in Figure 2.

The soil is modeled using 15-node triangular elements. Properties of locally available sand and clay are adopted in the analyses. The material properties adopted are outlined in Table 2.

3 RESULTS AND DISCUSSIONS

Vertical Stress vs Settlement curves for Rectangular Cross Section with flanges; obtained from finite element analyses are presented in Figure 3. It is seen that the presence of flanges below the edges of the strip footing improves the load-settlement behaviour. The improvement for d/B values up to 0.6 is less whereas for values of d/B > 0.6, improvement is more.

The load-settlement behaviour of footing with sloping cross section is presented in Figure 4. The optimum improvement is observed when d/B = 0.4. For higher values of d/B,

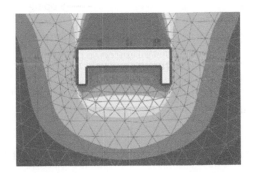

Figure 2.   Typical stress distribution after loading.

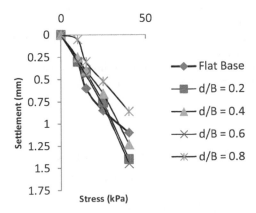

Figure 3.   Vertical stress vs settlement curves for rectangular cross section with flanges.

267

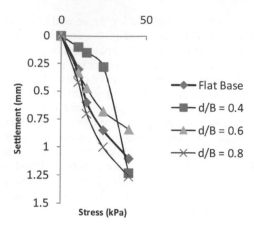

Figure 4.   Vertical stress vs settlement curves for footing with sloped cross section.

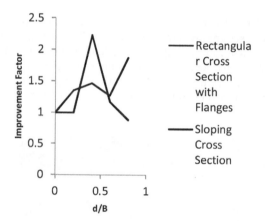

Figure 5.   Improvement factor.

the improvement reduces. This behaviour is quite contrast to that of Rectangular Cross Section with Flanges.

To quantify the improvement in load-settlement behaviour attained due to anchoring the reinforcement, an Improvement Factor (If) is defined as the ratio of stress with Various Shapes of Cross Section to that with Rectangular Cross Section; at 0.5 mm settlement.

It is seen that for rectangular cross section with flanges, the improvement factor initially increases and then reduces when d/B = 0.6. However for further higher values of d/B, the improvement factor increases.

The optimum value of d/B for footing with sloping cross section is observed to be 0.4. For higher values of d/B, the improvement factor is reducing.

4   CONCLUSIONS

From the finite element analyses carried out, the following conclusions are drawn.

- The load-settlement behaviour can be improved by altering the shape of cross section of footings
- The improvement factor depends of the geometrical parameters of the cross section

REFERENCES

Craig, R.F. (2004). Craig's soil mechanics, Taylor & Francis, New York.

Ericson Hans L, Drescher Anderw. Bearing capacity of circular footings. J Geotech Geoenviron Eng, ASCE 2002; 128(1):38–43.

Griffiths D.V., Fenton Gordon A., Manoharan N. Bearing capacity of rough rigid strip footing on cohesive soil: probabilistic study. J Geotech Geoenviron Eng, ASCE 2002; 128(9):743–50.

Hossain, M.S., and Randolph, M.F. (2009). "New mechanism-based design approach for spudcan foundations on single layer clay." J. Geotech. Geoenviron. Eng., 10.1061/(ASCE)GT.1943-5606.0000054, 1264–1274.

Hossain, M.S., and Randolph, M.F. (2010). "Deep-penetrating spudcan foundations on layered clays: centrifuge tests." Géotechnique, 60(3), 157–170.

Hu, Y., Randolph, M.F., and Watson, P.G. (1999). "Bearing response of skirted foundation on non homogeneous soil." J. Geotech. Geoenviron. Eng., 10.1061/(ASCE)1090-0241 (1999), 125:11(924), 924–935.

Lee J. and Salgado R. Estimation of bearing capacity of circular footing on sands based on CPT. J Geotech Geoenviron Eng, ASCE 2005; 131(4):442–52.

Merifield, R.S., Sloan, S.W., and Yu, H.S. (2001). "Stability of plate anchors in undrained clay." Géotechnique, 51(2), 141–153.

Meyerhof G. Shallow foundations. J Soil Mech Found Div, ASCE 1965; 91(SM2):21–31.

Michalowski Randoslaw L. and Dawson M.E. Three-dimensional analysis of limit loads on Mohr–Coulomb soil. Found Civil Environ Eng 2002; 1(1):137–47.

Ming Zhu and Radoslaw L. Michalowski, (2005), "Shape Factors for Limit Loads on Square and Rectangular Footings", Journal of Geotechnical and Geoenvironmental Engineering, Vol. 131, No. 2, ASCE, 223–231.

Prandtl L. Über die Eindringungsfestigkeit (Härte) plastischer Baustoffe und die Festigkeit von Schneiden. Zeit Angew Math Mech 1921(1):15–20.

Reissner H. Zum Erddruckproblem. In: Proceedings of first international congress of applied mechanics. Delft; 1924. p. 295–311.

Salgado R., Lyamin A.V., Sloan S.W., Yu H.S. Two-and three-dimensional bearing capacity of foundation in clay. Geotechnique 2004; 54(5):297–306.

Singh, S.K., and Monika, K., (2016), "Load carrying capacity of shell foundations on treated and untreated soils" Indian Geotechnical Conference 2016, IIT Madras Taiebat HA, Carter JP. Numerical studies of the bearing capacity of shallow foundation. Geotechnique 2000; 50(4):409–18.

Terzaghi K. Theoretical soil mechanics. New York: Wiley; 1943.

Wang, C.X., and Carter, J.P. (2002). "Deep penetration of strip and circular footings into layered clays." Int. J. Geomech.,10.1061/(ASCE)1532-641, (2002), 2:2(205), 205–232.

Zhang, Y., Bienen, B., Cassidy, M.J., and Gourvenec, S. (2012). "Undrained bearing capacity of deeply buried flat circular footings under general loading." J. Geotech. Geoenviron.Eng., 10.1061/(ASCE) GT.1943-5606.0000606, 385–397.

Zhang, Y., Wang, D., Cassidy, M.J., and Bienen, B. (2014). "Effect of installation on the bearing capacity of a spudcan under combined loading in soft clay." J. Geotech.Geoenviron. Eng., 10.1061/(ASCE) GT.1943-5606.0001126, 04014029.

Emerging Trends in Engineering, Science and Technology for Society,
Energy and Environment – Vanchipura & Jiji (Eds)
© 2018 Taylor & Francis Group, London, ISBN 978-0-8153-5760-5

Experimental study to determine the elastic wave velocities of marineclay

K. Anagha & M.N. Sandeep
Department of Civil Engineering, IES College of Engineering, Thrissur, Kerala, India

K.S. Beena
Department of Civil Engineering, Cochin University of Science and Technology, Kochi, Kerala, India

ABSTRACT: P-wave velocity and shear wave velocity are the dynamic parameters used to determine soil characteristics. Ultrasonic pulse velocity tests are usually used for p-wave velocity measurements. In this study, tests were conducted to investigate the effect of water content and dry densities on p-wave velocity and unconfined compressive strength and thus, to establish a correlation between p-wave velocity and unconfined compressive strength. Marine clay was the material used for the study. Experiments were conducted on a soil sample of diameter 3.75 cm and length 7.5 cm. Results show that p-wave velocity and unconfined compressive strength demonstrate a similar trend while varying the parameters such as water content and dry density. Theoretical correlations connecting elastic wave velocities and electrical resistivity were established for finding the shear wave velocities. Gassmann's and Archie's equations for porosity were used to derive the equation for shear wave velocity. Electrical resistivity was calculated experimentally.

1 INTRODUCTION

Elastic waves are generated in nature by the movement of tectonic plates, explosions, landslides and so on. Seismic waves are the vibrations generated at the interior of the earth, during rupture or explosions, which take energy away from the center of the earth. Vibrations which travel through the interior of the earth rather than surface are called body waves. Body waves are classified into two, primary waves or compression waves (p-waves), and secondary or shear waves (s-waves). The determination of elastic waves and their movement is an important parameter for different field applications such as insertion of deep foundations in soil, soil stabilization, compaction characteristics, anisotropic behavior of soils, stiffness evaluation of soils, sample quality determination, stratification of soils and so on.

These waves can also be modeled artificially by using different piezoelectric transducers in the field as well as in the laboratory. P-wave velocity is mainly determined using ultrasonic methods. Ultrasonic methods are usually used for assessing concrete quality and strength determination (Lawson et al., 2011). Similarly, ultrasonic methods can also be used for investigating the compaction characteristics of soils, finding that the variation of velocity with water content is similar to the variation of density with water content (Nazli et al., 2000). P-wave velocity, shear wave velocity and damping characteristics can be estimated by different calibrated ultrasonic equipment (Zahid et al., 2011). Elastic wave velocities are determined by piezo disk elements and electrical resistivity by electrical resistivity probes. A theoretical correlation connecting the elastic wave velocity and electrical resistivity was proposed by Jong et al. (2015).

The focus of this study was to find the primary wave velocity of soft soil using an ultrasonic pulse velocity test at different water contents and dry densities and to correlate the primary wave velocities to the unconfined compressive strength of the soil. The shear wave velocity of the soil was determined by proposing a new equation for shear wave velocity which connects elastic wave velocity and electrical resistivity.

Figure 1. Marine clay used for the study.

Table 1. Properties of soil.

Properties	Values
Specific gravity	2.5
Liquid limit (%)	70
Plastic limit (%)	24.4
Plasticity index (%)	45.6
Shrinkage limit (%)	12.7
Clay (%)	47
Silt (%)	33
Sand (%)	20
Max.dry density (g/cc)	1.62
Optimum moisture content (%)	20

2 MATERIALS USED

The soil used for the study was marine clay, which was collected from Kochi. It is blackish in color. Figure 1 shows the soil used for this study.

The soil properties were determined and are listed in Table 1.

3 TEST DESCRIPTION

To prepare specimens of diameter 3.75 cm and length 7.5 cm for the unconfined compressive test and ultrasonic pulse velocity test, three densities 1.62 g/cc, 1.5 g/cc and 1.4 g/cc were fixed from the compaction curve. The compaction curve is shown in Figure 2. Water content corresponding to these densities were taken from the wet and dry sides of the compaction curve and are tabulated in Table 2.

3.1 Ultrasonic pulse velocity test

A pair of piezo electric transducers, which are capable of producing stress waves with frequencies higher than 20 kHz were used for testing. Here, 60 kHz center frequency transducers were used. A battery controlled ultrasonic tester was used to obtain the time of travel through the soil. Through transmission method is used for the study, in which two opposite surfaces are available for testing. After calibrating the instrument with a glass prism, a sample was placed in between the two identical transducers. A suitable couplant was used on the surfaces of the transducer to transmit the signal effectively into the specimen. By dividing the path length or length of specimen by the time required to travel the length, ultrasonic pulse velocity or p-wave velocity can be calculated. The test setup for the ultrasonic pulse velocity test is shown in Figure 3.

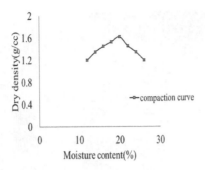

Figure 2.   Compaction curve.

Table 2.   Sample properties.

Sample	Dry density (g/cc)	Water content (%)
1	1.62	20
2	1.5	17.5
3	1.5	21.4
4	1.4	15.3
5	1.4	23

Figure 3.   Test setup for ultrasonic pulse velocity test.

3.2   *Electrical resistivity test*

An apparatus was constructed for measuring electrical resistivity, which consists of copper electrodes, cylindrical mold, ammeter, voltmeter, auto transformer and an AC Supply. It is grooved to fix a cylinder into it. A cylinder of length 8.7 cm and diameter 3.75 cm was used to prepare the mold. The cylinder used was a PVC pipe, which is a good electrical insulator. The mold was filled to the required moisture content and dry density. Then, it was connected to an alternating current source of 230 V. To regulate the voltage, an auto transformer with varying voltage between 0–200 V was used. A voltmeter of accuracy 10 V was connected in parallel to the mold and an ammeter in series to the mold. A voltage of 10 V was applied to the sample and then current flowing through the sample was measured from the ammeter and resistance was calculated from Ohm's Law. These electrical resistance values are used for calculating resistivity. Figure 4 shows the experimental arrangement for finding the electrical resistivity.

3.3   *Shear wave velocity from electrical resistivity values*

Archie developed (Jong et al., 2015), an equation for electrical resistivity of soil mixture in terms of porosity (n) and electrical resistivity of the electrolyte (E_{el}):

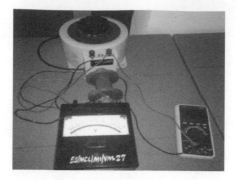

Figure 4.   Test setup for finding electrical resistivity of soil.

$$E_{mix} = \alpha \, E_{el} \, n^{-m} \tag{1}$$

where α is the cementation factor taking values between 0.6 and 3.5, 'm' is the shape factor indicating the shape of the particle, porous structure and specific surface. It takes values in the range of 1.4–2.2. The only term which connects the electrical resistivity and elastic wave velocity is the porosity. Movement of waves results volume change in soil. Gassmann suggested (Jong et al., 2015) an equation for bulk modulus in terms of bulk modulus of soil grain (B_g), bulk modulus of soil skeleton (B_{sk}), bulk modulus of pore fluid (B_f) and porosity:

$$B_{Gassmann} = Bsk + \frac{\left(1 - \dfrac{Bsk}{Bg}\right)}{\dfrac{n}{Bf} - \dfrac{Bsk}{Bg^2} + \dfrac{1-n}{Bg}} \tag{2}$$

Gassmann suggested another equation for bulk density in terms of the elastic wave velocities:

$$B_{Gassmann} = \rho \, (Vp^2 - 4/3 \; Vs^2) \tag{3}$$

By equating Equations 2 and 3, an equation for porosity is obtained as:

$$n_{Gassmann} = \frac{\dfrac{\dfrac{Bsk}{Bg} - 1}{Bsk - \rho\left(Vp^2 - \dfrac{4}{B}Vs^2\right)} - \dfrac{1}{Bg} + \dfrac{Bsk}{Bg^2}}{\dfrac{1}{Bf} - \dfrac{1}{Bg}} \tag{4}$$

From Equation 1:

$$n_{Archie} = E_{mix}^{\,1/-m}/\alpha \, E_{el} \qquad \text{(Jong et al., 2015)} \tag{5}$$

By equating Equations 4 and 5, a new equation for finding the shear wave velocity is derived.

The equation for shear wave velocity can be written as:

$$V_s = \frac{\left(\sqrt{3}\left[\sqrt{\left(E_{mix}^{\,1/(-m)}(\rho V_p^2 - B_{sk})(B_g - B_f)B_g^2\right.}\right.}{\left.\left.-\left((B_g - B_{sk})\rho V_p^2 + B_g^2 - B_{sk}^2\right) \times \alpha E_{el}B_g B_f\right)\right)\right)}{\left(2\sqrt{\rho}\sqrt{\left((E_{mix}^{\,1/(-m)}(B_g - B_f)B_g^2\right) - (B_g - B_{sk})\alpha E_{el}B_g B_f\right)}\right)} \tag{6}$$

274

where B_g, B_f, B_{sk} are the bulk moduli of the soil grain, fluid and the soil skeleton, respectively, and it takes the values as 14.5 GPa, 2.18 GPa and 7.78 GPa and E_{el} is the electrical resistivity of electrolyte and is taken as 0.312 Ωm (Jong et al., 2015).

4 RESULTS AND DISCUSSION

4.1 *P-wave velocity measurements*

4.1.1 *Variation of ultrasonic pulse velocity and unconfined compressive strength with dry densities*

Figures 5 and 6 show the variation of p-wave velocity and unconfined compressive strength with dry density at dry of optimum and wet of optimum water contents respectively.

For both wet of optimum and dry of optimum water contents, as the dry density increases both p-wave velocity and unconfined compressive strength increase.

The increase in p-wave velocity for an increase in dry density shows that wave propagation will be faster in the case of solids than through voids. Both p-wave velocity and unconfined compressive strength show a similar trend while varying dry density for different soils. Therefore, the ultrasonic pulse velocity can be used as a parameter to estimate unconfined compressive strength of soil indirectly.

4.1.2 *Variation of ultrasonic pulse velocity and unconfined compressive strength with varying water contents*

Table 3 shows obtained p-wave velocity and unconfined compressive strength.

As the water content increases from 17.5 to 21.4%, p-wave velocity and the unconfined compressive strength decreases, for the dry density of 1.5 g/cc. For 1.4 g/cc, water content increases from 15.3 to 23%, while both p-wave velocity and unconfined compressive strength decreases. Waves will propagate faster in solids than liquids for the same dry density.

4.2 *Shear wave velocity measurements*

The shear wave velocity soil is obtained from Equation 6, which connects the electrical resistivity with elastic wave velocities. Table 4 shows the obtained resistivity from the resistivity method and obtained shear wave velocity by using Equation 6 for soil.

For marine clay, the resistivity values range from 0.86 to 6.23 Ωm. The shear wave velocity values for marine clay vary between 255 and 337 m/s.

This method can be effectively used to determine the shear wave velocity of soft soils and it can provide a fast approach for finding the shear wave velocities.

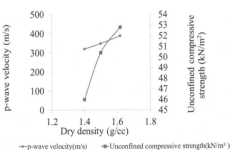

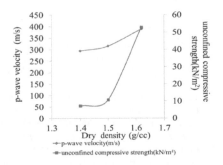

Figure 5. Variation of unconfined compressive strength and p-wave velocity for different dry density at dry of optimum water content.

Figure 6. Variation of unconfined compressive strength and p-wave velocity for different dry density at wet of optimum water content.

Table 3. Obtained p-wave velocity and unconfined compressive strength.

Dry density (g/cc)	1.4		1.5	
Moisture content (%)	15.3	23	17.5	21.4
P-wave velocity (m/s)	320.16	294.11	348.31	314.8
Decrease in velocity (%)		8.14		9.62
Unconfined compressive strength (kN/m^2)	46	7.3	50.4	10.8
Decrease in strength (%)		84.1		78.6

Table 4. Obtained electrical resistivity and shear wave velocity.

Dry density and water content	Electrical resistivity (Ωm)	Shear wave velocity (m/s) Vs
1.62 g/cc, 20%	6.23	336.58
1.5 g/cc, 17.5%	3.48	301.64
1.5 g/cc, 21.4%	1.53	272.63
1.4 g/cc, 15.3%	2.05	277.26
1.4 g/cc, 23%	0.86	254.71

5 CONCLUSIONS

P-wave velocity and shear wave velocity are important parameters in geotechnical engineering and can be used to predict the properties of soil without sampling and testing. The ultrasonic pulse velocity test is mainly used to determine the p-wave velocity.

As the dry density increases, both compression wave velocity and unconfined compressive strength increase for both dry of optimum and wet of optimum water contents. This is because the wave transmission through solids is faster than through voids. P-wave velocity and unconfined compressive strength decrease with an increase in moisture content for the same dry density. The decrease in p-wave velocity is due to the slower rate of wave propagation in liquids than in solids. Therefore, the ultrasonic pulse velocity can be used as a parameter to estimate unconfined compressive strength of soil indirectly.

A theoretical correlation connecting elastic wave velocities and electrical resistivity was used for finding the shear wave velocities. It was found that the shear wave velocity of the soil varies with changes in the dry density and water content of the soil.

REFERENCES

Jong, S.L. & Yoon, M.H. (2015). Theoretical relationship between elastic wave velocity and electrical resistivity. *Journal of Applied Geophysics, 116*, 51–61.

Lawson, K.A., Danso, H.C., Odoi, C.A. & Quashie, F.K. (2011). Non-destructive evaluation of concrete using ultrasonic pulse velocity. *Research Journal of Applied Sciences, Engineering and Technology, 3*(6), 499–504.

Nazli, Y. Inci, G. & Miller, C.J. (2000). Ultrasonic testing for compacted clayey soils. *Journal of Geotechnical and Geoenvironmental Engineering, ASCE, 287, 5*, 54–68.

Nazli, Y., James, L.H. & Mumtaz, A.U. (2003). Ultrasonic assessment of stabilized soils. *Journal of Geotechnical and Geoenvironmental Engineering, ASCE, 301, 14,170*–181.

Zahid, K., Cascante, G. & Naggar, M.H. (2011). Measurement of dynamic properties of stiff specimens using ultrasonic waves. *Canadian Geotechnical Journal, 48*, 1–15.

Emerging Trends in Engineering, Science and Technology for Society,
Energy and Environment – Vanchipura & Jiji (Eds)
© 2018 Taylor & Francis Group, London, ISBN 978-0-8153-5760-5

Behavior of a single pile under combined and uplift loads: A review

D. Divya, R.B. Jiniraj & P.K. Jayasree
Department of Civil Engineering, College of Engineering Trivandrum, Thiruvananthapuram, India

ABSTRACT: Pile foundations are deep foundations constructed to transfer loads from the superstructure to the hard strata beneath. The common loads acting on pile vertical, lateral and uplift loads. As the vertical load on the pile changes, variations are shown by the pile in its lateral behaviour. As the vertical load increases, the lateral capacity of the piles also tends to increase. The lateral behaviour of the pile under combined loading depends on various factors such as the order of loading, properties of the soil and pile. Even though the vertical force is the most common force acting on a pile, uplift forces are also seen, especially in foundations of structures in a harbor and in the case of submerged platforms of waterfront structures. Hydrostatic pressure, overturning moments, lateral force and swelling of surrounding soil cause uplift of piles. In the case of sandy soils, uplift resistance depends on the skin friction of piles and is determined by considering the shape of the failure surface, the shear strength of the soil and weight of the pile. This paper presents a review of the research works conducted on pile foundations subject to combined (vertical and lateral) loading conditions and foundations subject to uplift loads.

1   INTRODUCTION

Pile foundations are slender members provided to transfer a load from the super structure to the hard strata beneath. Lateral loads are also experienced in pile foundations of some structures. High velocity wind, dynamic earthquake pressure, soil pressure and so on, can cause lateral loads. Practically, when lateral loading is present on pile foundations, the actual condition occurring involves combined vertical and lateral loading on the structures. Hence, the behavior of piles under combined loading is an important matter to be considered during the structural design of piles. Many theoretical methods exist to evaluate the behavior of piles under different loads such asaxial-compressive, axial-uplift and lateral loading. To minimize the complexity in analyzing different loads simultaneously, loads are analyzed separately. Vertical loads on piles are analyzed to determine the bearing capacity and lateral loads to determine the elastic behavior of piles. However, in the field, lateral loads are of a large order, hence the effect of combined loading cannot be neglected. Foundations of retaining walls, anchors for bulkheads, bridge abutments, piers, anchorage for guyed structures and offshore structures, supported on piles are exposed to large inclined uplift loads. Most of the studies on piles are concentrated on vertical loads rather than uplift forces. Uplift forces may result either from a lateral force or direct pull out, hydrostatic pressure, overturning moments, lateral force and swelling of the surrounding soil. The effect of geometric properties like slenderness ratio on the lateral capacity of piles under combined lateral and vertical loading has also been a topic of study. This paper presents a review on the research works addressing both of these aspects.

2   INFLUENCE OF VERTICAL LOAD ON THE LATERAL RESPONSE OF THE PILE

Combined loading effects on piles have been a matter of study for the last ten decades. Trochanis et al. (1991) used 3D FE to study the effect of combined loading on piles and reported that a vertical load has a lesser effect on the lateral behavior piles. Anagnostopoulos and Georgiadis (1993) studied the lateral behavior of piles under axial load using experimental models and 2D FE analysis, and reported that FEM analysis is a useful tool to predict the stress developed and volume changes in soil mass under combined vertical and lateral loads when compared to other theoretical and 2D analytical methods. Karthigeyan et al. (2006, 2007) investigated the performance of piles under combined loading using 3D FE analysis and noticed that sandy soil experienced an increase in lateral capacity with an increase in vertical load, while in clayey soil the trend is reversed. Figure 1 shows the increase in lateral capacity with an increase in vertical load for sandy soil and Figure 2 shows the lateral behavior of piles under vertical load for various slenderness ratios. Achmus and Thieken (2010) used 3D FE methods to analyze the behavior of piles when acted upon by vertical and lateral loads in sandy soil and reported that the lateral loads mobilize passive earth pressure whereas vertical loads mobilize skin friction.

Hussien et al. (2012, 2014a, 2014b) used FE models to study the soil-pile interaction of free head piles installed in sandy soils and noticed a slight increase in lateral capacity of piles under vertical loads and inferred that the improvement in lateral capacity of piles may be due to the increase in confining pressure in the surrounding sand deposits. Maru and Vanza

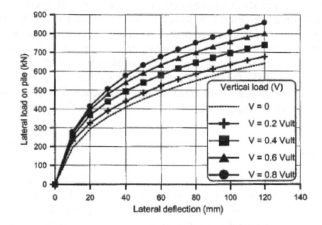

Figure 1.   Effect of vertical load on lateral capacity (Source: Karthigeyan et al., 2007).

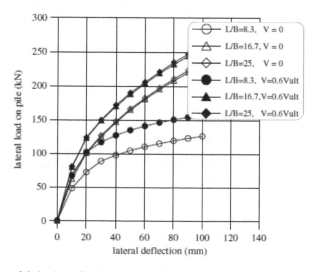

Figure 2.   The lateral behavior of piles under vertical load for various slenderness ratios (Source: Karthigeyan et al., 2006).

278

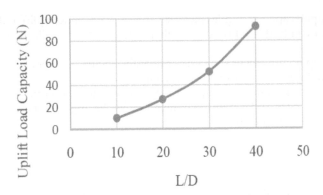

Figure 3.   Variation of uplift load capacity of a single pile with L/D ratio (Source: Parekh & Thakare, 2016).

(2017) conducted laboratory tests to investigate the lateral behavior of piles under axial-compressive loads and reported that the lateral load carrying capacity of piles improved with an increase in axial load and slenderness ratio. Hazzar et al. (2017) used 3D FD to study the effect of axial loads on the lateral behavior of pile foundations in sand and concluded that the behavior of piles in the lateral direction is not affected by the axial loads acting on the piles.

3 INFLUENCE OF UPLIFT LOAD ON PILES

The study of the behavior of a single pile under uplift loads and the factors affecting the uplift capacity of piles needs to be improved. Ayothiraman and Reddy (2015) investigated pile behavior under combined uplift and lateral loading in sand. The results concluded that the uplift load v/s axial displacement behavior is nonlinear for independent loading and in the case of combined loading. The behavior of a single pile, square pile group and hexago-nal pile group under independent lateral and uplift loading, and also under combined uplift and lateral loading were studied by Parekh and Thakare (2016). Based on their results, they concluded that, with an increase in L/D ratio of the pile, the uplift load capacity of a single pile increases linearly upto a certain point for both independent uplift loading and combined loading with a constant lateral load. Figure 3 shows the variation of uplift load capacity of single piles with an L/D ratio. To determine the ultimate uplift capacity of piles in sand, Chattopadhyay and Pise (1986) proposed an analytical method with an assumed curved fail-ure surface through the soil and determined the effects of factors like L/D ratio, the angle of shearing resistance and pile friction angle on the ultimate uplift capacity of the pile. Das and Seeley (1975) conducted some model tests in loose granular soil for determining the ultimate uplift capacity of vertical piles under axial pull. The results include the variation of unit uplift skin friction with the embedment depth. Rao and Venkatesh (1985) conducted labora-tory studies in uniform sands to find the uplift behavior of short piles and reported that the uplift capacity of piles was found to increases with the L/D ratio, the density of soil, particle size and pile roughness.

4 CONCLUSION

- In non-cohesive soils, the lateral behavior of piles is influenced by the vertical load acting on them which shows an increase in the lateral load carrying capacity.
- The effect of vertical load depends on the sequence of loading, geometric properties of the pile and material properties of the soil.
- Lateral load carrying capacity shows an increasing trend with an increase in vertical load as well as slenderness ratio.

- The ultimate uplift capacity of piles increases substantially under uplift and lateral loading.
- With an increase in both L/D ratio and relative density of the soil, the uplift capacity of piles increases.
- The pile head uplift movement shows a considerable increase under combined loading but very little on independent uplift and lateral loading.
- Longer piles are more resistant to uplift load compared to shorter piles.
- The resistance of piles under uplift load increases as the angle of shearing resistance and soil-pile friction angle increase.

REFERENCES

Achmus, M. & Thieken, K. (2010). On the behaviour of piles in non-cohesive soil under combined horizontal and vertical loading. *ActaGeotechnica, 5*(3), 199–210.

Anagnostopoulos, C. & Georgiadis, M. (1993). Interaction of axial and lateral pile responses. *Journal of Geotechnical Engineering, 119*(4), 793–798.

Ayothiraman, R. & Reddy, K.M. (2015). Experimental studies on behavior of single pile under combined uplift and lateral loading. *Journal of Geotechnical and Geo environmental Engineering*, ASCE, *141*, 1–10.

Chattopadhyay, B.C. & Pise, P.J. (1986). Uplift capacity of piles in sand. *Journal of Geotechnical Engineering,* ASCE, *112*(9), 888–904.

Das, B.M. & Seeley, G.R. (1975). Uplift capacity of buried model piles in sand. *Journal of GTE Div. ASCE, 10*, 1091–1094.

Hazzar, L., Hussien, M.N. & Karray, M. (2017). Influence of vertical load on lateral response of pile foundation in sands and clays. *Journal of Rock Mechanics and Geotechnical Engineering, 9*, 291–304.

Hussien, M.N., Tobita, T., Iai, S. & Karray, M. (2014a). Influence of pullout loads on the lateral response of pile foundation. *Proceedings of the 67th Canadian Geotechnical International Conference.*

Hussien, M.N., Tobita, T., Iai, S. & Karray, M. (2014b). On the influence of vertical loads on the lateral response of pile foundation. *Computers and Geotechnics, 55*, 392–403.

Hussien, M.N., Tobita, T., Iai, S. & Rollins, K.M. (2012). Vertical load effect on the lateral pile group resistance in sand response. *Geomechanics and Geoengineering, 7*(4), 263–282.

Karthigeyan, S., Ramakrishna, V.V.G.S.T. & Rajagopal, K. (2006). Influence of vertical load on the lateral response of piles in sand. *Computers and Geotechnics, 33*, 121–131.

Karthigeyan, S., Ramakrishna, V.V.G.S.T. & Rajagopal, K. (2007). Numerical investigation of the effect of vertical load on the lateral response of piles. *Journal of Geotechnical and Geoenvironmental Engineering, 133*(5), 512–521.

Maru, V. & Vanza, M.G. (2017). Lateral behaviour of pile under the effect of vertical load. *Journal of Information, Knowledge and Research in Civil Engineering, 4*(2), 482–485.

Parekh, B. & Thakare, S.W. (2016). Performance of pile groups under combined uplift and lateral loading. *International Journal of Innovative Research in Science, Engineering and Technology, 5*(6), 9219–9227.

Rao, K.S. & Venkatesh, K.H. (1985). Uplift behavior of short piles in uniform sand. *Journal of Soils and Foundations, 25*(4), 1–7.

Trochanis, A.M., Bielak, J. & Christiano, P. (1991). Three-dimensional nonlinear study of piles. *Journal of Geotechnical Engineering ASCE, 117*(3), 429–447.

Emerging Trends in Engineering, Science and Technology for Society,
Energy and Environment – Vanchipura & Jiji (Eds)
© 2018 Taylor & Francis Group, London, ISBN 978-0-8153-5760-5

Reported correlations on compaction characteristics of fine grained soils in the standard proctor test—a critical reappraisal

G. Sreelekshmy Pillai
N.S.S. College of Engineering, Palakkad, Kerala, India

P. Vinod
Government Engineering College, Thrissur, Kerala, India

ABSTRACT: The determination of compaction characteristics is very important in the case of earthworks construction and as such, attempts to develop empirical equations for their prediction has been greatly received. For preliminary design and assessment, correlations of compaction characteristics with index properties have been attempted by various investigators. The reported methods for prediction of compaction characteristics of fine grained soils in the standard Proctor compaction test are critically reviewed in this paper. Among the various correlations, those proposed by Dokovic et al. (2013) and Vinod and Sreelekshmy Pillai (2017), are observed to be the most accurate and precise ones. With the use of toughness limit, which effectively takes care of the combined effect of liquid limit (w_L) and plastic limit (w_p), a new set of equations for the prediction of compaction characteristics at the standard Proctor compactive effort, are also proposed.

1 INTRODUCTION

In the construction of many earth structures, such as embankments, it is essential to determine the compaction characteristics. The compaction characteristics of a soil, as obtained from laboratory compaction tests, are the Maximum Dry Unit Weight ($\gamma_{d\text{-max}}$) and the Optimum Water Content (OMC). The procedure for the determination of $\gamma_{d\text{-max}}$ and OMC, although simple, is time consuming and laborious since soil from highly variable sources need to be tested to assess their suitability for their desired purpose. Delay in testing results could adversely impact project timing and unnecessarily limit the search for sources of suitable material. Therefore, for the preliminary assessment of the suitability of borrow materials, predictive models can be used, especially when index properties are already known. Several attempts have been made in the past for the determination of γd-max and OMC of fine grained soils at the standard Proctor compactive effort. Most of them correlate compaction characteristics with the Liquid Limit (w_L) or the Plastic Limit (w_p) or a combination of both. In this paper, the reported methods for prediction of compaction characteristics of fine grained soils in terms of index properties are critically reviewed, and a more refined correlation is proposed.

2 CORRELATIONS OF COMPACTION CHARACTERISTICS WITH INDEX PROPERTIES AT THE STANDARD PROCTOR COMPACTIVE EFFORT

Al-Khafaji (1993) conducted studies on fine grained soils from Iraq and the USA, to find out the relationship between Atterberg limits and compaction characteristics at the standard Proctor compactive effort. The following equations are recommended for Iraqi soils:

$$OMC = 3.130 + 0.630\,w_p + 0.240\,w_L \qquad (1)$$
$$\gamma_{d\text{-max}} = 2.440 - 0.020\,w_p - 0.008\,w_L \qquad (2)$$

For soils of the USA, the equations are as follows:

$$OMC = 0.140\ w_L + 0.540\ w_P \tag{3}$$
$$\gamma_{d\text{-max}} = 2.270 - 0.019\ w_P - 0.003\ w_L \tag{4}$$

An empirical method was proposed by Blotz et al. (1998) for estimating $\gamma_{d\text{-max}}$ and OMC of fine grained soils at any rational compactive effort. They used the soil data from published literature and their experimental study. The following equations were recommended for compaction characteristics at the standard Proctor compactive effort:

$$OMC = -33.82\ \log w_L - 0.67\ w_L + 43.53 \tag{5}$$
$$\gamma_{d\text{-max}} = 6.29\ \log w_L - 0.16\ w_L + 14.41 \tag{6}$$

Gurtug and Sridharan (2004) brought out the effect of compaction energy on compaction characteristics of fine grained soils. The analysis carried out was based on data from published literature and experimental studies. The equations suggested were:

$$OMC = 0.92\ w_P \tag{7}$$
$$\gamma_{d\text{-max}} = 0.98\ \gamma_{d\text{-wP}} \tag{8}$$

Here, $\gamma_{d\text{-max}}$ was expressed in terms of $\gamma_{d\text{-wP}}$ which is the Maximum Dry Unit Weight at Plastic Limit Water Content, and is given by:

$$\gamma_{d}.w_P = (G\ \gamma_w)/(1 + w_P G) \tag{9}$$

Studies were carried out by Sridharan and Nagaraj (2005) to determine which index property correlated well with the compaction characteristics of fine grained soils at the standard Proctor compactive effort. The correlation of the compaction characteristics with w_P was found to be much better than that with w_L and the plasticity index. Using the data from their study and from the literature, the following equations were recommended:

$$OMC = 0.92\ w_P \tag{10}$$
$$\gamma_{d\text{-max}} = 0.23\ (93.30 - w_P) \tag{11}$$

Based on the experimental studies on fine grained soils from different regions in Turkey, and based on the published data, Sivrikaya (2008), also pointed out the importance of w_P in the prediction of compaction characteristics. The proposed equation for compaction characteristics at the standard Proctor compactive effort being:

$$OMC = 0.94\ w_P \tag{12}$$
$$\gamma_{d\text{-max}} = 21.97 - 0.27\ OMC \tag{13}$$

Gunaydin (2009) conducted a multiple regression analysis to determine the relationship between percentage of soil finer than 75 μm (F_C), Percentage of Sand (S_C), Percentage of Gravel (G_C), Specific Gravity (G), w_L, w_P and compaction characteristics at the standard Proctor compactive effort. The analysis was undertaken using the data of fine grained soils from Nidge, Turkey. For prediction of OMC and $\gamma_{d\text{-max}}$, equations using w_L and w_P were recommended:

$$OMC = 0.323\ w_L + 0.157\ w_P \tag{14}$$
$$\gamma_{d\text{-max}} = 0.078\ w_L - 0.062\ w_P \tag{15}$$

Dokovic et al. (2013) performed the standard Proctor tests on samples of fine grained soils from Serbia. Multiple regression analysis was carried out to determine the relationship between Atterberg limits and compaction characteristics, the empirical equation being:

$$OMC = 4.180 + 0.160\ w_L + 0.323\ w_P \tag{16}$$
$$\gamma_{d\text{-max}} = 2.140 - 0.007\ w_L - 0.005\ w_P \tag{17}$$

Varghese et al. (2013) proposed empirical equations to relate $\gamma_{d\text{-max}}$ and OMC at the standard Proctor compactive effort with soil index properties. Empirical equations were developed using multiple regression analysis on fine grained soil data from literature. The equations were:

$$OMC = 5.926 + 0.613\ w_P - 0.026\ w_L \qquad (18)$$
$$\gamma_{d\text{-max}} = 20.159 - 0.157\ w_P - 0.001\ w_L \qquad (19)$$

By defining a fine grained soil by its w_L and w_P, the interrelationship between compaction energy and compaction characteristics was brought within a definite framework by Sreelekshmy Pillai and Vinod (2016), using data from literature. The recommended equations for compaction characteristics of fine grained soils at the standard Proctor compactive effort were:

$$OMC = 0.172\ w_L + 0.563\ w_P \qquad (20)$$
$$\gamma_{d\text{-max}} = 0.314\ \gamma_{d\text{-wL}} + 0.742\ \gamma_{d\text{-wP}} \qquad (21)$$

where $\gamma_{d\text{-wL}}$ and $\gamma_{d\text{-wP}}$ are Maximum Dry Unit Weight at w_L and w_P Water Contents, respectively.

Through a review and analysis of reported literature on compaction characteristics of fine grained soils, Vinod and Sreelekshmy Pillai (2017) showed that Toughness Limit (w_T) (which is a function of w_L and w_P) bears a good correlation with $\gamma_{d\text{-max}}$ and OMC; the recommended equations for compaction characteristics at the standard Proctor compactive effort being:

Table 1. Summary of the reported empirical equations proposed by various researchers.

Source	Empirical equation	No. of data used for development of correlation	Source of data
Al-Khafaji (1993)	$OMC = 3.130 + 0.63\ w_P + 0.240\ w_L$ $\gamma_{d\ max} = 2.440 - 0.020\ w_P - 0.008\ w_L$	87	Fine grained soils of Iraq
	$OMC = 0.140\ w_L + 0.540\ w_P$ $\gamma_{d\text{-max}} = 2.270 - 0.019\ w_P - 0.003\ w_L$	15	Fine grained soils from the USA
Blotz et al. (1998)	$OMC = 34.32 - 33.82\ \log w_L + 0.67\ w_L + 9.21$ $\gamma_{d\text{-max}}\ 6.28\ w_L - 2.60 + 0.16\ w_L + 17.02$	22	Published data
Gurtug and Sridharan (2004)	$OMC = 0.92\ w_P$ $\gamma_{d\text{-max}} = 0.98\ \gamma_{d\text{-wP}}$	86	Experimental studies and published data
Sridharan and Nagaraj (2005)	$OMC = 0.92\ w_P$ $\gamma_{d\text{-max}} = 0.23\ (93.30 - w_P)$	64	Experimental studies and published data
Sivrikaya et al. (2008)	$OMC = 0.940\ w_P$ $\gamma_{d\text{-max}} = 21.97 - 0.276\ OMC$	156	Fine grained soils from Turkey and published data
Gunaydin (2009)	$OMC = 0.323\ w_L + 0.157\ w_P$ $\gamma_{d\text{-max}} = +0.078\ w_L + 0.062\ w_P$	126	Fine grained soils of Turkey
Dokovic et al. (2013)	$OMC = 4.180 + 0.160\ w_L + 0.323\ w_P$ $\gamma_{d\text{-max}} = 2.140 - 0.007\ w_L - 0.005$	72	Fine grained soils of Serbia
Varghese et al. (2013)	$OMC = 5.926 + 0.613\ w_P - 0.026\ w_L$ $\gamma_{d\text{-max}} = 20.159 - 0.157\ w_P - 0.001\ w_L$	95	Published data
Sreelekshmy Pillai and Vinod (2016)	$OMC = 0.172\ w_L + 0.563\ w_P$ $\gamma_{d\text{-max}} = 0.314\ \gamma_{d\text{-wL}} + 0.742\ \gamma_{d\text{-wP}}$	63 28	Published data
Vinod and Sreelekshmy Pillai (2017)	$OMC = 0.615\ w_T$ $\gamma_{d\text{-max}} = 1.134\ \gamma_{d\text{-wT}}$	137 102	Published data

$$OMC = 0.615 \, w_T \tag{22}$$
$$\gamma_{d\text{-max}} = 1.134 \, \gamma_{d\text{-wT}} \tag{23}$$

where,

$$w_T = w_P + 0.42 \, (w_L - w_P) \tag{24}$$

and $\gamma_{d\text{-wT}}$ represents the Maximum Dry Unit Weight at Toughness Limit Water Content.

Table 1 summarizes the empirical equations proposed by various researchers, source of data used by them in the development of correlations and the number of data points used.

A critical review of the reported methods for prediction of compaction characteristics of fine grained soils is presented in the following section.

3 RESULTS AND DISCUSSION

In order to determine the accuracy and precision of the above correlations, compaction characteristics of 493 fine grained soils reported in the literature review, along with their index properties, was used. The data was obtained from the following sources: McRae (1958); Wang and Huang (1984); Daniel and Benson (1990); Al-Khafaji (1993); Daniel and Wu (1993); Benson and Trast (1995); Blotz et al. (1998); Gurtug and Sridharan (2004); Sridharan and Nagaraj (2005); Horpibulsuk et al. (2008); Sivrikaya (2008); Gunaydin (2009); Roy et al. (2009); Patel and Desai (2010); Datta and Chattopadhyay (2011); Beera and Ghosh (2013); Varghese et al. (2013), Shirur and Hiremath (2014); Talukdar (2014); and Nagaraj et al. (2015).

According to Cherubini and Giasi (2000), a logical assessment of the validity of any empirical correlation can be made by an evaluation technique which simultaneously takes into consideration accuracy as well as precision. Accuracy can be estimated by the Mean Value (μ) and precision by means of Standard Deviation (s). A global evaluation of the accuracy of a correlation can then be made by two different indices. Ranking Distance (RD), (Cherubini & Orr, 2000) and Ranking Index (RI) (Briaud & Tucker, 1998) can be defined as follows:

$$RD = \sqrt{\left\{\left[1 - \mu\left(\frac{\text{predicted value}}{\text{observed value}}\right)\right]^2 + s^2\left(\frac{\text{predicted value}}{\text{observed value}}\right)\right\}} \tag{25}$$

$$RI = \mu\left|\ln\left(\frac{\text{predicted value}}{\text{observed value}}\right)\right| + s\left|\ln\left(\frac{\text{predicted value}}{\text{observed value}}\right)\right| \tag{26}$$

For a good correlation, both of these indices tend to zero. The μ, s, RD and RI of the predicted to observed values of $\gamma_{d\text{-max}}$ and OMC were calculated and used to compare the relative accuracy and precision of the above correlations (Equation 1 through Equation 8 and Equation 10 through Equation 23). A summary of the results are given in Tables 2 and 3.

It was seen that the correlations which used w_T, and those that used both w_L and w_P as input parameters, provided greater accuracy and precision than based on w_L and w_P alone. However, as far as the prediction of $\gamma_{d\text{-max}}$ is concerned, all the reported equations, except that of Blotz et al. (1998) and Gunaydin (2009), were seen to yield satisfactory results.

More accurate and precise empirical correlations for the prediction of compaction characteristics of fine grained soils was developed, taking into consideration of all the data points hitherto reported in the literature (493 in total). The subsequent multiple linear regression analyses resulted in the following equations:

$$OMC = 0.623 \, w_T \tag{27}$$
$$\gamma_{d\text{-max}} = 1.15 \, \gamma_{d\text{-wT}} \tag{28}$$

Table 2. Comparison of the μ, s, RD and RI values of predicted to observed values of OMC.

Source	μ	s	RD	RI
Dokovic et al. (2013)	0.99	0.19	0.19	0.22
Vinod and Sreelekshmy Pillai (2017)	1.02	0.23	0.23	0.21
Varghese et al. (2013)	0.99	0.214	0.21	0.24
Sreelekshmy Pillai and Vinod (2016)	1.06	0.23	0.25	0.25
Al-Khafaji (1993)	0.965	0.22	0.22	0.27
Blotz et al. (1998)	0.968	0.23	0.23	0.27
Sridharan and Nagaraj (2005)	1.08	0.28	0.29	0.30
Gunaydin (2009)	0.94	0.22	0.23	0.31
Gurtug and Sridharan (2004)	1.09	0.28	0.29	0.31
Sivrikaya et al. (2008)	1.11	0.28	0.30	0.32
Al-Khafaji (1993)	1.13	0.28	0.31	0.33

Table 3. Comparison of the μ, s, RD and RI values of predicted to observed values of $\gamma_{d\text{-max}}$.

Source	μ	s	RD	RI
Dokovic et al. (2013)	1.02	0.08	0.08	0.09
Varghese et al. (2013)	1.007	0.08	0.08	0.09
Al-Khafaji (1993)	1.019	0.08	0.08	0.09
Sridharan and Nagaraj (2005)	0.99	0.08	0.08	0.09
Vinod and Sreelekshmy Pillai (2017)	0.98	0.09	0.09	0.10
Sreelekshmy Pillai and Vinod (2016)	0.98	0.09	0.09	0.10
Gurtug and Sridharan (2004)	0.985	0.08	0.08	0.104
Al-Khafaji (1993)	0.97	0.10	0.11	0.14
Sivrikaya et al. (2008)	0.99	0.13	0.13	0.13
Blotz et al. (1998)	1.08	0.20	0.22	0.25
Gunaydin (2009)	0.29	0.10	0.71	1.6

The RD and RI values of the predicted to observed values of OMC (as per Equation 27) were 0.19 and 0.18, respectively. The RD and RI values were better than that corresponding to all reported equations in the literature review. For $\gamma_{d\text{-max}}$, RD and RI values of predicted to observed values (as per Equation 28) were 0.09 and 0.09, respectively. These values are highly satisfactory.

4 CONCLUSIONS

A critical review and analysis of the published literature on compaction characteristics of fine grained soils at the standard Proctor compactive effort has led to the following conclusions:

- The accuracy and precision of OMC prediction, at the standard Proctor compactive effort, is better when w_L along with w_P is used.
- Among the various reported correlations, those proposed by Dokovic et al. (2013) and Vinod and Sreelekshmy Pillai (2017) are the most accurate and precise. With the use of w_T, which effectively takes care of the combined effect of w_L and w_P, a new set of equations for the prediction of compaction characteristics at the standard Proctor compactive effort have also been proposed,and are as follows:
- $OMC = 0.623\ w_T$
- $\gamma_{d\text{-max}} = 1.15\ \gamma_{d\text{-wT}}$

REFERENCES

Al-Khafaji, A.N. (1993). Estimation of soil compaction parameters by means of Atterberg limits. *Quarterly Journal of Engineering Geology, 26*, 359–368.

Benson, C.H. & Trast, J.M. (1995). Hydraulics conductivity of thirteen compacted clays. *Clay Minerals, 4*(6), 669–681.

Beera, A.K. & Ghosh, A. (2011). Regression model for prediction of optimum moisture content and maximum dry unit weight of fine grained soil. *International Journal of Geotechnical Engineering, 5*, 297–305.

Blotz, R.L., Benson, C.H. & Boutwell, G.P. (1998). Estimating optimum water content and maxi dry unit weight for compacted clays. *Journal of Geotechnical and Geoenvironmental Engineering, 124*(9), 907–912.

Briaud, J.L. & Tucker, L.M. (1998). Measured and predicted axial load response of 98 piles. *Journal of Geotechnical Engineering, 114*(9), 984–1001.

Cherubini, C. & Giasi, C.I. (2000). Correlation equations for normal consolidated clays. A Nakase & T. Tschida (Ed(s))., *International coastal geotechnical engineering in practice* Vol. 1, pp. 15–20. Rotterdam, The Netherlands: A.A Balkema.

Cherubini, C. & Orr, T.L.L. (2000). A rational procedure for comparing measured and calculated values in geotechnics. A. Nakase & T. Tschida (Ed(s).), *International coastal geotechnical engineering in practice,* Vol. 1, pp. 261–265). Rotterdam, The Netherlands: A.A Balkema.

Daniel, D.E. & Benson, C.H. (1990). Water content—density criteria for compacted soil liners. *Journal Geotechnical Engineering, 116*(12), 1181–1190.

Daniel, D.E. & Wu, Y.K. (1993). Compacted clay liners and covers for arid sites. *The Journal of Geotechnical Engineering, 119*(2), 223–237.

Datta, T. & Chattopadhyay, B.C. (2011). Correlation between CBR and index properties of soil, D.K. Sahoo, T.G.S. Kumar, B.M. Abraham & B.T. Jose (Ed(s).), *Proceedings of Indian Geotechnical Conference, Kochi, India* (pp. 131–133).

Dokovic, K., Rakic, D. & Ljubojev, M. (2013). Estimation of soil compaction parameters based on the Atterberg limits. *Mining and Metallurgy Engineering Bor, 4*, 1–7.

Gunaydin, O. (2009). Estimation of soil compaction parameters by using statistical analysis and artificial neural networks. *Environmental Geology, 57*, 203–215.

Gurtug, Y. & Sridharan, A. (2004). Compaction behaviour and prediction of its characteristics of fine grained soils with particular reference to compaction energy. *Soils and Foundations, 44*(5), 27–36.

Horpibulsuk,S., Katkan,W. & Apichatvullop, A. (2008). An approach for assessment of compaction curves of fine grained soils at various energies using one point test. *Soils and Foundations. Japanese Geotechnical Society, 48*(1),115–126.

McRae, J.L. (1958). Index of compaction characteristics. *Symposium on application of soil testing in highway design and construction* (Vol. 239, pp. 119–127). Philadelphia; ASTM STP.

Nagaraj, H.B., Reesha, B., Sravan, M.V. & Suresh, M.R. (2015). Correlation of compaction characteristics of natural soils with modified plastic limit. *Transportation Geotechnics, 2*, 65–77.

Patel, R.S. & Desai, M.D. (2010). CBR predicted by index properties for alluvial soils of south Gujarat. R. Beri (Ed.), *Proceedings of Indian Geotechnical Conference, Mumbai, India* (Vol. 1, pp. 79–82).

Roy, T.K., Chattopadhyay, B.C. & Ro, S.K. (2009). Prediction of CBR from compaction characteristics of cohesive soil. *Highway Research Journal*, 7–88.

Shirur, N.B. & Hiremath, S.G. (2014). Establishing relationship between CBR value and physical properties of soil. *Journal of Mechanical and Civil Engineering, 11*, 26–30.

Sivrikaya, O. (2008). Models of compacted fine-grained soils used as mineral liner for solid waste. *Environmental Geology, 53*, 1585–1595.

Sreelekshmy Pillai, G.A. & Vinod, P. (2016). Re-examination of compaction parameters of fine grained soils. *Ground Improvement, 169*(3), 157–166.

Sridharan, A. & Nagaraj, H.B. (2005). Plastic limit and compaction characteristics of fine grained soils. *Ground Improvement, 9*(1), 17–22.

Talukdar, D.K. (2014). A study of correlation between California Bearing Ratio (CBR) value with other properties of soil. *International Journal of Emerging Technology and Advanced Engineering, 4*, 559–562.

Varghese, V.K., Babu, S.S., Bijukumar, R., Cyrus, S. & Abraham, B.M. (2013). Artificial neural networks: A solution to the ambiguity in prediction of engineering properties of fine grained soils. *Geotechnical and Geological Engineering, 31*, 1187–1205.

Vinod, P. & Sreelekshmy Pillai, G. (2017). Toughness limit: A useful index property for prediction of compaction parameters of fine grained soils at any rational compactive effort. *Indian Geotech Journal. 47*(1), 107–114.

Wang, M.C., ASCE, M. & Huang, C.C. (1984). Soil compaction and permeability prediction models. *Journal of Environmental Engineering, 110*(6), 1063–1082.

Emerging Trends in Engineering, Science and Technology for Society,
Energy and Environment – Vanchipura & Jiji (Eds)
© *2018 Taylor & Francis Group, London, ISBN 978-0-8153-5760-5*

A study on the correlation of the shear modulus of soil with the California bearing ratio and dynamic cone penetration value

S. Athira & S. Parvathy
College of Engineering, Trivandrum, India

V. Jaya
Department of Civil Engineering, Government Engineering College, Barton Hill, India

ABSTRACT: Stiffness of the base and sublayers is an important parameter in the design and quality assurance of pavements. In existing pavements, prior to resurfacing, it is essential to know the condition of the base that has been subject to traffic loading and environmental conditions. When failure of a pavement occurs, a quick and accurate measurement of the properties of the base layer is essential. The most popular methods are CBR, dynamic cone penetration and resilient modulus tests. The bender element technique, which is applicable to a wide variety of soils, can be used on existing pavements under construction in the field. To accept a method for design purposes, it should be validated with the conventional methods such as the penetration methods and CBR method. This study aims at developing a correlation between the Dynamic Penetration Index (DPI) and CBR values with the shear modulus obtained from this method. A correlation of these values with the shear modulus can be of use to our road sector.

Keywords: shear modulus, bender element method, subgrade CBR, dynamic cone penetration testing

1 INTRODUCTION

The performance of pavements depends on the properties of the subgrade soil and pavement materials. Accurate measurement of the properties of the base layer is essential to avoid failure of the subgrade. Shear modulus is a soil characteristic, which determines the strength of the subgrade and hence, its measurement is required for the design and construction of pavements.

The shear modulus of subgrade soil can be determined in a laboratory using the theory of wave propagation. Piezo ceramic sensors known as bender elements are currently used for the determination of the shear modulus of soil.

A pair of in-line bender elements is usually used, where one acts as the transmitter sending off the shear waves, while the other on the opposite end captures the arriving waves.

Dynamic Cone Penetrometer (DCP) results have never been used as an absolute indicator of the *in situ* strength or stiffness of a material in a pavement or subgrade. The California Bearing Ratio (CBR) value and DPI interpretation are important soil parameters for the design of flexible pavements and airstrips. It can also be used for determination of the subgrade reaction of soil by using correlation. It is essential to familiarize this simple wave propagation technique and correlate with CBR such that shear modulus can be used by practicing engineers for design and performance assessment of pavements.

This paper focuses on the measurement of the shear modulus of soil using bender elements and the separate development of a correlation with CBR and DPI.

The incorporation of bender elements in a triaxial apparatus is arguably the most common practice for the determination of shear modulus of soil samples in the laboratory, as demonstrated by Jovičić et al. (1996). The maximum shear modulus (Gmax) for saturated sand is estimated from shear wave velocity measurements using bender elements. The procedure for the selection of input parameters and the methods of signal interpretation are described by Jaya et al. (2007). The CBR test is laborious, time consuming, depends on the skill of the technician and the results are not accurate enough for a good quality design (Roy, Chattopadhyay & Roy, 2010). Direct evaluation of shear modulus in the laboratory and correlating it with CBR can lead to a more simple and reliable method for CBR evaluation. This paper describes a method of shear modulus determination using bender elements and its correlation with CBR.

2 MATERIALS AND METHODOLOGY

Experiments were conducted in the laboratory to determine the shear modulus, DPI and CBR value of selected subgrade soil. The properties of the materials, apparatus and methodology are explained in the following sections.

2.1 Materials

The soil samples used in this test were collected from selected subgrade locations in Thiruvananthapuram, Kerala. The physical and engineering properties of the soil were determined according to IS methods. The field density and specific gravity were 2.43 g/cc and 2.58, respectively.
    Based on IS:1498-1970, the soil samples were found to be gravelly sand.

2.2 Methodology

The bender element test was carried out on unconfined specimens in a triaxial cell. A pair of in-line bender elements were fixed in the test apparatus, where one acts as the transmitter sending off the shear waves, while the other on the opposite end captures the arriving waves. A wave generator was used to transmit sine waves into the soil samples and received signals as waveforms on an oscilloscope. These recorded waveforms were used for interpretation of travel time through the sample. The shear wave velocity (v_s) was derived by dividing the travel distance of the waves (between the transmitter and receiver) with the arrival time, which in turn is squared and multiplied with the specimen's bulk density to obtain the shear modulus.
    A mathematical description of the sets of variables is the best method of scientific explanation, as prior to this, in a graphical presentation there was always an element of bias or misleading presentation (Barua & Patgiri, 1996). The study of regression enables us to get a close functional relation between two or more variables (Kapur & Saxena, 1982).

3 RESULTS AND DISCUSSION

Compaction properties were determined by a standard Proctor test as per IS:2720 (Part VII). Unsoaked CBR values of the soil sample were determined as per the procedure laid down in IS: 2720 (Part XVI) (1979). Dynamic cone penetration tests were conducted on the samples according to IS:4968 (Part I) (1978).
    From the results obtained, we can see that the variation of OMC with respect to CBR values was found to be inversely proportional. Variations of compaction properties with respect to corresponding CBR values and shear modulus were determined.
    From Figure 1, it may be noted that when the value of the CBR decreases there is an increase in optimum moisture content. It was also observed that the CBR value has a significant correlation with OMC. The relation between CBR and OMC is $Y = 347.02x^{-1.243}$ with an R^2 value of 0.9503.
    From Figure 2, it may be noted that the shear modulus of soil shows a similar trend of variation with optimum moisture content. When the water content increases the shear stiffness of the soil reduces. The relation between CBR and OMC is $Y = 2.2604x^{-.65}$ with an R^2 value of 0.6708.

Table 1. Correlation of shear modulus with CBR value.

Sample no	Max dry density (g/cm³)	OMC (%)	Unsoaked CBR Value (%)	Shear modulus (GPa)	DPI (mm/blow)
1	1.82	15.75	7.41	0.253	12.06
2	1.65	21.25	11.11	0.216	12.26
3	1.76	15.55	11.05	0.322	13.685
4	1.76	15.55	13.96	0.252	13.86
5	1.97	13.43	10.56	0.347	11.51
6	2.05	17.6	9.71	0.347	11.01
7	1.85	17.88	12.57	0.347	13.92
8	1.95	14.9	14.57	0.346	11.13
9	1.93	12.33	12.75	0.346	9.84
10	1.93	14.81	10.93	0.346	13.54

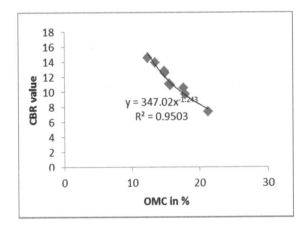

Figure 1. Variation of CBR with OMC.

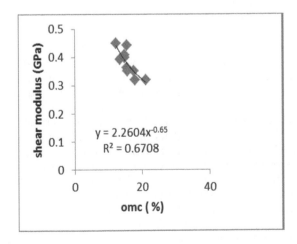

Figure 2. Variation of shear modulus with OMC.

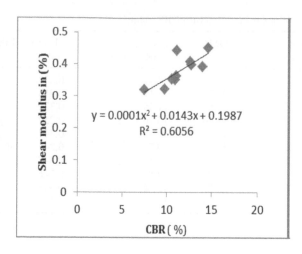

Figure 3. Variation of CBR with shear modulus.

a. Correlation between CBR and shear modulus

From Figure 3 it may be noted that the CBR and shear modulus values show a trend with an increase in OMC for all soil samples. The increase in water content reduces the stiffness of the soil samples. Hence, its shear modulus and penetration resistance decrease with an increasing water content. The failure of soil in pavements due to the presence of water is one of the major reasons for poor performance of pavements. Evaluation of shear modulus with maximum dry density and varying water content can predict the behavior of pavements under load with different moisture conditions. A correlation between CBR and shear modulus was obtained as $y = .0001x^2 + .0143x + 1987$ with $R^2 = 0.6056$.

Based on the relation obtained between CBR and shear modulus, we can predict the value of the shear modulus and compare it with obtained results of shear modulus.

The experimental and predicted shear modulus are almost similar for soil samples with a CBR of less than 10%. However, for soil samples with a CBR value greater than 10%, there is a large difference in experimental and predicted shear modulus. A comparison of shear modulus value (Figure 4), shows that in some soil samples, the laboratory and computed value of shear modulus are no different. The maximum difference is 7.3%, but in most cases the minimum differences are <3%.

b. Correlation of shear modulus with DPI

The correlation equation obtained is $DPI = -.026G + 20.37$, where DPI = Cone penetration index, G = shear modulus. The equation is in the form $y = mx + c$, which clearly implies a relationship between the experimental variables to be linear as shown in Figure 5.

The coefficient of correlation 'r', obtained from the linear regression analysis is 0.94. Its value being closer to one is a clear implication that there exists a strong functional relationship between the two variables and being a negative value clearly indicates that, as the value of small strain modulus increases, the dynamic cone penetration index decreases. Thus, that stiff soils will have a low penetration is clearly inferred from this result.

The coefficient of determination r^2, is determined to be 0.88 for the above data. This value, which is a measure of validity of the regression model, is only 88% as other factors that influence these two variables were not considered, thus, making the model only 88% correct.

The parity plot for the given data is shown in Figure 6, which depicts the quality of fit to be good. As the values lie close to the parity line, normal behavior of the data is predicted.

Experimental small strain shear modulus and predicted small strain shear modulus are plotted against the experimental penetration index as shown in Figure 7 and it can be seen that only less than 10% of the predicted values do not agree with the experimentally determined values, which clearly implies that the functional relationship between the variables is strong.

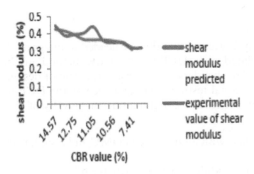

Figure 4.   Comparison of experimental and predicted value of shear modulus.

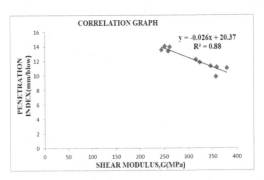

Figure 5.   Final correlation graph between small strain shear modulus and penetration index.

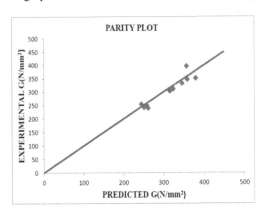

Figure 6.   Parity plot between experimental and predicted values.

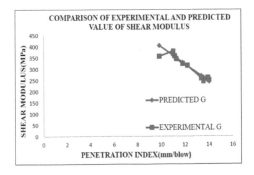

Figure 7.   Comparison between experimentally determined and predicted small strain shear modulus.

291

4   CONCLUSIONS

From the literature study, it was evident that we needed to integrate the engineering behavior of the hitherto uncorrelated tests (DCP testing, CBR method and bender element test). To extend the application of the bender element test in the field of design of pavement subgrade, validation with a conventional testing technique such as the DCP testing and CBR testing was necessary. The main conclusions, highlighting the novelty of the work, are briefly mentioned as follows:

- The CBR values decrease with an increase in OMC and a similar trend is also observed for shear modulus.
- The decrease of shear modulus with an increase of water content can predict the soil strength degradation due to the presence of moisture in pavements.
- The shear modulus of the soil samples increases with a decrease in DPI after the development of a linear regression model.
- The shear modulus of soil predicted using the derived relationship was found to be 90% in agreement with the experimental values when the DPI was considered as the parameter.

REFERENCES

Cahyadi, J. E.C. Leong & H. Rahardjo (2009). Measuring shear and compression wave velocities of soil using bender-extender elements. *Canadian Geotechnical Testing Journal, 46,* 792–812.

Dyvik, R. & Madshus, C. (1985). *Proceedings of ASCE Convention on Advances in the Art of Testing Soils under Cyclic Conditions.* Detroit, Michigan, *64,* 186–196.

Halsko, H.A. & Zeng, X. (2010). Piezoelectric probe for measurement of soil stiffness. *International Journal of Pavement Engineering, 11*(1).

Jaya, V., Dodagoudar, G.R. & Boominathan, A. (2007). Estimation of maximum shear modulus of sand from shear wave velocity measurements by bender elements. *Indian Geotechnical Journal, 37*(3), 159–173.

Lee, J.S. & Santa Marina, J.C. (2005). Bender elements: Performance and signal interpretation. *Journal of Geotechnical and Geoenvironmental Engineering, ASCE, 131,* 1090–0241.

Roy, T.K., Chattapadhyay, B.C. & Roy, S.K. (2010). *California bearing ratio, evaluation and estimation: A study of comparison.* Paper presented at the Indian Geotechnical Conference, Indian Institute of Technology, Mumbai, 19–22.

Sawangsuriya, A., Bosscher, P.J. & Edil, T.B. (2005). Alternative testing techniques for modulus of pavement bases and subgrades geotechnical applications for transportation infrastructure. *ASCE, Geotechnical Practice Publications, 3,* 108–121.

Emerging Trends in Engineering, Science and Technology for Society,
Energy and Environment – Vanchipura & Jiji (Eds)
© 2018 Taylor & Francis Group, London, ISBN 978-0-8153-5760-5

Load-settlement behaviour of footing on laterally confined soil

P.S. Sreedhu Potty, Mayuna Jeenu, Anjana Raj, R.S. Krishna, J.B. Ralphin Rose,
T.S. Amritha Varsha, J. Jayamohan & R. Deepthi Chandran
LBS Institute of Technology for Women, Thiruvananthapuram, India

ABSTRACT: The necessity of in-situ treatment of foundation soil to improve the bearing capacity has increased considerably due to non-availability of good construction sites. Soil confinement is one such method of soil improvement which can be economically adopted. The improvement in bearing capacity and reduction in settlement of footings resting on clay due to the addition of a laterally confined granular soil layer underneath, is investigated by carrying out a series of finite element analyses using the FE software PLAXIS 2D. The influence of parameters like radius, depth etc. of laterally confined granular soil layer is studied. It is observed that the load-settlement behaviour of isolated footings resting on clay can be considerably improved by providing a laterally confined granular soil layer underneath it.

Keywords: Laterally Confined Soil, Finite Element Analyses, Load-Settlement Behaviour

1 INTRODUCTION

The decreasing availability of good construction sites has led to increased use of sites with marginal soil properties. The necessity for in situ treatment of foundation soil to improve its bearing capacity has increased considerably. The soil confinement is a promising technique of improving soil capacity. This technique of soil confinement, though successfully applied in certain areas of soil engineering, has not received much attention in foundation applications. In the last few decades, great improvements in foundation engineering have occurred, along with the development of new and unconventional types of foundation systems through considerations of soil-structure interaction. In the past few decades many researches have been carried out to investigate the improvement in bearing capacity due to confining the underlying soil. It has been proved that by confining the soil there is a reduction in the settlement resulting in an increase in bearing capacity.

Much research has been carried out on soil reinforced with geosynthetics (Mahmoud and Abdrabbo (1989), Khing et al. (1993), Puri et al. (1993), Das and Omar (1993), Dash et al. (2001a&b), Schimizu and Inui (1990), Mandal and Manjunath (1995), Mahmoud and Abdrabbo (1989), Rajagopal et al. (1999)). Several authors have also studied strip foundations but reinforced with different materials (Verma and Char, 1986, Dawson and Lee, 1988).

Sawwaf and Nazer (2005) studied the behavior of circular footing resting on confined sand. They used confining cylinders with different heights and diameters to confine the sand. Krishna et al. (2014) carried out laboratory model tests on square footings resting on laterally confined sand. Vinod et al. (2007) studied the effect of inclination of loads on footings resting on laterally confined soil.

In this research, the beneficial effects of providing a laterally confined granular layer underneath a footing resting on clay are investigated by carrying out nonlinear finite element analysis using the FE software *PLAXIS 2D*. The influence of dimensions of the laterally confined granular soil layer on the load-settlement behaviour, axial force distribution in the confining walls, stress distribution at the interface between confining walls and surrounding soil etc. are particularly studied.

2 FINITE ELEMENT ANALYSES

Finite element analyses are carried out using the commercially available finite element software *PLAXIS* 2D. For simulating the behaviour of soil, different constitutive models are available in the FE software. In the present study Mohr-Coulomb model is used to simulate soil behaviour. This non-linear model is based on the basic soil parameters that can be obtained from direct shear tests; internal friction angle and cohesion intercept. Since circular footing is used, axi-symmetric model is adopted in the analysis. The settlement of the rigid footing is simulated using non zero prescribed displacements.

The displacement of the bottom boundary is restricted in all directions, while at the vertical sides; displacement is restricted only in the horizontal direction. The initial geostatic stress states for the analyses are set according to the unit weight of soil. The soil is modelled using 15 noded triangular elements.

Mesh generation can be done automatically. Medium mesh size is adopted in all the simulations. The size of the strip footing (B) is taken as one metre and the width and depth of soil mass are taken as 10B in all analyses.

The footing and the confining walls are modelled using plate elements. To simulate the interaction between the confining walls and surrounding soil, an interface element is provided on both outer and inner surfaces. The geometric model is shown in Fig. 1 and the typical deformed shape in Fig. 2. The soil is modelled using 15-node triangular elements. Properties of locally available sand and clay are adopted in the analyses. The material properties adopted are outlined in Table 1.

The influence of dimensions of laterally confined soil on the load-settlement behaviour and interaction between confining walls and surrounding soil are particularly studied. Various geometric parameters studied are indicated in Fig. 3. The diameter of footing is B, radius and thickness of confined soil are r and h respectively. The diameter of footing is taken as one metre for all the cases. Various values of r and h adopted in the analyses are indicated in Table 2.

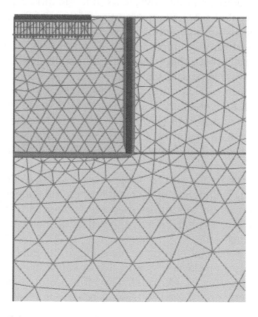

Figure 1.   Geometric model.

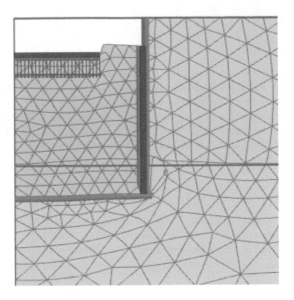

Figure 2.   Typical deformed shape.

Table 1.   Properties of soil.

Property	Sand	Clay
Dry Unit Weight (kN/m³)	17	16
Specific Gravity	2.3	2.1
Cohesion (kPa)	2	20
Angle of Shearing Resistance (°)	30	0
Modulus of Elasticity (kPa)	15000	7000
Poisson's Ratio	0.25	0.3

Table 2.   Values of r and h.

Parameter	r/B	h/B
Values	1, 1.5, 2, 2.5, 3	0.5, 1, 1.5, 2, 2.5, 3

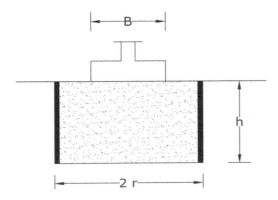

Figure 3.   Geometric parameters.

3   RESULTS AND DISCUSSIONS

3.1   *Load-settlement behaviour*

Vertical Stress vs Settlement curves obtained from FEA, for various values of (h/B) when r/B = are presented in Figure 4. It is seen that the load-settlement behaviour considerably improves due to the lateral confinement of underlying soil. The improvement, when (h/B) = 3 is almost ten times when compared with that of unconfined soil at 2% settlement. The influence of radius of the confining area, when the thickness is equal to 1.5B, is presented in Figure 5. It is seen that the load-settlement behaviour considerably improves when the radius of confinement increases.

3.2   *Interaction between confining walls and sand*

The interaction between confining walls and soil in the confined area is investigated by studying the distribution of normal and shear stresses at the interface. Figure 6 presents the distribution of normal stresses at the interface between confining walls and sand, obtained from FEA, for various values of (r/B) when (h/B) = 3.5.

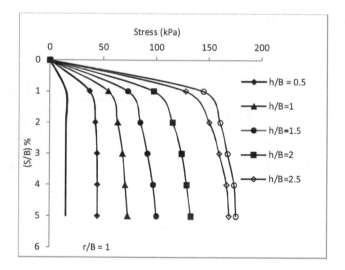

Figure 4.   Vertical Stress vs Settlement Curves for various values of (h/B); when (r/B) = 1.

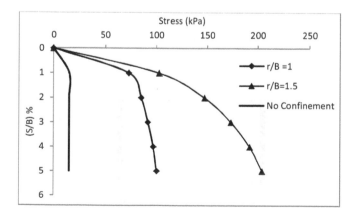

Figure 5.   Vertical Stress vs Settlement Curves for various values of (r/B); when (h/B) = 1.5.

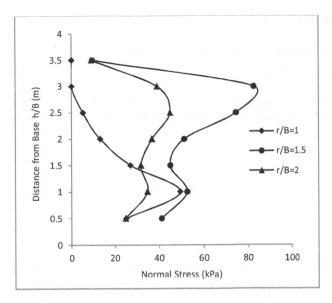

Figure 6. Distribution of normal stress at the interface between confining walls and Sand for various values of (r/B).

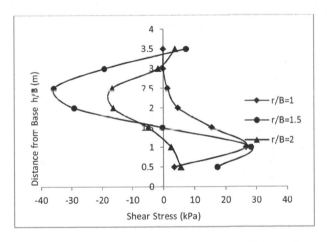

Figure 7. Distribution of shear stress at the interface between confining walls and Sand for various values of (r/B).

It is seen that the normal stress increases with the radius of confining area upto (r/B) = 1.5 and thereafter reduces. When (r/B) = 1, the peak stress occurs at a height of **B** from base. For higher values of (r/B), the point of peak stress shifts upwards and occurs at a height of **3B** from base.

Figure 7 presents the distribution of shear stress along the interface between confining walls and sand. It is seen that the distribution of shear stress drastically changes with the increase in radius of confining area. The pattern of soil movement within the confining area changes with the increase in radius which influences the shear stress at interface.

It is observed that the distribution of normal and shear stresses at larger cell widths is very different from that of smaller ell widths. This validates the observation of Vinod et al. (2007), that at smaller cell width the confining cell—soil and footing behaves as a single unit (deep foundation) and this behaviour changes as the radius of confinement is increased.

4 CONCLUSIONS

The following conclusions are drawn from the results of finite element analyses

1. The load-settlement behaviour considerably improves due to lateral confinement of under-lying soil
2. The distribution of normal and shear stresses at the interface between confining walls and the confined soil is considerably influenced by the radius of confinement.

REFERENCES

Binquet, J., and K.L. Lee (1975), "Bearing capacity tests on reinforced earth slabs." *Journal of Geotechnical Engineering Division*, 101(12), 1241–1255.

Das, B.M., and M.T. Omar (1993), "The effects of foundation width on model tests for the bearing capacity of sand with geogrid reinforcement." *Geotechnical and Geological Engineering*, 12, 133–141.

Dash, S., N. Krishnaswamy and K. Rajagopal (2001 a), "Bearing capacity of strip footing supported on geocell-reinforced sand." *Geotextile and Geomembrane*, 19, 535–256.

Dash, S., K. Rajagopal and N. Krishnaswamy (2001 b), "Strip footing on geocell reinforced sand beds with additional planar reinforcement." *Geotextile and Geomembrane*, 19, 529–538.

Dawson, A. and R. Lee (1988), "Full scale foundation trials on grid reinforced clay," *Geosynthetics for Soil Improvement*. 127–147.

Khing, K.H., B.M. Das, V.K. Puri, E.E. Cook and S.C. Yen (1993), "The bearing capacity of a strip foundation on geogrid-reinforced sand", *Geotextiles and Geomembranes*, 12, 351–361.

Krishna, A., Viswanath, B. and Nikitha, K. (2014), "Performance of Square footing resting on laterally confined sand", *International Journal of Research in Engineering and Technology*, Vol 3, Issue 6.

Mahmoud, M.A., and F.M. Abdrabbo (1989) "Bearing capacity tests on strip footing on reinforced sand subgrade." *Canadian Geotechnical Journal*, 26, 154–159.

Mandal, J.M., and V.R. Manjunath (1995), "Bearing capacity of strip footing resting on reinforced sand subgrades." *Construction and Building Material*, 9 (1), 35–38.

Puri, V.K., K.H. Khing, B.M. Das, E.E. Cook and S.C. Yen (1993), "The bearing capacity of a strip foundation on geogrid reinforced sand." *Geotextile and Geomembrane*, 12, 351–361.

Rajagopal, K., N. Krishnaswamy and G. Latha (1999), "Behavior of sand confined with single and multiple geocells", *Geotextile and Geomembrane*, 17, 171–184.

Sawwaf, M.E., and A. Nazer (2005), "Behavior of circular footing resting on confined granular soil." *Journal of Geotechnical and Geoenvironmental Engineering*, 131(3), 359–366.

Vinod, K.S., Arun, P., and Agrawal, R.K., (2007), " Effect of Soil Confinement on Ultimate Bearing Capacity of Square footing under eccentric-inclined load", *Electronic Journal of Geotechnical Engineering*, Vol. 12, Bund E.

Emerging Trends in Engineering, Science and Technology for Society,
Energy and Environment – Vanchipura & Jiji (Eds)
© 2018 Taylor & Francis Group, London, ISBN 978-0-8153-5760-5

Deformation behavior of sheet pile walls

H.S. Athira, V.S. Athira, Fathima Farhana, G.S. Gayathri, S. Reshma Babu,
N.P. Asha & P. Nair Radhika
Department of Civil Engineering, LBS Institute of Technology for Women, Trivandrum, India

ABSTRACT: Sheet pile walls are a common form of earth retaining structures. Earth pressures developed on either side of the sheet pile wall ensure its moment and force equilibrium. When the height of the earth that needs to be retained is rather high, the sheet pile walls are usually anchored near the top. On the other hand when the height is small, cantilever sheet pile walls are employed. Contrary to the conventional methods, the study takes in account the stiffness and structural capacity of sheet pile walls. The aim of the study is to analyse the deformation behaviour of cantilever and bulk head anchored sheet pile walls, for different depth of embedment to height ratio by Finite Element Analysis using **PLAXIS 2D** software. It is observed that wall deformation decreases when wall penetration depths are increased in cohesionless soils.

1 INTRODUCTION

Retaining walls are used to hold back soil and maintain a difference in the elevation of the ground surface. The retaining wall can be classified according to system rigidity into either rigid or flexible walls. A wall is considered to be rigid if it moves as a unit in rigid body and does not experience bending deformations. Flexible walls are the retaining walls that undergo bending deformations in addition to rigid body motion. Steel sheet pile wall is the most common example of the flexible walls because it can tolerate relatively large deformations.

Sheet pile walls are one of the oldest earth retention systems utilized in civil engineering projects. They consist of continuously interlocked pile segments embedded in soils to resist horizontal pressures. Sheet pile walls are used for various purposes; such as large and waterfront structures, cofferdams, cut-off walls under dams, erosion protection, stabilizing ground slopes, excavation support system, and floodwalls. The sheet pile walls can be either cantilever or anchored. The selection of the wall type is based on the function of the wall, the characteristics of the foundation soils, and the proximity of the wall to existing structures. While the cantilever walls are usually used for wall heights less than 6 m, anchored walls are

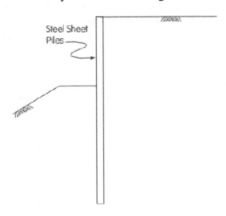

Figure 1. Sheet pile wall.

required for higher walls or when the lateral wall deformations are needed to be restricted. Typically the anchors are installed when the wall height exceeds 6 m or the wall supports heavy loads from a structure.

Contrary to conventional methods, aim of this study is to analyse the deformation behaviour of sheet pile walls taking into account its stiffness and structural capacity. Therefore, this research study was conducted to understand the steel sheet pile wall behavior in terms of deformations with different wall penetration depths.

2 NUMERICAL MODELLING AND PARAMETRIC STUDY

2.1 Numerical model

PLAXIS 2-D finite element analysis software package was used for the parametric study. PLAXIS program is a special purpose finite element program used to perform deformation and stability analysis for various types of geotechnical applications such as excavation, foundations, embankments and tunnels. PLAXIS, 2-D program consists of three main parts which are Model, Calculation and Output mode. In the Model mode geometry is built. When the geometry model is complete, the finite element model (or mesh) is generated. PLAXIS 2D allows for a fully automatic mesh generation procedure. In the calculation mode, a number of calculation phases will be defined.

2.2 Parametric study

The parametric study is focused primarily on studying the effect of varying wall penetration depths larger than the ones calculated by the conventional design methods on sheet pile wall behavior. Different types of the sheet pile walls (cantilever and anchored walls), were considered for the parametric study. The study was performed using the PLAXIS 2D version 8 Finite Element program employing 15-noded triangular elements for soil layers and 5-noded line elements for sheet pile walls. A total of seven cases have been modeled and analyzed in this parametric study.

The geometry of the finite element soil model adopted for the analysis is 20×30 m with depths to height ratios as 0.3, 0.45, 0.6 and 0.7. The properties of soil used in the analysis are listed in Table 1.

The material properties used in PLAXIS for the sheet pile walls are listed in Table 2

A. Cantilever sheet pile walls
The parametric study results were performed to investigate the effect of increasing wall penetration depth (D) on the cantilever sheet pile wall behavior by using medium dense sand soil ($\varphi = 30°$) for 6 m high wall (H = 6 m). Figures show the models of the sheet pile wall modelled for different D/H ratios. The total and horizontal wall displacements and also shear force for the lower and upper ranges of the wall penetration depth are analyzed for cantilever sheet pile wall, considered in this parametric study. The ratios adopted in this parametric study are 0.3, 0.45, 0.6 and 0.7.

Table 1. Material properties for the soil type studied.

Parameter	Name	Value	Unit
Type of material		Sand	
Soil saturated unit weight	γ_{unsat}	19	kN/m³
Soil unsaturated unit weight	γ_{sat}	16	kN/m³
Poisson's ratio	ν	0.3	
Young's modulus	E'	1.3×10^4	kN/m²
Cohesion	C'_{ref}	1	kN/m²
Friction angle	φ'	30	P
Dilatency angle	Ψ	0	P

Table 2. Material properties of the sheet pile wall.

Parameter	Name	Value	Unit
Type of behaviour	Material type	Elastic	
Normal stiffness	EA	2.738×10^6	kN/m
Bending stiffness	EI	2.3×10^4	kNm²/m
Equivalent thickness	D	0.3175	M
Weight	W	1.053	kNm²/m
Poisson's ratio	N	0.15	

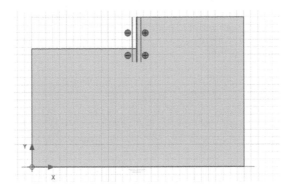

Figure 2. Cantilever sheet pile wall with D/H = 0.3.

B. Anchored sheet pile walls

The parametric studies were performed to investigate the effect of wall penetration depth (D) on deformation behavior of anchored sheet pile wall by using medium dense sand soil ($\varphi = 30°$) for 9 m high wall (H = 9 m). Figures show the models of the sheet pile walls modelled for different D/H ratios for different anchor block dimensions. The anchor was provided at a depth of 0.6 m from the ground level.

The total and horizontal wall displacements and also shear force for the lower and upper ranges of the wall penetration depth are analyzed for anchored sheet pile walls considered in this parametric study. The D/H ratios adopted in this parametric study are 0.3, 0.45, 0.6 and 0.7. The anchor block dimension used is 5 × 5 cm.

3 RESULTS

A. Cantilever wall

The analysis results in terms of, maximum horizontal displacements, and shear force with increasing wall penetration depth, for the 6 m cantilever sheet pile wall in medium dense sand soil are given in the following figures. The wall penetration depths D, were normalized by the wall height H, for all cases.

Wall displacements: Figures show the effect of increasing wall penetration depth on maximum horizontal wall displacements. The analyses results indicate that a significant decrease in the wall displacements is obtained with an increase in the wall penetration depth. These reductions are relative to the deformations of a wall designed using the conventional design methods. The results in these figures show that by increasing the wall penetration depth to height ratio in medium dense sand soils to 0.6 reduces the horizontal wall displacement to about 40% of the wall displacements observed when the ratio was 0.3. There is a considerable reduction in the wall displacements when ratio is increased to 0.7.

The graphs of D/H ratios versus total horizontal displacement were plotted for different D/H ratios.

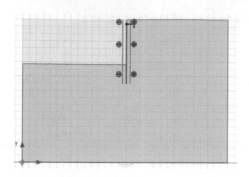

Figure 3.   Anchored sheet pile wall with D/H = 0.3.

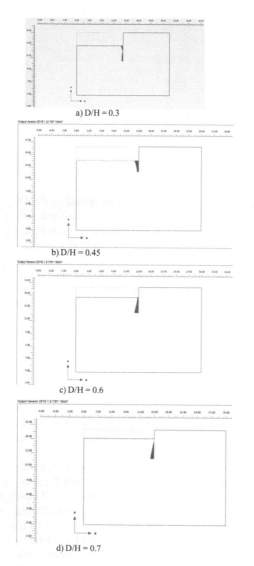

a) D/H = 0.3

b) D/H = 0.45

c) D/H = 0.6

d) D/H = 0.7

Figure 4.   Horizontal displacements of cantilever wall.

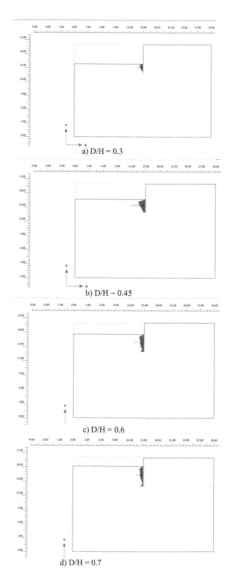

a) D/H = 0.3

b) D/H – 0.45

c) D/H = 0.6

d) D/H = 0.7

Figure 5.   Shear stress distribution for different D/H ratios.

302

B. Anchor wall

The analysis results in terms of, maximum horizontal displacements, and shear force with increasing wall penetration depth, for the 9 manchor sheet pile wall in medium dense sand soil are given in the following figures. The wall penetration depths D, were normalized by the wall height H, for all cases.

Wall displacements: Figures show the effect of increasing wall penetration depth on maximum horizontal wall displacements. The analyses results indicate that a slight decrease in the wall displacements is obtained with an increase in the wall penetration depth. The change in

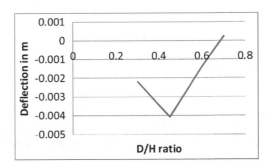

Figure 6.   Deflection versus D/H for cantilever wall.

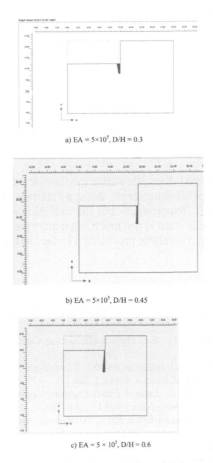

a) EA = 5×10⁵, D/H = 0.3

b) EA = 5×10⁵, D/H = 0.45

c) EA = 5 × 10⁵, D/H = 0.6

Figure 7.   Total horizontal displacement for different D/H ratios for anchored wall.

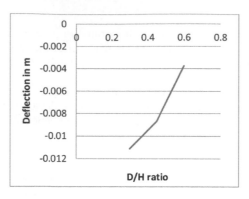

Figure 8.   Deflection versus D/H ratio.

wall displacements for all cases studied is minimal because the anchored wall is not displaced both at the bottom of the wall and also at the anchor position. Although the wall can bend between these positions, the overall wall displacements will be quite small with increasing the wall penetration depth due to these fixities.

4   CONCLUSIONS

A. Cantilever walls
Analysis of a 6 m cantilever sheet pile wall in medium dense sand was done using Plaxis 2D. As seen in the figure, increasing the wall penetration depth decreases the wall displacements. Maximum displacement was obtained when the D/H ratio was 0.3. This indicates that a sheet pile wall with D/H ratio 0.3 has a greater probability of failure. At the same time, minimum displacement was obtained for D/H ratio 0.7. But, providing a D/H ratio of 0.7 is not economical. Whereas a D/H ratio of 0.6 reduces the wall displacement by about 40%. Though this is not as much as that obtained from D/H ratio 0.7, it is still a significant reduction. Hence providing a D/H ratio of 0.6 is ideal.

B. Anchor walls
In the analysis of a 9 m anchored sheet pile wall in medium dense sand using Plaxis 2D, it was found that increase in wall penetration depth does in fact that increase in wall penetration depth does in fact reduce wall displacement but the reduction is not much. It was seen that the displacement in each case in itself is not much, no matter the D/H ratio. This is due to the fixity provided by the anchor. Therefore, providing a lower D/H ratio like 0.45 would suffice.

REFERENCES

Amer, Hetham A. Ramadan (2013). "Effect of wall penetration depth on the behaviour of sheet pile walls": A Liteature Review.
Bransby P.L.& G.W.E. Milligan (1975). "Soil Deformations near Cantilever Sheet pile walls", *Geotechnique 25*, No. 2 175–195.
GopalMadabhushi S.P. & V. S. Chandrasekaran (2008). "Centrifuge Testing of a Sheet Pile Wall with Clay Backfill". *Indian Geotechnical Journal*, 38(1), 1–20.
Omer Bilgin, P.E. & M. Asce (2012). "Lateral Earth Pressure Coefficients for Anchored Sheet Pile Walls".*International Journal of Geomechanics* Vol 12: 584–595.
Omer Bilgin, P.E., M. Asce & M. Bahadir Erten 2009)."Analysis of Anchored Sheet Pile Wall Deformations". *International Foundation Congress and Equipment Expo.*
Prakash Kumar Gupta et al., (2017). "A study relation between soil and cantilever sheet pile. A model of theory and designing". *International journal of engineering sciences & research technology.*

Progressive Developments in Mechanical Engineering (PDME)

Fluid flow and fluid power systems

Emerging Trends in Engineering, Science and Technology for Society,
Energy and Environment – Vanchipura & Jiji (Eds)
© 2018 Taylor & Francis Group, London, ISBN 978-0-8153-5760-5

On the role of leading-edge tubercles in the pre-stall and post-stall characteristics of airfoils

V.T. Gopinathan & R. Veeramanikandan
Department of Aeronautical Engineering, Hindusthan College of Engineering and Technology, Coimbatore, India

J. Bruce Ralphin Rose
Department of Mechanical Engineering, Anna University Regional Campus, Tirunelveli, India

V. Gokul
Department of Aerospace Engineering, Madras Institute of Technology, Chennai, India

ABSTRACT: Biologically inspired designs provide different set of strategies and tools to deal with engineering problems. The humpback whale is one such bio-inspiring species, being the most acrobatic of the baleen whales and capable of performing good maneuvers. The presence of large rounded protuberances or tubercles along the Leading Edge (LE) of the humpback whale flipper highlights its uniqueness. The LE tubercles act as passive flow control devices that improve the performance and maneuverability of the flipper. The aerodynamic performance of NACA airfoils such as NACA 0015 and NACA 4415, and modified airfoils with leading-edge tubercles (BUMP 0015, 4415) are numerically investigated at a Reynolds number (Re) of 1.83×10^5. The popular commercial Computational Fluid Dynamics (CFD) tool FLUENT was used. The post-stall and pre-stall characteristics are analyzed in terms of coefficients of lift (C_L) and drag (C_D) with respect to various Angles of Attack (AoAs). Both airfoils, with and without tubercles, are investigated. Comparisons of streamline distribution, pressure coefficient (C_P), C_L and C_D between the baseline airfoils and the airfoils with tubercles help to explain the momentum transfer characteristics of tubercles and hence how a stall is delayed.

Keywords: Bio inspired, Humpback Whale Flipper, Tubercles, Flow separation, Stall

1 INTRODUCTION

Tubercles are rounded protuberances of the Leading Edge (LE) that alter the flow field characteristics around an airfoil. It has been suggested that tubercles on the flipper of the humpback whale function as lift-enhancement devices, allowing the flow to remain attached for a larger Angle of Attack (AoA), and thus delaying stall. The protuberances found along the LE of the humpback whale flipper vary in amplitude and wavelength across the span. Further, the amplitude of the protuberances ranges from 2.5% to 12% of the chord and the wavelength varies from 10% to 50% of the chord. From a more morphological point of view, Fish and Battle (1995) proved that the geometrical properties of the tubercles could influence the aerodynamic performance of a wing.

The wind tunnel experiments of Miklosovic et al. (2004) demonstrated drastic enhancements in lift for a post-stall AoA, and also showed a delay in the stall angle of up to 40%. These experiments performed at the Reynolds number (Re) in the range of 10^5. One of the mechanisms of performance enhancement is believed to be the generation of stream-wise vortices, which improve the momentum exchange in the boundary layer. The potential benefits of tubercles on the aerodynamic performance of an airfoil were addressed by Bushnell and Moore (1991). Over the last two decades, several studies have been performed experimentally and numerically to assess the influence of tubercles. The first numerical study was performed by Watts and Fish (2001) using a 3-dimensional numerical panel method. They observed a slight increase in coefficient of lift (C_L) of about 4.8%, and a reduction in coefficient of drag (C_D) of approximately 11% in the pre-stall AoA ranges.

Johari et al. (2007) carried out a systematic experimental study on a NACA 63_4-021 airfoil with tubercles of different combinations of amplitude and wavelength. The cross section of the humpback flipper has a profile similar to the NACA 63_4-021. The results from these experiments showed that incorporating LE protuberances into the airfoil generated ~50% higher lift without any drag penalty in the post-stall region as compared to the baseline airfoil. While amplitude of the protuberance exhibited a notable effect on the performance of the airfoil, it was observed that the wavelength had meager influence.

The effect of a wavy leading edge on a NACA 4415 airfoil was investigated experimentally at a low Re of 1.2×10^5, using 2D Particle Image Velocimetry (PIV) measurement. The size of recirculating zone and wake width was investigated experimentally by Karthikeyan et al. (2014). A numerical investigation into the effects of a sinusoidal LE fitted to NACA 0015 and NACA 4415 profiles was analyzed with OpenFOAM software and the aerodynamic characteristics discussed by Corsini et al. (2013). Hansen et al. (2011) experimentally investigated the influence of tubercles on NACA 63_4-021 and NACA 0021 airfoils. They found that for both airfoil profiles, reducing the tubercle amplitude led to a higher maximum lift coefficient and larger stall angle.

Zhang et al. (2013) examined the effects of sinusoidal LE protuberances on two-dimensional full-span airfoil aerodynamics at a low Re of 5.0×10^4. In terms of controlling the boundary-layer separation, the function of protuberances is very similar in some respects to a low-profile vortex generator. The increase in airfoil thickness causes aerodynamic deterioration at pre-stall regimes for a wavy LE airfoil. The effect of variation in the airfoil thickness in the wavy LE phenomenon was examined for NACA 0020 and NACA 0012 profiles by De Paula et al. (2016).

2 GEOMETRY

NACA 0015 and NACA 4415 profiles are used as the wing cross section. The leading-edge tubercles design being formulated through unequal chord length. The ratio of amplitude (A) to wavelength (λ), expressed as ($\eta = A/\lambda$), retains high priority in the design perspective of tubercle research. The variable mean chord length (C), along with span-wise ordinates (z), are elucidated by the wave equation:

$$C(z) = A\cos(2\pi z /\lambda) + C$$

The chord length variation (Δc) along the span-wise direction is computed and is presented in Figure 1.

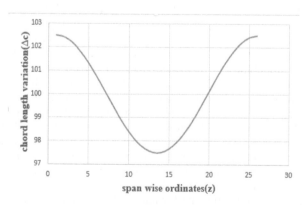

Figure 1. Plot of chord length variation vs. span-wise ordinates.

Figure 2. BUMP 0015 airfoil with η = 0.05.

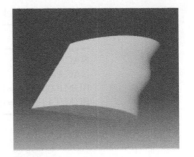

Figure 3. BUMP 4415 airfoil with η = 0.05.

2.1 *Details of the cases analyzed*

Airfoil profiles   :   NACA 0015 & NACA 4415
Leading-edge lines :   Straight & Bump
Mean chord length :   100 mm
Amplitude   :   2.5 mm
Wavelength   :   50 mm
η = A/λ = 0.05

3   MESHING

A boundary is created for airfoil design in which the computational domain extends 10C upstream and 10C downstream. The top and bottom boundaries of the domain are located 10C away from the foil. The whole computational domain is discretized with an unstructured grid with triangular mesh, as shown in Figures 4a and 4b.

3.1 *Numerical methodology*

Numerical analysis is done using the FLUENT tool. The steady state compressible Reynolds-averaged Navier–Stokes (RANS) equation was solved using a K-ω model:

$$\rho\left(\frac{\partial U_i}{\partial t} + U_k \frac{\partial U_i}{\partial x_k}\right) = -\frac{\partial P}{\partial x_i} + \frac{\partial}{\partial x_i}\mu\left(\frac{\partial U_i}{\partial x_j}\right) + \left(\frac{\partial R_{ij}}{\partial x_j}\right) \tag{1}$$

where R_{ij} is Reynolds stresses; U, p, ρ represents the velocity, pressure, density. The y⁺ values for all the simulations are maintained below 3. Boundary conditions for numerical analysis are pressure far field and wall.

4   RESULTS AND DISCUSSION

To examine the influence of tubercles, the viscous fluid flow over a baseline airfoil is first simulated. Then the airfoils with tubercles are simulated with the same Reynolds number. The differing values in these simulations show the influence of the tubercles. The numerical experiments are carried out at angles of attack of 0° to 21°. Favorable effects of tubercles are reported on aerodynamic coefficients (C_L and C_D) and stalling angle.

4.1 *Streamline patterns*

The two-dimensional streamlines along the peak, trough and mid-span planes are determined for all the airfoils at AoAs of 0°, 10° and 15°. For the baseline airfoil, the streamline patterns

Table 1.   Mesh element counts.

Airfoil series	Total number of elements
NACA 0015	1,323,205
NACA 4415	1,425,762
BUMP 0015 ($\eta = 0.05$)	1,136,096
BUMP 4415 ($\eta = 0.05$)	1,135,394

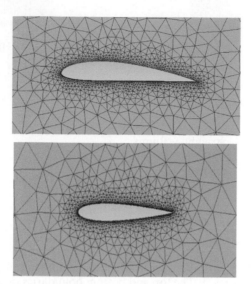

Figure 4.   (a) Mesh generated for cambered airfoil. (b) Mesh generated for symmetrical airfoil.

are visualized along the mid-span plane. The suction surface of the baseline airfoil is in a separated condition when the AoA is increased, as highlighted in Figures 5a, 5b and 5c, and Figures 9a, 9b and 9c. This applies to both NACA 0015 and NACA 4415 airfoils.

The streamlines of the peak and trough regions are captured at x/b values of 0.5 and 0.25, respectively. For the BUMP 0015 airfoil at 0° angle of attack, the separation effect is not significant. For BUMP 0015 at an AoA of 15°, in the peak region the attached flow covers the suction surface, forming counter-rotating vortex structures, and the flow at the trough region is detached for the same AoA, as shown in Figures 8a and 8b. By contrast, for the BUMP 4415 airfoil at an AoA of 15°, the peak region has attached flow but the trough region has a more separated flow than the BUMP 0015 airfoil, as highlighted in Figure 12b.

The streamline patterns over peak and trough regions at AoAs of 0° and 10° indicate that the flow is close to the surface, as shown in Figures 6, 7, 10 and 11, with meager amounts of separation.

4.2   *Effects of tubercles on aerodynamic coefficients*

The effect of aerodynamic coefficients with different angles of attack is plotted for both baseline and modified airfoils, as shown in Figure 13. The projected plan form area are used to calculate the C_L and C_D. At low angles of attack, the difference between the baseline airfoil and modified airfoil is not significant. The NACA 0015 airfoil's C_L is increasing and stalls at 14°. The BUMP 0015 airfoil stalls at 19°. This result indicates the stall-delaying effect of tubercles. In particular, when considering the pre-stall operations of the NACA 0015 airfoil, the lift coefficient is lower than for the corresponding BUMP 0015 airfoil after 10°. On the other hand, the drag coefficient difference is not significant. The use of leading-edge bumps for cambered airfoils results in an increase in lift over

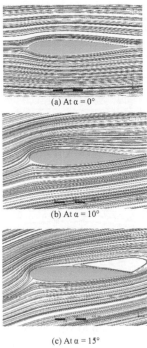

(a) At $\alpha = 0°$

(b) At $\alpha = 10°$

(c) At $\alpha = 15°$

Figure 5.   treamlines of NACA 0015 airfoil at mid-span location.

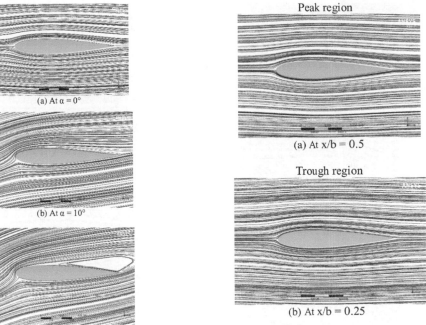

Peak region

(a) At x/b = 0.5

Trough region

(b) At x/b = 0.25

Figure 6.   Streamlines of BUMP 0015 airfoil at $\alpha = 0°$.

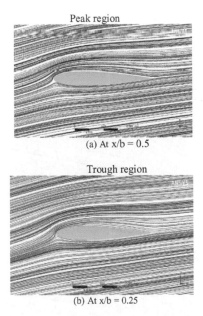

Peak region

(a) At x/b = 0.5

Trough region

(b) At x/b = 0.25

Figure 7.   Streamlines of BUMP 0015 airfoil at $\alpha = 10°$.

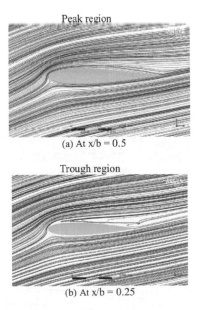

Peak region

(a) At x/b = 0.5

Trough region

(b) At x/b = 0.25

Figure 8.   Streamlines of BUMP 0015 airfoil at $\alpha = 15°$.

angles of attack ranging from 11° to 15°. The cambered NACA 4415 airfoil stalls at 12° and the BUMP 4415 airfoil stalls at 15°. The drag coefficient decreased for the BUMP 4415 airfoil in the post-stall condition.

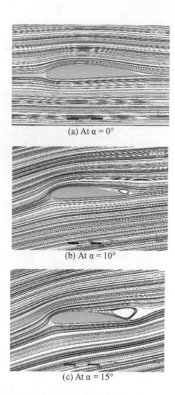

(a) At $\alpha = 0°$

(b) At $\alpha = 10°$

(c) At $\alpha = 15°$

Figure 9.   Streamlines of NACA 4415 airfoil at mid-span location.

Peak region

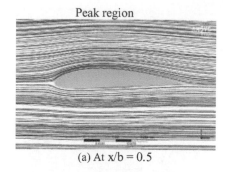

(a) At x/b = 0.5

Trough region

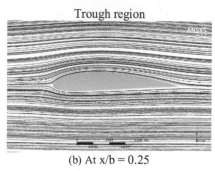

(b) At x/b = 0.25

Figure 10.   Streamlines of BUMP 4415 airfoil at $\alpha = 0°$.

Peak region

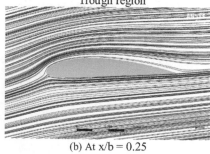

(a) At x/b = 0.5

Trough region

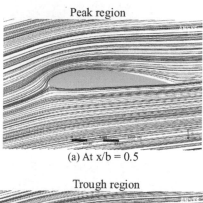

(b) At x/b = 0.25

Figure 11.   Streamlines of BUMP 4415 airfoil at $\alpha = 10°$.

Peak region

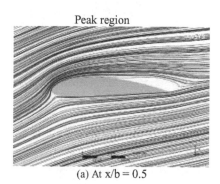

(a) At x/b = 0.5

Trough region

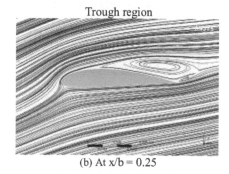

(b) At x/b = 0.25

Figure 12.   Streamlines of BUMP 4415 airfoil at $\alpha = 15°$.

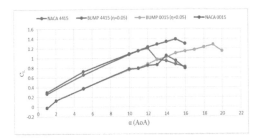

Figure 13. Comparison of lift coefficient with angle of attack.

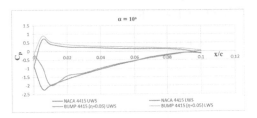

Figure 14. Pressure distribution plot for baseline and modified cambered airfoils at $\alpha = 10°$.

4.3 Effect of tubercles on C_p distribution

The chord-wise pressure (C_p) distributions for baseline and modified airfoils are plotted in Figure 14. For the baseline airfoil, C_p achieves a minimum value near the leading edge and, after the peak value, C_p increases along the downstream creating a strong adverse pressure gradient. This causes the boundary layer to separate from the top surface. The most negative values of the peak pressure developed over the suction surface of the BUMP 4415 airfoil at $\alpha = 10°$ are associated with the trough cross section.

4.4 Flow field induced by tubercles

The implementation of tubercles at the leading edge of airfoil leads to the formation of counter-rotating pairs of stream-wise vortices between the tubercle peaks (i.e. trough region). The row of tubercles redirects the flow of air into the scalloped valley between each tubercle, causing swirling vortices that roll up and over the airfoil, which actually enhances lift properties. The swirling vortices exchange momentum with the flow and this exchange keeps the flow attached to the suction side of the airfoil and delays stall to higher angles of attack. The benefits of tubercles act in a similar way to the vortex generators that are often used in conventional aircraft. Determination of the most effective ratio of height to boundary-layer thickness for the tubercles and optimization of tubercle geometries could serve as a potential replacement for vortex generators.

5 CONCLUSION

The effect of leading-edge tubercles is studied on the pre-stall and post-stall behavior of an airfoil. The formation of counter-rotating pairs of stream-wise vortices is said to enhance the airfoil characteristics in comparison to baseline airfoils. During pre-stall, the presence of a laminar separation bubble increases the lift of the airfoil with minimum drag effect. The implication is that tubercles on airfoils and wings helps to increase the lift beyond that of the baseline airfoils or wings. The C_L is improved from 20% to 30% for modified airfoils, as presented in Figure 13. The bursting of the laminar separation bubble at the leading edge of the baseline airfoil during the stall increases the drag with a significant loss in lift. In an airfoil employing tubercles, the separation bubbles are restricted to trough regions between the tubercles. Because the row of tubercles redirects the flow of air into the scalloped valley between each tubercle, it causes swirling vortices that roll up and over the airfoil, which actually enhances lift properties. Therefore, the tubercled airfoil does not stall so quickly. At post-stall angles of attack, wings experience a reduction in lift due to flow separation. Post-stall lift can be attained in larger amounts with the aid of tubercles with larger amplitude. The structure of the vortices increases with the increase in amplitude of the tubercles. The stream-wise vortices convert the momentum into flow, and the circulation of flow in the downstream direction increases. As a result, the flow tends to attach to the upper surface of the wing, yielding a large increase in lift.

NOMENCLATURE

Re   = Reynolds number
PIV   = Particle Image Velocimetry
A   = Amplitude
λ = Wavelength
z   = Span-wise ordinate
C   = Mean chord
η = A/λ
c   = Chord
C_L = Lift coefficient
C_D = Drag coefficient
α = Angle of attack
UWS   = Upper Wing Surface
LWS   = Lower Wing Surface

REFERENCES

Bushnell, D.M. & Moore, K.J. (1991). Drag reduction in nature. *Annual Review of Fluid Mechanics, 23*, 65–79.

Cai, C., Zuo, Z., Liu, S. & Wu, Y. (2015). Numerical investigations of hydrodynamic performance of hydrofoils with leading-edge protuberances. *Advances in Mechanical Engineering, 7*(7), 1–11.

Corsini, A., Delibra, G. & Sheard, A.G. (2013). On the role of leading-edge bumps in the control of stall onset in axial fan blades. *Journal of Fluids Engineering, 135*, 081104.

De Paula, A.A., Padilha, B.R.M., Mattos, B.D. & Meneghini, J.R. (2016). *The airfoil thickness effect on wavy leading edge performance.* Paper presented at the 54th AIAA Aerospace Sciences Meeting, AIAA SciTech, San Diego, CA. doi:10.2514/6.2016–1306.

Edel, R.K. & Winn, H.E. (1978). Observations on underwater locomotion and flipper movement of the humpback whale Megaptera novaeangliae. *Marine Biology, 48*, 279–287.

Fish, F.E. & Battle, J.M. (1995). Hydrodynamic design of the humpback whale flipper. *Journal of Morphology, 225*, 51–60.

Hansen, K.L., Kelso, R.M. & Dally, B.B. (2011). Performance variations of leading-edge tubercles for distinct airfoil profiles. *AIAA Journal, 49*(1), 185–194.

Johari, H., Henoch, C.W., Custodio, D. & Levshin, A. (2007). Effects of leading-edge protuberances on airfoil performance. *AIAA Journal, 45*(11), 2634–2642.

Karthikeyan, N., Sudhakar, S. & Suriyanarayanan, P. (2014). *Experimental studies on the effect of leading edge tubercles on laminar separation bubble.* Paper presented at the 52nd Aerospace Sciences Meeting, AIAA SciTech 2014, 13 Jan 2014, National Harbor, MD. doi:10.2514/6.2014–1279.

Lohry, M.W., Clifton, D. & Martinelli, L. (2012). *Characterization and design of tubercle leading-edge wings.* Paper presented at the Seventh International Conference on Computational Fluid Dynamics (ICCFD7–4302), Big Island, Hawaii.

Miklosovic, D.S., Murray, M.M., Howle, L.E. & Fish, F.E. (2004). Leading edge tubercles delay stall on humpback whale flippers. *Physics of Fluids, 16*(5), L39–L42.

Rostamzadeh, N., Hansen, K.L., Kelso, R.M. & Dally, B.B. (2014). The formation mechanism and impact of stream wise vortices on NACA 0021 airfoil's performance with undulating leading edge modification. *Physics of Fluids, 26*, 107101.

Skillen, A., Revell, A., Pinelli, A., Piomelli, U. & Favier, J. (2015). Flow over a wing with leading edge undulations. *AIAA Journal, 53*(2), 464–472.

Watts, P. & Fish, F.E. (2001). The influence of passive leading edge tubercles on wing performance. In *Proceedings 12th UUST, Durham, NH, August 2001.* Lee, NH: Autonomous Undersea Systems Institute.

Zhang, M.M., Wang, G.F. & Xu, J.Z. (2013). Aerodynamic control of low-Reynolds number airfoil with leading-edge protuberance. *AIAA Journal, 51*(8), 1960–1971.

Emerging Trends in Engineering, Science and Technology for Society,
Energy and Environment – Vanchipura & Jiji (Eds)
© 2018 Taylor & Francis Group, London, ISBN 978-0-8153-5760-5

Aerodynamic investigation of airfoil inspired HALE UAV

R. Saravanan, V. Vignesh, V. Venkatramasubramanian, R. Vaideeswaran &
R. Naveen Rajeev
Department of Aeronautical Engineering, Hindusthan College of Engineering and Technology,
Coimbatore, India

ABSTRACT: The present investigation focuses on structure and shape modification for a High Altitude Long Endurance (HALE) Unmanned Aerial Vehicle (UAV). Aerodynamics and performance parameters are defined during the conceptual design. Accordingly, the fuselage shape is designed by NACA 2410 (low – Reynolds number) and the wing modification is carried out using Gurney flaps and winglets. The objectives of this investigation are fully stated as a consequence of the design, and the wind tunnel model is briefly described. This project intended to get some coherent results between theoretical calculations and both the experimental and Computational Fluid Dynamics (CFD) results. Two methods for calculating the pressure coefficient are used, one is experimental low speed subsonic wind tunnel testing and the other is a numerical solution from CFD ANSYS R15.0. The Spallart-Allmaras turbulence model was used for solution initialization. The lift and drag coefficients of the airfoil at different angles of attack were observed for both computational and experimental study. The pressure and velocity distribution were obtained using ANSYS R15.0. The aerodynamic characteristics of the UAV model have been carried out at different angles of attack. The aerodynamic parameters, such as the lift coefficient, drag coefficient and (L/D) ratio are improvised by optimizing the design.

1 INTRODUCTION

An Unmanned Aerial Vehicle (UAV) is a flying machine without a human pilot on board. The flight of UAVs as a rule work with different degrees of self-rule. Development of an advanced unmanned aerial vehicle (UAV) for civil and military applications has driven the development of modern aviation. In this study, a structure and shape modification procedure for a high altitude long endurance (HALE) unmanned aerial vehicle (UAV) is presented. The improvement and utilization of cutting edge remote detection of UAV, for example, HALE, that could more suitably address the necessities of the crisis administration group, requires extensive consideration and assessment. State-of-the-art craftsmanship in HALE stage innovation is exemplified by the US Air Force's Global Hawk Unmanned Endurance Air Vehicle, which is in framework trials, is intended to go for completely independently operation (Goraj et al., 2004) with a most extreme flight continuance of 40 hours. Worldwide Hawk could fly up to 3,000 nautical miles (5,556 km) at up to 67,000 ft. (20, 422 m), saunter over an objective territory for 24 hours while utilizing sensors or other hardware, and afterwards come back to base, all without human guidance.

The Gurney flaps (Zerihan & Zhang, 2001) are a simple device consisting of a short strip fitted perpendicular to the pressure surface along the trailing edge of a wing. With a typical size of 1–15% of the wing chord (Bechert et al., 2000), it can exert a significant effect on the lift (down force), with a small change in the stalling incidence, leading to a higher C_L max, as documented by (Liebeck, 1978). Although the device was named after Gurney in the 1960s, mechanically similar devices were employed earlier, for example, by Gruschwitz, Schrenk and Duddy. It is anything but difficult to outline an air ship on the off chance that have information's about officially existing flying machines of comparative sort. It gives more fulfillments and maintains a strategic distance from perplexity while picking some plan parameters for our flying machine (Parezanovic et al., 2005). In this definite overview some numerous essential outline drivers like viewpoint proportion, wing stacking, general measurements and motor determinations are resolved for our reference. It helps with proposing another plan and changes in our outline, which will enhance the execution of the proposed air ship. This guarantees the execution of the flying machine according to the outline figuring's and a simple method for planning an airplane inside specific timeframe.

2 METHODOLOGY

2.1 Computational details

Computational Fluid Dynamics (CFD) is attractive to industry as it is more financially viable than the physical testing. In any case, one must note that the mind boggling stream reproduction is testing and blunder inclined and it takes a great deal of building ability to get approved arrangements. Applying the basic laws of mechanics to a liquid gives the overseeing conditions for a liquid. The protection of mass condition is

$$\frac{\partial p}{\partial t} + \Delta.\left(\rho u\right) = 0$$

where ρ = density; u = velocity and t = time

The above conditions along with the protection of vitality condition frame an arrangement of coupled and nonlinear fractional differential conditions. The design of the (Hendrick et al., 2008) HALE UAV is undertaken using Solidworks software for the surface design. A model was imported to an ANSYS ICEM CFD 15 in meshing tool. The model is imported as Para solid format for getting a fine meshing a model with 160,000 elements and 260,000 nodes, the interval counts are 500,250 and 100. The model is placed in a rectangular domain.

2.2 Exploratory set-up

Four noteworthy parts of the HALE UAV display are wing, fuselage, flat stabilizer, vertical stabilizer, winglets and Gurney folds. Mid wing sort show has been chosen for manufacture. NACA 2410 cambered airfoil has been utilized for creation of wing, fuselage, even stabilizer and vertical stabilizer of the said display.

2.3 Subsonic breeze burrow

Shut circuit subsonic breeze burrow has been utilized to test the manufactured HALE UAV. The measurements of the working segment of the wind burrow are 35 cm (length) x 35 cm (stature). Air enters the breeze burrow through an efficiently planned diffuser (cone).

A control handle could be pivoted to control the velocity from 0 to 25 m/s. The 'L-area' is fitted with the working segment of the breeze passage to gauge the lift, drag, weight and their coefficients. A photograph of the subsonic breeze passage and HALE UAV fitted with the working area of the subsonic breeze burrow are shown in Figure 3 and Figure 4 individually.

Table 1. Boundary conditions and input parameters in fluent 14.

S. no	Flow parameters	Values
1	Pressure (P) in Pascal	1.0132
2	Density (ρ) in kg/m^3	1.225×10^3
3	Velocity at inlet (m/s)	25
4	Reynolds number	3×10^4

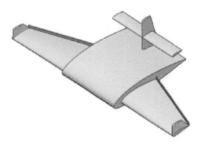

Figure 1.   Designed HALE UAV.

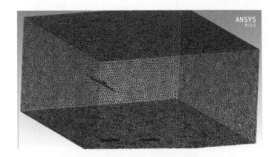

Figure 2.   Wire frame mesh of designed HALE UAV.

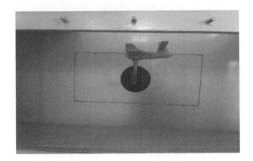

Figure 4.   Model fitted in a test section.

Figure 3.   Low speed subsonic wind tunnels.

3   RESULTS AND DISCUSSION

The lift and drag coefficients of the airfoil at different angles of attack were observed for both computational and experimental study. Data is taken by varying the angle of attack and by keeping the free stream velocity constant for both tests.

3.1   *Comparison between computational and experimental data*

Experiment results are compared with computational results and it illustrates the similarities and minimum contrast behavior between computational and experimental data. Results are calculated by keeping free stream velocity as a constant and varying the angle of attack. The plot is drawn between locations of pressure tapings and pressure coefficient. The plots shown will give the pressure coefficient at various angle of attack for both numerical and analytical results. Figure 5, the pressure at the suction region decreases because of decreasing the lift coefficient value and this scenario is similar for both experimental and numerical calculations and are become evident through the plot. In Figure 7 the pressure at suction zone tends to increase by simultaneously increasing the lift coefficient. Both CFD and experimental results indicate the similar traits, thus in turn to improve the aerodynamic parameters (Jahangir Alma et al., 2013). In Figure 10, the pressure difference value reaches a peak point and the lift coefficient value drops rapidly such that the stall region is approached in both experimental and CFD results. The computational results for the HALE UAV are obtained by using the Sparlart Allmaras turbulence model. Keeping the inlet velocity at 25 m/s and by varying the angle of attack, the velocity distribution and pressure distribution is obtained as contour diagrams. Thus the values of the lift coefficient, drag coefficient and pressure coefficient is obtained.

The value of pressure coefficient is drawn in the graph against the location of pressure tapings, where the location is plotted by using the solver. The lift coefficient plotted with angle of attack gives the lift curve and the drag coefficient is plotted against angle of attack

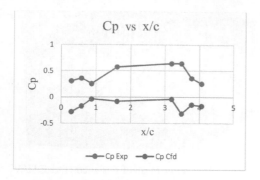

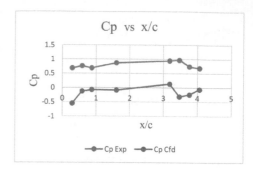

Figure 5. Comparison of pressure coefficient vs. locations between computational and experimental data at $\alpha = 0°$ @ V = 25 m/s.

Figure 6. Comparison of pressure coefficient vs. locations between computational and experimental data at $\alpha = 8°$ @ V = 25 m/s.

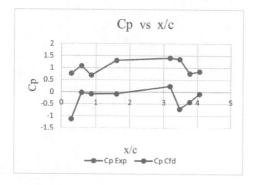

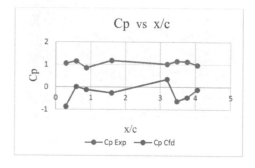

Figure 7. Comparison of pressure coefficient vs. locations between computational and experimental data at $\alpha = 12°$ @ V = 25 m/s.

Figure 8. Comparison of pressure coefficient vs. locations between computational and experimental data at $\alpha = 16°$ @ V = 25 m/s.

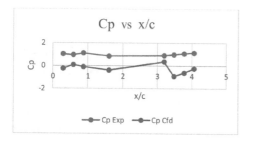

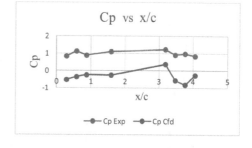

Figure 9. Comparison of pressure coefficient vs, locations between computational and experimental data at $\alpha = 24°$ @ V = 25 m/s.

Figure 10. Comparison of pressure coefficient vs. locations between computational and experimental data at $\alpha = 28°$ @ V = 25 m/s.

to visualize its effect at various levels. The drag variation is represented in the polar diagram in which C_L is plotted against C_D. In this plot, lift coefficient increases until a particular angle of attack after which it drops due to subsequent increase in the drag coefficient. The graph between pressure coefficient and location of pressure tapings is plotted for both experimental as well as computational results. The obtained results are compared and validated. It is found that the optimized lift coefficient and the stall region of the designed HALE UAV is similar in both experimental and CFD results.

4 CONCLUSIONS

The primary goal of this paper is to investigate the aerodynamic characteristics of HALE unmanned aerial vehicle (UAV) in order to make a UAV with optimized aerodynamic performances. To that end, it explains how improvisation of aerodynamic parameters such as coefficient of lift (C_L) and coefficient of drag (C_D) is done. The coefficient of pressure (C_P) for the model is calculated from the experimental results. The report concludes the result by using the methodology that has been implemented in it. The desired result is achieved by modifying the shape of the fuselage. The presence of a Gurney flap on the wing served to increase the lift generated on the body. The presence of winglets makes the flow laminar by preventing the creation of vortices. The induced drag has been counterattacked by using the winglets and the vortex drag has been counterattacked by Gurney flaps. For three dimensional analysis, favorable results were obtained with the ANSYS, FLUENT software. The C_L Vs. α and C_D Vs. α graphs give the satisfactory results with values of C_L and C_D. Different data is obtained by keeping the free stream velocity constant and by varying the angle of attack for both the experimental and investigational study. Since HALE UAVs fly at high altitude, and in low density conditions at lower speeds, it is difficult to locate the transition point. This problem is overcome by using a low Reynolds number airfoil shaped fuselage structure. Manometric readings are obtained using a subsonic wind tunnel at a constant velocity V = 25 m/s. From the manometric readings, free stream static pressure and total pressures were calculated. These manometric readings are calculated with the aid of pressure tapings. Computational investigations have been performed to examine the effectiveness of the aerodynamic parameters such as C_L, C_D and C_P for the HALE UAV. The stall angle for the designed HALE UAV is 28°. The coefficient of lift is found to be maximum at the angle of attack $\alpha = 16°$. Experimental results compared with computational simulations are next in relation to the aerodynamic coefficients. The pressure contour at stall angle of attack is found to be maximum at the pressure side and minimum at the suction side. The velocity contour and pressure distribution exhibits the same result. Finally, the result concludes that the aerodynamic parameter C_P has been efficient for experimental and computational results. This project achieved some coherent results between theoretical calculations and both the experimental and CFD results. This analysis yields better results for a HALE UAV and can be implemented on future UAV projects to get the optimized coefficient of lift (C_L) value. Using this design optimization, the HALE UAVs will be able to achieve higher ranges and longer endurance.

REFERENCES

Bechert, D.W., Meyer, R. & Hage, W. (2000). *Drag reduction of air-foils with mini flaps—AIAA*. Berlin, Germany.

Goraj, Z., Frydrychewicz, A., Ašwitkiewicz, R., Hernik, B., Gadomski, J., Goetzendorf-Grabowski, T., Figat, M. & Chajec, W. (2004). High altitude long endurance unmanned aerial vehicle of a new generation-a design challenge for a low cost, reliable and high performance aircraft. *Bulletin of the Polish Academy of Sciences Technical Sciences, 52*(3), 173–194.

Hendrick, P., Verstraete, D. & Coatanea, M. (2008). Preliminary design of a joined wing HALE UAV. *26th International Congress of the Aeronautical Sciences, (ICAS2008)*.

Jahangir Alam, G.M., Md. Mamun, Md Abu, T.A, Md. Quamrul Islam, Md. & Sadrul Islam, A.K.M. (2013). Investigation of the aerodynamic characteristics of an airfoil shaped fuselage UAV model, *International Conference on Mechanical Engineering*.

Liebeck, R.H. (1978) Design of subsonic air-foils for high lift. *Journal of Aircraft, 15*(9), 547–561.

Parezanovic, V., Rasuo, B. & Adzic, M. (2005). Design Airfoils for wind turbine blades. Research Gate Publications. Retrieved from https://www.researchgate.net/publication/228608628_DESIGN_OF_AIRFOILS_FOR_WIND_TURBINE_BLADES.

Zerihan, J., & Zhang, X. (2001). Aerodynamics of Gurney flaps on a wing inground effect, *AIAA JOURNAL, 39*(5), 772–780.

Industrial engineering and management

Emerging Trends in Engineering, Science and Technology for Society,
Energy and Environment – Vanchipura & Jiji (Eds)
© 2018 Taylor & Francis Group, London, ISBN 978-0-8153-5760-5

The usability of road traffic signboards in Kottayam

R. Rajesh, D.R. Gowri & N. Suhana
Rajiv Gandhi Institute of Technology, Kottayam, Kerala, India

ABSTRACT: Traffic signboards provide important information, directions and warnings on the road; they are designed and placed to provide assistance to drivers. In this study, the signboard usability issues of drivers in Kottayam district was studied. The significant factors that affect usability are determined from a literature survey, and a preliminary observational field study was conducted. Next, based on the brainstorming sessions, a cause and effect diagram for the poor usability of signboards was developed. A questionnaire survey was conducted. A majority of the driver respondents (58%) do not use the signboards. The reasons for non-usage or poor usage are attributed to operational factors, environmental factors, or human factors. The respondents perceived that poor signs would have the largest negative impacts on driving control and confusion. Numerous factors relating to driver, road, signboard, and environment affect the usability of traffic signs. Signboard factors such as readability, font size, language, position, long-distance visibility, and multi-sign configuration, road factors such as advertisements, greenery, buildings, and speed, driver factors such as driver age, eyesight, road familiarity, and driver experience, and environmental factors such as weather, light, and police presence, all affect traffic sign usability. Traffic signs could be made better by applying ergonomic principles and a number of suggestions to improve traffic sign usability are provided. The study is limited by its sample size and a larger sample is needed to draw a generalized conclusion. A field-based observational study or a naturalistic driver study could provide for more quantitative analysis

Keywords: Traffic signboards, warnings on the road, multi-sign configuration

1 INTRODUCTION

Traffic signboards provide important information, directions and warnings on the road; they are designed and placed as assistance to drivers. They keep traffic flowing freely by helping drivers reach their destinations and letting them know entry, exit, and turn points in advance. Pre-informed drivers will naturally avoid committing mistakes or taking abrupt turns and causing bottlenecks. Comprehension of traffic signboards is crucial to safety, but they are not always detected or recognized correctly. Signboards present issues in terms of detection and recognition due to poor visibility, bad weather conditions, the color combinations used, their height and position, vehicle speed, and driver's age and vision.

Usability indicates ease or convenience of use. In this study, we consider signboard usability issues. The objective of this paper is to assess the usability problems of signboards faced by drivers and to study the effectiveness of signboards in Kottayam.

2 LITERATURE REVIEW

Numerous studies have indicated that regulatory and warning signs help to improve the flow of traffic, reduce accidents and ensure that pedestrians can safely use designated crosswalks. Traffic sign usability is influenced by numerous factors.

It is essential for increased road safety that drivers can understand signs correctly and quickly (Cristea & Delhomme, 2015). Meaningfulness, recognition, simplicity, and symbol design affect sign comprehension (Yuan et al., 2014). Familiarity, standardization, simplicity, and the symbol-concept compatibility of traffic signs also influence comprehension (Ou & Liu, 2012; Shinar & Vogelzang, 2013). For better usability, Jamson and Mrozek (2017) emphasized color and shape, while Khalilikhah and Heaslip (2016) stressed sign condition. Traffic signs, traffic factors, and driver factors affect the driver and their safety performance in relation to accuracy, correctness of answers, and reaction time (Ng & Chan, 2008; Di Stasi et al., 2012; Shinar & Vogelzang, 2013; Yuan et al., 2014; Kazemi et al., 2016; Domenichini et al., 2017). A study of signboard comprehension by Ben-Bassat and Shinar (2015) indicated that younger drivers performed significantly better than older drivers on both accuracy and response time.

3 METHODOLOGY

The significant factors that affect usability have been determined from the literature survey. A preliminary observational field study was carried out on three 15-kilometer stretches of road in the Kottayam district of Kerala state in India. The field study provided an opportunity to comprehend the signboard types and the factors highlighted in the literature that affect their usability. Figures 1 and 2 show some of the signboards on the roads observed. Subsequently, a brainstorming session was undertaken with four student project members and two experts to focus on the road signboards' attributes and their usability. Figure 3 shows a cause and effect diagram for the poor usability of signboards according to these brainstorming sessions. Next, a four-part questionnaire was prepared and a pilot survey was conducted in Kottayam district. Convenience sampling was used for the survey. Where possible, respondents were approached individually using hard copy; others were surveyed using an online method. A total of 236 responses were obtained. These were filtered for missing values and, subsequently, descriptive statistics were used for data analysis.

The cause and effect diagram (Figure 3) highlights four components that affect usability, that is, driver, signboard, road, and environment. Driver factors include age, gender, experience, familiarity with signs, familiarity with roads, and driver behavior; road factors include

Figure 1.   Signboards with two languages, color combination, variable font size, and graphics.

Figure 2.   Damaged, vegetation-obstructed, and poor-visibility signboards.

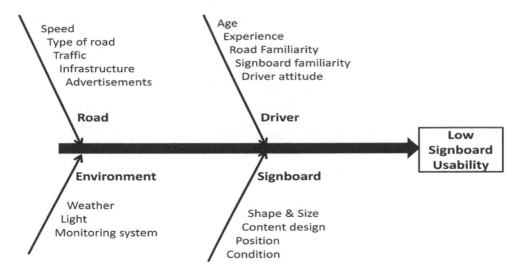

Figure 3.   Cause and effect diagram for signboard usability.

speed of vehicle, type of road, traffic conditions, road infrastructure, and advertisements; signboard design attributes include shape, size, color, content, position, and condition; environmental factors include weather, light, and monitoring system. Numerous studies have provided evidence of the differing impacts of various factors on traffic sign effectiveness. For example, sign layout, shape, and familiarity would improve driver comprehension and hence driver response, while damaged signs, obstructing vegetation, and advertisements would make signs unusable for the driver. Higher speed, driver age, and traffic volumes would impair sign usability as a result of the shorter driver response times necessitated by them.

4   RESULTS

Kottayam contains 406 km of state highways and a total road length of 3,449 km. The three main types of signs, that is, mandatory signs, cautionary signs, and information signs, have been used regularly along the roads (see, for example, Figures 1 and 2). The languages used are predominantly Malayalam and English. The color combination for the text on information signs is white text on a green background.

The sample was relatively young (mean = 25.19 years; SD = 8.863 years) with 91% male respondents. A proportion (50%) of respondents used two-wheelers. A majority of drivers (58%) reported that they do not use signboards or only use them sometimes (36%). The reasons for non-usage or low usage could be attributed to operational factors, environmental factors or human factors. Some of these aspects are examined in the following sections.

4.1   Impact of poor traffic signboard usability

Poor and improper use of signboards causes numerous issues to drivers, for example, confusion, traffic congestion, time loss, penalties, loss of driving control, and accidents. In this survey, it has been found that poor signboards lead to traffic congestion (50%), time loss (24%) and confusion (13%). Two probable reasons for non-usage or poor usage of signs are: (i) drivers are familiar with the roads and don't perceive a need to use the signboards; (ii) drivers have difficulty in using the signboards. The two major reasons for poor signboard usability are damaged signage (49%) and fading colors (33%). Interestingly, 'sign knowledge', 'user behavior' and 'time pressure' have not been indicated by the respondents as the primary reason for poor signboard use.

325

Among the six impacts assessed, that is, 'driving control', 'confusion', 'time loss', 'traffic congestion', 'incidents', and 'traffic compliance', respondents indicated that poor traffic signs cause the most negative impact on 'driving control' (88%) and 'confusion' (83%); 68% of respondents felt that poor traffic signs cause 'time loss', 66% that they would lead to 'traffic congestion', and 51% indicated they would lead to 'incidents'. In terms of 'traffic compliance', respondents were divided: 49% indicated that poor traffic signs would have a negative impact on this, while 31% indicated there would be no effect. A statistical T-test found significantly greater negative effects from poor traffic signs on 'congestion', 'time loss', and 'driving control' for those with less than five years' driving experience.

4.2 Factors affecting signboard usability

Figure 4 shows the signboard attribute rating. Shape attributes (color combination, height, size, layout) have been favorably rated, while readability (inside text, font size, language) and visibility (weather, distance, night-time, position and configuration) attributes have been negatively rated. Readability and visibility attributes are among the likely causes of poor signboard usability. In addition, vehicular (speed), environmental (greenery, buildings, advertisements) and human (age and eyesight) factors produce a further negative effect on usability. As far as other human factors (driving experience and road familiarity) are concerned, they are perceived to produce a positive effect on usability. This implies that there could be a positive trade-off in the attitude to compliance of drivers regarding road traffic signboard use.

There are numerous external factors affecting traffic signboard usability, such as high vehicle speed, greenery and vegetation, presence of buildings, presence of flags/advertisements/placards, increase in age of drivers, driver's eyesight, driver's compliance attitude, driver's familiarity with road, driver's experience, and police presence (see Figure 5). Among these, high vehicle speed, greenery and vegetation, presence of buildings, presence of flags/advertisements/placards, increase in age of drivers, and driver's poor eyesight have a negative effect on signboard usability, while driver's familiarity with the road, driver's experience, and police presence have a positive effect on signboard usability. During the field study, it was observed that the roads had significant bends, were densely populated with buildings, and had significant vegetation along them. Some of this is visible in Figures 1 and 2. Details of the field study would be made available in the part-2 of this article. Contrary to the expectation that a driver's attitude to compliance might have a positive effect on signboard usability, it was found that only 27% of respondents perceived that compliance attitude had a positive effect on signboard usability. This is a behavior issue that needs to be further examined.

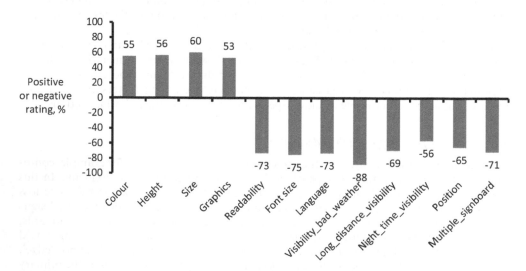

Figure 4. Signboard attribute rating.

326

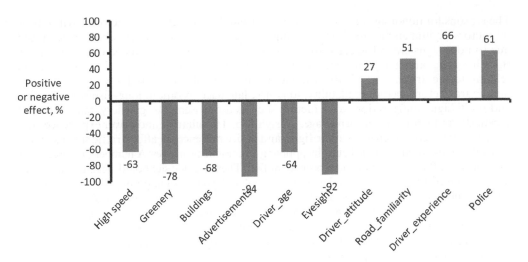

Figure 5.   Effects of road conditions and human factors on road signboard usage.

4.3   *Discussion*

Numerous factors relating to driver, road, signboard, and environment affect the use of traffic signs (see Figure 3). Readability, font size, language, position, long-distance visibility, and multi-sign configuration are prominent attributes that affect sign usability. Advertisements, greenery, buildings, and high speed are road factors that reduce traffic sign usability. Driver factors such as driver age and eyesight are perceived to have a negative effect, but road familiarity and driver experience are perceived to have a positive effect on traffic sign usability. The environmental factors of weather, light, and police presence affect traffic sign usability. For better use and compliance with road signage there is a need to redesign traffic signs. A user-centered design of traffic signs that considers driver, road, signboard, and environmental factors would enable better comprehension of the signs, their more effective usability, and better driver responses. Ultimately, this would lead to better traffic flow and safety performance. Traffic signs could be made better by using ergonomic principles (Jamson & Mrozek, 2017).

From the survey carried out, 32% of the respondents felt that an on-board recognition system could improve sign usability. A good traffic sign could help in better design of intelligent driving-aid systems (Amditis et al., 2010; Wen et al., 2016). The respondents emphasized the use of reflective signs (30%) and improved road infrastructure (20%) for enhanced traffic sign usability.

We make the following specific suggestions:

– Signboards should be cleaned and maintained every six months.
– The font size, line spacing and sign size should be as per the Indian standards of motor vehicle legislation.
– Focus on better content layout in signs for better readability.
– Position and align signboards to maximize long-distance visibility.
– Remove greenery regularly from the lines of sight of signs.
– Provide for night-time sign visibility through reflective signs or night lighting.
– Examine the possibility of providing user-compatible in-vehicle traffic sign devices that help to address human age or eyesight issues.
– Increase monitoring efforts for traffic sign compliance.

5   CONCLUSION

The road traffic sign usability issues of drivers in Kottayam district were studied. A majority of the drivers surveyed (58%) do not use the signboards or only use them sometimes (36%).

The reasons for non-usage or poor usage are attributable to operational factors, environmental factors or human factors. The two major issues that the respondents have experienced due to poor signs are traffic congestion and time loss. The respondents perceived that poor signs have the largest negative impact on driving control and confusion. Numerous factors relating to the driver, road, signboard, and environment affect the usability of traffic signs. Signboard factors such as readability, font size, language, position, long-distance visibility, and multi-sign configuration, road factors such as advertisements, greenery, buildings, and speed, driver factors such as driver age, eyesight, road familiarity, and driver experience, and environment factors such as weather, light, and police presence, all affect traffic sign usability.

Traffic signs could be made better by applying ergonomic principles. A number of suggestions to improve traffic sign usability have been identified. These include signboard cleaning and maintenance, use of Indian traffic standards, better content layout design, positioning of signboards for maximum visibility, removal of obstructions and greenery along lines of sight, improved reflectivity, the provision of in-vehicle traffic sign devices, and increased monitoring efforts.

This study was limited by its sample size and a larger sample is needed to allow a more generalized conclusion. A field-based observational study or a naturalistic driver study could provide for more quantitative analysis.

ACKNOWLEDGMENTS

We thank Mrs. Asha Lakshmi, Ms. Harsha Surendranand, Dr. Vinay V. Panikar, Dr. Basha S.A, and students of M. Tech 2016 for helping us carry out this study.

REFERENCES

Amditis, A., Pagle, K., Joshi, S. & Bekiaris, E. (2010). Driver Vehicle environment monitoring for on-board driver support systems: Lessons learned from design and implementation. *Applied Ergonomics, 41*(2), 225–235.

Ben-Bassat, T. & Shinar, D. (2015). The effect of context and drivers' age on highway traffic signs comprehension. *Transportation Research Part F: Traffic Psychology and Behaviour, 33*, 117–127.

Cristea, M. & Delhomme, P. (2015). Factors influencing drivers' reading and comprehension of onboard traffic messages. *European Review of Applied Psychology, 65*(5), 211–219.

Di Stasi, L.L., Megías, A., Cándido, A., Maldonado, A. & Catena, A. (2012). Congruent visual information improves traffic signage. *Transportation Research Part F: Traffic Psychology and Behaviour, 15*(4), 438–444.

Domenichini, L., La Torre, F., Branzi, V. & Nocentini, A. (2017). Speed behaviour in work zone crossovers. A driving simulator study. *Accident Analysis & Prevention, 98*, 10–24.

Jamson, S. & Mrozek, M. (2017). Is three the magic number? The role of ergonomic principles in cross country comprehension of road traffic signs. *Ergonomics, 60*(7), 1024–1031.

Kazemi, M., Rahimi, A.M. & Roshankhah, S. (2016). Impact assessment of effective parameters on drivers' attention level to urban traffic signs. *Journal of the Institution of Engineers (India): Series A, 97*(1), 63–69.

Khalilikhah, M. & Heaslip, K. (2016). The effects of damage on sign visibility: An assist in traffic sign replacement. *Journal of Traffic and Transportation Engineering, 3*(6), 571–581.

Ng, A.W. & Chan, A.H. (2008). The effects of driver factors and sign design features on the comprehensibility of traffic signs. *Journal of Safety Research, 39*(3), 321–328.

Ou, Y.K. & Liu, Y.C. (2012). Effects of sign design features and training on comprehension of traffic signs in Taiwanese and Vietnamese user groups. *International Journal of Industrial Ergonomics, 42*(1), 1–7.

Shinar, D. & Vogelzang, M. (2013). Comprehension of traffic signs with symbolic versus text displays. *Traffic & Transportation Research, 44*(1), 3–11.

Wen, C., Li, J., Luo, H., Yu, Y., Cai, Z., Wang, H. & Wang, C. (2016). Spatial-related traffic sign inspection for inventory purposes using mobile laser scanning data. *IEEE Transactions on Intelligent Transportation Systems, 17*(1), 27–37.

Yuan, L., Ma, Y.F., Lei, Z.Y. & Xu, P. (2014). Driver's comprehension and improvement of warning signs. *Advances in Mechanical Engineering, 6*, 582–606.

Emerging Trends in Engineering, Science and Technology for Society, Energy and Environment – Vanchipura & Jiji (Eds)
© *2018 Taylor & Francis Group, London, ISBN 978-0-8153-5760-5*

Development and analysis of robust neighbourhood search for flow-shop scheduling problems with sequence dependent setup times

V. Jayakumar & Rajesh Vanchipura
Department of Mechanical Engineering, Government Engineering College, Thrissur, Kerala, India

ABSTRACT: This paper focuses on the problem of determining a permutation schedule for n jobs in an m-machine flow shop environment. It operates in a Sequence Dependent Setup Time (SDST) environment. The objective function considered is minimization of makespan. A new heuristic algorithm called, 'Robust Neighbourhood Search'—a local search algorithm with added global search properties is used for minimization of makespan, the objective. The proposed heuristic involves an explorative search with random variables converge to near optimal solution. More than one cycle is used to ensure achieving results close to global minima. For purpose of experimentation, 1080 SDST benchmark problem instances are used. That involves nine problem sizes at eight different levels of setup times. Graphical analysis, relative performance index analysis, and statistical analysis are carried out on the makespan results obtained for all the benchmark problem instances. The analysis reveals that for all problem instances the RNS works superior to the Genetic Algorithm.

1 INTRODUCTION

The production industries generally involve various processes such as design, manufacturing, marketing. And the manufacturing process consists of operations like, welding, milling etc. In reality there can be numerous of operations to be carried out on a raw material before it comes out as a final product. Many times other materials were added to or removed from the original material, some totally different in nature. Each of this operations consume different processing times. In addition to that, each of these operations need separate machines to perform each action. Hence, in the shop floor each machine is used for a particular job at a time. Once a job is done, the setup on the machine needs to be changed to do another job. There will be some time required to change the setup done for doing one job to another job. The setup times are invariably involved in all scheduling situations. However, they are added to processing times in many of the situations. Certainly, this procedure will reduce the complexity of the problem solved. On the other hand it will affect the quality of solutions obtained. Hence, there is need for explicit consideration setup time, which is addressed in the present study.

In shop floor there are many configurations such as single machine, parallel machine, flow shop, job shop etc. However, real manufacturing situations encounter numerous variations of these basic shop configurations. It is observed that more than one-third of production systems follow flow shop configuration (Foote and Murty, 2009). A flow shop is characterised by the flow of work that is unidirectional i.e., there are n jobs to be processed on m machines. The order jobs are processed on machines is assumed to be same i.e., permutation flow shop. If there are n jobs, there are $n!$ total number of solutions. The present research considers a realistic variation of the general flow shop, i.e. a flow shop operating in a sequence dependent setup time environment. When the setup time is added, the complexity of the problem becomes NP complete in nature (Jatinder & Gupta, 1986).

The flow shop scheduling problems are widely used in industry. Reducing the makespan is the main objective that is desired in most of the situations. Exact solution methodologies such as branch-and-bound, dynamic programming explored by researchers but are limited small size problems involving two or three machines. Since there are no solution methodologies giving exact solution for these problems, heuristic methods are widely employed. The heuristics can be constructive or improvement in nature. Constructive heuristic gives a single solution all the time while improvement heuristics give different solutions every time. The advantage is that the algorithms can be easily changed to meet some specific requirements. Both have their own merits and demerits.

Improvement heuristics are applied more than constructive ones owing to their flexibility. Majority of the improvement heuristics are meta-heuristics such as genetic algorithm, particle swarm optimization algorithm, and simulated annealing algorithms etc. By observing the nature or otherwise, scientists and industrialists try to make new heuristics to solve flow shop scheduling problems. Such an experiment is what we try to do in this paper. The present paper proposes a novel variation of neighborhood improvement heuristic for Flow shop scheduling problems with setup times.

2 LITERATURE REVIEW

The flow shop scheduling problems have been an intense subject of study over past few decades. The literature review carried out in the present paper focuses on the works done on flow shop scheduling with and without set up times for jobs. The earlier works are done without taking the setup times into consideration. Both constructive and improvement heuristics are worked on it quite well. It can be seen that make span minimization is given the premium importance in all of them. For practical size problems, the researchers have used either constructive or improvement heuristics.

Hence the literature review can be divided into two sections.

2.1 *Constructive heuristics*

For the SDST flow shop scheduling problems, the work of Rios-Mecardo and Bard (1998) stand a good constructive algorithm. They present a constructive algorithm known as NEHRB, which is an extension of another well known constructive algorithm of Nawas Enscore Ham. The same NEH is again modified by adopting idea of job insertion to it by Chakraborty & Laha (2009). Another significant work done is by Vanchipura & Sridharan (2012) where the idea of fictious jobs which are pair of jobs with highest processing time taken as one job and added into the sequence in decreasing order.

2.2 *Improvement heuristics*

Rios-Mecardo and Bard (1998) has also given an improvement heuristic called the greedy randomised adaptive search procedure (GRASP). Das et al. considers problem instances while setup times are quite higher than the processing time and focuses on saving time. The well known meta heuristic Simulated Annealing is worked on by Parthasarathy & Rajendran (1997). Ruiz et al. (2005) does the GA analysis for hybrid flow shop with sequence dependent set up times with four new cross over operators. Bat intelligence approach was developed for problems without setup times by Malakooti et al. (2007). Liu (2007) does the Particle swarm optimisation technique. In particular, the PSO applies the evolutionary searching mechanism which is characterized by individual improvement, population cooperation, and competition to effectively perform exploration. Marichelvan (2002) does an improved hybrid Cuckoo Search metaheuristic. He found that the results are good but as the problem size increased, it was necessary to change search characteristics. Yagmahan & Yenisey (2007) do the Ant-colony optimization method. They study the multi objectives of makespan, total flow time and total machine idle time. Taillard (1989) gives benchmark problems for permutation flow shop, job shop and open shop scheduling problems.

3   PROBLEM FORMULATION

3.1   *Experimental problem description*

There are altogether 1080 problem instances which are tested with the new heuristic method developed. For example one sample problem can be a 20 job problem with 5 machines having setup times of job equal to 0% of processing time. These problems are made from Taillard Benchmark problems.

For example one sample problem can be a 20 job problem with 5 machines having setup times of job equal to 0% of processing time. These problems are made from Taillard Benchmark problems.

There are 120 problem instances of Taillard are available for general flow shop. But these have only processing time only. To add setup times also to it, the setup time matrix is so formed at nine levels as percentages of processing time as described above.

4   SOLUTION METHODOLOGY

The present study proposes a novel variation of neighbourhood search for scheduling SDST flow shop. Generally, all neighbourhood search procedures use only one type of neighbourhood. In order to provide more intensity to the search procedure two different types of neighbourhood are searched in the same local search. These are swap and insertion neighbourhoods. In the swap neighbourhood, two job numbers are randomly generated and their positions are mutually exchanged. In the insertion neighbourhood procedure, a job number is randomly generated. The randomly generated job is inserted in the random position generated. The advantage of this procedure is that both these neighbourhoods generate mutually exclusive set of solution sequences, which result in intensified search. The proposed heuristic method involves local and global search parameters which are optimized to make it a robust algorithm. The working procedure of the algorithm is as shown.

The local search

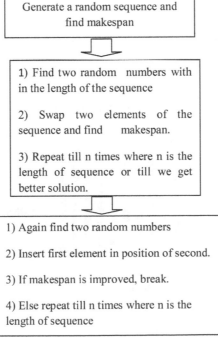

Repeat the local search

5 RESULTS AND DISCUSSION

The makespan results are obtained for all the 1080 problem instances. They are compared with results of Genetic algorithm obtained from Vanchipura R and Sridharan R. The various analysis methods used are Graphical Analysis, Relative performance index analysis and Statistical analysis. It is found that the Robust Neighbourhood search performs superior to GA for all problem instances and as the problem size increases in terms of both setup times and number of jobs, the difference become more and more evident.

5.1 *Graphical analysis*

The RNS is tested for altogether 1080 problem instances. They are grouped into 9 groups based on set up times. In each group, there are twelve different problem instances. Each problem instance has 10 sample problems. Makespan values obtained using RNS are compared with that of GA. Setup times taken are 1%, 5%, 10%, 25%, 50%, 75%, 100%, 125%, 150%.

5.2 *The relative performance index analysis*

The relative performance of these two algorithm is the difference of Robust Neighbourhood Search based on the Genetic Algorithm. It can be mathematically stated that

$$RPI = (RNS - GA/GA)$$

For the two sets of 200 jobs problems, ie the 200*10 and 200*20 the RPI values are respectively −0.02706 and −0.044. The RPI value for 500*20 problem is −0.06411. These results clearly shows that RNS performs better than GA.

Out of this 100% setup time level is taken as the base case and graphs are plotted. There are 12 graphs corresponding to 12 different size problems at 100%. Due to space limitations, 5 graphs are shown (Figs. 1 to 2) corresponding to 20*10, 50*10, 100*10, 200*10 and 500*20 size problems.

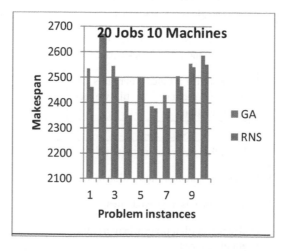

Figure 1.   SDST 100, 20 Jobs & 10 Machines.

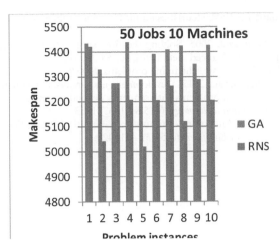

Figure 2. SDST 100, 50 Jobs & 10 Machines.

The result clearly shows that the new algorithm, ie the Robust Neighbourhood Search performs better than GA for all problem instances.

6 CONCLUSION

The proposed algorithm, namely Robust Neighbourhood Search, is tested for altogether 1080 problem instances with a varying number of jobs, machines and setup times. The parameters of the heuristics are optimized using design of experiment methodology. Further, experimentations were carried out and makespan results were obtained. Three stages of analyses; graphical analysis, relative performance index analysis and statistical analysis are performed on the results obtained. In the analyses, the proposed heuristic is compared with GA and are found superior.

The main advantage of RNS is that the algorithm is very flexible and robust. It gives improved result in successive iterations without getting trapped in local optima. The other advantage is that the search techniques can be easily changed so as to adapt for real life situations. The local and global search parameters can also be changed. The statistical analysis reveals that for some problem groups, the proposed algorithm is found to be not superior in spite of the better makespan results obtained, which can be identified as shortcoming of the proposed heuristic.

REFERENCES

Jatinder N.D & Gupta. 1986. The two machine sequence dependent flow-shop scheduling problem. *European Journal of Operational Research. 2*4(3): 439–446.
Laha D & Chakraborty UK .2009. A constructive heuristic for minimizing makespan in no-wait flow shop scheduling. *Int. j. Adv. Manuf. Technol.* 41(1–2): 97–109.
Liu B. 2007. An effective PSO based algorithm for scheduling. *IEEE* 37(1): 45–52.
Malakooti B, Kim H & Shikh S .2011. Bat intelligence search with application to multi objective multiprocessor scheduling optimization. *International journal for advanced manufacturing technology.* 60(9): 1071–1086.
Marichelvan M. 2002. An improved hybrid Cuckoo Search (IHCS) meta heuristic algorithm for permutation flow shop scheduling problems. *International journal for bio inspired computation* 4(4): 116–128.
Parthasarathy S, Rajendran CA .1997. A simulated annealing heuristic for scheduling to minimise weighted tardiness in a flow shop with sequence dependent setup times of jobs—a case study, *Produ Plan Control* 8(5): 475–483.

Rios-Mecardo R.Z. & Bard J.F. 1998. Heuristics for the flow line problem with setup cost. *EJOR* 110(1): 76–78.

Ruiz R., Maroto C. & Alcatraz J. 2005. Solving the flow shop scheduling problems with sequence dependent setup times using advanced meta heuristic. *EJOR,* 165(1): 34–54.

Taillard E. 1989. Benchmark for basic scheduling problems. *EJOR* 64(2): 278–285.

Vanchipura, Rajesh & R. Sridharan, 2012. Development and analysis of constructive heuristics algorithms for flow shop scheduling problems with sequence dependent setup times. *International journal for advanced manufacturing and technology.* 67(5-8): 1337–1353.

Yagmahan B., Yensiey M. 2007. Ant colony optimization for multi-objective flow shop scheduling problem. *Computers and Industrial Engineering.* 54(3): 411–420.

Emerging Trends in Engineering, Science and Technology for Society,
Energy and Environment – Vanchipura & Jiji (Eds)
© 2018 Taylor & Francis Group, London, ISBN 978-0-8153-5760-5

A real-world analysis of the impact of knowledge management on cost of quality in construction projects

Reshma Chandran & S. Ramesh Krishnan
Department of Mechanical Engineering, Rajiv Gandhi Institute of Technology, Kottayam, Kerala, India

ABSTRACT: Knowledge Management (KM) is vital in the case of the construction industry because of its impact on integrating knowledge within and outside the industry. Knowledge management implementation strategies play an important role in increasing project performance. The need to reduce Cost of Quality (CoQ) is clear, but the effect of KM in lowering CoQ is uncertain. This paper reviews the factors contributing to CoQ and the effect of KM on this. A literature survey and qualitative inquiries were adopted to address the research aims. Organizations should understand that there is a native CoQ problem that needs to be addressed. The main factors contributing to CoQ are design changes, errors and omissions, and poor skills. Here, the logic is based on the desire to reduce CoQ and the need to tackle as well as integrate knowledge across personal, project, organizational, and industry boundaries. Knowledge management was found to have a positive impact on lowering the cost of quality.

Keywords: Real-world analysis, Knowledge Management, Cost of Quality

1 INTRODUCTION

Knowledge Management (KM) is the process of efficient handling of information and resources within a commercial organization. Knowledge in an organization can be broadly classified as explicit knowledge and tacit knowledge. The major region of focus is actually the cost associated with the unwanted problem of redoing processes that have been inaccurately implemented, which is often referred to as the cost of poor quality. This consists of costs involving errors and omissions, poor skills, design changes, and the substantial costs often associated with client disappointment.

The study focuses on KM aspects such as repetition of mistakes and an absence of lessons learned, which may directly give rise to Cost of Quality (CoQ) issues.

2 LITERATURE REVIEW

2.1 *Knowledge*

Knowledge can be defined as the theoretical or practical understanding of a subject. A grouping of information, background, and understanding can be captured, utilized and shared for business purposes (Wibowo & Waluyo, 2015). Explicit knowledge is solitary that is able to be calculated, taken into custody, examined, and can effortlessly be passed onto others in a codified layout. It is that type of knowledge that can be uttered in statements of words and numbers. It can be supplementary, transferred, dispersed and transformed in a methodical and prescribed way into facts (Wibowo & Waluyo, 2015). Tacit knowledge, on the other hand, comes from one's experience. It can be considered as human knowledge that can be an intuition, finding, talent, experience, a form of body language, or a value or belief. It is highly personal and context-specific, which is very complicated to create, exchange in a few words, or distribute onwards to a community.

2.2 *Knowledge management*

Knowledge management was actually introduced more than two decades ago to help companies generate, categorize, and utilize knowledge in an orderly manner. KM can be defined as the classification, optimization and dynamic management of logical property to create worth, increase efficiency, and gain benefit (Wibowo & Waluyo, 2015).

2.3 *Issues associated with knowledge management in construction*

Knowledge management is very useful to the construction industry as it is significant for construction organizations to bind together knowledge on the way to improve effectiveness and amplify profitability. It is predominantly significant because of the unique nature of projects, such as the difficult character of operations, professions and organizations, short-term team members, heavy reliance on experience, rigid schedules, and restricted budgets (Nonaka & Konno, 1998).

3 RESEARCH METHODOLOGY

This paper reviews the factors contributing to the cost of quality and the effect of knowledge management on the cost of quality. A literature survey and qualitative inquiries were adopted to address the research aims.

The main factors contributing to the cost of quality are established using a literature survey. Qualitative inquiries are used to identify the effect of knowledge management on the cost of quality. These are carried out using semi-structured interviews with open-ended questionnaires. For the research, ten experts (see Table 1) were selected from construction companies across Thiruvananthapuram with knowledge management strategies. The questionnaire has three sections. The first consists of general information, the second deals with the factors contributing to the cost of quality, and the third section inquires into the effect of knowledge management on CoQ.

4 DATA COLLECTION AND ANALYSIS

Ten respondents were asked to rate the intensity of effect of KM processes on the component elements of CoQ, that is, design changes, poor skills, and errors and omissions, based on their own experience on construction projects. A four-point Likert scale was used to rate the effect as follows: 1 = Strong Negative Impact; 2 = Negative Impact; 3 = Positive Impact; 4 = Strong Positive Impact. The mean values of the ratings were considered and the process with most impact on CoQ in practice is determined.

Most of the respondents rated the KM processes as having a positive impact or strong positive impact, that is, as 3 and 4. The data collected were analyzed and the leading five

Table 1. Profile of interviewees.

ID	Years of experience	Project experience*	Organization experience*
A	36	1	1
B	27	1	1,2
C	24	1,2	1,2,3
D	12	1	1,2
E	14	1	1,2
F	12	1	1
G	10	1	1
H	13	1,2	1,2
I	12	1,2	1,2
J	14	1	1

*Key – Project experience: 1 – Building construction, 2 – Highway; Organization experience: 1 – Client organization, 2 – Consultancy, 3 – Design.

Table 2.   The impact of KM on errors and omissions.

	N	Min.	Max.	Mean
Knowledge champions	10	3	4	3.5
Knowledge capture	10	3	4	3.41
Knowledge creation	10	3	4	3.36
Knowledge transfer	10	3	4	3.34
Knowledge sharing	10	3	4	3.29

Table 3.   The impact of KM on design changes.

	N	Min.	Max	Mean
Knowledge sharing (early)	10	3	4	3.55
Knowledge creation	10	3	4	3.41
Knowledge sharing (team)	10	3	4	3.37
Knowledge capture	10	3	4	3.22
Knowledge dissemination	10	3	4	3.21

Table 4.   The impact of KM on poor skills.

	N	Min.	Max.	Mean
Knowledge identification	10	3	4	3.66
Knowledge transfer	10	3	4	3.51
Knowledge capture	10	3	4	3.42
Knowledge champions	10	3	4	3.39
Knowledge creation	10	3	4	3.32

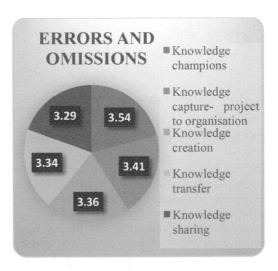

Figure 1.   Mean values of KM processes impacting errors and omissions.

KM processes that affect cost of quality were found by taking the mean of the values. This is captured in Tables 2, 3 and 4, and also represented in the form of pie charts (see Figures 1, 2, and 3). In the first case, that is, for the effect of KM processes on omissions and errors, the responses obtained from the ten interviewees were 3, 3, 3, 3, 4, 4, 4, 4, 4 and 3. The average value was 3.5. All other means were similarly calculated.

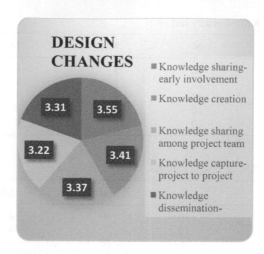

Figure 2.   Mean values of KM processes impacting design changes.

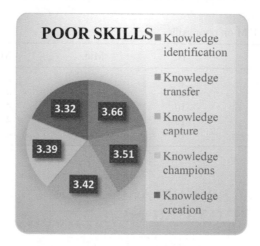

Figure 3.   Mean values of KM processes impacting poor skills.

5   FIVE HIGHEST-RANKING KM PROCESSES IMPACTING CoQ

The cost of quality mainly involves design changes, errors and omissions, and poor skills. The major KM processes reported to affect errors and omissions are knowledge champions, knowledge capture, knowledge creation, knowledge transfer, and knowledge sharing. Those affecting design changes are knowledge sharing, knowledge creation, knowledge capture, and knowledge dissemination. On the other hand, poor skills are mainly affected by knowledge identification, knowledge transfer, knowledge capture, knowledge champions, and knowledge creation.

6   RESULTS AND DISCUSSION

The main contributing factors to the cost of quality are errors and omissions, design changes, and poor skills. These factors are summarized in Tables 5, 6 and 7.

According to the results, knowledge management has most impact in reducing errors and omissions. The unifying process across all three CoQ contributing factors is knowledge capture.

Table 5. Factors contributing to the cost of errors and omissions.

Factors contributing to the cost of errors and omissions	Description
Knowledge-related mistakes	Insufficient idea about a task results in wrong solution
Rule-related mistakes	Selecting an improper solution for a problem
Lack of lessons learned	Due to insufficient learning from the past
Organizational culture	Beliefs and norms which may lead to conflict or opposition
Time constraints	Errors made because workers are under time pressures
Budget constraints	Limited budget leads to absence of good quality resources
Poor communication	This may include language barriers, improper transfer of knowledge

Table 6. Factors contributing to the cost of design changes.

Factors contributing to the cost of design changes	Description
Client change	Client alteration to product requirements, definition and project scope
Design change inevitable	Design is actually an iterative practice, and is expected to change during construction
Unforeseen site conditions	Unpredicted ground conditions, weather, etc.
Poor client expertise	Improper understanding of design scope and cost
Errors and omissions in the site	Non-reversible mistakes and omissions resting on site results in alteration of obtainable plan
Procurement strategy	Those which do not involve design innovation

Table 7. Factors contributing to the cost of poor skills.

Factors contributing to the cost of poor skills	Description
High personnel turnover	There is a cycle of recruitment
Lack of dedication	This is mainly with younger people, who are not prepared to continue for a long time
Lack of training	Industry lacks adequate training
Time constraints	This may lead to inadequate transfer of knowledge because workers are under time pressures
Organizational culture	Insufficient knowledge distribution and knowledge conveyance between project personnel

7 CONCLUSION

The main aims of this paper were to determine the factors contributing to the cost of quality and the effect of knowledge management on CoQ. It was found that the main factors contributing to CoQ are errors and omissions, design changes, and poor skills. From the pie charts, it is clear that KM processes such as knowledge sharing, knowledge creation, knowledge capture, knowledge dissemination, knowledge transfer, and knowledge champions have the highest impacts on cost of quality. Through qualitative inquiries, it was found that knowledge management has a positive impact on reducing the cost of quality. Therefore, it is crucial that every organization implement its own knowledge management strategies.

REFERENCES

Furcada, N. & Marcarulla, M. (2013). Knowledge management perceptions in construction and design companies. *Automation in Construction, 29*, 83–91.

Gromoff, A., Kazantsev, N. & Bilinkis, J. (2016). An approach to knowledge management in construction service-oriented architecture. *Procedia Computer Science, 96*, 1179–1185.

Nonaka, I. & Konno, N. (1998). The concept of "Ba": Building a foundation for knowledge creation. *California Management Review, 40*(3), 40–54.

Wibowo, M.A. & Waluyo, R. (2015). Knowledge management maturity in construction companies. *Procedia Engineering, 125*, 89–94.

Emerging Trends in Engineering, Science and Technology for Society,
Energy and Environment – Vanchipura & Jiji (Eds)
© 2018 Taylor & Francis Group, London, ISBN 978-0-8153-5760-5

Maintenance strategies for realizing Industry 4.0: An overview

A.A. Sambrekar, C.R. Vishnu & R. Sridharan
Department of Mechanical Engineering, NIT Calicut, Kerala, India

ABSTRACT: This paper presents an overview of different maintenance strategies widely discussed in the industry by consolidating the results and inferences from a wide range of research articles. This information will be useful to maintenance personnel during the selection and execution of maintenance activities to realize the concept of Industry 4.0. The paper follows a combined meta-analytic and descriptive procedure for reviewing articles that have provided path-breaking contributions in maintenance engineering and management. Details of top-cited research articles related to maintenance are downloaded from the Web of Science database to carry out the meta-analysis. A text analytic tool called BibExcel is utilized for fetching and analyzing textual information. Subsequently, a close examination of significant papers that describes different types of maintenance strategies is analyzed by following the descriptive review procedure to report the advantages and limitations of each strategy. Accordingly, it is found that predictive maintenance strategy is more aligned with the objectives of Industry 4.0. Hence, an emphasis is placed on predictive maintenance techniques and tools since this approach is gaining more attention as a result of advancements in soft computation, cloud technology, data analytics, machine learning and artificial intelligence.

1 INTRODUCTION

Generally, maintenance involves the set of activities carried out in the industry to make sure all the machineries and other physical assets are available for production. The main purpose of industrial maintenance is to achieve minimum breakdown and to maintain efficiency of the production facilities at the lowest possible cost. Maintenance activities and its execution depend on the manufacturing system and layout of the plant. However, in any case, maintenance should not be considered as a cost-centric activity, but a profit-generating function (Alsyouf 2007).

Maintenance helps in adding value to the organization through better utilization of production facilities, enhancing product quality as well as reducing rework and scrap. ISO 55000: 2014 Asset Management System upholds, "Assets exist to provide value to the organization and its stakeholders". Unfortunately many companies still consider maintenance activities as a "necessary evil", due to the blurred perception about its role in attaining company's objectives and goals (Duffuaa et al. 2002). For those companies, the first step is to change the corporate mindset such that the role of maintenance in achieving customer oriented performance parameters such as quality, on-time delivery, etc., is significant.

Unexpected failures affect three key elements of competitiveness—quality, cost, and productivity. In the modern world, all firms are striving hard to elevate these key features to develop a strategic advantage against their competitors. Simply, waiting for the failure to occur is not affordable in today's business operations scene. Hence, companies have to adopt different maintenance strategies suitable for their businesses.

The concept of Industry 4.0 originated from Germany, but its vision has caught the attention of organizations across the globe (Zezulka et al. 2016). Industry 4.0 has spawned a new wave of technology revolutionizing manufacturing environment through "smart factory" in which machines cooperate with humans in real time via the cyber-physical systems, thus creating a customer-oriented production field utilizing the technologies of Internet-of-Things (IoT). The machine rather than working as an independent entity will be able to collect data, analyze it and act accordingly. The use of advanced IoT sensors on the machines has made it possible to detect any faults that go unseen by the human eye. Instead of reacting to the failure, predictive maintenance will provide early warnings well in advance so that machines or humans can take the necessary action to minimize the frequency of failures (Tupa et al. 2017).

This paper presents an overview of different maintenance strategies widely discussed in the literature. The paper has two main objectives. The first objective is to realize a systematic review of literature in maintenance which is described in section 2. The second objective is to explore how each of these maintenance strategies approaches toward maintenance decision making which is described in section 3. Since predictive maintenance is widely being adopted in today's industry, section 4 explores the developments in this approach in maintenance.

2 META-ANALYSIS OF LITERATURE

A comprehensive amount of research papers have been published in the area of plant maintenance. It would be non-exhaustive process to review all the papers in the literature. The Web of Science (WoS) database (after proper refinement) suggests 12,576 articles connected with the keywords-Preventive Maintenance, Predictive Maintenance, Corrective Maintenance and Reliability Centered Maintenance as on October, 2017. The collection includes 7,736 journal articles, 4,744 conference proceedings and 609 reviews.

This huge volume of articles makes it difficult to provide a simple narration and critical review of each research article. To overcome this common issue in conducting a literature review, Glass et al. (1981) introduced a new approach known as Meta-analysis. Meta-analysis is a methodology used to integrate research findings from a large body of articles using statistical analysis and sophisticated measurement techniques (Krishnaswamy ct al. 2007). This methodology is widely accepted in the research community to obtain firsthand information from a large pool of articles from which the current direction of research can be projected and significant articles in the domain can be shortlisted for further study.

To achieve the objective of this review paper, statistical techniques are employed to the data retrieved from a pool of research papers extracted by data mining using BibExcel software tool. Bibexcel is used to carry out bibliographic and statistical analysis by extracting the data of textual nature such as title of the paper, author names, journal name, keywords, etc. This free software tool also allows modifying and/or adjusting data that can be imported from various databases including Scopus, WoS, Mendeley among others (Fahimnia et al. 2015).

Initially, a WoS outfile is created in plain text format that contains relevant information of top 250 cited research papers in the maintenance domain, to be used as input for BibExcel. The result of the analysis discloses the major contributors to the domain, major journals publishing top quality articles, tools used in maintenance studies, etc.

2.1 Keyword frequency analysis

The occurrence of different research aspects can be related to the frequency of keywords appeared in maintenance related articles as presented in the Figure 1.

It is evident from the figure that reliability is the most related aspect in industrial maintenance. Reliability has always been an important aspect in the assessment of industrial equipment's health. The choice of maintenance strategy highly influences the reliability of a system. This is followed by optimization, preventive maintenance, fault diagnosis, condition monitoring, etc. It is also inferred that genetic algorithm is the most used meta-heuristics to solve optimization problems in maintenance. The title term frequency analysis conducted on the same set of papers also substantiates similar results.

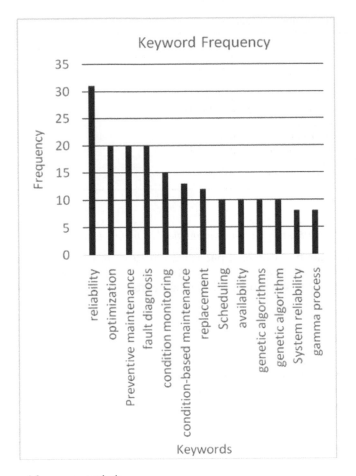

Figure 1.   Keyword frequency analysis.

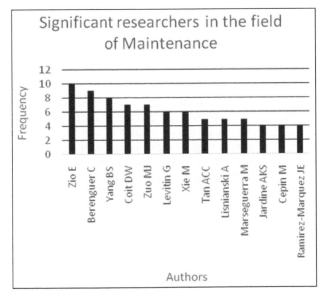

Figure 2.   Significant researchers.

Table 1. Distribution of papers across journals.

Journal	Number of papers
Reliability Engineering & System Safety	112
Expert Systems with Applications	28
European Journal of Operational Research	21
International Journal of Production Economics	13
IEEE Transactions	13
Computers & Operations Research	9
Journal of the Operational Research Society	8
Naval Research Logistics	8
Computers & Industrial Engineering	7

2.2 *Analysis of significant researchers in the field of maintenance*

The significant researchers who have published articles in the field of industrial maintenance are represented in the Figure 2. It can be noted that Zio E., Berenguer C., Yang B.S., Coit D.W. and Zuo M.J. have published substantial number of articles in the area of industrial maintenance.

2.3 *Journal frequency analysis*

The majority of the top cited articles in industrial maintenance appeared in the journals such as Reliability Engineering & System Safety, Expert Systems with Applications and European Journal of Operational Research (Table 1). Therefore, these journals could be acknowledged as leading ones in the areas of industrial maintenance.

3 MAINTENANCE STRATEGIES

In general, maintenance strategies can be generally classified into two categories: Reactive and Proactive. Reactive maintenance focuses on repairing an asset once failure occurs. Proactive maintenance focuses on avoiding repairs and asset failure through preventive and predictive methods. These strategies meet the objectives laid out in the philosophy of Total Productive Maintenance (TPM).

3.1 *Corrective maintenance*

Corrective maintenance (CM) also known as reactive maintenance refers to activities that are done when equipment has already broken down, in order to restore the equipment to its normal operating condition. The occurrence of sudden failure of the asset leads to high levels of machine downtime. There is no initial cost associated with corrective maintenance and it requires no planning. CM is suitable for non-critical components where the failure neither compromises employee safety nor the production objectives.

3.2 *Scheduled maintenance*

Scheduled maintenance (or preventive maintenance) refers to the set of activities that are performed at regular intervals on a system to reduce the likelihood of it failing. Usually, preventative maintenance is carried out with minimal disruption in production activities such that unexpected breakdowns will not occur. The maintenance is scheduled either time based or usage based. An air-conditioner which is serviced every year, before summer is an example of time based preventative maintenance whereas, the maintenance of motor vehicles are usually carried out on a usage based schedule (e.g. every 10,000 km).

3.3 Predictive maintenance

Machines usually undergo degradation before failure occurs. It is possible to monitor the trend of degradation so that any faults can be corrected before they cause any failure and machine breakdown. Predictive maintenance (PdM) is one such strategy that helps us to predict failures before they actually occur. It is a more efficient strategy since maintenance is only performed on machines when it is required. This strategy requires condition monitoring of the asset that detects signs of decreasing performance or upcoming failure by utilizing sensor technologies.

The asset is monitored for various parameters such as vibration, temperature, lubricating oil, contaminants, and noise levels. The general classification of maintenance strategies is provided in Figure 3.

Maintenance decision making under the PdM can be analyzed in two perspectives namely: diagnosis and prognosis. Diagnosis is the process of identifying the root-cause of a fault (Jeong & Villalobos 2007), while prognosis is the process of predicting failure that may occur in future (Lewis & Edwards 1997). Diagnostics comprises (a) fault detection, (b) fault isolation and (c) fault identification, while prognostics include (a) remaining useful life (RUL) prediction and (b) confidence interval estimation (Efthymiou et al. 2012). Even if the asset is running in a degraded state, it cannot be concluded that the asset has failed. It can probably still be utilized for certain duration before failure occurs. To deal with these circumstances, prognosis is required. The key purpose of prognosis is to provide early warning to facilitate better maintenance planning. Therefore, appropriate maintenance activities for the equipment can be planned to prevent failures.

3.4 Reliability-centered maintenance

Reliability-Centered Maintenance (RCM) is a systematic approach for determining the most effective approach for maintenance. Effectiveness is determined by considering either reliability (or probability of failure) and overall cost. RCM is one of the best known and most used techniques to preserve the operational efficiency in critical sectors like power plants, artillery system, aviation industry, railway networks, oil and gas industry and ship maintenance (Carretero et al. 2003).

Generally, RCM is conceived as the optimum mix of all the traditional maintenance strategies in order to utilize the advantages of those maintenance strategies. While preventive maintenance is generally considered to be worthwhile, there are some disadvantages such as huge cost and need of specialist labor. Hence, preventive maintenance need not be the cost effective strategy for every machinery/component especially for the non-critical assets that every industry possesses.

Thus, for proper maintenance of the plant, it is better to adopt an integrated method of breakdown and preventive maintenance strategies to make use of the respective strength

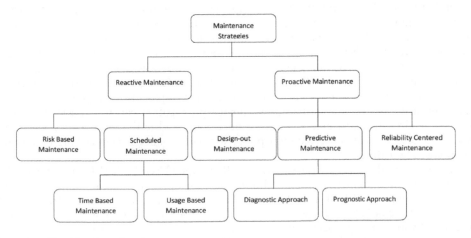

Figure 3. Classification of maintenance strategies.

alone for which RCM was introduced. Accordingly, RCM selects the most appropriate and tailor made maintenance strategy for all the equipment in the plant based on its criticality score and reliability parameters (Vishnu & Regikumar 2016).

3.5 Risk based maintenance

Risk Based Maintenance (RBM) prioritizes maintenance resources toward assets that carry the most risk if they were to fail. Risk-based maintenance framework comprises two major phases: (Selvik & Aven 2011, Arunraj & Maiti 2010)

i. Risk assessment
ii. Maintenance planning based on risk

Risk is adopted as an index for clarifying priority in risk maintenance technologies. Researchers have also proposed risk based models for optimizing maintenance activities mainly using the tools of simulation (Nielsen & Sørensen 2011, Sørensen 2009). As per Khan & Haddara (2003), risk is be calculated as the product of failure probability and consequences of failure.

Generally, it is conceived that 20% of the components in the system will comprise 80% of the total risk of the system. In that case, it would be irrational to inspect all devices with the same level of priority. It is thus important to identify the 20% of devices and raise their priority level in the inspection program. This is the basic concept of risk based maintenance strategy. This concept is called the 80–20 rule (Pareto principle).

3.6 Design-out maintenance

Design-out maintenance focuses on the vulnerabilities in the machines that lead to frequent failures and aims to minimize the regular maintenance activities by redesigning those machines and facilities. This will raise machine performance from the maintenance point of view, especially to those machines having longer repair time or huge replacement cost.

4 DEVELOPMENTS IN PREDICTIVE MAINTENANCE

Predictive Maintenance is the current excitement in the field of maintenance. This strategy is based on maintenance predictive models working on the principle of machine learning. When building predictive models, we use historical data to train the model which can then recognize hidden patterns and further identify these patterns in the future data. These models are trained with examples described by their features and the target of prediction. The trained model is expected to make predictions on the target by analyzing the features of the new examples. It is crucial that the model captures the relationship between features and the target of prediction. In order to train an effective machine learning model, we need training data which includes features that actually have predictive power towards the target of prediction.

Therefore, quantitative data is of foremost importance for training the model. Since failures are very rare to occur and the availability of such data is difficult in real time, there are a few data repositories like Prognostics and Health Management (PHM) that facilitate the availability of run-to-failure data to encourage development of prognostic algorithms. Saxena et al. (2008) have described how damage propagation can be modeled in various modules of aircraft gas turbine engines for developing and testing prognostics algorithms. This paper also presents an evaluation policy for performance benchmarking of different prognostics algorithms.

In Prognostics, the objective is to predict the Remaining Useful Life (RUL) before a failure occurs, given the current machine condition and past operation profile. Since the dataset was made available by PHM in 2008, researchers have built different prognostics methods. Ramasso & Saxena (2014) reviewed different approaches to PHM dataset and analyzed to understand why some approaches worked better than others. This paper presents three top winning approaches described in the order of their rank as follows:

- Wang et al. (2008) present a similarity-based approach for estimating the RUL in prognostics, however, it was only suitable when abundant amount of run-to-failure data was available.
- Heimes (2008) utilizes advanced recurrent neural network architecture to estimate the RUL of the system,
- Peel (2008) utilizes ensemble regression modelling for determination of RUL.

Different other approaches proposed by Fagogenis et al. (2015) and Bucknam (2017) have also been utilized for predicting the RUL of the aircraft engine for the PHM dataset after 2008. The results of these algorithms are significantly closer to the three winning approaches.

The value of predictive maintenance is already well recognized by the biggest players in the industry. Rolls Royce adopted predictive maintenance strategy under its Total Care Program which resulted in transforming them from a loss making aircraft manufacturer to a leading aircraft engine manufacturer. The change of maintenance strategy highly impacted the business decisions for Rolls-Royce and the company has now moved to leasing its engines in the monthly subscription model. They have added complex system monitoring sensors in its engines that sends real-time engine performance metrics when the plane is mid-air to one of the company's R&D centers. The data is then analyzed immediately and any concerns will be conveyed to the aircraft to ensure its safety.

5 SCOPE FOR FUTURE WORK

The previous discussion reveals that although the application of PdM is more beneficial compared to other strategies from a practical point of view, further research on PdM is necessary in order to realize the concept of Industry 4.0. The application of PdM is more complex and expensive because PdM heavily relies on the implementation of complex sensor systems to monitor the health conditions of the equipment in real time. As discussed, research based on data repository provided by Prognostic and Health Management (PHM) seems the right direction for future research on condition based maintenance of sensitive and complex equipment/systems. As observed from the literature, around 70 publications have been found that have utilized the PHM dataset for the development of prognostic algorithms. However, the PHM dataset exhibits an exponential degradation pattern. No comparison has been made to check whether the same prognostic algorithms can perform effectively on different datasets.

6 CONCLUSION

This paper presented a systematic review of different maintenance strategies. Although many papers have been published in this area, only a few papers present the advantages and challenges associated with implementing each maintenance strategy. This paper identifies certain 'core' articles which may prove beneficial for the people seeking to research the area. We identified some of the recent path-breaking research papers that contribute towards the advancements in maintaining production facilities and other complex machinery to achieve efficiency and effectiveness.

Furthermore, it can be seen that there have been tremendous excitements all over the world to implement predictive maintenance strategy for monitoring and executing maintenance activities. This motivation is associated with the advancements in the field of systems engineering especially in machine learning, data analytics, and instrumentation technologies. Hence, PdM can be seen as the most promising maintenance strategy for realizing the objectives of Industry 4.0. However, a single maintenance strategy cannot be the most economical strategy for all the equipment in the plant. Especially for non-critical items, it is always better to follow breakdown maintenance. Therefore, it can be concluded that an optimum mix of the above strategies such as RCM, with more emphasis on predictive maintenance, is the most suited maintenance strategy in the emerging industrial scenario.

REFERENCES

Alsyouf, I. 2007. The role of maintenance in improving companies' productivity and profitability. *Int. J. Prod. Eco.* 105(1): 70–78.

Arunraj, N.S. & Maiti, J. 2010. Risk-based maintenance policy selection using AHP and goal programming. *Safety Science*, 48(2): 238–247.

Bucknam, J.S. 2017. Data analysis and processing techniques for remaining useful life estimations.

Carretero, J., Pérez, J.M., García-Carballeira, F., Calderón, A., Fernández, J., García, J.D., Lozano, A., Cardona, L., Cotaina, N. & Prete, P., 2003. Applying RCM in large scale systems: a case study with railway networks. *Relia. Engng. & System Safety 82*(3): 257–273.

Duffuaa, S.O., Al-Ghamdi, A.H. & Al-Amer, A. 2002. Quality function deployment in maintenance work planning process. In *Proc. of the 6th Saudi engineering conference KFUPM, Dhahran, Kingdom of Saudi Arabia*.

Efthymiou, K., Papakostas, N., Mourtzis, D. & Chryssolouris, G. 2012. On a predictive maintenance platform for production systems. *Procedia CIRP, 3*: 221–226.

Fagogenis, G., Flynn, D. & Lane, D., 2014. Novel RUL prediction of assets based on the integration of auto-regressive models and an RUS Boost classifier. *Prognostics and Health Management (PHM), 2014 IEEE Conference:* 1–6.

Fahimnia, B., Sarkis, J. & Davarzani, H. 2015. Green supply chain management: A review and bibliometric analysis. *Int. J. Prod. Eco.*162: 101–114.

Glass, G.V., McGaw, B., Smith, M.L. 1981. Meta Analysis in Social research. Beverly Hills, CA: *Sage Publications*, New York, NY.

Heimes, F.O., 2008. Recurrent neural networks for remaining useful life estimation. In *Prognostics and Health Management (PHM), 2008 IEEE Conference:* 1–6.

Jeong, I.J., Leon, V.J. & Villalobos, J.R. 2007. Integrated decision-support system for diagnosis, maintenance planning, and scheduling of manufacturing systems. *Int. J. Prod. Research*, 45(2): 267–285.

Khan, F.I. & Haddara, M.M. 2003. Risk-based maintenance (RBM): a quantitative approach for maintenance/inspection scheduling and planning. *J. of loss prevention in the process industries*, 16(6): 561–573.

Kim, N.-H., Joo-Jo, C. & An, D. 2017. *Prognostics and Health Management of Electronics.*

Krishnaswamy, K.N., Sivakumar, A.I., Mathirajan, M. 2007. Management Research Methodology, *Dorling Kindersley: Pearson India Publications*, India.

Kuhn, M. & Johnson, K. 2013. *Applied predictive modeling* (Vol. 810). New York: Springer.

Lewis, S.A. & Edwards, T.G. 1997. Smart sensors and system health management tools for avionics and mechanical systems. *16th DASC. AIAA/IEEE Digital Avionics Systems Conference. Reflections to the Future. Proceedings* 2: 8.5–1–8.5–7. doi: 10.1109/DASC.1997.637283.

Nielsen, J.J. & Sørensen, J.D. 2011. On risk-based operation and maintenance of offshore wind turbine components. *Relia. Engng. & System Safety*, 96(1): 218–229.

Peel, L. 2008. Data driven prognostics using a Kalman filter ensemble of neural network models. *2008 Int. Conference on Prognostics and Management* 1–6, doi: 10.1109/PHM.2008.4711423.

Ramasso, E. & Saxena, A. 2014. Performance Benchmarking and Analysis of Prognostic Methods for CMAPSS Datasets. *Int. J. Prognostics and Health Management 5*(2):1–15.

Saxena, A., Goebel, K., Simon, D. & Eklund, N. 2008. Damage Propagation Modeling for Aircraft Engine Prognostics. *Response.*

Selvik, J.T. & Aven, T. 2011. A framework for reliability and risk centered maintenance. *Relia. Engng. and System Safety* 96(2):324–331.

Si, X.S., Zhang, Z.X. & Hu, C.H., 2017. *Data-Driven Remaining Useful Life Prognosis Techniques: Stochastic Models, Methods and Applications.* Springer.

Sørensen, J.D. 2009. Framework for risk-based planning of operation and maintenance for offshore wind turbines. *Wind energy*, 12(5): 493–506.

Tupa, J., Simota, J. & Steiner, F. 2017. Aspects of risk management implementation for Industry 4.0, *Procedia Manufacturing* 11, 1223–1230.

Vishnu, C.R. & Regikumar, V. 2016. Reliability Based Maintenance Strategy Selection in Process Plants: A Case Study. *Procedia Technology* 25:1080–1087.

Wang, T., Yu, J., Siegel, D. & Lee, J., 2008. A similarity-based prognostics approach for remaining useful life estimation of engineered systems. In *Prognostics and Health Management (PHM) 2008. IEEE Conference:*1–6.

Zezulka, F., Marcon, P. Vesely, I. & Sajdl, O. 2016. Industry 4.0 – An Introduction in the phenomenon. *IFAC-Papers Online*, 49(25), 8–12.

Manufacturing technology and material science

Emerging Trends in Engineering, Science and Technology for Society,
Energy and Environment – Vanchipura & Jiji (Eds)
© 2018 Taylor & Francis Group, London, ISBN 978-0-8153-5760-5

Characterization of nanoliposomes and their modification for drug delivery

K.S. Athira
Department of Materials Engineering, Indian Institute of Science, Bangalore, Karnataka, India

K.W. Ng
School of Materials Science and Engineering, Nanyang Technological University, Singapore

ABSTRACT: Atherosclerosis causes heart disease and stroke and is a main cause of death in many countries. Nanoliposomes are a potential candidate for use in drug targeting for its treatment. The size of the nanoliposomes affects their cellular uptake. An optimal size range is necessary for effective site-specific drug delivery. Such optimum-sized nanoliposomes were developed. Their sizes were characterized in this study using dynamic light scattering and nanoparticle tracking analysis. The sizes of the nanoliposomes were found to be in the ranges of 79–128 nm via the former method and 86–99 nm via the latter. Liposomes grafted with polyethylene glycol show an improved stability and increased circulation time, while those grafted with fluocinolone acetonide help to reduce inflammation. It was observed that normal nanoliposomes had a larger size than these grafted liposomes, which each had a larger size than those grafted with both polyethylene glycol and fluocinolone acetonide.

Keywords: nanoliposomes, polyethylene glycol, fluocinolone acetonide, drug delivery

1 INTRODUCTION

Atherosclerosis is a chronic inflammatory disease in which endothelial stress or injury causes low-density lipoproteins (LDLs) to infiltrate into the tunica intima of the artery. These LDLs may undergo oxidization, and in oxidized form they are toxic to the endothelial cells. The inflamed cells will then recruit monocytes (Wick et al., 2004). They migrate into the tunica intima where they differentiate into macrophages. The macrophages take up oxidized LDLs that permeate into and transform to foam cells (Gerrity, 1981). They aggregate to form an atheromatous core, which later becomes necrotic (Ross, 1993).

Liposomes are small artificial vesicles of spherical shape that can be created from cholesterol and phospholipids (Akbarzadeh et al., 2013). They have one or more concentric spherical lipid bilayers with aqueous phases inside and between lipid bilayers. Nanoliposomes are a new technology for the encapsulation and delivery of bioactive agents. They have potential applications in nanotherapy (diagnosis, cancer and atherosclerosis therapy, gene and drug delivery) because of their biocompatibility, biodegradability and nanosizing. They can upgrade the functioning of bioactive agents by enhancing their dissolvability and bioavailability, *in vitro* and *in vivo* stability, and, in addition, blocking their undesirable interactions with other molecules. Another preferred application of nanoliposomes is cell-specific targeting, which is essential to achieving the concentrations of drugs required for ideal therapeutic value at the target site while limiting antagonistic impacts on healthy cells and tissues. Water-soluble compounds can be trapped in the aqueous phase, and lipophilic agents can be trapped between liposomal bilayers (Torchilin, 2005). Liposomes naturally target the Mononuclear Phagocytic System (MPS), that is, monocytes, macrophages and dendritic cells, and thus they can be used to deliver drugs to the MPS. The size, charge and lipid composition affect the efficiency of liposomes targeting MPS cells

(Kelly et al., 2010). Characterization of the liposomes in terms of their size is very important because the size of the drug will affect cellular uptake. Liposomes can be modified by grafting with Polyethylene Glycol (PEG) or Fluocinolone Acetonide (FA).

Because PEG is an exceptionally hydrophilic polymer and has low toxicity, PEG and its subsidiaries have been generally used to enhance the stability and pharmacokinetics of drug carriers and parent drugs (Harris et al., 2001). In liposomal drug delivery, PEG has been broadly utilized for liposome surface alteration (PEGylation), and this procedure has been utilized for making liposomal drug delivery systems, which are known as PEGylated liposomes (Santos et al., 2007). PEG also increases the circulation time (Bergstrand, 2003), which is necessary for an efficient site-specific drug delivery.

Fluocinolone acetonide is an anti-inflammatory drug. The results of Vafaei et al. (2015) reveal that the liposomes have promising potential as an effective delivery system for incorporation of fluocinolone acetonide following the formation of inclusion complex to release the drug in a sustained manner for the treatment of ocular inflammatory disease. It can thus be used to control inflammation in atherosclerosis.

Fluocinolone acetonide-loaded PEGylated liposomes have a strong pharmacological effect and low toxicity (Vafaei et al., 2015). Hence, liposomes modified by grafting with PEG or FA or both also need to be characterized as they provide a better drug delivery system.

It has been shown that liposomes in the size range of 50 to 800 nm show a trend in which the small liposomes are optimal for internalization by the MPS cells, while the large ones induce toxicity and cytokine activation, which leads to inflammation (Epstein-Barash et al., 2010; Takano et al., 2003). Increasing liposome size resulted in increased inhibitory effect of alendronate liposomes, liposomal clodronate and with h-monocytes and J774 macrophage cell lines (Epstein-Barash et al., 2010). Reactive oxygen species generation was decreased by PEG coating as the association with macrophage-like RAW264.7 cells and the induction of apoptosis were reduced (Takano et al., 2003).

In this article, we have used two different methods to analyze the size characterization of nanoliposome materials, together with their PEG coating material, the loaded drug and, finally, the combined material, as effective surface modifications of nanoliposomes for site-specific drug delivery in the treatment of atherosclerosis: the results obtained are compared and discussed.

2 MATERIALS AND METHODS

Four types of nanoliposome were procured from the collaborative research program between the School of Materials Science and Engineering, Nanyang Technological University, Singapore, and other international universities. The four types of liposomes were:

a. Normal liposomes (blank);
b. Liposomes grafted with PEG;
c. FA-loaded liposomes;
d. Liposomes grafted with PEG-FA.

2.1 Dynamic light scattering

The size of particles suspended in a solution can be determined by measuring the changes in the intensity of light scattered from the solution; this method is called Dynamic Light Scattering (DLS).

The principle of dynamic light scattering is that fine particles that are in Brownian motion (constant and random thermal motion) diffuse with a speed that is related to their size. Smaller particles diffuse faster than larger ones. To measure this diffusion speed, the pattern produced by illuminating the particles with a laser is observed.

The four samples were diluted to 10X, 100X, 1000X and 100,000X. The size of each sample was measured by the DLS method using a Zetasizer Nano ZS90 (Malvern Instruments Ltd, UK) at 25°C.

2.2 *Nanoparticle tracking analysis*

In Nanoparticle Tracking Analysis (NTA), the size of the suspended particles in a solution is found by measuring the rate of Brownian motion, which is correlated to the size of these particles.

Similarly to DLS, the particles of the sample can be visualized by the light they scatter when illuminated by a laser. The light scattered by the particles is captured using a camera and the motion of each particle is tracked frame to frame by software. This rate of particle movement is related to a sphere-equivalent hydrodynamic radius as calculated through the Stokes–Einstein equation.

The four samples were diluted to 10,000X & 100,000X and the size was measured using a Nanosight NS300 (Malvern Instruments Ltd, UK) at 25°C.

3 RESULTS AND DISCUSSION

Representative size data for the four types of nanoliposomes from the DLS measurement is shown in Figure 1. It can be seen from the figure that the size of the normal (blank) liposome is approximately 100 nm. Similarly, data were obtained for all the different liposomes in each and every dilution. The variation in the sizes of the different types of liposomes at varying dilutions found by the DLS method is shown in Figure 2. The size of the nanoliposomes can be seen to be in the range of 79–140 nm. The blank liposomes have a size range of 94–128 nm, with an average of 111 nm. The PEG-coated liposomes have a size range of 96–124 nm, with an average of 110 nm. The FA-loaded liposomes have a size range of 91–97 nm, with an average of 94 nm. The PEG-FA-incorporated liposomes have a size range of 79–140 nm, with an average of 110 nm. As the dilution of 10,000X is at the limit of the sensitivity of the DLS equipment, there is a large amount of error in the data obtained at this dilution. So, excluding this, the average sizes of the liposomes are plotted in Figure 3. Thus, the size of different types of nanoliposomes ranges from 79 to 128 nm, with an average of 104 nm. The trend of sizes from DLS measurement can be described as: blank liposomes > PEG liposomes > FA liposomes > PEG-FA liposomes.

Representative size data from the NTA measurement is shown in Figure 4, in which the concentration of particles of a particular size is plotted. In the given data, most of the particles have a size of 100 nm. Further, the small peaks of higher sizes can be ignored because of the very small number of particles in that size range.

The average of the size values thus obtained from the dilutions 10,000X and 100,000X is shown in Figure 5. The size of the nanoliposomes were found to be in the range of 86–99 nm. The trend of sizes obtained from NTA measurement is: blank liposomes > PEG liposomes ≈ FA liposomes > PEG-FA liposomes.

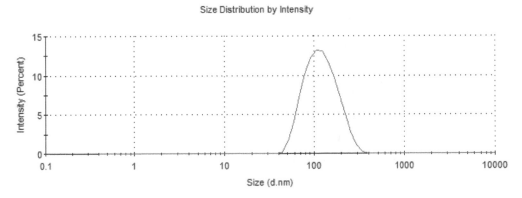

Figure 1.   Representative data of dynamic light scattering: Intensity of scattered light with size of nanoliposomes (blank).

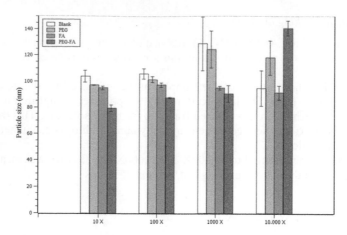

Figure 2.   Size of nanoliposomes by dynamic light scattering.

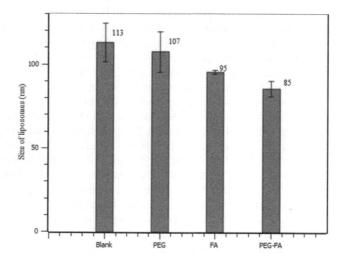

Figure 3.   Average sizes of the different types of nanoliposomes obtained by the dynamic light scattering method.

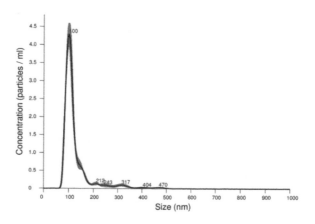

Figure 4.   Representative data for nanoparticle tracking analysis: Concentration of nanoliposomes versus size.

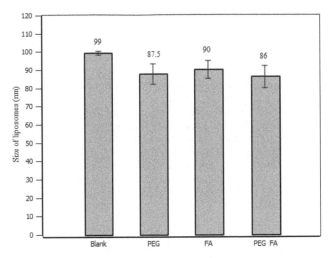

Figure 5. Size of nanoliposomes by nanoparticle tracking analysis: Average of 10,000X and 100,000X.

Because NTA is a more accurate method, the overall trend obtained by combining the results from DLS and NTA measurements can be summarized as: blank liposomes > PEG liposomes ≈ FA liposomes > PEG-FA liposomes.

Larger liposomes exhibited increased internalization by h-monocytes (Epstein-Barash et al., 2010), but with a higher degree of apoptosis induction (Takano et al., 2003). A significant reduction in activity was also found after treatments with very small liposomes (55 ± 15 nm) (Epstein-Barash et al., 2010). Hence, a size range between these two extremes, such as 75–150 nm, would be ideal for a successful site-specific drug delivery.

It has been observed by Nicholas et al. (2000) that the addition of PEG or FA brought the weight ratio of liposomes down due to permeabilities, reaction temperatures, and phase transition between the mushroom and brush regimes. This might be the reason for the reduction in size observed in the liposomes grafted with PEG, FA, and both.

4   CONCLUSIONS

The size characterization study of the different types of nanoliposomes procured from the collaborative research program between Nanyang Technological University, Singapore, and other international universities was done using DLS and NTA methods.

The size of the nanoliposomes alone, as well as in forms grafted with different materials, were found to be in the range of 79–128 nm via the DLS method and 86–99 nm via the NTA method adopted in the present study.

Because all the samples were monodispersed, both techniques are appropriate and the results are comparable. The standard deviation in the DLS method was 30.5 and that with NTA was only 6.5. Hence, the NTA method is adjudged as the better method as it produces less deviation in the data.

The trend of size variations in the materials used in the current study can be described as: blank liposomes > PEG liposomes ≈ FA liposomes > PEG-FA liposomes. Thus, the nanoliposomes grafted with the two different materials necessary for drug delivery were found to be better than blank nanoliposomes with respect to size. The PEG-FA grafted nanoliposomes characterized with a size of 85.5 nm procured from this laboratory are found to be superior in site-specific drug delivery for atherosclerosis with respect to size.

REFERENCES

Akbarzadeh, A., Rezaei-Sadabady, R., Davaran, S., Joo, S.W., Zarghami, N., Hanifehpour, Y. & Nejati-Koshki, K. (2013). Liposome: Classification, preparation, and applications. *Nanoscale Research Letters, 8*(1), 102.

Bergstrand, N. (2003). Liposomes for drug delivery: From physico-chemical studies to applications (Doctoral dissertation, Acta Universitatis Upsaliensis, Sweden).

Epstein-Barash, H., Gutman, D., Markovsky, E., Mishan-Eisenberg, G., Koroukhov, N., Szebeni, J. & Golomb, G. (2010). Physicochemical parameters affecting liposomal bisphosphonates bioactivity for restenosis therapy: Internalization, cell inhibition, activation of cytokines and complement, and mechanism of cell death. *Journal of Controlled Release, 146*(2), 182–195.

Gerrity, R.G. (1981). The role of the monocyte in atherogenesis: I. Transition of blood-borne monocytes into foam cells in fatty lesions. *American Journal of Pathology, 103*(2), 181–190.

Harris, J.M., Martin, N.E. & Modi, M. (2001). Pegylation. *Clinical Pharmacokinetics, 40*(7), 539–551.

Kelly, C., Jefferies, C. & Cryan, S.A. (2010). Targeted liposomal drug delivery to monocytes and macrophages. *Journal of Drug Delivery, 2011*, 727241.

Nicholas, A.R., Scott, M.J., Kennedy, N.I. & Jones, M.N. (2000). Effect of grafted polyethylene glycol (PEG) on the size, encapsulation efficiency and permeability of vesicles. *Biochimica et Biophysica Acta (BBA) - Biomembranes, 1463*(1), 167–178.

Ross, R. (1993). The pathogenesis of atherosclerosis: A perspective for the 1990s. *Nature, 362*(6423), 801–809.

Santos, N.D., Allen, C., Doppen, A.M., Anantha, M., Cox, K.A., Gallagher, R.C., ... Webb, M.S. (2007). Influence of poly (ethylene glycol) grafting density and polymer length on liposomes: Relating plasma circulation lifetimes to protein binding. *Biochimica et Biophysica Acta (BBA) - Biomembranes, 1768*(6), 1367–1377.

Takano, S., Aramaki, Y. & Tsuchiya, S. (2003). Physicochemical properties of liposomes affecting apoptosis induced by cationic liposomes in macrophages. *Pharmaceutical Research, 20*(7), 962–968.

Torchilin, V.P. (2005). Recent advances with liposomes as pharmaceutical carriers. *Nature Reviews Drug Discovery, 4*(2), 145–160.

Vafaei, S.Y., Dinarvand, R., Esmaeili, M., Mahjub, R. & Toliyat, T. (2015). Controlled-release drug delivery system based on fluocinolone acetonide–cyclodextrin inclusion complex incorporated in multivesicular liposomes. *Pharmaceutical Development and Technology, 20*(7), 775–781.

Wick, G., Knoflach, M. & Xu, Q. (2004). Autoimmune and inflammatory mechanisms in atherosclerosis. *Annual Review of Immunology, 22*, 361–403.

Emerging Trends in Engineering, Science and Technology for Society,
Energy and Environment – Vanchipura & Jiji (Eds)
© 2018 Taylor & Francis Group, London, ISBN 978-0-8153-5760-5

Investigation of compressive strength of impact damaged hybrid composite laminate

A. Madhan Kumar, V. Kathiresan, K. Ajithkumar, A. Hukkim Raja &
D. Balamanikandan
Department of Aeronautical Engineering, Hindusthan College of Engineering and Technology,
Coimbatore, India

ABSTRACT: The objective of this research work is to investigate the compressive strength of Glass-Carbon/epoxy hybrid laminate subjected to impact damage. Glass-Carbon/epoxy hybrid laminate was fabricated using vacuum assisted compression molding and a novel arrangement of quasi-isotropic sequence was followed. Coupon specimens were prepared according to ASTM standard for low velocity drop impact and Compression After Impact (CAI) to assess its compressive strength after the impact. Results showed that the stacking sequence has minimized the impact damage area. The failure of the laminate after CAI was majorly due to the buckling of the sub laminate. The Hybridized effect played a vital role in the performance of the laminate.

1 INTRODUCTION

Low velocity impact induced damage on composite structures is a dangerous phenomenon, since it barely leaves any indication on the surface. Propagation of this internal damage can lead to catastrophic failure of the complete structure. Moreover, the compressive strength of the structure is reduced significantly. The internal damage, induced by the impact causes matrix cracks, fiber breakage and delamination in the structure. Brittle nature of the matrix is a crucial factor for delamination. Compression after Impact (CAI) strength is affected by matrix cracks, matrix and fiber micro cracking, weakening of interlaminar strength as shown by Hao Yan et al. (2010). Slattery, P.G. et al. (2016), Hao-Ming Hasiao. (2012), Yuichiro et al. (2008), V. Kostopoulos et al. (2010), Xinguang Xu et al. (2014), Daniele Ghelli & GiangiacomoMinak. (2011), Samuel Rivallant et al. (2013). Hakim Abdulhamid et al. (2016), Jun-Jiang Xiong et al. (2008) carried out investigations with Carbon fiber [2–10] and Hao Yan et al. (2010), Jefferson Andrew J. et al. (2015). Mehmet Aktas et al. (2012), Mandar D. Kulkarni et al. (2011), Mehmet Aktas et al. (2009) investigated with Glass fiber. As Hybridization has evolved in recent years, it is important to study the hybridized effect of Glass and Carbon.

Many researches were done on CAI tests. CAI simulation works were carried out by Hakim Abdulhamid et al. (2016), Panettieri E. et al. (2016), Gonzalez E.V. et al. (2012), Thomas E. Lacy & Youngkeun Hwang. (2003), Wei Tan et al. (2015). Experimental CAI works were done by Slattery, P.G. et al. (2016), Hao-Ming Hasiao. (2012), Yuichiro et al. (2008), V. Kostopoulos et al. (2010), Xinguang Xu et al. (2014), Tan K.T. et al. (2012), Mannov E. (2013), Hakim Abdulhamid et al. (2016) and Sanchez Saez S. et al. (2005) on FRPs. CAI on sandwich panels were carried out by Thomas E. Lacy & Youngkeun Hwang. (2003), Bin Yang et al. (2015), Guoqi Zhang et al. (2013), Davies G.A.O. et al. (2004), Bruno Castanie et al. (2008), Gilioli A. et al. (2014), Vaidya U.K. et al. (2000), Michael W. Czabaj et al. (2010) and Andrey Shipsha & Dan Zenkert. (2005). Modifications were brought in through repair by Slattery, P.G. et al. (2016) and Jefferson Andrew J. et al. (2015), stitch by Alaattin Aktas et al. (2014), Aymerich F. & P. Priolo. (2008), Tan K.T. et al. (2015) and Tan K.T. et al. (2012), pinning by Zhang X. et al. (2006) and Vaidya U.K. et al. (2000), fillers by V. Kostopoulos et al. (2010), Xinguang Xu et al. (2014) and Mannov E. (2013), Tapered laminates byHakim Abdulhamid et al. (2016), repeated impact by Cao D F et al. (2015) and Hakim Abdulhamid et al. (2016) and through hygrothermal environment by Yuichiro et al. (2008), Mehmet Aktas et al. (2009), Berketis K. & D. Tzetzis. (2010) and Michael Dale et al. (2012). However CAI on Hybrid FRPs is meager. This paper is focused on the hybrid laminate.

2 MATERIALS AND METHOD

2.1 *Fabrication*

Seven layers of Plane weave Carbon (warp)-Glass (weft) hybrid (C-G) and six layers of Plane weave Glass (G) (warp and weft) fiberswere selected for the laminate fabrication. Fibers were cut to a size of 550 mm*400 mm. A total of 13 layers were taken for the laminate. A novel quasi isotropic sequence was selected as shown in Figure 1 (a). Epoxy resin LY556 and hardener HY952 were used in the ratio 10:1. Vacuum assisted compression molding was used to fabricate the laminate. A pressure of 600 Pa was applied over the laminate for 4 hour and cured at a temperature of 120°C for 12 hours and was left for ambient cure for 2 hours. Figure 1 (b) shows the fabricated laminate of thickness 4.2 mm. Coupon specimens were prepared according to ASTM standards to determine the basic properties and CAI properties.

2.2 *Experiments*

2.2.1 *Basic properties*
Uniaxial tensile test was performed to determine the basic properties. Samples were prepared according to ASTM D3039. Table 1, shows the basic properties of Carbon-Glass/ epoxy hybrid laminate.

2.2.2 *Low velocity impact and CAI test*
Samples were prepared according to ASTM D7137. Coupon specimen were prepared for the impact and CAI test. The impact test setup consists of a drop tower with hemispherical impactor of mass 3.5 kg and 15.3 mm diameter. The impactor, when released from a height,

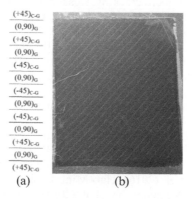

(a)   (b)

Figure 1.   (a) Stacking sequence. (b) Fabricated laminate.

Table 1.   Basic Properties of Glass-Carbon/epoxy hybrid laminate.

	E_1, GPa	E_2, GPa	G_{12}, GPa	v_{12}
Glass-Carbon/epoxy hybrid laminate	14.32	11.55	4.493	0.5

impacts the specimen which was clamped below, with the corresponding energy. Energy level can be varied by varying the height of the impactor on the drop tower. Three samples for each of 25J, 35J and 45J were chosen. Compression tests were carried out for the impacted samples in CAI fixture. The fixture consists of anti-buckling plates to prevent the buckling of the sample during CAI test. The CAI was linked with the data acquisition system, from which results were obtained.

3 RESULTS AND DISCUSSION

Figures 2–4 shows the samples of impact energies of 25J, 35J and 45J respectively. For 25J and 35J, visible damage area was observed. However,the sample of 45J energy,perforation was observed. Matrix cracking has taken place at 25J. A reduction in the area of impact damage was observed as compared to the work of Cartie D.D.R. & P.E. Irving. (2002) on pure Carbon.

The matrix crack that formed because of the impact usually propagates and ends at a place where it meets a stiffer fiber. Since the adopted stacking sequence has stiffer fibers covering all the directions, the crack propagation was terminated. In comparison with the damaged

Figure 2.   Impacted sample at energy 25J.

Figure 3.   Impacted sample at energy 35J.

Figure 4.   Impacted sample at energy 45J.

samples in the literature, it is evident that the adopted method of stacking sequence was effective using the plain weave hybrids.

Hybridization and stacking sequence tailored the brittle nature of carbon. The plain wave of fibers prohibited the propagation of the crack and minimized the damage area,as observed in Figure 4.

Figure 5 shows the CAI damage of Glass-Carbon/epoxy laminate for all three energies. The crack formed by the compressive force propagates in the direction perpendicular to loading. The intensity of the crack and sub laminate buckling increased with the increase of impact energy.

The Force-Time history of Glass-Carbon/epoxy was given in Figure 6. The maximum force required for the sample to fail decreased with the increase in impact energy. The time to failure for the 35J and 45J were near. However, the impact of 25J energy took a longer time to failure.

Figure 7, shows the Force vs Displacement curve for all the three energies. The curves were linear and hence the displacement decreased with the increase of impact energy. The Displacement vs Time curve was exactly linear and proportional to each other for all the energies as shown in Figure 8. However, the time to failure and the displacements were inversely proportional. Higher energies require a lesser time and lesser energy requires a longer time.

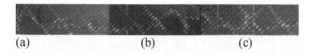

(a)   (b)   (c)

Figure 5.   CAI crack formation for (a) 25J (b) 35J (c) 45J.

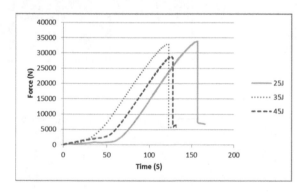

Figure 6.   CAI result of Force vs Time.

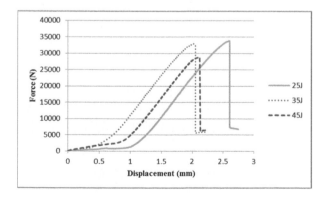

Figure 7.   CAI result of Force vs Displacement.

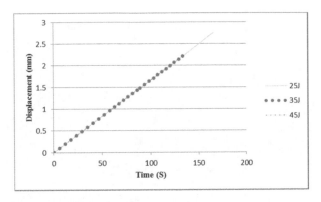

Figure 8. CAI result of Displacement vs Time.

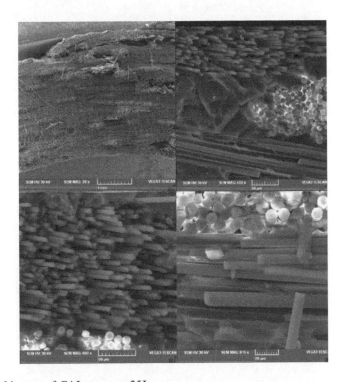

Figure 9. SEM images of CAI at energy 25J.

Figures 9–11 shows the SEM image of CAI specimen for energies 25J, 35J and 45J respectively. From the figures, it is evident that failure occurred because of the sub laminate buckling. Matrix crack was dominant for the energy at 25J. The dominant parameters for the laminates of energy greater than 25J were fiber pullout, fiber micro cracking and Delamination.

The damage of the CAI happens in the direction perpendicular to loading. The tensile or the shear load that occurs as a result of impact initiates the matrix crack, which propagates through the thickness of the laminate.

Figures 10 and 11 shows the SEM images of energy greater than 25J. These images witness the fiber pullout, fiber fracture, matrix fiber de-bonding and results in severe plastic deforma-

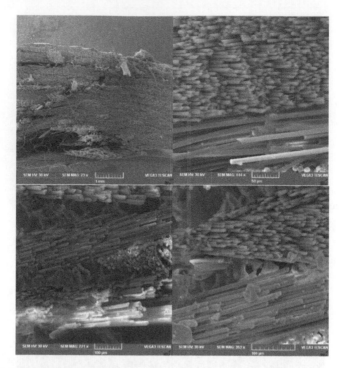

Figure 10.   SEM images of CAI at energy 35J.

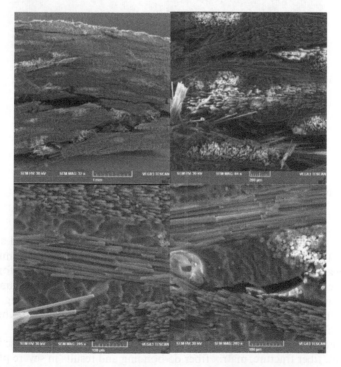

Figure 11.   SEM images of CAI at energy 45J.

tion. The larger value of in-plane properties of Carbon and the higher impact property of Glass complimented each other to result in an increased performance.

Dent in the matrix caused by the fiber pull out is clearly visible in Figure 11. Micro cracking of the fiber is visible only in carbon because of its brittle nature.

Quasi isotropic stacking sequence with plane weave fabrics has an added advantage especially to the impact response. Stacking sequence is also a main parameter to determine the CAI strength of the laminate.

4 CONCLUSION

From the above experimentation, the authors found out the following:

- Crack propagation due to impact can be terminated by stiffer fibers in the path of crack.
- For energies less than 25J, matrix crack is predominant.
- For energies greater than 25J other parameters such as fiber pullout, fiber micro cracking and Delamination.
- Delamination occurs as a result of sub laminate buckling.
- The contact force and time were inversely proportional.
- The time to failure and the displacement were inversely proportional.
- Stacking sequence is also a main parameter to determine the CAI strength of the laminate.
- Hybridization results in an increased performance, where two or three fibers in a stacking sequence, as mentioned in this research, can complement each other in the performance as a single laminate.

REFERENCES

Alaattin Aktas et al. 2014. Impact and post impact (CAI) behavior of stitched woven-knit hybrid composites.*Composite Structures*. 116, 243–253.

AndreyShipsha & Dan Zenkert. 2005. Compression after impact strength of sandwich panels with core crushing damage. *Applied Composite Materials*. 12, 149–164.

Aymerich F. & P. Priolo. 2008. Characterization of fracture modes in stitched and unstitched cross-ply laminates subjected to low-velocity impact and compression after impact loading. *Int. J. Impact Engineering*. 35, 591–608.

Berketis K. & D. Tzetzis. 2010. The compression after impact strength of woven and non-crimp fabric reinforced composites subjected to long term water immersion ageing. *J. Mater Sci*. 45, 5611–5623.

Bin Yang et al. 2015. Study on the low velocity impact response and CAI behavior of foam filled sandwich panels with hybrid facesheet. *Composite Structures*. 132, 1129–1140.

Bruno Castanie et al. 2008. Core crushing criterion to determine the strength of sandwich composite structures subjected to compression after impact, *Composite Structures*, 86, 243–250.

Cao D.F. et al. 2015. Compressive properties of SiC particle reinforced aluminum matrix composites under repeated impact loading. *Strength of Materials*. 47, 61–67.

Cartie D.D.R. & P.E. Irving. 2002. Effect of resin and fibre properties on impact performance of CFRP. *Composites: Part A*. 33, 483–493.

Daniele Ghelli & Giangiacomo Minak. 2011. Low velocity impact and compression after impact tests on thin carbon/epoxy laminates. *Composites: Part B*. 42, 2067–2079.

Davies G.A.O. et al. 2004. Compression after impact strength of composite sandwich panels. *Composite Structures*. 63, 1–9.

Gilioli A. et al. 2014. Compression after impact test (CAI) on NOMEX honeycomb sandwich panels with thin aluminum skins. *Composites: Part B*. 67, 313–325.

Gonzalez E.V. et al. 2012. Simulation of drop weight impact and compression after impact tests on composite laminates. *Composite Structures*. 94, 3364–3378.

Guoqi Zhang et al. 2013. The residual compressive strength of impact damaged sandwich structures with pyramidal truss cores. *Composite Structures*. 105, 188–198.

Hakim Abdulhamid et al. 2016. Experimental study of compression after impact of asymmetrically tapered composite laminate. *Composite Structures*. 149, 292–303.

Hakim Abdulhamid et al. 2016. Numerical simulation of impact and compression after impact of asymmetrically tapered laminated CFRP. *International J. Impact Engineering*. 95, 154–164.

Hao Yan et al. 2010. Compression-after-impact response of woven fiber-reinforced composites. *Composite Science and Technology*. 70, 2128–2136.

Hao-Ming Hasiao. 2012. Compression after impact strength and surface morphology in toughned composite materials. *Int J. Fract*. 176, 229–236.

Jefferson Andrew J. et al. 2015. Compression after impact strength of repaired GFRP composite laminates under repeated impact loading. *Composite Structures*. 133, 911–920.

Jun-Jiang Xiong et al. 2008. A strain based residual strength model of Carbon/epoxy composites based on CAI and fatigue residual strength concepts. *Composite Structures*. 85, 29–42.

Kostopoulos V. et al. 2010. Impact and compression after impact properties of carbon fibre reinforced composites enhanced with multi wall carbon nanotubes. *Composites science and technology*. 70, 553–563.

Mandar D. Kulkarni et al. 2011. Effect of back pressure on impact and compression after impact characteristics of composites. *Composite Structures*. 93, 944–951.

Mannov E. 2013. Improvement of compressive strength after impact in fibre reinforced polymer composites by matrix modification with thermally reduced graphene oxide. *Composite Science and Technology*. 87, 36–41.

Mehmet Aktas et al. 2009. Compression after impact behavior of laminated composite plates subjected to low velocity impact in high temperatures. *Composite Structures*. 89, 77–82.

Mehmet Aktas et al. 2012. Impact and post impact behaviour of layered fabric composites. *Composite Structures*. 94, 2809–2818.

Michael Dale et al. 2012. Low velocity impact and compression after impact characterization of woven carbon/vinylester at dry and water saturated conditions. *Composite Structures*. 94, 1582–1589.

Michael et al. 2010. Compression after impact of sandwich composite structures: Experiments and Modelling. *AIAA*. 2867, 1–16.

Panettieri E. et al. 2016. Delaminations growth in compression after impact test simulations: Influence of cohesive elements parameters on numerical results. *Composite Structures*. 137, 140–147.

Samuel Rivallant et al. 2013. Failure analysis of CFRP laminates subjected to compression after impact: FE simulation using discrete interface elements. *Composites: Part A*. 55, 83–93.

Sanchez Saez S. et al. 2005. Compression after impact of thin composite laminates. *Composites Science and Technology*. 65, 1911–1919.

Slattery, P.G. et al. 2016. Assessment of residual strength of repaired solid laminate composite materials through testing, *Composite Structures*. 147, 122–130.

Tan K.T. et al. 2012. Effect of stitch density and stitch thread thickness on compression after impact strength and response of stitched composites. *Composites Science and Technology*. 72, 587–598.

Tan K.T. et al. 2015. Finite element model for compression after impact behavior of stitched composites. *Composites: Part B*. 79, 53–60.

Thomas E. Lacy & Youngkeun Hwang. 2003. Numerical simulation of impact damaged sandwich composites subjected to compression after impact loading.*Composite Structures*. 61, 115–128.

Vaidya U.K. et al. (2000). Low velocity impact and compression after impact response of Z-pin reinforced core sandwich composites. *Transactions of the ASME*. 122, 434–442.

Wei Tan et al. (2015). Prediction of low velocity impact damage and compression after impact (CAI) behavior of composite laminates. *Composites: Part A*. 71, 212–226.

Xinguang Xu et al. (2014). Improving compression after impact performance of carbon fiber composites by CNTs/thermoplastic hybrid film interlayer. *Composites science and technology*. 95, 75–81.

Yuichiro et al. (2008). Effect of hygrothermal condition on compression after impact strength of CFRP laminates. *Composites science and technology*. 68, 1376–1383.

Zhang X et al. (2006). Improvement of low velocity impact and compression after impact performance by z-fibre pinning.*Composites Science and Technology*. 66, 2785–2794.

Emerging Trends in Engineering, Science and Technology for Society,
Energy and Environment – Vanchipura & Jiji (Eds)
© 2018 Taylor & Francis Group, London, ISBN 978-0-8153-5760-5

Effect of polarity in micro-electrical discharge machining

Jibin Boban, Arun Lawrence & K.K. Manesh
Department of Mechanical Engineering, Government Engineering College, Thrissur, India

Leeba Varghese
Department of Mechanical Engineering, Viswajyothi College of Engineering and Technology, Ernakulam, India

ABSTRACT: Micro Electric Discharge machining can be used to generate micro features and micro level dimensions on the work-piece irrespective of the hardness of the material. This paper discusses the effect of polarity in tool wear during micro-EDM drilling of stainless steel work-piece (SS 304). An experimental investigation has been carried out to understand the effect of change in polarity in tool wear using three different tool electrodes (Cu, Brass and W). Direct polarity has significant impact over reverse polarity in reducing tool wear for all the three electrodes. Further, observations indicated that material removal rate for stainless steel is maximum in case of direct polarity.

Keywords: Tool wear rate, Tool electrodes, Material removal rate, Polarity

1 INTRODUCTION

EDM is a non-traditional machining process which involves the removal of electrical conductive material by a series of electric sparks between two electrodes submerged in a dielectric fluid. The material removal mechanism involves the melting and vaporization of the work-piece material caused by these electric sparks.

In the current scenario, micromachining of materials has become essential to make precise and accurate components (Yuangang et al. 2009). Micro-EDM is a recently developed method that can be used for producing micro-parts within the range of 50 μm–100 μm. It is an efficient machining process for the fabrication of miniaturized products, micro channels, micro-metal holes and micromold cavities with a lot of merits resulting from its characteristics of non-contact and thermal metal removal process (Yeakub Ali & Mohammed 2009).

The tool wear in micro-EDM directly affects the machining precision and efficiency (Jingyu et al. 2017). Hence the minimization of tool wear is of great importance in micro-edm

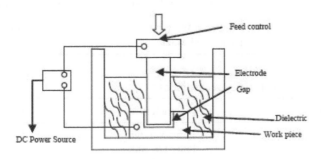

Figure 1. Schematic diagram for micro-EDM (Equbal et al. 2009).

process to achieve the required accuracy of machined features (Bissacco et al. 2010). In macro-EDM, the workpiece is made positive (anode) and the tool electrode is made the cathode. This is referred to as direct or straight polarity (Cyril et al. 2017). Similarly, if the tool electrode is made positive, it is referred to as negative polarity. The influence of change in polarity in micro-EDM is less studied by researchers. Thus, experiments are conducted by changing the polarity of tool electrode or work-piece in order to study its effect on tool wear in micro-EDM. In the study of Yoshiyuki et al. (1991), change in polarity has significant effect on the electrical discharge machining performance. Also, Lee & Li (2001) have reported that negative polarity of tool offers less tool wear.

In micro-EDM, the machinability of the work-piece material mainly depends on the thermal conductivity and melting point (Yu et al. 2014). Stainless steel (SS 304) has high melting point in the range of about 1400–1455°C. But the tool electrodes used for machining have melting point lower than the work-piece material. So chance for tool wear is high which makes the study important. Experimental investigation is carried out to determine how the change in polarity affects the tool-wear. Also the effect of polarity in the material removal rate of stainless steel is also checked for both the polarities.

2 EXPERIMENTAL SETUP DETAILS

An In House built Micro-EDM was used for the experimental investigation. Stainless steel specimens ($30 \times 20 \times 5$ mm) for experiment were cut using abrasive cutters. Copper, Brass & Tungsten are the tool electrodes used for study. The details of the experiment are given in Table 1:

The major input parameters used in micro-EDM are Input voltage, Input current, Pulse On time and Pulse off-time. By conducting a lot of pilot experiments, optimum values of process parameters which gives good machining is identified and selected for the study. The experiment levels are presented in Table 2:

The properties of the tool electrodes and work-piece used are given in Tables 3 and 4.

Experiment is carried out using separate tools and work-pieces for both direct and reverse polarities. The weights of both tool and work-piece are noted using precision weighing balance, before and after machining in each case.

Material Removal Rate (MRR) and Tool Wear Rate (TWR) are calculated using the equation:

$$MRR = \frac{\text{Weight before machining} - \text{Weight after machining}}{\text{Machining Time}}$$

$$TWR = \frac{\text{Weight before machining} - \text{Weight after machining}}{\text{Machining Time}}$$

Table 1. Experimental details.

Work-Piece	Stainless Steel (SS 304)	30 mm × 20 mm × 5 mm
Tool	Cu, Brass, W	0.8 mm ϕ, 40 mm length
Dielectric	De-ionized water	–

Table 2. Experimental conditions.

SI No	Parameters	Values
1	Input Voltage	50 V
2	Pulse On-time	110 μS
3	Pulse Off-time	30 μS

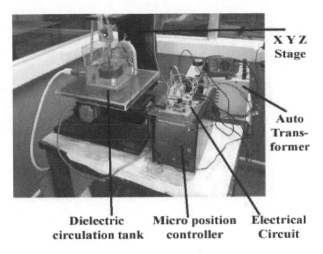

X Y Z Stage

Auto Transformer

Dielectric circulation tank

Micro position controller

Electrical Circuit

Figure 2.   In-house built micro-EDM experimental setup.

Table 3.   Properties of tool electrodes.

	Specifications of electrodes		
Properties	Copper	Brass	Tungsten
Density (kg/m³)	8910	8490	19.25
Thermal Conductivity (W/m-K)	392	158	173
Melting Point (°C)	1083	900	3422

Table 4.   Properties of stainless steel (SS 304).

Properties	Values
Brinell Hardness	123
Density	8000 kg/m³
Thermal Conductivity	16.2 W/m-K
Melting Point	1400–1455°C

3   RESULTS AND DISCUSSIONS

The effect of change in polarity on the tool wear rate and Material removal rate are discussed. The results are based on the experimental investigation performed by machining stainless steel (SS 304) specimen.

The response table for TWR and MRR is shown in Table 5:

3.1   *Effect of polarity on tool wear*

Figure 3 indicates the effect of change in polarity on tool wear rate. It can be noticed that the tool wear is minimum in straight or direct polarity compared to reverse polarity for all the three electrodes. In positive polarity, the tool is made the cathode and work piece is made the anode. Electrons are lighter in mass and hence they get accelerated faster from the cathode (tool electrode). Therefore electrons bombarding on the anode (work-piece material) will generate more heat energy than positively charged particles hitting the tool electrode. This helps to reduce the

Table 5. TWR and MRR for different tool electrodes.

Tool electrode	TWR (g/min)		MRR (g/min)	
	Direct	Reverse	Direct	Reverse
Copper	0.0028	0.0149	0.0036	0.0024
Brass	0.0032	0.0135	0.0028	0.001
Tungsten	0.0007	0.0017	0.0025	0.0012

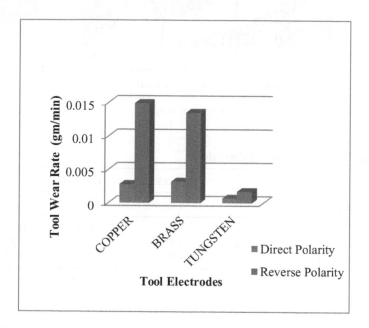

Figure 3. Effect of polarity on tool wear rate.

tool wear rate. The tool wear is reduced by about 81.2% for copper tool, 76.3% for brass tool and 58.8% for Tungsten tool when direct polarity is used instead of reverse polarity.

3.2 *Effect of polarity on MRR*

Figure 4 indicates the effect of change in polarity on material removal rate for stainless steel. It can be noticed that the MRR is minimum in reverse polarity compared to direct polarity for all the three electrodes. This is due to the high kinetic energy of electrons as compared to that of a positively charged ion. The high motion energy of electrons may be attributed to the light weight of electron and the high acceleration of the electric field. Thus, electron bombards the anodic surface and gives more energy to the work-piece, thereby increasing the material removal rate.

3.3 *Comparison of tool electrodes*

The three tool electrodes show different tool wear characteristics during micro electrical discharge machining of the work-piece. From Figure 3, it can be observed that, in both polarities, tool wear is minimum for Tungsten electrode. This is due to the high melting point of tungsten electrode (3422°C) compared to that of stainless steel work-piece (1450°C). Also maximum MRR is obtained by copper tool electrode irrespective of the polarities. This can be inferred from Figure 4.

3.4 *Physical evaluation of work-piece surface*

It is also observed from visual inspection that large deposition of tool material on the work-piece surface occurs in reverse polarity compared to that of direct polarity. Figure 5 shows the images of work-piece surface obtained by means of image acquisition system. The amount of tool deposition has to be further analyzed using spectrometric analysis or using Scanning Electron Microscopy.

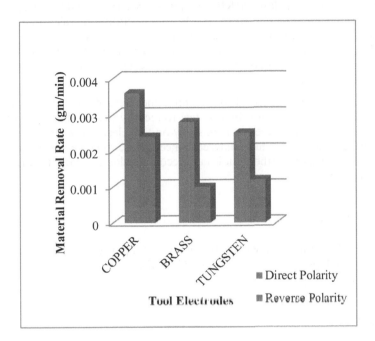

Figure 4.   Effect of polarity on material removal rate.

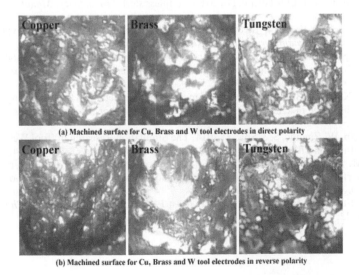

(a) Machined surface for Cu, Brass and W tool electrodes in direct polarity

(b) Machined surface for Cu, Brass and W tool electrodes in reverse polarity

Figure 5.   Tool deposition on the machined surface for three electrodes.

4 CONCLUSIONS

In this study, the effect of change in polarity is evaluated for various tool electrodes by measuring TWR and MRR. The following conclusions are derived from the experimental investigation:

1. In direct polarity, tool wear is found to be minimum compared to that of reverse polarity on account of more heat generation at the work-piece surface as in case of direct polarity.
2. Direct polarity also offers large MRR than reverse polarity for all the three electrodes.
3. The comparative study of Cu, Brass and W tool electrodes revealed that Tungsten tool offers minimum tool wear in both direct and reverse polarities.
4. High MRR is provided by the copper tool electrode while machining stainless steel.
5. From physical observation, deposit of tool electrode material on the work-piece surface is higher in case of reverse polarity.

The study can be further extended to determine the effect of change in polarity on other work-piece materials rather than stainless steel to identify the suitable polarity that can be used in micro-EDM. In the present paper, study on tool electrode wear is given more focus and hence surface roughness is not analyzed. Since surface roughness is too a major output parameter in micro-EDM, further studies are recommended to study the effect of polarity on surface roughness.

ACKNOWLEDGEMENT

The authors would like to acknowledge the financial support of the Centre for Engineering Research and Development (Proceedings no. C3/RSM86/2013, dated 09/02/2015 of Kerala Technological University), Government of Kerala, India.

REFERENCES

Bissacco, G. Valentincic, J. Hansen H.N. & Wiwe. B.D. 2010. Towards the effective tool wear conrol in micro-EDM milling. *Int. J. Adv. Manuf. Technol.*, 47: 3–9.

Cyril Pilligrin, J., Asokan, P. Jerald, J. Kanagaraj G., J.M. Nilakantan & Nielsen, I. 2017. Tool speed and polarity effects in micro EDM drilling of 316 L stainless steel. *Production & Manufacturing Research*, 5(1): 99–117.

Equbal, A. & Sood, A.K. 2014. Electric Discharge Machining; An overview on various areas of research. *Journal of Manufacturing and Industrial Engineering*, 13: 1–6.

Jingyu, P., Zhang, L., Du, J., Zhuang, X., Zhou, Z., W. Shunkun & Zhu, Y. 2017. A model of tool wear in electrical discharge machining process based on electromagnetic theory. *International Journal of Machine Tools & Manufacture*, 117: 31–41.

Lee, S.H. & Li, X.P. 2001. Study of the effect of machining parameters on the machining characteristics in electrical discharge machining of tungsten carbide, *Journal of Materials Processing Technology*, 115: 344–358.

Liua, Y., Zhangb, W., Zhangc, S. & Sha, Z. 2014. The Simulation Research of Tool Wear in Small Hole EDM Machining on Titanium Aloy, *Applied Mechanics and Materials*, 624:249–254.

Yeakub Ali, M. & Mohammad, A.S. 2009. Effect of Conventional EDM Parameters on the Micro machined Surface Roughness and Fabrication of a Hot Embossing Master Micro tool, *Materials and Manufacturing Processes*, 24: 454–458.

Yuangang, W. Fuling, Z. & Jin, W. 2009. Wear-resist Electrodes for Micro-EDM, *Chinese Journal of Aeronautics*, 22: 339–342.

Emerging Trends in Engineering, Science and Technology for Society,
Energy and Environment – Vanchipura & Jiji (Eds)
© 2018 Taylor & Francis Group, London, ISBN 978-0-8153-5760-5

Characterization and Taguchi based modeling and analysis of dissimilar TIG welded AISI 316L austenitic stainless steel-HSLA steel joints

P.V. Shaheer, Anwar Sadique & K.K. Ramachandran
Department of Mechanical Engineering, Government Engineering College, Thrissur, Kerala, India

ABSTRACT: In this study, 3 mm thick plates of AISI 316L austenitic stainless steel and High Strength Low Alloy (HSLA) steel were dissimilar GTA welded and the tensile strength and microstructural properties were investigated. The experimental trials were carried out as per the Taguchi design and the welding current, welding speed, wire feed rate and filler material were selected as the parameters. The results showed that the highest joint strength of 610 MPa was obtained at welding current of 100 A, welding speed of 9 cm/min and wire feed rate of 1.6 m/min with 304 steel as the filler material. ANOVA revealed that the welding current and wire feed rate are the most significant and least significant parameters, respectively. The regression model showed that the welding speed and filler material and the welding current and wire feed rate have interaction effect on the tensile strength of the joints.

1 INTRODUCTION

Austenitic stainless steel, AISI316L is known for its inherent superior properties like high strength at elevated temperatures, increased resistance to pitting and general corrosion, high creep strength etc. High strength low alloy (HSLA) steels possess very high strength, high strength to weight ratio and increased corrosion resistance together with relatively low cost. 316L austenitic stainless steel and HSLA steels have wide range of combined application in the marine, automotive and locomotive sectors. In spite of having excellent qualities, individually, dissimilar fusion welded 316L/HSLA steel joints (by conventional techniques) are prone to complications such as grain coarsening, sensitization, stress corrosion cracking, hot cracking etc. Thus, accurate control of welding process parameters is essential to regulate the heat input and hence to reduce the associated problems.

With regard to fusion welding of austenitic stainless steels, Anand Rao V & Daivanathan B (2015) conducted detailed experimental study on TIG welding of 310 steel joints. In their research, a total of 9 welded joints were fabricated and tested with the objectives of analysis and optimization of the TIG welding process. The results showed that welding current of 120 A with 309L steel filler metal had produced the highest joint tensile strength of 454.6 MPa while a welding current of 80 A with 316L filler metal produced the lowest tensile strength of 517.9 MPa. Navid Moslemi et al. (2009) conducted a study on the effect of welding current on the mechanical and microstructural characteristic of TIG welded 316 steel joints. The mechanical characteristics of the welded joints such as tensile strength and microhardness were evaluated in the study. Microstructural studies confirmed the presence of secondary sigma phase that caused embrittlement in the weld zone.

Bharatha et al. (2013) reported the process optimization and joint analysis of 316 steel TIG welded joints using Taguchi technique. With regard to dissimilar fusion welding of stainless steels and HSLA steels, only a very few works are reported in the literature. Anant et al. (2017) have developed a special nozzle for the GMAW and successfully welded 25 mm thick plates of dissimilar 304L stainless steel and SA543 HSLA steel. The authors have used 308 L steel as filler metal and the weld was completed by multiple passes.

To the best of knowledge of the authors, studies on dissimilar fusion welding of 316L and HSLA steels are not reported in the literature. Also, implementation of the dissimilar joining techniques needs the knowledge of the effect of operating parameters on the performance of the joints and optimization of the welding process. Therefore, in this work, austenitic stainless 316L and HSLA steel are dissimilar TIG welded and the influence of process parameters on the joint performance are investigated.

2 MATERIALS AND METHODS

The base materials selected for this research were 3 mm thick rolled sheets of HSLA steel, IRS-M42-93 and austenitic stainless steel, AISI 316L. The composition of HSLA steel and 316L steel is given in Table 1. The optical micrographs of the base metal are shown in Figure 1. The base 316L steel consists of mostly coarse austenite with minor amount of ferrite. Annealed twins crossing the grain boundary interface were observed. The base HSLA steel is with approximately equiaxed and fine grained ferritic–pearlitic microstructure. Work pieces of 150 mm × 50 mm × 3 mm size were cut by cold shearing. Three types of filler wires were used in this investigation; AISI 304 steel, 309L steel and 316L steel. The composition of each of the filler material is shown in Table 3. The welding trials were performed using automated TIG welding equipment and the joints were finished in a single pass welding.

The Taguchi – L9 orthogonal design is selected as the design of experiment (DOE) technique (Krishnaiah, K. & Shahabudeen, P. 2012). The most important stage in the DOE is the very selection of DOE technique and the selection of the control factors. The travel speed (welding speed), welding current, wire feed rate and filler material were selected as the factors (variables) of the design. The range and upper and lower bounds of the selected variables were fixed by trial and error with the criterion of visually defect free joints. The four factor, three level; L-9 design matrix given in Table 2 was developed using the statistical software, Minitab 17. Two sets of joints were fabricated at each set of factor setting and the average of the response was considered for modeling and analysis.

Table 1. Chemical composition of the base materials.

Material	Composition (wt%)														
	C	Mn	Si	P	S	Al	Cu	Cr	Mo	Ni	Nb	Ti	V	N	Fe
HSLA Steel; IRSM-42-97	0.11	0.42	0.32	0.10	0.01	0.029	0.31	0.54	0.001	0.22	0.001	0.002	0.002	–	Bal
AISI 316L steel	0.03	2.0	0.75	0.045	0.03	–	–	17	2–3	10–14	–	–	–	0.1	Bal

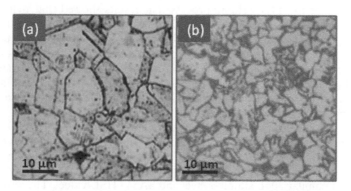

Figure 1.   Optical micrographs of base materials; (a) 316L steel (b) HSLA steel.

Table 2. The L-9 Taguchi DOE matrix.

Sl. No.	Filler material	Current (A)	Travel speed (cm/min)	Feed (m/min)
1	SS 304	80	8	1.2
2	SS 304	90	8.5	1.4
3	SS 304	100	9	1.6
4	SS 316L	80	8.5	1.6
5	SS 316L	90	9	1.2
6	SS 316L	100	8	1.4
7	SS 309L	80	9	1.4
8	SS 309L	90	8	1.6
9	SS 309L	100	8.5	1.2

Table 3. Chemical composition of the filler materials materials.

Material	Composition (Wt. %)										
	Ni	Cr	Si	Mn	C	P	S	Mo	N	Cu	Fe
AISI 304	9.25	19	1	2	0.08	0.045	0.03	–	–	–	Bal
AISI 316L	10–14	17	0.75	2	0.03	0.045	0.03	2–3	0.10	–	Bal
AISI 309L	12–14	23.5	0.3–.65	1–.5	0.03	0.03	0.03	0.75	–	0.75	Bal

The specimens for the tensile test were cut by conventional milling process with the geometry as per **ASME SEC IX** (2015): Boiler and Pressure Vessel Code (QW-462 Test specimen). The tensile test was done on a universal testing machine with 50 kN capacity at crosshead speed of 4 mm/min. The specimen for optical microscopy was prepared by using standard metallographic procedures. The etched specimens were observed under optical microscope model: BX51.

2 RESULTS AND DISCUSSION

Visual examination of welded joints showed that in seven out of nine welding conditions, no visible surface defects were observed. But, for the sixth and seventh runs, the joints were observed with visual defects such as excessive deposition and lack of fusion on the HSLA side, respectively. The observed surface defects may be caused due to excessive/low heat generation as a result of the combined effect of welding current, welding speed, and filler metal feed rate.

The average tensile strength of the joints is shown in Table 4. The highest tensile strength of 610 MPa and the least strength of 415 MPa is obtained for the fifth and third trail run, respectively. The third trial that resulted the highest joint strength is corresponding to welding current of 100 A, welding speed of 9 cm/min and feed rate of 1.6 m/min. Whereas the joint with the least strength correspond to welding current of 80 A, welding speed of 8.5 cm/min and wire feed rate of 1.6 m/min. For the best joint (highest strength), the filler material used was 304 steel whereas for the joint with the least strength, the filler material was 316L steel.

Figures 2 and 3 portray typical optical micrographs of the weld nugget region of the joints that possess the lowest and highest joint strength, respectively. Figure 2 reveals the formation of coarse grain in the nugget zone. Also, it seems that the grain boundaries are characterized by the precipitation of secondary phase. The precipitation may be sigma phase as a result of the very high heat generation and the use of 316L filler metal during welding. Under high heat input, there will be increased distribution of sigma phase in nugget zone with a wider

Table 4. Coded and actual values of the variables and values of tensile strength.

Sl. No.	Coded values of variables				Actual values of variables				TS (Mpa)	S/N Ratio
	M	I	U	F	M (Grade)	I (A)	U (cm/min)	F (m/min)		
1	1	1	1	1	304	80	8	1.2	420	52.4650
2	1	2	2	2	304	90	8.5	1.4	549	54.7914
3	1	3	3	3	304	100	9	1.6	610	55.7066
4	2	1	2	3	316L	80	8.5	1.6	415	52.3610
5	2	2	3	1	316L	90	9	1.2	596	55.5049
6	2	3	1	2	316L	100	8	1.4	471	53.4604
7	3	1	3	2	309L	80	9	1.4	530	54.4855
8	3	2	1	3	309L	90	8	1.6	533	54.5345
9	3	3	2	1	309L	100	8.5	1.2	459	53.2363

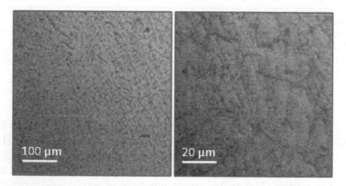

Figure 2. Optical micrographs of the weld nugget of the lowest strength joint.

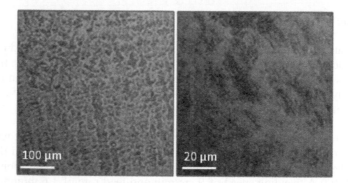

Figure 3. Optical micrographs of the weld nugget of the highest strength joint.

heat affected zone (HAZ). Under tensile load, cracks can easily propagate through the grain boundaries causing failure at low loads. Thus the observed microstructure clearly substantiates the lowest joint strength resulted for the joint.

Referring to Figure 3, it seems that the weld nugget is with low concentration of secondary phase formations. The marginally low heat generation and the use of 304 steel filler wire would be the probable causes for the low concentration of secondary phases. Vitek, J. M. & David, S. A. (1984) reported that the sigma phase reaction is accelerated by the large scale atomic movement (advance of re-crystallization front) rather than the chromium content in steel. Therefore, though 304 steel is having higher chromium content, the higher welding speed could be the probable reason for the higher joint tensile strength.

3 MODELING AND ANALYSIS

The response, tensile strength (TS) of the joint is modeled as a function of the four factors that were selected for the DOE. The generalized form of the regression model is given by;

$$TS = f\ (I,\ U,\ F,\ M) \tag{1}$$

where I is the welding current, U is the welding speed, F is the wire feed rate and M is the filler material. The coded and actual values of the variables and the values of TS are given in Table 4. The regression model developed using the proprietary statistical software; Minitab17 is given in equation (2).

$$TS = 248.1 - (28.7 \times M) + (283.7 \times I) + \\ (7.3 \times U) - (77 \times F) - (86.33 \times I^2) + \\ (22.33 \times M \times U) + (51 \times I \times F) \tag{2}$$

The adequacy of the developed model is verified by the analysis of variance (ANOVA) and the actual and adjusted coefficients of determination (R^2). The results of the ANOVA are shown in Table 5. The larger values of F-ratio indicate that the model is adequate and capable to predict the TS accurately. The actual and adjusted R^2 values of the model are 95.96% and 99.65%, respectively. Considering the nature of variability of the welding process, the above values of R^2 shows adequacy of the developed model. Conformity experimental trials agree well with the TS values predicted by the model with an average error of 0.82%.

3.1 *Analysis of the developed model*

From equation (2) it can be seen that only the factors, current and wire feed rate and filler material and welding speed have mutual interaction effect on the TS of the joints. The developed model is analyzed to illustrate the individual and combined effects of the factors (parameters) on the response (TS) of the joints. The individual and combined effects of parameters on TS are plotted by keeping the other parameters at the middle values.

Referring to Figure 4 (a), it can be seen that for the mid values of current and wire feed rate (90 A and 1.4 m/min), at welding speeds of 8.5 and 9 cm/min, 309L filler material is giving the highest joint strength. But, at low welding speed of 8 cm/min, the best filler material is 304 steel. Also, at low heat input, the suitable filler materials are 316L and 309L. This may be

Table 5. ANOVA of the developed model.

Source	Adjusted sum of squares	Adjusted mean squares	F-Value	P-Value
Regression	41069.8	5867.11	330.03	0.042
M	99.4	99.41	5.59	0.255
I	3613.8	3613.78	203.27	0.045
U	6.6	6.59	0.37	0.652
F	2615.7	2615.74	147.14	0.052
I^2	7453.4	7453.44	419.26	0.031
M*U	249.4	249.39	14.03	0.166
I*F	6502.5	6502.50	365.77	0.033
Error	17.8	17.78	–	–
Total	41087.6	–	–	–

due to the evolution of approximately equal amount of austenitic and ferritic phases in the weld nugget, rather than the formation of chromium carbide.

Figure 4(b) show the influence of welding current on the tensile strength of the joints at different wire feed rates. The plots clearly show that the welding current and wire feed rate have significant interaction on the joint strength. Also, the highest and lowest joint strength is resulted for wire feed rate of 1.6 m/min. The peak joint strength corresponds to 95 A and 1.6 m/min. The proper material deposition under moderate heat input and wire feed rate and the resulted better microstructure could be the probable reason for the higher joint strength.

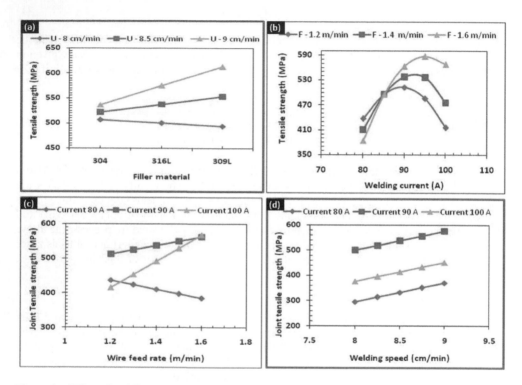

Figure 4.   Effect of welding parameters on joint strength.

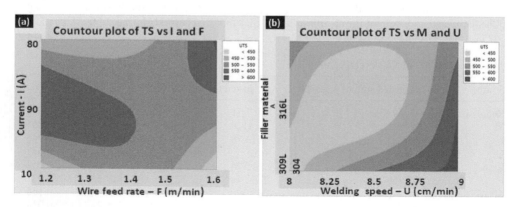

Figure 5.   Contour plot of Tensile Strength (TS) vs current & feed rate and TS vs filler metal & welding speed.

The influence of wire feed rate on joint strength at different welding currents is shown in Figure 4(c). Consistently higher joint strength is resulting at welding current of 90 A. At higher current of 100 A, though the strength is very low at low feed rate, it sharply increases and reaches the peak value at 1.6 m/min. This indicates that the higher heat input at 100 A is compensated by the increase wire feed rate and probably resulting better microstructure with less secondary phase precipitation. Figure 4(d) portrays the influence of welding speed on joint strength. The joint strength shows an increasing trend as the welding speed increases, at all the currents. But, the highest strength is obtained at 90 A, welding current. This implies that the heat input corresponding to 90 A is near optimal and results better nugget microstructure with less precipitation of secondary phases. Figure 5 shows the contour plot of tensile strength versus current and feed rate and TS vs filler material and welding speed. Other parameters do not have any significant interaction effect on the joint strength.

4   CONCLUSION

In the present research, dissimilar AISI316L stainless steel and HSLA steel were successfully TIG welded and the joint tensile strength and microstructure were investigated. The following conclusions were made.

- The highest joint strength of 610 MPa is obtained for the joint fabricated at welding current of 100 A, welding speed of 9 cm/min and wire feed rate of 1.6 m/min with 304 steel as filler metal.
- The micrographs revealed signs of secondary phase evolution at the grain boundaries in the weld nugget region.
- ANOVA of the results showed that the welding current and wire feed rate are the most significant and least significant parameter, respectively.
- Analysis of the developed model suggest that the welding current and wire feed rate have significant interaction effect on the joint strength and the welding speed and filler metal have moderate interaction effect on the joint strength.

REFERENCES

Anand Rao, V. & Deivanathan, R. 2014. Experimental Investigation for welding aspects of stainless steel 310 for the Process of TIG welding. *Procedia Engineering,* 97, 902–908.
Bharath, P. Sridhar, V.G. & Senthil Kumar, M. 2014. Optimization of 316 stainless steel weld joint characteristics using Taguchi technique, *Procedia Engineering*, 97, 881–891.
Krishnaiah, K. & Shahabudeen, P. 2012. Applied Design of Experiments and Taguchi Methods.
Nabendu Ghosh, Pradip Kumar Pal. & Goutam Nandi 2016. Parametric optimization of MIG welding on 316L austenitic stainless steel by Grey-based Taguchi method, *Procedia Technology* 25, 1038–1048.
Navid Moslemi, Norizah Redzuan, Norhayati Ahmad & Tang n Hor. (2015) Effect of current on characteristic for 316 stainless steel welded joint including microstructure and mechanical properties, *Procedia CIRP.* 26. 560–564.
Ramachandran, K.K. Murugan, N. & Shashi Kumar, S. 2015. Influence of tool traverse speed on the characteristics of dissimilar friction stir welded aluminium alloy, AA5052 and HSLA steel joints. *Archives of civil and mechanical Engineering*, 15, 822–830.
Ramkishor Anant & Ghosh, P.K. 2017. Ultra-narrow gap welding of thick section of austenitic stainless steel to HSLA steel, *Journal of Materials Processing Technology*, 239, 210–221.
Vitek, J.M. & David, S.A. 1984. The sigma phase transformation in austenitic stainless steels, 65th Annual AWS Convention in Dallas, Tex.

Mechanical design, vibration and tribology

Emerging Trends in Engineering, Science and Technology for Society,
Energy and Environment – Vanchipura & Jiji (Eds)
© 2018 Taylor & Francis Group, London, ISBN 978-0-8153-5760-5

Behavior of stress intensity factor of semi-elliptical crack at different orientations subjected to thermal load

M.B. Kumaraswamy & J. Sharana Basavaraja
Department of Mechanical Engineering, BMS College of Engineering, Bengaluru, Karnataka, India

ABSTRACT: Fracture mechanics is a very important tool used for improving the life cycle of mechanical components. During manufacturing, flaws or cracks will be formed in all metal structures. Studying and monitoring the propagation of the crack in a component forms the core of fracture study. In the present work, a parametric study on the propagation of semi elliptical crack in a turbine blade is carried out. A turbine blade with its hub is modeled using CATIA software. The model is then imported into ANSYS Workbench and a Finite Element analysis is performed. Rotational velocity is applied on cracks at different orientations ranging from 0° to 90° at different crack depth to half crack length ratios and stress intensity factor (K) is determined for two cases with thermal load and without thermal load i.e. static load case. The Finite Element results are validated by an empirical solution of Raju-Newman solution using the MATLAB software. The results will be useful in the assessment of structural integrity of the component.

1 INTRODUCTION

Fracture mechanics analysis forms the basis of damage tolerant design methodology. Its objectives are the determination of stress intensity factor (K), energy release rate (G), path independent integral (J), Crack Tip Opening Displacement (CTOD) and prediction of mixed mode fracture, residual strength and crack growth life.

Solution for Stress Intensity Factor (SIF) in mode I for a surface crack in a plate is presented in empirical form by Newman & Raju (1981). This empirical equation is presented as a function of parametric angle, depth and length of the crack as well as thickness and width of the plate for tension and bending loads. Witek (2011) discusses the failure of compressor blade due to due to bending fatigue loads, and the calculation of SIF is performed using the Raju-Newman solution for a semi-elliptical crack. Barlow & Chandra (2005) discuss the fatigue crack growth rates at the fan blade attachment in an aircraft engine due to centrifugal and aerodynamic loads. Song et al. (2007) deliberate on the failure of a jet-engine turbine blade due to improper manufacturing techniques. One of the observations that can be made from the above-mentioned discussions is that the cracks may originate in any form and direction due to improper design or manufacturing methods. Hence it becomes imperative to perform analysis to obtain SIF values of cracks at various orientations, for growth analysis.

In this work, fracture analysis is performed on the turbine blade of third stage turbine bucket of a gas turbine. Initially, static analysis is carried out to obtain the region of crack nucleation, and then semi-elliptical cracks at various orientations with respect to the rotor axis is analyzed using the finite element technique to obtain the values of stress intensity factors. A parametric study varying the crack parameters is conducted to analyse their effects on the stress intensity factor in three modes.

2 FINITE ELEMENT MODELLING OF GAS TURBINE BLADES

A turbine blade constitutes a part in the turbine region of a gas turbine engine, and functions as the component responsible for absorbing energy from the gas at high temperature and pressure created in the combustor of the engine. The turbine blades very often are the limiting components of gas turbines since they are subjected to extreme thermal and fluid stresses. Due to this reason, turbine blades are often made out of materials like alloys of Titanium containing exotic additives. They also use different and ingenious techniques of cooling, such as air channels inside the blade itself, boundary layer cooling, and thermal barrier coatings.

2.1 Geometric model

Geometric modeling of the turbine blade has been carried out in CATIA. For the present study, turbine blade of third stage turbine bucket of a gas turbine is considered. To create the geometric model, aerofoil profile of the blade has been taken from Ref (2005). This data provides the coordinates of points on the aerofoil that have been imported into CATIA V5 CAD software. The blade profile is attached to a cylindrical hub (Figure 1) with dimensions $r_i = 110$ mm, $r_o = 137.5$ mm.

2.2 Material properties

Turbine blades operate under very tough conditions of pressures and temperatures, and hence the blade material should have the capacity to resist failing under corrosion, fatigue and impact loading. An alloy of Titanium (Ti-6Al-2Sn-4Zr-2Mo) has been selected for the turbine blade material (2016). Table 1 specify the physical properties of Titanium alloy.

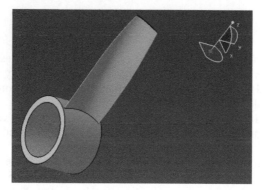

Figure 1.   Turbine blade attached to the hub.

Table 1.   Physical properties: Titanium alloy (2016).

Physical property	Value
Density	4540 kg/m³
Young's modulus	120 GPa
Shear modulus	45.5 GPa
Poisson's ratio	0.32
Ultimate tensile strength	1010 MPa
Yield strength	990 MPa
Fracture toughness	148 $MPa\sqrt{m}$
Coefficient of thermal expansion	$8.1 \times 10^{-6}/°C$
Operating temperatures	T1 = 500°C
	T2 = 550°C
	T3 = 601°C

2.3 Validation of finite element model

Finite Element Modelling of the turbine blade has been carried out using ANSYS-Workbench using 3D tetrahedral elements. In order to validate the FE model, the benchmark solution from Raju & Newman (1981) has been considered. The geometric dimensions for the benchmark plate geometry [9] is as shown in Figure 3: h = 250 mm; b = 250 mm; t = 200 mm; Crack dimensions: Width 2c = 80 mm; Depth a = 80 mm. The plate and crack dimensions meet the criteria mentioned for the application of the Raju-Newman solution, which is,

$$0 < a/c \leq 1; 0 \leq a/t < 1; c/b < 0.5$$

This problem involves a plate with a semi-elliptical crack subjected to tensile load with the opposite face being fixed. The FE model is shown in Figure 2.

The benchmark solution (1981) is an empirical equation for the stress-intensity factors for a surface crack. The empirical equation covers a wide range of configuration parameters. The ratios of crack depth to plate thickness and the ratios of crack depth to crack length range from 0 to 1. The stress-intensity factor equation (1981) for combined tension and bending loads is given by,

$$K_I = (S_t + HS_b)\sqrt{\Pi\left(\frac{a}{Q}\right)}F\left[\frac{a}{t},\frac{a}{c},\frac{c}{b},\phi\right] \tag{1}$$

For, $0 < a/c \leq 1; 0 \leq a/t < 1; c/b < 0.5$ and $0 \leq \varnothing \leq \pi$.
Where Q is an approximation given by,

$$Q = 1 + 1.464\left[\frac{a}{c}\right]^{1.65} \tag{2}$$

The term HS_b can be ignored because only the tensile loading is considered. Hence, the equation for stress intensity factor becomes

$$K_I = (S_t)\sqrt{\Pi\left(\frac{a}{Q}\right)}F\left[\frac{a}{t},\frac{a}{c},\frac{c}{b},\phi\right] \tag{3}$$

The Raju-Newman solution for the stress intensity factor is for tensile and bending stress. And since only the rotational loading is considered for analysis, the bending load part of

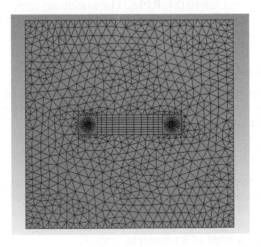

Figure 2. Finite element model of the plate with a semi-elliptical crack.

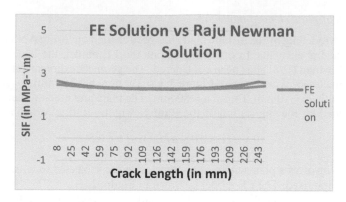

Figure 3. ANSYS solution vs Raju-Newman solution.

the solution is ignored. Hence, coming to the boundary conditions applied to the plate, the bottom face of the plate is considered as a fixed support and the on the top face, a tensile pressure is applied to simulate the centrifugal load. This simulates the crack opening mode.

Figure 3 shows a comparison of mode 1 stress intensity factor values obtained from ANSYS Workbench and the target solution of Raju-Newman empirical equation given by (3).

The present solution is in good agreement with empirical solution of (1981). The maximum variation is found to be 2.457%. These values can be considered satisfactory and hence the Finite Element model developed is considered to be validated.

3 FRACTURE MECHANICS ANALYSIS

3.1 *Location of the crack on blade*

It is essential to determine the critical location of crack on the turbine blade where the crack may nucleate. To determine the location, a static analysis is performed on the blade by applying rotational load on it. By finding the region where the maximum von-Mises stress is induced, the location of crack can be approximated.

To simulate the centrifugal loading on the turbine blade the central hub is fixed by constraining all the degrees of freedom of the nodes present in its mesh. A rotational velocity is imparted at the center of the hub along its axis by creating a new coordinate system. Static analysis is carried out for 10,000 RPM. The maximum stress is found approximately at bottom/near root of the turbine blade. Considering the above result, the crack for the analysis has been introduced in the region of maximum von-Mises stress.

3.2 *Finite element model of semi-elliptical crack*

The base mesh around the Fracture Affected Zone comprises of tetrahedral elements. The convergence study was carried out and the final details of the finite element model and crack mesh is given in Figure 4.

3.3 *Evaluation of SIF*

Once the meshing is completed as shown in Figure 4 and boundary conditions are applied and the analysis is performed. Here, the stress intensity factors in mode 1, mode 2 and mode 3 are calculated. Parametric study done with the crack length and depth as the varying parameters is discussed following paragraphs. Figure 5 shows the plots of stress intensity factor in mode-1 along the crack front of an example analysis.

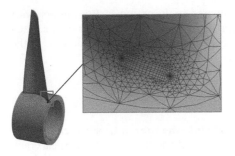

Figure 4.   Blade and crack mesh.

Figure 5.   SIF plot for opening mode.

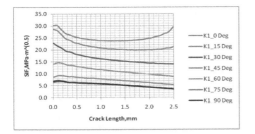

Figure 6.   K1 for crack angles 0° to 90° at the interval of 15°.

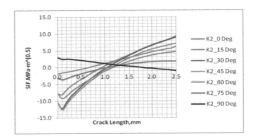

Figure 7.   K2 for crack angles 0° to 90° at the interval of 15°.

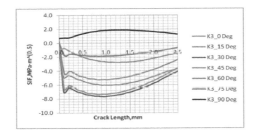

Figure 8.   K3 for crack angles 0° to 90° at the interval of 15°.

The crack mesh has 7 solution contours for each a/c ratio. The Post-Processing of the solution involves exporting the values to Microsoft Excel for further analysis. For the parametric studies, the average of the values obtained from all seven solution contours is considered.

Seven different analyses of the blade are carried out with the crack at 0°, 15°, 30°, 45°, 60°, 75° and 90° orientation with respect to the longitudinal axis of the turbine rotor. The crack length and depth in all the seven cases are kept constant at 2.0 mm and 0.80 mm (2002) respectively. The plots of SIF along the crack front for the crack angles 0° to 90° at the interval of 15° are shown in Figures 6, 7 and 8.

The behavior of the values of K1 is shown in Figure 6. It is observed that the values of K1 are highest for the 0° crack which indicates that cracks at 0° angle to load have the lowest chance of propagation of crack compared to others. The SIF values decrease with the increase in crack orientation, and minimum for 90° crack which has highest chance of propagation of crack. It is observed in Figure 7 that the values of SIF in sliding mode vary from a positive number to a negative number between the crack front for all the angles except for 90° angle where in it reverses its behavior. The value of K3 plotted in the graph is shown in Figure 8. For the cracks of all the orientations, the stress intensity factor values are observed to be lower at the center of the crack front when compared to the extremes. But this behavior reverses for crack of 90° orientation, with the values being higher at the center of the crack front when compared to the ends. Also, the values become negative with the increase in the crack length from extremes to center of crack front.

4 PARAMETRIC STUDIES

The parametric studies of this project provide an insight into the stress intensity factor behavior at the crack front for various thermal loads and static load condition. Semi-elliptical crack length, crack depth and crack orientations are the parameters considered for the analysis. In the current work, how the stress intensity factor varies in three modes for cracks at different crack orientations with different a/c ratios and different loading conditions are studied.

Thermal-Static loads: For thermal-static load case, Titanium alloy of yield strength 990 MPa, fracture toughness 148 MPa $(m)^{1/2}$ and coefficient of thermal expansion $8.1 \times 10^{-6}/°C$ with temperature T3 = 601°C are considered.

Mode-1: The behavior of mode-1 (opening mode) stress intensity factor of thermal and static load conditions are very similar for cracks of all orientations. From the graphs (Figures 12–17), the values of K1 for cracks of different a/c ratios proportionally decrease with the increase in crack orientations. There is a notable variation in the proportional decrease of K1 value with the increase in crack orientation i.e. less reduction at one end of crack front in comparison with the other end where the variation is large. This can be attributed to the curvature and twist along the span of the blade, leading to K1 values which result in crack closure due to compression. Maximum value of K1 for thermal load case is reduced nearly 5 times in comparison with static load case. The value of K1 becomes maximum for larger value a/c ratio in static load case where as in thermal load case the value of K1 increase from crack angle 0° to 30° and keeps on decreasing for higher crack angle. This shows that cracks with a larger length when compared to the depth have a higher tendency of propagation near the crack tip and cracks which are comparatively smaller in length have a higher tendency of propagation near the Centre of crack front, leading to component failure.

The Figures 9 to 13 show that the values of K1 changes from along the crack length at 0° crack angle to diagonal way continuously with the increase in crack angle. The extreme values of K1 are positive in static load case where as in thermal load case, the value of K1 changes

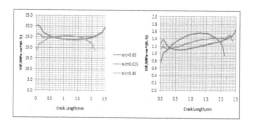

Figure 9.   K1 values for 0° crack angle.

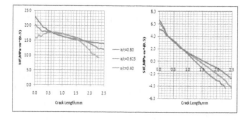

Figure 10.   K1 values for 30° crack angle.

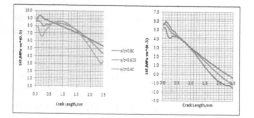

Figure 11.   K1 values for 60° crack angle.

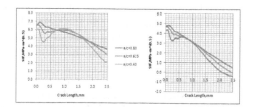

Figure 12.   K1 values for 75° crack angle.

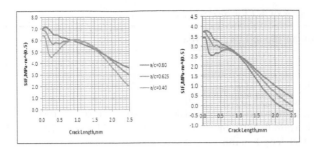

Figure 13.   K1 values for 90° crack angle.

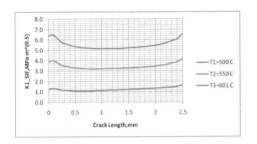

Figure 14.   K1 values for 0° crack.

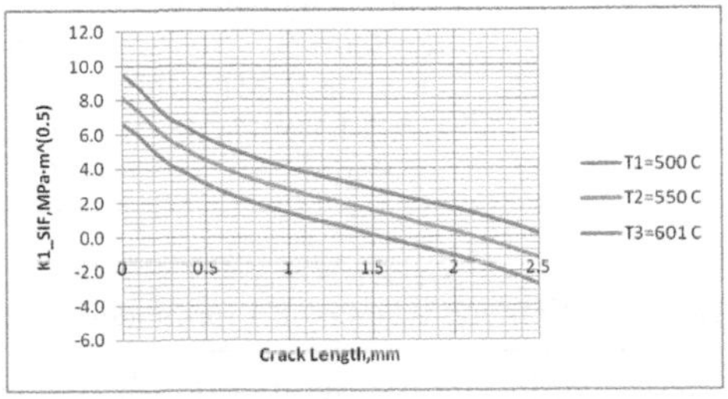

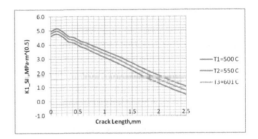

Figure 15.   K1 values for 30° crack.

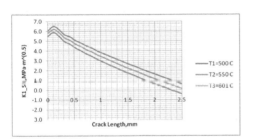

Figure 16.   K1 values for 60° crack.

Figure 17.   K1 values for 75° crack.

from a positive value to a negative value between the two crack fronts except for a crack angle 0°. Two important parameters; semi-elliptical crack length and depth are considered for the analysis, and variation in the stress intensity factors in three modes for cracks at different orientations and different a/c ratios are studied in the present work.

The Figures 9 to 13 show that cracks with a larger length when compared to the depth have a higher tendency of propagation, leading to component failure.

The behavior of K1 values for cracks of orientations 60° and 75° is observed to be different from the rest. The K1 values, from Figures 9–13, are negative for these two cracks. This can be attributed to the curvature and twist along the span of the blade, leading to K1 values which result in crack closure due to compression.

As seen from the Table 2 for both static and thermal load case, extreme values of K1 for all the orientation of the crack is located near the proximity of crack tip except for crack angle 0° where the maximum value of K1 is located at the centre of the crack length. For static load case, (K1) max = 30 Mpa √m at a/c ratio 0.80 and the crack orientation 0°. Similarly (K1) min = 3.6 Mpa √m at a/c ratio 0.40 and the crack orientation 90°. Extreme values of K1 are decreasing proportionally with the increase of crack orientation.

Thermal load case, (K1) max = 6.6 Mpa √m at a/c ratio 0.80 and the crack orientation 30°. Similarly (K1) min = −3.2 Mpa √m at a/c ratio 0.40 and the crack orientation 30°. Maximum value of K1 initially increases from crack angle 0° to 30° and decreases with the further

387

Table 2. Extreme values of K1 for mode-1 case.

Stress intensity factor for mode-1 (K_1) in Mpa √m

Crack orientation in deg		0	15	30	45	60	75	90
Static	(K_1) max in Mpa √m	30	28	23	14.8	9.2	6.8	7.2
	Crack length in mm	0	0	0	0.10	0.15	0.15	0.10
	a/c Ratio	0.80	0.80	0.80	0.80	0.80	0.80	0.80
	(K_1) min in MPa √m	18	16	11.5	7.6	5.0	3.7	3.6
	Crack length in mm	0	2	2	2	2	2	2
	a/c Ratio	0.40	0.40	0.40	0.40	0.40	0.40	0.40
Thermal $(T_3 = 601°C)$	(K_1) max in Mpa √m	1.6	4.4	6.6	6.4	5.8	4.8	3.8
	Crack length in mm	1.3	0	0	0.10	0.10	0.10	0.10
	a/c Ratio	0.40	0.80	0.80	0.80	0.80	0.80	0.80
	(K_1) min in MPa √m	0.7	−2.0	−3.2	−2.2	−0.8	0.15	0.2
	Crack length in mm	0	2	2	2	2	2	2
	a/c Ratio	0.40	0.40	0.40	0.40	0.40	0.40	0.40

Table 3. Extreme values of K1 for mode-1 case.

Stress intensity factor for mode-1 (K_1) in Mpa √m

Thermal load →				
Parameters ↓		T1 = 500°C	T1 = 550°C	T1 = 601°C
Maximum value	(K_1) max in MPa √m	9.6	8.2	6.6
	Crack length in mm	0	0	0
	a/c Ratio	0.80	0.80	0.8
	Crack orientation in deg	30	30	30
Minimum value	(K_1) min in MPa √m	−0.6	−2.0	−3.2
	Crack length in mm	2	2	2
	a/c Ratio	0.40	0.40	0.40
	Crack orientation in deg	30	30	30

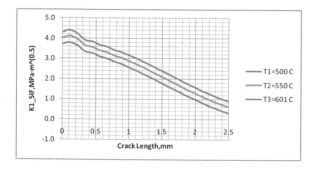

Figure 18. K1 values for 90° crack.

increase of crack angle and minimum value of K1 initially decreases from crack angle 0° to 30° and increases with the further increase of crack angle.

As seen from the Table 3 for both static and thermal load case, extreme values of K1 for all the orientation of the crack is located near the proximity of crack tip except for crack angle 0° where the maximum value of K1 is located at the centre of the crack length. For static load case, (K1) max = 30 Mpa √m at a/c ratio 0.80 and the crack orientation 0°. Similarly

(K1) min = 3.6 Mpa √m at a/c ratio 0.40 and the crack orientation 90°. Extreme values of K1 are decreasing proportionally with the increase of crack orientation.

Maximum value of K1 initially increases from crack angle 0° to 30° and decreases with the further increase of crack angle and minimum value of K1 initially decreases from crack angle 0° to 30° and increases with the further increase of crack angle as shown in the Figures 14–18.

5 CONCLUSIONS

The following conclusions are drawn from the above work.

Thermal-static loads

Maximum value of K1 for thermal load case is reduced nearly 5 times in comparison with static load case. From the FEA results, for static load case, it can be concluded that fracture by mode-1 is likely to occur because the values of K1 and it is found at 0° crack orientation. The maximum values K1 and K2 for thermal load case are found at crack angle 30° and 0° respectively.

For the temperature from T1 = 500°C to T2 = 550°C, maximum value of K1 decrease by 14.6% and further increase of temperature from T2 = 550°C to T3 = 601°C, maximum value of K1 decrease by 19.5%. From FEA results thermal load case, it is observed that with the increase of temperature, the value of K1 and K3 decreases in proportion along the crack length whereas the value of K2 increases with the increase of temperature.

ACKNOWLEDGMENT

We would like to express my sincere gratitude to TEQIP III and Management B.M.S. College of Engineering, Bengaluru, for extending financial support in publishing and presenting the paper.

REFERENCES

ASM Aerospace Specification Metals Inc., *13 May 2016*. Available: http://asm.matweb.com/search/SpecificMaterial.asp?bassnum=MTP641.

Barlow, K.W. & Chandra, R. 2005. *Fatigue crack propagation simulation in an aircraft*: International Journal of Fatigue 27, pp. 1661–1668.

Kyo-Soo Song, Seon-Gab Kim, Daehan Jung & Young-Ha Hwang, 2007. *Analysis of the fracture of a turbine blade on a turbojet engine*: Engineering Failure Analysis 14, pp. 877–883.

Lucjan Witek. 2011. *Stress intensity factor calculations for the compressor blade with half-elliptical surface crack using Raju-Newman solution*: Fatigue of Aircraft Structures, Vol. 1, pp. 154–165.

Nalla, et al., R.K. 2002. *Mixed-mode, high cycle fatigue-crack growth thresholds in Ti-6Al-4V: Role of small cracks*: International Journal of Fatigue.

Newman, J.C. & Raju, I.S. 1981. *An empirical stress-intensity factor equation for the surface crack*: Engineering Fracture Mechanics Vol. 15, No. 1–2, pp. 185–192.

Pierre Ladevèze, T. 2016. *Advanced Modeling and Simulation in Engineering Sciences*: Springer Open, 3.30.

Wilson Frost, et al., 2003. *Fourth-stage turbine bucket airfoil*: U.S. Patent 6503059 B1.

Numerical models and computational methods

Emerging Trends in Engineering, Science and Technology for Society,
Energy and Environment – Vanchipura & Jiji (Eds)
© 2018 Taylor & Francis Group, London, ISBN 978-0-8153-5760-5

Computational Fluid Dynamics (CFD) simulation of gas-liquid-solid three-phase fluidized bed

G.P. Deepak & K. Shaji
Government Engineering College, Kozhikode, Kerala, India

ABSTRACT: Three-phase fluidized beds have much significance since they offer excellent heat and mass transfer rates. Hence they are utilized in major industries such as biotechnology, pharmaceuticals, food, chemicals, and environmental and refining plants. Computational Fluid Dynamics (CFD) is an economical method by which to study the hydrodynamic properties of three-phase fluidized beds, because experimental and theoretical methods have their own limitations. A study of the hydrodynamics of a gas-liquid-solid (three-phase) fluidized bed has been made using ANSYS Fluent simulation software. The simulation has been carried out on a cylindrical column 1.8 m tall and 0.1 m diameter. Glass particles of diameter 2.18 mm and 3.05 mm were used for initial bed heights of 0.267 m and 0.367 m. The hydrodynamic properties, such as bed expansion, holdup of all three phases, and pressure drop across the column, were studied by varying inlet water velocity and inlet air velocity. Finally, comparison was made between the results obtained from the simulation and the experimental results. The CFD simulation result shows excellent agreement with the experimental results.

Keywords: Computational Fluid Dynamics, three-phase fluidized beds, simulation

1 INTRODUCTION

Fluidization is achieved by a continuous supply of fluid on a bed of solid particles. The upward drag force exerted on the solid particles by the fluid becomes exactly the same as the gravitational force at a particular value of fluid velocity and thus keeps the particles in a suspended state inside the fluid. After reaching this condition the bed has achieved a fluidized state and exhibits fluid-like behavior. The velocity of the fluid where fluidization is achieved is considered as the superficial fluid velocity.

Gas-liquid-solid fluidization uses three different phases where a bed of solid particles are held in the medium of two different fluids, as shown in Figure 1. This leads to excellent contact between different phases in the system.

Saha et al. (2016) presented a 2-dimensional Computational Fluid Dynamics (CFD) study of a three-phase fluidized bed. They studied holdups of liquid and gas in the column. They found that the gas holdup varies inversely with liquid velocity. By contrast, holdup of gas was found to vary directly with air velocity. Witt et al. (1998) implemented a multiphase Eulerian–Eulerian approach in estimating the transient performance of a fluidized bed. A body-fitted grid system was used to prevent the problem of exact geometrical representation. They studied the isothermal flow in a 3-D bubbling fluidized bed. Li et al. (1999) proposed a 2-dimensional model to study three-phase fluidization using a Eulerian–Lagrangian approach. The flow of gas was represented by a volume of fraction method, the flow of solid was represented by the Eulerian method, and the dispersed particle method was used to represent the liquid phase. They studied bubble wake behavior in the fluidized bed. Zhang and Ahmadi (2005) proposed

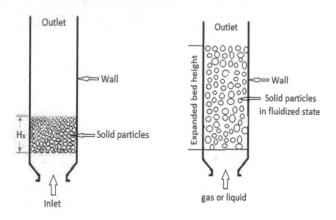

Figure 1. Three-phase fluidization.

a 2-D model for a three-phase slurry reactor by using the Eulerian–Lagrangian approach. They investigated the transient characteristics of flow of all the phases.

Sivalingam and Kannadasan (2009) carried out an experiment on a three-phase fluidized bed. They made an effort to study the relationship between fluid flow rates and hydrodynamic characteristics. The conclusion drawn from the experiment was that the gas flow rate influences the design of a fluidized bed. Sau and Biswal (2011) conducted an experimental study and a 2-dimensional CFD study. They made an attempt to study and compare the hydrodynamic properties of a two-phase tapered fluidized bed using the Eulerian approach. They concluded that 3-dimensional study would provide better results than 2-dimensional study. Mohammed et al. (2014) presented an experimental study on a three-phase fluidized bed by considering water, kerosene and spherical plastic particles as different phases. From the experiment, they found that the holdup of the dispersed phase increases with its velocity as well as particle size. However, it decreases with continuous phase velocity. Jena et al. (2008) presented an experimental study on a three-phase fluidized bed. They considered glass beads of different diameters as the solid phase. Liquid was taken as continuous phase. They found that the holdup of gas increased with gas velocity and it increased when particle size was increased.

From the brief review above of the literature, it is clear that several experimental studies have been done on three-phase fluidized beds. It was concluded that only a few works have been done on three-phase fluidized beds using CFD. For better comparison with experimental results, 3-dimensional is mostly preferred over 2-dimensional simulation. This is because 3-dimensional simulation gives more realistic results. Studying the three-phase fluidized bed using CFD is found to be a promising approach, with reduced cost and effort when compared to experiments.

In this work, the important hydrodynamic properties of a three-phase fluidized bed have been studied using CFD. For this, ANSYS Fluent simulation software is used to model and solve the problem using a Eulerian approach. The simulation results are validated by comparison with experimental results.

2 COMPUTATIONAL MODEL AND BOUNDARY CONDITIONS

A cylindrical column of 1.8 m height and 0.1 m diameter has been considered for the study. Glass, water and air are taken as solid, liquid and gas phases, respectively. Water is treated as a continuous phase. Secondary phases are glass particles and air. A uniform velocity inlet and pressure outlet boundary condition with mixture gauge pressure 0 Pa has been used. At the wall, no slip condition has been used for water. X = 0, Y = 0 specified shear was used for air and glass. The parameters used in the simulation are tabulated in Table 1.

Table 1. Important parameters.

Parameter	Value
Diameter of the column	0.1 m
Height of the column	1.8 m
Density of glass particles	2470 kg/m^3
Density of air	1.225 kg/m^3
Density of water	998.2 kg/m^3
Viscosity of air	1.789×10^{-5} kg/ms
Viscosity of water	0.001003 kg/ms
Mean particle size	2.18 mm, 3.05 mm
Initial bed height	0.267 m, 0.367 m
Inlet air velocity	0.02–0.08 m/s
Inlet water velocity	0.08–0.16 m/s
Initial solid holdup	0.6

2.1 Interphase momentum exchange

Interphase momentum exchange terms consist of different forces. Added mass force is negligible when compared to drag force. In most cases, lift force is insignificant. Hence, simulation was carried out by assuming that drag force acts as the only interphase momentum exchange term.

2.1.1 Drag models

Schiller–Neumann correlation has been used for liquid–air interaction and a Gidaspow model has been used for both solid–air and solid–liquid interactions (ANSYS, 2003).

2.2 Solid pressure

The pressure arising through the collision of solid particles was calculated using the Kinetic Theory of Granular Flow (KTGF) (ANSYS, 2003).

2.3 Turbulence models

Transport equations for the turbulent kinetic energy k and its dissipation rate ε have been used for the study. The turbulence has been modeled using a realizable k-ε model. A high Reynolds number is assumed in this model (ANSYS, 2003).

2.4 Equations

2.4.1 Continuity equation

$$\frac{\partial}{\partial t}(\varepsilon_k \rho_k) + \nabla(\varepsilon_k \rho_k \vec{u}_k) = 0 \tag{1}$$

where

$$\varepsilon_g + \varepsilon_l + \varepsilon_s = 1 \tag{2}$$

2.4.2 Momentum equation
For the liquid phase:

$$\frac{\partial}{\partial t}(\rho_l \varepsilon_l \vec{u}_l) + \nabla(\rho_l \varepsilon_l \vec{u}_l \vec{u}_l) - \varepsilon_l \nabla P + \nabla(\varepsilon_l \mu_{\mathrm{eff},l}(\nabla \vec{u}_l + (\nabla \vec{u}_l)^\mathrm{T}) + \rho_l \varepsilon_l g + \mathrm{M}_{i,l} \tag{3}$$

For the gas phase:

$$\frac{\partial}{\partial t}\left(\rho_g \varepsilon_g \vec{u}_g\right) + \nabla\left(\rho_g \varepsilon_g \vec{u}_g \vec{u}_g\right) = -\varepsilon_g \nabla P + \nabla(\varepsilon_g \mu_{eff,g}\left(\nabla \vec{u}_g + \left(\nabla \vec{u}_g\right)^T\right) + \rho_g \varepsilon_g g - M_{i,g} \tag{4}$$

For the solid phase:

$$\frac{\partial}{\partial t}(\rho_s \varepsilon_s \vec{u}_s) + \nabla\,(\rho_s \varepsilon_s \vec{u}_s \vec{u}_s) = -\varepsilon_s \nabla P + \nabla\,(\varepsilon_s \mu_{eff,s}(\nabla \vec{u}_s + (\nabla \vec{u}_s)^T) + \rho_s \varepsilon_s g + M_{i,s} \tag{5}$$

3 GRID INDEPENDENT STUDY

Grid independent study has a major role in determining the time required for the simulation. As the number of elements/cells decreases, simulation time reduces. But a decreased number of elements leads to variation from the expected results. Hence it is important to find the number of elements which leads to accurate results with a minimum number of iterations. Simulations were performed for different numbers of elements. The values of pressure drop obtained are shown in Figure 2.

From Figure 2 it is clear that as the number of cells increases the value of pressure drop also increases. With 55390 cells, the pressure drop obtained is 1380 Pa, which is close to the experimental result with the same boundary conditions. When the number of cells was increased further, the variation in the value of pressure drop is smaller. Hence the mesh with 55390 cells was chosen for the calculation to minimize the simulation time.

4 RESULTS AND DISCUSSION

A fluidized bed which uses liquid as primary phase, and gas and solid particles as secondary phases has been modeled and solved using ANSYS Fluent. Bed heights of 0.267 m and 0.367 m were used to study the fluidized bed with 2.18 mm and 3.05 mm glass beads. The fluidized bed has been studied by varying inlet superficial velocity of water and inlet air velocity.

During the simulation of the fluidized bed, variation in the profile of the bed is observed, as shown in Figure 3. But no major change in the profile is seen after some time. This is the proof that the bed is in a fluidized condition.

4.1 Dynamics of individual phases

Figure 4 shows the contours of phase dynamics of individual phases, that is, glass, water and air, after reaching quasi-steady state.

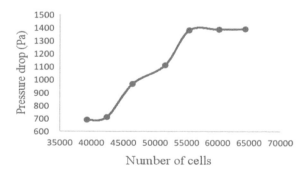

Figure 2. Values of pressure drop for different numbers of cells (V_l = 0.04 m/s, V_g = 0.02 m/s).

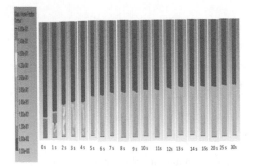

Figure 3. Contours showing the variation of volume fraction of glass particles with time (H_s = 0.267 m, V_w = 0.12 m/s, V_g = 0.02 m/s, D_p = 3.05 mm).

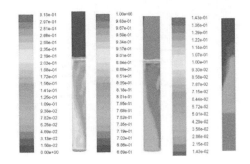

Figure 4. Contours of volume fraction of glass, water and air (V_w = 0.12 m/s, V_g = 0.02 m/s, H_s = 0.267 m, D_p = 3.05 mm).

It is clear from the contour that the bed has reached a fluidized state. The water volume fraction is relatively smaller in the fluidized region and that of air is greater when compared to the remaining part.

4.2 Bed expansion

The change in bed expansion with inlet air velocity at various inlet water velocities is shown in Figure 5. Only slight variation in bed expansion is observed with inlet air velocity. Bed expansion increases as the inlet water velocity increases. This is because the force exerted by liquid at lower inlet water velocities is not sufficient enough to overcome the gravitational pull on the glass particles.

The plot of change in bed expansion with inlet water velocity at different heights of bed is shown in Figure 6. Bed expansion increases when bed height is increased.

Figure 7 is the plot of the behavior of bed expansion with inlet velocity of water for inlet air velocity 0.02 m/s for 2.18 mm and 3.05 mm particles at bed height 0.367 m. It can be seen that the bed expansion is lower for higher particle size.

Figure 8 is the plot of comparison of experimental and simulation results of bed expansion ratio for glass particles of diameter 3.05 mm at initial bed height of 0.267 m for air velocity of 0.02 m/s. From the comparison curve, it can be seen that the values of the simulation make a satisfactory match with experimental results.

4.3 Phase holdups

Holdup of individual phase is the mean area weighted average of volume fraction in the fluidized region. Average value is taken as holdup because the values differ at different locations of the fluidized part.

Figure 9 shows the representation of variation of air holdup with water velocity at an initial bed height of 0.267 m for various air velocities for particles of 3.05 mm. It is clear that air holdup decreases when water velocity is increased, and it increases with increased air velocity.

Figure 10 shows the behavior of glass holdup with inlet water velocity for glass particles of diameter 3.05 mm by keeping 0.267 m bed height. Glass holdup decreases when water velocity is increased by keeping inlet air velocity constant. When air velocity is increased, glass holdup decreases slightly.

The variation of water holdup with water velocity for glass particles of diameter 3.05 mm at 0.267 m bed height is plotted in Figure 11. Water holdup increases when the inlet water velocity is increased. And water holdup decreases when inlet air velocity is increased.

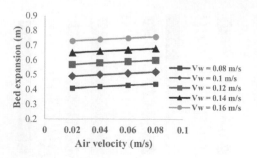

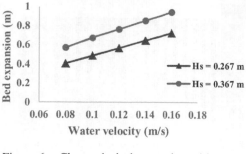

Figure 5. Variation of bed expansion with inlet air velocity ($H_s = 0.267$ m, $D_p = 3.05$ mm).

Figure 6. Change in bed expansion with water velocity at altered bed heights ($V_g = 0.02$ m/s, $D_p = 3.05$ mm).

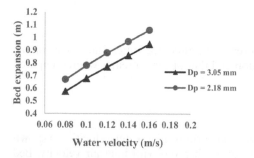

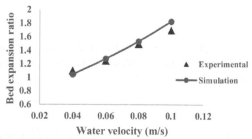

Figure 7. Change in bed expansion with water velocity for various particle sizes ($H_s = 0.367$ m, $V_g = 0.02$ m/s).

Figure 8. Variation of bed expansion ratio with water velocity ($D_p = 3.05$ mm, $H_s = 0.267$ m, $V_g = 0.02$ m/s).

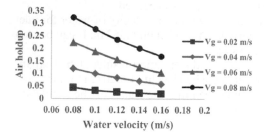

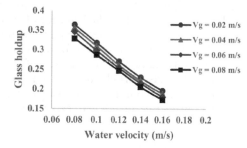

Figure 9. Change in air holdup with water velocity at various air velocities ($D_p = 3.05$ mm, $H_s = 0.267$ m).

Figure 10. Variation of glass holdup with water velocity at various air velocities ($D_p = 3.05$ mm, $H_s = 0.267$ m).

The evaluation of experimental and simulation results of air holdup is made in Figure 12 for glass particles of 3.05 mm at 0.267 m bed height for inlet air velocity of 0.02 m/s. The match between simulation result and experimental result is excellent.

4.4 *Pressure drop*

The difference of pressures at inlet and outlet is known as pressure drop. The pressure drop variation with inlet water velocity for various inlet air velocities for glass particles of diameter 3.05 mm and 0.267 m bed height is shown in Figure 13.

The pressure drop varies directly with water velocity. At lower air velocities, the operation becomes liquid-fluid because the volume fraction of air is too low and hence the pres-

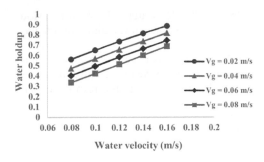

Figure 11. Variation of water holdup with water velocity at various air velocities (D_p = 3.05 mm, H_s = 0.267 m).

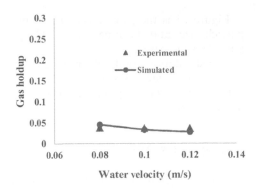

Figure 12. Variation of air holdup with water velocity (V_g = 0.02 m/s, D_p = 3.05 mm, H_s = 0.267 m).

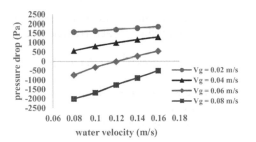

Figure 13. Change in pressure drop with water velocity for various air velocities (D_p = 3.05 mm, H_s = 0.267 m).

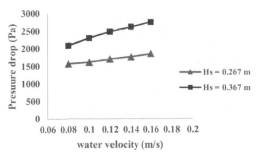

Figure 14. Change in pressure drop with water velocity at various bed heights (D_p = 3.05 mm, V_g = 0.02 m/s).

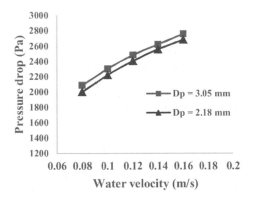

Figure 15. Change in pressure drop with water velocity for various particles (H_s = 0.367 m, V_g = 0.02 m/s).

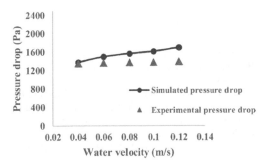

Figure 16. Variation of pressure drop with water velocity (V_g = 0.02 m/s, D_p = 3.05 mm, H_s = 0.267 m).

sure drop does not increase much. Drop in pressure varies inversely with air inlet velocity. This is because as the inlet air velocity increases gas holdup increases, and water holdup in the column decreases. Because the density of air is much lower than that of water, as water holdup decreases pressure drop also decreases.

Figure 14 shows the behavior of pressure drop across the column with inlet water for 3.05 mm glass particles at 0.267 m and 0.367 m bed heights. The pressure drop increases when the initial bed height increases.

Figure 15 is the plot of changes in pressure drop with inlet water velocity for various particle sizes at 0.367 m bed height. The pressure drop increases when the diameter of glass particle is increased.

The simulation results of pressure drop are validated with experimental results for glass particles of diameter 3.05 mm at 0.267 m bed height for 0.02 m/s air velocity. The comparison curves are shown in Figure 16. The simulation results differ by 12% or less from the experimental results.

5 CONCLUSIONS

A study of a gas-liquid-solid three-phase fluidized bed has been made using a Eulerian–Eulerian approach. A 3-dimensional model having 1.8 m height and 0.1 m diameter was developed. The hydrodynamic properties, such as bed expansion, air holdup, water holdup, and pressure drop, were studied by varying inlet water velocity, inlet air velocity, particle diameter and bed height. The major conclusions from the study are summarized as follows:

- Bed expansion is directly proportional to water velocity. It is not much affected by inlet air velocity. It increases when static bed height is increased and reduces when particle size is increased.
- Gas holdup is directly proportional to inlet air velocity. It decreases when inlet water velocity is increased.
- Glass holdup decreases when inlet water velocity is increased and only slight variation is observed in glass holdup when inlet air velocity is increased.
- Water holdup varies directly with inlet water velocity and it decreases with an increase in air velocity.
- Pressure drop varies directly with inlet water velocity. It varies inversely with inlet air velocity. Pressure drop increases when initial bed height and particle diameters were increased.
- The simulation results show excellent agreement with experimental results.

REFERENCES

ANSYS. (2003). *Fluent 6.1 User's Guide* (pp. 1–5). Canonsburg, PA: ANSYS.
Blazek, J. (2001). *Computational fluid dynamics: Principles and applications* (1st ed.). Oxford, UK: Elsevier.
Jena, H.M., Sahoo, B.K., Roy, G.K. & Meikap, B.C. (2008). Characterization of hydrodynamic properties of a gas-liquid-solid three-phase fluidized bed with regular shape spherical glass bead particles. *Chemical Engineering Journal, 145*, 50–56.
Li, Y., Zhang, J. & Fan, L.S. (1999). Numerical simulation of gas-liquid-solid fluidization system using a combined CFD-VOF-DPM method: Bubble wake behaviour. *Chemical Engineering Science, 54*, 5101–5107.
Mohammed, T.J., Sulaymon, A.H. & Abdul-Rahmun, A.A. (2014). Hydrodynamic characteristic of three phase (liquid-liquid-solid) fluidized beds. *Journal of Chemical Engineering and Process Technology, 5*, 188.
Saha, S.N. Dewangan, G.P. & Gadhewal, R. (2016). Gas-liquid-solid fluidized bed simulation. *International Journal of Advanced Research in Chemical Science, 3*, 1–8.
Sau, D.C. & Biswal, K.C. (2011). Computational fluid dynamics and experimental study of the hydrodynamics of a gas–solid tapered fluidized bed. *Applied Mathematical Modelling, 35*, 2265–2278.
Sivalingam, A. & Kannadasan, T. (2009). Effect of fluid flow rates on hydrodynamic characteristics of co-current three phase fluidized beds with spherical glass bead particles. *International Journal of ChemTech Research, 1*, 851–855.
Witt, P.J., Perry, J.H. & Schwarz, M.P. (1998). A numerical model for predicting bubble formation in a 3D fluidized bed. *Applied Mathematical Modelling, 22*, 1071–1080.
Zhang, X. & Ahmadi, G. (2005). Eulerian–Lagrangian simulations of liquid-gas-solid flows in three-phase slurry reactors. *Chemical Engineering Science, 60*, 5089–5104.

Emerging Trends in Engineering, Science and Technology for Society,
Energy and Environment – Vanchipura & Jiji (Eds)
© 2018 Taylor & Francis Group, London, ISBN 978-0-8153-5760-5

Fault diagnosis of self-aligning troughing rollers in a belt conveyor system using an artificial neural network and Naive Bayes algorithm

S. Ravikumar
GKM College of Engineering and Technology, Tamilnadu, India

S. Kanagasabapathy
National Engineering College, Kovilpatti, Tamilnadu, India

V. Muralidharan
B.S. Abdur Rahman University, Chennai, Tamilnadu, India

R.S. Srijith & M. Bimalkumar
GKM College of Engineering and Technology, Tamilnadu, India

ABSTRACT: The belt conveyor system is used for conveying large volumes of materials from one location to another. The Self-Aligning Troughing Roller (SATR) is one of the critical components in the belt conveyor; it is quite influential in riding the belt conveyor in fault free conditions. SATR has to operate under heavy axial and shear forces, which lead to frequent failures. Hence continuous monitoring and fault diagnosis of SATR becomes essential. The self-aligning troughing idler arrangement has a long roll to support the belt and handle maximum loads per cross section. The self aligning troughing roller has machine elements, including the ball bearing, a central shaft and the external shell. In the belt conveyor system certain faults, such as Bearing Flaws (BF), Central Shaft Faults (CSF), combined Bearing Flaws and Central Shaft Faults (BF & CSF) occur frequently. A prototype investigational model has been made with the above mentioned faults and the vibration signals were attained from the set-up. The vibration data acquired was fed as algorithm input into Artificial Neural Networks (ANN) and Naive Bayes (NB) algorithms, which are used for classification of acquired signals. In the present effort, the artificial neural networks and Naive Bayes algorithms were found to achieve 82.1 and 90% classification accuracy, which acknowledges that the Naive Bayes algorithm has a better gain over artificial neural networks in the field of fault diagnosis applications.

1 INTRODUCTION

The self-aligning troughing roller (SATR) is an essential element of the belt conveyor system. It may fail due to multidimensional forces, derisory lubrication, culpable sealing, uneven loading and improper training of belt. The critical elements that fail periodically in the self-aligning troughing roller are the groove ball bearing and the central shaft.

The malfunction of these parts directly affects the efficiency of the SATR, which can hinder the proper functioning of the belt conveyor system. In these circumstances, to avoid overwhelming damage of the belt conveyor, a failure prediction system is a major requirement. The various conditions for this research are SATR running in Fault Free Condition (FFC), Bearing Fault Condition (BFC), Central Shaft Fault (CSF) and Bearing Fault and Central Shaft Fault (BFC & CSF). The malfunction of these components affects the functioning of SATR which in turn leads to under-performance of the belt conveyor system.

The conventional and Fast Fourier Transform (FFT) methods work well when signals are static, but in continuous varying conditions they are ineffective. Here the components of SATR generate vibration signals with significant variation. These signals are acquired and the selection features are extracted from it. The selection of these features is based on their impact in fault prediction, which is the consequent stage of SATR condition monitoring. Apart from this, a good quality fault diagnosis algorithm tool has to be utilized for classification. At present there are a fair number of classification algorithms, each having their own pros and cons.

In the present case artificial neural networks (ANN) and Naive Bayes (NB) algorithms are selected for analysis. The classification correctness differs among algorithms, likewise ANN and NB were found to achieve different performance levels. Thus, it is necessary to select a relevant algorithm that can be used to assess the condition monitoring and fault diagnosing of SATR.

2 RELATED WORK

Murru (2016) presented an original algorithm for initialization of weights in back propagation neural net with application to character recognition. The initialization method was mainly based on a customization of the Kalman filter, translating it into Bayesian statistics terms. A metrological approach was used in this context considering weights as measurements, modeled by mutually dependent normal random variables. The algorithm performance was demonstrated by reporting and discussing results of simulation trials. Results were compared with random weights initialization and other methods. The proposed method showed an improved convergence rate for the back propagation training algorithm.

Wong et al. (2016) proposed a Probabilistic Committee Machine (PCM), which combines feature extraction, a parameter optimization algorithm and multiple Sparse Bayesian Extreme Learning Machines (SBELM) to form an intelligent diagnostic framework. Results showed that the proposed framework was superior to the existing single probabilistic classifier. Zhang et al. (2016) developed a Bayesian statistical approach developed for modal identification using the free vibration response of structures. The results indicated that a frequency-domain Bayesian framework was created for modal identification of Most Probable Values (MPVs) and modal parameters Mori & Mahalec (2016) introduced a decision tree structured conditional probability representation that can efficiently handle a large domain of discrete and continuous variables. Experimental results indicated that our method was able to handle the large domain discrete variables without increasing computational cost exponentially.

Hu et al. (2016) developed the framework of Non-Negative Sparse Bayesian Learning (NNSBL). The algorithm obviated pre-setting any hyper parameter, where the Expectation Maximization (EM) algorithm was exploited for solving this NNSBL problem. Without a prior knowledge of the source number, the proposed method yielded performances in the underdetermined condition illustrated by numerical simulations.

Kiaee et al. (2016) utilized the concept of random effects in the Extreme Learning Machine (ELM) framework to model inter-cluster heterogeneity, provided the inherent correlation among the samples of a particular cluster is taken into account, as well. The proposed random effect model includes additional variance components to accommodate correlated data. Inference techniques based on the Bayesian evidence procedure were derived for the estimation of model weights, random effect and residual variance parameters as well as for hyper parameters. The proposed model is applied to both synthesis and real-world clustered datasets. Experimental results showed that our proposed method can achieve better performance in terms of accuracy and model size, compared with the previous ELM-based models.

Wang et al. (2016) used the Gaussian kernel function with smoothing parameter to estimate the density of attributes. A Bayesian network classifier with continuous attributes was established by the dependency extension of Naive Bayes classifiers. The information provided to a class for each attributes as a basis for the dependency extension of Naive Bayes classifiers is analyzed. Experimental studies on UCI datasets showed that Bayesian network classifiers using Gaussian kernel function provided good classification accuracy compared to the approaches when dealing with continuous attributes. Magnant, et al. (2016) proposed

Bayesian non-parametric models. It showed the possible functional forms of the state noise covariance matrices, which eliminate the number of time switching hyper parameters in many applications. Results showed that the online estimation of the state noise precision matrix used by DP methods allowed improvisation of tracking accuracy.

Muralidharan & Sugumaran (2016) have introduced a more logical algorithm for fault diagnosis of the centrifugal pump. The extracted features were classified and analyzed by SVM and Extreme Learning Machine (ELM). Finally it is concluded that ELM has higher accuracy than SVM. Muralidharan et al. (2015) elucidated, that the centrifugal pump fault can be identified using the pattern detection method. The input signal is extracted through Stationary Wavelet Transform (SWT) and classification is made through Bayes net. Finally it is concluded that the Bayes net is better for fault diagnosis of the centrifugal type monoblock pump.

3 EXPERIMENTAL SET-UP

SATR fault diagnosis involves several steps regarding the conveyor set-up: (i) design and fabrication with multiple fault conditions, (ii) acquisition of signals, (iii) feature extraction and (iv) feature classification. The procedure of the process can be clearly understood from Figure 1. Initially, the belt conveyor model is allowed to run with parts working in a fault free condition and the signals were acquired. One set of shaft and bearing are prefabricated with faults. As given in Table 1, the outer ring thickness of 4.5 mm and 4.51 mm respectively were ground for 4 mm and 3.90 mm respectively for developing faults in the groove bearings.

The bearing was attached to a self-aligning troughing roller set-up in the conveyor system. Similarly the central shaft was ground to create shaft fault as shown in Table 2.

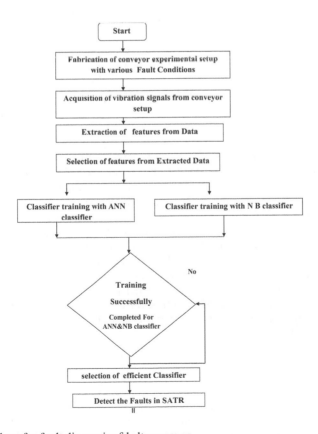

Figure 1.   Flowchart for fault diagnosis of belt conveyor.

Table 1. Groove bearing diameter readings.

| Sl. No. | Bearing outer ring thickness mm | |
	Before grinding	After grinding
1	4.50 (LHS side)	4.0
2	4.51 (RHS side)	3.90

Table 2. Central shaft diameter readings.

| Sl. No. | Diameter of shaft mm | | |
	Before grinding	After grinding	Side
1	10.01	9.88	Left Side
2	10.00	9.90	
3	10.00	9.98	Right Side
4	9.99	9.99	

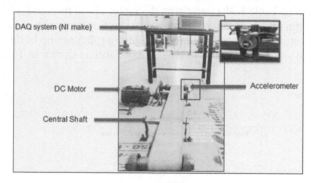

Figure 2. Experimental set-up.

The roller bearing (Model No KG6200Z) and central shaft were prefabricated with a fault to acquire the vibration readings as given in Tables 1 and 2. The different fault conditions, like SATR having Bearing Fault (BF), central shaft fault (CSF) and combined central shaft and bearing fault (CSF & BF) were subsequently set up one by one and the corresponding vibration signals were acquired. Figure 2 shows a schematic arrangement about the SATR vibration analysis experimental set-up. A piezoelectric accelerometer sensor (Model No 3055B1) was mounted over the vibration zone to absorb the vibrations generated due to faults.

A signal conditioning unit was connected to the accelerometer sensor which in turn was connected to an Analog-to-Digital Converter (ADC). The digital vibration signals acquired from the ADC were fed to the computer for further processing through relevant tools. The LabVIEW software was used to record the vibration signals in the digital form and store them in the computer hard disk memory. It is further processed different features were extracted by add-on in Microsoft Excel.

4 FEATURE EXTRACTION

Different statistical parameters are used to show the fault conditions in the fabricated model. The various statistical parameters include mean, median, mode, standard error, standard deviation, kurtosis, skewness, minimum value, maximum value, sample variance and range.

Table 3. Individual feature classification accuracy.

Individual feature classification accuracy (%)	
Standard deviation	71
Skewness	63.45
Sample variance	50
Standard error	44.66
Range	38.01

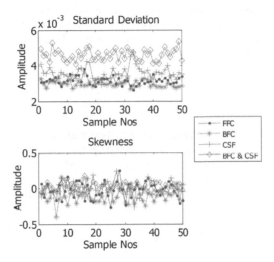

Figure 3. Sample plots of selected features with different conditions.

4.1 Selection of important features

In this stage, eight common features were chosen from the acquired vibration signals in order to see the difference between fault and fault free conditions. The classification accuracy of each feature was noted down. The feature with high accuracy of 71% was found be with standard deviation. The next highest value of 63.45% was found to be with skewness. Hence, when the second feature was combined with the first one, then the classification accuracy followed an increasing trend. Similarly, the next highest value was found to be with sample variance and hence it was combined with the other two higher accuracy features.

The trend was found to be increasing. The procedure was repeated for all the features separately and the features with more competence to classify the faults were identified.

The selected features are standard deviation, skewness, sample variance, and standard error, which have high classification accuracy over the remaining statistical features as can be seen in Figure 3.

5 CLASSIFIER

5.1 Artificial neural network

Artificial neural network (ANN) is a method that has been developed with the concept of genetic neurons of the nervous systems. It is capable of self-interpreting the signals that are provided as inputs.

It has hidden layers between the input and output layers. The input signals are learned and interpreted by means of activation functions, which are associated with every hidden layer. Usually, the processing is performed by multilayer perceptrons.

5.2 Multi-Layer Perceptron (MLP)

An MLP can be viewed as a logistic regression can learn nonlinear transformation. This transformation projects the input data into a space where it becomes linearly separable. This intermediate layer is referred to as a hidden layer. A single hidden layer is sufficient to make MLPs a universal approximator. However, we will see later on that there are substantial benefits to using many such hidden layers, i.e. the very premise of deep learning. Naive Bayes.

5.3 Naive Bayes

Naive Bayes method is a powerful algorithm for the classification of sample by a set of supervised learning algorithms. It is applied based on Bayes' theorem with the "naive" assumption of independence between every pair of features. Given a class variable y and a dependent feature vector x 1 through x n, Bayes' theorem states the following relationship:

$$P(y/x_1,...x_n) = P(y)P\left(\frac{P(y/x_1,...x_{n/y})}{P(x_1,...x_n)} \right)$$

(1)

It can be used for a Maximum A Posteriori (MAP) estimation to estimate P(y) and P(x$_i$\y); the former is then the relative frequency of class *y* in the training set. The Naive Bayes classifiers differ mainly by the assumptions they make regarding the distribution of P(x$_i$\y) in spite of their apparently over-simplified assumptions. Naive Bayes classifiers have worked quite well in many real-world situations, particularly for document classification and spam filtering. They require a small amount of training data to estimate the necessary parameters. Naive Bayes learners and classifiers can be extremely fast compared with more sophisticated methods. The decoupling of the class conditional feature distributions means that each distribution can be independently estimated as a one dimensional distribution. This in turn helps to alleviate problems stemming from the curse of dimensionality.

6 RESULTS AND DISCUSSION

The vibration signals were acquired for the various faults, like fault free condition, central shaft fault, bearing fault condition and combined central shaft and bearing faults. A total of 250 data points were taken for each fault condition. Furthermore, the sample was split into two equal parts with the first phase for training followed by testing. In each phase 125 signals were taken for classification.

In section 4 and 4.1, feature extraction, feature selection has been discussed. Of the existing statistical eleven features were suggested for classification. However, the best ones were selected for the following reasons:

- To avoid unnecessary computation and poor results (dimensionality reduction).
- To save time and build the robust model.

For the classification and validation Naive Bayes and ANN algorithm have been utilized and the results were discussed. The effectiveness of the sample quality is understood by the True Positive (TP) rate and False Positive (FP) rate in Table 4. A quality classification has a value approaching '0' for the false positive (FP) and for true positive the value has to be near to 1. In Table 4 the TP value is very significant as it approaches 1 and the FP value is near 0, which highlights the quality of this classification. In addition, the classified data may be exhibited in the form of a confusion matrix as indicated in Table 5.

The important features that are highly participative in deciding the various faults of the SATR are standard deviation, skewness, standard variance, and standard error is vital in deciding the faults. The standard deviation determines how much variability is sorted in a coefficient estimate. A coefficient is significant if it is non-zero. Standard deviation is used to measure the number of faults around and non-faulty conditions. The higher the standard

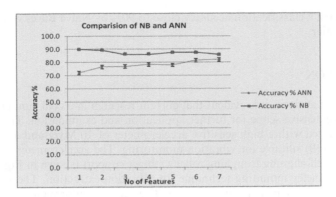

Figure 4.   Comparison of NB and ANN for different number of features.

Table 4.   Individual feature classification accuracy.

TP rate	FP rate	Precision	Recall	F measure	ROC area	Class
0.945	0.062	0.738	0.745	0.853	0.861	FFC
0.854	0.073	0.053	0.654	0.672	0.836	CSF
1	0	1	1	1	1	BFC
0.841	0.036	0.891	0.843	0.824	0.885	BFC & CSF
Weight Avg.	0.882	0.047	0.784	0.782	0.782	−

Table 5.   Confusion matrix.

	FFC	CSF	BFC	BFC & CSF
FFC	212	38	0	0
CSF	33	193	0	22
BFC	0	2	248	0
BFC & CSF	0	14	0	236

deviation value the larger the gap between faulty conditions and good condition. Skewness measures the symmetry in the samples.

When the skewness value reaches zero, it indicates an error free condition. The critical features selected for classification has shown a clear margin from each other, which substantiate the selection of these features (Figure 3). The results of the 1,000 samples were reviewed with the help of a confusion matrix. The understanding of the confusion matrix is essential before exposing the same which is presented in Table 5.

The fault free condition (FFC) is represented in the first row, followed by central shaft fault (CSF) subsequently the bearing fault condition (BFC) and the combined fault condition (CSF & BFC) was shown in the third and fourth rows in the table.

It was obvious from the confusion matrix (Table 5), that 250 samples were taken for the different conditions of the self-aligning training roller. Those diagonal elements of the confusion matrix speak about the effectively ordered information and the incorrectly classified data points are positioned as non-diagonal elements. This is how the classification accuracies are predicted from the confusion matrix.

In this case, 212 fault free condition (FFC) data has been effectively ordered and the remaining 36 data demonstrate as central shaft fault (CSF). Similarly, 193 data points of central shaft fault (CSF) have been effectively ordered and 38 ineffectively ordered as fault free conditions. In this manner, the confusion matrix is inferred and the classification precision obtained was 90%. These results are for a particular dataset hence, the classification accuracy of 90% may guarantee a similar performance for all similar feature data. Furthermore it is possible to expect a classification accuracy close to 90% by Naive Bayes when compared to the ANN classifier.

7 CONCLUSION

From the current analysis, it is evident that current research in a coal handling belt conveyors SATR has extensive scope for further application and examination. The prototype setup has been created with a high sensitive accelerometer of 10 Mv/g and a frequency range of 0–2000 Hz (3dB) suitable for vibration monitoring. The accelerometer was hermetically mounted at the self-aligning troughing roller's stinger support (shown in Figure 2), which is an ideal accelerometer mounting technique for vibration extraction. The vibration signals were acquired using a data acquisition system. The acquired information has been preprocessed to extract those measurable features. The superlative features were distinguished utilizing the Naive Bayes algorithm and the faults were classified with them.

Since the information has been acquired in a particular working state, the end result may be comprehensive for similar cases. Aiming at the shortcomings of the conventional failure analysis of self-aligning troughing roller, the methodology adopted would definitely serve as a guideline for the future research work on this area. However, the classification accuracy of 90% is significant in this application. It can be concluded that the statistical features and Naive Bayes algorithms are the best options for fault diagnosis of self-aligning troughing roller in a bulk material handling belt conveyors.

REFERENCES

Hu, B., Sun, J., Wang, J., Dai, C. & Chang, Localization for sparse array using nonnegative sparse Bayesian learning. *Signal Processing, 127,* 37–43.

Kiaee, F., Sheikhzadeh, H. & Eftekhari Mahabadi, S. (2016). Sparse Bayesian mixed-effects extreme learning machine, an approach for unobserved clustered heterogeneity. *Neurocomputing 175,* 411–420.

Magnant, C., Giremus, A., Grivel, E., Ratton, L. & Joseph, B. (2016). Bayesian non-parametric methods for dynamic state-noise covariance matrix estimation: Application to target tracking. *Signal Processing, 127,* 135–150.

Mori, V. Mahalec, (2016). Inference in hybrid Bayesian networks with large discrete and continuous domains. *Expert Systems with Applications, 49,* 1–19.

Muralidharan V., Sugumaran, V. & Sakthivel, N. (2014). Fault diagnosis of monoblock centrifugal pump using stationary wavelet features and Bayes algorithm. *Asian Journal of Science and Applied Technology, 3,* 1–4.

Muralidharan, V. & Sugumaran, V. (2017). A comparative study between Support Vector Machine (SVM) and Extreme Learning Machine (ELM) for fault detection in pumps. *Indian Journal of Science and Technology, 9,* 1–4.

Murru, N. & Rossini, R. (2016). A Bayesian approach for initialization of weights in backpropagation neural net with application to character recognition. *Neurocomputing, 193,* 92–105.

Wang, R., Gao, L.-M. & Wang, (2016). Bayesian network classifiers based on Gaussian kernel density. *Expert Systems with Applications, 51,* 207–217.

Wong, P.-K., Zhong, J., Yang, Z.-X. & Vong, C.-M. (2016). Sparse Bayesian extreme learning committee machine for engine simultaneous fault diagnosis. *Neurocomputing, 174,* 331–343.

Zhang, J. Chen, Z. Cheng, P. & Huang, X. (2016). Multiple-measurement vector based implementation for single-measurement vector sparse Bayesian learning with reduced complexity, *Signal Processing, 118,* 153–158.

Other topics related to mechanical engineering

Emerging Trends in Engineering, Science and Technology for Society,
Energy and Environment – Vanchipura & Jiji (Eds)
© 2018 Taylor & Francis Group, London, ISBN 978-0-8153-5760-5

Structural analysis of midship section using finite element method

Sirosh Prakash
Department of Naval Architecture and Shipbuilding Engineering, Sree Narayana Gurukulam College of Engineering, Kolenchery, Ernakulam, India

K.K. Smitha
Department of Civil Engineering, Sree Narayana Gurukulam College of Engineering, Kolenchery, Ernakulam, India

ABSTRACT: The design used in the present work is from the conceptual design of a container ship and includes preliminary analysis of the structural design of the midship section. The main objective of the work is to study the structural response of the midship section to static loading. The present work is carried out in ANSYS, which is well-known finite element modeling software. The first step is to produce the required model in ANSYS, which we can either model in ANSYS or can import from any CAD software. For this analysis, the model has been developed in ANSYS Parametric Design Language (APDL). As it is a preliminary analysis, there are many assumptions, such as reducing the number of girders and longitudinal stiffeners. The different static forces taken into account are the hydrostatic pressure acting on the ship's hull, the self-weight of structural members, and the loads from containers. There are two approaches for load application: the load combination method and the resultant force method. In the first method, we apply the forces acting on each structural member. In the second method, we find the resultant forces and these are applied on some respective structural members.

Keywords: structural analysis, finite element method, midship

1 INTRODUCTION

Ships when exposed to sea will undergo different kind of forces, including deformation, so it is necessary do a preliminary analysis to find the response of ship structures. Generally, there are two types of loads acting on a ship, dynamic loads and static loads. Ship structural design is a challenging task during the shipbuilding process. The structural design should fulfill two main objectives. One is to design the ship structure to withstand the loads acting on it; the other is to design the structural members economically.

The next step in structural design is the evaluation of loads, nature of loads, and so on. The initial structural dimensions are fixed according to stress analysis of beams, plates and the shell under hydrostatic pressure, bending and concentrated loads. The loads that strongly affect the deformation of hull girders are hydrostatic pressure and cargo loads (Eyres, 1988; Taggart, 1980; Souadji, 2012).

2 MODEL DEVELOPMENT IN ANSYS SOFTWARE

The modeling procedure using ANSYS software can be divided into three parts, that is, the modeling of the side shell, the modeling of side girders, and the modeling of the bulkhead. The following steps illustrate the process of model development:

1. First, it is necessary to specify the type of analysis, whether it is thermal, structural, magnetic or electric. For our model, we have chosen structural analysis.
2. The next step is to specify the element. This software tab includes different types of elements: solid, beam, shell and link. For 3D modeling, an 8-noded brick element has been chosen.
3. For modeling the midship structure, the side shell must first be created, and the key point must be defined. The key point is positioned according to the coordinate values of the side shell model.
4. After modeling the side shell, the bottom structure, including center girders and side girders, has to be modeled. The inputs for modeling the girders are the thickness of the plate and the height of the girders. The four key points of the girders are provided. These four key points are joined using the *Line* command in modeling. The plane obtained is extruded to the required dimensions.
5. The modeling of the bulkhead starts by modeling a plate, which is obtained by defining key points that are joined by lines to create a default area. This area is extruded about an axis to model the plate. Once the plate for the bulkhead is created, modeling of vertical frames and horizontal stiffeners are carried out (Souadji, 2012).

3 MESHING IN ANSYS

In meshing, the surface and volumes are divided into a number of elements using nodes. The accuracy of the meshed model varies according to the element selected (e.g. 10-node, 20-node and 4-node quad). The accuracy also depends on the type of meshing—coarse or fine. A coarse mesh gives poor results, because the effect of continuity is reduced. Hence, to obtain near-perfect results it is necessary to use a fine mesh. The mesh can be modified using the *Refine* option. The mesh is refined with respect to the nodes, elements, area, and volumes. It is refined near to the edges and the joints to obtain better results. Figure 1 shows a meshed model of a structure.

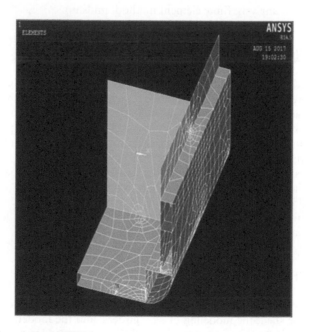

Figure 1. Meshed model in ANSYS.

4 APPLYING LOADS IN ANSYS

Once the meshing is completed we can apply various loads on the structures of ship. The side shell of the ship will be subjected to hydrostatic pressure. This pressure can be calculated using the formula ρgh, where ρ is the density of water, g is the acceleration due to gravity, and h is the height of the water. This can be applied as a pressure load to the bottom and side shell up to the draft level. Using ANSYS, we can apply pressure loads, forces, moments, uniformly distributed loads (UDLs), and so on. The boundary conditions can be applied by specifying the displacements UX, UY and UZ, and the rotations RX, RY and RZ. The upper side of the bulkhead will be under the loads from decks. These loads can be applied as a uniformly distributed load. The forces are applied to selected elements by selecting the nodes, areas, and elements.

5 APPLYING LOADS IN ANSYS

Analysis of a container ship carried out by Souadji (2012) has been taken as the basis for the present analysis. The details of ship cross section are shown in Table 1 and Figure 2.

Structural analysis in ANSYS involves various steps, including modeling, meshing, load application, obtaining the solution, and understanding the output. The various steps used for the present structural analysis were as follows:

a. Assigning the type of elements used; for this work, shell type element SHELL 4 NODE 181 was used.
b. Assigning the material properties, the modulus of elasticity, Poisson's ratio, and density. Defining these properties helps the software to select the appropriate material.
c. Providing thickness value of the shell element. There is a separate library called *Sections* to provide this thickness value.
d. In modeling, the coordinate values are provided. As the model becomes more complex, the number of key points also increases. The key points are joined using lines, which will give us the required area.
e. Once the modeling is completed the next process is to develop a meshed model. This is the process in which the geometry is converted into a number of elements. Meshing is necessary before the application of loads.
f. Various loads are then applied to the structure. The different static loads acting are the hydrostatic pressure, loads from containers, and the self-weight of structural members. The resultant load is calculated and applied to the respective parts of the midship section. Load calculations:
  i. Calculation of loads from containers:
    Payload from one container   = volume of the container × density of the material stored = 0.454 × 80350 kg
    No. of containers in a compartment = 70
    Total load from containers   = 2000 N/compartment
    Total load including self-weight = 3000 N/compartment
  ii. Hydrostatic pressure   = 105581 kg/m^2

Table 1. Ship particulars.

Length overall, LOA (m)	220.5
Length between perpendiculars, LBP (m)	210.2
Breadth, B (m)	32.24
Depth to main deck, D (m)	18.70
Draft scantling, T (m)	12.15
Deadweight capacity (t)	41850
Block coefficient (CB)	0.67
Speed at design draft V (kn)	22.30
Tonnage about (dt)	35881

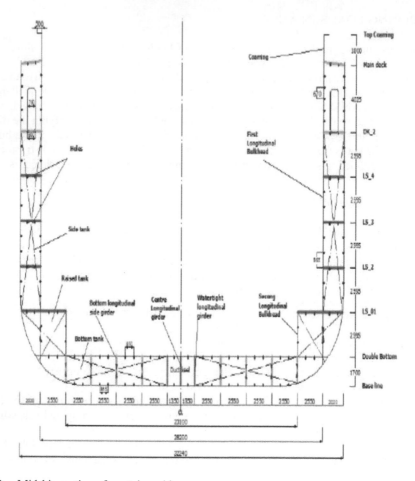

Figure 2. Midship section of container ship.

g. Application of boundary conditions is an important criterion. This is done using the *Apply loads* tab.
h. The next step is to solve the geometry. The solution gives the stress and displacement components in x, y and z directions.

6 DISCUSSION OF RESULTS

Figure 3 shows the deflected profile of the structure. The maximum value of deflection is derived as 0.84 m. Due to the complexity of the structure and the high computational time necessary to solve the model, some of the longitudinal members are omitted from the ANSYS model. This reduces the stiffness of the modeled structure compared to the actual physical structure and hence a higher value of deflection is obtained in the finite element analysis. In order to obtain a value for deflection that is closer to that of the physical model, the full structure must be modeled mathematically in ANSYS, which shows the importance of precise modeling of physical structures. Figure 3 also shows the stress in the x direction. The maximum stress value is 0.301×10^9 N/m²; this value occurs at one or two points where the stress concentration comes into the picture. This can be avoided by refining the mesh near the stress concentration points. The stresses in x, y and z directions can be obtained from the ANSYS software.

414

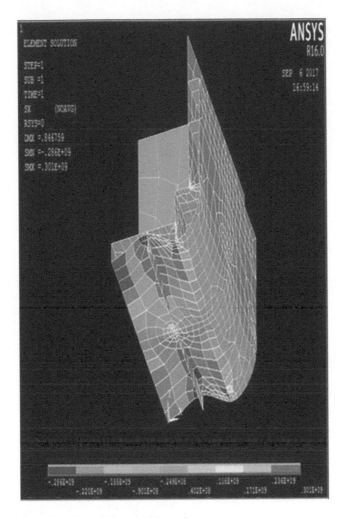

Figure 3. Stresses in X direction of the midship section.

7 CONCLUSION

Structural analysis of a midship section can be carried out using ANSYS finite element software. The deflection and stresses in the x direction can be evaluated.

REFERENCES

Eyres, D.J. (1988). *Ship construction* (3rd ed., pp. 201–320). Oxford, UK: Butterworth-Heinemann.

Larsson, R. (1988). *Ship structures—Basic course (MMA130)*. Gothenburg, Sweden: Department of Applied Mechanics, Chalmers University of Technology. Retrieved from http://www.am.chalmers.se/~ragnar/ship_structures_home/lectures/L1.pdf.

Souadji, W. (2012). Structural design of a containership approximately 3100 TEU according to the concept of general ship design B-178 (Master's thesis, Western Pomeranian University of Technology, Szczecin, Poland). Retrieved from http://m120.emship.eu/Documents/MasterThesis/2012/Wafaa%20Souadji%20.pdf.

Taggart, R. (Ed.). (1980). *Ship design and construction* (pp. 130–224). New York, NY: The Society of Naval Architects & Marine Engineers.

REFERENCES

Emerging Trends in Engineering, Science and Technology for Society,
Energy and Environment – Vanchipura & Jiji (Eds)
© *2018 Taylor & Francis Group, London, ISBN 978-0-8153-5760-5*

Experimental analysis of properties of a biolubricant derived from palm kernel oil

K. Sandeep
Department of Mechanical Engineering, Karunya University, Coimbatore, Tamil Nadu, India

M. Sekar
GMR Institute of Technology, Srikakulam, Andhra Pradesh, India

ABSTRACT: Lubricants with natural origin are known for their biodegradability and hence are called biolubricants. This study examined the tribological, physical and chemical properties of a biolubricant derived from Palm Kernel Oil (PKO). Zinc dialkyldithiophosphate (ZDDP) is used as an additive and comparison of the properties of this newly developed oil with pure PKO and SAE 20W40 engine oil were conducted. Friction and wear tests were performed in a four-ball tribo tester as per ASTM D4172 standards. Test results reveal that pure PKO has good tribological and physical properties (with the exception of its melting point) compared to other pure vegetable oils. Modification of PKO with ZDDP made the results even better— values of wear scar diameter and coefficient of friction were lower than for SAE 20W40 oil. The melting point also reduced to 9°C, which can be further reduced by chemical modification.

Keywords: Experimental analysis, biolubricant, palm kernel oil

1 INTRODUCTION

Lubricating oils are used in domestic and industrial processes to increase the life of machinery. They also make the provision of energy easier and at lower cost. Growing consumption of different lubricant types that are mostly mineral-based or synthetic leads to accidental but unavoidable inflow of considerable quantities of non-biodegradable lubricants into the environment. The increase in ecological concern inspires research in the lubricant industry into raw materials from renewable sources.

Biodegradability is the ability of a substance to be decomposed by microorganisms. Vegetable oils have biodegradability of 97% to 99%, but that of mineral oils is only 20% to 40%, according to Rudnick (2006). However, vegetable oils have failed to meet the demands of industrial lubricants by not having acceptable physical and tribological properties. Researchers are seeking methods and additives to improve these properties to an acceptable level so that there can be considerable reductions in the discharge of non-biodegradable oils into the environment.

There are various methods of improving the tribological and physical properties of vegetable oils. Additives and chemical modification are the most frequently adopted methods. Various classes of additives, such as extreme pressure additives, pour-point depressors, viscosity modifiers, corrosion inhibitors, and nanoparticles, are used nowadays to improve such properties. One of the most commonly used additives is zinc dialkyldithiophosphate (ZDDP).

2 EXPERIMENTAL DETAILS

2.1 *Equipment used*

The equipment used for the experiments is described below.

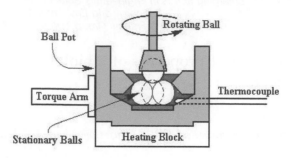

Figure 1. Schematic of four-ball tribo tester.

2.1.1 *Ducom TR-30L-PNU-IAS Four-ball tester*
A four-ball tribo tester has three balls in the ball pot at the bottom that are stationary, and one ball at the top, attached to a rotating spindle (see Figure 1). The ball pot is filled with the oil sample to be tested. There is provision to heat the oil inside the ball pot. Load is applied to the rotating ball as per ASTM (1999) standard during rotation for a specified time (3,600 seconds here).

During the experiment, a real-time graph of Coefficient of Friction (CoF) can be observed. After the experiment, the stationary balls are examined under a microscope to measure the Wear Scar Diameter (WSD).

2.1.2 *Brookfield DV2T extra viscometer*
Viscosity indexes were determined using a procedure based on international standard ASTM D2270-04. In order to calculate these, the kinematic viscosity of oils at 40°C and 100°C were experimentally measured using a viscometer.

2.2 *Materials used*

CoF, WSD, and viscosity tests were conducted on various oil samples as follows:

a. pure Palm Kernel Oil (PKO)
b. SAE 20W40 oil
c. PKO + ZDDP

In the latter sample, the weight percentage of ZDDP was varied between 0.5, 1.0, 1.5, 2.0, and 2.5%. CoF and WSD were determined using a Ducom four-ball tester with test conditions of 392 N load, 75 °C, 1200 RPM and 3,600 seconds. For this testing purpose, chromium alloy steel balls as per ASTM D2783 (IP 239) were used. Three balls at the bottom were fixed in a ball pot and one ball was attached to the motor spindle. Load is given via the output from the air compressor. The chemical properties of the oils were analyzed as per American Oil Chemists' Society methods.

The main challenge in using vegetable oil is the inconsistency of the properties of the oil samples. The properties of each oil sample depend on environmental factors wherein the mother plant grows, including soil quality, humidity, temperature, and water availability. Because of this, test results can vary from sample to sample. Samples were sourced from three different places; all tests were repeated on these three samples and average results taken if the difference between maximum and minimum values was within 1%. In tests for which this could not be achieved, worst values were taken.

3 RESULTS AND DISCUSSION

3.1 *Effect of ZDDP on anti-wear properties*

The wear scar diameter of PKO was found to decrease with increasing concentration of ZDDP additive up to a particular limit. PKO + 1.5% ZDDP showed the lowest value of

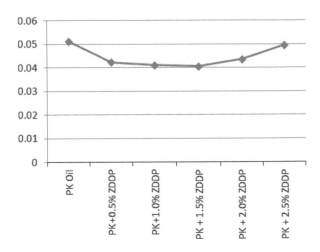

Figure 2.   Coefficient of friction vs oil samples tested.

Table 1.   CoF and WSD values of various oil samples tested.

Oil name	CoF	WSD Micrometer	Viscosity index
PKO	0.0513	460.97	146
SAE 20W40	0.0475	498.39	122
PKO + 0.5% ZDDP	0.0424	449.33	151
PKO + 1.0% ZDDP	0.0411	421.55	156
PKO + 1.5% ZDDP	0.0406	401.37	161
PKO + 2.0% ZDDP	0.0437	429.82	163
PKO + 2.5% ZDDP	0.0496	487.94	169

WSD. Azhari et al. (2015) explained the mechanism behind this as the reaction of ZDDP with the metal surface to form a solid protective film and the reaction layer. When metal is immersed in ZDDP solution in a lubricant or other non-polar solvent, a thermal film rapidly forms at the metal surface.

The WSD and CoF values of pure PKO, PKO with additives, and SAE 20W40 oil are shown in Figure 2 and Table 1.

3.2   *Effect of ZDDP on anti-friction properties*

Initially, it was observed that as the concentration of ZDDP increases in PKO, the CoF decreases. But this trend continues only up to 1.5 Wt.% of ZDDP, after which CoF increases. According to Mahipal et al. (2014), the former observation is due to a hydrodynamic boundary film being formed at this optimal concentration. At this concentration, frictional torque on the contacting surfaces will be reduced and the coefficient of friction at the contact surfaces will be reduced. However, at higher concentrations, the excess ZDDP adversely affects boundary film formation, due to excess zinc adsorption on the contact surfaces, leading to an increase in the frictional torque. The CoF values of pure PKO and PKO with additives are as shown in Figure 2 and Table 1.

3.3   *Effect of ZDDP on chemical properties*

The chemical properties of PKO are affected by ZDDP additive. It improved the saponification value and reduced the acid value and iodine value. The saponification, iodine, acid and

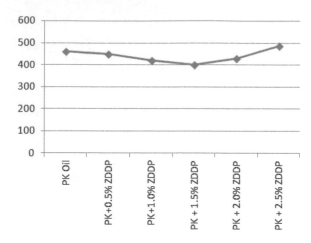

Figure 3. Wear scar diameter (in micrometers) vs oil samples tested.

Table 2. Chemical properties of PKO and PKO + 1.5% ZDDP.

Oil name	Saponification value mg KOH/g	Iodine value g/100g	Acid value mg KOH/g	Ester value mg KOH/g
PKO	244	21	3.2	277.8
PKO + 1.5% ZDDP	256	16	2.9	292.3

Table 3. Thermal properties of servo oil, palm kernel oil, and palm kernel oil + 1.5% ZDDP.

Oil name	Flash point °C	Fire point °C	Pour point °C	Cloud point °C
Servo oil	230	260	−6	−3
PKO	242	251	22	23
PKO + 1.5% ZDDP	256	264	12	14

ester values of PKO and PKO + 1.5% ZDDP, which gave the best results in wear and friction tests, are shown in Table 2.

3.4 *Effect of ZDDP on thermal properties*

Palm kernel oil exhibits a good flash point and fire point. The Achilles' heel of PKO is its melting point. Pour point and cloud point are high for a lubricant.

As the ZDDP additive has better thermal properties, as explained by Azhari et al. (2014), it can improve the thermal properties of the oil. The thermal properties of PKO, PKO + 1.5% ZDDP, and servo oil are shown in Table 3.

4 CONCLUSION

Tribological tests show that palm kernel oil has better anti-wear and anti-friction properties than SAE 20W40 oil. PKO samples with ZDDP additive resulted in minimum wear scar diameter and coefficient of friction. The viscosity index was also comparable with commercial 20W40 oil. The chemical properties of palm kernel oil are better than all other vegetable oils we have tested, including coconut oil, soybean oil, sunflower oil, mustard oil and karanja oil. In terms of thermal properties, flash point and fire point are on a par with servo oil. Properties that show poorer results are pour point and cloud point. Palm kernel oil is in a semi-solid state at 20°C. Though the melting point value is depressed considerably after adding ZDDP, PKO is still poor compared to commercial SAE 20W40 oil. According to Mahipal et al. (2014), this can be improved by esterification, which has been proven to suppress the melting pour point of such oils.

REFERENCES

American Society for Testing Materials (ASTM). (1999). *D4172-94: Standard Test Method for Wear Preventive Characteristics of Lubricating Fluid (Four-Ball Method)*. West Conshohocken, PA: ASTM International.

Azhari, M.A., Fathe'li, M.A., Aziz, N.S.A., Nadzri, M.S.M. & Yusuf, Y. (2015). A review on addition of zinc dialkyldithiophosphate in vegetable oil as physical properties improver. *ARPN Journal of Engineering and Applied Sciences*, *10*(15), 6496–6500.

Azhari, M.A., Suffian, Q.N. & Nuri, N.R.M. (2014). The effect of zinc dialkyldithiophosphate addition to corn oil in suppression of oxidation as enhancement for bio lubricants: A review. *ARPN Journal of Engineering and Applied Sciences*, *9*(9), 1447–1449.

Balamurugan, K., Kanagasabapathy, N. & Mayilsamy, K. (2010). Studies on soya bean based lubricant for diesel engines. *Journal of Scientific & Industrial Research*, *69*, 794–797.

Barnes, A.M., Bartle, K.D. & Thibon, V.R. (2001). A review of zinc dialkyldithiophosphates (ZDDPs): Characterisation and role in the lubricating oil. *Tribology International*, *34*, 389–395.

Dinda, S., Patwardhan, A.V., Goud, V. & Pradhan, N.C. (2008). Epoxidation of cotton seed oil by aqueous hydrogen peroxide catalyzed by liquid inorganic solids. *Bio Resource Technology*, *99*, 3737–3744.

Erhan, S.Z., Sharma, B.K. & Perez, J.M. (2006). Oxidation and low temperature stability of vegetable oil-based lubricants. *Industrial Crops and Products*, *24*, 292–299.

Luna, F.M.T., Cavalcante, J.B., Silva, F.O.N. & Cavalcante, C.L., Jr. (2015). Studies on biodegradability of bio-based lubricants. *Tribology International*, *92*, 301–306.

Mahipal, D., Krishnamurni, P., Mohammed Rafeekh, P. & Jnyndan N.H. (2014). Analysis of lubrication properties of zinc-dialkyl-dithio-phosphate (ZDDP) additive on karanja oil (Pongamia pinnatta) as a green lubricant. *International Journal of Engineering Research*, *3*(8), 494–496.

Rudnick, L.R. (Ed.). (2006). *Synthetics, mineral oils, and bio-based lubricants: Chemistry and technology*. Boca Raton, FL: CRC Press.

Zulkifli, N.W.M., Azman, S.S.N., Kalam, M.A., Masjuki, H.H., Yunus, R. & Gulzar, M. (2014). Lubricity of bio-based lubricant derived from different chemically modified fatty acid methyl ester. *Tribology International*, *93*, 555–562.

Emerging Trends in Engineering, Science and Technology for Society,
Energy and Environment – Vanchipura & Jiji (Eds)
© 2018 Taylor & Francis Group, London, ISBN 978-0-8153-5760-5

Influence of perforated tabs on subsonic set control

Dharmahinder Singh Chand & D.L. Vasthadu Vasu Kannah
Department of Aeronautical Engineering, Tagore Engineering College, Chennai, India

S. Thanigaiarasu
Department of Aerospace Engineering, MIT Campus, Anna University, Chennai, India

S. Elangovan
Department of Aerospace Engineering, Bharath University, Chennai, India

ABSTRACT: An experimental analysis has been carried out to examine the mixing encouraging effectiveness of tabs with circular perforation of different perforation diameters, 1.5 mm, 2 mm and 2.5 mm. The geometrical blockage offered by the perforated tabs, placed diametrically opposite, at the nozzle exit, were 8.42%, 7.55% and 6.42%, respectively, for perforation diameters 1.5 mm, 2 mm and 2.5 mm. The Mach along the jet central axis and Mach profiles in the directions along and tangential to the tabs were calculated at various axial locations. The results of the Mach 0.4, 0.6 and 0.8 jets studied show that the tab with 2 mm perforation is a better mixing promoter than 1.5 mm and 2.5 mm perforations, effecting in a core length decrement of 62% for Mach 0.6 jet. The corresponding reduction in core length for 1.5 mm and 2.5 mm holes are only 47.61% and 42% respectively.

Keywords: Perforated tab; passive control; potential core; mixing enhancement

NOMENCLATURE

X = Co-ordinate jet central axis
Y = Co-ordinate along the tabs
Z = Co-ordinate tangential to the tabs
NPR = Nozzle Pressure Ratio
φ_p = Diameter of perforation
M_j = Jet Mach number
M = Localized Mach number
P = Pitot measured pressure
P_0 = Settling chamber pressure

1   INTRODUCTION

Restricting of jet has become an active area of analysis owing to its application potential, such as improvement of stealth capabilities, minimization of base heating and reduction of aero acoustic,. Most of these applications require mixing improvement of the jet, i.e. the mass from the adjoining region entrained by the jet has to be mixed with the jet fluid mass as rapid as possible (Reeder & Zaman, 1996).

To achieve mixing enhancement, mixing promoting small scale vortices needs to be originate and commence at the nozzle exit (Reeder & Samimy, 1996), with this aim considerable number of passive and active techniques have been identified by the researchers over the recent few decades (Rathakrishnan, 2010). Among these, passive control in the form of tab has become popular due to its simple geometry and efficient mixing, promoting capability (Chiranjeevi. & Rathakrishnan, 2010).

The effect of wedge-shaped protrusions on the progress of a low-speed jet was examined by Bradbury and Khadem (1975). This tab was helpful to distort the jet propagation. Bohl and Foss (1996), Wishart et al. (1993) and Zaman et al. (1994) have determined that the tab creates a pair of counter-rotating stream-wise vortices. Zaman (1993) reported that the relative volume of the crest stream-wise vortices was observed to be about 20% of that of the crest azimuthal vortices for a tabbed circular jet at a Mach number of 0.3. Zaman et al. (1992) established that the distortion introduced by a mechanical tab is due to combination of duo stream-wise vortices. These vortices were found to be accountable for the phenomenal entrainment. Bohl and Foss (1996) reported that the two achievable sources of stream-wise vortices for the flow above a tab are the pressure gradients, which flux stream wise vortices into the flow and the well-known 'necklace' or 'horseshoe' vortices owing to boundary layer reorientation.

A rapid development of the mixing layer was observed by Reeder and Samimy (1996) when tabs were inserted. The consequence of the stream-wise tab position within the nozzle was investigated by Reeder and Zaman (1996). They observed that tabs placed at the exit plane increased entrainment into the jet, whereas tabs located further upstream in the nozzle caused an ejection of core fluid. Sreejith and Rathakrishnan (2002) studied a wire across a diameter (cross-wire) as a passive control to improve the jet mixing. The stream-wise eddied inducted by the cross-wire lead to an additional and rapid decay of the centerline pitot pressure. Also, it was noticed that the cross-wire deteriorated the shocks in the jet core significantly. The authors confirmed that the limit for tab length is the nozzle exit radius and not the boundary layer thickness. This limit of tab length is termed as Rathakrishnan limit.

Vortices emitted from the edges of flat and arc plates, kept normal to a consistent water flow, were investigated by Takama et al. (2008). It was found that the lookalike vortex in the rear of plate was much bigger than that for a circular cylinder. This indicated the major role played by the reverse flow on the vortex formation. Extending this effect Thanigaiarasu et al. (2013) studied the effect of arc tabs on the mixing behavior of subsonic and correctly expanded sonic jets. Two tabs in the form of a semi-circular arc of diameter 1.5 mm and length 2 mm were placed at the nozzle exit of 10 mm in diameter. The mixing in close proximity of the jet was studied for two designs of the tab; namely concave surface of the arc tab in front of the flow exiting the nozzle (arc tab facing in) and the convex surface of the front of the flow (arc tab facing out). Mixing promotion caused by these tabs was evaluated with a plain rectangular tab of the same blockage ratio of 7.64%. It was discover that a core length reduction of 80% was achieved in arc tab facing in, whereas the corresponding reduction in core length with arc tab facing out was just 50%. Mach profiles in the planes tangential to the jet axis showed that the arc tab facing in deforms the jet effectively by forcing the jet to spread faster and wider in the plane normal to the tab than that along the tab.

The proficiency of corrugated tabs for high speed jet control was studied by Chiranjeevi and Rathakrishnan (2010). They employed two tabs of rectangular shape with a 4.2% blockage; the edges were corrugated and located diametrically opposite at the convergent-divergent nozzle exit of a Mach 1.8. It was observed that a better mixing efficiency was achieved compared with the uncontrolled jet. More than 78% reduction in core length was obtained with corrugated tabs for the jet operated at a nozzle pressure ratio of 7; the corresponding reduction with plain tab was noticed at only 54%. It was found that the mixing capability of the corrugated tabs improved consistently with increase in nozzle pressure ratio.

Arun Kumar and Rathakrishnan (2013a) studied the influence of triangular tabs on supersonic jets. They demonstrated superiority of triangular tabs in promoting jet mixing in all three jet zones with Mach 2, in the presence of slightly favorable and unfavorable pressure gradients. Also, they investigated triangular corrugated tabs for jet control. It was demonstrated that among triangular tabs, the tab with a truncated apex is found to be a better mixing enhancer than the one with a sharp apex. The reason for the superiority is envisaged as the fact that the vortices at the tab tip do not interrelate themselves and drop energy, as in the case of the pointed apex. Corrugations at the edges of the tabs greatly augment the jet mixing.

Truncated corrugated triangular tabs were also found to be effective in boosting the mixing capability by Arun Kumar and Rathakrishnan (2013b). Semi-circular as well as square corrugated tabs were beneficial to bifurcate the jet into two fractions (lobes) at X/D ≤ 1. Also they established that the shape of corrugation plays a major role in the mixing enhancement. Tab geometries, such as rectangular, triangular, circular, a d arc have been studied extensively. However, these solid tabs shed mixing, encouraging small eddies of homogeneous size only from a specified height of the tab. But it may be more beneficial from a mixing point of view, if the vortices themselves possess varied sizes. To generate such small vortices of varied size, tabs with perforation over their flat faces are explored in the present investigation.

2 EXPERIMENTAL DETAILS

The experiments were carried out in the High Speed Jet Laboratory at MIT, Anna University, Chennai. The test facility consists of an air delivering system (compressors and reservoir) and an open jet testing facility as shown in Figure 1.

2.1 Open jet

Exit diameter (D) 20 mm of convergent nozzle prepared from gunmetal was used in this study. The nozzle was fixed at the end of the settling chamber with an o-ring sealing to minimize the leakage. Required stagnation pressures were protected in the settling chamber to generate the desired subsonic jet Mach number. A pitot tube made of stainless steel tube of inside diameter 0.4 mm and outside diameter 0.6 mm, fixed on a traverse mechanism, was used for pressure (P) measurements along and across the jets. The pitot probe has to be small enough such that there is insignificant disturbance to the incoming flow because of its existence. Thus the proportion of nozzle exit area to the probe area is $(10/0.6)^2 = 1111.11$, which is higher than the limit of 64, for concerning the probe blockage as minor[4]. The pitot tube was linked to a pressure transducer for pressure measurements.

2.2 Tab jets

Perforated tabs of stainless steel were used in the present study. The length and breadth of the tab were 5 mm and 3 mm, respectively. Two identical tabs were placed, at diametrically opposite positions, at the nozzle exit as shown in Figure 2. The fraction of the tab area normal to the flow nozzle exit area studied were 8.42%, 7.55% and 6.42%. The pitot pressures measured were translated into the Mach number using isentropic relations, using the ambient pressure in the laboratory as the static pressure, because subsonic jets are always considered to be correctly expanded.

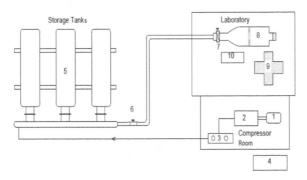

Figure 1.   An artistic view of high speed laboratory.
1. 80 HP induction motor, 2. Reciprocating compressors, 3. Activated charcoal filter, 4. Water cooling unit and silica gel dryer units, 5. Storage tanks, 6. Gate valve, 7. Pressure regulating valve, 8. Settling chamber, 9. Traversing system, 10. Instrumentation desk.

3 RESULTS AND DISCUSSION

3.1 *Centerline Mach number*

Centerline jet decay can be taken as a genuine measure of promoting jet mixing (Rathakrishnan, 2009). The centerline jet reduction of uncontrolled and controlled Mach 0.4 are compared in Figure 5. It is seen that the potential core span of the jet is reduced by 58.33% with the tab with 7.55% blockage (2 mm circular perforation), while reduction is 47.91% for the tab with 8.42% blockage (1.5 mm circular perforation) and 42.7% for the tab with 6.42% blockage (2.5 mm circular perforation) compared with uncontrolled jet. It may be due to non-uniform

Figure 2.   A pictorial view of tab arrangement at nozzle exit.

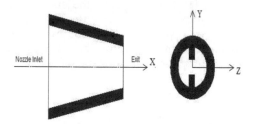

Figure 3.   Schematic diagram of tab at nozzle exit and co–ordinates implemented.

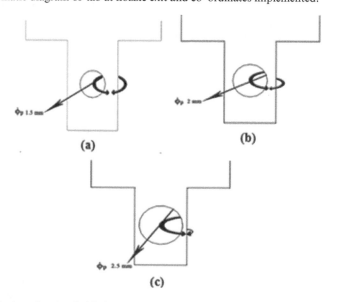

Figure 4.   Physical mechanism behind vortex generation between perforation and the tab.

426

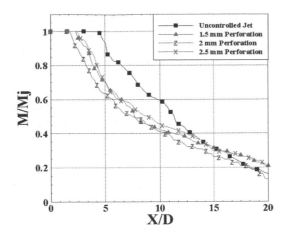

Figure 5.   Centerline Mach decay at Mach 0.4 jet.

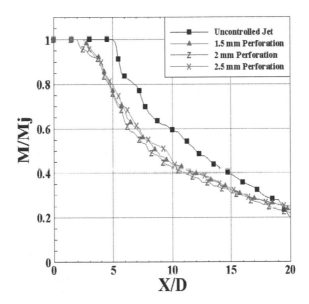

Figure 6.   Centerline Mach decay at Mach 0.6 jet.

vortices emitted by web and perforation. Also it is observed that decay in transition region is very rapid beyond 11D compared with uncontrolled jet. This is due to the fact that small scale vortices are stronger than large scale vortices and travels longer distance 20D.

Figure 6 depicts the variation in the potential core area at Mach 0.6 for uncontrolled jet, tabs with 8.42% blockage (1.5 mm circular perforation), 7.55% blockage (2 mm circular perforation) and 6.42% blockage (2.5 mm circular perforation). Potential core for uncontrolled jet extends almost up to X/D = 5.25. The core span extends to X/D = 2 for the tab with 7.55% blockage (2 mm circular perforation), whereas core length extends to X/D = 2.75 and X/D = 2.78 for the tab arrangements of 8.42% blockage (1.5 mm circular perforation) and 6.42% blockage (2.5 mm circular perforation), respectively. The tab with 7.55% blockage (2 mm circular perforation) is more effectual for mixing enrichment than other perforation tab configurations studied.

This perhaps may be because of the varying radii of curvature created by perforation inside the tab, which is responsible for manipulating the dimensions of vortices (Quinn 1995).

It is apparent from Figure 7 that the potential core span is reduced by 50% with the tab with 7.55% blockage (2 mm circular perforation) and the core reduction with tabs of 8.42% blockage (1.5 mm circular perforation) and 6.42% blockage (2.5 mm circular perforation) is 40% and 38% respectively, compared with uncontrolled jet. The jet stirring efficiency with atmospheric air is the maximum for the tab with blockage of 7.55% (2 mm diameter circular perforation) among other perforated tabs because of 'necklace' or 'horseshoe' vortices due to the boundary layer rearrangement over a tab (Chiranjeevi & Rathakrishnan, 2010).

When the perforation diameter is 1.5 mm the mass, or more precisely energy, of the jet through the perforation is less, resulting in lower shear layer interaction between core jet and perforated jet. Also the eddies liberated by the perforation and web are of the same size as in Figure 4 (a). This combination results in lesser mixing enhancement. Similarly energy through the perforation of diameter 2.5 mm as in Figure 4 (c) is greater compared with 1.5 mm and 2 mm perforated tabs. But the combination of shear layer interaction (between core jet and perforation jet) and vortices exited by the web (w) (shortest distance between the outer edge of tab and inner edge of the perforation) to promote mixing are insufficient. The reason may be due tp the reduced web width owing to the increased perforation diameter.

3.2 *Mach number profile*

Figure 8 shows that more dip and sudden rise in regard to tabs with 7.55% blockage (2 mm circular perforation) due to combination of vortices and perforation at $X/D = 0.15$ for $M_j = 0.4$, which in turn entrains mass from atmosphere to promote mixing efficiently. This may be due to the fact that non-circular jets entrain more ambient fluid than their circular counterparts, and also to the non-uniform self-induction brought about by azimuthal curvature variation of the initial vortices generated at the nozzle exit plane (Mrinal et al., 2006).

Figure 8 shows the shrunk area compared with other controlled and uncontrolled flows studied for the outcome of perforation size as an indication of the scattering of vortices along the tab applicable to the tab with 7.55% blockage (2 mm circular perforation) at axial location of $X/D = 1$ for jet Mach 0.4.

Figure 9 exhibits that for all perforated tabs, Mach number variation in the perpendicular plane is almost identical to the uncontrolled jet at $X/D = 0.15$ for $M_j = 0.4$. This implies that the thrust loss caused by the tabs is only marginal[7]. Also from Figure 11 it is clear that the existence of core is observed up to $Z/D = 0.4$, for the tab with 7.55% blockage (2 mm circular

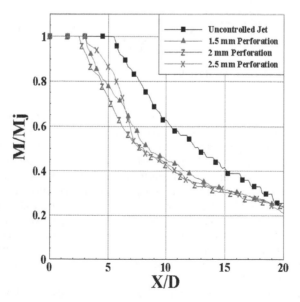

Figure 7.   Centerline Mach decay at Mach 0.8 jet.

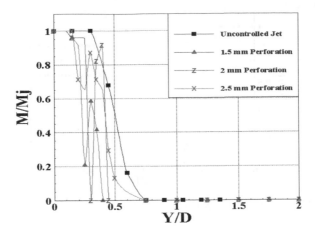

Figure 8.   Y-Mach number profiles at X/D = 0.15 for M_j = 0.4.

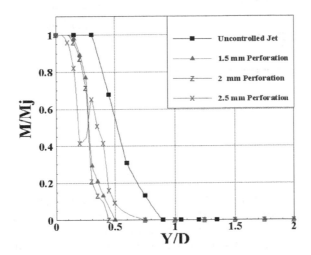

Figure 9.   Y-Mach number profiles at X/D = 1 for M_j = 0.4.

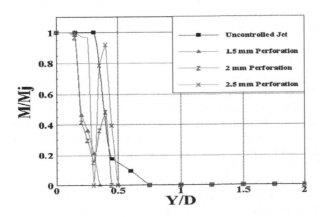

Figure 10.   Z-Mach number profiles at X/D = 0.15 for M_j = 0.4.

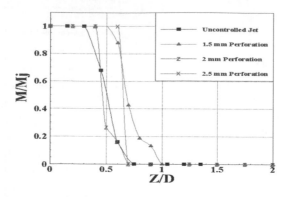

Figure 11. Y-Mach number profiles at X/D = 0.15 for $M_j = 0.8$.

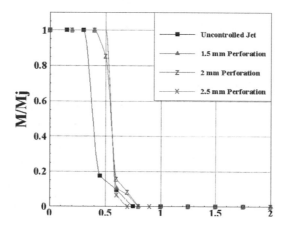

Figure 12. Z-Mach number profiles at X/D = 0.15 for $M_j = 0.8$.

perforation), whereas it is about Z/D = 0.7 and Z/D = 0.6 for tabs with 8.42% blockage (1.5 mm circular perforation) and 6.42% blockage (2.5 mm circular perforation) respectively, at axial position X/D = 0.15 and 0.25. This shows that the tab with 7.55% blockage (2 mm circular perforation) alters jet cross-sectional area more effectively and entrains more air from the nearby compared to other perforated tabs studied.

From Figure 10, it is perceived that the mixing enhancement performance in XY-plane is better for tab with 7.55% blockage (2 mm circular perforation) due to more shrunk in area.

Figure 12 proves that the jet develops faster along Z/D for tab with 7.55% blockage (2 mm circular perforation) at X/D = 0.15, for Mach 0.8. This may be connected to the drift of small eddies sidewise from the nozzle outlet.

3.3 *Iso-mach contour*

The pitot pressure variations in the jet field with control is measured in the YZ-plane and converted into Mach number using isentropic relation at axial locations, as illustrated in Figure 13. Three pairs of vortices can be noticed from the Figure 13 (b), each at root, tip and web. This may be credited to suitable combination of web and perforation for shedding varying sizes of vortices. In the case of 1.5 mm perforation only two pairs can be seen from Figure 13 (a). Again from Figure 13 (c) a pair of vortices is visible at root only. This means that the tab with 7.55% blockage (2 mm circular perforation) is more capable mixing promoter among other circular perforation tabs, namely, 8.42% blockage (1.5 mm circular perforation) and 6.42% blockage (2.5 mm circular perforation). Also there are dual pressure mountains on both sides of the jet axis, as seen in Figure 13 (b). It may be taken as an indication of jet bifurcation[13], which is a vital factor for mixing enhancement.

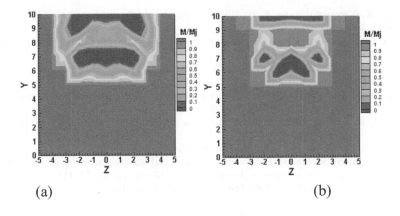

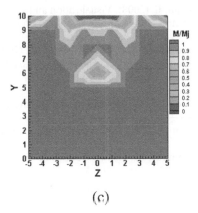

(c)

Figure 13.   Iso-Mach contour for tabs, at X/D = 0.1, for M_j = 0.6. (a) 1.5 mm circular perforation (b) 2 mm circular perforation (c) 2.5 mm circular perforation.

4   CONCLUSION

The outcome of the current study on subsonic jet control with perforated tabs show that all the perforations studied are profound and effective in promoting mixing close to jet field. However, among them, perforation of 2 mm diameter is a highly effective stirring promoter, resulting in a core span reduction of 62%. The corresponding core span reduction for 1.5 mm and 2.5 mm holes are only around 47%.

REFERENCES

Arun Kumar, P. & Rathakrishnan, E. (2013a). Corrugated triangular tabs for supersonic jet control. *Journal of Aerospace Engineering, 1–15.*
Arun Kumar, P. & Rathakrishnan, E. (2013b). Corrugated truncated triangular tabs for supersonic jet control. *Journal of Fluid Mechanics, 135,* 1–11.
Arun Kumar, P. & Rathakrishnan, E. (2013c). Truncated triangular tabs for supersonic—jet control. *Journal of Propulsive Power, 29,* 50–65.
Bohl, D. & Foss, J.F. (1996). Enhancement of passive mixing tabs by the addition of secondary tabs *AIAA paper,* 96–054.
Bradbury, L.J.S. & Khadem, A.H. (1975). The distortion of a jet by tabs. *Journal of Fluid Mechanics, 70* (4), 801–813.

Chiranjeevi, P.B. & Rathakrishnan, E. (2010). Corrugated tabs for supersonic jet control. *Journal of AIAA, 48*(2), 453–465.

Elangovan, S. Solaiappan, A. & Rathakrishnan, E. (1996). Studies on twin elliptic jets. *Journal of Aerospace, 100*(997), 295–296.

Mrinal, K., Pankaj, S.T. & Rathakrishnan, E. (2006). Studies on the effect of notches on circular sonic jet mixing. *Journal of Propulsive Power, 22*, 211–214.

Quinn, W.R. (1995). Turbulent mixing in a free jet issuing from a low aspect ratio contoured rectangular nozzle. *Journal of Aerospace, 99,* (988), 337–342.

Rathakrishnan, E. (2009). Experimental studies on the limiting tab. *Journal of AIAA, 47*(10), 2475–2485.

Rathakrishnan, E. (2010). *Applied gas dynamics.* New Jersey: Wiley.

Reeder, M.F. & Samimy, M. (1996). The evolution of a jet with vortex-generating tabs: real-time visualization and quantitative measurements. *Journal of Fluid Mechanics 311,* 73–118.

Reeder, M.F. & Zaman, K.B.M.Q. (1996). Impact of tab location relative to the nozzle exit on jet distortion. *Journal of AIAA, 34*(1), 197–199.

Sreejith, R.B. & Rathakrishnan, E. (2002). Cross-wire as passive device for supersonic jet control. *AIAA paper,* 2002–4052.

Takama, Y., Suzuki, K. & Rathakrishnan, E. (2008). Visualization and size measurement of vortex shed by flat and arc plates in a uniform flow. *International Journal of Review of Aerospace Engineering, 1*(1), 55–60.

Thanigaiarasu, S., Jayaprakash, S., Elangovan, S. & Rathakrishnan, E. (2008). Influence of tab geometry and its orientation on under expanded sonic jets. *Journal of Aerospace Engineering, 222,* 331–339.

Wishart, D.P., Krothapalli, A. & Mungal, M.G. (1993). Supersonic jet control disturbances inside the nozzle. *Journal of AIAA, 31*(7), 1340–1341.

Zaman, K.B.M.Q. (1993). Streamwise vorticity generation and mixing enhancement in free jets by delta-tabs. *AIAA paper,* 93–3253.

Zaman, K.B.M.Q., Reeder, M.F. & Samimy, M. (1992). Supersonic jet mixing enhancement by delta-tabs. *AIAA paper,* 92–3548.

Zaman, K.B.M.Q., Reeder, M.F. & Samimy, M. (1994). Control of an axi-symmetric jet using vortex generators. *Physics of Fluids, 6,* 778–793.

Renewable energy and alternate fuels

Emerging Trends in Engineering, Science and Technology for Society,
Energy and Environment – Vanchipura & Jiji (Eds)
© 2018 Taylor & Francis Group, London, ISBN 978-0-8153-5760-5

Modification of a parabolic trough collector and its exergy analysis

O. Arjun, M.C. Nikhil & B. Sreejith
Government Engineering College, Kozhikode, Kerala, India

ABSTRACT: Extraction of thermal energy from solar energy is a challenge in harvesting solar energy. When higher temperatures are required, it becomes necessary to concentrate solar radiation. Parabolic Trough Collectors (PTCs) are formed by a cylindrical surface of mirrors having high reflectivity with a parabolic shape that concentrates solar radiation on a receiver tube located at the focal point of the parabola. Working fluid, which is heated up by the solar radiation, circulates inside the receiver tubes. Energy losses at higher temperatures are the main cause of efficiency loss in PTCs. In previous work, the collector was exposed to the atmosphere and thus thermal losses were high. Some modifications to the design of PTCs for reduction of energy losses was proposed in this work. In order to reduce convective losses, an evacuated tube collector was introduced instead of copper tube. A modified PTC was fabricated and its performance evaluated. Exergy analysis was performed to study the effects of operational and environmental parameters on the performance of PTCs.

Keywords: Parabolic Trough Collectors, Exergy analysis, solar radiation

1 INTRODUCTION

The sun is the head of the family of planets, and is the most abundant source of renewable energy for our earth. Reserves of other energy sources, such as coal and fossil fuel, will eventually diminish. Solar energy contains radiant heat and light energy from the sun, which can be harnessed with modern technologies like photovoltaic (PV) cells, solar heating, artificial photosynthesis, and solar thermal electricity. Solar thermal collectors gain energy through radiation, conduction and convection. A flat plate collector will lose energy through conduction as well as convection; thus, it reduces the amount of energy that can be transferred to working fluid. The evacuated tube collector is a new harnessing technology. It is ideal for high-temperature applications such as boiling water, pre-heating, and steam production.

Exergy is the ability of a system to do useful work before it has been brought into thermal, mechanical and chemical equilibrium with the environment. It is derived from both the first and second laws of thermodynamics. When a system and its surroundings are not in equilibrium with each other, then we can extract work. This means that if there is any difference in temperature between a system and its surroundings, it will be in unstable equilibrium. This situation can be used to produce work. On the basis of the second law of thermodynamics, it is impossible to convert low-grade energy completely into shaft work. The part of low-grade energy that can be converted into useful work is termed available energy or exergy. The performance of Parabolic Trough Collectors (PTCs) can be explained in terms of exergy, which provides a useful basis for the design and optimization of PTCs.

The present work includes modification of an existing PTC with substitution of an evacuated tube. The main drawback of the existing system is that the outside surface of the receiver tube is exposed to the atmosphere; thus convective energy loss will be dominant, which reduces the performance of the PTC. In order to reduce this loss, evacuated tubes are introduced instead of copper tubes. Exergy analysis was performed to assess the performance of the modified PTC.

Yadav et al. (2013) conducted an experiment on parabolic troughs with various reflectors, such as stainless-steel sheet, aluminum foil, and aluminum sheet, and obtained temperatures of 42.1°C, 48.2°C and 52.3°C, respectively. A buffed aluminum sheet will produce even better results in comparison to an untreated sheet.

Previous work by Nikhil and Sreejith (2016) used acrylic glass and its performance using Al_2O_3 nanofluid as the working fluid was studied. The main limitation of the model is that the collector is exposed to ambient air and thus heat loss to the surroundings is higher. In the current work, the collector is replaced with an evacuated tube collector and an exergy analysis is performed.

Jafarkazemi and Ahmadifard (2013) conducted an energy and exergetic analysis of a flat plate collector; based on their theoretical results, the maximum energy and exergy efficiency of a flat plate collector are close to 80% and 8%, respectively. Energy efficiency is always higher than that of exergy efficiency due to irreversible work associated with flow.

A study was performed by Padilla et al. (2014) to simulate the heat transfer model for a PTC. They made an exergy balance of the receiver in the control volume. They also calculated exergy efficiency and destruction rates.

Tyagi et al. (2007) performed a parametric study of concentrating-type solar collectors for different mass flow rates and concentration ratios using hourly solar radiation. They concluded that exergy output and thermal efficiency are an increasing function of the mass flow rate for a given value of solar intensity.

The performance and efficiency of PTCs have increased significantly during the last three decades. This is due to the development of new technologies and advances in material science. New collectors, such as evacuated tubes and Apercus evacuated tubes, can absorb solar energy more effectively than flat plate collectors. However, combining these collectors in a concentrating technology like a parabolic trough, with dishes and Fresnel lenses will enhance the overall performance of the system.

2   THERMAL PERFORMANCE OF PTC

The performance of a PTC can be estimated by the efficiency factor η, which is defined as the ratio of the net heat gain to the solar radiation energy, based on the diffuse reflection area of the solar collector:

$$\eta = \frac{\dot{m}C_p\left(T_{out} - T_{in}\right)}{I\left(A_a - A_s\right)\rho}$$

Calculation of the thermal performance of concentrating collectors is similar to that of flat plate collectors. Because concentrating collectors operate at higher temperatures than flat plate collectors, thermal loss will be greater. In order to calculate thermal loss, we must know the overall heat loss coefficient, U_L, which is given by:

$$U_L = \frac{Q_{loss}}{A_r\left(T_r - T_a\right)} \tag{2}$$

2.1   Design of parabolic trough

In the present study, the material used for the PTC reflector is acrylic glass, whose reflectivity is nearly the same as MIRO-SUN® (Alanod GmbH, Ennepetal, Germany), which is used in reflectors in concentrating collectors. The dimensions and specification of the parabolic trough are shown in Figure 1 and Table 1, respectively.

2.2   Exergy analysis

The exergy balance equation must be applied on the control volume of the receiver. According to Padilla et al. (2014), the partial differential equation of the exergy balance is given by:

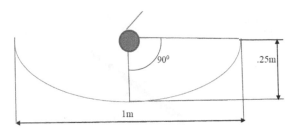

Figure 1.   Dimensions of the trough.

Table 1.   Specifications of PTC.

Parameter	Value
Width of the concentrator, W	1 m
Length of the concentrator, L	1.8 m
Rim angle of the trough, ϕ	90°
Concentration ratio, C	5.2
Aperture area, A_a	1.8 m²
Focal length	0.25 m
Half acceptance angle	1.5°

$$\frac{dEcv}{dt} = \sum_j \dot{E}_{qj} - \dot{W}_{cv} + \sum_i \dot{m}_i e_{fi} - \sum_e \dot{m}_e e_{fe} - \dot{E}_d - \dot{E}_{loss}$$

It is assumed that steady state flow, and kinetic and potential energy were negligible, and that the specific heat and other properties remain constant during operation.

Exergy efficiency is defined as the ratio of exergy gain to maximum possible solar radiation exergy, that is:

$$\eta_{ex} = \frac{\dot{E}_{gain}}{\dot{E}_{sr}} = \frac{\dot{E}_i - \dot{E}_e}{\dot{E}_{sr}}$$

$$\eta_{ex} = \frac{\dot{m}\left[\int_{T_i}^{T_e} C_p(T) dT - T_o \int_{T_i}^{T_e} \frac{C_p(T)}{T} dT \right]}{I_b(A_a - A_s)\rho\psi}$$ (3)

$$\eta_{ex} = \frac{\dot{m}C_p\left[(T_e - T_i) - T_o \ln\frac{T_e}{T_i} \right]}{I_b(A_a - A_s)\rho\psi}$$

where ψ is the relative potential of the maximum useful work extracted from radiation and is calculated with Petela's (2003) formula:

$$\psi = 1 - \frac{4}{3}\left(\frac{T_o}{T_s}\right) + \frac{1}{3}\left(\frac{T_o}{T_s}\right)^4$$

3   EXPERIMENTAL FACILITY

In order to avoid confusion in terminology and ensure consistency between terms we use 'concentrating collector' to represent the entire setup. The term 'concentrator' is used for the optical subsystem that directs the solar radiation onto the absorber, and the term 'receiver' represents the subsystem consisting of the absorber, coating and cover, as shown in Figure 2.

437

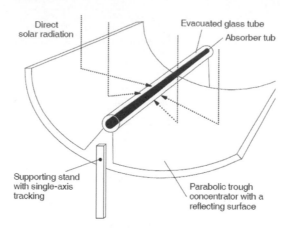

Figure 2.   Schematic diagram of PTC.

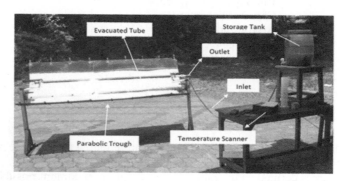

Figure 3.   Experimental setup.

3.1   Experimental setup

The experiment was conducted for six hours per day. Figure 3 shows the experimental setup of the PTC. It consists of a storage tank, inlet pipe, outlet pipe, K-type thermocouple, temperature scanner, parabolic trough, lux meter, measuring jar and anemometer. The pipes and all fittings are made from Chlorinated Polyvinyl Chloride (CPVC), which is used in solar heating and high-temperature applications instead of ordinary PVC because of its ability to withstand high temperatures. A lux meter and anemometer are used to measure solar intensity and wind velocity, respectively, at a particular instant. The flow rate was measured with a measuring jar in which the time for a 200 ml rise was taken as the reference. Thus, we obtain the mass flow rate in kg/s. Minerally insulated K-type (nickel-chromium/nickel-alumel) thermocouples are used in this work. The thermocouples and all the readings of temperature described here are calibrated. The storage tank is placed above the receiver so that water will flow using gravitational force, which eliminates a pumping system. There are two valves, located at the inlet and outlet of the receiver; thus, we can control the flow rate.

4   RESULTS AND DISCUSSION

The experiment was carried out during February to March 2017 at the Government Engineering College, Kozhikode (11.2858° N, 75.7703° E), between the hours of 9 a.m. and 3 p.m. The experiment was conducted in differing climatic situations, such as cloudy and sunny skies. Further, the study was carried out in different solar intensities, varying from 600 to 1000 W/m². The efficiency of the PTC was calculated by using Equation 1.

Figures 4 and 5 represent the variation of outlet temperature and efficiency with respect to mass flow rate. As the mass flow rate increases, the outlet temperature is reduced. That is, for lower mass flow rates the outlet temperature was high and vice versa. In addition, when the mass flow rate increases the heating efficiency increases.

If the mass flow rate remains fixed, the effect of solar intensity on outlet temperature can be studied. Solar intensity is varied from 600 to 1000 W/m². It can be clearly seen from Figure 6 that even under cloudy conditions, or when the intensity of radiation is at a minimum (600 W/m²), we can obtain temperatures between 70°C and 92°C depending upon the mass flow rate. Thus, the PTC is a promising technology for future energy needs. Furthermore, because of the wide range of outlet temperature, it can be used in a wide variety of applications, such as process heating, desalination, and pasteurization.

As illustrated in Figure 7, the maximum outlet temperature was obtained between 12 noon and 2 p.m. This is due to the fact that solar radiation falls perpendicularly on the trough at this time and almost all radiation is collected via the entire length of the receiver. Furthermore, the higher the intensity of radiation, the higher the outlet temperature. On the other hand, efficiency steadily decreases until 2 p.m. This is because of temperature rise as solar intensity rises. Then efficiency increases because the high outlet temperature drops as compared to reduction in intensity of solar radiation.

The overall heat loss coefficient, U_L, can be computed using Equation 2. The variation in heat loss coefficient with receiver temperature is shown in Figure 8. One can clearly observe that as the temperature increases the heat loss coefficient also increases. This is due to the fact that radiation loss will be higher at higher temperatures.

4.1 Comparison with previous work

The variation in efficiency with mass flow rate for different receivers, such as evacuated tube, copper tube and aluminum tube, is shown in Figure 9. The efficiency of the PTC can be

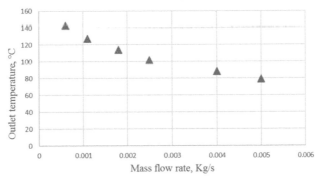

Figure 4. Variation of outlet temperature with mass flow rate (for constant 1000 W/m² solar intensity).

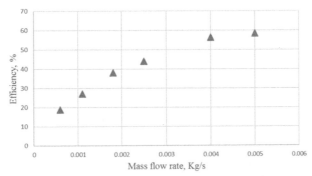

Figure 5. Variation of efficiency with mass flow rate (for constant 1000 W/m² solar intensity).

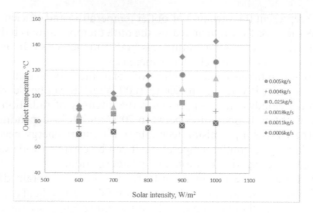

Figure 6. Variation of outlet temperature with solar intensity (for fixed mass flow rate).

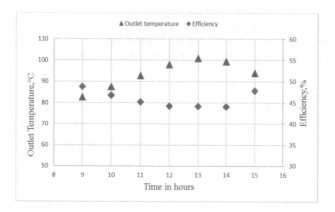

Figure 7. Variation of outlet temperature and efficiency with time (for constant $\dot{m} = 0.0025$ kg/s).

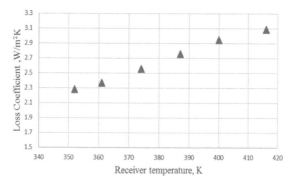

Figure 8. Variation of heat loss coefficient with receiver temperature.

calculated using Equation 1. The evacuated tube collector shows the greatest efficiency, nearly twice that of the copper tube collector with Al_2O_3 nanofluid as the working fluid. The maximum efficiencies obtained were 58.3%, 29.3%, 18.3%, and 11% for evacuated tube, copper tube with nanofluid, copper tube, and aluminum tube, respectively.

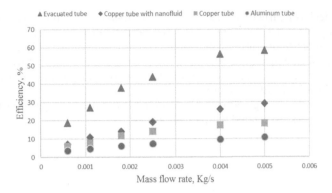

Figure 9. Variation of efficiency with mass flow rate.

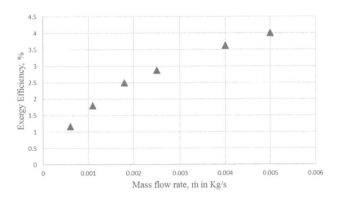

Figure 10. Variation of exergy efficiency with mass flow rate.

4.2 *Exergy efficiency*

Figure 10 depicts the variation in exergy efficiency with mass flow rate. Exergy efficiency can be calculated by using Equation 3. The variation in exergy efficiency was similar to that of energy efficiency. However, energy efficiency will always be higher than that of exergy efficiency. According to the second law of thermodynamics, we cannot convert low-grade energy completely into high-grade energy because of the heat transfer irreversibility between a system and its surroundings. This is because of the irreversible work associated with any real process. The exergy losses include exergy flowing out to the surroundings, but in exergy destruction indicates the loss of exergy within the system boundary due to its irreversibility.

5 CONCLUSIONS

Some modifications were performed in the design of a PTC and experiments were conducted with different mass flow rates, various climates and different solar intensities to compare its performance. The study demonstrated the effectiveness of the proposed modification to the design of the PTC. Exergy analysis was performed in order to establish various parameters that affect the performance of the PTC. This was a cost-effective way of obtaining high temperatures in response to future energy demands. The following conclusions were drawn upon the completion of the work:

- The maximum temperature obtained was 143°C, which is 33°C (~30%) higher than that of the previous model using copper tube with water/Al_2O_3 nanoparticles.
- The maximum temperature obtained for the previous model without using nanoparticles was 96°C, which is 49% lower than that of the new model.

- The new model has nearly twice the efficiency compared to the previous one.
- Solar intensity plays a significant role in the performance of the PTC. At higher solar intensity, the outlet temperature will be higher, thus giving higher efficiency.
- As the mass flow rate increases, efficiency increases and outlet temperature decreases.
- Efficiency will decrease at higher temperatures. This is because as the temperature increases the radiative heat loss will also increase.
- The results show that exergy efficiency was much lower than energy efficiency because of the irreversible work associated with a real process.

NOMENCLATURE

A_a	Aperture area	m^2
A_s	Shaded area	m^2
C	Concentration ratio	
D	Diameter	m
U_L	Overall heat loss coefficient	$W/m^2\,K$
W	Aperture	m
$\dot{E}_e$	Exergy output	W
$\dot{E}_i$	Exergy input	W
$\dot{m}$	Mass flow rate	kg/s
η	Energy efficiency	
η_{ex}	Exergy efficiency	
$\theta\alpha$	Acceptance angle	degree
ρ	Reflectivity	
σ	Stefan–Boltzmann constant	$W/m^2/K^4$

Subscripts
a	Atmosphere
r	Receiver

REFERENCES

Jafarkazemi, F. & Ahmadifard, E. (2013). Energetic and exergetic evaluation of flat plate solar collectors. *Renewable Energy*, 56, 55–63.

Nikhil, M.C. & Sreejith, B. (2016). Performance evaluation of a modified parabolic trough concentrator using nanofluid as the working fluid. *NET*, 192–201.

Padilla, R.V., Fontalvo, A., Demirkaya, G., Martinez, A. & Quiroga, A.G. (2014). Exergy analysis of parabolic trough solar receiver. *Applied Thermal Engineering*, 67, 579–586.

Petela, R. (2003). Exergy of undiluted thermal radiation. *Solar Energy*, 74, 469–488.

Tyagi, S.K., Wang, S., Singhal, M.K., Kaushik, S.C. & Park, S.R. (2007). Exergy analysis and parametric study of concentrating type solar collectors. *International Journal of Thermal Sciences*, 46, 1304–1310.

Yadav, A., Kumar, M. & Balram. (2013). Experimental study and analysis of parabolic trough collector with various reflectors. *International Journal of Energy and Power Engineering*, 7(12), 1659–1663.

Emerging Trends in Engineering, Science and Technology for Society,
Energy and Environment – Vanchipura & Jiji (Eds)
© *2018 Taylor & Francis Group, London, ISBN 978-0-8153-5760-5*

Analytical computation of GWP, ODP, RF number and TEWI analysis of various R134a/R1270/R290 blends as R22 alternatives

Sharmas Vali Shaik & T.P. Ashok Babu
NIT Karnataka, Surathkal, Karnataka, India

ABSTRACT: The principal objective of the present work is to compute the GWP, ODP, RF number and TEWI analysis of various ternary R134a/R1270/R290 blends as alternatives to R22. In this study thirteen refrigerant blends consists of R134a, R1270 and R290 at different compositions are taken. GWP and ODP of refrigerant blends are computed by using various simple correlations. The estimation of emission of greenhouse gases and flammability study of refrigerants are done by using TEWI and RF analysis respectively. Analytical results revealed that all the thirteen studied fluids are ozone friendly in nature. The GWP of refrigerant M6 (651) is lower than that of GWP of R22 (1760). RF analysis exhibited that all the thirteen refrigerant blends are categorized as ASHRAE A2 flammability category. Thermodynamic analysis revealed that COP of M6 (3.608) is higher that of COP of R22 (3.534). TEWI of M6 is lower among the R22 and thirteen studied fluids. Hence refrigerant M6 (R134a/R1270/R290 50/5/45 by mass %) is an alternative to R22.

Keywords: COP, GWP, ODP, RF number, R22 alternatives, TEWI

NOMENCLATURE

HFCs	Hydrofluorocarbons
HOC	Heat of combustion (kJ/mol)
L	Lower flammability limit (kg/m^3)
MW	Molecular weight (kg/kmol)
RF	Refrigerant flammability (kJ/g)
U	Upper flammability limit (kg/m^3)
C_i	Composition of ith component
$\dot{m}$	Mass flow rate of refrigerant (kg/min)
m_i	Mass fraction of ith component
W_c	Compressor work (kJ/kg)
W_{CP}	Compressor power (kW)

1 INTRODUCTION

Refrigerant R22 has adverse environmental impacts like high ozone depletion potential (ODP) and high global warming potential (GWP) (Mohanraj et al. 2009). Therefore an international Montreal protocol decided to phase out R22 by the year 2030 (UNEP 1987). Currently global warming has become very significant issue and hence Kyoto protocol was recommended to resolve this problem, for which hydrofluorocarbons (HFCs) were classified as one among the targeted global warming refrigerants (GECR 1997). Hence in this study an attempt was made to develop the refrigerants which will meet the requirements of both the Montreal and Kyoto protocol respectively. Formerly various performance studies were carried out to find suitable alternative for the refrigerant R22. Theoretical analysis revealed that R444B was a suitable candidate to replace R22 (Atilla G.D. & Vedat O. 2015). Experimental studies reported that R134a requires a larger size of compressor in order to replace

R22 (Devotta et al. 2001). The ODP and GWP values of various pure refrigerants were found from the literature (ASHRAE 2009, IPCC 2014). Lorentzen suggested that R290 was an appropriate substitute to R22 (Lorentzen G. 1995). Experimental studies exhibited that R410A was an alternative refrigerant to replace R22 (Chen 2008). Theoretical performance investigation results revealed that the performance of R134a/R1270/R290 (50/5/45 by mass percentage) was 2.10% higher than the R22 (Sharmas Vali S. and Ashok Babu T.P. 2017). The present work focuses on the computation of GWP, ODP, RF number and TEWI values of various R134a/R1270/R290 blends for which authors are not covered in the above literature. Hence a significant work is identified to compute the above considered parameters. And also from the above literature COP (3.608) of R134a/R1270/R290 (50/5/45 by mass percentage) was higher than the COP of R22 (3.534).

2   COMPUTATION OF GWP AND ODP OF R134A/R1270/R290 BLENDS

GWP and ODP values of pure refrigerants (R22, R134a, R1270 and R290) are required to compute the GWP and ODP of various refrigerant blends. The values of GWP and ODP are taken from the literature and they are listed in Table 1 (ASHRAE 2009, IPCC 2014). In the present study total thirteen ternary refrigerant blends (R134a/R1270/R290) at various compositions are considered and their corresponding designation followed for the blends are given in Table 1. The correlations used to compute the GWP and ODP of various refrigerant blends are taken from literature and they are given below (Ahamed J.U. 2014, Arora R.C. 2010).

$$GWP_{mix} = m_1 GWP_1 + m_2 GWP_2 + \ldots = \sum m_i GWP_i \tag{1}$$

$$ODP_{mix} = m_1 ODP_1 + m_2 ODP_2 + \ldots = \sum m_i ODP_i \tag{2}$$

The GWP and ODP of thirteen investigated blends are computed by using equations (1) and (2) respectively. The values of GWP and ODP of thirteen studied fluids are also given in Table 1.

Table 1.   Environmental properties of refrigerants.

Refrigerant designation	Composition by mass %	GWP (100 years)	ODP
R22	Pure fluid	1760	0.055
R134a	Pure fluid	1300	0
R290	Pure fluid	3	0
R1270	Pure fluid	3	0
M1 (R134a/R1270/R290)	80/10/10	1040.6	0
M2 (R134a/R1270/R290)	60/20/20	781.2	0
M3 (R134a/R1270/R290)	65/5/30	846.05	0
M4 (R134a/R1270/R290)	60/5/35	781.2	0
M5 (R134a/R1270/R290)	55/5/40	716.35	0
M6 (R134a/R1270/R290)	50/5/45	651.5	0
M7 (R134a/R1270/R290)	65/30/5	846.05	0
M8 (R134a/R1270/R290)	60/35/5	781.2	0
M9 (R134a/R1270/R290)	55/40/5	716.35	0
M10(R134a/R1270/R290)	50/45/5	651.5	0
M11(R134a/R1270/R290)	50/15/35	651.5	0
M12(R134a/R1270/R290)	50/20/30	651.5	0
M13(R134a/R1270/R290)	50/25/25	651.5	0

3   COMPUTATION OF REFRIGERANT FLAMMABILITY NUMBER

Study of flammability is most crucial for the investigators while developing alternative refrigerants from the view point of safety. Flammability gives the range of fuel concentration within which the refrigerant blends can burn or ignite. These limits are important while computing the hazards of liquid or gaseous fuel mixtures. Hence Jones proposed the correlation to estimate the upper and lower flammability of gases and vapors (Jones, G.W. 1938, Zabetakis, M.G. 1965). An index named refrigerant flammability (RF) number is used for indicating the hazards of combustion of the refrigerants. It is reliable to express the hazards of combustion with respect to limits of flammability of each refrigerant by using RF number. An empirical correlation used for computing the RF number is given below (Shigeo Kondo et al. 2002).

$$RF = \left[\left(\frac{U}{L}\right)^{1/2} - 1\right] \times \frac{HOC}{MW} \tag{3}$$

To compute the limits of upper and lower flammability of refrigerant blends, Le Chaterier's rule can be used (Shigeo Kondo et al. 2002).

$$\frac{1}{U_{mix}} = \frac{C_1}{U_1} + \frac{C_2}{U_2} + \ldots = \sum \frac{C_i}{U_i} \tag{4}$$

$$\frac{1}{L_{mix}} = \frac{C_1}{L_1} + \frac{C_2}{L_2} + \ldots = \sum \frac{C_i}{L_i} \tag{5}$$

From the above literature RF number of R290 and R1270 are 52.2 and 62.1 kJ/g respectively. Flammability limits of R22 and R134a are not available in literature and hence Jones correlations can be used to compute the flammability limits of R22 and R134a. However from ASHRAE design safety standard 34, the refrigerants R22 and R134a are classified as nonflammable ASHRAE A1 category (ASHRAE 34 2007). Based on RF number, refrigerants are classified into various groups. If RF number is below 30 then it is considered as slightly flammable (ASHRAE A2) group and in between 30 to 150 classified as flammable (ASHRAE A3) group. To compute the RF number of thirteen R134a/R1270/R290 blends equations from (3 to 5) is used and corresponding values are shown in Table 2.

Table 2.   RF number of various studied refrigerant blends

Refrigerants	RF number (kJ/g)	ASHRAE flammability group
M1	14.90	A2*
M2	20.52	A2*
M3	18.35	A2*
M4	19.79	A2*
M5	21.35	A2*
M6	23.05	A2*
M7	19.57	A2*
M8	21.29	A2*
M9	23.14	A2*
M10	25.15	A2*
M11	23.55	A2*
M12	23.81	A2*
M13	24.07	A2*

*Computed values.

4 TEWI ANALYSIS

To compute TEWI of various R134a/R1270/R290 blends the compressor power is required and it is taken from the literature (Sharmas Vali S. and Ashok Babu T.P. 2017).

Totale quivalent warming index is a measure of impact of global warming based on emission of greenhouse gases during the working of the refrigeration equipment and the dumping of the operating refrigerants at the end of life (AIRAH 2012). TEWI takes into an account of both direct CO_2 emissions due to direct emission of refrigerant when the device is operating with particular refrigerant and indirect CO_2 emissions from the various fuels used to generate electrical energy to operate the apparatus during its lifetime (Thomas W.D. 2004).

$$TEWI = Direct\ CO_2\ emissions + Indirect\ CO_2\ emissions$$
$$= \left(GWP_{100} \times m \times L \times S_{Life} \right) + \left(E_{An} \times S_{Life} \times C \right) \tag{6}$$

were GWP_{100} = GWP of a given fluid for a time period of 100 years, m = Charge of the given fluid (kg), L = Leakage rate of the refrigerant (%), S_{Life} = Service lifetime of the device (years), E_{An} = Energy consumption per annual (kWh), C = Indirect emission factor (kg CO_2/kWh).

Table 3. Compressor power of investigated blends.

Refrigerants	$\dot{m}$ (kg/min)	W_c (kJ/kg)	W_{CP} (kW)
R22	1.226	38.949	0.7958
M1	1.028	47.366	0.8115
M2	0.894	54.919	0.8182
M3	0.907	52.487	0.7934
M4	0.875	54.032	0.7879
M5	0.847	55.432	0.7825
M6	0.819	57.072	0.7790
M7	0.945	53.754	0.8466
M8	0.919	54.795	0.8392
M9	0.895	56.47	0.8423
M10	0.871	58.215	0.8450
M11	0.829	57.524	0.7947
M12	0.835	57.675	0.8026
M13	0.841	57.828	0.8105

Table 4. TEWI of investigated blends.

Refrigerant	TEWI (kg CO_2/kWh)
R22	21586
M1	21164
M2	21037
M3	20499
M4	20290
M5	20084
M6	19925
M7	21809
M8	21554
M9	21558
M10	21551
M11	20312
M12	20506
M13	20701

The assumptions made while computing the TEWI of various studied refrigerants are taken from the literature (Vincenzo L.R. 2011). The charge of refrigerant considered in the device is = 1.5 kg, dischargelevel of refrigerant per annum = 5%, service life of the device = 15 years, operating period of air conditioner = 10 h/d and indirect emission factor = 0.45 kg CO_2/kWh. TEWI values of various refrigerant blends are tabulated in the Table 4.

5 RESULTS AND DISCUSSIONS

5.1 GWP of various refrigerant blends

Figure 1 shows the global warming potential of various refrigerant blends. From the fig 1 it is observed that the GWP of four fluids M6, M11, M12, and M13 are lower among the R22 and thirteen studied fluids. Since all the four refrigerant blends (M6, M11, M12, and M13) are blended with higher composition of very low GWP refrigerants like R290 and R1270.

5.2 ODP of various refrigerant blends

From the Table 1 it is observed that all the thirteen investigated refrigerants are ozone friendly in nature. This is due to zero ozone depletion potential of refrigerants (R134a, R1270, and R290). Since R134a, R1270 and R290 are free from ozone layer depleting substance called chlorine.

5.3 RF number of various refrigerant blends

Figure 2 shows the refrigerant flammability number of various refrigerants considered in the present study. Referring to Figure 2 it is evident that RF number of all the thirteen studied refrigerants are less than 30 kJ/g. Refrigerants with RF number less than 30 are categorized into mildly flammable (ASHRAE A2). Therefore all the thirteen investigated refrigerant blends are mildly flammable.

5.4 TEWI of various refrigerant blends

Figure 3 shows the TEWI of various refrigerants investigated in the present study. Refrigerants with a low value of TEWI are desirable from the view point of greenhouse effect.

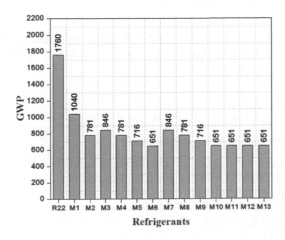

Figure 1. GWP of various refrigerant blends.

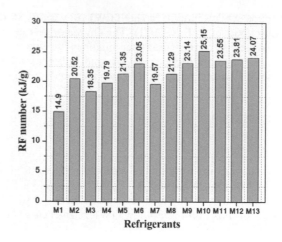

Figure 2. RF number of various refrigerant blends.

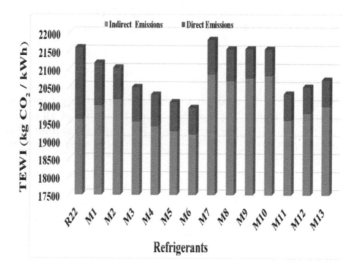

Figure 3. TEWI of various refrigerant blends.

Referring to Figure 3 it is observed that TEWI of M6 is lower among the R22 and thirteen studied refrigerant blends. This is due to lower compressor power and lower global warming potential of M6.

6 CONCLUSIONS

From the present study the following can be drawn.

- The COP of M6 was 2.10% higher among the R22 and thirteen studied refrigerants.
- The GWP of M6, M10, M11, M12 and M13 refrigerant blends were lower among the R22 and thirteen studied refrigerants.
- The ODP of all the thirteen studied fluids were zero compared to ODP of R22 (0.055).
- RF analysis exhibited that all the thirteen investigated refrigerant blends were classified into weakly flammable (ASHRAE A2) group. Since RF number of all the thirteen fluids were less than 30.

- TEWI of ternary blend M6 was lower among the R22 and thirteen studied blends.
- Over all the performance of refrigerant blend M6 (R134a/R1270/R290 50/5/45 by mass percentage) is better than R22 from the stand point of COP, GWP, ODP and TEWI among the thirteen investigated fluids and hence it could be an appropriate alternative candidate to Replace R22 used in air conditioners.

REFERENCES

Ahamed, J.U., Saidur. R, & Masjuki M.M. 2014. Investigation of Environmental and Heat Transfer Analysis of Air Conditioner Using Hydrocarbon Mixture Compared to R22. *Arabian Journal for Science and Engineering*, 39: 4141–4150.

AIRAH. 2012. Methods of calculating Total Equivalent Warming Impact Best Practice Guide lines: 2–20.

ANSI/ASHRAE 2007. Standard 34. Designation and Safety Classification of Refrigerants.

Arora R.C. 2010. *Refrigeration and Air conditioning*. New Delhi: PHI learning Private Limited.

ASHRAE. 2009. *Handbook Fundamentals* (SI) chapter 29. Refrigerants: 29.1–29.10.

Atilla Gencer Devecioğlu & Vedat Oruça. 2015. Characteristics of Some New Generation Refrigerants with Low GWP. *Journal of Energy Procedia*, 75: 1452–1457.

Chen, W. 2008. A comparative study on the performance and environmental characteristics of R410 A and R22 residential air conditioners. *Applied Thermal Engineering*, 28: 1–7.

Devotta, S., Waghmare, A.V., Sawant, N.N., & Domkundwar, B.M. 2001. Alternatives to HCFC-22 for air conditioners. Applied Thermal Engineering, 21: 703–715.

Global Environmental Change Report, 1997. *A brief analysis of the Kyoto protocol.* 9(24).

IPCC. 2014. Fifth Assessment Report chapter 8. *Anthropogenic and Natural Radiative Forcing*: 659–740.

Jones, G.W. 1938. Inflammation Limits and Their Practical Application in Hazardous Industrial Operations. *Chemical. Reviews*, 22, 1–26.

Lorentzen, G. 1995. The use of natural refrigerants: a complete solution to the CFC/HCFC predicament. *International Journal of Refrigeration*, 18 (3): 190–197.

Mohanraj, M., Jayaraj, S. & Muraleedharan, C. 2009. Environment friendly alternatives to halogenated refrigerants-A review. *International Journal of Greenhouse Gas Control*, 3(1): 108–119.

Sharmas Vali, S. & Ashok Babu T.P. 2017. Theoretical Performance Investigation of Vapour Compression Refrigeration System Using HFC and HC Refrigerant Mixtures as Alternatives to Replace R22. *Journal of Energy Procedia*, 109: 235–242.

Shigeo Kondo, Akifumi Takahashi, Kazuaki Tokuhashi, Akira Sekiya. 2002. RF number as a new index for assessing combustion hazard of flammable gases. *Journal of Hazardous Materials*, A93: 259–267.

Thomas W.D. & Ottone C. 2004. A low carbon, low TEWI refrigeration system design. Applied Thermal Engineering 24: 1119–1128.

United Nations Environmental Programme, (1987). *Montreal Protocol on substances that deplete the ozone layer*, Final act. New York: United Nations.

Vincenzo L.R. & Giuseppe P. 2011. Experimental performance evaluation of a vapour compression refrigerating plant when replacing R22 with alternative refrigerants. Applied Energy 88: 2809–2815.

Zabetakis, M.G. 1965. Flammability Characteristics of Combustible Gases and Vapors. *Bulletin 627, US Bureau of Mines*.

Emerging Trends in Engineering, Science and Technology for Society,
Energy and Environment – Vanchipura & Jiji (Eds)
© 2018 Taylor & Francis Group, London, ISBN 978-0-8153-5760-5

Indoor performance evaluation of a Photovoltaic Thermal (PVT) hybrid collector

K.K. Janishali & C. Sajith Babu
Government Engineering College, Calicut, Kerala, India

Sajith Gopi
Kerala Water Authority, Thrissur, Kerala, India

ABSTRACT: A PVT system combines a solar cell, which produces electricity with a solar thermal collector, which extracts thermal energy from the sun so that it can extract both energies simultaneously. A solar panel is capable of achieving a maximum electrical efficiency of 15 to 25%, while a PVT hybrid collector produces combined energy efficiency in the range of 55 to 70%. This ability of the PVT collector to trap a large amount of the sun's energy makes them superior over the conventional solar panels. This study is aimed to analyse the performance of a PVT collector under a solar simulator. A 100 watt solar panel is used for the construction of the PVT collector. Thermal energy is extracted by water flowing through the system, which is in direct contact with the rear side of the PV panel. Water enters through the inlet pipe, heats up by absorbing thermal energy from the panel and leaves through the exit pipe. The experiment was conducted for three light intensities of 600, 800 and 1000 W/m². The performance analysis of the PVT system shown that, these systems are four to five times more efficient than normal PV systems working at outdoor conditions. The results also shown that, the increase in PV panel temperature results a reduction in electrical efficiency, while the overall efficiency of PVT system remained almost maximum at all temperatures for a particular light intensity.

1 INTRODUCTION

In this mechanized world, life will be very easy if sufficient amount of energy is available for us. Every automated system exploits energy to give the desired work output. Therefore, if energy is not available, every machine is a waste. The overexploitation of non-renewable energy sources such as fossil fuels will lead to energy crisis. It's our duty to conserve the available energy for our upcoming generations by depending on renewable energy sources and also by utilizing them in an efficient way. While considering the renewable energy sources, solar energy is considered to be the best among them. This is because, it is sufficiently available and its tapping is cheaper and easy compared to other energy forms. A photovoltaic thermal (PVT) system is used to extract energy from sun to a maximum extend. The PV module will generate electrical energy and the heat developed on the PV module is absorbed by water or air flowing through the system.

APVT system will bring the advantage of both solar thermal collector and PV panel to extract thermal as well as electrical energy. This make PVT panels higher efficient when compared to a solar thermal collector or a PV panel individually. A PVT system is an integration of a PV panel module with an air or water heating system. It produces electrical energy along with thermal energy. A PVT system consists of a PV panel below which a heat transfer fluid (air or water) will be flowing to extract the thermal energy from the panel. Since the heat generated on the PV module is taken away by the heat transfer fluid (HTF), the operating temperature of the panel will be kept lower than normal conditions so that the panel efficiency is improved.

Indoor simulation is used for testing air based PVT collectors (Solanki et al. 2009). From the test results, an overall energy efficiency of 50% was obtained, out of which electrical efficiency obtained was 8.4% and thermal efficiency was 42%. Thermal efficiency obtained was considerably higher compared to the electrical efficiency. Sixteen halogen lampseach having a rating of 500 was used for indoor solar simulation. A small 12v DC fan is used for the circulation of air through the PVT/air system.

A study was conducted (Bahaidarah et al. 2013) to evaluate the performance of PV module with back surface water cooling. Both numerical and experimental studies of the water based PVT system were conducted for the hot climate of Saudi Arabia. The working temperature of the photovoltaic module was significantly reduced about 20%. At the same time the electrical efficiency of the PV module also shown a hike of 9% due to the water cooling provided. This study was conducted at a light intensity of 900 W/m^2 in which about 80% of total energy was captured by the PVT/water system whereas the solar cell was able to extract only 20% of the total irradiance.

In another work conducted on a water based PVT system in Saudi Arabia (Harbi, Eugenio & Zahrani 1998), a higher thermal efficiency is obtained during the summer climate conditions. However the higher temperature of PV panel leads to a loss of 30% in the available electrical energy. But the overall efficiency of the system remained good due to higher thermal efficiency. In winter, the PV power output increased even though the thermal side performance was reduced compared to summer climate.

Researches in the field of PVT collectors started in early of 1940s. Towards the end of 20th century, researches on PVT systems kept on increasing and the area of interest changed to building integrated photovoltaic options. This is because PVT systems are aesthetically better when compared to a separate solar thermal and PV systems.

PVT systems are different type on the basis of coolants used. They are air, water, refrigerant and heat-pipe based systems. Air based PVT systems are the most common type used and they are commercially applied in several engineering practices. These PVT systems can achieve a maximum of 8% electrical efficiency and 39% of thermal efficiency. Lower heat removal rate of air based systems is a major disadvantage compared to other systems. Water based systems are the second popular PVT collector. These systems can achieve electrical efficiency of about 9.5% and thermal efficiency of 50%. When compared to air based systems, water based PVT collector will provide increased thermal as well as electrical efficiency.

1.1 *Indoor simulation*

Indoor simulation of solar energy refers to the process of producing illumination in laboratory conditions and it facilitates in getting approximate natural light in a controllable condition. That is, different light intensities can be obtained easily. This method is used in testing of devices such as photovoltaic panels and sun screens. Light sources like xenon arc lamp and tungsten halogen lamps are generally used in indoor solar simulation. The solar simulator for this work consists of 21 halogen lamps connected in series. Each halogen lamp is having a rating of 150 W. Different light intensity can be obtained using a voltage regulator connected to rear side of the halogen lamps system.

2 EXPERIMENTAL SETUP

A photovoltaic thermal system is developed with water as the heat transfer fluid. The performance evaluation of the developed system is conducted with the help of indoor simulation system.

The experimental setup for the testing of PVT system is shown in Figure 1. It consists of a PVT system, an indoor solar simulator, voltage, current and temperature measuring setup and a water circulating system consisting of a pump and storage tank.

The PVT system consists of a PV panel of 100 Wand an attachment to the rear side of the panel to extract thermal energy from it. Energy extraction is achieved by creating a passage through which water flows and keep a direct contact with the rear side of the panel. The water enters the system through the inlet pipe and heats up by extracting the heat from the PV module. This will provide hot water and also help to reduce the module temperature so that its efficiency is increased. The model of the PVT panel showing plastic frame and mica sheet designed in Creo parametric 2.0 is shown in Figure 2.

Thermocouple are attached on the back side of PV panel at four different places and paralleled to get the average value of temperature. The terminal of the PV module is taken outside by drilling hole to the mica sheet. After fixing Mica sheet and connecting the thermocouple wires, the gaps are scaled with silicone sealant. Polystyrene foam (Thermocol) sheet is used to insulate the back side of PVT system. Then GI sheet is screwed on the frame of the PV panel to complete the design of PVT system.

Power output from the solar panel is obtained using digital multimeter. A variable resistor is used to study the IV characteristics and obtain maximum power point. A mini-pump is used to force the water to flow through the PVT system which is controlled with the help of a temperature controller. A temperature controller is used to control the pump by turning it on or off according to the pre-set value of temperature. There will be terminals to connect the sensing thermocouple and a relay switch which will control the pump. The temperature at which relay turns on is called set point and it can be adjusted with the help of button switches provided on the temperature controller. It is important to measure temperature at various areas to study the performance of the developed system. Temperature of the PV module, inlet and exit temperature of water, and ambient temperatures are measured using K-type thermocouples. A temperature indicator is used to read temperatures from the thermocouples.

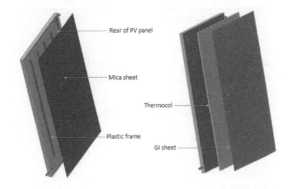

Figure 1.   Experimental setup.

Figure 2.   Creo design of PVT.

453

2.1 Test procedure

The PV panel and the PVT system are tested in indoor conditions. For testing the PV system, the PV panel and halogen lamp system are arranged as shown in Figure 1 and the connections required for temperature and power measurement are also made. Required light intensity is obtained with the help of voltage regulator provided on the halogen lamp system and measuring with a lux meter. The initial temperature of the PV panel and power output are measured and noted. Temperature and power variations are noted in a regular interval of one minute. The testing of the system is continued up to a maximum panel temperature of 70°C so that higher temperature may cause damage to the PV panel. The experiment is repeated for different light intensities.

For testing the PVT system, an additional water circulating system is required. After doing all arrangements similar to PV testing setup, the pump is made on to fill the PVT system fully with water. Required light intensity is obtained with the help of voltage regulator provided on the halogen lamp system and a lux meter. The initial temperature of the PV panel and power output are measured and noted. Temperature and power variations are noted in a regular interval of one minute. The experiment is continued up to a panel temperature of 55°C so that the efficiency of the PV panel is not reduced much. The experiment is repeated for different light intensities.

3 DATA REDUCTION

3.1 PV system

To analyze a PV system, it is very simple compared to a PVT system as it includes only electrical energy output. The power output at the maximum power point is used to find the electrical efficiency of the PV system. The equation for electrical efficiency of the system is given by,

$$\eta_e = \frac{PV\ power\ output}{Energy\ in} = \frac{Imp \times Vmp}{G \times A} \tag{1}$$

where, Imp and Vmp are the current and voltage at the maximum power point, G is the incident solar irradiance normal to surface and A is the collector aperture area.

3.2 PVT system

A PVT system provides electrical energy as well as thermal energy as output. The analysis of the electrical performance of the PVT system is similar to the PV system. Electrical efficiency of the PV system is given by equation 1 and thermal efficiency of the PVT system is given by,

$$\eta_t = \frac{Thermal\ energy\ output}{Energy\ in} = \frac{\dot{m}C(Tout - Tin)}{G \times A} \tag{2}$$

where, $\dot{m}$ is the mass flow rate of the water through the PVT collector, T_{in} and T_{out} are the temperatures of water at inlet and outlet of PVT panel. The overall performance of the PVT system is given by the sum of electrical and thermal efficiency.

$$\eta_{overall} = \eta_{electricall} + \eta_{thermal} \tag{3}$$

4 RESULTS

The PVT and PV systems were tested using indoor solar simulation and the results obtained at different illuminations are plotted below.

The test results of PV and PVT systems at a light intensity of 600 W/m² are shown in Figure 3 and Figure 4 respectively.

Efficiency of PV panel decreases with increase in the module temperature. With an increase of 34°C in PV panel temperature, the PV efficiency shown a relative decrease of 13.2%.

In the case of PVT system, the decrease in overall efficiency is really small. With an increase of 20°C in PV panel temperature, only a reduction of 0.18% in overall efficiency is obtained. A comparison of PVT and PV efficiencies at different panel temperatures with an illumination of 600 W/m² is shown in Table 1. The PV panel efficiency is reducing with increase in temperature while PVT efficiency variation is very low.

Performance of PV and PVT systems at 800 W/m² is shown in Figure 5 and Figure 6. The PV and PVT performance at 800 W/m² is shown a similar trend to that at 600 W/ m².

The relative PV efficiency is reduced by 17.8% for an increase of 34°C in PV temperature. But in the case of PVT system, only a reduction of 1.14% is noted for an increase of 20°C in PV temperature.

Table 2 gives the comparison of PV and PVT at different panel temperatures. At 800 W/m² also PV efficiency is reduced largely compared to the PVT.

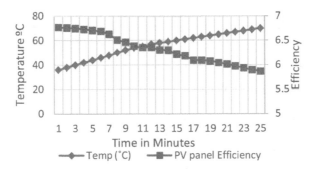

Figure 3. PV performance at 600 W/m².

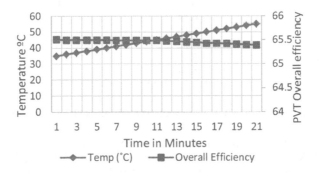

Figure 4. PVT performance at 600 W/m².

Table 1. Comparison of PV an PVT at 600 W/m².

Temperature	PV Efficiency	PVT Efficiency
35	6.78	65.51
40	6.75	65.50
45	6.67	65.49
50	6.51	65.43
55	6.36	65.39
60	6.22	–
65	6.05	–
70	5.87	–

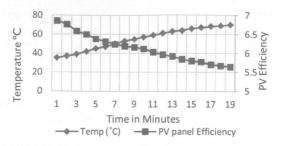

Figure 5. PV performance at 800 W/m².

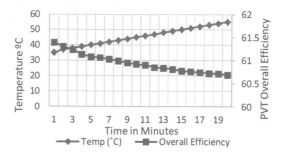

Figure 6. PVT performance at 800 W/m².

Table 2. Comparison of PV an PVT at 800 W/m².

Temperature	PV Efficiency	PVT Efficiency
35	6.86	61.39
40	6.55	61.08
45	6.38	60.92
50	6.24	60.77
55	6.16	60.69
60	5.99	–
65	5.82	–
70	5.64	–

The performance of PV and PVT system at 1000 W/m² is shown in Figure 7 and Figure 8 respectively.

It is clear from figures that the performance of both systems shown a similar trend as of 800 W/m² and 600 W/m². There was a reduction of 16.95% in relative performance of the PV system for a temperature rise of 34°C. In the case of PVT system, the reduction is only 1.17% of available energy output for a temperature rise of 19°C.

PV and PVT efficiency at different panel temperature is tabulated below for 1000 W/m² illumination. In this case also, the PV efficiency is significantly affected by temperature whereas PVT efficiency is not affected much.

From the performance studies of PV and PVT systems at three different light intensities, it is clear that the PV efficiency increases with increase in light intensity. But the temperature increase of PV module resulted in reduction of efficiency for all the three light intensities.

In the case of PVT system, the overall efficiency is showing slight decrease with increase in light intensity. It is because, the corresponding rate of increase in thermal energy is less compared to rate of increase in light energy obtained with the halogen lamp simulator, when luminance is increased from 600 W/m² to 800 W/m² and then to 1000 W/m².

Even the PV panel temperature reached up to 55°C, the temperature attained for water is only 50°C to 51°C only. This is due to the heat transfer loss from PV panel to water.

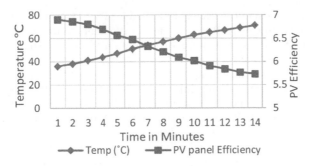

Figure 7.   PV performance at 1000 W/m².

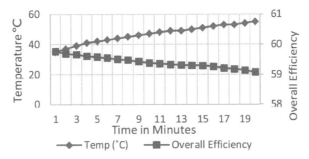

Figure 8.   PVT performance at 1000 W/m².

Table 3.   Comparison of PV an PVT at 1000 W/m².

Temperature	PV Efficiency	PVT Efficiency
35	6.92	59.77
40	6.82	59.63
45	6.66	59.48
50	6.54	59.30
55	6.30	59.07
60	6.09	–
65	5.91	–
70	5.75	–

Table 4.   Flow rate and T_{out} at different irradiation.

Irradiation intensity W/m²	Flowrate Kg/minute	Outlet temperature °C
600 W/m²	0.2143	51
800 W/m²	0.2609	50
1000 W/m²	0.3158	50

The Table 4 will show the flow rates and outlet temperatures of water obtained at three different light intensities for indoor testing.

5   CONCLUSIONS

Performance of the developed PVT system was studied in indoor condition for light intensities of 600, 800 and 1000 W/m². The variation of electrical efficiency of PV system and overall efficiency of PVT system were analysed at all three intensities.

457

From the analysis, the overall efficiency obtained for PVT system is very high compared to the PV panel electrical efficiency. With increase in the temperature, PVT system performed well whereas the normal PV system suffers to maintain efficiency. That is, PVT overall efficiency maintained almost constant at all temperatures whereas PV efficiency reduced at higher temperatures. While considering the PVT system, electrical efficiency of PV module reduced slightly due to increase in module temperature. But a small compromisation in the electrical efficiency is very well compensated with higher overall efficiency of a PVT system. The cooling in PVT system will help to maintain the PV panel temperature at a lower value which will prevent reduction in electrical efficiency.

PV panels normally give an electrical efficiency of 12 to 15%. When compared to that, a PVT system is 4 to5 times more efficient. It is utilizing less space and is cost effective as compared to separate water heating and PV system. The PVT systems are more visually appealing and can be integrated to buildings.

NOMENCLATURE

A	Collector aperture area	m^2
C	Specific heat of water	J/Kg-K
G	Incident sola rirradiance	W/m^2
Imp	Maximum powerpoint current	A
Isc	Short circuit current	A
$\dot{m}$	Mass flow rate of water	Kg/s
Pmax	Maximum Power	W
T_{in}	Inlet temperature of water	°C
T_{out}	Outlet temperature of water	°C
Vmp	Maximum power point voltage	V
Voc	Open circuit voltage	V
ηe	Electrical efficiency	%
$\eta power$	Electrical power generation efficiency	%
$\eta saving$	Energy saving efficiency	%
ηt	Thermal efficiency	%

REFERENCES

Adnan Ibrahim & Goh Li Jin. 2009. Hybrid Photovoltaic Thermal (PV/T) Air and Water Based Solar Collectors Suitable for Building Integrated Applications. *American Journal of Environmental Sciences*. 5: 614–624.

Al Harbi, Y., Eugenio, N.N. & Al Zahrani, S. 1998. Photovoltaic-thermal solar energy experiment in Saudi *Arabia. Renewable Energy*. 15: 483–486.

Bahaidarah, H., Abdul Subhan., Gandhidasan., & Rehman, S. 2013. Performance evaluation of a PV (photovoltaic) module by back surface water cooling for hot climatic conditions. *Energy*. 59: 445–453.

Chegaar, M., Hamzaoui, A., Namoda, A., Petit, P., Aillerie, M., & Herguth, A. 2013. Effect of illumination intensity on solar cells parameters. Advancements in Renewable Energy and Clean Environment, *Energy Procedia*. 36: 722–729.

Chow, T.T. 2010. A review on photovoltaic/thermal hybrid solar technology. *Applied Energy*. 87: 365–379.

Garg, H. & Agarwal, R. 1995. Some aspects of a PV/T collector/forced circulation at plate solar water heater with solar cells. *EnergyConverse Management*. 36: 87–99.

Garg, H.P. & Adhikari, R.S. 1997. Conventional hybrid photovoltaic/thermal (PV/T) air heating collectors: steady state simulation, *Renewable Energy*. 11: 363–85.

Huang, B., Lin, T., Hung, W. & Sun, F. 2001. Performance evaluation of solar photovoltaic/thermal systems. *Solar Energy*. 70: 443–448.

Kumar, K., Sharma, S.D. & Jain, L. 2007. Standalone Photovoltaic (PV) Module Outdoor Testing Facility for UAE Climate. CSEM-UAE Innovation Center LLC 2007.

Solanki, S.C., Swapnil Dubey & Arvind Tiwari. 2009. Indoor simulation and testing of photovoltaic thermal (PV/T) air collectors. Centre for Energy Studies, *Applied Energy*, IIT Delhi. 86: 2421–2428.

Xingxing Zhanga., Xudong Zhaoa., Stefan Smitha., Jihuan Xub. & XiaotongYuc. 2012. Review of R&D progress and practical application of the solar photovoltaic/thermal (PV/T) technologies. Renewable and Sustainable *Energy Reviews*. 16: 599–617.

Emerging Trends in Engineering, Science and Technology for Society,
Energy and Environment – Vanchipura & Jiji (Eds)
© *2018 Taylor & Francis Group, London, ISBN 978-0-8153-5760-5*

Properties of biodiesel and blends: An investigative and comparative study

P.P. Yoosaf, C. Gopu & C.P. Sunil Kumar
Department of Mechanical Engineering, Government Engineering College, Thrissur, Kerala, India

ABSTRACT: Biodiesel is a renewable energy source and an alternate fuel for compression ignition engines as an alternative for diesel. Biofuel satisfies the physical and chemical standards of diesel. Hence it can be used as an alternative for diesel in compression ignition engines. Compared with normal diesel, biodiesels cause less pollution. The main objective of this experimental study is to compare the various properties of pure diesel, fish biodiesel, coconut testa biodiesel, coconut testa Biodiesel-Ethanol-Diesel blend (BED), which are made up of 5 vol% of ethanol and 10 vol% of biodiesel to 85 vol% diesel fuels, and coconut testa Biodiesel-Ethanol blend (BE), which are made up of 5% ethanol and 95% biodiesel. Properties such as flash-fire point, density, viscosity, Acid Value (AV), Saponification Value (SV), Iodine Value (IV), and calorific value of different biodiesels were analyzed.

Keywords: Biodiesel, Transesterification and Properties

1   INTRODUCTION

Recent development in the applications of alternative fuels for compression ignition engines have attracted attention in the automobile domain due to the depletion of fossil fuels and increasing air pollution problems caused by the emissions from various engines. The plastic pyrolysis oil is another alternative fuel for some engine application in certain operation conditions. Biodiesel consists of a mixture of ethyl or methyl esters of fatty acids derived from vegetable oils or animal fats, which are obtained from the transesterification reaction with short-chain alcohol, methanol or ethanol, respectively and in the presence of a catalyst (Parente, 2003). The properties, namely density, viscosity, flash point and fire point of fish oil biodiesel are higher and the calorific value is 0.92 times that of diesel (Shivraj et al., 2014). Flash and fire points increase with an increase in the amount of biodiesel in the blend. The cetane number of Fish Oil Biodiesel (FOB) is higher; this ensures the complete combustion of FOB. The calorific value of B100 is less and increases with the increase in the amount of diesel fuel in the blend, and the flash and fire points also increase with increase in the amount of biodiesel in the blend (Pavan & Venkanna, 2014).

Fuel stability related properties, acid value and iodine value of testa biodiesel is within the range, which shows it has good storage stability (Swaroop et al., 2016). The best proposed solution for reducing diesel engine pollutants is using biofuels that consist of a combination of diesel, biodiesel and ethanol (Hoseini, 2017). A mixture of biodiesel-diesel-ethanol blend is utilized to increase the poor cold-flow properties of biodiesel as the cetane number and lubricity of ethanol-diesel blends is too low (Hatkard et al., 2015). The blended fuels reduced PM emissions, while increased NO x emissions, but reduce smoke and CO emissions (Çelikten, 2011). Using ethanol as fuel or a fuel additive in diesel engines is limited by their miscibility problems with diesel fuel. Other problems are low their cetane number, low lubricity and reduced heating value (Altun et al., 2011). In comparison with the diesel fuel, biodiesel blends produced lower sound levels due to many factors, including an increase in oxygen content, reduction in the ignition delay, higher viscosity and lubricity (Liaquat et al., 2013). The use of different vegetable oils affects production processes and costs, and the resulting biodiesel characteristics used cooking oil require pre-treatment prior to traditional alkali-catalyzed transesterification (Titipong & Ajayk, 2014). The thin brown colored layer outside the fleshy white kernel of coconut when peeled out and pressed produces the coconut testa oil. The testa oil thus produced is saturated and non-edible. Coconut testa is a waste product in the coconut processing industries. It is after peeling out the testa the industries producing products like coconut milk, virgin oil etc. It is also used as a cattle feed. Oil produced from testa is used in paints in industries. This work also aims at producing biodiesel from coconut testa. Due to the wide availability and non-edible nature of the oil, biodiesel produced from testa has enormous potential in meeting the future fuel demands. Being saturated and non-edible, it is one of the best feedstocks available for biodiesel production.

2 MATERIAL AND METHODS

2.1 Production of biodiesel

Arbee Biomarine is the largest producer and manufacturer of biodiesel and glycerin in Kerala, India. The company's biodiesel refinery employs the latest refining technology and equipment and quality assurance procedures. Arbee biodiesel can be directly used in pure form, or in combination with petrodiesel. With virtually no sulphur content, biodiesel has better lubricating properties and much higher cetane ratings. Biodiesel is also proven to reduce fuel system wear. Arbee Biomarine's biodiesel are ideally suited for diesel engines. Biodiesels have much more applications in boilers.

2.2 Transesterification reaction

Transesterification is the process of the reaction of triglyceride (oil/fats) with an alcohol in the presence of an acidic or alkaline catalyst, and it requires the reaction temperature to be below the boiling point of the alcohol used The reaction time should not be less than 30 minutes or more than two hours to form mono alkyl ester that is biodiesel and glycerol. It is a widely employed procedure to reduce the high viscosity of triglycerides.

Triglycerides + Methanol → Methyl ester + Glycerol (Catalyst – KOH).

This is the method used for the preparation of biodiesel from testa. 1 liter of coconut testa oil was taken in a conical flask. 250 mL methanol taken in a beaker and 8 g of KOH was added to it. KOH was allowed to dissolve in methanol to form methoxide. The coconut testa oil was heated to 60°C with constant stirring. The methoxide solution was poured into the oil. The solution was stirred continuously for 45 minutes. The resulting mixture was poured into a separating flask and kept for 24 hours. After 24 hours the glycerol settled at the bottom of the flask with biodiesel on the top. The glycerol settled at the bottom of the flask was separated out to obtain the biodiesel.

3 EXPERIMENTAL PROCEDURES

3.1 Density

Density was measured using the standard method (BIS, 1972). A capillary stopper relative density bottle of 50 ml capacity was used to determine the density of the biodiesel. Density was calculated using the following equation.

$$Density = \frac{W_3 - W_1}{W_2 - W_1} \times \rho H_2O \tag{1}$$

3.2 Kinematic viscosity

Viscosity is a measure of resistance to flow of a liquid due to internal friction caused by one part of a fluid moving over another. Viscosity of biodiesel is measured using a Brookfield viscometer, which satisfies the ASTM standard. The Brookfield viscometer consists of a cylindrical container with a metallic spindle rotating within. For the measurement of viscosity, the biodiesel is poured into the container up to the mark. The spindle rotates within the cylinder at a constant rpm. The instrument measures the torque and thus calculates the viscosity.

3.3   Iodine value

Iodine value or iodine number was introduced in biodiesel quality standards for evaluating its stability to oxidation, and is expressed as the g of iodine that will react with 100 g of biodiesel. A higher iodine value indicates more unsaturated content. Iodine value is a measurement of total unsaturation of fatty acids. Biodiesel having high iodine value is easily oxidized in contact with air. The iodine value largely depends on the nature and ester composition of the feedstocks used in biodiesel production.

B = Volume of sodium thiosulphate for blank
S = Volume of sodium thiosulphate for sample
W = Weight of sample, g

$$Iodine\,value = \frac{(B-S) \times N \times 12.69}{W} \qquad (2)$$

3.4   Acid value

The acid number or neutralization number is a measure of the amount of free fatty acids contained in a fresh fuel sample and of free fatty acids and acids from degradation in aged samples. Acid value determination is used to quantify the presence of acid moieties in a biodiesel sample. The acid number determines the degree of degradation of biodiesel when the fuel is used. The acid value of biodiesel is defined as the number of milligrams of NAOH required to neutralize free acid present in one gram of biodiesel.

$$Acid\,value = \frac{N \times V \times 56.1}{W} \qquad (3)$$

N = Normality = 0.1
V = Volume of NAOH required in mL
W = Weight of sample in g

3.5   Saponification value

The process formation of soap is called saponification. The saponification value is defined as the milligrams of KOH required to saponify 1 g of fat or oil.

$$Saponification\,Value = \frac{N \times (b-s) \times 56.1}{W} \qquad (4)$$

W = Weight of the sample
B = Titre value for blank
S = Titre value of sample

3.6   Flash point and fire point

The temperature at which the vapor of a liquid flash when subjected to a naked flame is known as the flash point of the liquid.
  The flash point is used to classify fuel for transport, storage and distribution according to its hazard level.

3.7 *Calorific value*

The calorific value is the amount of heat energy released by the combustion of unit mass of a fuel. Fuel has higher calorific value and lower calorific value. The amount of heat released as H_2O after the combustion is condensed to liquid state is called higher calorific value. The amount of heat released as H_2O after combustion is in the vapor state is called lower calorific value. The calorific value of fuel was determined by using IKA C200 bomb calorimeter.

4 RESULTS AND DISCUSSION

4.1 *Density*

The density of different fuels is shown in Figure 1 and it is different for each fuel sample. The value of density of BED blend is 828.37 kg/m³, and the density of testa biodiesel is 832.3 kg/m³, both of which are comparable with the density of diesel (833 kg/m³). The density of BE blend is 875 kg/m³ and for fish biodiesel the density is found to be 787.5 kg/m³. All these values satisfy ASTM standards (575–900 kg/m³).

4.2 *Kinematic viscosity*

The kinematic viscosity of different fuels are shown in Figure 2 and it is different for each fuel sample. The value of kinematic viscosity of BED blend is 2 cP, and the density of testa biodiesel is 5.9 cP. The kinematic viscosity of fish biodiesel is comparable with the kinematic viscosity of diesel (2.95 cP). All these values satisfy ASTM standards (1.9–6).

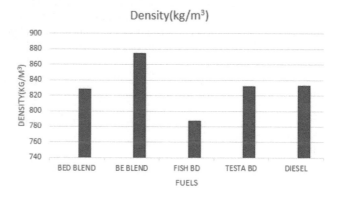

Figure 1.   Density of different fuels.

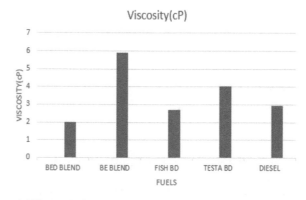

Figure 2.   Viscosity of different fuels.

4.3 *Flash point and fire point*

The flash point and fire point of different fuels are shown in Figure 3 and it is different for each fuel sample. The flash point and fire point were determined for all investigated fuels and it is found that biodiesel blended with ethanol has the lowest flash and fire points. Lower values indicate that those fuels are difficult to handle and transport. The flash and fire point of pure biodiesels, such as fish biodiesel and testa biodiesel, are much higher than pure diesel.

4.4 *Iodine value*

The iodine value of different fuels are shown in Figure 4 and it is different for each fuel sample. It is found that diesel has the lowest iodine value at 2.66 mg I/g. All biodiesels have higher iodine values than the pure diesel, which indicates the fuel samples have more unsaturated contents.

4.5 *Acid value*

The acid value of different fuels are shown in Figure 5 and it is different for each fuel sample. The maximum value of acid value for biodiesels based on ASTM is 0.5 mgKOH/g. All the investigated fuels meet the standard value. BED and BE blend have almost equal and higher values. The acid value is not applicable for pure diesel since it has a lower degree of degradation due to the low amount of carboxylic acid groups.

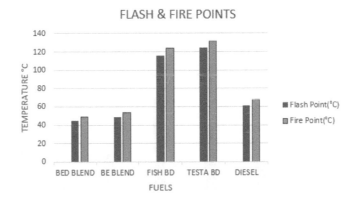

Figure 3.   Flash and fire point of different fuels.

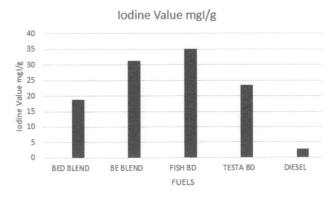

Figure 4.   Iodine value of different fuels.

4.6 *Saponification value*

Saponification of different fuels are shown in Figure 6 and it is different for each fuel sample. Saponification is minimal for pure diesel and much higher for all other biodiesel samples. The maximum value is found in fish biodiesel of the samples. All these values satisfy ASTM standards (max 120).

4.7 *Calorific value*

The calorific value of different fuels are shown in Figure 7 and it is almost equal for each fuel sample. Diesel fuel has the maximum value and all other biodiesel samples have a low and almost equal values.

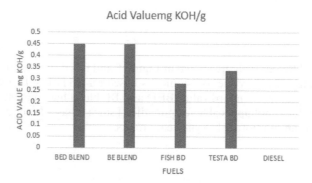

Figure 5.   Acid value of different fuels.

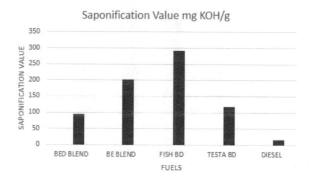

Figure 6.   Saponification value of different fuels.

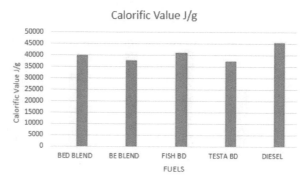

Figure 7.   Calorific value of different fuels.

464

5 CONCLUSION

The proper constant quality of biodiesel can only be promising by analyses the biodiesel quality standards like ASTM, EN and BIS. To attain this aim it is very important to check the quality throughout the biodiesel production, such as trans-esterification, emulsification or any other production techniques, and from the feedstock to the distribution units. The physical and chemical properties of different biodiesels are mainly influenced by the composition of feedstock used in their production process, and the nature of the feedstock and its storage conditions, such as air, sunlight and humidity. Furthermore, different area markets need different quality requirements. The main differences are found in viscosity, iodine value, density and acid value. Various other reasons encountered for these variations are the performance describing properties at very low temperature, such as density at 15°C, as per ASTM standards, and the exposed conditions. It is not possible to devise a formula for standards of biodiesels due to these differences. This would be a major disruption for both biodiesel imports and exports among different countries of the world, and also for the automotive industry.

REFERENCES

Altun, S., C. Öner, F. Yaşar, & Firat, M. (2011). Effect of a mixture of biodiesel-diesel ethanol as fuel on diesel engine emissions. *International Advanced Technologies Symposium (IATS'11)*, Elazığ, Turkey, 16–18.

Çelikten, I. (2011). The effect of biodiesel, ethanol and diesel fuel blends on the performance and exhaust emissions in a DI diesel engine. *Gazi University Journal of Science, 24*(2), 341–346.

Hatkard, N., Salunkeg, B. & Lawande, V.R. (2015). The impact of biodiesel-diesel-ethanol blends. *volume 2, Issue 5.*

Hoseini, S. (2017). The effect of combustion management on diesel engine emissions fueled with biodiesel-diesel blends; *Renewable and Sustainable Energy Reviews, 73*, 307–331.

Liaquat, A.M, Masjuki, H.H., Kalam, M.A., Rizwanul Fattah, I.M., Hazrat, M.A., Varman, M., Mofijur, M. & Shahabuddin, M. (2013). Effect of coconut biodiesel blended fuels on engine performance and emission characteristics. *Procedia Engineering, 56.*

Parente, E.J. (2003). Biodiesel: uma aventura tecnológica num país engraçado (1st ed). Fortaleza: Unigráfica.

Pavan, P. & Venkanna, B.K. (2014). Production and characterization of biodiesel from mackerel fish oil. *International Journal of Scientific & Engineering Research, 5*(11).

Shivraj, H., Astagi, V. & Omprakash, D.H. (2014). Experimental investigation on performance, emission and combustion characteristics of single cylinder diesel engine running on fish oil biodiesel. *International Journal for Scientific Research & Development (IJSRD), 2*(7), *ISSN (online): 2321-0613.*

Swaroop, C., Tennison, K. Jose, & Ramesh, A. (2016). Property testing of biodiesel derived from coconut testa oil and its property comparison with standard values, *ISSN, 2394–6210, 2*(2).

Titipong, I. & Ajayk, D. (2014). Biodiesel from vegetable oils. *Renewable and Sustainable Energy Reviews, 31*, 446–471.

Thermal sciences and transport phenomena

Emerging Trends in Engineering, Science and Technology for Society,
Energy and Environment – Vanchipura & Jiji (Eds)
© 2018 Taylor & Francis Group, London, ISBN 978-0-8153-5760-5

Computational study on two phase flow of liquid nitrogen with different internal coatings

S.S. Bindu, Sulav Kafle, Godwin J. Philip & K.E. Reby Roy
TKM College of Engineering, Kollam, Kerala, India

ABSTRACT: One of the most important application of cryogenics includes the transfer of cryogenic fluid from storage site to its utilization. To optimize the initial phase of cryogenic heat transfer, twisted channels are coated with different coating materials, which increases the chill down efficiency of cryogenic systems. Applications of coating materials like graphene, CNT, polyurethane and Teflon on twisted channels has significant time saving in cool down compared to conventional channels surfaces. Computational study was performed to evaluate the enhancement of heat transfer and coating effectiveness on chill down time. The chill down of two surfaces uncoated and coated is compared and latter one is found to be more efficient.

Keywords: Chill down, Cryogenics, Liquid Nitrogen, Nucleate boiling, Polyurethane coatings, Twisted channels

NOMENCLATURE

Ta	External temperature
Ti	Coil initial temperature
T	Inlet temperature
qc, qu	heat load on coated and uncoated panels.
Uc, Uu	overall heat transfer coefficient for coated and uncoated panels.
$(\Delta\theta)c, (\Delta\theta)u$	temperature difference between skin temperature and LN2 for coated and uncoated panels.
Ac, Au	panel area for the coated and uncoated panels.

1 INTRODUCTION

Cryogenic liquids are used in many technological applications: such as propulsion systems, cooling of superconducting magnets etc. It is also being widely adopted for various clinical applications. Cryogen transfer involving two-phase flow is an indispensable procedure before the operation of these systems. This transfer process is characterized by their highly transient nature. Cryogenic chill down refers to the process by which the temperature of the transfer line is lowered to the saturation temperature of cryogen. This process is highly unstable and characterized by large pressure fluctuations accompanied by transient boiling heat transfer.

A team of researchers at MIT (Preston, Mafra, Miljkovic, Kong, & Wang, 2015) studied the usefulness of ultrathin grapheme coatings on conducting materials using CVD process and reported that it promotes drop wise condensation. CFD analysis of single-phase flows through coiled tubes was (Jayakumarar et al. 2010) and found that the fluid particles undergo oscillatory motion inside the pipe causing fluctuation in heat transfer rates.

Numerical investigation on heat transfer from hot water in shell to cold water flowing in a helical (Neshat, HossainpourF, & Bahiraee, 2014) coil was made and identified that the mass flow rate and specific heat of fluids are dependent on the shell side fluid temperature and geometric parameters of coil.

Goli et al. (2013) conducted an experiment and demonstrated that chemical vapor deposition of graphene on Copper films strongly enhance their thermal diffusivity and thermal conductivity. The enhancement is primarily due to changes in Cu morphology during graphene deposition and associated with temperature treatment. Allen et al. conducted an experiment in which (Allen, 1965) they observed a sudden drop in test section wall temperature at the start of the chilldown process. They concluded it as the result of higher rate of cryogenic heat transfer between transfer line wall and the cryogen being transported. After a period of time, the rate of decrease in wall temperature remains approximately continual until the entire test line is cooled down. In coated panels, the initial heat transfer rate is reduced due to the thermal resistance offered by the Teflon coating and the gas film to the cryogen. As the fluid volume increases, there arises a thermal gradient on each side of the coating leading to the faster attainment of nucleate boiling on the surface in contact with the cryogen. This results in shorter chill down time.

Equations 1 and 2 helps in the comparison of overall heat transfer coefficients under steady state for coated and uncoated panels.

$$qc = UcAc(\Delta\theta)c \tag{1}$$

$$qu = UuAu(\Delta\theta)u \tag{2}$$

The coatings on good conductors provide rapid lowering in surface temperature which, in turn permits rapid attainment of liquid solid contact with good heat removal rates (Maddox, 1966). Comparison of cool down time between internally coated and uncoated propellant lines (Reed, Fickett, & T, 1967) was conducted and found that Kel-F coated pipe cool down rate is faster than that of uncoated line. Sheffer et al. conducted two phase flow of LN2 with pulsated flows and arrived at a conclusion that (Shaeffer et al. 2013) flows with higher Reynolds number will results in higher efficiency. Reed et al. (1967) reported that unlike in uncoated transfer lines where, immediately after cryogen entry into the transfer line a thin layer of vapor is established between the cryogen and the pipe wall causing an increase in chill down time, a non-conductive coating on a conducting pipe can reduce chilldown time by creating a thermal gradient between the fluid-fluid interface.

2 NUMERICAL STUDY USING COMSOL

Heat transfer analysis in twisted channel and effect of coating materials are analyzed by varying coating material, thickness of coating and fluid flow rates. Coating materials used are Polyurethane, Teflon, Graphene and CNT. Flow rates are varied from 0.1 m/s to 0.01 m/s. Coating thickness are varied from 0.025 mm to 0.1 mm for Polyurethane and Teflon, and 0.1 mm for Teflon and CNT. The properties of coating materials are shown in Table 1.

Table 1. Properties of coating materials.

Material/properties	Thermal conductivity (k) W/m.K	Specific heat (Cp) J/kg.K	Density kg/m³
Polyurethane	0.026	1.76	30
Teflon	0.25	970	2200
Graphene	2000	730	1500
CNT	3000	450	1300

2.1 Heat transfer in twisted coil

Steps involved are as follows:

- First model is drawn by drawing two right handed helices giving axial pitch as 50 mm, major and minor radius as 100 mm and 10 mm with a single number of turn.
- Coil material used is copper and a thin layer is added to the inner wall of the coil. The properties of the thin layer are given that of different coating material with all the following properties.

  - The flow of fluid is taken as Turbulent with time dependent under k-epsilon model.
  - Mesh used is physics controlled coarse and mesh statistics are as follows:
    The number of Tetrahedral, Prism, Triangular, Quadrilateral, Edge and Vertex elements are 19240, 15250, 5494, 160, 863 and 16 respectively.
  - Study is done using surface average to find coil surface temperature.

2.2 Boundary conditions

For all case LN2 is used as a fluid, the other boundary conditions used are as follows:

Inlet temperature is taken as 77 K with air as external convection medium. External and initial Coil temperature was 300 K. The Coating thickness for Polyurethane and Teflon was taken as 0.025 mm and 0.1 mm respectively. The corresponding values for Graphene and CNT was 0.1 mm.

3 ANALYSIS

3.1 Case 1: (a) Velocity = 0.01 m/s, Coating of polyurethane having thickness 0.025 mm and 0.1 mm

Figure 1 shows surface temperature variation along the coil. It depicts that the temperature goes on increasing as very cold liquid nitrogen is passed through inlet, lower right end. Temperature increases as there is heat transfer from outside hot region to inner cold fluid.

Figures 2 and 3 depicts the variation of average surface temperature with time when LN2 is flowing through the coil at the flow rate of 0.01 m/s. Average total surface temperature of coated tube is decreased as cold fluid LN2 is passed through it. In case of coating thickness 0.025 mm, it is seen that the tube surface temperature is 267 K at 56.7 second for uncoated coil and same temperature is attained at 56.3 second for coated one. Coated one takes 0.7% lesser time to reach that temperature. For a coating thickness of 0.1 mm, the tube surface temperature is 268 K at 56.7 seconds for uncoated coil and same temperature is attained at 56 seconds for coated one. Coated one takes 1.25% lesser time to reach the same temperature.

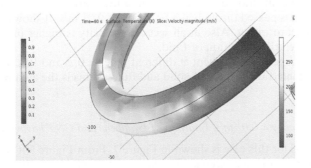

Figure 1. Close view of temperature contour.

471

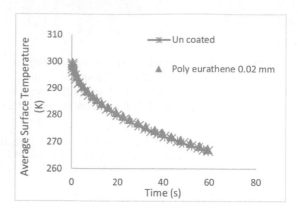

Figure 2.   Variation of surface temperature with time when inlet velocity is 0.01 m/s, coating thickness 0.025 mm.

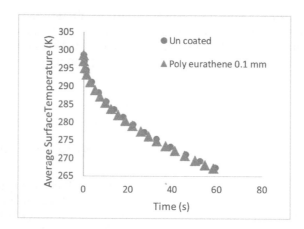

Figure 3.   Variation of surface temperature with time when inlet velocity is 0.01 m/s, coating thickness 0.1 mm.

3.2   *Case 1:(b) Velocity = 0.01 m/s, Coating of Teflon having thickness 0.025 mm*

The result obtained is:

In this case, from Figure 4, it is seen that the tube surface temperature is 272.88 K at 40 seconds for uncoated coil and same temperature is at 39.3 seconds for coated one. So, coated coil takes 0.7 seconds i.e. 1.75% lesser time to reach that temperature.

Figure 5 depicts the variation of velocity magnitude at various section slices along the coil when liquid nitrogen is passed through at a flow rate of 0.01 m/s. It shows velocity is goes on decreasing from inlet to outlet. Also at all sections, velocity is higher in outward direction, this may be due to centrifugal action of fluid.

Figure 6 shows temperature gradient at different time t = 0, 5 and 10 s. Temperature gradient is greater along interface surface of solid and fluid as this is the area where heat transfer interaction occurs.

3.3   *Case 1: (c) Velocity = 0.01 m/s, Coatings of Graphene and CNT having thicknesses 0.1 mm*

The geometric model for this case is shown in Figure 7. From Figure 8 it is seen that the tube surface temperature is 277 K at 28 seconds for uncoated coil and same temperature is at 23.1 seconds for graphene coated. Similarly, 277 K is at 20.9 seconds for CNT coated coil. So, graphene coated coil takes 18.2% lesser time and CNT coated takes 25.35% to reach that temperature.

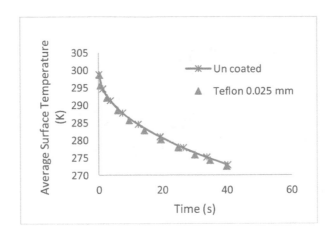

Figure 4.   Variation of surface temperature with time.

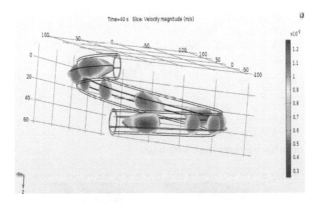

Figure 5.   Contour of velocity magnitude.

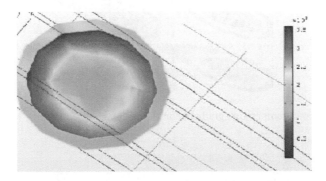

Figure 6.   Temperature gradient at inlet section when inlet velocityis 0.01 m/s.

Figure 9, depicts temperature gradient variation at inlet section slice of the coil at different time, t = 1, 10, 20, 30, 40, and 50 s. Temperature gradient is greater along interface surface of solid and fluid as this is the area where heat transfer interaction occurs. Figure 10 shows surface temperature variation along the coil. It depicts that the temperature goes on increasing as very cold liquid nitrogen is passed through inlet (lower right end in above figure). Temperature increases as there is heat transfer from outside hot region to inner cold fluid. Figures 11 and 12 depicts the variation of average surface temperature with time when LN2

473

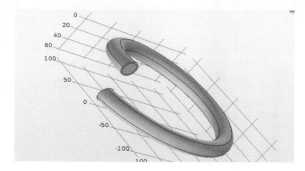

Figure 7.   Geometric model.

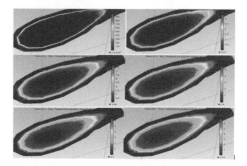

Figure 8.   Variation of surface temperature with time.

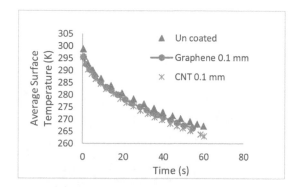

Figure 9.   Close view of temperature gradient contour.

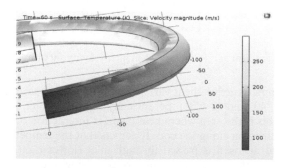

Figure 10.   Close view of temperature contour.

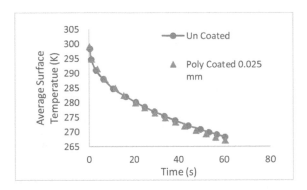

Figure 11.   Variation of surface temperature with time when inlet velocity is 0.1 m/s, coating thickness 0.025 mm.

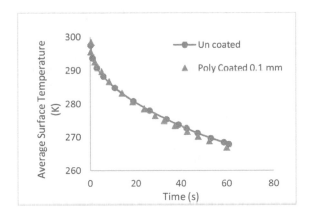

Figure 12.   Variation of surface temperature with time when inlet velocity is 0.1 m/s, coating thickness 0.1 mm.

is flowing through the coil at the flow rate of 0.1 m/s. Average total surface temperature is decreased when cold fluid LN2 is passed through the tube.

3.4   Case 2: (a) Velocity = 0.1 m/s, Coating of polyurethane having thickness 0.025 mm and 0.1 mm

In case of coating thickness 0.025 mm, the tube surface temperature is 268.1 K at 59.9 seconds for uncoated coil and same temperature is at 56 seconds for coated one. So, coated coil take 3.9 seconds i.e. 6.5% lesser time to reach that temperature. Again, in case of coating thickness of 0.1 mm, it is seen that the tube surface temperature is 268.1 K at 55.7 seconds for uncoated coil and same temperature is at 59.9 seconds for coated one. So, coated coil take 4.3 seconds i.e. 7.17% lesser time to reach that temperature.

Figure 13 depicts the variation of velocity magnitude at various section slices along the coil when liquid nitrogen is passed at flow rate of 0.1 m/s. It shows velocity goes on decreasing from inlet (lower part in above figure) to outlet. Also at all sections, velocity is higher in outward direction, may be due to centrifugal action of fluid.

Figure 14 shows temperature gradient at different time t = 1, 5 and 10 s. Temperature gradient is greater along interface surface of solid and fluid as this is the area where heat transfer interaction occurs.

Figure 15 shows surface temperature variation along the coil. It depicts that the temperature goes on increasing as very cold liquid nitrogen is passed through inlet (lower right end in above figure). Temperature increases as there is heat transfer from outside hot region to

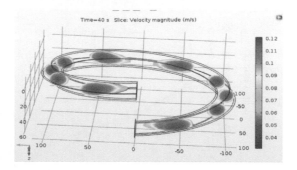

Figure 13. Contour of velocity magnitude.

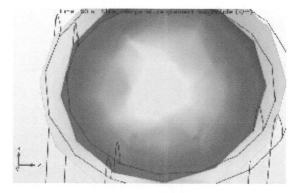

Figure 14. Temperature gradient at inlet section when inlet velocity is at 0.01 m/s.

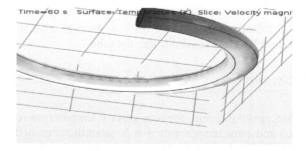

Figure 15. Close view of temperature contour.

inner cold fluid. Figure 16 depicts the variation of average surface temperature with time when liquid nitrogen is flowing through the coil at the flow rate of 0.1 m/s. Average total surface temperature is decreased when coated as cold fluid LN2 is passed through the tube. In case of coating thickness 0.025 mm, it is seen that the tube surface temperature is 268.1 K at 59.9 seconds for uncoated coil and same temperature is at 56.4 seconds for coated one. So, coated coil take 3.5 seconds i.e. 5.85% lesser time to reach that temperature. Again, in case of coating thickness 0.1 mm, it is seen that the tube surface temperature is 268.1 K at 57.6 seconds for uncoated coil and same temperature is at 59.9 seconds for coated one. So, coated coil take 2.3 seconds i.e. 4% lesser time to reach that temperature.

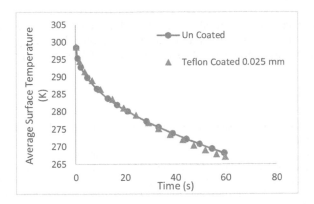

Figure 16.   Variation of surface temperature with time when inlet velocity is 0.1 m/s, coating thickness 0.025 mm.

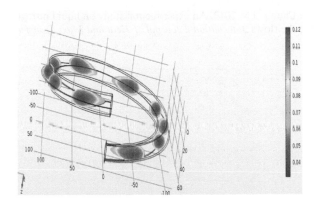

Figure 17.   Contour of velocity magnitude.

3.5   *Case 2: (b) Velocity = 0.1 m/s, Coating of Teflon having thickness 0.025 mm and 0.1 mm*

Figure 17 depicts the variation of velocity magnitude at various section slices along the coil when liquid nitrogen is passed at a flow rate of 0.1 m/s. It shows that velocity goes on decreasing from inlet (lower part in above figure) to outlet. Also at all sections, velocity is higher in outward direction, may be due to centrifugal action of fluid.

4   CONCLUSION AND SCOPE OF FUTURE WORK

From numerical study, by using polyurethane, Teflon, CNT and graphene, the cool down rate is significantly lesser for coated helical channels than uncoated ones. These results should prove useful in the design of future transfer lines. Future work can employ different coating materials to cool down transfer lines made of stainless steel and other suitable materials with different coil shapes.

ACKNOWLEDGEMENT

The authors would like to acknowledge the Space Technology laboratory of TKM college of Engineering and also the Kerala State Council for Science Technology and Environment (KSCSTE) for providing facilities for the successful completion of the project.

REFERENCES

Allen, L.D. 1965. Advances in Cryogenic Engineering. Texas: Cryogenic Engineering Conference Rice University Houston.

Goli, P., Ning, H., Li, X., Lu, C.Y., Novoselov, K.S., & Balandin, A.A. 2013. Strong Enhancement of Thermal Properties of Copper Films after Chemical Vapor Deposition of Graphene.

Jayakumarar, J.S.S.M. Mahajania, J.C. MandalaKannan, N. Iyera, & P.K. Vijayan. 2010. CFD analysis of single-phase flows inside helically coiled tubes. *Computers & Chemical Engineering, 34*:430–446.

Maddox, J.P. 1966. Advances in Cryogenic Engineering. (pp. 536–546). New York: Plenum Press.

Neshat, E., Hossainpour F.S., & Bahiraee. 2014. Experimental and numerical study on unsteady natural convection heat transfer in helically coiled tube heat exchangers. *Heat and Mass Transfer, 50* (6): 877–885.

Preston, D.J., Mafra, D.L., Miljkovic, N., Kong, J., & Wang, E.N. 2015. Scalable Graphene Coatings for Enhanced Condensation Heat Transfer. *American Chemical Society, 15*: 2902–2909.

Reed, R.P., Fickett, F., & T, L. 1967. Advances in Cryogenic Engineering. *12*, pp. 331–339. Plenum Press.

Shaeffer. R, Hu. H, & Chung. J.N. 2013. An experimental study on liquid nitrogen pipe chilldown and heat transfer with pulse flows. *International Journal of Heat and Mass Transfer, 67*: 955–66.

Emerging Trends in Engineering, Science and Technology for Society,
Energy and Environment – Vanchipura & Jiji (Eds)
© 2018 Taylor & Francis Group, London, ISBN 978-0-8153-5760-5

An experimental investigation of cryogenic chilldown time in a polyurethane-coated helical coil

Jesna Mohammed, Pranav Shyam Praveen & K.E. Reby Roy
Thangal Kunju Musaliar (TKM), College of Engineering, Kollam, Kerala, India

ABSTRACT: To optimize the cryogenic chilldown of transfer lines and to improve the efficiency of cryogenic systems, chilldown time should be reduced. Such time saving is associated with reduced consumption of cryogenic fluids. Experiments were performed on helical channels with a polyurethane coating for different inlet pressures, and the performance of an untreated helical surface and one with a coating were compared at corresponding pressures. The results indicated that there are substantial savings in chilldown times with coated surfaces compared to non-coated ones. Liquid nitrogen was used as the cryogen and was passed through helical coils made of copper. The significant reduction in chilldown time is observed only after the onset of a nucleate boiling regime, as indicated by a graph of average surface temperature versus time.

Keywords: Chilldown, Cryogenics, Helical channels, Liquid nitrogen, Polyurethane coating

1 INTRODUCTION

The scope of cryogenic fluid use is typically in industries, space exploration, cooling of electronic components, and in the medical field. Transfer of cryogens to their associated installations is very important prior to their operation. Cryogen transfer is accompanied by phase changes in flow, pressure surges and flow reversal. When cryogens are introduced into a transfer line that is in thermal equilibrium with the ambient temperature, uncontrolled evaporation occurs. In order to establish a steady flow of fluid during this initial phase, cooling down of equipment is a prerequisite, which is termed cryogenic chilldown. To design a cryogenic transfer line, phenomena such as heat transfer, fluid flow, and changes in pressure across the test line need to be identified and managed. Observing chilldown processes, Yuan et al. (2008) indicated that the cryogenic liquid encounters three boiling regimes during chilldown: film, transition, and nucleate boiling. This differs from boiling experiments in that no external heat is provided in the test section.

Berger et al. (1983) found that helically coiled tubes are superior to straight tubes for heat transfer applications. Because of the curvature of helical coils, centrifugal force is introduced resulting in the development of secondary flows, as reported by Dravid et al. (1971). Thus, the movement of the outermost fluid tends to be faster than that at the inside of the coil, which increases the turbulence and thereby increases heat transfer. As reported by Cowley et al. (1962), a reduction in the time taken for cooldown of cryogenic equipment was seen when metallic components were coated with materials of poor thermal conductivity. Allen (1966) suggested that heat transfer by virtue of forced convection between the entering liquid and the transfer line wall results in a sudden temperature drop during the initial phases of the chilldown process. After a brief time period, the rate of temperature drop reduces to a minimum and is then maintained until chilldown is attained. During this period, the flow encountered is film boiling with relatively low-velocity gas flow. When low-conducting coating materials such as Teflon are introduced between the transfer line wall and the cryogen, a thermal gradient is developed resulting in the early attainment of a temperature corresponding to a nucleate boiling regime. This eventually results in higher rates of heat transfer, leading to faster chilldown of the line. Maddox and Frederking (1966) reported that heat removal rates can be improved by providing a coating on good conductors. Leonard et al. (1967), through a comparative study of the chill down times of transfer lines with and without coatings under low-velocity fluid flow, concluded that the coating can enhance heat transfer. In their study on Critical Heat Flux (CHF) in helically coiled tubes, Jensen and Bergles (1981) concluded that CHF initially increases with mass velocity, but then decreases after achieving a maximum value; they found that CHF occurs at the inside surface first. Because of higher centrifugal force and secondary flow, a coil with smaller radius was found to have a higher critical value for the same heat flux, mass velocity and tube length.

In a comparison of heat transfer rates between a straight tube and a helically coiled heat exchanger, Prabhanjan et al. (2002) concluded that helical coils have better heat transfer characteristics and that geometry and flow rate play significant roles in elevating the temperature of the fluid.

In a study of the thermochemical characteristics of refrigerant R134a flow boiling in helically coiled tubes at low mass flux and low pressure, Chen et al. (2010) found that the heat transfer coefficient increases with rises in mass flux, vapor quality, and pressure. Their study also provided many correlations for helical flow.

By conducting transient cryogenic chilldown processes in horizontal and inclined pipes, Johnson and Shine (2015) proposed an upward inclination for effective cryogen transport. In reviewing the two-phase heat transfer characteristics of helically coiled tube heat exchangers, Fsadni and Whitty (2016) suggested some practical correlations for two-phase heat transfer coefficients.

Research by Hardik and Prabhu (2017) on the critical heat flux in helical coils at low pressure done on film boiling crisis concluded that CHF increases as the mass flux increases and decreases with increase in quality and, for the same quality, CHF may decrease with increase in tube and coil diameter.

Thus, the present experiment combines two chilldown strategies by creating a helical geometry and coating it with polyurethane, and the objective of this research is to prove that coated coils are more efficient for use in cryogenic system installations than uncoated ones.

2 EXPERIMENTAL APPARATUS

2.1 *Basic experimental facility*

The experimental setup consisted of a gaseous nitrogen cylinder, a liquid nitrogen Dewar vessel, pressure regulator, valves, a Data Acquisition (DAQ) system, pipes and brass fittings, electric heater, flow meter and a test section consisting of a polyurethane-coated helical coil with T-type thermocouples. Figure 1 shows a schematic of this. The test section was made

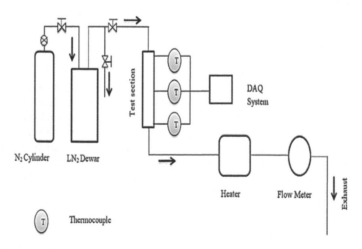

Figure 1. Schematic of experimental system setup.

480

of Oxygen-Free High thermal Conductivity (OFHC) copper tubing (UNS C10100) with an internal diameter of $5/_{16}$ inch (0.0079 m). The test section had helical channels with major diameter of 106 mm and was coated with polyurethane (PU) using a Chemical Vapor Deposition (CVD) process. The thickness of the PU coating was 0.1 mm. The wall temperature measurements were taken at five equidistant points along the curve length by using T-type thermocouple wires. The leak of external heat into the test section was minimized by using a thick layer of polyurethane foam (thermal conductivity of 0.02 W/m K, density of 11 kg/m^3), which is a type of expanded foam insulation. The test section was also covered by yarn and then by nitrile rubber insulation. Liquid nitrogen was supplied to the test section through ½-inch 304-grade stainless-steel pipes and brass fittings from a 55 L Dewar vessel (TA-55) made by IBP Co. Ltd. (Kolkata, India). The liquid nitrogen was pumped from the Dewar vessel by external pressurization using a gaseous nitrogen cylinder having 47 L capacity, which could supply a maximum pressure of 15 bar. The flow rate of liquid nitrogen was manually regulated by a pressure regulator mounted on the external pressurization line. Initially, the entire test section was purged with gaseous nitrogen. A bypass line was introduced to ensure the entry of saturated liquid N$_2$ into the test section. Temperature measurement of the test section was done using T-type thermocouples connected to a Keysight 34972 A data acquisition/data logger switch unit with scanning frequency of 30 milliseconds, which was also connected to a personal computer for data analysis. The average mass flux was measured using a volume flow meter having an accuracy of 0.05 mm^3/s at the exit line from the test section. A mercury thermometer was mounted at the outlet to measure the average temperature of the outgoing gas. A single-phase flow meter was provided to measure flow rate. To ensure that a single-phase gas flow entered the flow meter, the outlet line was placed in a hot water bath heated by electric heaters. A pressure gauge was provided to measure and regulate initial pressure. Different flow control valves were provided to control the flow.

2.2 Procedure

Straight copper tube was first coated with polyurethane to a thickness of 0.1 mm. From this, a helical test section with pitch angle of 8° and the above-mentioned dimensions was prepared. The thermocouples were fixed on the surface of the test section circumferentially (120° apart) at five equally spaced locations with three thermocouples in each section (in one pitch length). The test section was covered first by yarn and then by nitrile rubber insulation. Polyurethane foam was sprayed onto it for further insulation. The experiment was conducted at two different inlet flow pressures of 6.89 kN/m^2 and 8.61 kN/m^2. The equivalent mass flow rates were calculated to be 93 kg/m^2 s and 116 kg/m^2 s, respectively.

3 RESULTS AND DISCUSSIONS

3.1 Temperature profile in chilldown process

The temperature profile determines how fast the cryogenic fluid can cool down the tube. Comparison of temperature profiles is useful in identifying the chilldown effectiveness of coated and uncoated surfaces. Three main boiling regimes, film boiling, transition boiling, and nucleate boiling, are the focus while analyzing the chilldown phenomenon. The initial phase is film boiling where because of a larger temperature difference between the cryogen and the tube wall, the fluid undergoes rapid evaporation and the test section is filled entirely with vapor. The point at which liquid comes into contact with the wall for the first time is termed the Leidenfrost point and is the point of minimum heat flux. This represents the initiation of the transition boiling regime. Because of the presence of liquid contact with the wall, heat transfer greatly increases. This maximum heat transfer point is called the critical heat flux point, which signals the onset of the nucleate boiling regime. Because the heat transfer coefficient is large, the nucleate boiling regime has a higher heat transfer rate and results in faster cooling. On uncoated surfaces, the solid–fluid thermal resistance forms the basis for the heat transfer at the interface. The vapor film developed immediately between the wall and fluid during the initial flow period causes an increase in the time required for chilldown because it takes time to break this boiling envelope. In the case of coated surfaces, the low conductivity of the coating material results in a thermal gradient on either side causing a reduction in temperature between fluid–fluid interfaces, thus resulting in a reduction in chilldown time.

3.2 *Comparison of temperature profile between coated and uncoated surfaces*

The experiments were conducted on two helical test sections with coated and uncoated surfaces for two different mass flow rates. The average of all working thermocouples was deemed as the chilldown time, and the results are shown in Figure 2. It can be seen that the coated channel attains chilldown more quickly than the uncoated one; here, in the case of coated channels, the time required to cover film boiling region is reduced to 28%. The uncoated test section shows a gradual attainment of nucleate boiling while the coated test section produces a sudden attainment of nucleate boiling, thereby resulting in a substantial reduction in chilldown time. This significant chilldown time reduction is seen after 275 seconds, that is, only after attaining the nucleate boiling regime. Figure 2 describes the variation of average surface temperature with time for the mass flow rate of 93 kg/m^2 s. From Figure 3, it is clear that, at the average chilldown temperature of 117 K, the uncoated tube requires 352 seconds to reach this temperature, while the coated one requires just 298 seconds, thereby resulting in a time saving of 54 seconds. Figure 4 illustrates the variation of average surface temperature with time for the mass flow rate of 116 kg/m^2 s. It can be concluded that the significant chill down time reduction for this condition is only seen after 165 seconds, that is, only after the attainment of the nucleate boiling regime. However, for the uncoated test section the transition occurs at around 72 seconds. The high specific heat of the polyurethane coating could have led to the time delay in completing the transition regime. From Figure 5, it can be seen that to reach an average chilldown temperature of 126 K, the uncoated tube requires 307 seconds, while the coated tube took 251 seconds, thereby resulting in a time saving of 18%. The high specific heat of polyurethane keeps the temperature difference higher for longer, which improves heat transfer and helps to reduce the chilldown time compared to an uncoated coil. Figures 6 and 7 show the variation of temperature along the tube from inlet to outlet at 200 and 300 seconds, respectively. At 200 seconds, the temperature varies between 118.9 K and 129.8 K for the coated tube and between 123.7 K and 119.3 K for the uncoated one. Similarly, at 300 seconds, the temperature varies from 135.2 K to 135 K for the coated tube and 131.07 K to 141.60 K for the uncoated tube. From Figure 7, it can be seen that the temperature towards the end of the tube is increasing for both cases, which is due to reduced temperature difference between

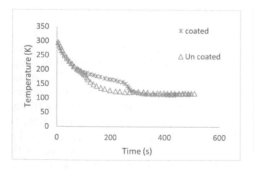

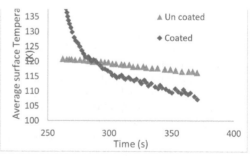

Figure 2. Average surface temperature vs. time for mass flow rate of 93 kg/m^2 s.

Figure 3. Average surface temperature vs. time after 270 seconds for mass flow rate of 93 kg/m^2 s.

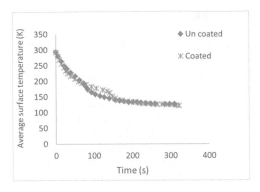

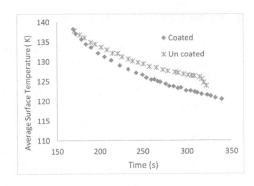

Figure 4. Average surface temperature vs. time for mass flow rate of 116 kg/m^2 s.

Figure 5. Average surface temperature vs. time after 170 seconds for mass flow rate of 116 kg/m^2 s.

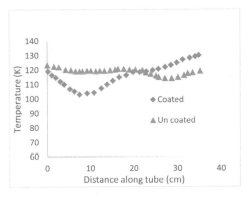

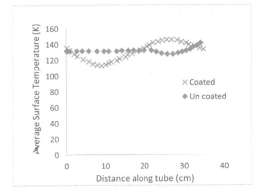

Figure 6. Variation of temperature along tube from inlet to outlet at 200 seconds.

Figure 7. Variation of temperature along tube from inlet to outlet at 300 seconds.

the coil and the fluid. This in turn reduces heat transfer to the fluid, which causes a lower reduction in temperature compared to the inlet temperature drop.

According to Cowley et al. (1962), chilldown time can be decreased by using a thin layer of insulating material on the surface of the metal, and they also postulated that it can be applied to the internal surface of cryopanels. The narrow region in which the maximum heat transfer exists, after which the nucleate boiling begins, can be widened with the addition of a low thermal conductivity coating material on the conducting surface. When enough fluid is present, a thermal gradient is developed because of this layer that helps to attain the temperature in the critical maximum heat transfer region, resulting in the shortening of chill down time. From our observations, with an increase in mass flow rate, the chilldown time decreases as the quantity of fluid flowing through the tube increases. Because the film boiling regime has a lower heat transfer rate than the nucleate boiling regime, we have focused our comparison on the nucleate boiling regimes of the coated and uncoated surfaces. On the basis of the higher slope of the nucleate boiling regime of the coated tubes and the temperature profiles, it can be inferred that heat transfer would be enhanced by coating the inner walls of the tube, resulting in further reduction of chilldown time.

4 CONCLUSION AND SCOPE OF FUTURE WORK

In our experiment on the chilldown of polyurethane-coated and uncoated helical coils, it can be concluded that polyurethane coating of the helical coil has increased the time required for transition from film boiling to nucleate boiling as a result of the high specific heat of polyurethane.

Because of this high specific heat, the temperature drop of the polyurethane-coated tube takes longer than the uncoated one. This maintains the flow in a film boiling regime for longer, but after transition the temperature drop was rapid and for a longer duration. Here, the temperature drop of the polyurethane is considerably lower, which keeps the temperature difference higher, causing higher heat transfer compared to the uncoated tube.

Because of the sudden increase in heat transfer after the transition, it was found that the polyurethane-coated coil has a shorter chilldown time for different mass flow rates. These results should prove useful in the design of transfer lines. Future work can employ different combinations of materials for transfer lines and coating materials with lower conductivity. The effectiveness of different geometries can also be investigated. An in-depth understanding of this phenomenon can be obtained by considering the heat transfer coefficient variation in the three flow regimes during the chilldown process.

ACKNOWLEDGMENTS

The authors would like to thank the Space Technology Laboratory of TKM College of Engineering, and also the Technical Education Quality Improvement Programme Phase II (TEQIP-II), promoted by the National Project Implementation Unit, Ministry of Human Resource Development, Government of India, for their support.

REFERENCES

Allen, L. (1966). A method of increasing heat transfer to space chamber cryo panels. *Advances in Cryogenic Engineering, 11*, 547–553.

Berger, S., Talbot, L. & Yao, L. (1983). Flow in curved pipes. *Annual Review of Fluid Mechanics, 15*, 461–512.

Chen, C.-N., Han, J.-T., Jen, T.-C. & Shao, L. (2010). Thermo-chemical characterstics of R134a flow boiling in helically coiled tubes at low mass flux and low pressure. *Thermochimica Acta, 512*, 1–7.

Cowley, C., Timson, W. & Sawdye, J. (1962). A method for improving heat transfer to cryogenic fluid. *Advances in Cryogenic Engineering, 7*, 385–390.

Dravid, A., Smith, K. & Merrill, E. (1971). Effect of secondary fluid motion on laminar flow heat transfer in helically coiled tubes. *AIChE Journal, 17*, 1114–1122.

Fsadni, A.M. & Whitty, J.P. (2016). A review on the two-phase heat transfer characteristics in helically coiled tube heat exchangers. *International Journal of Heat and Mass Transfer, 95*, 551–565.

Hardik, B. & Prabhu, S. (2017). Critical heat flux in helical coils at low pressure. *Applied Thermal Engineering, 112*, 1223–1239.

Jensen, M.K. & Bergles, A.E. (1981). Critical heat flux in helically coiled tubes. *Journal of Heat Transfer, 103*, 660–666.

Johnson, J. & Shine, S. (2015). Transient cryogenic chill down process in horizontal and inclined pipes. *Cryogenics, 7*, 7–17.

Leonard, K., Getty, R. & Franks, D. (1967). A comparison of cooldown time between internally coated and uncoated propellant lines. *Advances in Cryogenic Engineering, 12*, 331–339.

Maddox, J. & Frederking, T. (1966). Cooldown of insulated metal tube to cryogenic temperature. *Advances in Cryogenic Engineering, 11*, 536–546.

Prabhanjan, D.G., Raghavan, G.S.V. & Rennie, T.J. (2002). Comparison of heat transfer rates between a straight tube heat exchanger and helically coiled heat exchangers. *International Communications in Heat and Mass Transfer, 29*, 185–191.

Yuan, K., Chung, Y.J.N. & Shyy, W. (2008). Cryogenic boiling and two-phase flow during pipe chilldown in earth and reduced gravity. *Journal of Low Temperature Physics, 150*, 101–122.

Emerging Trends in Engineering, Science and Technology for Society,
Energy and Environment – Vanchipura & Jiji (Eds)
© 2018 Taylor & Francis Group, London, ISBN 978-0-8153-5760-5

Effect of concentrated solar radiation on reduction of pressure drop in oil pipeline

V.C. Midhun, K. Shaji & P.K. Jithesh
Department of Mechanical Engineering, Government Engineering College, Kozhikode, India

ABSTRACT: Oil transportation through pipeline is considered as an effective and economical method. Heating is an effective method used for the reduction of crude oil viscosity and pressure drop. In the present work, 3D steady state CFD analysis carried out in an oil pipeline which is heated by applying concentrated solar radiation using a Parabolic Trough Collector (PTC) where the pipeline acts as an absorber pipe. The analysis focuses on the effect of concentrated solar radiation for reducing pressure drop in the oil pipeline. This reduction in pressure drop helps to reduce the pumping power requirement. The influence of concentrated solar radiation and Reynolds number on pressure drop in heated oil pipeline is investigated. Results from the CFD analysis indicates that considerable pressure drop reduction can be achieved by heating the pipe with concentrated solar radiation using PTC.

1 INTRODUCTION

The solar energy is considered as the most essential, clean and inexhaustibly accessible renewable energy. Its two main applications can be classified into heating and generating electrical energy. Solar energy utilisation has a great scope in the present situation. This work is related to the utilization of solar energy to solve the transportation problem faces in oil pipelines.

It is observed that, heating is one of the usual methods used to reduce the crude oil viscosity for reducing pressure drop when compared with other chemical treatment methods. In (Midhun et al. 2015), authors mentioned a novel method of heating crude oil pipelines using Parabolic Trough Collector (PTC) for reducing pumping power by applying concentrated solar radiation on pipe surface. Here the authors tried to investigate the pressure drop in heated oil pipeline and adiabatic pipe and its comparison is made to show the pressure drop reduction. The hydrodynamic and thermal characteristics of the flow were also investigated to explain the nature of flow and heat transfer inside the pipe. Also the relevance of a three

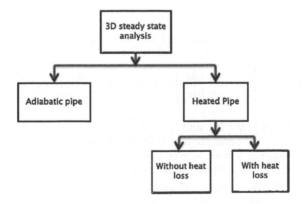

Figure 1. Oil pipeline at different conditions.

dimensional steady state numerical analysis of pipe line was discussed. But the effect of different concentrated radiation and different flow condition were not investigated. In this paper, a 3D steady state CFD analysis on oil pipeline was carryout by applying different level of concentrated solar radiation its surface. Here a detailed analysis is carried out to find the effect of different concentrated solar radiation (heat flux) and different Reynolds number flow in pressure drop reduction in oil pipeline. In practical situation, there will be convective and radiation losses occur from pipeline surface to surrounding. So a comparative analysis between the heated oil pipeline with and without heat loss is also included. Through this the effect of heat loss in pressure drop reduction is investigated in this work. Figure 1 shows the different conditions that an oil pipeline confronts, which are analysed in this work.

2 METHOD

In this work, the main motive is to find the effect of Reynolds number of the flow and different level of concentrated solar radiation on pressure drop reduction in oil pipelines. Thus it is necessary to conduct a single phase three dimensional analysis of the flow through an oil pipeline to determine the pressure drop inside the pipeline along with hydrodynamic and thermal characteristics of the flow. So this analysis mainly focuses on the impact of variation of Reynolds number of the flow and heat flux on reduction of pressure drop in oil pipelines heated by PTC. Numerical analysis is done by using CFD software tool ANSYS FLUENT 14.5.

The steady state equation for conservation of mass or can be written as follows:

$$\frac{\partial}{\partial x_i}(\rho u_i) = 0 \tag{1}$$

Momentum Equation:

$$\frac{\partial}{\partial x_i}(\rho u_i u_j) = -\frac{\partial P}{\partial x_i} + \frac{\partial}{\partial x_i}\left[(\mu + \mu_t)\left(\frac{\partial u_i}{\partial x_j} + \frac{\partial u_j}{\partial x_i}\right) - \frac{2}{3}(\mu + \mu_t)\frac{\partial u_i}{\partial x_l}\delta_{ij}\right] + \rho g_i \tag{2}$$

Energy equation:

$$\frac{\partial}{\partial x_i}(\rho u_i u_j) = \frac{\partial}{\partial x_i}\left[\left(\frac{\mu}{Pr} + \frac{\mu_t}{\sigma_T}\right)\frac{\partial T}{\partial x_i}\right] + S_R \tag{3}$$

The most frequently adopted turbulence model is the k – ε models. Transport equation for Realizable k-ε Model:

$$\frac{\partial}{\partial x_j}(\rho k u_j) = \frac{\partial}{\partial x_i}\left[(\mu + \mu_t)\frac{\partial k}{\partial x_j}\right] + G_k + G_b - \rho\varepsilon - Y_M + S_k \tag{4}$$

$$\frac{\partial}{\partial x_i}(\rho \varepsilon u_i) = \frac{\partial}{\partial x_i}\left[\left(\mu + \frac{\mu_t}{\sigma_\varepsilon}\right)\frac{\partial \varepsilon}{\partial x_i}\right] + \rho C_1 S\varepsilon - \rho C_2 \frac{\varepsilon^2}{k + \sqrt{\vartheta\varepsilon}} + C_3\frac{\varepsilon}{k}C_{3\varepsilon}G_b + S_\varepsilon \tag{5}$$

where $C_1 = \max\left[0.43, \frac{\eta}{\eta+5}\right]$, $\eta = S\frac{k}{\varepsilon}$ and $S = \sqrt{2S_{ij}S_{ij}}$

G_k and G_b corresponds the generation of turbulence kinetic energy due to the mean velocity gradient, and turbulence kinetic energy generation due to buoyancy, Y_M stands for the contribution of the fluctuating dilatation in compressible turbulence to the overall dissipation rate. $C_{1\varepsilon}$ and C_2 are the constants. σ_k and σ_ε stands for turbulent Prandtl numbers for k and ε. S_R, S_k and S_ε represents the user-defined source terms.

Single phase steady state fluid flow without gravitational effect is taken for the analysis. It is assumed that heat flux around the pipe surface by direct and concentrated solar radiation on the absorber tube is taken in approximated manner. Inner surface of the pipe is considered to be smooth.

2.1 Computational domain

Inner diameter and outer diameter of pipe are 0.02 m and 0.025 m. This small diameter pipe is normally used in boilers for carrying furnace oil. The Figure 2 describes the practical condition that an oil pipeline heated by concentrated solar radiation along with the heat loss from pipe surface to surroundings. The computational domain is created by using ANSYS Design Modeler. The fluid domain is for crude oil and solid domain for pipe material. The pipe wall of 1 m from the inlet side is considered as adiabatic wall in all the three case. The concentrated radiation is applied on the base of pipe surface by using a Parabolic Trough Collector (PTC) of rim angle 160° where the pipeline acts as absorber pipe. In heated region of the oil pipe, the concentrated solar radiation falls on the bottom surface of the pipe and direct solar radiation falls on the top surface.

2.2 Mesh details

Meshing of the computational domain explained above was done by using ANSYS ICEM. The total no of elements in the domain are 435774 with 454608 nodes. The fluid and solid domain are created by using sweep method. Inflation is provided in the fluid domain closer to the inside pipe wall so that fluid domain nearer to the pipe domain will be finely meshed which is necessary to incorporate the effect of boundary layer. The surface of the pipe is divided in to some parts for providing proper boundary conditions.

Average fluid outlet temperature is monitored with different levels of meshing for Reynolds number 12000 and heat flux of 80000 W/m². The variation of fluid temperature decreases with increase in number of mesh elements and temperature approaches a value of 299.93 K. So further increase in mesh elements have no considerable effect in temperature effect. The temperature approaches a tolerance value of 0.001 K and the domain becomes grid independent at 435774 mesh elements.

2.3 Material properties

In this analysis fluid domain is Iranian Light Dead crude oil. Aluminium is taken as the pipe material because of its reasonable mechanical and thermal properties. Properties of Aluminium as $\rho = 2719$ kg/m³, $c_p = 871$ J/kg K, $k = 202.4$ W/m K respectively. The heat capacity and thermal conductivity of crude oil are 1887 J/kg K and 0.1483 W/m K. The density is taken as a function of temperature:

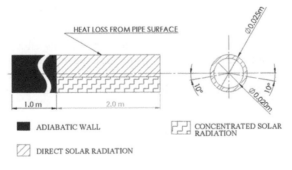

Figure 2.   Heating of oil pipeline by applying concentrated solar radiation on base surface (with heat loss).

$$\rho = -0.732T + 869.3 \tag{6}$$

from (Tavakoli and Baktash 2012) where temperature T is in °C. The correlation used for dynamic viscosity (μ) from (Sattarin et al, 2007) for light dead crude oil is,

$$\mu = a \times \frac{e^{\frac{b}{API}}}{API} \tag{7}$$

where $a = 0.00735T^2 - 4.3175T + 641.3572$, $b = -1.51T + 568.84$.

Here the dynamic viscosity μ is in centipoise (cP) and temperature T is in Kelvin (K).

2.4 Solution methodology

The governing equations were solved using the commercial software package of FLUENT. The second order upwind scheme was employed for convective term discretization of momentum, energy, turbulent kinetic energy and turbulent dissipation; whereas least square cell based scheme for gradients and standard for pressure equation quantities at cell faces are computed using a multidimensional linear reconstruction approach. SIMPLE algorithm was chosen for pressure and velocity coupling. The solution was considered to be converged when the difference was limited to the third decimal point for solution of velocity terms and the sixth decimal point for energy and continuity solutions.

2.5 Boundary conditions

The pipe inlet is given with velocity inlet boundary condition with oil temperature of 298 K and outlet with pressure outlet condition. The inlet velocity is taken on the basis of Reynolds number at inlet with property values based on the temperature of the crude oil. The operating pressure is taken as 101.3 kPa and a gauge pressure of 0 Pa is given as the outlet pressure. For the analysis of insulated oil pipeline, the pipe wall is given with adiabatic wall condition whereas in the case of heated pipeline, the pipe domain surface is split up in to two basic regions—adiabatic region and heated region. 1 m length of pipe wall surface is given with adiabatic condition. The remaining part of the pipe surface is split in to four regions (top and bottom surface in each 1 m length). The top surface of 2 m length of pipe is provided with wall boundary condition for applying the effect of direct solar radiation. The corresponding bottom portion of the pipe is also provided with wall boundary condition to account the effect of concentrated solar radiation. Heat generation (q_g) equivalent to constant heat flux of 1000 W/m² is given to the top surface. In addition to that the effect of concentrated solar radiation is applied on the pipe surface in the form of heat generation on pipe surface equivalent to the heat flux. This equivalent value of heat generation is calculated by using the relation:

$$q_g''' \times \frac{\pi}{4}\left(\left(D_0 + 10^{-6}\right)^2 - D_0^2\right) \times L = q'' \times \pi D_0 \times L \tag{8}$$

By using mixed type of wall boundary condition, the heating effect (by heat generation) along with convection and radiation losses from the pipe surface can be considered. The combined heat transfer coefficient for forced convection and natural convection loss from the pipe to the surrounding can be calculated by a correlation against wind speed as from (Duffie and Beckman 2013):

$$h_0 = 5.7 + 3.8u_W \tag{9}$$

where u_w is the velocity of wind. In this work h_0 is taken as approximately 10 W/m².

The calculation for convection loss with atmospheric is based up on equation 9 assuming the atmospheric temperature $T_\infty = 300$ K, convection coefficient $h_0 = 10$ W/m² K. Radiation losses from pipe surface to the sky are accounted by calculating the sky temperature by using the relation given in (Sharma and Mullick 1991).

$$T_{sky} = 0.0552 T_\infty^{1.5} \qquad (10)$$

from that $T_{sky} = 286.82$ K and the atmospheric temperature is taken as $T\infty = 300$ K. The emissivity of sky ε_{sky} is calculated by using Trinity equation:

$$\varepsilon_{sky} = 0.787 + 0.0028 T_{dp} \qquad (11)$$

where $\varepsilon_{sky} = 0.83$ and 15.1°C as dew point temperature.

3 RESULTS AND DISCUSSIONS

The main objective of the present work is to find the effect of Reynolds number of the flow and different level of concentrated solar radiation on pressure drop reduction in oil pipelines. The heat loss from the pipe to the surroundings is also considered in this analysis, so it can be considered as almost a practical case. Due to this heat loss, naturally the effect of heating in reducing pressure drop is less when compared to the ideal case (without heat loss) except at higher level of heating. Here the hydrodynamic and thermal characteristics of the flow are analysed and compared with ideal case in-order to find the effect of heat loss. The pipe line is provided with non-uniform heating across the section, so average value of temperature, friction factor and Nusselt numbers at each section of the oil pipeline are considered here.

3.1 Effect of Reynolds number at constant heat flux (with heat loss)

Pressure distribution in the adiabatic pipe and heated pipe at different Reynolds number can be compared from the Figure 3. The pressure drop in adiabatic pipe is higher than the heated pipe for each Reynolds number. This pressure drop reduction in heated pipe is due to decrease in viscosity.

Figure 3 shows the comparison of pressure variation with axial length at constant heat flux and adiabatic condition under different Reynolds number. The effect of heat loss is not much reflected in the reduction in pressure drop. The nature of pressure curve is similar to that of the ideal case (without heat loss). The difference between the pressure drop in heated oil pipe having thermal leakage and heated oil pipe at ideal case is very small. It is observed that

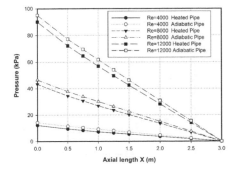

Figure 3. Pressure distribution along the heated pipe (at constant concentrated solar radiation of 80000 W/m²) and adiabatic pipe under different Reynolds number.

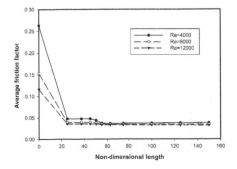

Figure 4. Comparison of average friction factor of the heated pipe along dimensionless lengthat concentrated solar radiation = 80000 W/m² under different Reynolds number.

pressure drop in ideal case is higher when compared with pressure drop in heated pipe with heat loss. Because, here the increase of friction factor due to increase in velocity gradient at heated pipe wall near to the outlet is lower than the ideal case.

The variation of friction factor (f) with dimensionless length (X/D) in a heated pipe at different Re is shown in the Figure 4. Friction factor will be higher at the entrance region due to higher velocity gradient and attains a constant value at fully developed region but here the pipe is heated by direct and concentrated solar radiation, so the viscosity and density of fluid are reduced. After the sudden decrease in friction factor there is a gradual increase in friction factor along the axial length in the heated region. This rise in friction factor is due to the increase in velocity gradient at the pipe wall. Decrease in fluid density near to the heated pipe wall leads to increase in fluid velocity near the pipe wall. Thus the velocity gradient increases and overcomes the effect of decrease in viscosity.

Figure 5 shows the pressure drop variation with Reynolds number in heated pipe and adiabatic pipe where the concentrated solar radiation (heat flux) is maintained as 80000 W/m². From the plot it is observed that the difference between the pressure drop in adiabatic pipe and heated pipe is increasing with increase in Reynolds number. While increasing Reynolds number there will be a decrease in friction factor. Pressure drop difference between adiabatic pipe and heated pipe is increasing with increase in Reynolds number. Thus at higher Reynolds number this heating method is more effective in reducing the pressure drop for this corresponding heat flux.

3.2 *Effect of different heat flux (concentrated solar radiation) at constant Reynolds number*

In this analysis the Reynolds number at the inlet is maintained at 3000 while the concentrated radiation (heat flux) is varied around the bottom of the oil pipe surface. Here also the heat loss from the pipe is considered.

Sudden decreases in friction factor due to decrease in viscosity by heating can be observed from Figure 6, but after that f attains a very small rise along the axial length. This is because the density of fluid near to the wall decreases further while the viscosity variation is very small. The velocity of fluid near to the wall increases because of decrease in density. This increase in fluid velocity leads to the increase in wall shear stress because of high velocity gradient at the wall surface. Thus the friction factor began to increase gradually up to outlet of the pipe. It leads to an adverse effect on pressure drop reduction with increase in heat flux which is analysed further in detail by obtaining plot for pressure drop versus heat flux.

Figure 7 shows the comparison of pressure drop between oil pipes at ideal case and with thermal leakage at constant Re under different heat flux. Pressure drop in heated pipe without heat loss and with heat loss are plotted against heat flux. The pressure drop increase is

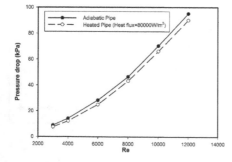

Figure 5. Comparison of pressure drop variation with Reynolds number in adiabatic pipe and heated pipe.

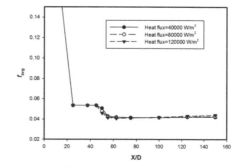

Figure 6. Comparison of average friction factor along the pipe at $Re = 3000$ and different concentrated solar radiation (heat flux).

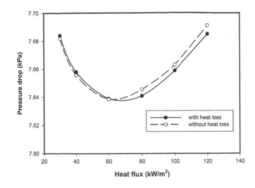

Figure 7. Comparison of pressure drop variation with heat flux ($Re = 3000$).

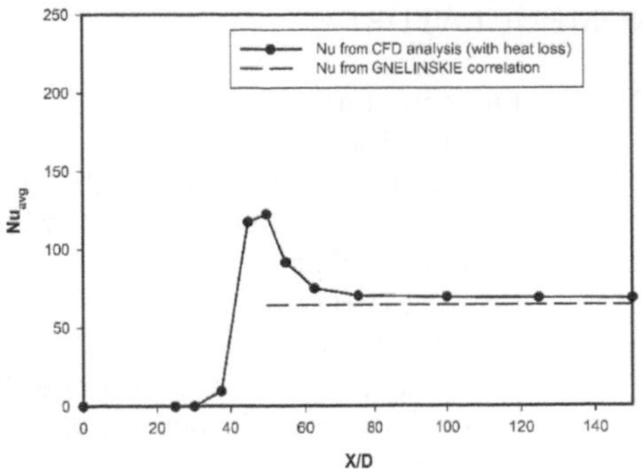

Figure 8. Comparison of Nussult number from CFD analysis with Gnelinskie's correlation.

due to the increase in friction factor due to increase in velocity gradient. This friction factor increases with heat flux after a particular value of heat flux. This may depend on the length of heating, heat loss and fluid properties. It is clear that the heat loss has some effect in the pressure drop reduction. The pressure drop curve is shifted towards right side. Pressure drop in ideal case (without heat loss) is lesser than the pressure drop in pipe with heat loss up to certain limit of heat flux after that the pressure drop in pipeline with heat loss is found to be higher than ideal case.

3.3 Validation

Average Nusselt number obtained from the CFD analysis are used for validation purpose. Nusselt number is compared with the value obtained from Gnelinskie's correlation. Here the $Re = 3000$ at inlet and the concentrated solar radiation as 40000 W/m² with heat loss condition.

Fluid properties were calculated based on this bulk mean temperature. While the average Nusselt number for oil pipeline with heat loss obtained from analysis agrees with Gnelinski's correlation (Eq. 11) with an error of 6.48%. This may be due to the effect of severe property variation and large difference between wall temperature and fluid temperature.

$$Nu = \frac{\left(\frac{f}{8}\right)(Re - 1000) \times Pr}{1 + 12.7\left(\frac{f}{8}\right)^{0.5}\left(Pr^{\frac{2}{3}} - 1\right)}, \qquad (11)$$

4 CONCLUSIONS

The utilisation of PTC for pressure drop reduction in oil pipelines by applying concentrated solar radiation on the pipe surface, in order to reduce pumping power was analysed. Pressure variation in heated oil pipeline was compared with adiabatic oil pipeline for corresponding Reynolds number of flow to determine the effect of heating in pressure drop reduction. The analysis shows that with increase in Reynolds number, the effect of heating in pressure drop reduction is getting significantly increased. Heating effect is found to be more effective in higher Reynolds number. The pressure drop versus different concentrated solar radiation was plotted. Also the effect of pressure drop reduction varies with different concentrated solar radiation in a peculiar manner. At first the pressure drop gets reduced with increase in heat flux but after that pressure drop curve showed a reverse trend due the increase in friction factor.

NOMENCLATURE

D Diameter (m)
f Friction Factor
h Convective heat transfer coefficient (W/m^2 K)
k   Thermal conductivity of fluid (W/mK)
Nu Nusselt number
P Pressure (Pa)
Pr Prandtl number
q_g Volumetric heat   generation rate (W/m^3)
q'' Rate of heat transfer per unit area (W/m^2)
Re Reynolds number
T Temperature (K)
u Velocity (m/s)
X Axial length (m)

Greek symbols

ε Emissivity
μ Dynamic viscosity (Ns/m^2)
v Kinematic viscosity (m^2/s)
ρ density (kg/m^3)

REFERENCES

Cengel Y.A. 2013. *Heat Transfer A Practical Approach*. Cambridge: Cambridge University Press.

Duffie, J.A., & Beckman, W.A. 2013. *Solar Engineering of Thermal Processes*. Hoboken, NJ, USA: John Wiley & Sons, Inc.

Forristall, R. 2003. Heat Transfer Analysis and Modeling of a Parabolic Trough Solar Receiver Implemented in Engineering Equation Solver. *Golden, Colo.: National Renewable Energy Laboratory*, NREL/TP; 550-34169.

Hart, A. 2014. A Review of Technologies for Transporting Heavy Crude Oil and Bitumen via Pipelines. *Journal of Petroleum Exploration and Production Technology* 4(3): 327–36.

Mammadov, F.F. 2006. Application of Solar Energy in the Initial Crude Oil Treatment Process in Oil Fields. *Journal of Energy in Southern Africa* 17(2): 27–30.

Martínez-Palou, R., Mosqueira, M. de L., Zapata-Rendón, B., Mar-Juárez, E., Bernal-Huicochea, C., de la Cruz Clavel-López, J., & Aburto, J. 2011. Transportation of Heavy and Extra-Heavy Crude Oil by Pipeline: A Review. *Journal of Petroleum Science and Engineering* 75(3–4): 274–82.

Midhun V.C., Shaji, K. & Jithesh, P.K. 2015. Application of Parabolic Trough Collector for Reduction of Pressure Drop in Oil Pipelines. *International Journal of Modern Engineering Research* 5(3): 40–48.

Price, H., Lüpfert, E., Kearney, D., Zarza, E., Cohen, G., Gee, R., & Mahoney, R. 2002. Advances in Parabolic Trough Solar Power Technology. *Journal of Solar Energy Engineering*, 124(2), 109.

Matthew Roesle, Volkan Coskun, & Aldo Steinfeld. 2011. Numerical Analysis of Heat Loss From a Parabolic Trough Absorber Tube With Active Vacuum System. *Journal of Solar Energy Engineering* 133(3): 31015.

Saniere, A., Hénaut, I. & Argillier, J.F. 2004. Pipeline Transportation of Heavy Oils, A Strategy, Economic and Technological Challenger: Oil and Gas. *Science and Technology-Rev. IFP* 59(5): 455–466.

Sattarin, M., Modarresi, H. & Teymori, M. 2007. New Viscosity Correlations for Dead Crude Oils. *Petroleum & Coal* 49(2): 33–39.

Sharma, V. B & Mullick, S.C. 1991. Estimation of Heat-Transfer Coefficients, the Upward Heat Flow, and Evaporation in a Solar Still. *Journal of solar energy engineering* 113(1): 36–41.

Tavakoli, A., & Baktash, M. 2012. Numerical Approach for Temperature Development of Horizontal Pipe Flow with Thermal Leakage to Ambient. *International Journal of Modern Engineering Research* 2(5): 3784–94.

Emerging Trends in Engineering, Science and Technology for Society,
Energy and Environment – Vanchipura & Jiji (Eds)
© 2018 Taylor & Francis Group, London, ISBN 978-0-8153-5760-5

Effects of air jet on bluff body stabilized flame: Validation by simulation

S. Parvathi, V.P. Nithin, S. Nithin, N. Nived, P.A. Abdul Samad & C.P. Sunil Kumar
Government Engineering College, Thrissur, Kerala, India

ABSTRACT: A burner having a conical bluff body with a central air injector is considered. In this paper, the effects of the central air jet on the heat load of the bluff body are investigated. The flame structures and the flame blowoff temperatures were compared with corresponding simulated outcomes. Simulation results show that the considerable reduction in the heat load to the bluff body by the central air jet determined experimentally is quite valid. Thus the problem caused by the high heat load in practical applications has a solution. The addition of central air jet alters the flame structures and blowout temperatures, as shown in simulation as well as experiment. Various blowout behaviors caused by the air jet observed experimentally also match those that were simulated. It is evident from simulation and experimentation that the center air injection could cool down the bluff-body. However, the flame stability could not be accomplished.

1   INTRODUCTION

In a multitude of applications, such as afterburners, gas turbine combustors, industrial furnaces and heat recovery steam generators, flame stabilization in premixed fuel–air streams is of technological interest (Chaparro & Cetegen, 2006).

For stabilising a diffusion flame, the bluff-body with a central jet is used. This is because the use of a bluff body flame holder improves the mixing properties and also increases the control over combustion. The recirculation of hot gas behind a bluff-body enhances mixing and reignites gas mixtures (Guo et al. 2010). But the studies on the flame holder's temperature distribution were minimal. Bluff-body stabilized premixed flame has been studied before in a few works by Zukoski & Marble (1955a), Zukoski & Marble (1955b), Longwell (1953), Longwell et al. (1953), Wright (1959) and Pan et al. (1991). The effects of some properties and geometry of the bluff body on the blowoff performance of flame holders supplied with mixtures of gaseous propane and air were summarized in Lefebvre et al. (2010). Shanbhogue et al. (2009) and Chaudhuri et al. (2009) investigated the dynamics of two-dimensional bluff-body stabilized flames. Roquemore et al. (1986) tested the reaction behavior of flows in an axisymmetric bluff-body burner. Esquiva-Dano et al. (2001) concluded that six regimes of non-premixed bluff-body stabilized flames existed, based on his experiment. Tang et al. (2013) studied the effects of the Reynolds number of central fuel and annular air jet on the flame extinction and its structure. Clearly none of the existing works the effects of central air jet on bluff body stabilized flames by simulation. The high heat load and temperature of the bluff-body surface were the challenges of the bluff body in many practical applications. To understand the influence of the central jet on the bluff-body surface temperature and the premixed flame stabilization, few efforts have been identified. In this work, the investigated effects of central air jet on the bluff-body stabilized premixed methane-air flame is compared with the results arrived at through simulation and thus validated. The flame blowoff temperatures, bluff-body surface temperature and flame structures were mainly considered.

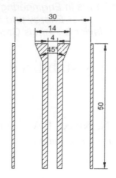

Figure 1. Schematic of the nozzle (all dimensions are in mm).

2 METHODOLOGY

The schematic of the conical bluff-body burner is shown in Figure 1. After the literature survey, the simulation model was decided. Since the bluff body burner was symmetrical, a 2-D simulation was chosen. The model for the same was created in Ansys Fluent software, which is based on the finite volume method. The general transport equations for mass, momentum, energy etc. are applied to each cell and discretized. All equations are then solved to render the flow field. As in the experiment carried out by Tong et al. (2017), a 45° conical bluff-body was placed in the center of the burner. The inner diameter of the circular pipe for the methane-air flow is 30 mm. The bluff-body has top and inner diameters of 14 mm and 4 mm respectively. The thickness of the pipe wall is 2 mm. Premixed methane-air is fed through the annular channel. Air is injected through the central pipe. The mass flow rate of the methane-air mixture is carried by varying the equivalence ratio of the same. The equivalence ratio is set between $\Phi_{annular} = 0.64$ to blowoff limit.

The boundary conditions are applied as per the experiment. The velocities are varied accordingly. The combustion equation of methane is chosen as given in the Fluent software.

3 APEX TEMPERATURE VS. TIME

As cited in the literature (Euler et al. 2014), the temperature distribution is highest at the center of the bluff-body. T_{apex} is taken as the temperature at the apex of the bluff-body surface. For the boundary condition, the emissivity (ε) of the bluff-body is set as 0.58 (ε of stainless steel varies 0.54 to 0.63). Taking central air jet velocity $U_{-jet} = 0$, annular velocity $U_{-annular} = 2.77$ m/s and the annular flow equivalence ratio $\Phi_{annular} = 0.64$, the temperature at the apex of the injection hole is approximately 480 K, which is referred as T_0 as the reference. *Fluent* is used for simulation throughout the study. Figure 2 shows temperature changes of T_{apex} when shutting down the fuel supply in simulation as well as in the experiment.

The fuel supply is shut down by reducing the mass flow rate of the methane-air flame gradually to zero. Because of the weak flame attached to the bluff-body, the temperature decreases with a sharp slope at first. Thereafter, due to the heat convection to the environment, when the effect of the flame completely disappears, the surface temperature changes slowly over time. After the blowoff, the rate of decrease of temperature is less than 3 K/s in both cases. This temperature is taken as T_{apex}.

In the simulation as well as the experimental result, the temperature before shutting down the fuel supply is 513 K at 1 sec. After shutting down fuel supply, at 0 sec, the temperature drops down to 481 K, in both cases.

In both cases, we have selected the temperature of the bluff body surface at times within 1 second after the flame is totally blown off. Thus the simulated pattern matches the experimental results.

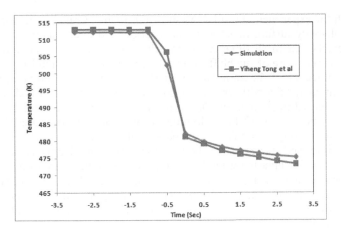

Figure 2.　Changes of apex temperature when shutting down the fuel supply in experiment as well as simulation.

4　GOVERNING EQUATIONS USED IN SIMULATION

Continuity:　$\dfrac{\partial}{\partial x}(\rho u)+\dfrac{\partial}{\partial y}(\rho v)+\dfrac{\partial}{\partial z}(\rho w)=0$

Momentum:　$\dfrac{\partial(\rho uu)}{\partial x}+\dfrac{\partial(\rho uv)}{\partial y}+\dfrac{\partial(\rho uw)}{\partial z}=-\dfrac{\partial p}{\partial x}+\dfrac{\partial \tau_{xx}}{\partial x}+\dfrac{\partial \tau_{xy}}{\partial y}+\dfrac{\partial \tau_{xz}}{\partial z}+\rho f_x$

$\dfrac{\partial(\rho vu)}{\partial x}+\dfrac{\partial(\rho vv)}{\partial y}+\dfrac{\partial(\rho vw)}{\partial z}=-\dfrac{\partial p}{\partial y}+\dfrac{\partial \tau_{yx}}{\partial x}+\dfrac{\partial \tau_{yy}}{\partial y}+\dfrac{\partial \tau_{yz}}{\partial z}+\rho f_y$

$\dfrac{\partial(\rho wu)}{\partial x}+\dfrac{\partial(\rho wv)}{\partial y}+\dfrac{\partial(\rho ww)}{\partial z}=-\dfrac{\partial p}{\partial z}+\dfrac{\partial \tau_{zx}}{\partial x}+\dfrac{\partial \tau_{zy}}{\partial y}+\dfrac{\partial \tau_{zz}}{\partial z}+\rho f_z$

ρ – density, P – pressure, ζ – shear stress, f – body force, U,v,w – velocity components in x,y,z directions.

Energy:　$\dfrac{\partial(\rho uh)}{\partial x}+\dfrac{\partial(\rho vh)}{\partial y}+\dfrac{\partial(\rho wh)}{\partial z}=\dfrac{\partial(k\partial T)}{\partial x^2}+\dfrac{\partial(k\partial T)}{\partial y^2}+\dfrac{\partial(k\partial T)}{\partial z^2}+$

$\sum_i \left[\dfrac{\partial}{\partial x}\left(h_i\rho D_{i,m}\dfrac{\partial Y_i}{\partial x} \right)+\dfrac{\partial}{\partial y}\left(h_i\rho D_{i,m}\dfrac{\partial Y_i}{\partial y} \right)+\dfrac{\partial}{\partial z}\left(h_i\rho D_{i,m}\dfrac{\partial Y_i}{\partial z} \right) \right]-\sum_i h_i R_i$

Species:　$\dfrac{\partial(\rho Y_i u)}{\partial x}+\dfrac{\partial(\rho Y_i v)}{\partial y}+\dfrac{\partial(\rho Y_i w)}{\partial z}=-\left[\dfrac{\partial}{\partial x}\left(\rho D_{i,m}\dfrac{\partial Y_i}{\partial x} \right)+\dfrac{\partial}{\partial y}\left(\rho D_{i,m}\dfrac{\partial Y_i}{\partial y} \right) \right.$

$\left. +\dfrac{\partial}{\partial z}\left(\rho D_{i,m}\dfrac{\partial Y_i}{\partial z} \right) \right]+R_i$

h – Enthalpy of species, k – Thermal conductivity, T – Temperature, Y – Mass fraction of species, R – Production or consumption of species, D – Mass diffusivity.

5　RESULTS AND DISCUSSION

5.1　Bluff-body surface temperature

Taking the premixed annular flow velocity $U_{annular}$ = 2.77 m/s and equivalence ratio $\Phi_{annular}$ = 0.64, while changing $U_{jet}/U_{annular}$ = 0~8.8, T_{apex}/T_0 variation in simulation, the results compared with the experimental results. As shown in Figure 3, the temperature of the inner apex of the bluff-body surface varies with the injection of the central air jet.

Figure 3 depicts that in all three cases, namely simulation, experimentation and steady condition, the temperature of the bluff body surface drops to less than 81% of T_0, even with a small amount injection of the central air jet. In the experiment, the temperature ratio comes back to approximately 85% with an increase of the central air flow, which clearly matches with the 86% achieved in simulation. Deviation to 95% in steady conditions is noteworthy. The annular flow near the bluff-body dominates the local flow structures. That is why the central air jet with $U_{jet}/U_{annular} \approx 1$ causes the peak cooling effect.

A layer of air is formed above the bluff-body surface when central air jet is introduce. Thus the heat convection from the flame to the bluff-body is prevented. When the velocity ratio is increased to 1, neither flow dominates, making the effect of cooling weaker. Thus, to reduce the heat load to the bluff-body, the central injection of air can be practically employed, taking into account T_{apex}.

Obviously with regard to bluff body temperature, simulation results validate the achieved experimental outcomes.

5.2 Flame blowoff temperature

To evaluate the performance of the bluff-body stabilized flame, bluff body face temperature variation with respect to face radial distance is considered (Figure 4). It is seen that variation of blow-

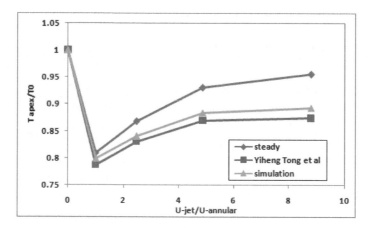

Figure 3. Comparison of effects of central air jet on the temperature of bluff-body surface in experiment. simulation and steady state.

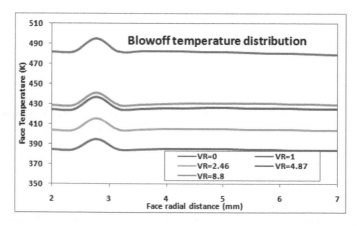

Figure 4. Blowoff temperature distribution with respect to bluff body face radial distance.

off temperature occurs between 2.5 mm and 3.25 mm face radial distance in all velocity ratios $U_{jet}/U_{annular}$. The highest blowoff temperature occurs without the central air jet. As the velocity ratio $U_{jet}/U_{annular}$ increases, the blowoff face temperature also increases. For all velocity ratios, the maximum face temperature occurs at a face radial distance of 2.9 mm. When there is no central air jet, the maximum blowoff face temperature reaches 495 K. Furthermore it becomes 393 K, 416 K, 438 K and 440 K as the velocity ratios change to 1, 2.46, 4.87 and 8.8 respectively.

Taking $U_{annular} \approx 1.85$ m/s and $\Phi_{annular} = 0.64$, steady state face temperature distribution with respect to face radial distance are also found (Figure 5). Obviously for both cases, the overall flame blowoff temperature increases with the introduction of the central air jet, which is one of the main drawbacks.

At the beginning, the increase in flame blowoff temperature is due to the annular dominated flow structure. A fresh air layer is formed by the central air jet upon the bluff body surface. When the annular flow velocity is higher, the recirculation zone becomes small. This results in detached flame from the bluff-body, which makes it easy to blowoff.

When the velocity ratio is equal to 1, the flame may get reattached to the bluff-body's surface. Also, when central air jet dominates the flow field, it results in a linear increase in the flame blowoff temperature.

5.3   Flame structures

Taking $U_{annular} = 1.85$ m/s and $\Phi_{annular} = 0.64$, Figure 6 shows the contour of temperature along the flame structures. Evidently the flame structures change with the injection of the central air jet.

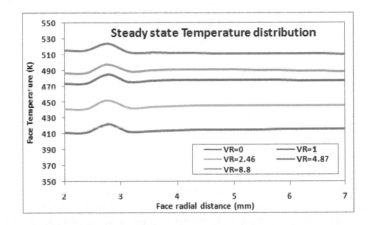

Figure 5.   Steady state temperature distribution with respect to bluff body face radial distance.

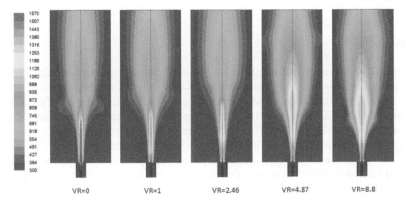

Figure 6.   Contour of temperature (K).

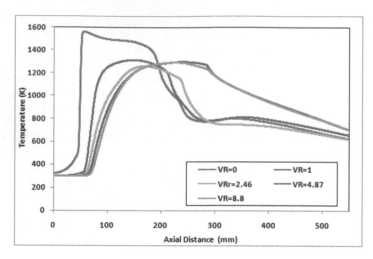

Figure 7.   Temperature vs. axial distance.

With conditions of $U_{jet}/U_{annular}$ = 0, 1, 2.46, 4.87 and 8.8 (from left to right), variations of temperature in K with respect to axial distance are shown in Figure 7.

From Figure 6, it is obvious that the temperature downstream of the bluff-body is the highest without the introduction of the central air jet. When the central air jet is injected, it creates a layer separating the main heat release zone and the flame, thereby appearing to get attached to the surface of the bluff-body. The fresh cold air from the central jet fills the non-luminous recirculation zone. With the addition of central air jet, the fuel air ratio decreases creating leaner. The flame becomes weaker due to the reduction in size of the heat release zone. That is why the bluff-body temperature with central air jet is smaller than T_0. With small velocity ratios of $U_{jet}/U_{annular}$, the annular flow dominates the flow field making the temperature with $U_{jet}/U_{annular}$ ~ 1 the lowest, as is evident in Figure 7. When the recirculation zones downstream of the bluff-body are dominated by the annular flow, to avoid the flame getting attached to the bluff-body, the small amount of central air jet may form a fresh air layer. Also the heat convection of burnt products to the bluff-body is reduced by the cold central air jet layer.

As is evident from Figure 7, the peak temperatures occur at axial distances of 150 mm, 200 mm, 260 mm and 300 mm as velocity ratios change from 1, 2.46, 4.87 and 8.8 respectively. This means that the heat releasing zone becomes larger and travels to farther distances with increases in central air jet velocity. Obviously in the absence of central air jet, the peak temperature of 1,550 K occurs at 50 mm distance and the heat releasing zones cling together in small region.

6   CONCLUSION

In this paper, the effects of central air jet on the bluff body stabilized premixed methane-air flame, namely bluff-body surface temperature, flame blowoff temperature and flame structures are studied by simulation which, in turn, resulted in the validation the corresponding experimental outcomes given by Tong et al. In both cases, it can be seen that the central air jet reduces the heat load on the bluff-body surface. But on further addition of the air jet, the flame becomes unstable. The flame blows off easily with the central air jet. Variation of apex temperatures with different velocity ratios, variation of flame blowoff temperature with face radial distance and variation contour of temperature in flame structure with different velocity ratio are thus validated by simulation study.

REFERENCES

Chaparro, A.A. & Cetegen, B.M. (2006). Blowoff characteristics of bluff-body stabilized conical premixed flames under upstream velocity modulation. *Combustion and Flame, 144*(1), 318–335.

Chaudhuri, S. & Cetegen, B.M. (2009). Blowoff characteristics of bluff-body stabilized conical premixed flames in a duct with upstream spatial mixture gradients and velocity oscillations. *Combustion Science and Technology, 181*(4), 555–569.

Esquiva-Dano, I., Nguyen, H.T. & Escudie, D. (2001). Influence of a bluff-body's shape on the stabilization regime of non-premixed flames. *Combustion and Flame, 127*(4), 2167–2180.

Euler, M., Zhou, R., Hochgreb, S. & Dreizler, A. (2014). Temperature measurements of the bluff body surface of a Swirl Burner using phosphor thermometry. *Combustion and Flame, 161*(11), 2842–2848.

Guo, P., Zang, S. & Ge, B. (2010). Technical brief: predictions of flow field for circular-disk bluff-body stabilized flame investigated by large eddy simulation and experiments. *Journal of Engineering for Gas Turbines and Power, 132*(5), 054503.

Lefebvre, A.H. & Ballal, D.R. (2010). *Gas turbine combustion. CRC Press.*

Longwell, J.P., Frost. E.E. & Weiss, M.A. (1953). Flame stability in bluff body recirculation zones. *Industrial & Engineering Chemistry, 45*(8), 1629–1633.

Longwell, J.P. (1953). Flame stabilization by bluff bodies and turbulent flames in ducts. *Symposium (International) on Combustion. Elsevier, 4*(1).

Pan, J.C., Vangsness, M.D. & Ballal, D.R. (1991). Aerodynamics of bluff body stabilized confined turbulent premixed flames. *ASME 1991 International Gas Turbine and Aeroengine Congress and Exposition. American Society of Mechanical Engineers.*

Roquemore, W.M., Tankin, R.S., Chiu, H.H. & Lottes, S.A. (1986). A study of a bluff-body combustor using laser sheet lighting. *Experiments in Fluids, 4*(4), 205–213.

Shanbhogue, S.J., Husain, S. & Lieuwen, T. (2009). Lean blowoff of bluff body stabilized flames: Scaling and dynamics. *Progress in Energy and Combustion Science, 35*(1), 98–120.

Tang, H., Yang, D., Zhang, T. & Zhu, M. (2013). Characteristics of flame modes for a conical bluff body burner with a central fuel jet. *Journal of Engineering for Gas Turbines and Power, 135*(9), 091507.

Wright, F.H. (1959). Bluff-body flame stabilization: blockage effects. *Combustion and Flame, 3*, 319–337.

Tong, Y., Li. M., Thern, M., Klingmann, J., Weng, W., Chen, S. & Li, Z. (2017). Experimental investigation on effects of central air jet on the bluff body stabilized premixed methane-air flame, *Energy Procedia, 107*, 23–32.

Zukoski, E.E., Marble, F.E. (1955a). The role of wake transition in the process of flame stabilization on bluff bodies. *AGARD Combustion Researches and Reviews*, 167–180.

Zukoski E.E., Marble F.E. (1955b). Gas dynamic symposium on aerothermochemistry. Northwestern University, Evanston, IL.

Emerging Trends in Engineering, Science and Technology for Society,
Energy and Environment – Vanchipura & Jiji (Eds)
© 2018 Taylor & Francis Group, London, ISBN 978-0-8153-5760-5

Identification of energy-intensified equipment for reliability analysis

C. Sanjeevy & Jacob Elias
Department of Mechanical Engineering, School of Engineering, CUSAT, Cochin, India

ABSTRACT: A methodology is developed for the identification of Energy-Intensified Equipment (EIE) for reliability analysis in the chemical processing industry. There are several methods based on classification that can be used for identifying such equipment, such as Always Better Control/ABC, Vital/Essential/Desirable (VED), Scarce/Difficult/Easily available (SDE), High/Medium/Low (HML), and Fast/Slow/Non-moving (FSN), but these selective inventory control methods do not indicate the criticality of an item. The damage grounds from failure and the failure modes to plan an optimum maintenance program. The method is applicable in understanding the equipment in the operational phase where there is only limited data available. When available data is scarce or generic, critical data is retrieved from some related selective inventory data banks. In this method, based on physical factors, the situation under which the equipment is working, such as external/internal load/pressure, is used in modeling the equipment. Pareto's 80/20 principle is employed to identify its criticality and calculate its risk factor. The current methodology applies criticality importance analysis and criticality allocation to optimize the maintainability correlated with Reliability-Centered Maintenance (RCM) models. Evaluating the reliability of life-threatening equipment in reverse engineering of the (competitive) operational phase is one of the applications of this method. As a case study, EIE is used for assessment of the proposed method and the results identify the equipment and sub-systems that are critical elements from a reliability and maintenance perspective. A benchmark of the results indicates the effectiveness and quality of the method in identification of energy-intensified equipment for reliability analysis.

Keywords: Energy-intensified equipment, reliability analysis, reliability-centered maintenance

1 INTRODUCTION

For any item of equipment that has a vital role in a production sequence, a hazard in such critical equipment will affect the entire production and create greater damage, so requires significant care. The identification of Energy-Intensified Equipment (EIE) is one of the crucial phases of Reliability-Centered Maintenance (RCM), through combination of quantitative analysis with qualitative analysis (Barabady & Kumar, 2008). This research has been carried out in a chemical company called Travancore Cochin Chemicals (TCC) Limited, situated in Ernakulam district, Kerala, India. The chemical processing industry plays an essential role in the manufacture of many chemicals, such as caustic soda, sodium chloride, chlorine, sulphuric acid, hydrochloric acid and bleaching powder. There are more than 600 types of equipment involved in their production. The industry provides a tremendous variety of materials to other manufacturers, such as textiles, rayons, plastics, aluminum, detergents, drugs, fertilizers, food preservatives, and paper-producing industries. It also produces chemical products that benefit people directly. Several changes in equipment have been taking place in the processing of chemicals, and chemical processing industries in India are facing certain challenges that need to be addressed for their survival in the era of globalization. This analysis focuses on the implementation of a rational methodology to detect reliability-related issues in the chemical processing industry. An intense literature survey has been conducted in relation to the equipment-related problems of chemical processing industries. The survey makes clear that several aged chemical processing industries face severe problems in equipment identification. The most significant problem is non-adoption of innovative technologies in production. Because the technology in use therefore remains ancient, criticality-related difficulties become common. The investigation of criticality and its combination with the reliability of equipment are used to identify the related problems of analytical research.

2 LITERATURE REVIEW

This literature review is a brief review of research conducted in the areas of reliability and maintenance programs of equipment, in order to identify the equipment and tools to be adopted for the present study. Initially, the literature reveals several methods that can be used to identify equipment, based on classifications such as ABC/Always Better Control, Vital/Essential/Desirable (VED), Scarce/Difficult/Easily available (SDE), High/Medium/Low (HML), and Fast/Slow/Non-moving (FSN). methods have not found the criticality, mutilation causing from failure and the failure modes as well as to plan an optimum maintenance program (Sanjeevy & Thomas, 2014).

Today, the methods most commonly applied in this field are Failure Mode Effect and Criticality Analysis (FMECA) and Reliability-Centered Maintenance (RCM). Reliability can be expressed as the possibility of process or equipment which perform its function or task under stated environment for a definite surveillance period. Reliability analysis methods have been increasingly accepted as typical tools for the development and management of regular and intricate processing methods since the mid-1980s. The occurrence of failure cannot be prevented completely, but it is important to reduce both its chance of happening and the impact of failures when they occur (Barabady & Kumar, 2008). To sustain the intended reliability, availability, and maintainability features and to attain expected performance, a valid maintenance plan is essential. Both corrective and preventive maintenance have direct consequences on the reliability of equipment and, consequently, its performance. Hence the identification of energy-intensified equipment is critical for reliability evaluation procedures. To overcome the limitations of the traditional classifications such as the selective inventory controls of ABC, VED, SDE, HML, and FSN, and evaluation using FMECA methodology, the present methodology is adopted. Ben-Daya and Raouf (1996) noted that the economic model proposed by Gilchrist (1993) addresses a problem that differs from the problem FMECA is intended to address. They combined the expected cost model proposed by Gilchrist with their improved Risk Priority Number (RPN) model in order to provide a quality improvement technique at the production stage. They also confirmed that if the assessment of the factor scores on a 1 to 9 scale is not appropriate then the treatment of identical significance is not practical. According to their model, the probability of an event should be more significant and their model suggests the probability of an event (with scale 1–9) is increased to the power of 2 (Tang et al., 2017; Puthillath & Sasikumar, 2012).

There are two kinds of criticality analysis: quantitative and qualitative. To use the quantitative criticality analysis method, the investigation group has to identify the dependability/unpredictability for every element, in a specified working period, to recognize the part of the element's unpredictability that can be attributed to each probable failure mode, and rate the possibility of loss (or severity) that will result from each failure mode that can occur (Sachdeva et al., 2009). Several authors make use of fuzzy set theory to tackle uncertainties in maintenance decision-making, Chang et al. (1999) argued for the use of gray theory to obtain critical valuations. The use of fuzzy logic theory for maintenance-critical inquiry is also suggested in the literature (Eti et al., 2006; Teng & Ho, 1996; Jayakumar & Asgarpoor, 2004).

3 RESEARCH METHODOLOGY

In reality, all items of equipment cannot be controlled with equal attention. An effective critical equipment identification calls for an understanding of the nature of care. Some equipment may be very important while some is too small or too unimportant to call for a rigorous and intensive mechanism. Criticality analysis means variances in the method of control from equipment to equipment, founded on the basis of physical factors. The criterion used for this purpose may be criticality, risk, maintenance difficulties, or something else. Controlling the area of operation for good performance involves the time, money and effort required to conduct operations that take less time and avoid sudden damage. Therefore, to achieve this objective it need not be necessary to control the entire area of operations but only that area of operation that is not controlled and is likely to cause damage. Thus, criticality analysis means selecting the areas of mechanism so that the required objective is achieved as early as possible without the loss of time that would be involved in taking care of the full area.

3.1 *Critical analysis*

When monitoring equipment, a company may be using many types in various numbers and configurations. Equipment monitoring needs rational and clear classification. Thus, equipment identification is needed to ensure objectivity in monitoring that equipment for maintenance processes. The critical ranking number of a system or piece of equipment is a measure of the system's or equipment's influence on the production when the system or equipment fails to function, no matter how frequently the malfunction occurs. Clear critical ranking rules are needed to help in allocating critical rankings to systems or equipment through the analysis. The procedure is established by considering the combined scores for all evaluation criteria. The critical ranking numbers, number range, and the model for assigning weighted scores to the systems or equipment under assessment are defined before conducting the analysis. Critical ranking numbers are allocated to systems or equipment based on the procedures developed. This is accomplished by comparing the equipment's criteria with the important factors in the critical ranking number rules. If the equipment matches the rules, the equipment is assigned that critical ranking number. The equipment is always allocated the highest critical ranking number that it scores. The 24 most appropriate critical factors selected for equipment grouping are as listed below:

1. Percentage of utilization of machine
2. External/internal load/pressure acting on the equipment
3. Availability of machine
4. Weight of machine
5. Availability of substitute equipment
6. Impact on other equipment of breakdown
7. Cost of machine
8. Power consumption of machine
9. Age of equipment
10. Speed of repair
11. Equipment material
12. Availability of spares
13. Manufactured or purchased externally
14. Quality of work performed on machine
15. Ease of purchase
16. Effect of corrosion
17. Machine handling
18. Repaired or disposed of
19. Re-usability
20. Reliability
21. Equipment life cycle

22. Recyclability
23. Environmentally friendly
24. Maintenance history.

From these 24 basic factors, the nine most common factors suitable for the critical analysis are selected and a questionnaire is prepared, based on a five-point "Likert" scale. This is constructed and tested for reliability. The equipment data was collected from the Ashagi Glass Corporation (AGC) plant using the opinions of experts in the operations and maintenance department. The nine factors which are most important and relevant are given weightings and allocated overall scores from 1 to 9, which is 100%. In the questionnaire, each question is assigned a five-point Likert score by assigning I–V weightings to maintain consistency of degree: I – 20%; II – 40%; III – 60%; IV – 80%; V – 100%. To check whether the questionnaire measure the same latent variable of Cronbach's alpha is run on a sample selected as explained in the analysis section. The utmost critical issues are allotted the upper weightings as related to less significant issues are shown in Table 3. Moreover, using the survey data all of the equipment in the plant is categorized according to the criticality score, which is classified as follows: Critical – degree score above 80%; Very important – degree score between 60% and 80%; Important – degree score between 40% and 60%; Less important – degree score below 40%.

Based on the above, a well-planned questionnaire was prepared and circulated to the heads of important departments such as Operations, Maintenance, Inventory Control, Purchasing, Quality Control, Fire and Safety, Planning, Electrical Maintenance, Instrumentation and Control, Troubleshooting, Emergency Operations, Training, Computer Control Section, Chemical Section, Hazard Control Section, and Research and Development Section to gather their feedback. The results obtained are described in the following section.

4 RESULT ANALYSIS AND DISCUSSION

4.1 *Reliability according to questionnaire*

A group of 63 experts was selected from the total of 75. The chosen group included personnel from service, production, and maintenance working in the department. Of the 63 experts, 59 replied to the questionnaire, which is shown in Appendix I. Only 56 responses were selected as valid and were tested for internal consistency at a 95% confidence level using SPSS software, which found that the Cronbach's alpha was 0.82. The data represents presence of same latent variable and found good. Respondents responded differently. The reliability statistics are shown in Table 1.

Table 1. Reliability statistics.

Cronbach's alpha[a]	Cronbach's alpha based on standardized items	No. of items
0.82	0.78	9

Table 2. Critical analysis output.

Class	Total no. of items	% of total no. of items	Annual usage value (in lakhs)	Critical scores
Critical	62	26.38	9.5	80–100%
Important	121	51.48	6.9	40–80%
Less important	52	22.12	0.35	<40%

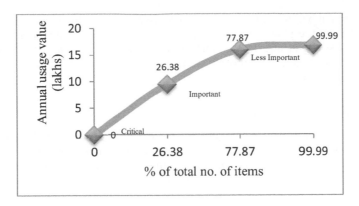

Figure 1. Trend in criticality of equipment.

Table 3. Criticality scores for equipment.

No	Factor description	Weighting	Degree I (score)	Degree II (score)	Degree III (score)	Degree IV (score)	Degree V (score)
1	% utilization of m/c	10	M/c loaded for 8 hours (2)	1 shift (4)	2 shifts (6)	3 shifts (8)	Overload 10
2	External/internal pressure acting on m/c	10	Very low (2)	Moderate (4)	Average (6)	High (8)	Extremely high (10)
3	Availability of substitute	10	4 or more substitutes (2)	3 substitutes (4)	2 substitutes (6)	1 substitute (8)	No substitute (10)
4	Impact on other machines of breakdown	15	10% of machines or men idle (3)	10–20% idle (6)	20–40% idle (9)	40–60% idle (12)	>80% idle (15)
5	Age of m/c	10	Up to two years (2)	2–4 years (4)	4–6 years (6)	6–8 years (8)	>8 years (10)
6	Ease of repair	15	Mechanical fault easy to repair (5)	Electrical and hydraulic fault difficult to diagnose (10)	Electronic fault difficult to find (15)	–	–
7	Quality of work done on m/c	5	Wide tolerance; no effect on quality (1)	Tolerance 0.05–0.10; no effect on quality (2)	Tolerance 0.025–0.05; some effect on quality (3)	Tolerance 0.010–0.0250; major loss to production (4)	Tolerance 0.01; complete production loss (5)
8	Ease of procurement in terms of lead time/manufacture in the company	10	1 month; can be manufactured in the company (2)	1–3 months; can be manufactured in the company (4)	3–6 months; in-house manufacture not possible (6)	Over 6 months; in-house manufacture not possible (8)	To be imported; uncertain procurement period (10)
9	Maintenance history of m/c in terms of repair orders received	15	0–1 (3)	1–3 (6)	3–5 (9)	5–10 (12)	>10 (15)

Source: P & IC Hand Book.

4.2 Results of criticality analysis

Criticality analysis was carried out for the Ashagi Glass Corporation plant (a manufacturing plant installed in a collaboration with Japan), at TCC Ltd. At first the criticality of machine to which equipment belongs found the critical score of equipment. This was on the basis of data collected from the plant with the help of the survey. They were classified as Critical, Important or Less important. It was found that 62 items of equipment were Critical, 121 items were Important, and 52 items Less important, as shown in Table 2. A graph of the trend is shown in Figure 1.

4.3 Discussion of criticality analysis

Damage of equipment in the Critical category will result in production stoppages or threats to workers. Consequently, such equipment has never to be maintained freely and strategies that will offer adequate safety must be implemented for these types. Important items must be reasonably checked because their damage will have an effect on production for a shorter period. Safe controls can be maintained for Less important items that will not create hazards or impact production.

5 CONCLUSION

The identification of EIE is one of the key phases in elimination of accidents and severe damage to production systems. The current investigation demonstrates that a methodical approach has been implemented to distinguish the EIE and address problems arising in the chemical processing industry. A number of proposals are suggested in our discussion of criticality analysis for caring for such types of equipment and improving the existing maintenance strategy. In addition, it is very clear from the literature survey that there are many methods available for analyzing the reliability and maintenance program of equipment and lacks in to identify the equipment with hazard involved in failure. This literature review is intended to provide an idea of the preceding works that facilitated this work, helping to identify various significant methodologies used in this field. Moreover, it assists in selecting the appropriate procedure to identify the energy-intensified equipment for reliability analysis and the tool for developing the maintenance program. The present work extends the scope of detailed analysis in energy-intensified equipment.

REFERENCES

Barabady, J. & Kumar, U. (2008). Reliability analysis of mining equipment: A case study of a crushing plant at Jajarm bauxite mine in Iran. *Reliability Engineering and System Safety*, *93*, 647–653.
Ben-Daya, M. & Raouf, A. (1996). A revised failure mode and effect analysis model. *International Journal of Quality & Reliability Management*, *13*(1), 43–47.
Eti, M.C., Ogaji, S.O.T. & Probert, S.D. (2006). Development and implementation of preventive-maintenance practices in Nigerian industries. *Applied Energy*, *83*, 1163–1179.
Gilchrist, W. (1993). Modeling failure modes and effect analysis. *International Journal of Quality and Reliability Management*, *10*, 16–23.
Jayakumar, A. & Asgarpoor, S. (2004). Maintenance optimization of equipment by linear programming. *Probability in Engineering and Information Science*, *20*, 183–193.
Puthillath, B. & Sasikumar, R. (2012). Selection of maintenance strategy using failure mode effect and criticality analysis. *International Journal of Engineering and Innovative Technology*, *1*(6), 73–79.
Sachdeva, A., Kumar, D. & Kumar, P. (2009). Multi-factor failure mode critically analysis using TOPSIS. *Journal of Industrial Engineering International*, *5*(8), 1–9.
Sanjeevy, C. & Thomas, C. (2014). Use and application of selective inventory control techniques of spares for a chemical processing plant. *International Journal of Engineering Research & Technology*, *3*(10), 301–306.
Tang, Y., Liu, Q., Jing, J., Yang, Y. & Zou, Z. (2017). A framework for identification of maintenance significant items in reliability centered maintenance. *Energy*, *118*, 1295–1303.
Teng, S.-H. & Ho, S.-Y. (1996). Failure mode and effects analysis: An integrated approach for product design and process control. *International Journal of Quality & Reliability Management*, *13*, 8–26.

APPENDIX I

TCC Limited, Ernakulam.
Questionnaire for Criticality Analysis

1. Machine/Equipment..
2. Code...
3. Specification..
   [Put tick mark ( ) for the appropriate options]
4. Percentage utilization of machine
   a. M/c loaded for <8 hours b. 1 shift c. 2 shifts d. 3 shifts e. Overload
5. External/internal pressure acting on machine
   a. Low b. Moderate c. Average d. High e. Extremely high
6. Availability of substitute
   a. 4 or more substitutes b. 3 substitutes c. 2 substitutes d. 1 substitute e. No substitute
7. Impact on other machines of breakdown
   a. 10% m/c or men idle b. 10–20% c. 20–40% d. 40–60% e. >80%
8. Age of machine
   a. Up to two years b. 2–4 years c. 4–6 years d. 6–8 years e. >8 years
9. Ease of repair
   a. Mechanical fault easy to repair
   b. Electrical and hydraulic difficult to diagnose
   c. Electronic fault difficult to find
10. Quality of work done on machine
    a. Wide tolerance; no effect on quality
    b. Tolerance 0.05–0.10; no effect on quality
    c. Tolerance 0.025–0.05; some effect on quality
    d. Tolerance 0.010–0.025; major loss to production
    e. Tolerance <0.01; complete production loss
11. Ease of procurement in terms of lead time/manufacture in the company
    a. 1 month/can be manufactured in the company
    b. 1–3 months/can be manufactured in the company
    c. 3–6 months/in-house manufacture not possible
    d. Over 6 months/in-house manufacture not possible
    e. To be imported; uncertain procurement period
12. Maintenance history of machine in terms of repair orders received
    a. 0–1 b. 1–3 c. 3–5 d. 5–10 e. >10

International Conference on Advances in Chemical Engineering (ICAChE)

Emerging Trends in Engineering, Science and Technology for Society,
Energy and Environment – Vanchipura & Jiji (Eds)
© 2018 Taylor & Francis Group, London, ISBN 978-0-8153-5760-5

Design of Fractional Filter Fractional Order Proportional Integral Derivative (FFFOPID) controller for higher order systems

R. Ranganayakulu, G. Uday Bhaskar Babu & A. Seshagiri Rao
Department of Chemical Engineering, National Institute of Technology, Warangal, India

ABSTRACT: This article presents a Fractional Filter Fractional Order Proportional Integral Derivative (FFFOPID) controller design method for higher order systems, approximated as Non-Integer Order Plus Time Delay (NIOPTD) models. The design uses an Internal Model Control (IMC) scheme and the resulting controller has a series form of Fractional Order Proportional Integral Derivative (FOPID) term, in series with a fractional filter. An analytical tuning method is then used to identify the optimum controller settings. Simulation results on different systems show that the proposed method gives better output performance for set point tracking, disturbance rejection, parameter variations, and for measurement noise in the output. The robust stability of the system regarding process parametric uncertainties is verified with robustness analysis. Controllability index analysis is also undertaken to ascertain the closed loop system performance and robustness.

Keywords: Internal Model Control, robust stability, closed loop system, Fractional Order Proportional Integral Derivative

1 INTRODUCTION

Higher order models describe the process dynamics more accurately than lower order models (Isaksson & Graebe, 1999; Malwatkar et al., 2009). However, they complicate the controller design and tuning for quality control. There are several controllers tuning rules for higher order models, approximated as First Order Plus Time Delay (FOPTD) models. Most of these rules are to tune a controller having a Proportional Integral Derivative (PID) structure, which has been widely used to date (Aström & Hägglund, 1995; Skogestad, 2003). The controller designed for such FOPTD models may not give the satisfactory performance, as the dynamics are compromised during the approximation. An alternative to preserve the dynamics, while ensuring satisfactory control, is to approximate them as Non-Integer Order Plus Time Delay (NIOPTD) models (Pan & Das, 2013).The major advantage of NIOPTD models is that they represent the process behavior more compactly than integer order systems (Podlubny, 1999).

Fractional order control for fractional order systems has been in focus in the last two decades (Shah & Agashe, 2016). Several fractional order controller structures have been proposed and the widely accepted one is the Fractional Order Proportional Integral Derivative (FOPID) controller (Monje et al., 2008; Luo & Chen, 2009; Tavakoli-Kakhki & Haeri, 2011; Padula & Visioli, 2011; Vinopraba et al., 2012; Das et al, 2011; Valério & da Costa, 2006). The FOPID controller has the ability to enhance the closed loop performance, but the tuning is complex as it has more tuning parameters than the PID controller. Recently, there was work found in the literature where the five FOPID parameters are identified, based on the stability regions of a closed loop system (Bongulwar & Patre, 2017). Further, the simulation results were shown only for the servo response.

In this paper, a Fractional Filter Fractional Order Proportional Integral Derivative (FFFOPID) controller is proposed using Internal Model Control (IMC). The present work uses a series form of a FOPID controller (Hui-fang et al., 2015). The resulting controller has a structure consisting of a FOPID term along with a fractional filter, with only two parameters to be tuned (Ranganayakulu et al., 2017). The tuning parameters are identified such that the Integral Absolute Error (IAE) and Total Variation (TV) are minimal. The selection is proved to be optimum by observing the trends of IAE and TV through an analysis, after varying the tuning parameters in the range of +10% and −10%. The simulations have been performed for different inputs and the measures used to assess the system performance are % overshoot (%OS), Settling Time (ST), IAE and TV. Also, the applicability of the proposed method for apparent changes in time delay is verified by varying the L/T ratio called as controllability index (Ranganayakulu et al., 2016; Lin et al., 2008). The performance of the proposed method is validated with three different examples.

This article is sectioned as follows: the IMC scheme is described in Section 2. Section 3 presents the controller design and tuning, and robust stability analysis is provided in Section 4. The simulation results for different systems are presented in Section 5. The conclusion is given in Section 6, and the paper ends with references.

2 PRELIMINARIES

2.1 *Internal model control*

The IMC method block diagram and feedback loop are shown in Figure 1. The IMC based controller design procedure is briefly given below.

1. Identify the non-invertible (all time delays and unstable zeros) and invertible (minimum phase elements) parts of the process model:

$$\tilde{G}(s) = \tilde{G}^+(s)\tilde{G}^-(s) \tag{1}$$

2. The IMC controller is:

$$C_{IMC}(s) = \frac{f(s)}{\tilde{G}^-(s)} \tag{2}$$

where f(s) is the IMC filter
3. The feedback controller is:

$$C(s) = \frac{C_{IMC}(s)}{1 - C_{IMC}(s)\tilde{G}(s)} \tag{3}$$

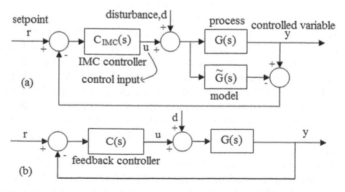

Figure 1.   (a) IMC scheme; (b) feedback loop.

3 DESIGN OF FFFOPID CONTROLLER

The structure of the controller used here is:

$$C(s) = (fractional\ filter)K_p\left(\frac{\tau_i s^\lambda + 1}{\tau_i s^\lambda}\right)(1 + \tau_d s^\mu) \qquad (4)$$

To design the controller, the higher order system approximated as a NIOPTD model is considered and is given by Equation 5:

$$\tilde{G}(s) = \frac{Ke^{-Ls}}{Ts^\alpha + 1} \qquad (5)$$

where K, L, T and α are gain, time delay, time constant and fractional order of the model, respectively. The feedback controller C(s), according to the procedure in Section 2.1 by using the IMC filter f(s) = $1/(\gamma s^p+1)$ and first order Pade's approximation for time delay $e^{-Ls} = (1-0.5\ Ls)/(1+0.5\ Ls)$ is:

$$C_{proposed}(s) = \left(\frac{1}{0.5\gamma Ls^p + \gamma s^{p-1} + L}\right) \times \left(\frac{L}{2K}\right)\left(\frac{1+0.5Ls}{0.5Ls}\right)(1 + Ts^\alpha) \qquad (6)$$

Comparing Equations 4 and 6, the settings are:

$$\left.\begin{array}{l} K_p = \dfrac{L}{2K}; \tau_i = 0.5L; \lambda = 1; \tau_d = T; \mu = \alpha \\[2mm] fractional\ filter = \dfrac{1}{0.5\gamma Ls^p + \gamma s^{p-1} + L} \end{array}\right\} \qquad (7)$$

where γ and p are two tuning parameters of the controller.

3.1 Tuning

The tuning parameters γ and p are chosen in a way that the measures IAE and TV are minimal. The optimum values are identified through the behavior of IAE and TV by varying γ and p in the range of (−10%, +10%). Finally, γ is chosen for the minimum of both IAE and TV, and also p.

4 ROBUSTNESS ANALYSIS

The stability of a closed loop system should always be analyzed for process parameter uncertainties because the process model is an approximation of the real plant. The robust stability condition (Morari & Zafiriou, 1989) is:

$$\|l_m(s)T(s)\| < 1; s = j\omega; \forall\ \omega \in (-\infty, \infty) \qquad (8)$$

where $1_m(s) = G(s) - \tilde{G}(s)/\tilde{G}(s)$ is the bound on multiplicative uncertainty and T(s) = C(s) G(s)/1+C(s)G(s) is the complementary sensitivity function. For uncertainty in both L and K, the following condition must be satisfied:

$$\|T(j\omega)\|_\infty < 1 \bigg/ \left\|\left(\frac{\Delta K}{K} + 1\right)e^{-\Delta L} - 1\right\| \qquad (9)$$

Another constraint to be satisfied for robust control performance is:

$$\|T(s)1_m(s)+(1-T(s)w_m(s))\|<1 \qquad (10)$$

where $1-T(s)$ is the sensitivity function and $w_m(s)$ is the uncertainty bound on the sensitivity function.

5 SIMULATION STUDY

Three higher order systems approximated as NIOPTD models are simulated in MATLAB and the system performance is compared with the Bongulwar and Patre (2017) method (hereafter addressed as the Patre (2017) method). The effectiveness of the proposed method is verified with the performance measures %OS, ST, IAE and TV, which are defined in Table 1.Settling time is defined as the time taken for the response to settle within 2% to 5% of its final value.

The closed loop system's unit step response is observed, with step change in disturbance of magnitude 0.5 applied at a later time. Also, the step response is observed for perturbations of +10% in L and K and for measurement noise with a variance of 0.0001. The system robustness for uncertainty is illustrated in the following sections through stability analysis. The frequency used for Oustaloup approximation of fractional order is (0.01,100)rad/s. In addition, the trend of closed loop behavior is interpreted for variation in controllability index (i.e. L/T ratio in the range of 0.1 to 2). This analysis demonstrates the difficulty in control for large changes in time delay.

5.1 Example 1

Consider the higher order system (Shen, 2002) and its equivalent NIOPTD (Patre, 2017) model:

$$G_1(s)=\frac{1}{(s+1)^4}=\frac{0.99149}{2.8015s^{1.0759}+1}e^{-1.6745s} \qquad (11)$$

The proposed and Patre (2017) controllers are:

$$C_{proposed}(s)=\left(\frac{1}{0.7535s^{1.01}+0.9s^{0.01}+1.6745}\right)\times 0.99149\left(\frac{0.8372s+1}{0.8372s}\right)\left(1+2.8015s^{1.0759}\right) \qquad (12)$$

$$C_{old}(s)=1.5129+\frac{0.3432}{s^{1.1}}+0.1733s^{1.05} \qquad (13)$$

The optimum values of γ and p for the proposed controller are identified as 0.9 and 1.01 (Figure 2). The performance measures for set point tracking are given in Table 2. The proposed method is superior in performance compared to the Patre (2017) method with lower values of %OS, ST, IAE and TV. The servo response for a disturbance applied at t = 30 s is illustrated in Figure 3. Betterment is observed, even with disturbance with lower values of performance measures, which is clear from Table 3. It is evident from Figure 3 that a satisfactory performance is observed in terms of disturbance rejection. Figure 4 shows the step response for a perturbed model, and the response for white noise in the measured output is illustrated in Figure 5. It is observed that there is enhanced performance (Table 2) for both

Table 1. Expressions for %OS, IAE and TV.

%OS	IAE	TV
$\dfrac{y_{peak}-y_{ss}}{y_{ss}}\times 100$	$\displaystyle\int_0^\infty |e(t)|\,dt$	$\displaystyle\sum_{i=0}^{\infty}|u_{i+1}-u_i|$

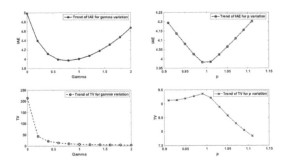

Figure 2. Identification of optimum γ and p for example1.

Table 2. Servo performance comparison for the three different systems.

System	Method	%OS	ST	IAE	TV
$G_1(s)$	Proposed	18.2	8.7	2.87	8.45
	Patre (2017)	22.3	15	3.57	18.7
$G_2(s)$	Proposed	2.4	1.54	0.78	3.86
	Patre (2017)	5.02	2.97	0.83	4.75
$G_3(s)$	Proposed	21.2	0.94	0.42	12.74
	Patre (2017)	40.7	1.95	0.53	12.35

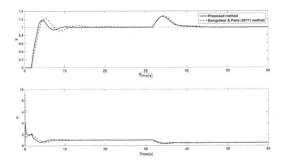

Figure 3. Closed loop response of $G_1(s)$ for step input.

Table 3. Comparison of IAE and TV for the three examples.

System	Method	Perfect case		Perturbed case		Noise case	
		IAE	TV	IAE	TV	IAE	TV
$G_1(s)$	Proposed	3.984	9.2133	4.837	10.6533	4.281	38.7359
	Patre (2017)	4.907	19.567	5.988	21.1318	5.02	137.5908
$G_2(s)$	Proposed	1.176	4.393	1.263	4.992	1.207	7.2042
	Patre (2017)	1.243	5.3571	1.353	6.1191	1.248	10.4974
$G_3(s)$	Proposed	0.5663	13.5002	0.6612	16.185	0.6131	20.1658
	Patre (2017)	0.6224	13.506	0.8196	20.2817	0.6767	20.6138

the cases with the proposed method, as compared to the Patre (2017) method. Also, there is significantly less control effort with the proposed method for all the possible input changes.

The magnitude plot is shown in Figure 6 for +10% uncertainty in K; +10% and +50% uncertainty in L. Robust stability condition in Equation 9 is violated by both the complementary sensitivity functions for +50% uncertainty in time delay. The proposed method violates the

515

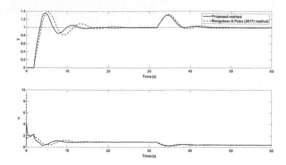

Figure 4.   Closed loop step response of $G_1(s)$ for perturbations.

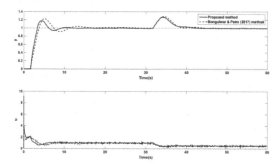

Figure 5.   Step response in presence of measurement noise.

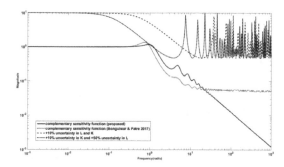

Figure 6.   Magnitude plot for example1.

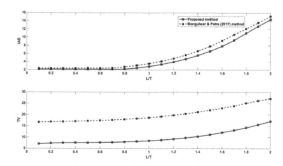

Figure 7.   L/T ratio versus IAE, TV for step change in set point.

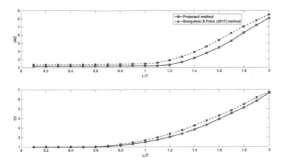

Figure 8.   L/T ratio versus IAE, TV for step change in disturbance.

condition a bit earlier than the old method. Figure 7 and Figure 8 show the trends of IAE and TV for servo and regulatory response with increasing L/T ratio. It is evident that increasing trends are observed with the Patre (2017) method, compared to the proposed method. Hence, the proposed method can be considered for enhanced closed loop performance of processes with large changes in time delay.

5.2   Example 2

The second example (Chen et al., 2008; Patre, 2017) considered for performance comparison is:

$$G_2(s) = \frac{9}{(s+1)(s^2+2s+9)} = \frac{1.0003}{0.8864s^{1.0212}+1} e^{-0.4274s} \tag{14}$$

The proposed controller and the controller with Patre (2017) method are:

$$C_{proposed}(s) = \left(\frac{1}{0.07479s^{1.01} + 0.35s^{0.01} + 0.4274} \right) \times 0.2136 \left(\frac{0.2137s+1}{0.2137s} \right) (1 + 0.8864s^{1.0212}) \tag{15}$$

$$C_{old}(s) = 1.4996 + \frac{1.2203}{s^{1.05}} + 0.0409s^{1.05} \tag{16}$$

The optimum values of γ and p are identified as 0.35 and 1.01. The closed loop system gives good servo response with the proposed method. This is true with the lower values of %OS, ST, IAE and TV given in Table 2. The unit step response with disturbance applied at t = 6 s is shown in Figure 9 and the corresponding performance measures are given in Table 3. The proposed method gives better servo response, which is evident with lower values of performance measures, while the regulatory performance is almost the same for both the controllers. The system response for perturbations is presented in Figure 10. Figure 11 presents the closed loop response for white noise in the output. The proposed method continues to give the superior performance compared to the Patre (2017) method, which is clear with the lower values of IAE and TV (Table 3).

The closed loop robust stability for uncertainties in K and L is illustrated through the magnitude plot in Figure 12. The closed loop system gives robust performance up to +100% uncertainty in time delay and +10% uncertainty in gain with the proposed controller, whereas the stability condition fails for +90% uncertainty in time delay with the Patre (2017) method. Figure 13 and Figure 14 show the trends of IAE and TV for servo and regulatory response with increase in L/T ratio. The proposed method shows less control effort for servo and regulatory response for the entire variation of L/T ratio. The trend followed by IAE for set point tracking is almost the same up to L/T ratio of 1 for both the methods; after that, it starts increasing with the old method. In the case of disturbance rejection, the IAE values are lower up to L/T ratio of 1.3 with the old method, and then it increases. Hence, the proposed method is a good choice to have a better control for increasing L/T ratio, compared to the old method (Patre, 2017).

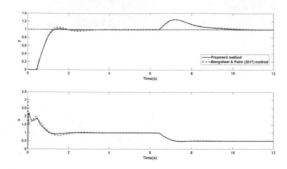

Figure 9.   Closed loop response of $G_2(s)$ for step input.

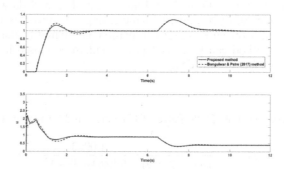

Figure 10.   Closed loop step response of $G_2(s)$ for perturbations.

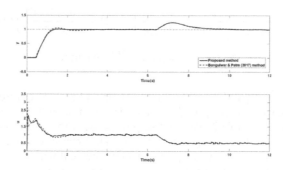

Figure 11.   Step response in presence of measurement noise.

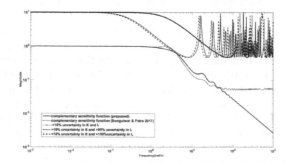

Figure 12.   Magnitude plot for example2.

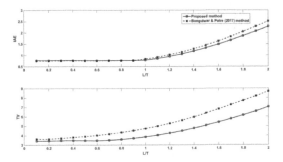

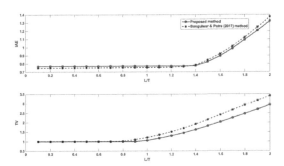

Figure 13.   L/T ratio versus IAE, TV for step change in set point.

Figure 14.   L/T ratio versus IAE, TV for step change in disturbance.

5.3   *Example 3*

The higher order system studied in Panagopoulos et al. (2002) is considered as the third example:

$$G_3(s) = \frac{1}{(s+1)(0.2s+1)(0.04s+1)(0.008s+1)} = \frac{0.99932}{1.0842s^{1.0132}+1}e^{-0.1922s} \qquad (17)$$

The proposed and old (Patre, 2017) controllers are given as follows:

$$C_{proposed}(s) = \left(\frac{1}{0.0096s^{1.1}+0.1s^{0.1}+0.1922} \right) \times 0.0961 \left(\frac{0.0961s+1}{0.0961s} \right)(1+1.0842s^{1.0132}) \qquad (18)$$

$$C_{old}(s) = 5.0034 + \frac{6.14}{s^{1.1}} + 0.0163s^{1.1} \qquad (19)$$

The values of γ and p for the proposed method are 0.1 and 1.1. The performance measures shown in Table 2 for servo response indicates that the %OS, ST and IAE values are lower than with the proposed method, but that the TV value is slightly higher compared to the old (Patre, 2017) method. Similarly, the step response for a change in disturbance applied at t = 5 s is shown in Figure 15. The closed loop step response for process parameter variations and for output noise is illustrated in Figure 16 and Figure 17. The corresponding perform-ance measures for all the above cases are presented in Table 3. It is evident from all these Figures that the proposed method gives superior servo performance but is a bit slow in reject-ing the disturbance compared to the old method.

The proposed method gives robust performance up to an uncertainty of +70% in L and +10% in K, while the Patre (2017) method fails for less than +50% uncertainty in L (Figure 18). Figure19 and Figure 20 show the performance for variation of L/T ratio. For servo response, the variation of IAE is low with the proposed method, while the control effort is slightly high

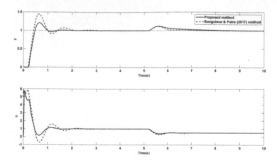

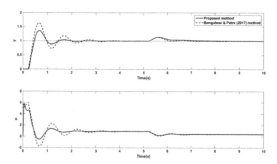

Figure 15. Closed loop response of $G_3(s)$ for step input.

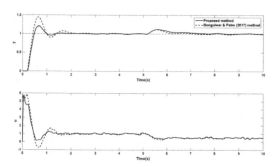

Figure 16. Closed loop step response of $G_3(s)$ for perturbations.

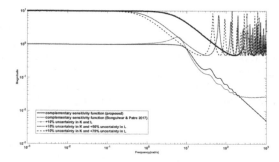

Figure 17. Step response in presence of measurement noise.

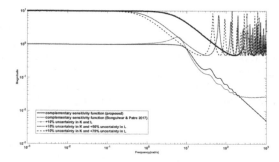

Figure 18. Magnitude plot for example 3.

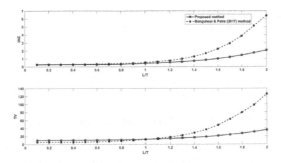

Figure 19.   L/T ratio versus IAE, TV for step change in set point.

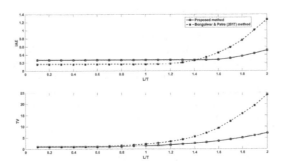

Figure 20.   L/T ratio versus IAE, TV for step change in disturbance.

up to L/T = 0.9, and then it increases drastically with the old method. In the case of regulatory control, the IAE values are higher with the proposed method and the variation of TV is low. Hence, there is a trade-off between IAE and TV for increasing L/T, and the proposed method is recommended for servo response, while it can be used for disturbance rejection at higher values of L/T.

6   CONCLUSIONS

In this paper, a FFFOPID controller is proposed for higher order systems approximated as NIOPTD models using IMC method. Analytical method is followed for identifying the tuning parameters by minimizing IAE and TV. Enhanced closed loop performance is observed with the proposed method for changes in set point and disturbance. In particular, the proposed method is effective in terms of there being less control effort. The closed loop system is robust with the proposed method for high uncertainty in process parameters. Also, the proposed method assures better control for large changes in time delay, which is proved through the variation of L/T ratio.

REFERENCES

Aström, K.J. & Hägglund, T. (1995). PID controllers: Theory, design, and tuning. Research Triangle Park, NC: ISA.

Bongulwar, M.R. & Patre, B.M. (2017). Stability regions of closed loop system with one non-integer plus time delay plant by fractional order PID controller. *International Journal of Dynamics and Control*, 5(1), 159–167.

Chen, Y.Q., Bhaskaran, T. & Xue, D. (2008). Practical tuning rule development for fractional order proportional and integral controllers. *ASME Journal of Computational Nonlinear Dynamics, 3*(2), 021403.

Das, S., Saha, S., Das, S. & Gupta, A. (2011). On the selection of tuning methodology of FOPID controllers for the control of higher order processes. *ISA Transactions, 50*(3), 376–388.

Hui-fang, W., Qiu-sheng, H., Zhi-cheng, Z & Jing-gang, Z.(2015). A design method of fractional order $PI^{\lambda}D^{\mu}$ controller for higher order systems. In *34th Chinese Control Conference (CCC), Hangzhou, China* (pp. 272–277).IEEE.

Isaksson, A. & Graebe, S. (1999). Analytical PID parameter expressions for higher order systems. *Automatica, 35*(6), 1121–1130.

Lin, M.G., Lakshminarayanan, S. & Rangaiah, G.P. (2008). A comparative study of recent/popular PID tuning rules for stable, first-order plus dead time, single-input single-output processes. *Industrial & Engineering Chemistry Research, 47*(2), 344–368.

Luo, Y. & Chen, Y. (2009). Fractional order [proportional derivative] controller for a class of fractional order systems. *Automatica, 45*(10), 2446–2450.

Malwatkar, M., Sonawane, S. & Waghmare, L. (2009). Tuning PID controllers for higher-order oscillatory systems with improved performance. *ISA Transactions, 48*(3), 347–353.

Monje, C.A.,Vinagre, B.M., Feliu, V. & Chen, Y. (2008). Tuning and auto-tuning of fractional order controllers for industry applications. *Control Engineering Practice, 16*(7), 798–812.

Morari, M. & Zafiriou, E. (1989). *Robust process control.* Englewood Cliffs, New Jersey: Prentice Hall.

Padula, F. & Visioli, A. (2011). Tuning rules for optimal PID and fractional order PID controllers. *Journal of Process Control, 21*(1), 69–81.

Pan, I. & Das, S. (2013). Model reduction of higher order systems in fractional order template. In *Intelligent Fractional Order Systems and Control* (pp. 241–256). Berlin, Heidelberg: Springer.

Panagopoulos, H., Aström, K.J. & Hägglund, T. (2002). Design of PID controllers based on constrained optimisation. *IEEProceedings-Control Theory and Applications, 149*(1), 32–40.

Podlubny, I. (1999). Fractional-order systems and $PI^{\lambda}D^{\mu}$ controllers. *IEEE Transactions on Automatic Control, 44*(1), 208–214.

Ranganayakulu, R., Babu, G.U.B., Rao, A.S. & Patle, D.S.(2016). A comparative study of fractional order $PI^{\lambda}/PI^{\lambda}D^{\mu}$ tuning rules for stable first order plus time delay processes. *Resource-Efficient Technologies, 2*, S136–S152.

Ranganayakulu, R., Babu, G.U.B. & Rao, A.S. (2017). Fractional filter IMC-PID controller design for second order plus time delay processes. *Cogent Engineering, 4(1)*, 1366888.

Shah, P. & Agashe, S. (2016). Review of fractional PID controller. *Mechatronics, 38*, 29–41.

Shen, J.C. (2002). New tuning method for PID controller. *ISA Transactions, 41*(4), 473–484.

Skogestad, S. (2003). Simple analytic rules for model reduction and PID controller tuning. *Journal of Process Control, 13*(4), 291–309.

Tavakoli-Kakhki, M. & Haeri, M. (2011). Fractional order model reduction approach based on retention of the dominant dynamics: Application in IMC based tuning of FOPI and FOPID controllers. *ISA Transactions, 50*(3), 432–442.

Valério, D. & da Costa, J.S. (2006).Tuning of fractional PID controllers with Ziegler–Nichols-type rules. *Signal Processing, 86*(10), 2771–2784.

Vinopraba, T., Sivakumaran, N., Narayanan, S. & Radhakrishnan, T.K. (2012). Design of internal model control based fractional order PID controller. *Journal of Control Theory and Applications, 10*(3), 297–302.

Emerging Trends in Engineering, Science and Technology for Society,
Energy and Environment – Vanchipura & Jiji (Eds)
© *2018 Taylor & Francis Group, London, ISBN 978-0-8153-5760-5*

Effect of parameters on electro-Fenton process for removal of oil and grease from refinery wastewater

Mikhila Vikramaraj, A.M. Manilal & P.A. Soloman
Department of Chemical Engineering, Government Engineering College, Thrissur, Kerala, India

ABSTRACT: In the petroleum industry, oily water occurs in the stages of production, transportation and refining, as well as during the use of derivatives. Crude oil is one of the major components of wastewater from the petroleum industry. The project focuses on the use of the Electro-Fenton (EF) technique for the removal of oil content from wastewater from the petroleum industry. Hydrogen peroxide is used as Fenton's reagent and Fe^{2+} is provided from sacrificial cast iron anodes. The crude oil samples are taken from the Cochin refinery. Experimental study has been conducted to investigate the influence of various factors, for instance current density, time of electro-Fenton, feed pH, and H_2O_2 concentration in the electro-Fenton process.

Keywords: Electro-Fenton technique, wastewater, Hydrogen peroxide

1 INTRODUCTION

Petroleum products play an unavoidable role in our daily lives and our demand is increasing day by day. Petroleum products include transportation fuels, fuel oils for heating and electricity generation, and feed stocks for making the chemicals, plastics, and synthetic materials that are in nearly everything we use. Of the approximate 7.19 billion barrels of total US petroleum consumption in 2016, 48% was motor gasoline (including ethanol), 20% was distillate fuel (heating oil and diesel fuel), and 8% was jet fuel. So, these products have become essential in our routine lives.

These petroleum products are made from crude oil through a refining process. During the refining process, petroleum refineries unavoidably generate large amounts of oily wastewater. These wastes occur at different stages of oil processing, such as during production, transportation and refining. However, during the production phase large amounts of oily wastes are generated, which become mixed with the sea water and cause pollution. Coelho et al. (2006) reported that the quantity of water used in the oil refinery processing industry during the production stage ranges from 0.4 to 1.6 times the volume of processed oil, and this wastewater may, if untreated, cause serious damage to the environment.

The presence of oil and grease in the water bodies accounts for a major part of water pollution. Alade *et al.* (2011) explains the effects of these to the economy. The oil and grease will form a thin oily layer above the water medium, which reduces the light penetration into the water medium, thereby decreasing photosynthesis. Thus, it affects the survival of aquatic life in water since the amount of dissolved oxygen in the water is less. It also affects the aerobic and anaerobic wastewater treatment process due to the reduction in oxygen transfer rates, and also due to the reduction in the transport of soluble substrates to the bacterial biomass.

So, more attention must be given to the treatment of oily wastewater. There are various methods available for oil removal from wastewater, such as physical treatment, chemical treatment, biological treatment, membrane treatment and advanced oxidation processing (Krishnan et al., 2016).

The motivation of this project is the application of the Electro-Fenton (EF) process for large-scale industries and wastewater treatment plants. Electro-Fenton treatment is regarded as being a better mechanism for water treatment units.

In electro-Fenton, organic substances are removed at two stages of oxidation and coagulation. Oxidation of organic substances is due to -OH radicals and coagulation is ascribed to the formation of ferric hydroxo complexes (Nidheesh & Gandhimathi, 2012).

The degradation mechanism of organic pollutants by Fenton reaction is given in the following Equations.

$$Fe^{2+} + H_2O_2 \rightarrow Fe^{3+} + OH^- + HO^{\cdot} \tag{1}$$

$$RH + HO^{\cdot} \rightarrow R^{\cdot} + H_2O \tag{2}$$

where, RH denotes organic pollutants

$$R^{\cdot} + Fe^{3+} \rightarrow R^+ + Fe^{2+} \tag{3}$$

$$Fe^{2+} + HO^{\cdot} \rightarrow Fe^{3+} + OH^- \tag{4}$$

The viability of electro-Fenton is studied by conducting batch experiments for investigating the influencing parameters. These four parameters were identified viz. current density, initial pH, hydrogen peroxide concentration, and reaction time.

2 MATERIALS AND METHODS

The project mainly focuses on the study of different parameters that affect the electro-Fenton process for removal of oil and grease content from crude oil wastewater. The electro-Fenton process has two different configurations. In the first, Fenton reagents is added to the reactor from outside and inert electrodes with high catalytic activity are used as anode material. While in the second configuration, only hydrogen peroxide is added from outside and Fe^{2+} is provided from sacrificial cast iron anodes (Nidheesh & Gandhimathi, 2012). Here we are using the second configuration. In this project we are using mild steel/iron as the anode and stainless steel as the cathode. Therefore, an additional supplement of Fe^{2+} is not needed.

2.1 Experimental setup

Figure 1 shows the experimental setup for the electro-Fenton process for the removal of oil and grease from crude oil processed wastewater.

The reactor is designed for a volume of 2 liters of water to be treated. Therefore the dimensions of the reactor selected are 25 cm × 10 cm × 10 cm, with a volume of 2.5 liters. The reactor is made of plastic. The crude oil feed is fed into the reactor. The pH of the feed is measured using a pH meter and the pH is adjusted by adding dilute HCl and NaOH. The electro-Fenton process needs anodes and cathodes for the removal of crude oil. The anodes are made of mild steel and the cathodes of stainless steel. The anodes and cathodes are designed in such a way that they fit well in the reactor. The electrodes are placed in a parallel arrangement. Here five anodes and four cathodes are used. The dimensions of the anodes and cathodes are 8 × 5. The total area of ten electrodes was 4 dm².

The hydrogen peroxide is added from outside as per required. Current and potential is passed through the electrodes using a regulated DC power supply with the anode connected to the positive, and the cathode to the negative ends of the DC power supply. The input voltage specification of the Aplab regulated power supply unit is 0–230V AC, and output voltage specification is 0–30V DC. A magnetic stirrer was provided inside the reactor with varying rotational speeds. The magnetic stirrer is an integral part of the mechanism because proper contact of the electrodes with the wastewater is required to reduce the internal resistance (IR) drop of the system. Once the experiment is completed, output samples are pippeted out from the middle of the reactor. The samples are allowed to settle for half an hour and then analyzed using Chemical Oxygen Demand (COD) and fluorescence spectrometry.

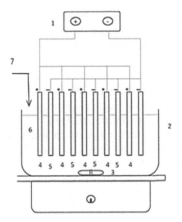

Figure 1.   Electro-Fenton experimental setup.
Schematic of EF reactor – 1. Power supply with indication; 2. Reactor; 3. Magnetic stirrer; 4. Anodes (mild steel); 5. Cathodes (stainless steel); 6. Water to be treated; 7. H_2O_2 added from outside.

2.2   Experimental procedure

The crude oil feed preparation procedure and electro-Fenton experimental procedure is explained below.

2.2.1   Feed preparation

The crude oil from the desalter unit of Bharath Petroleum Corporation Limited (BPCL), Kochi is taken for conducting the experiments. To prepare the feed, take 2,000 ml of water from the reactor. Then 4 g crude oil, 1.5 g salt (NaCl) and 0.5 g SDS (Sodium Dodecyl Sulfate) are added. Stir well using a magnetic stirrer at 500rpm for 10 mins. This is the feed. The pH of the feed is measured using a pH meter. The pH is maintained according to experimental requirement. It is maintained by adding dilute HSO_4 or dilute NaOH. The SDS, a surfactant, was added to help the oil particles better dissolve in the water. The characteristics of feed are different for each set of experiments.

2.2.2   Experiment

Before moving to the experiment, first the electrodes are cleaned using dilute HCl. For that, 10 ml dilute HCl is used to clean the electrodes and the weights of the anodes and cathodes are noted. Then we move on to the Fenton process. For that, 2 liters feed is taken from the reactor. The electrodes (cathodes and anodes) are arranged bipolar. Electrodes are connected to the DC supply. Current and voltage are adjusted. The experiment is run according to the values of the parameters (current density, H_2O_2 concentration, pH, time) in the experiment list. The salt concentration of the feed is 1.5 g.

   After the experiment, the samples are collected and analyzed using COD value, fluorescence spectrometry and gravimetric analysis.

2.3   Experimental analysis

The experimental results are analyzed using the COD analyzer, fluorescence spectroscopy, and gravimetric analysis.

2.3.1   Chemical oxygen demand

The samples are analyzed using COD. The COD value indicates the amount of oxygen needed for the oxidation of all organic substances in water in mg/l. The COD is closely related to the laboratory standard dichromate-method. The method involves using a strong oxidizing chemical, potassium dichromate, to oxidize the organic matter in solution to carbon dioxide and water under acidic conditions. Often, the test also involves a silver compound to encourage oxidation of certain organic compounds, and mercury to reduce the interference from oxidation of chloride ions. Based on this method, the COD has become a commonly used method in wastewater analysis.

2.3.2 *Fluorescence spectroscopy*

The Hitachi Fluorescence Spectrophotometer Model F-4600 is also used for the analysis of the samples. It is a research-grade fluorescence spectrophotometer with high sensitivity and high scanning speeds, capable of multi-mode analysis.

In this project fluorescence analysis for the oil in water was conducted. Fluorescence is the emission of light by a substance that has absorbed light. In general, the emitted light has a longer wavelength, and therefore lower energy, than the absorbed radiation. A light source is provided inside the instrument. A monochromatic filter is provided for selecting the excitation wavelength. Then the spectral distribution of the light emitted from the sample is analyzed. This is detected by a light detector. The sample was placed inside the fluorescence spectrometer in a quartz cuvette for the analysis.

3 RESULTS AND DISCUSSION

The project focuses on removal of oil and grease from crude oil wastewater by the electro-Fenton process. COD and analysis of oil and grease using a fluorescence spectroscope and gravimetric analysis are used to estimate the percentage oil removal in each experiment. The samples from each experiment are analyzed using COD and fluorescence spectroscope and gravimetric method and the percentage removal of oil and grease is found by comparing with the feed values. The COD value gives the amount of organic substances found in the water. Therefore, it is an important measure of water quality.

3.1 *Calibration of fluorescence spectroscopy*

Fluorescence spectroscopy is used to find the oil content from the fluorescence values of different emission wavelengths which were supplied with a constant excitation wavelength.

The working parameters for conventional spectra of crude oil were an excitation wavelength of 380 nm, an emission range of 200–900 nm (5 nm intervals) with 5.0 nm excitation, and 5.0 nm emission slits with a scan rate of 1,200 nm/min (Sotelo et al., 2008). Fluorescence values were calibrated in terms of concentration in ppm. The calibration was done with known concentrations. The calibration curve is shown in Figure 2.

From the calibration curve:
Fluorescence $= 0.0788 \times$ Concentration $+ 3.2801$

From this the concentration was calculated as:
Concentration (ppm) $= 12.488$ (Fluorescence $- 37.765$)

The excitation wavelength of crude oil is given as 380 nm. The fluorescence spectrometer generates the spectrum. The emission wavelength is given in the range of 200–900 nm. The height at emission wavelength apex 418.2 nm is taken for noting the crude oil concentration.

3.2 *Effect of parameters on electro-Fenton process*

The preliminary stage experiments were conducted to analyze the effect of different parameters that influence the electro-Fenton process. The individual influence of a parameter was studied by keeping all other parameters constant.

3.2.1 *Influence of pH*

The pH value will affect the oxidation and coagulation of the electro-Fenton process. The impact of initial pH value on percentage removal was studied. The initial pH was adjusted using 0.1 N HCl and 0.1 N NaOH. The initial pH value was found using a digital pH meter.

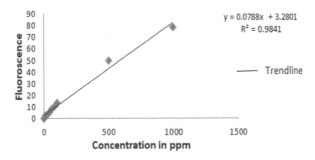

Figure 2.   Calibration curve for fluorescence spectroscopy – fluorescence v/s concentration.

The experiment was conducted at current density 0.5 A/dm^2, reaction time of 20 mins, and H$_2$O$_2$ concentration 0.275 g/l.

The initial pH plays an important role in the electro-Fenton process. Figure 3 shows the percentage of oil removal using COD and fluorescence analysis. The study was conducted at a pH range from 2–7. Generally, the Fenton process was conducted at a pH below 7 (i.e. at acidic medium).

From the results it was clear that the maximum removal was obtained at a pH around 4. The removal becomes less effective at a pH < 3. It is due to regeneration of Fe^{2+} ions, through reaction between Fe^{3+} and H$_2$O$_2$. At higher pH the removal also decreases rapidly: at a pH > 5, the % removal was found to decrease to a value of 50%. It is due to the fact that H$_2$O$_2$ is unstable in basic solution (Nidheesh & Gandhimathi, 2012).

3.2.2   *Influence of current density*
Current density is the electric current per unit area of cross section of the anode. As the current increases, current density also increases.

The maximum current density taken was 0.75 A/dm^2 and the oil removal was found to increase with an increase in current density. The experiment was conducted at pH 4, reaction time 20 mins and H$_2$O$_2$ concentration 0.275 g/l. As current density increases, removal percentage also increases. The maximum removal is obtained at maximum current density (i.e. at a current density of 0.73 A/dm^2). The applied current is the driving force for the reduction of oxygen, leading to the generation of hydrogen peroxide at the cathode.

3.2.3   *Influence of H$_2$O$_2$ concentration*
Hydrogen peroxide is used as Fenton's reagent, which plays a major role in the removal of oil. The removal efficiency increases with the increase in hydrogen peroxide concentration. For the experimental analysis, the amount of hydrogen peroxide taken for a 2 liter crude oil feed is 0.42 g, 0.51 g, 0.55 g, 0.6 g, and 0.68. All other parameters were kept constant (i.e. current density 0.5 A/dm^2, reaction time 20 mins, and at pH 4).

Figure 5 shows that as the amount of hydrogen peroxide increases, initially the removal increases and then decreases. The maximum efficiency was obtained at 0.55 g hydrogen peroxide for 2 liters of crude oil feed. As the H$_2$O$_2$ concentration increases, the increase in removal efficiency is due to the increase in hydroxyl radical concentration. At high dosage of H$_2$O$_2$, the decrease in removal efficiency was due to the scavenging effect of H$_2$O$_2$ and the recombination of hydroxyl radical (Yu et al., 2013).

3.2.4   *Influence of reaction time*
The last parameter analyzed was reaction time. Time increment resulted in a sharp increment in the removal efficiency.

This might be due to the Farady's law of electrolysis, which gives a direct relationship between time and the amount of anode material released. A saturation point was observed at around 20 mins for the electro-Fenton process. There was not much increase in removal after the saturation value, as can be seen in Figure 6.

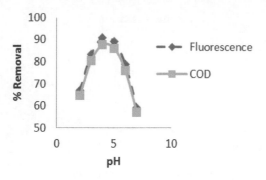

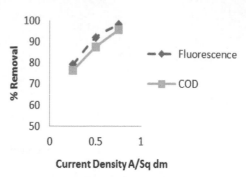

Figure 3. Variation of % oil removal with change in pH.

Figure 4. Variation of % oil removal with change in current density.

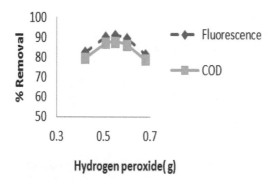

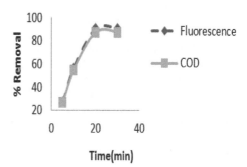

Figure 5. Variation of % oil removal with change in H_2O_2 concentration.

Figure 6. Variation of % oil removal with change in time.

4 CONCLUSION

The work focused on the removal of oil and grease from crude oil processed wastewater releasing from refineries. Crude oil wastewater from BPCL, Kochi was selected as the sample for treatment purpose. For the removal of oil and grease from crude oil processed wastewater, the electro-Fenton technique was used. Both oxidation and coagulation contributed to COD removal through Fenton treatment of wastewater.

COD test and fluorescence spectrometry analysis were the major analyses conducted for the treated water. Electro-Fenton experiments were conducted with mild steel and stainless steel electrodes. The important parameters affecting the EF process were analyzed, –viz. pH, current density, H_2O_2 concentration and time. At pH 4 percentage removal was maximum, and decreased with an increase and decrease in the pH of water. The oil removal increases with increase in current density. At highest current density, the maximum removal efficiency was obtained. In the case of H_2O_2 concentration, the maximum removal was obtained at 0.55 g/l with a reaction time of 30 mins.

REFERENCES

Alade, A.O., Jameel, A.T., Muyibi, S.A., Karim, M.I.A., & Alam, Z. (2011). Application of semifluidized bed bioreactor as novel bioreactor system for the treatment of palm oil mill effluent (POME). *African Journal of Biotechnology*, *10*(81), 18642–18648.

Al-Harbawi, A.F.Q., Mohammed, M.H. & Yakoob, N.A. (2013). Use of Fenton's reagent for removal of organics from Ibn Al-Atheer Hospital wastewater in Mosul City. *Al-Rafidain Engineering Journal, 21*(5), 127–135.

Atmaca, E. (2009). Treatment of landfill leachate by using electro-Fenton method. *Elsevier Journal of Hazardous Materials, 163*(1), 109–114.

Coelho, A., Castro, A.V. Dezotti, M. & Sant'Anna Jr, G.L. (2006). Treatment of petroleum refinery sourwater by advanced oxidation processes. *Elsevier Journal of Hazardous Materials, 137*(1), 178–184.

Gordon Jr, D.C., Keizer, P.D. & Dale, J. (1974). Estimates using fluorescence spectroscopy of the present state of petroleum hydrocarbon contamination in the water column of the northwest Atlantic Ocean. *Elsevier Scientific Publishing Company. Amsterdam. Marine Chemistry, 2*(4), 251–261.

Jameel, A.T., Muyubi, S.A., Karim, M.I.A & Alam, M.Z. (2011). Removal of oil and grease as emerging pollutants of concern (EPC) in wastewater stream. *IIUM Engineering Journal, 12*(4), 161–169.

Kalra, S.S., Mohan, S., Sinha, A. & Singh, G. (2011). Advanced oxidation processes for treatment of textile and dye wastewater: A review. In *2nd International Conference on Environmental Science and Development IPCBEE, 4*, 271–275.

Krishnan, S., Chandran, K., & Sinnathambi, C.M. (2016). Wastewater treatment technologies used for the removal of different surfactants: A comparative review. *International Journal of Applied Chemistry, 12*(4), 727–739.

Nidheesh, P.V. & Gandhimathi, R. (2012). Trends in electro-Fenton process for water and wastewater treatment: An overview. *Elsevier Desalination, 299*, 1–15.

Patil, A.D. & Raut, P.D. (2014). Treatment of textile wastewater by Fenton's process as an advanced oxidation process. *IOSR Journal of Environmental Science, Toxicology and Food Technology (IOSR-JESTFT)*. e-ISSN: 2319–2402. p-ISSN: 2319-2399, 8, 29–32.

Soloman, P.A., Basha, C.A., Velan, M. & Balasubramanian, N. (2009). Electrochemical degradation of pulp and paper industry waste-water. *Journal of Chemical Technology and Biotechnology, 84*(9), 1303–1313.

Sotelo, F.F., Pantoja, P.A., López-Gejo, J., Le Roux, G.A.C., Quina, F.H. & Nascimento, C.A.O. (2008). Application of fluorescence spectroscopy for spectral discrimination of crude oil samples. *Brazilian Journal of Petroleum and Gas, 2*(2), 63–71.

Yu, R.F., Lin, C.H., Chen, H.W., Cheng, W.P. & Kao, M.C. (2013). Possible control approaches of the electro-Fenton process for textile wastewater treatment using on-line monitoring of DO and ORP. *Elsevier Chemical Engineering Journal, 218*, 341–349.

Zhang, J. & Dong, H. (2009). Study of treatment of oil pollution from water with electro-Fenton technology. *IEEE*. 978-1-4244-2902-8/09.

Emerging Trends in Engineering, Science and Technology for Society,
Energy and Environment – Vanchipura & Jiji (Eds)
© 2018 Taylor & Francis Group, London, ISBN 978-0-8153-5760-5

Conversion studies of methanol to olefin on boric acid treated Al/MCM-41

D. Kumar, N. Anand & A. Kedia
University School of Chemical Technology, GGS IP University, Delhi, India

ABSTRACT: The use of a large-pore catalyst with a high surface area, MCM-41, loaded with alumina, is explored for the production of ethylene and propylene from methanol conversion. MCM-41 is a silica-based catalyst with low Lewis acid site strength. Al/MCM-41 was treated with boric acid at three different concentrations. A modified catalyst was characterized using BET, chemisorptions, XRD and SEM. The total surface area was observed to reduce after boric acid treatment with the treatment of MCM-41. The maximum decrease in surface area was obtained with the treatment of Al/MCM-41 with boric acid (1M). An N_2 adsorption-desorption plot shows a change in the porous structure of the catalyst after treatment with boric acid. The conversion studies were performed at different temperatures between 250–450°C and liquid flow rate in the range of 30 to 120 ml/min. The effect of the catalyst on the selectivity of ethylene and propylene was studied with Al/MCM-41 and B-Al/MCM-41. Results showed that the boric acid treated Al/MCM-41 helps to increase the selectivity of the catalyst toward propylene production. Gas yield was also observed to increase after using the boric acid treated catalyst. A 20.3% decrease in the coke yield was observed when the experiments were performed with the boric acid treated catalyst as compared to an untreated catalyst. However, no significant effect on the coke production was obtained in the presence of three boric acid treated Al/MCM-41 catalysts.

Keywords: large-pore catalyst, boric acid, propylene production, coke production

1 INTRODUCTION

In the recent past, the scientific community has been putting more effort into looking for alternative routes for the production of feedstock materials required in the petrochemical industries. At present, olefins are mainly produced from methanol conversion. The main source of methanol is methane, obtained from the petroleum refineries (Tian et al., 2015). However, as petroleum refineries are now being modified to process the gas obtained from the natural resources, new and effective technologies are needed to convert methane to methanol and further feedstock materials such as DME and formaldehyde. In coming years, more methane will be obtained from shale gas or gas hydrates (Lefevere et al., 2014).

Zeolites such as ZSM-5 has been widely used throughout the world, due to its shape selectivity, durability and reusability for a wide range of reactions in the petroleum refinery and petrochemical industries (Khare et al., 2017). However, the small size of cage and small pore size of ZSM-5 are its major drawbacks in the Methanol to Olefin (MTO) process, wherein large-size products cannot escape from the small cage opening, thereby deactivating the catalyst. To overcome the fast deactivation issue of catalysts, a large-pore silica-based mesoporous catalyst has been used for different reactions at various laboratories in the last decade, and is a subject of interest (Wu et al., 2012). These catalysts have both two-dimensional and three-dimensional structures with large-pore diameters and high surface area (e.g. SBA-15, MCM-41, MCM-22, FDU-13 and MFU). However, these catalysts lack mechanical strength and have low acid site concentrations required for reactions (Olsbye et al., 2012; Li et al.,

2011; Almutairi et al., 2013; Xu et al., 2013). These catalysts can be tailored for their pore size as well as acid site concentration. The surface functionalization of the hydroxyl groups with alumina has affected the acidity of these silicious catalysts. Studies on alumina and other metals like Pd, Ga, Cu, and Ge impregnated MCM-41, have been reported. These metals help in modifying the Lewis as well as Bronsted acid sites, for various reactions on mesoporous as well as microporous catalyst, including ZSM-5, for various reactions (Linares et al., 2014; Naik et al., 2010; Bhattacharyya et al., 2003; Sang et al., 2013). As mentioned by a few authors, faster deactivation of catalyst might occur due to loading of alumina. But is reported that the loading of alumina might be helpful in increasing the mechanical strength and increasing the number of strong acid sites. The modification of the acid site concentration by the treating of ZSM-5 with boric acid, or oxalic acid, or phosphoric acid have been reported (Du et al., 2006; Murthy et al., 2010; Sang et al., 2013; Epelde et al., 2015). This modification might be applied for the conversion of methanol to gasoline hydrocarbons.

This work reports on the modification of alumina-loaded MCM-41 with boric acid to study the effect of change in the type of acid site concentration, and thus the conversion of methanol and selectivity for ethylene and propylene in a fixed-bed quartz reactor.

2 MATERIALS AND METHODOLOGY

2.1 Catalyst modification

Alumina-loaded MCM-41 was purchased from ACS Material LLC and was modified with boric acid. The solid powder was mixed in 0.5, 1.5 and 2.5M H_3BO_3 with solution/catalyst ratio of 20 cm^3/g at 80°C for 8h, and calcined at 550°C for 6h.

2.2 Catalyst characterization

The purchased catalyst and the modified catalyst were characterized for their properties, such as surface area, surface structure, crystalline structure and mesoporous structure. Fourier Transform Infrared Spectroscopic (FT-IR) studies of the catalysts were obtained on a Nicolet Model: 6700 in the wavelength range from 4,000 to 500 cm^{-1}. The catalysts, before and after boric acid treatment, were first dried at 110°C, and then 0.1 g of each catalyst was mixed with KBr and pressed in a pelletizer to form pellets of 1 mm thickness. X-ray diffraction spectra were obtained on a Philips X'Pert diffractometer, equipped with a monochromatic CuK$_\alpha$, with radiation having an angular range (2θ) from 0.5 to 5°; preliminary phase analysis was performed with a scan speed of 1 s per step and a step size of 0.05°. N$_2$ adsorption-desorption studies were performed on a Quantachrome Autosorb iQ instrument at 77°K. Both the catalysts were out-gassed in vacuum for 3h at 300°C. Both the mesoporous and the overall surface area were obtained with the help of using the BET method. NH$_3$ temperature-programmed desorption (NH$_3$-TPD) was performed with a temperature-programmed desorption system. The ammonia desorption amount was measured as a function of temperature. The distribution of type of hydrocarbons in the liquid product was measured using a PerkinElmer GC/Mass Spectrometer (GC/MS) Clarus 500, fitted with an Elite-5MS column (30 m length, 0.25 mm i.d. and 0.5 μm film thickness) with helium as the carrier gas. The oven program was started at 35°C for 8 minutes, 100°C for 5 minutes, 150°C for 2 minutes, and 250°C for 40 minutes. The mass spectra were obtained at the ionization energy of 70 eV from m/z 12/400.

2.3 Experimental setup

The performance studies of the purchased and modified catalysts were performed in a quartz reactor (2 cm diameter and 40 cm length). In a typical run 0.5 g of catalyst was packed in the middle of reactor, with the help of glass beads and glass wool packed at the two ends, with a N$_2$ flow at 50 mL/min. The studies were conducted at a temperature range of 250–450°C and a liquid flow rate of 30–120 ml/min. Methanol with water (in a ratio of 1:4) was made to first enter a pre-heater fixed at 150°C with the help of a syringe pump, and the vapors were then made to pass through the catalyst bed. The vapors from the reactor were then passed via a condenser maintained at 4°C, with the help of a chiller using water and isopropyl alcohol. The condensed liquid and uncondensed vapors from the condenser were further sent to a separator, in which liquid was collected at different time intervals and gases were injected into the GC for the analysis. The conversion and ethylene/propylene selectivity were evaluated as follows:

$$X_{Methanol}(\%) = \frac{F_{Methanol, fed} - F_{Methanol, out}}{F_{Methanol, fed}} \qquad (1)$$

$$S_i(\%) = \frac{\text{yield of i}^{th} \text{ product}}{\text{total gas products}} \qquad (2)$$

2.4 Product analysis

Uncondensed gases obtained from the outlet of the separator were injected into the GC TCD, as well as the FID equipped with a Porapak column. The estimation of the gases such as carbon dioxide, carbon monoxide, methane, ethane, ethylene, propane, propylene, butane, butane, butylenes, pentane and pentene was performed, and then its quantification was obtained using standards available. The analysis of liquid products was performed on the GC/MS for estimation of distribution of the components, and on the GC equipped with capillary column for the quantification estimation of conversion of methanol and yield of other compounds such as DME.

3 RESULTS AND DISCUSSIONS

3.1 Characterization of catalyst

FT-IR spectra of both the treated and untreated catalysts are shown in Figure 1. As shown in the Figure, the intense band at 1,050 cm^{-1} confirms asymmetric stretching of the Al-O-Al or Si-O-Si groups, whereas the peak at 810 cm^{-1} confirms the symmetric stretching of Al-O-Al or Si-O-Si groups. The band at 955 cm^{-1} shows the presence of defective Si-OH groups, and the band at 1,640 cm^{-1} is the bending vibration absorption of OH (Bhattacharyya et al., 2003). These bands confirm that Al/MCM-41 and B-Al/MCM have a skeleton similar to that of MCM-41.

The XRD of both the catalysts Al/MCM-41 and B-Al/MCM-41 were performed and the peaks were obtained at different theta values of 2.30, 3.94 and 4.54, resulting from the (100), (110) and (200) crystal faces of MCM-41. These results confirmed that even after alumina loading and boric acid treatment, the ordered hexagonal-lattice mesopore structure is sustained (Du et al., 2006). However, a decrease in the intensity of the diffraction peak at (100) and (200) was observed after boric acid treatment. The N$_2$ adsorption-desorption isotherms and the pore size distributions obtained from chemisorption for treated and untreated Al/MCM-41 is shown in Figure 2.

As shown in Figure 1, all the N$_2$ adsorption-desorption isotherms were observed to be similar to those of type IV, indicative of the existence of mesopores. BET analysis showed that the Al-MCM-41 has higher surface areas (890 m^2/g) with pore diameter of 3.15, which was observed to decrease to 782, 693 and 601 m^2/g on treatment with boric acid 0.5, 1.5 and 2.5M, respectively. However, no significant decrease in the pore diameter was observed after boric acid treatment. For studying variations in the surface acidity and strength of acid sites, NH$_3$-TPD characterization was used for the Al/MCM-41 before and after boric acid treatment. The peaks at 210 and 330°C were obtained for weak acidic sites, whereas the peak

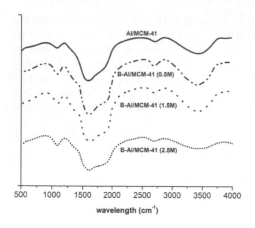

Figure 1. FT-IR spectra of the catalyst.

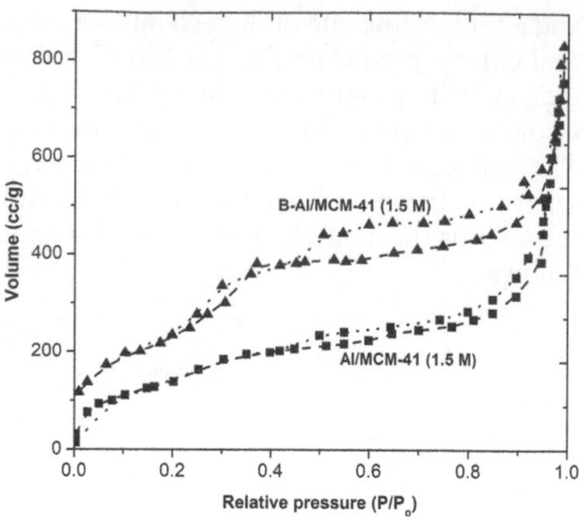

Figure 2. N$_2$ adsorption-desorption curves of Al/MCM-41 and boric acid treated Al-MCM-41.

obtained at 500°C indicates the strong acidic sites. In general, Al impregnated MCM-41 has a higher strength of strong acid site, which was observed to decrease after boric acid treatment.

A slight enhancement was observed after boric acid treatment. Based on the strength of weak acid site concentration and strong site concentration, Al/MCM-41 treated with boric acid (1.5 M) was selected for further studies.

3.2 Catalyst performance studies

3.2.1 Methanol conversion and gaseous product yield at different temperature

Conversion of methanol was performed at 250, 300, 350, 400 and 450°C in the fixed-bed quartz reactor for Al/MCM-41 and B-Al/MCM-41 (1.5M) and is shown in Figure 3. Lower conversions, 56.3% and 67.1%, were obtained at 250°C. However, as the temperature was increased from 250–450°C, above 90% conversions were obtained for Al/MCM-41 and B-Al/MCM-41 (1.5M), respectively. Higher change (approximately 12%) in the yield of liquid products was observed (from 19.2% to 7.2%) with increase of temperature from 250°C to 450°C in the presence of boric acid treated Al/MCM-41, whereas the gas yield was increased from 78.3% to 91.4%. A similar pattern was observed with Al/MCM-41; however, the percentage changes were different.

3.2.2 Methanol conversion and gaseous product yield at different flow rate

The conversion studies were performed with 0.5 g catalyst but at a flow rate range from 30 ml/hr to 120 ml/min, keeping the nitrogen flow rate at 50 ml/min and a methanol-to-water ratio of 1:2 (Figure 4). The maximum methanol conversion (89.2%) was obtained at 90 ml/hr with B-Al/MCM-41, whereas with Al/MCM-41 catalyst the maximum conversion was 83.4%. A similar pattern in gaseous products yield was observed (90.3%), whereas maximum gas yield with Al/MCM-41 was 85.3%. Lower liquid yield was obtained in both the cases.

3.2.3 Selectivity of ethylene and propylene

The gases coming from the outlet of the reactor were analyzed and found to include methane, ethylene, propylene, butane and pentane in the presence of Al/MCM-41 and B-Al/MCM-41. Table 1 shows the various conversion and selectivity of various hydrocarbons. The selectivity of ethylene was 4.3% with Al/MCM-41 and 7.1% with B-Al/MCM-41, indicating a 7.2% decrease. However, the selectivity for propylene was increased from 8.1% to 15.3%. No major change was observed in the selectivity of C4 hydrocarbons between 7.3% and 10.4%. C5 was approximately 10.2% and was observed to decrease with boric acid treatment. It has been

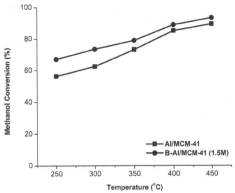

Figure 3. Methanol conversion at different temperatures.

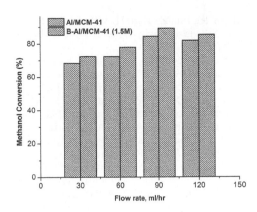

Figure 4. Methanol conversion at different temperatures.

Table 1. Catalytic performance for MTO reaction.

Catalyst	Conv (%)	Selectivity (%)				
		C_{1-4}	C_2H_4	C_3H_6	C_4H_8	C_{5+}
Al/MCM-41	92.1	13.2	4.3	8.1	7.3	10
B-Al/MCM-41 (1.5M)	94.1	6.3	7.1	15.3	10.4	3.7

mentioned by a few reporters that a moderate decrease in strong acid site presence due to decrease in alumina sites is responsible for increase in selectivity for propylene.

4 CONCLUSIONS

The conversion of methanol for the production of ethylene and propylene was performed in a fixed-bed reactor. The highest conversions were obtained at 450°C and at a flow rate of 90 ml/min. Alumina-loaded MCM-41 was treated with boric acid and the selectivity of propylene was observed to increase from 8.1% to 15.3%. The selectivity of ethylene was observed to decrease and was mainly observed to promote the propylene yield.

REFERENCES

Abrokwah, R.Y., Deshmane, V.G. & Kuila, D. (2016). Comparative performance of M-MCM-41 (M: Cu, Co, Ni, Pd, Zn and Sn) catalysts for steam reforming of methanol. *Journal of Molecular Catalysis A: Chemical*, *425*, 10–20.
Almutairi, S.M.T., Mezari, B., Pidko, E.A., Magusin, P.C.M. & Hensen, E.J.M. (2013). Influence of steaming on the acidity and the methanol conversion reaction of HZSM-5 zeolite. *Journal of Catalysis*, *307*, 194–203.
Bhattacharyya, K.G., Talukdar, A.K., Das, P. & Sivasanker, S. (2003). Al-MCM-41 catalysed alkylation of phenol with methanol. *Journal of Molecular Catalysis A: Chemical*, *197*(1–2), 255–262.
Du, G., Lim, S., Yang, Y., Wang, C., Pfefferle, L. & Haller, G.L. (2006). Catalytic performance of vanadium incorporated MCM-41 catalysts for the partial oxidation of methane to formaldehyde. *Applied Catalysis A: General*, *302*(1), 48–61.
Epelde, E., Santos, J.I., Florian, P., Aguayo, A.T., Gayubo, A.G., Bilbao, J. & Castañoa, P. (2015). Controlling coke deactivation and cracking selectivity of MFI zeolite by H3PO4 or KOH modification. *Applied Catalysis A: General, 505*, 105–115.

Khare, R., Liu, Z., Han, H. & Bhan, A. (2017). A mechanistic basis for the effect of aluminum content on ethene selectivity in methanol-to-hydrocarbons conversion on HZSM-5. *Journal of Catalysis*, *348*, 300–305.

Lefevere, J., Mullens, S., Meynen, V. & Van Noyen, J. (2014). Structured catalysts for methanol-to-olefins conversion: A review. *Chemical Papers*, *68*(9), 1143–1153.

Li, J., Wei, Y., Liu, G., Qi, Y., Tian, P., Li, B., ... Liu, Z. (2011). Comparative study of MTO conversion over SAPO-34, H-ZSM-5 and H-ZSM-22: Correlating catalytic performance and reaction mechanism to zeolite topology. *Catalysis Today*, *171*(1), 221–228.

Linares, N., Silvestre-Albero, A.M., Serrano, E., Silvestre-Albero, J. & García-Martínez, J. (2014). Mesoporous materials for clean energy technologies. *Chemical Society Reviews*, *43*(22), 7681–7717.

Murthy, K.V.V.S.B.S.R., Kulkarni, S.J., Chandrakala, M., Mohan, K.K., Pal, P. & Rao, T.P. (2010). Alkylation of 1-naphthol with methanol over modified Silicoaluminophosphate and MCM-41 molecular sieves. *Journal of Porous Materials*, *17*(2), 185–196.

Naik, S.P., Bui, V., Ryu, T., Miller, J.D. & Zmierczak, W. (2010). Al-MCM-41 as methanol dehydration catalyst. *Applied Catalysis A: General*, *381*(1–2), 183–190.

Olsbye, U., Svelle, S., Bjorgen, M., Beato, P., Janssens, T.V.W., Joensen, F., & Lillerud, K.P. (2012). Conversion of methanol to hydrocarbons: How zeolite cavity and pore size controls product selectivity. *Angewandte Chemie International Edition*, *51*(24), 5810–5831.

Sang, Y., Li, H., Zhu, M., Ma, K., Jiao, Q. & Wu, Q. (2013). Catalytic performance of metal ion doped MCM-41 for methanol dehydration to dimethyl ether. *Journal of Porous Materials*, *20*(6), 1509–1518.

Sang, Y., Liu, H., He, S., Li, H., Jiao, Q., Wu, Q. & Sun, K. (2013). Catalytic performance of hierarchical H-ZSM-5/MCM-41 for methanol dehydration to dimethyl ether. *Journal of Energy Chemistry*, *22*(5), 769–777.

Tian, P., Wei, Y., Ye, M. & Liu, Z. (2015). Methanol to olefins (MTO): From fundamentals to commercialization. *American Chemical Society Catalysis*, *5*(3), 1922–1938.

Wu, L., Degirmenci, V., Magusin, P.C.M.M., Szyja, B.M. & Hensen, E.J.M. (2012). Dual template synthesis of a highly mesoporous SSZ-13 zeolite with improved stability in the methanol-to-olefins reaction. *Chemical Communications*, *48*(76), 9492–9494.

Xu, A., Ma, H., Zhang, H., Ying, W. & Fang, D. (2013). Conversion of methanol to propylene over a high silica B-HZSM-5 catalyst. *International Journal of Chemical, Molecular, Nuclear, Materials and Metallurgical Engineering*, *7*(4), 175–184.

*Emerging Trends in Engineering, Science and Technology for Society,
Energy and Environment – Vanchipura & Jiji (Eds)*
© 2018 Taylor & Francis Group, London, ISBN 978-0-8153-5760-5

Degradation of diphenamid by UV/hydrogen peroxide advanced oxidation process

M.S. Manju & V. Nishan Ahammed
Department of Chemical Engineering, Government Engineering College, Thrissur, Kerala, India

K.P. Prasanth Kumar
Department of Chemistry, Maharaja's College, Ernakulam, India

ABSTRACT: Degradation and mineralization of diphenamid by hydrogen peroxide in the presence of UV was investigated. The results indicate that a diphenamid sample treated with hydrogen peroxide along with UV irradiation has the maximum degradation in the given time. Hydrogen peroxide alone can give considerable degradation, but hydrogen peroxide together with UV can maximize the extent as well as providing a better scheme of degradation. The study indicates that UV irradiation with the injection of 80 microliters of 20 v/v hydrogen peroxide to 1 liter of 0.1 ppm sample solution can completely disintegrate the pesticide content in two hours.

Keywords: diphenamid, UV, hydrogen peroxide, Degradation, mineralization

1 INTRODUCTION

Recently, Advanced Oxidation Processes (AOPs) have emerged as promising methods for the removal of organic pollutants from water. AOPs are based on the use of hydroxyl radicals for oxidative disintegration of organic pollutants into environmentally benign substances such as CO_2 and H_2O. Diphenamid (DPA) is a herbicide used for controlling annual grasses and weeds in tomato, potato, peanut, and soybean plants (Schultz & Tweedy, 1972; Sirons et al., 1981). As in the case of other pesticides and herbicides, DPA also enters into the water bodies and poses a threat to the environment in general, and to the aquatic organisms in particular. Therefore, the development of effective methods for remediation of polluted water containing even trace amounts of DPA is significant. Researchers have established that toxic pollutants impact on the health of the ecosystem and present a threat to humans through the contamination of drinking water supplies (Eriksson et al., 2007).

Several researchers had studied the photochemical degradation of DPA in aqueous solution. Rosen (1967) studied the homogeneous photodegradation of DPA by UV and sunlight irradiation. Rahman et al. (2003) investigated the photocatalytic degradation of DPA in aqueous P25 TiO_2 suspension under the illumination of a medium-pressure mercury lamp. Liang et al. (2010) studied the homogeneous and heterogeneous degradations of DPA in aqueous solution by direct photolysis with UVC (254 nm) and by photocatalysis with TiO2/UVA (350 nm). H_2O_2-based AOP studies on the degradation of diphenamid have not been reported by researchers so far.

The objective of the present work is to study the application of UV/H_2O_2-based AOP for the removal of DPA from water. Degradation of the pesticide is not the only concern for us, but the compounds and the intermediates which are formed during the course of these reactions are also of utmost importance. Hence, we have employed the most advanced analytical techniques in this work, viz. the application of High-Performance Liquid Chromatography (HPLC) and a Total Organic Carbon (TOC) analyzer. Designing a method with a safe degradation pathway is always a challenging task. It is equally important to develop an economically viable and time-bound method which easily blends in with the existing purification methods. This makes this work exciting and socially relevant.

2 MATERIALS AND METHODS

2.1 *Materials*

The diphenamid we used was 99.99% pure, purchased from Sigma Aldrich.
The hydrogen peroxide solution was standard lab quality, which is 20 v/v.

2.2 *Methods*

2.2.1 *Extent of degradation analysis*
2.2.1.1 Preparation of standard solution for extent of degradation analysis
DPA solution of 1,000 ppm (1 g of diphenamid dissolved in 1,000 ml solution) was magnetically stirred for 60 min. All reactions were performed at room temperature.

2.2.1.2 TOC analysis of sample with UV degradation
About 200 ml of the DPA standard solution was taken in a beaker and placed in a UV reactor. The solution was thoroughly stirred using a magnetic stirrer. The reactor started, and the samples were collected from the solution at regular intervals of time.

2.2.1.3 TOC analysis of sample with H_2O_2 degradation
About 200 ml of the DPA standard solution was taken in a beaker, placed on a magnetic stirrer and thoroughly stirred. To this solution 10 microliters of 20 v/v H_2O_2 was added and the initial sample was collected. Then the timer was started, and the samples were collected from the solution at regular intervals.

2.2.1.4 TOC analysis of sample with H_2O_2 and UV degradation
About 200 ml of the DPA standard solution was taken in a beaker and placed in a UV reactor. The solution was thoroughly stirred using a magnetic stirrer. To this solution 10 microliters of 20 v/v H_2O_2 was added and the initial sample was collected. The timer was started, and the samples were collected from the solution at regular intervals.

2.2.1.5 Analytical method
The samples obtained are analyzed using a TOC analyzer. Initially the analyzer was calibrated using a blank sample. After that, each sample was analyzed using a suitable method. Here, the method used was the Non-Purgeable Organic Carbon (NPOC) measurement method.

2.2.2 *Effect of H_2O_2 loading*
2.2.2.1 Preparation of standard solution for degradation time analysis
Prior to photolytic reaction, the DPA solution of 0.1 ppm (0.1 mg of diphenamid dissolved in 1,000 ml solution) was magnetically stirred for 60 min. All reactions were performed at room temperature.

2.2.2.2 Methodology for reaction with UV irradiation: Standard sample without H_2O_2
About 250 ml of the DPA standard solution was pipetted into a beaker and placed in a UV reactor. The samples were collected from the solution at different intervals. The DPA concentration was determined using HPLC.

2.2.2.3 Methodology for reaction with UV irradiation: Sample with H_2O_2
About 250 ml of the DPA standard solution was pipetted into a beaker and placed in a UV reactor. To this solution 10 microliters of 20 v/v H_2O_2 was added and the initial sample was collected. Samples were collected from the solution at different intervals. The DPA concentration was determined using HPLC. Then the procedure was repeated for 20 microliters of H_2O_2.

2.2.2.4   Analytical method

Samples obtained were analyzed using HPLC. Initially HPLC was calibrated using a standard solution. Thus, retention time was obtained along with the peak area for different samples. Later, each sample was analyzed. The peak areas obtained were consolidated and standardized to the scale from 0 to 1 and plotted.

3   RESULTS AND DISCUSSIONS

3.1   *Extent of degradation*

3.1.1   *Result sample 1 – Diphenamid sample irradiated with UV*

The reaction of diphenamid was conducted in the immersion type photo reactor as five-hour long runs, resulting in the following observations:

i. Degradation rate in the first hour was very low and only 0.1% of the total diphenamid present in the solution was degraded.
ii. The degradation rate in the second hour was comparatively faster, but there was only 0.5% of degradation.
iii. The degradation rate in the third hour was the fastest in the entire run, and 1.2% of the total diphenamid present in the sample was degraded.
iv. The degradation pattern in the fourth hour seems similar to the degradation pattern in the second hour. This period showed a degradation of 0.4% and was considered to be the slowest rate in the entire run.
v. The degradation pattern in the fifth hour seemed similar to the degradation pattern in the third hour, with a degradation of 1%.
vi. In the five hours, UV irradiation had given a total degradation of 3.1% of the total diphenamid present in the solution.

Figure 1 shows the fraction of diphenamid in the sample vs time plot. Overall, degradation rate in this sample run is very low. The degradation pattern is non-uniform, showing a wide variation in the degradation rate. Irradiation caused only a total degradation of 3.1% in five hours.

3.1.2   *Result sample 2 – Diphenamid sample mixed with hydrogen peroxide*

The experimental run of diphenamid sample mixed with hydrogen peroxide was conducted for five hours as per the standard procedure. The observations are noted below:

i. Degradation rate exhibited a drastic increase to 18% in the first hour. This was the maximum degradation rate in the entire run.
ii. The degradation rate in the second hour seemed to be very low. In this period the degradation was only 0.95%.

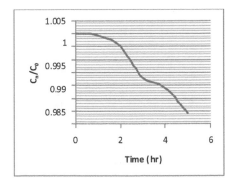

Figure 1.   TOC analysis data of sample irradiated with UV.

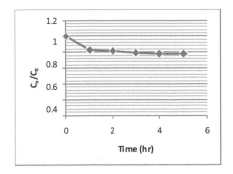

Figure 2.   TOC analysis data of sample treated with hydrogen peroxide.

539

iii. The degradation rate in the third hour was lower compared to the degradation in the second hour. In this period the degradation was only 0.29%.
iv. As time passed, the degradation pattern showed a reduction in the rate, as in the fourth hour the degradation was further reduced to 0.11%. In the fifth hour the degradation was as low as 0.06%.
v. In five hours, there was a total degradation of 23% of the diphenamid present in the solution.

Figure 2 shows the fraction of diphenamid in the sample vs time plot. The degradation rate was very high in the beginning, but quickly reduced with time. The degradation rate was nearly zero in the fifth hour. The degradation pattern shows that the initial hike in the degradation rate may be attributed to the fast and spontaneous reaction due to the addition of hydrogen peroxide, which is reduced with increase in time. The five hours of stirring after the addition of hydrogen peroxide brought about around 23% degradation of the diphenamid initially present in the solution.

3.1.3 *Result sample 3 – Diphenamid sample treated with H_2O_2 irradiated by UV*

The photo reaction of diphenamid solution mixed with hydrogen peroxide was conducted as a continuous run for five hours, as per the standard procedure. The observations are as given below:

i. The degradation in the first hour seemed to be moderate with degradation of 13% of the initial concentration of the diphenamid in the solution.
ii. The degradation rate was lower in the second hour. In this period the extent of degradation was 6.5%.
iii. In the third hour the degradation pattern showed a further reduction behavior with a degradation of 3.1%.
iv. The reduction in degradation with respect to time was observed in the fourth hour too, with a degradation of 1.5%.
v. However, in the fifth hour there was a variation with a slight increase in the degradation rate, giving a degradation of 1.7%.
vi. In five hours, UV irradiation had given a total degradation of 26.8% of the total diphenamid present in the solution.

Figure 3 shows the fraction of diphenamid in the sample vs time plot. The degradation pattern shows a uniform reduction of degradation rate with time. Initial hours of reaction show the maximum degradation rate with 13% degradation. However, degradation rates in the fourth to fifth hours show an increase. The sample exhibits a non-uniform degradation pattern with alternate increase and decrease. Overall degradation was 26.8% in five hours.

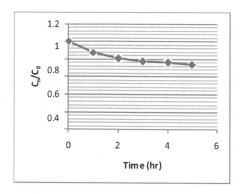

Figure 3. TOC analysis data of sample treated with hydrogen peroxide/UV.

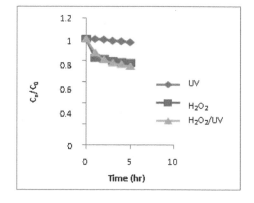

Figure 4. C_n/C_0 data plot for extent of degradation analysis.

A diphenamid sample with added hydrogen peroxide and irradiated by UV has a degradation pattern which moreover shows a uniform reduction of degradation rate with time. Initial hours of reaction showed the maximum degradation rate with 13% degradation. However, degradation seems to be active after five hours with an increase in the degradation rate. Also, the pattern exhibits a non-uniform degradation rate reduction behavior with alternative increase and decrease. An overall degradation of 26.8% happened in five hours.

Figure 4 shows a comparison plot of degradation rate of all the above samples. It is clear that UV irradiation gave the least extent of degradation in the given time, showing that UV is incapable for total degradation of the compound. It can be only used for sterilization in order to eliminate microorganisms. A diphenamid sample treated with hydrogen peroxide shows a high degradation of 18% in the beginning, but the total degradation percentage was limited to 23%, while a diphenamid sample with hydrogen peroxide and irradiated/UV treatment shows comparatively less degradation of 13% initially, but made a total degradation of 26.8%.

The observations from the extent of degradation analysis confirm that a diphenamid sample with added hydrogen peroxide and irradiated by UV has the maximum degradation in the prescribed time of five hours. The addition of hydrogen peroxide can give considerable degradation, but hydrogen peroxide together with UV can give the maximum degradation as well as a better degradation scheme, since hydrogen peroxide can degrade pesticides and UV can destroy microorganisms.

3.2 *Effect of amount of H_2O_2 on extent of degradation*

Figure 5 shows the effect of H_2O_2 loading on the rate of degradation of diphenamid. The degradation curve for diphenamid solution UV irradiated without hydrogen peroxide shows a very slow degradation pattern. Degradation rate continuously decreases with time. About 95% of the total diphenamid in the solution degrades in the first four hours. However, total degradation time was found to be five hours.

When the diphenamid solution is UV irradiated after adding 10 microliters of hydrogen peroxide, the sample shows a fast degradation pattern. The degradation rate is very high in the first hour, degrading 79% of the diphenamid present. The curve shows a stagnation rate in-between one to two hours. The total degradation time is around 4 hours.

The degradation curve for the diphenamid solution UV irradiated after adding 20 microliters of hydrogen peroxide shows a drastic degradation pattern, with 91% of diphenamid in the sample degraded in one hour. The degradation rate shows a decreasing pattern with time in general. The total degradation time is around two hours.

The results show that UV irradiation followed by the addition of hydrogen peroxide leads to a much faster degradation of diphenamid. Also, degradation time decreases with increase in the volume of hydrogen peroxide. Only 20 microliters of hydrogen peroxide was sufficient to degrade 250 ml of 0.1 ppm diphenamid solution in two hours. That means that 80 ml of

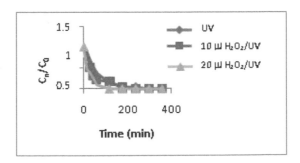

Figure 5.   Cn/C0 data plot for effect of H_2O_2 loading.

20 v/v hydrogen peroxide is sufficient to degrade 1,000l of water with 0.1 ppm diphenamid concentration.

4 CONCLUSIONS

The results of the present study confirm that a diphenamid sample with added hydrogen peroxide and irradiated by UV has the maximum degradation in the prescribed time of five hours. The addition of hydrogen peroxide can give considerable degradation, but hydrogen peroxide together with UV can give maximum degradation as well as a better degradation scheme, since hydrogen peroxide can degrade pesticides and UV can destroy microorganisms.

However, the knowledge about the compounds formed due to the degradation of diphenamid and the behavior and properties of the intermediate compounds in the reaction pathway are beyond the scope of this work, which will definitely serve as a scope for future investigations in this topic.

ACKNOWLEDGMENT

One of the authors (Manju M.S.) acknowledges with thanks the financial assistance received from Centre for Engineering Research (CERD), Government of Kerala.

REFERENCES

Eriksson, E., Baun, A., Mikkelsen, P.S. & Ledin, A. (2007). Risk assessment of xenobiotics in stormwater discharged to Harrestup Å, Denmark. *Desalination, 215*(1–3), 187–197.

Liang, H.C., Li, X.Z., Yang, Y.H. & Szc, K.H. (2010). Comparison of the degradations of diphenamid by homogeneous photolysis and heterogeneous photocatalysis in aqueous solution. *Chemosphere, 80*(4), 366–374.

Rahman, M.A., Muneer, M. & Hahnemann, D. (2003). Photocatalysed degradation of a herbicide derivative, diphenamid in aqueous suspension of titanium dioxide. *Journal of Advanced Oxidation Technologies, 6*(1), 100–108.

Rosen, J.D. (1967). The photolysis of diphenamid. *Bulletin of Environmental Contamination and Toxicology, 2*(6), 349–354.

Schultz, D.P. & Tweedy, B.G. (1972). Effect of light and humidity on absorption and degradation of diphenamid in tomatoes. *Journal of Agriculture and Food Chemistry, 20*(1), 10–13.

Sirons, G.J., Zilkey, B.F., Frank, R. & Paik, N.J. (1981). Residues of diphenamid and its phytotoxic metabolite in flue-cured tobacco. *Journal of Agriculture and Food Chemistry, 29*(3), 661–664.

Emerging Trends in Engineering, Science and Technology for Society,
Energy and Environment – Vanchipura & Jiji (Eds)
© 2018 Taylor & Francis Group, London, ISBN 978-0-8153-5760-5

Assessment of trihalomethanes in drinking water using gas chromatography

S.P. Aravind & P.A. Soloman
Department of Chemical Engineering, Government Engineering College Thrissur, Kerala, India

ABSTRACT: Water is an essential element to sustain life. To ensure this requires a safe, adequate and accessible supply. Therefore, efforts should be made to achieve a standard drinking water quality. This is achieved through the use of water treatment plants, where the major objective is the removal of pathogenic microorganisms to prevent the spread of water-borne diseases. It is important that water treatment works be equipped with adequate disinfection systems. Disinfection processes can result in the formation of both organic and inorganic Disinfection By-Products (DBPs). The most well-known of these are the organo-chlorine by-products such as Trihalomethane (THM) compounds. THM concentrations in drinking water are measured using Gas Chromatography (GC) at various places in Thrissur City, including at the Government Engineering College, which has its water treated at the Peechi water treatment plant. A study has been conducted on parameters that affect the formation of THMs. Response surface designs have been created for each THM using the design of an experiment tool in Minitab 17, with the most influencing parameters being observed from parameter study.

Keywords: Disinfection, Trihalomethane, THM, RSM

1 INTRODUCTION

There are many sources of contamination in drinking water, ranging from natural substances leaching from soil to harmful chemical discharges from industrial plants. In developing countries, nearly half of the population is suffering due to lack of potable water, or due to contaminated water (WHO, 1992). Since disinfection is the most popular step in the treatment of water, the cheapest method is preferred. Hence, chlorination is the most common disinfection method as it remains in water until it has been consumed (Sadiq & Rodriguez, 1999). However, this poses a chemical threat to human health as it reacts with organic matter available in the water to produce harmful products.

During the chlorination of water containing organic matter, different Disinfection By-Products (DBPs) are formed and more than 300 different varieties have been identified (Becher, 1999). In the range of 37–58% of the total measured halogenated by-products are trihalomethanes. Trihalomethanes (THMs) are a group of four volatile compounds that are formed when chlorine reacts with organic matter present in the water (Frimmel & Jahnel, 2003).

The THMs include: Trichloromethane (Chloroform-CF), Bromodichloromethane (BDCM), Dibromochloromethane (DBCM) and Tribromomethane (Bromoform-BF). These are classified as possible human carcinogens by the US Environment Protection Act (USEPA) (US Environmental Protection Agency, 1990). In the USEPA guidelines, the maximum contaminant level specified in the DBP Stage I Rule is 80 µg/L for total THMs (US Environmental Protection Agency, 1998). The formation of these compounds depends on several other factors such as temperature, pH, disinfectant dose, contact time, inorganic compounds and organic matter present in the drinking water supply (Bull et al., 1995; Wu et al., 2001; Bach et al., 2015).

In India, surface water, which can be divided into river, lake, and complex types, provides more than 90% of the country's drinking water. Accordingly, as the water sources are different, their characteristics differ and eventually the amount of THMs also differs. Therefore, we measured THM forming according to the quality of the raw water and the possible factors favoring the formation.

The objective of this study was to analyze trihalomethanes present in the available drinking water using Gas Chromatography (GC), and to study the parameters that influence the formation of THMs, to find optimum values and to model them.

2 MATERIALS AND METHODS

2.1 Chemicals

- Certified reference material of THM 2,000 µg mL^{-1} in methanol was purchased from Supelco, USA.
- n-hexane of HPLC grade (99%) was purchased from Sigma-Aldrich.
- Hypochlorite solution (13%) was purchased from suppliers.
- Working standards were prepared using n-hexane.
- Other chemicals (e.g. NaOH, H_2SO_4, NH_4Cl) were analytical grade.

2.2 Extraction and analysis

The standalone method for the extraction is liquid-liquid extraction, and for analysis is gas chromatography with electron capture detector (GC-ECD) or gas chromatography mass spectrometry (Frimmel & Jahnel, 2003; Siddique et al., 2015; Kim et al., 2002). Here THMs are extracted with n-hexane and analysis done using a Perkin Elmer Clarus 580 gas chromatograph, equipped with a headspace injection system and an electron capture detector (9 in. × 9 in. × 9.8 in. = 794 in^3. Its maximum usable depth = 6.3 in. It accepts 1/8 in. o.d. stainless steel, 6 mm o.d. glass and all fused silica, packed or capillary columns 6.5 in. diameter coil).

The method was developed by tuning the GC, based on the data from the supplier of the certified reference material and from the handbook of water treatment technologies (WHO, 2008; The Environmental Protection Agency, 1998). The tuned GC operating parameters for the analysis are given in Table 1. The GC is calibrated using five working standards.

2.3 Sample analysis

Water samples were collected three times at two-week intervals from various parts of Thrissur City to determine the concentration of THMs. Seven of the samples were from water treated at the Peechi water treatment plant, and two were from the pump houses at the Government Engineering College, Thrissur. Water from well 1 of the college was used as raw water for the parameter study.

Table 1. Tuned GC operating parameters for THM analysis.

Parameter	Conditions
Oven Temperature	80°C for 4 minutes
Injector Temperature	150°C
Detector Temperature	300°C
Carrier Gas Flow	Nitrogen gas (99.998%) @ 30 mL/min
Injection Volume	1 µL

2.4 Experimental set up for parameter study

Water samples from well 1 at Government Engineering College, Thrissur were collected for the parameter study. 100 mL of sample water was taken in a 500 mL beaker, which was then placed over a temperature-programmed magnetic stirrer. Experiments were carried out by varying one parameter but keeping the other parameters constant. This scheme was repeated for all parameters. Parameters chosen were chemical oxygen demand, HOCl concentration, temperature, pH, reaction time and mixing time.

2.5 Modeling using Minitab 17

There were 27 sets of experiments suggested by the response surface methodology tool of Minitab 17 for four sets of parameters that have more influence on THM formation. A response surface design refers to a set of advanced Design of Experiments (DOE) techniques, which help in experimental analysis and response optimization. Box-Behnken design was selected for the design of experiments.

The response Minitab optimization tool helps to identify the combination of input variable settings that jointly optimize a single response or a set of responses. Minitab calculates an optimal solution and draws a plot.

3 RESULTS AND DISCUSSION

3.1 Determination of THMs in samples

THM concentrations of the samples were analyzed three times. Their average concentrations in ppb are given in Table 2. The observed concentrations are far below permissible limits.

3.2 Parameter study

The concentrations of each THM present in the raw water is given in Table 2, and the levels seem to be very low. Chlorination experiments were carried out and the influence of all parameters in the formation of THMs were studied. It is observed that THM concentration increases with HOCl concentration, reaction time and temperature. However, the nature fluctuated for pH and mixing time. Also, THM increases to a particular level for COD, after which it decreases, except in the case of chloroform.

3.3 Response surface design

Four parameters were selected that have more influence in THM formation, as is seen from parameter study. Parameters and their domains are given in Table 3.

Table 2. Results of water sample analysis.

Location	CF	BDCM	DBCM	BF
Swaraj Round	3.5574	2.1885	0.6818	0.0208
Arimboor Panchayath	0.8597	0.0341	0.0101	0.0025
Koorkenchery	10.1506	4.0765	1.6565	0.1391
Ramavarmapuram	9.5372	3.5249	0.2425	0.0168
Peringavu	0.9445	0.6641	1.3561	0.0480
Poonkunnam	5.0526	3.3114	1.2292	0.0971
Pump House 1, GEC Thrissur	0.1004	0.0081	0.0003	0.0000
Pump House 2, GEC Thrissur	3.3620	2.9177	0.0019	0.0215
Raw Water	0.1004	0.0081	0.0003	0

Table 3. The level and range of variables.

| Parameters | Factor | Domain | |
		−1	1
COD (ppm)	A	40	800
HOCl Concentration (ppm)	B	0.2	3
Temperature (°C)	C	20	50
Reaction Time (min)	D	30	90

Table 4. Response surface design data.

Run order	COD (ppm)	HOCl (ppm)	Temperature (°C)	Reaction time (min)	CF (ppb)	BDCM (ppb)	DBCM (ppb)	BF (ppb)
1	40	1.6	35	90	0.9915	0.1036	0.02610	0.006880
2	420	1.6	35	60	0.3701	0.0101	0.00220	0.001000
3	420	3.0	20	60	0.3770	0.1502	0.01790	0.000000
4	420	1.6	20	90	0.3230	0.0219	0.00580	0.002200
5	420	1.6	35	60	0.3701	0.0101	0.00220	0.001000
6	40	1.6	35	30	0.3727	0.0349	0.00720	0.004700
7	800	1.6	35	90	0.4557	0.0071	0.00350	0.001700
8	420	0.2	50	60	0.1197	0.0004	0.00110	0.000100
9	420	0.2	20	60	0.3684	0.0035	0.00000	0.002700
10	40	1.6	50	60	0.4482	0.0527	0.01850	0.006500
11	420	1.6	50	30	0.4464	0.0668	0.01280	0.006400
12	420	1.6	50	90	0.1971	0.0189	0.00440	0.005500
13	800	1.6	50	60	0.2357	0.0221	0.00300	0.001500
14	420	3.0	35	90	0.5922	0.1611	0.02650	0.007332
15	40	1.6	20	60	0.3665	0.0591	0.01630	0.004490
16	420	0.2	35	90	0.2404	0.0072	0.00320	0.001700
17	800	0.2	35	60	0.4394	0.0038	0.00370	0.002200
18	800	3.0	35	60	0.5700	0.0305	0.00160	0.000400
19	420	3.0	35	30	0.3051	0.0548	0.00480	0.002700
20	420	3.0	50	60	0.4553	0.1231	0.01860	0.008800
21	420	1.6	20	30	0.2232	0.0736	0.01250	0.005200
22	40	0.2	35	60	0.3850	0.0001	0.00100	0.000300
23	800	1.6	35	30	0.8377	0.1234	0.01862	0.007700
24	800	1.6	20	60	0.2812	0.0193	0.00090	0.005490
25	40	3.0	35	60	0.4790	0.0924	0.02430	0.008800
26	420	1.6	35	60	0.3701	0.0101	0.00220	0.001000
27	420	0.2	35	30	0.2909	0.0038	0.00140	0.003200

The three level second-order designs demand comparatively less experimental data to enable precise prediction. In the Box-Behnken method a total number of 27 experiments, including three center points, are carried out to estimate the formation of THM. The quality of the fit of this model is expressed by the coefficient of determination R^2. The concentration of each THM after the chlorination experiment is subtracted from its level before chlorination, and the concentrations are given in Table 4.

Models have been created to determine the concentration or quantity of each compound based on the data. Regression equations for each component in uncoded units are given by:

a. Chloroform
CF formed = −0.435 + 0.000427 a −0.201 b + 0.0388 c + 0.00419 d + 0.000001 a*a −0.000480 c*c + 0.000079 d*d −0.000022 a*d + 0.00389 b*c + 0.00201 b*d −0.000194 c*d.

b. Bromodichloromethane

BDCM formed = 0.182 + 0.000192 a −0.0250 b −0.00646 c −0.00313 d + 0.000000 a*a + 0.01142 b*b + 0.000089 c*c + 0.000031 d*d −0.000031 a*b −0.000004 a*d + 0.000612 b*d.

c. Dibromochloromethane

DBCM formed = 0.0226 + 0.000031 a −0.00001 b −0.000786 c −0.000513 d + 0.000000 a*a + 0.00095 b*b + 0.000013 c*c + 0.000006 d*d −0.000012 a*c −0.000000 a*c −0.000001 a*d −0.000005 b*c + 0.000118 b*d −0.000001 c*d.

d. Bromoform

BF formed = 0.01870 + 0.000015 a −0.00385 b −0.000526 c −0.000302 d + 0.000000 a*a + 0.000007 c*c + 0.000003 d*d −0.000005 a*b −0.000000 a*c −0.000000 a*d + 0.000136 b*c + 0.000037 b*d.

where a is COD, b is HOCl concentration, c is temperature and d is reaction time. R^2 values obtained are 80.78%, 78.64%, 84.15%, and 89.46% respectively.

Response optimization has given COD 646.46 ppm, HOCl concentration 0.20 ppm, temperature 50°C, and reaction time 90 minutes, for obtaining minimum concentration values of all THMs. The minimum values obtained for BF, DBCM, BDCM and CF are −0.0011, −0.006, −0.0216 and −0.0968 respectively. Figure 1 shows the results given by the response optimizer tool of RSM for obtaining minimal formation of THMs.

Since the RSM optimized the system to a possible minimum combination, slightly negative values are obtained. In order to find the combination where zero THM are formed, the optimization is performed using the Microsoft Excel 2016 Solver tool.

The combinations observed are given in Table 5. Thus, the combination of parameters were obtained for zero production of CF, BDCM, DBCM and a near to zero value for BF.

Figure 2 shows some of the contour plots of different THMs obtained using the Minitab 17 tool. It gives the trend of compound formation by considering two different varying parameters and two hold parameters.

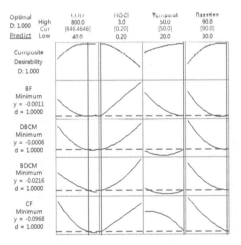

Figure 1.   Results of response optimizer for THM.

Table 5.   Optimized combinations given by solver.

Analytes	a (ppm)	b (ppm)	c (°C)	d (min)	THM Conc. (ppb)
CF	438.53	0.2	50	84.85	0
BDCM	800	0.2	34.55	45.76	0
DBCM	556.21	3	50	90	0
BF	40	0.2	35.63	49.10	0.0024

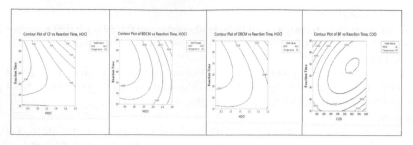

Figure 2.   Contour plots of THMs.

4   CONCLUSION

Disinfection is a crucial and necessary step in the drinking water treatment process. Trihalom-ethanes are the major disinfection by-products formed in drinking water when organic matter present in the water reacts with chlorine. It is very toxic and carcinogenic and causes severe health effects. THM concentrations are determined from water samples taken at two-week intervals from various parts of Thrissur City. The deviations from mean concentration values are not at all high. The concentrations of all the components are below that of the highest permissible level.

COD, HOCl concentration, temperature and reaction time are observed to be more influ-encing parameters from the parameter study. Response surface designs were created using Minitab 17 for all components. By knowing the parameters' values, uncoded coefficients and regression equation, it is possible to calculate the formation of individual THM components with around 83% reliability. The Minitab response optimizer tool gave optimized values of: COD 646.46 ppm, HOCl concentration 0.20 ppm, temperature 50°C, and reaction time 90 minutes, for the minimal formation of THMs. The Microsoft Excel solver tool optimized the values of parameters to achieve zero THM.

REFERENCES

Bach, L., Garbelini, E.R., Stets, S., Peralta-Zamora, P. & Emmel, A. (2015). Experimental design as a tool for studying trihalomethanes formation parameters during water chlorination. *Microchemical Journal, 123*, 252–258.

Becher, G. (1999). Drinking water chlorination and health. *CLEAN–Soil, Air, Water, 27*(2), 100–102.

Bull, R.J., Birnbaun, L.S., Cantor, K.P., Rose, J.B., Butterworth, B.E., Pegram, R. & Tuomisto, J. (1995). Water chlorination: Essential process or cancer hazard? *Toxicological Sciences, 28*(2), 155–166.

Frimmel, F.H. & Jahnel, J.B. (2003). Formation of haloforms in drinking water. In *Haloforms and Related Compounds in Drinking Water, 5(Part G)* (pp. 1–19). Springer, Berlin: Heidelberg.

Kim, J., Chung, Y., Shin, D., Kim, M., Lee, Y., Lim, Y. & Lee, D. (2002). Chlorination by-products in surface water treatment process. *Desalination, 151*(1), 1–9.

Sadiq, R., & Rodriguez, M.J. (2004). Disinfection by-products (DBPs) in drinking water and predictive models for their occurrence: A review. *Science of the Total Environment, 321*(1–3), 21–46.

Siddique, A, Saied, S., Mumtaz, M., Hussain, M.M. & Khwaja, H.A. (2015). Multipathways human health risk assessment of trihalomethane exposure through drinking water. *Ecotoxicology and Environmental Safety, 116*, 129–136.

The Environmental Protection Agency (1998). Water Treatment Manual: Disinfection.

US Environmental Protection Agency. (1990). *Risk assessment, management and communication of drinking water contamination, EPA/600/4-90/020.* Washington, DC.

US Environmental Protection Agency. (1998). *National primary drinking water regulations; disinfectants and disinfection by-products; final rule, fed. regist., 63(241),* 69389–69476.

World Health Organization. (1992). *Our planet our health: report of the WHO commission health and environment.* Geneva: World Health Organization.

World Health Organization. (2008). Guidelines for drinking-water quality [electronic resource]: 1st and 2nd addenda, vol. 1, Recommendations.

Wu, W.W., Benjamin, M.M., & Korshin, G.V. (2001). Effects of thermal treatment on halogenated disinfection by-products in drinking water. *Water Research, 35*(15), 3545–3550.

Emerging Trends in Engineering, Science and Technology for Society,
Energy and Environment – Vanchipura & Jiji (Eds)
© *2018 Taylor & Francis Group, London, ISBN 978-0-8153-5760-5*

Adsorption of perchlorate using cationic modified rice husk

Ann M. George
Department of Chemical Engineering, University of Kerala, Kerala, India

K.B. Radhakrishnan
Department of Chemical Engineering, TKM Engineering College, Kollam, India

A. Jayakumaran Nair
Department of Biotechnology, Kariavattom Campus, Thiruvananthapuram, India

ABSTRACT: Perchlorates are highly soluble anions and are used as ingredients in solid rocket fuels, fireworks, missiles, batteries etc. The potential human risk of perchlorate exposures includes effects on nervous system, inhibition of thyroid activity and mental retardation in infants. Various materials and techniques have been used to remove perchlorate from drinking water. For light polluted water of perchlorate, adsorption seems to be one of the most attractive, easiest, safest and cost effective physio-chemical treatment methods especially for drinking water. Rice husk, one of the major bi-products of rice milling industry can be used as a low cost adsorbent for perchlorate removal.

Present study deals with adsorption of perchlorate using cationic modified rice husk by optimizing various parameters like pH, adsorbent mass, adsorbate concentration, temperature and time of adsorption. The surface charge is the major governing factor for perchlorate removal compared to surface area. To enhance the adsorption capacities, modifications with cationic surfactants were made. Powdered rice husk was surface modified with Cetyl Trimethyl Ammonium Bromide (CTAB). The adsorption of perchlorate was studied experimentally after surface modifications and the different parameters including pH, adsorbent mass, adsorbate concentration, temperature and time were optimized. The performance of adsorption was found at different conditions and it has been observed that more than 97% adsorption efficiency was achieved in the perchlorate removal.

1 INTRODUCTION

Perchlorate (ClO_4^-) is highly soluble anion that consists of a central chloride atom surrounded by four oxygen atoms (John. D. Coates et al. 2004). Perchlorate salts have been manufactured and used as ingredients in solid rocket fuels highway safety flares, airbag inflators, fireworks, missile, fuels, batteries, matches (Mamie N.I. et al. 2014). It mostly exists as ammonium perchlorate, sodium perchlorate, potassium perchlorate, magnesium perchlorate and lithium perchlorate (Yali Shi et al. 2007, Urbansky 1998). The perchlorate ion is similar in size to an iodide ion and can therefore be taken up in place of iodide ions by the thyroid gland. Thus the perchlorate ions disturb the production of thyroid hormones and may disrupt metabolism in the human body and the effects can be significant in case of pregnant women and fetuses (Urbansky 2002). The potential human risk of perchlorate exposures include effects on nervous system, inhibition of thyroid activity and mental retardation in infants (Z. Li, et al. 2000) Various materials and techniques have been used to remove perchlorate from drinking water. These technologies are classified physical removal by the sorption on materials, chemical reduction by metal, biodegradation by bacteria and electrochemical reduction on metal electrodes and integrated techniques. Now a days the better method for the treatment of wastewater is considered as adsorption due to its universal nature, inexpensiveness and ease of operation (Guillame Darracq et al. 2014).

For light polluted water of perchlorate adsorption seems to be one of the most attractive, easiest, safest, versatile and cost effective physio-chemical treatment methods, especially for drinking water. Adsorption can also remove soluble and insoluble organic pollutants. The removal capacity by this method may be up to 99.9%. Due to these facts, adsorption has been used for the removal of a variety of organic pollutants from various contaminated water sources.

The biggest barrier in the application of this process by the industries is the high cost of adsorbents presently available for commercial use. The cost of adsorption technology can be reduced, if the adsorbent is in expensive. So there is a need to develop low cost and easily available adsorbents for the removal of these pollutants from the aqueous environment. An abundant source of potentially anionic compounds adsorbing biomass is cellulosic agricultural waste. They are widely available, inexhaustible and inexpensive material that exhibit specificity towards targeted ions (Umesh. K.G. et al, 2009).

The natural adsorbent can improve their adsorption capacity through various modifications. The treatment with surfactant is one of the effective modifications. The cationic surfactant modified adsorbent alters the surface property of the adsorbent from hydrophilicity to hydrophobicity or organophilicity. The modified adsorbent enhances their adsorption capacity for perchlorate in water (Umpuch et al, 2013). Present work focused on modification of rice husk using cationic surfactant CTAB for the better adsorption for perchlorate (Dipu Borah et al, 2009).

Recently several materials have been tested for perchlorate removal. A common adsorbent used for the adsorption of perchlorate was granular activated carbon. To enhance the adsorption capacities of granular activated carbon several modifications with cationic surfactants was made (Parette & Cannon, 2005). It was found that perchlorate adsorption on 10 types of commercial activated carbon followed Langmuir type of adsorption isotherm. The surface charge was the most important factor governing the perchlorate removal rather than specific surface area (Rovshan Mahmudov et al. 2010).

Rice husk is one of the major bye-product of the rice milling industry, is one of the most commonly available lingo-cellulosic materials that can be converted to different types of fuels and chemical feed stocks through a variety of thermochemical conversion processes. Chemical content of rice husk consists of 50% cellulose, 25–30% lignin and 15–20% silica (Humayatul Ummah et al. 2015).

2 MATERIAL AND METHODS

2.1 *Adsorbent collection and preparation*

Rice husk was used in the present work were collected from Kariavattom, Thiruvananthapuram – Kerala – GPRS 8° 33' 51" N, 76° 53' 11" E. The collected Rice husk were washed with distilled water several times to remove dirt particles and water soluble materials. The washed materials were dried in an air oven at 40°C for 24 h. The dried Rice husk then powdered and sieved to the desired particle size (40 μm). Finally the product was stored in a vacuum desiccator until required.

2.2 *Surface modification of rice husk using CTAB*

2 g of adsorbent (rice husk) was treated with 20 ml of 1% CTAB (1 g in 100 ml) and shaken in a temperature controlled shaker at 100 rpm for 24 h. The modified rice husk was then filtered and washed with distilled water several times. It was then dried in hot air oven at 60°C overnight.

2.3 *Adsorption experiment and analysis*

2.3.1 *Estimation of optimum pH for the experiment*
The optimum pH for the adsorption, was found out by taking 2 g of modified rice husk in a series of conical flask containing 10 mg/l of adsorbate. The pH was adjusted to 2–10 and

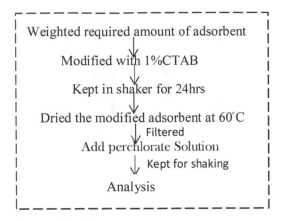

Figure 1. Overall procedure for measuring the residual perclorate.

kept for shaking. After 1 hr incubation the samples were filtered and analysed the residual perchlorate.

2.3.2 *Estimation of optimum amount of adsorbent*

The optimum amount of rice husk powder for the adsorption was found out by weighing the adsorbent ranging from 1 to 5 g in a series of conical flask containing 10 mg/l adsorbate. The solution was under optimum pH condition and kept on the shaker for 1 hr. After incubation the samples were filtered and residual perchlorate was analysed.

2.3.3 *Estimation of optimum temperature*

Optimum amount of modified rice husk was taken in a series of conical flask containing 10 mg/L adsorbate, and the solution was kept for shaking at temperatures ranging from 30°C to 70°C for 1 h. After incubation the samples were filtered and analysed the residual perchlorate.

2.3.4 *Estimation of optimum perchlorate concentration*

In order to obtain the optimum perchlorate concentration, the experiment was carried out at perchlorate concentration ranging from 5 to 25 ppm respectively; in a series of conical flask containing optimum amount of modified adsorbent. The other parameters like temperature, mass of adsorbent and pH condition were kept constant. After incubation, the samples were taken for perchlorate analysis using the ion sensitive electrode.

2.3.5 *Estimation of optimum level of perchlorate adsorption*

In order to obtain the optimum level of perchlorate adsorption, the experiment was carried out under optimum conditions of pH, adsorbent mass, adsorbate concentration and temperature. The samples were kept for shaking and after 1 hr of incubation it was filtered and residual perchlorate was analysed.

3 RESULT AND DISCUSSION

3.1 *Adsorption at different pH*

The effect of pH solutions on removal of perchlorate from aqueous solutions with range of pH varying from 2 to10 is studied. Figure 2 shows the effect of pH on percentage removal of

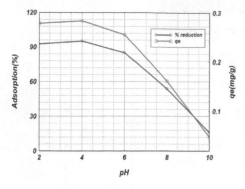

Figure 2. Effect of pH on the removal of perchlorate by rice husk at $37 \pm 2°C$.

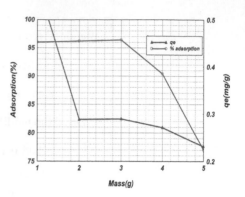

Figure 3. Effect of adsorbent mass on removal of perchlorate by rice husk at $37 \pm 2°C$ and pH4.

perchlorate. It can be noted that the percentage removal of perchlorate increases and reaches maximum at pH-4 and then decreases. This behavior can be due to the effect of pH solution on the charge of functional group of rice husk, and thus become more effective to adsorption in acidic pH. The modified rice husk containing a positive charge in this acidic pH is more effective for the adsorption of maximum amount of perchlorate.

3.2 *Adsorption at different adsorbent mass*

The effect of adsorbent dosage on the removal percentage of perchlorate increases and reaches maximum at 3 g of adsorbent and then decreased. Figure 3 shows the results obtained from this investigation. This increase in the removal percentage with increase amounts of adsorbent may be due to the availability of more effective sites of adsorption due to increased surface area.

The decrease of adsorption with increasing the amounts of adsorbent can be attributed to unsaturated absorption sites residual through adsorption procedure.

3.3 *Adsorption at different adsorbate concentration*

The effect of concentration of adsorbate on the removal percentage of perchlorate from aqueous solutions with perchlorate concentrations ranging from 5 to 25 mg/L is studied. Figure 4 shows the effect of adsorbate concentration on removal of perchlorate by rice husk at $37 \pm 2°C$ and keeping other parameters (pH-4, Adsorbent mass-3 g, adsorbate) constant. The adsorption of perchlorate in rice husk is maximum at 5 ppm and the percentage reduction was also high at this level. The percentage adsorption rate increases with increase in perchlorate concentration but the adsorption per unit gram of adsorbent reaches maximum at 20 ppm and reaches equilibrium. The effect of the adsorbed concentration factor depends on the relationship between the concentration of the adsorbed and the binding sites available on surface of the adsorbent.

3.4 *Adsorbtion at different temperatures*

The effect of temperature on the removal percentage of perchlorate from aqueous solutions with temperature varying between from 30°C to 70°C was studied. Figure 5 shows the effect of temperature on removal of perchlorate by rice husk at (30–70°C) by keeping the parameters (pH-4, Adsorbent mass-3°g, adsorbate concentration-20 ppm) constant. The adsorption of perchlorate in rice husk is maximum at 60°C and the percentage reduction was also high at this temperature. According to Le-Chatelier's Principle, adsorption occurs more readily at lower temperature and decreases with increase in temperature.

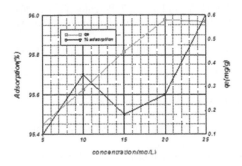 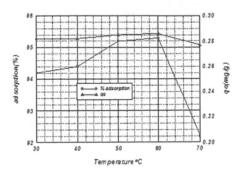

Figure 4. Effect of adsorbate concentration on removal of perchlorate by rice husk.

Figure 5. Effect of temperature on removal of perchlorate by rice husk at (30–70°C).

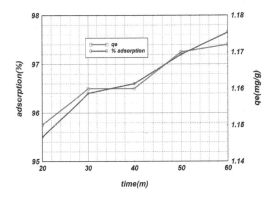

Figure 6. Effect of optimum parameters.

3.5 *Adsorption at optimum parameters*

The removal of perchlorate was investigated at time interval of 10 minutes. Figure 6: shows the effect of optimum parameters (pH-4, temperature (60°C), adsorbent mass-3 g, concentration 20 ppm) on removal of perchlorate by rice husk. Results showed that the removal percentage increased and reached the equilibrium point at 60 minutes.

4 CONCLUSION

This study deduces that the cationic surfactant modified rice husk powder can be used as a very low cost adsorbent for the removal of perchlorate contaminated water. From the results obtained it can be concluded that rice husk is a good adsorbent due to its ability to remove the perchlorate in water even at low concentrations. The experiment showed that at low pH the adsorption level was maximum because CTAB modified rice husk containing a positive charge which increases the adsorption rate of negatively charged.

REFERENCES

[1] John, D.C. & Laurie, A.A. 2004. Microbial perchlorate Reduction, Rocket fuelled metabolism. *Nature Reviews/Microbiology* 2:569–580.
[2] Mamie, N.I., Kate, M.S. & Dennis, C.R. 2005. Reduction of Perchlorate and Nitrate by Microbial Communities in Vadose Soil. *Applied and Environmental Microbiology*, 71(7):3928–3934.

[3] Yali, S., Ping, Z., Yawei, W., Jianbo, S., Yaqi, C., Shifen, M. & Guibin. 2007. Perchlorate in sewage sludge, rice, bottled water and milk collected from different areas in China. *Environment International* 33:955–962.

[4] Sridhar, K. 2013. Organoclays of high-charge synthetic clays and alumina pillared natural clays: Perchlorate uptake. *Applied Clay Science* 80:340–345.

[5] Urbansky, E.T. 2002. Perchlorate as an environmental contaminant. *Environmental Science and Pollution Research* 9:187–192.

[6] Z. Li, F.X. Li, D. Byrd, G.M. Deyhle, D.E. Sesser, M.R. Skeels, S.H. Lamm. 2000 Neonatal thyroxine level and perchlorate in drinking water. *Occup. Environ. Med.* 42:200–205.

[7] Robert, P. & Fred, S.C. 2005. The removal of perchlorate from ground water by activated carbon tailored with cationic surfactants. *Water Research* 39:4020–4028.

[8] Rovshan, M. & Chin, P.H. 2010. Perchlorate removal by activated cabon asorption. *Separation and purification technology* 70:329–337.

[9] Weifang, C., Fred, C. & Joserene, R.M. 2005. Ammonia-tailoring of Gac to enhance perchlorate removal II Perchlorate Adsorption. *Carbon* 43(3):581–590.

[10] Yanhua, X., Shiyn, L., Fei, W. & Guangli, L. 2010. Removal of perchlorate from aqueous solution using protonated cross-linked chitosan. *The Chemical Engineering Journal* 156(1):56–63.

[11] Muquing, Q., Chu, X.H., Xiaojie, L. 2016. The influence of the co-existing anions on the adsorption of perchlorate from water by the modified orange peels. *Mature Environment and Pollution Technology* 15 (4):1359–1362.

[12] Joo, Y.K., Sridhar, K., Robert, P., Fred, C., Hiroaski, K. 2011. Perchlorate uptake by synthetic layered double hydroxides and organo-clay materials. *Applied Clay Science* 51:158–164.

[13] Umesh, K.G., Kaur, D.S. & K. Garg. 2009. Removal of hexavalent chromium from aqueous solution by adsorption on treated sugarcane bagasse using response surface methodological approach. *Desalination* 249:475–479.

[14] C. Umpuch & B. Jutarat. 2013. Adsorption of Organic Dyes from Aqueous solution by Surfactant Modified Corn Straw. *International Journal of Chemical Engineering and Applications* 4(3).

[15] Dipu, B., Shigeo, S., Shigeon, K. & Toshinri, K. 2009. Sorption of As(V) from aqueous solution using acid modified Carbon black. *Journal of Hazardous Materials* 162:1269–1277.

[16] Humayatul, U., Dadang, A.S., Mary, S. & Abdul, W. 2015. Analysis of Chemical Composition of ice Husk Used as Absorber plates sea water into clean water, ARPN *Journal of Engineering and Applied Sciences* 10(14):6046–6050.

[17] Farai, M., Olga, K. & Pardon, K.K. 2014. Removal of Cr(VI)from aqueous solutions using Powder of Potato peelings as a low cost sorbent, *Hindawi Publishing Corporation*.

Emerging Trends in Engineering, Science and Technology for Society,
Energy and Environment – Vanchipura & Jiji (Eds)
© 2018 Taylor & Francis Group, London, ISBN 978-0-8153-5760-5

Biosorption of methyl orange from aqueous solution using cucurbita pepo leaves powder

M. Tukaram Bai, K. Latha, P. Venkat Rao & Y.V. Anudeep
Department of Chemical Engineering, Andhra University, Visakhapatnam, India

ABSTRACT: This paper investigates the biosorption of methyl orange from aqueous solution on cucurbita pepo leaves powder. The batch studies were carried out for the contact time, (5–60 min), biosorbent dosage (0.05–0.5 g), pH (2–9), initial concentration of dye (10–50 mg/L) and temperature, (283–323°K). The isotherms for the present study are Freundlich, Langmuir and Temkin. Of these isotherms, the Freundlich isotherm was best fitted. The kinetics were studied with pseudo first and second order. Biosorption kinetics of cucurbita pepo leaves were well correlated with pseudo second order.

Keywords: biosorption, methyl orange, cucurbita pepo leaves, effluent treatment, isotherms, kinetics

1 INTRODUCTION

The development of science and technology provides many benefits to human life, but for the effects of a negative impact on the surrounding environment, such as industrial waste problems. Many industries use dyes in order to color their products, and then dispose of a lot of colored wastewater as effluent. The major sources of the dyes are industries like carpet, leather, printing, textile. The wastewater from these industries causes hazardous health effects to aquatic and human life.

Several processes have been applied for the treatment of dyes from wastewater, such as chemical, biological and physical processes. Even though chemical and biological treatments are effective for removing dyes, they still require special equipment and are considered to be quite energy intensive in terms of the addition and large amounts of by products often generated. In recent years, a physical method through adsorption process based on activated carbon material has been considered to be a superior technique as compared to others. However, the commercial activated carbon is quite expensive and has limited its application. Due to economic reasons, the discovery toward alternative adsorbents to replace the costly activated carbon is highly recommended. Many investigators have studied the feasibility of using inexpensive alternative materials like chitosan beads (Negrulescu et al., 2014), calcined Lapindo (Jalil et al., 2010), chitosan intercalated montmorillonite (Umpuch & Sakaew, 2013), activated carbon coated monolith in a batch system (Darmadi & Thaib, 2010), sawdust and sawdust-fly ash (Lucaci & Duta, 2011), cork as a natural and low-cost adsorbent (Krika & Benlahbib, 2015), thermally treated eggshell (Belay & Hayelom, 2014), modified activated carbon from rice husk (Qiu et al., 2015), tree bark powder (Egwuonwu, 2013), and banana trunk fiber (Prasanna et al., 2014), as carbonaceous precursors for the preparation of activated carbons and for the removal of dyes from water and wastewater. The present investigation is an attempt to explore the possibility of using cucurbita pepo leaves powder to remove methyl orange in aqueous solution, since the raw material is harmless, cheaper, and plentiful.

Methyl orange is an anionic azo dye with a molecular formula $C_{14}H_{14}N_3NaO_3S$. The wide usage of methyl orange was as a pH indicator. It is used in titrations because it changes color at the pH of a mid-strength acid. Its anion form is yellow, and its acidic form is red.

The entire color change occurs in acidic conditions. It is partially soluble in hot water, very slightly soluble in cold water, and insoluble in diethyl ether and alcohol. It is soluble in pyrimidine. At room temperature it appears as a solid, odorless, orange/yellow powder that yields an orange solution when dissolved in water. The IUPAC name is Sodium 4-[(4-dimethylamino) phenyldiazenyl] benzenesulfonate.

Its oral consumption is toxic, hazardous and carcinogenic due to the presence of nitrogen. When discharged into the running streams it will affect aquatic life, causing detrimental effects in the liver, gill, kidney, intestine, and pituitary gonadotropic cells. In humans, it may cause irritation to the respiratory tract if inhaled and causes irritation to the gastrointestinal tract upon ingestion. This dye may enter into the food chain and could possibly cause carcinogenic, mutagenic, and tetragenic effects on humans.

2 MATERIALS AND METHODS

2.1 *Preparation of biosorbent*

The cucurbita pepo leaves used in the present study were collected from the plants near Gopalapatnam, Visakhapatnam. The collected leaves were washed twice with distilled water to remove dirt, completely dried in sunlight for 20 days, and then dried in an oven to remove the excess moisture content. The dried leaves were then cut into small pieces and powdered. The powder was screened to 150 μm size and directly used as a biosorbent.

2.2 *Procedure*

The methyl orange was obtained from Merck laboratories limited, Mumbai, India. A stock solution of 1,000 mg/L was prepared by dissolving 1g of methyl orange in 1,000 mL of distilled water, which was later diluted to required concentrations. All the solutions were prepared using distilled water. Solution pH for pH studies was adjusted by adding HCl and NaOH, as required. Concentrations of the dye solutions were determined from the absorbance spectrum of the solution at the characteristic wavelength of dye using a double beam UV-Visible spectrophotometer. Final concentrations were determined from the calibration curve. The absorption wavelength of methyl orange (λ_{max}) = 464 nm. Variables studied and their range: Contact time: 5–60 min., Aqueous dye solution pH: 2–9, Initial concentration of the dye: 10–50 mg/L, Biosorbent dosage: 0.05–0.2 g, Temperature: 283–323°K.

The effect of contact time was determined by shaking 0.1 g of adsorbent in 100 ml of synthetic solutions methyl orange of initial dye concentration 10 mgL^{-1}. Shaking was provided for different time intervals like 5, 10, 15, 20, up to 60 min at a constant agitation speed of 230 rpm. The effect of pH of the dye solution was determined by agitating 0.1 g of biosorbent and 100 ml of synthetic dye solutions of initial dye concentration 10 mg/L at different pH values of the solution, ranging from 2 to 9, by adding 0.1N HCl or 0.1 N NaOH. To study the effect of concentration, 100 ml of aqueous solution, each of methyl orange of different dye concentrations of 10 mg/L, 20 mg/L, 30 mg/L, 40 mg/L and 50 mg/L were taken in 250 ml conical flasks. 0.1 g/L of cucurbita pepo leaves powder was added to each of the flasks. The total dye concentration in solution was analyzed with double beam a UV Spectrometer at a wavelength of 464 nm for methyl orange dye solution.

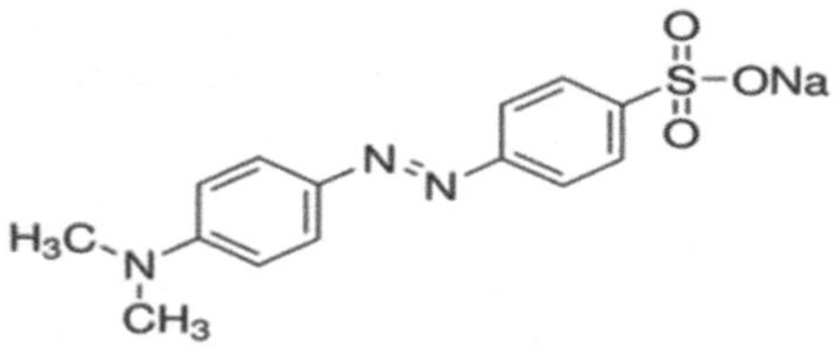

Figure 1.   Molecular structure of methyl orange.

3 RESULTS AND DISCUSSION

Experimental data was generated in a batch mode of operation to study the effect of various parameters for the removal of methyl orange from the aqueous solution (prepared in the laboratory) using cucurbita pepo leaves powder as the biosorbent. Various experimental runs were conducted in the present study. The parameters studied include: Contact time, t (min), pH of the solution, initial concentration of the solution, C_0 (mg/L), biosorbent dosage, w (g) and temperature, T (°K).

3.1 *Effect of contact time, t*

The effect of contact time on biosorption of methyl orange onto cucurbita pepo was studied at 10 mg/L. From Figure 2, it can be seen that the rate of biosorption was very rapid in the initial period of contact time. Thereafter, it gradually increased with contact time until biosorption was reached at the equilibrium point. This trend of biosorption kinetics was due to the biosorption of dye on the exterior surface of biosorbent at the initial period of contact time. When the biosorption on the exterior surface reached saturation point, the dye diffused into the pores of the biosorbent and was adsorbed by the interior surface of the biosorbent. The equilibrium time for the cucurbita pepo-methyl orange system was 40 min. No further biosorption occurred, even after an increase in the contact time from 40 min to 60 min. The % removal and dye uptake were calculated, as given in Equations 1 and 2 respectively:

$$\% \text{ Removal} = \frac{\left(C_0 - C_t\right)}{C_0} \times 100 \qquad (1)$$

$$\text{Dye uptake}\left(Q\right) = \frac{\left(C_0 - C_t\right) \times v}{w \times 1000} \qquad (2)$$

where
   C_0 = Initial concentration of the dye, mg/L
   C_t = Final concentration of the dye after time, t
   V = Volume of aqueous dye solution, ml
   W = Weight of biosorbent, g

3.2 *Effect of initial concentration of aqueous dye solution (C_0)*

Experiments were undertaken to study the effect of the initial dye concentration on the removal of methyl orange from the solution. The variations of % dye removal and dye uptake

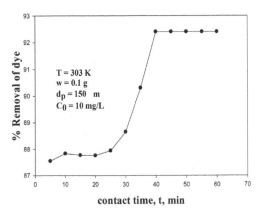

Figure 2. Effect of contact time on % removal of methyl orange dye.

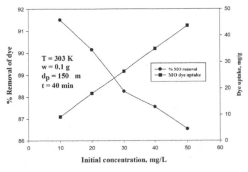

Figure 3. Effect of initial concentration of methyl orange dye on % removal and dye uptake.

with initial dye concentration are presented in Figure 3. The obtained curves show that the dye uptake increased with an increase in initial concentration of dye, while the percentage removal of dye decreased with an increase in initial dye concentration. The increase in dye uptake is a result of the increase in the driving force; that is, the concentration gradient, with an increase in the initial dye concentration (from 10 to 50 mg/L). However, the percentage removal of dye on cucurbita pepo leaves powder was decreased from 91.50 to 86.54% for methyl orange. Though an increase in dye uptake was observed, the decrease in percentage removal may be attributed to lack of sufficient surface area to accommodate much more dye available in the solution. The percentage removal at higher concentration levels shows a decreasing trend whereas the equilibrium uptake of dye displays an opposite trend.

3.3   *Effect of pH*

The pH of an aqueous dye solution is an important monitoring parameter in biosorption, as it affects the surface charge of the biosorbent material and the degree of ionization of the dye molecule. It is also directly related to the competition ability of hydrogen ions with biosorbate molecules to active sites on the biosorbent surface. In the present study methyl orange dye biosorption data was obtained in the pH range of 2 to 9 of the aqueous solution ($C_0 = 10$ mg/L) using 0.1 g of 150 μm size biosorbent. The effect of pH of aqueous solution on % biosorption of methyl orange dye is shown in Figure 4. The % biosorption of methyl orange dye was increased from 87.90 to 93.68% as pH increased from 2 to 6, and beyond the pH value of 6 it decreased. As the pH of the system decreased, the number of negatively charges surface sites decreased and the number of positively charged surface sites increased and this favors the biosorption of dye anions due to electrostatic attraction. When the acidity increased due to concentration of H+ that will decrease the negative charge for methyl orange, then the adsorption increased. However, when the basicity of solution was increased, the amount of biosorption decreased due to the increase of concentration of OH. Hence the optimum pH for methyl orange is taken as 6.

3.4   *Effect of biosorbent dosage (w)*

The percentage biosorption of methyl orange dye is drawn against a biosorbent dosage for 150 μm biosorbent size is shown in Fig. 5. The % biosorption of methyl orange dye increased from 88.10 to 95.80 and dye uptake decreased from 17.62 to 4.765 mg/g, with an increase in biosorbent dosage from 0.05 to 0.2 g. The number of active sites of biosorbent increase with an increase in biosorption dosage. This is due to the binding of almost all the dye molecules to adsorbent surfaces and the establishment of equilibrium between the dye molecules on the adsorbent surfaces and in the solution.

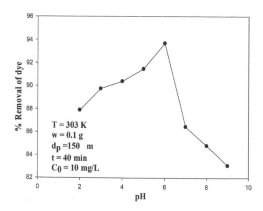

Figure 4.   Effect of pH on % removal of methyl orange dye.

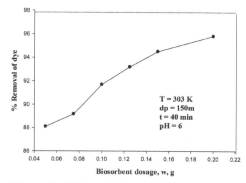

Figure 5.   Effect of biosorbent dosage on % removal of methyl orange dye.

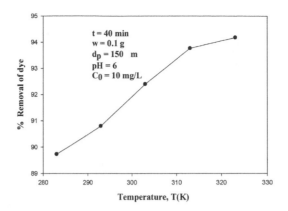

t = 40 min
w = 0.1 g
$d_p = 150$ m
pH = 6
$C_0 = 10$ mg/L

Figure 6.   Effect of temperature on % removal of methyl orange dye.

3.5   *Effect of temperature (T)*

The effect of change in the temperature on the methyl orange dye uptake is shown in Figure 6. The biosorption of methyl orange by cucurbita pepo leaves powder at different temperatures showed an increase in the biosorption capacity when the temperature was increased. The temperature has two main effects on the biosorption process. An increase in temperature is known to increase the diffusion rate of the biosorbate across the external boundary layer and within the pores. Furthermore, changing the temperature will modify the equilibrium capacity of the biosorbent for a particular biosorbate. The effect of temperature was investigated from batch experiments carried out at five constant temperatures: 283, 293, 303, 313 and 323°K. With an increase in temperature, the % removal was increased from 89.73 to 94.18% for methyl orange for the initial concentration of 10 mg/L, as shown in Figure 6. This indicates that the biosorption reaction is endothermic in nature because of the chemical interaction between biosorbate and the biosorbent, and due to the increased rate of intra-particle diffusion of dyes into the pores of the biosorbent at higher temperatures. However, the dye uptake capacity was increased with an increase in temperature.

4   BIOSORPTION KINETICS

The kinetics of the biosorption data was analyzed by two models, namely pseudo first order and pseudo second order. These models correlate solute uptake, which is important in the prediction of reactor volume.

4.1   *Pseudo first order kinetics*

The order of biosorbate—biosorbent interactions have been described using kinetic models. Traditionally, the Lagergren first order model finds wide application. The Lagergren first order rate Equation is:

$$(dQ/dt) = K_1 (Q_{eq} - Q) \tag{3}$$

where Q_{eq} and Q are the amounts of dyes adsorbed at equilibrium time and any time t, and K_1 is the rate constant of the pseudo first order biosorption.

The above Equation can be presented as:

$$\int (dQ/(Q_{eq} - Q)) = \int K_1 \, dt \tag{4}$$

Applying the initial condition $q_t = 0$ at t = 0, we get:

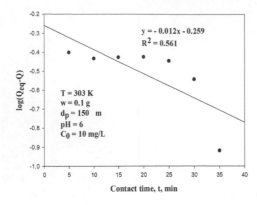

Figure 7. First order kinetics for % biosorption of methyl orange dye.

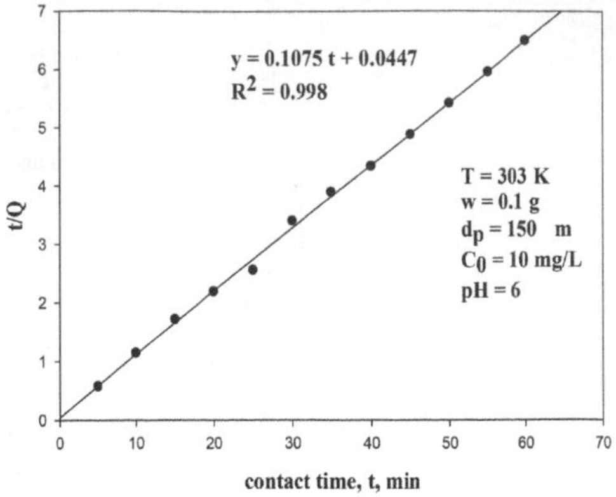

Figure 8. Second order kinetics for % biosorption of methyl orange dye.

$$\log (Q_{eq} - Q) = \log Q_{eq} - (K_1/2.303)\, t \qquad (5)$$

The plot of time t versus $\log (Q_{eq}-Q)$ gives a straight line for first order kinetics, facilitating the computation of biosorption first order rate constant (K_1).

In the present study, the kinetics were investigated with 100 ml of aqueous solution ($C_0 = 10$ mg/L) for the contact time of 5 to 60 min. The Lagergren first order plot is drawn in Figure 7. The first order model Equation obtained for the present study is:

$$\log (Q_{eq}-Q) = -0.012t - 0.259 \qquad (6)$$

4.2 Pseudo second order kinetics

The pseudo second order model is represented by the following Equation:

$$(t/Q) = (1/K_2 Q_{eq}^2) + (1/Q_{eq})\, t \qquad (7)$$

If the pseudo second order kinetics is applicable, the plot of time t versus (t/Q) gives a linear relationship that allows computation of K_2.

In the present study, the kinetics are investigated with 100 ml of aqueous solution ($C_0 = 10$ mg/L) in the agitation time intervals of 5 min to 60 min. The pseudo second order plot of time 't' versus (t/Q) is drawn in Figure 8. The second order kinetics obtained for the present study is given as:

$$t/Q = 0.1075\, t + 0.0447 \qquad (8)$$

5 ADSORPTION ISOTHERMS

In the present study, the isotherms studies are Langmuir, Temkin and Freundlich. The linear forms of these isotherms are obtained at room temperature and are shown in the Figures below.

5.1 Langmuir isotherm

The Langmuir isotherm has been successfully applied to many pollutant biosorption processes and has been the most widely used isotherm for the biosorption of a solute from a liquid solution.

A basic assumption of the Langmuir theory is that biosorption takes place at specific homogeneous sites within the biosorbent. It is then assumed that once a solute particle occupies a site, no further biosorption can take place at that site. The rate of biosorption to the surface should be proportional to a driving force. The driving force is the concentration in the solution, and the area is the amount of bare surface.

The Langmuir relationship is hyperbolic, and the Equation is shown as:

$$Q_{eq} = \frac{Q_{max} b C_{eq}}{1 + b C_{eq}} \tag{9}$$

The above Equation can be rearranged into the following linear form as:

$$\frac{C_{eq}}{q_{eq}} = \frac{1}{b Q_{max}} + \frac{1}{Q_{max}} C_{eq} \tag{10}$$

where

C_{eq} is the equilibrium concentration (mg/L)
Q_{eq} is the amount of dye ion adsorbed (mg/g)
Q_{max} is Q_{eq} for a complete monolayer (mg/g)
b is sorption equilibrium constant (L/mg)

Figure 9 is the plot of $[C_{eq}]$ versus $[C_{eq}/Qeq]$, which is a straight line with slope $1/Q_{max}$ and an intercept of $1/bQ_{max}$.

The Correlation coefficient $R^2 = 0.9671$ and the Langmuir Equation obtained for the present study is:

$$(C_{eq}/Q_{eq}) = 0.0107 C_{eq} + 0.0874 \tag{11}$$

5.2   Freundlich isotherm

In 1906, Freundlich studied the sorption of a material onto animal charcoal. He found that if the concentration of solute in the solution at equilibrium, C_{eq}, was raised to the power of m, the amount of solute adsorbed being Q_{eq}, then $C_{eq}{}^m/Q_{eq}$ is a constant at a given temperature. This fairly satisfactory empirical isotherm can be used for non-ideal sorption and is expressed by the following Equation:

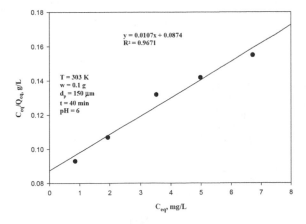

Figure 9.   Langmuir isotherm for % biosorption of Methyl Orange dye.

$$Q_{eq} = K_f C_{eq}^{1/m} \qquad (12)$$

The Equation is conveniently used in the linear form by taking the logarithm of both sides as:

$$\log (Q_{eq}) = (1/m)^* \log (C_{eq}) + \log (K_f) \qquad (13)$$

The Freundlich isotherm is derived assuming a heterogeneity surface. K_f and m are indicators of biosorption capacity and biosorption intensity respectively. The value of m should lie between 1 and 10 for favorable biosorption.

Figure 10 is a plot of $\log [C_{eq}]$ versus $\log [Q_{eq}]$, which is a straight line with a slope of 1/m and an intercept of $\log (K_f)$.

From the value of biosorption intensity, it can be concluded that the Freundlich isotherm indicates for favorable biosorption. The Freundlich Equation obtained for the present study is shown by:

$$\log Q_{eq} = 0.7454\log C_{eq} + 1.0227 \qquad (14)$$

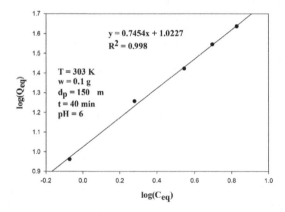

Figure 10.   Freundlich isotherm for % biosorption of methyl orange dye.

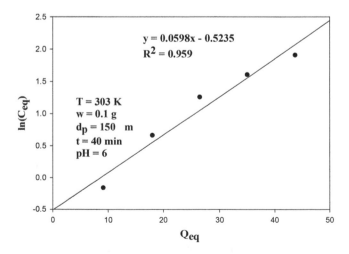

Figure 11.   Temkin isotherm for % biosorption of methyl orange dye.

562

5.3 Temkin isotherm

The derivation of the Temkin isotherm assumes that the fall in the heat of sorption is linear rather than logarithmic, as implied in the Freundlich Equation (Aharoni & Ungarish, 1977). The Temkin isotherm has generally been applied in the following form:

$$Q_{eq} = \frac{R*T}{b_T} \ln[A_T * C_{eq}] \tag{15}$$

where
   R = Universal gas constant (8.314 J/mol.K)
   T = Temperature of dye solution, K
   A_T, b_T = Temkin isotherm constants
This can be written as:

$$Q_{eq} = \frac{R*T}{b_T} \ln\left[C_{eq} \right] + \frac{R*T}{b_T} \ln\left[A_T \right] \tag{16}$$

Figure 11 shows a plot of $\ln[C_{eq}]$ versus Q_{eq}, which is a straight line with slope of RT/b_T and intercept of $RT/b_T \ln[A_T]$. The Temkin Equation obtained for the present study is:

$$Q_{eq} = 0.0598 \ln C_{eq} - 0.5235 \tag{17}$$

The isotherm constants obtained for various isotherm models are shown in Table 1.

The correlation coefficients obtained from the Langmuir, Freundlich, and Temkin models were 0.9671, 0.9981, 0.959 respectively for methyl orange, and the Freundlich Equation was observed to be more suitable, followed by Langmuir and Temkin for the experimental data of methyl orange dye.

The maximum metal uptakes for the biosorption of the methyl orange by using various other biosorbents are tabulated in Table 2.

Table 1. Isotherm constants for various isotherm models.

ISOTHERM	CONSTANT	R^2
Langmuir	Q_{max}, mg/g = 93.45 b, L/mg = 0.1224	0.9671
Freundlich	K_f = 10.536 m = 1.342	0.9981
Temkin	$A_T = 1.57 \times 10^{-4}$ b_T = 42,126	0.959

Table 2. Methyl orange uptake capacities for different biosorbents.

AUTHOR	BIOSORBENT	Q_{max}, mg/g
Jalil et al. (2010)	Calcined Lapindo volcanic mud	333.3
Darmadi and Tahib (2010)	Carbon coated monolith	36.72
Krika and Benlahbib (2015)	Cork powder	16.66
Faith Deniz (2013)	Prunus amygdalus (almond shell)	41.34
Danish et al. (2013)	Acacia mangium wood	7.54
Gong et al. (2013)	Finger citron residue	934.58
Su et al. (2014)	Wheat straw	50.4
Chaidir et al. (2015)	Durio zibethinus (murr seed)	6.352
Present study	**Cucurbita Pepo**	**93.45**

6 CONCLUSION

The following conclusions can be drawn from the above discussion. Methyl orange is removed efficiently by the cucurbita pepo leaves. The optimum contact time for the process is 40min at room temperature. The optimum pH is 6 and the optimum dosage is 0.1g. The biosorption favors the increase in the temperature. The adsorption kinetics are better described with pseudo second order kinetics. The isotherm studies are best fitted for the Freundlich isotherm, followed by the Langmuir and Temkin isotherms.

REFERENCES

Belay, K. & Hayelom, A. (2014). Removal of methyl orange from aqueous solutions using thermally treated egg shell (locally available and low cost biosorbent). *International Journal of Innovation and Scientific Research*, 8(1), 43–49. ISSN 2351-8014.

Chaidir, Z., Sagita, D.T., Zein, R & Munaf, E. (2015). Bioremoval of methyl orange dye using durian fruit (durio zibethinus) murr seeds as biosorbent. *Journal of Chemical and Pharmaceutical Research*, 7(1), 589–599.

Danish, M., Hashim, R., Ibrahim, M.N.M. & Sulaiman, O. (2013). Characterization of physically activated acacia mangium wood-based carbon for the removal of methyl orange dye. *BioResources*, 8(3), 4323–4339.

Darmadi, D. & Thaib, A. (2010). Adsorption of anion dye from aqueous solution by activated carbon coated monolith in a batch system. *Jurnal Rekayasa Kimia dan Lingkungan*, 7(4), 170–175. ISSN 1412-5064.

Deniz, F. (2013). Adsorption properties of low-cost biomaterial derived from *Prunus amygdalus* L. for dye removal from water. *The Scientific World Journal, 961671*. Egwuonwu, P.D. (2013). Adsorption of methyl red and methyl orange using different tree bark powder. *Academic Research International*, 4(1), 330.

Gong, R., Ye, J., Dai, W., Yan, X., Hu, J., Hu, X., & Huang, H. (2013). Adsorptive removal of methyl orange and methylene blue from aqueous solution with finger-citron-residue-based activated carbon. *Industrial & Engineering Chemistry Research*, 52(39), 14297–14303.

Jalil, A.A., Triwahyono, S., Adam, S.H., Rahim, M.D., Aziz, M.A.A., Hairom, N.H.H., & Mohamadiah, M.K.A. (2010). Adsorption of methyl orange from aqueous solution onto calcined Lapindo volcanic mud. *Journal of Hazardous Materials*, 181(1–3), 755–762.

Krika, F. & Benlahbib, O.E.F. (2015). Removal of methyl orange from aqueous solution via adsorption on cork as a natural and low-cost adsorbent: Equilibrium, kinetic and thermodynamic study of removal process. *Desalination and Water Treatment*, 53(13), 3711–3723.

Lucaci, D. & Duta, A. (2011). Removal of methyl orange and methylene blue dyes from wastewater using sawdust and sawdust-fly ash as sorbents. *Environmental Engineering and Management Journal*, 10(9), 1255–1262.

Negrulescu, A., Patrulea, V., Mincea, M., Moraru, C. & Ostafe, V. (2014). The adsorption of tartrazine, congo red and methyl orange on chitosan beads. *Digest Journal of Nanomaterials and Biostructures*, 9(1), 45–52.

Prasanna, N, Manivasagan, V., Pandidurai, S., Pradeep, D. & Leebatharushon, S.S. (2014). Studies on the removal of methyl orange from aqueous solution using modified banana trunk fibre. *International Journal of Advanced Research*, 2(4), 341–349.

Qiu, M.Q., Xiong, S.Y., Wang, G.S., Xu, J.B., Luo, P.C., Ren, S.C. & Wang, S.B. (2015). Kinetic for adsorption of dye methyl orange by the modified activated carbon from rice husk. *Advance Journal of Food Science and Technology*, 9(2), 140–145. ISSN: 2042-4868, e-ISSN: 2042-4876.

Su, Y., Jiao, Y., Dou, C. & Han, R. (2014). Biosorption of methyl orange from aqueous solutions using cationic surfactant-modified wheat straw in batch mode. *Desalination and Water Treatment*, 52(31–33), 6145–6155.

Umpuch, C. & Sakaew, S. (2013). Removal of methyl orange from aqueous solutions by adsorption using chitosan intercalated montmorillonite. *Songklanakarin Journal of Science & Technology*, 35(4), 451–459.

Emerging Trends in Engineering, Science and Technology for Society,
Energy and Environment – Vanchipura & Jiji (Eds)
© 2018 Taylor & Francis Group, London, ISBN 978-0-8153-5760-5

Catalytic hydrodechlorination of 1,4-dichlorobenzene from wastewater

C. Megha
Government Engineering College, Kozhikode, India

K. Sachithra
Government Engineering College, Thrissur, India

Sanjay P. Kamble
National Chemical Laboratory, Pune, India

ABSTRACT: Dechlorination is one of the promising methods to convert more toxic chlorinated aromatics into less toxic environmentally friendly value-added products. In this work catalytic hydrodechlorination is achieved with the aid of low-cost hydrogenation catalysts such as Raney nickel, bimetallic catalyst and palladium on activated carbon catalyst. Among them, Raney nickel shows the best result as it takes only a few hours to completely dechlorinate 1,4-dichlorobenzene to benzene. The detailed study on dechlorination using Raney nickel was made in this work. It is an economically feasible method for treating chlorinated pollutants in wastewater. Experiments were done with varying parameters like temperature, concentration, and pH. Dechlorination shows the best result at high temperature and lower pH. The product was confirmed using HPLC analysis and UV spectroscopy. Recycling of Rancy nickel catalyst was also performed in this study. It was not possible to do more than two recycles due to poisoning of the catalyst.

Keywords: Dechlorination, hydrodechlorination, 1,4-dichlorobenzene, wastewater

1 INTRODUCTION

Chlorinated aromatic compounds have at least one chlorine atom covalently attached to an aromatic ring. Due to the presence of a halogen group in the aromatic ring, it is highly resistant for biodegradability, and so its ubiquitous presence can be seen in every ecosystem. Chlorinated organic, such as Trichloroethylene (TCE), Carbon Tetrachloride (CT), chlorophenols, and Polychlorinated Biphenyls (PCBs) are among the most common contaminants. Most of these chloroorganics were widely used in industry during the past half-century as solvents, pesticides, and electric fluids.

Many chlorinated organic chemicals (COCs) have been detected in many surface waters and groundwater, in sewage, and in some biological tissues (Pearson, 1982). The observed levels are, in general, too low to cause immediate acute toxicity to mammals, birds, and aquatic organisms (Cheng et al., 2007). Treatment processes can and do reduce the concentrations of COCs in water However, the degree of efficacy is often a function of chemical structure, cost, and energy. All treatment processes have some degree of side effects, such as a generation of residuals or by-products. Among the different methods of treatment, catalytic hydrodechlorination is emerging as an effective way to reduce toxicity of COCs. This method reduces the toxicity and increases the biodegradability at low cost. It is a method of recycling compounds from which they have originally formed with low emissions.

COCs, including 1,4-dichlorobenzene, have been reported as being primary pollutants by Environmental Protection Agency. Several research studies revealed reducing the toxicity of compound by breaking carbon chlorine bond is not effectively done. The greatest challenge in implementing this strategy is the adoption of low-cost technologies using different metal and bimetallic catalysts.

Various metal, bimetallic, hydrogenation catalysts were used. New bimetals were prepared (Sopoušek et al., 2014) and dechlorination was done to convert 1,4-dichlorobenzene into benzene. Screening of 1,4-dichlorobenzene was done using various metal and bimetallic catalysts. Hydrogenation catalysts like Raney nickel, and palladium on activated carbon, were used to dechlorinate 1,4-dichlorobenzene at room temperature and pressure (Xia et al., 2009).

2 EXPERIMENT

Experiments were conducted for studying the effect of initial concentration of 1,4-dichlorobenzene, catalyst concentration, pH, and synergy of salts for Raney nickel. Dechlorination experiments were conducted with various metal and bimetallic catalysts. A bimetallic catalyst was prepared using its corresponding metal precursor and sodium borohydride was used as a reducing agent. All experiments were carried out at room temperature ($30 \pm 2°C$). Stock solution of 1,000 ppm 1,4-dichlorobenzene was prepared in a 100 mL standard flask with acetonitrile. The desired concentration of 1,4-dichlorobenzene for experiments was prepared by micro-pipetting from this stock solution into deionized water.

The solutions of 1,4-dichlorobenzen were prepared in the concentrations of 10, 20, 50, and 60 mg/L. Batch experiments were conducted with 150 mL of solution taken in a 250 mL conical flask with a tight lid. These bottles were kept in a rotary shaker at a constant shaking speed of 150 rpm for 24 hrs. After 24 hrs, 10 mL of sample was taken out and syringe filtered. The resulting 1,4-dichlorobenzene concentration was determined using HPLC. From this final 1,4-dichlorobenzene concentration, the percentage dechlorination was calculated,

$$\% \text{ Dechlorination} = (C-C0)/C0 *100 \qquad (1)$$

where C (mgL^{-1}) is the amount of 1,4-dichlorobenzene per liter at time t., and C_0 is the initial concentration (mg L^{-1}). The dechlorination efficiency or conversion percentage of 1,4-dichlorobenzene was calculated using the expression:

2.1 *Kinetic experiment study*

In order to estimate the rate constant and order of reaction for the dechlorination of 1,4-dichlorobenzene, time-dependent studies were conducted in a 250 mL round-bottom flask. Stirring was done continuously. The particular concentration of 1,4-dichlorobenzene was transferred into the flask and optimized catalyst loading (3 g/L) of the catalyst was added to it. Samples were withdrawn from the vessel at frequent time intervals and analyzed for the concentration of 1,4-dichlorobenzene using HPLC. The kinetics were checked, mainly for first-order reaction as most of the catalytic dechlorination reactions followed it.

$$\ln C/C0 = -K*t \qquad (2)$$

where
    C: concentration of 1,4-dichlorobenzene in ppm
    C0: initial. Concentration of 1,4-dichlorobenzene in ppm
    K: reaction rate constant
    t:time
    After a time interval, solutions were filtered and analyzed for 1,4-dichlorobenzene content using HPLC.

3   RESULTS AND DISCUSSIONS

3.1   *Calibration of 1,4-dichlorobenzene*

1,4-dichlorobenzene is calibrated in HPLC before any kind of analysis is done. 5, 10, 20, and 50 ppm 1,4-dichlorobenzene solution is prepared and kept for analysis.

Linear fit:
$ax + b$
$a = 5.98309 \times e\text{-}0.005$
$b = 0$
Goodness of fit is 0.994

3.2   *Catalysts screening*

The dechlorination experiments of 1,4-dichlorobenzene were conducted with various metals, bimetal, and metal nanoparticle with optimized catalyst loading 3g/L. Sodium borohydride was used as a reducing agent.

The results show that hydrogenation catalysts like Raney nickel and palladium on activated carbon give the best results at room temperature and pressure. Fe/Cu bimetal also shows

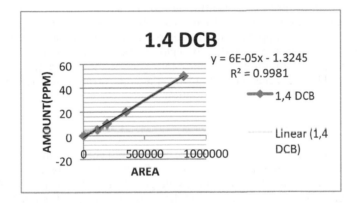

Figure 1.   Calibration curve of 1,4-dichlorobenzene.

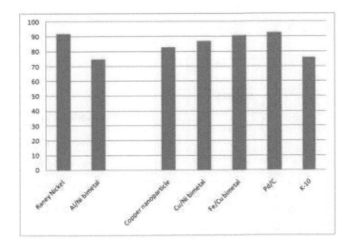

Figure 2.   Percentage dechlorination of various catalysts.

better dechlorination results. Economically Raney nickel is the feasible catalyst. Low-cost bimetals were prepared, and dechlorination of 1,4-dichlorobenzene. using catalysts like Fe/Cu, Al/Ni, Cu/Ni, and K-10 were reported for the first time. However, it gives comparatively poorer results than the hydrogenation catalyst due to leaching and blockage of the bimetal surface.

3.3 Effect of catalyst loading

Dechlorination of 1,4-dichlorobenzene was done using a Raney nickel catalyst of various loadings of 0.1 g, 0.3 g, 0.5 g, 0.7 g, and 0.9 g, keeping other parameters like initial concentration, pH, and temperature constant. The catalyst loading is the most important parameter in optimization of reaction conditions, which shows that as catalyst loading increases, the dechlorination also increases. The activity of the catalyst is increased in an increased amount as hydrogenation increases with loading. The screening was done with these loadings in a rotary shaker for six hours and samples were analyzed in HPLC. At 3.3 gm/l of catalyst loading, 99.9% dechlorination takes place.

3.4 Concentration screening

The concentration of reactant is an important parameter in dechlorination. The concentration is varied, such as 10 ppm, 20 ppm, 50 ppm, and 60 ppm, keeping the catalyst loading at 3 g/L at room temperature. As concentration is increased the dechlorination is decreased. Lowering the concentration too much also decreases the dechlorination. It has been observed that between 20 ppm and 50 ppm dechlorination reach 98%. When the concentration is high dechlorination is reduced; it may be due to agglomeration of more compounds in catalyst. Therefore, the reaction would not takes place in a good manner. The only reactant in this reaction is 1,4-dichlorobenzene. Thus, lowering the concentration will reduce the rate of reaction; in effect, it decreases the dechlorination.

3.5 Effect of salts on dechlorination

Salts like ammonium sulfate, sodium carbonate, magnesium chloride, and sodium nitrate are very common in ground water and other sources of water. It was observed that, except for sodium nitrate salt, other salts have not much synergic effect on dechlorination. As observed, dechlorination kinetics is dominant in the first hour. So, the sample after one hour was taken to check the effect of various salts on dechlorination. The samples without salts showed better dechlorination of the same initial concentration than with salts. The salts may affect

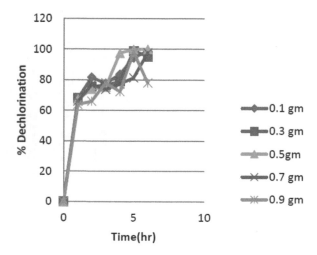

Figure 3.   Catalyst loading.

the hydrogenation in such a way that the ions of the salts may block the catalyst sites, and dechlorination efficiency is reduced comparatively. Sodium nitrate shows better dechlorination compared to without salt. This may be due to the presence of nitrate group in it, which enhances the reduction reaction.

3.6 *Effect of pH*

pH is an important parameter in dechlorination because the ground water for treatment will have different pH, due to different treatment and the presence of several salts and other chemicals. Most of dechlorination studies show that acidic pH favors the dechlorination. The experiments had been done with pH 2, 7, and 11 to know whether acidic, neutral or basic pH is good for dechlorination. pH was changed using 0.1 N HCl and 0.1 N NaOH. Figure 6 shows that acidic medium favors dechlorination in the first ten minutes as the supply of hydrogen ions from hydrochloric acid helps the fast formation of benzene, more than in a basic condition where it is difficult for the replacement of hydrogen ions from water to the aromatic ring. The result shows that acidic and neutral pH does not show much difference in dechlorination after four hours of reaction time. However, basic pH does not show comparatively good results. Therefore, optimization of dechlorination can be done at neutral pH, which is economically and chemically effective.

3.7 *Effect of temperature*

Studies based on the effect of temperature on dechlorination reaction were conducted at various temperatures, such as at 40°C and 50°C. The studies show that as reaction temperature increases the dechlorination also increases. At 50°C, 86.77% dechlorination is achieved while at 40°C only 37.54% dechlorination is achieved. The reaction was conducted only to a temperature of 50°C because mild conditions are referred for COCs.

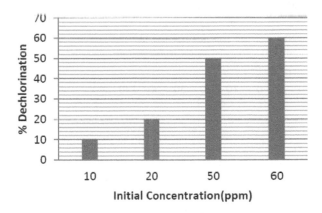

Figure 4.   Concentration screening.

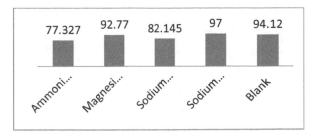

Figure 5.   Effect of salts.

569

3.8 Effect of reducing agent

Reducing agent in a reaction mixture is a provider of hydrogen. It enhances the hydrodechlorination reaction to a great extent. The dechlorination reaction is conducted without reducing agent and with reducing agent. Sodium borohydride is used as a reducing agent and Raney nickel is used as a catalyst at a loading of 3 gm/L. Reaction is conducted for two hours and samples are analyzed for the different time intervals. While using Raney nickel as a catalyst, usage of reducing agent can be avoided. From Figure 8 we can interpret that dechlorination is almost the same for with and without reducing agent.

3.9 Reducing agents

Reducing agent is a proton donor or hydrogen donor, so hydrodechlorination reaction will occur in a good manner. Different reducing agents like sodium borohydride, oxalic acid, 2-propanol, sodium nitrite, and activated charcoal were used for the comparative study. Reaction was kept for 24 hours. Sodium nitrite showed 99.5% dechlorination and sodium borohydride showed 96.45%. Considering the economic side, sodium borohydride was best.

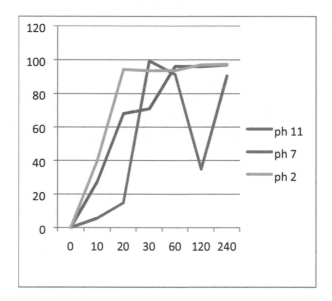

Figure 6.   Effect of pH.

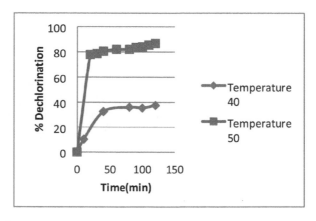

Figure 7.   Effect of temperature.

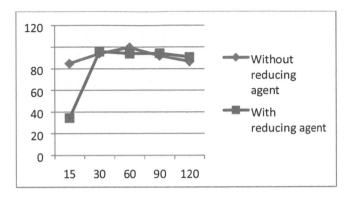

Figure 8.   Effect of reducing agent.

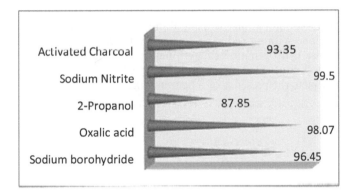

Figure 9.   Reducing agents.

Figures 10 and 11.   SEM image of Raney nickel before and after reaction.

Table 1.   Recycling of Raney nickel.

Fresh			Recycle 1			Recycle 2		
Initial conc	Final conc	%Dechlorination	Initial conc	Final conc	%Dechlorination	Initial conc	Final conc	%Dechlorination
25.2	6.68	73.55	54.383	18.8	65.43	77.46	29.598	61.79

3.10 Recycling of Raney nickel

The Raney nickel catalyst is checked for recyclability. The product formed is benzene. It can poison the catalyst and recyclability is reduced. The second recycle itself reduced the dechlorination by 8% and in a further recycle by more than 10%. This shows that the sites are blocked by chloride ions, and agglomeration may occur. Recycling is possible because contamination is less in the catalyst. This is evident from the SEM images of the Raney nickel catalyst before and after dechlorination, as shown in Figure 10 and Figure 11.

3.11 Kinetics of Raney nickel

Hydrodechlorination reaction is conducted for 24 hours and 90.78% dechlorination is achieved. Reaction follows first-order kinetics with a rate constant of 0.0136 min^{-1}. To enhance the reaction sodium borohydride is added as a reducing agent.

4 ANALYSIS OF PRODUCTS

Dechlorination of 1,4-dichlorobenzene using Raney nickel and palladium on activated carbon (0.5wt%) was done without reducing agents like hydrogen. It was observed that when palladium is used as a hydrogenation catalyst, nearly 100% dechlorination is attained. Formation of benzene was confirmed by UV spectroscopy. Figure 12 shows a sharp peak visible at 267 nm, which depicts the presence of a benzene ring.

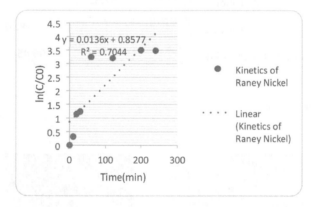

Figure 12. Kinetics of Raney nickel.

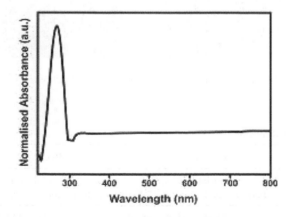

Figure 13. UV spectroscopy.

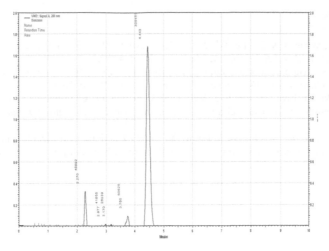

Figure 14.   HPLC peak for benzene.

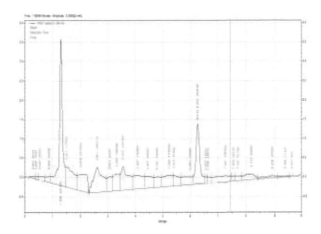

Figure 15.   HPLC analysis after reaction.

Benzene was separately injected into the HPLC column and the peak appeared at 4.5 min. Afterwards, the reaction sample was analyzed in HPLC. The HPLC analysis after reaction result is shown in Figure 15. A small peak appears at 4.457 min; it represents the presence of benzene.

5   CONCLUSIONS

Catalytic hydrodechlorination of 1,4-dichlorobenzene from wastewater was the aim of this work, in which a hydrogenation catalyst like Raney nickel, palladium on activated carbon, and bimetals were used. A hydrogenation catalyst such as Raney nickel and palladium showed fast dechlorination compared to bimetallic catalysts. For Raney nickel, the hydrogenation activity was comparable to that of palladium and the cost is also very much less. The Raney nickel catalyst was further studied in detail by changing parameters like catalyst loading, pH, initial concentration, the effect of salts, and the effect of temperature. Various reducing agents were also tried for catalytic hydrodechlorination reaction. UV spectroscopy analysis was done for various reaction mixture samples after the reaction, and the absorbance peak matched to that of benzene.

REFERENCES

Cheng, R.O.N.G., Wang, J. & Zhang, W. (2007). Reductive dechlorination of p-chlorophenol by nano-scale iron. *Biomedical and Environmental Sciences, 20*(5), 410–413.

Pearson, C.R. (1982). Halogenated aromatics. In O. Hutzinger (Ed.), *Anthropogenic Compounds: Volume 3, Part B (The Handbook of Environmental Chemistry)* (pp. 89116). New York: Springer-Verlag.

Sopoušek, J., Pinkas, J., Brož, P. Buršík, J., Vykoukal, V., Škoda, D, & Šimbera, J. (2014). Ag-Cu colloid synthesis: Bimetallic nanoparticle characterisation and thermal treatment. *Journal of Nanomaterials, 2014*, 1.

Xia, C., Liu, Y., Xu, J., Yu, J., Qin, W. & Liang, X. (2009). Catalytic hydrodechlorination reactivity of monochlorophenols in aqueous solutions over palladium/carbon catalyst. *Catalysis Communications, 10*(5), 456–458.

Emerging Trends in Engineering, Science and Technology for Society,
Energy and Environment – Vanchipura & Jiji (Eds)
© 2018 Taylor & Francis Group, London, ISBN 978-0-8153-5760-5

Controller tuning method for nonlinear conical tank system using MATLAB/Simulink

S. Krishnapriya & R. Anjana
Government Engineering College, Thrissur, India

ABSTRACT: In the process industries the control of liquid level is mandatory. But the control of nonlinear process is difficult. Many process industries use conical tanks because their nonlinear shape provides better drainage for solid mixtures, slurries and viscous liquids. Conical tanks are extensively used in the process industries, petrochemical industries, food process industries and wastewater treatment industries. So, control of conical tank level is a challenging task due to its nonlinearity and continually varying cross section. This is due to the relationship between the controlled variable level and manipulated variable flow rate, which have a square root relationship. The system identification of the nonlinear process is made and studied using mathematical modeling with Taylor series expansion, and the real time implementation is done in Simulink using MATLAB.

1 INTRODUCTION

Every industry faces the flow control and level control problem and have plentiful features such as nonlinearity, time-delay, and time invariants. These features cause difficulties in obtaining the exact model. Conical tanks are extensively used in the process industries, petrochemical industries, food process industries and wastewater treatment industries. The conical tank is generally nonlinear in nature due to its varying cross-sectional area. Although many innovative methodologies have been devised in the past 50 years to handle more complex control problems and to achieve better performances, the great majority of industrial processes are still controlled by means of simple Proportional-Integral-Derivative (PID) controllers. This seems to be because PID controllers, despite their simple structure, assure acceptable performances for a wide range of industrial plants, and their usage (the tuning of their parameters) is well known among industrial operators. Hence, PID controllers are simple and easy if the process is linear. Since the process considered is a nonlinear process, various other techniques are being implemented, which include Internal Model Control (IMC) and fuzzy logic control. The IMC design procedure is exactly the same as the open loop control design procedure. In addition, the IMC structure compensates for disturbances and model uncertainty. The filter parameters in IMC are considered and are used to tune the model of the given system to get the desired output. The use of fuzzy logic controllers seems to be particularly appropriate, since it allows us to make use of the operator's experience and therefore to add some sort of intelligence to the automatic control. Firstly, a PID controller has been designed by using the Ziegler-Nichols frequency response method, and its performance has been observed. The Ziegler-Nichols tuned controller parameters are fine-tuned to get satisfactory closed-loop performance. Secondly, it has been proposed for the same system to use IMC and fuzzy logic controllers. A performance comparison between the PID controller, IMC-based PID controller, and fuzzy logic controller is presented using MATLAB/Simulink. Simulation results are studied, and finally the conclusion is presented.

2 SYSTEM IDENTIFICATION

The system used is a conical tank, which is highly nonlinear due to the variation in area of cross section. The controlling variable is inflow of the tank. The controlled variable is the level of the conical tank. A level sensor is used to sense the level in the process tank and is fed into the signal conditioning unit. The required signal is used for further processing. The level process station is used to perform the experiments and to collect the data. One of the computers is used as a controller. It consists of the software which is used to control the level process station. The process consists of a process tank, reservoir tank, control valve, I to P (I/P) converter, level sensor and pneumatic signals from the compressor.

When the setup is switched on, the level sensor senses the actual level. Initially the signal is converted to a current signal in the range between 4 to 20 mA. This signal is then given to the computer through a data acquisition cord. Based on the controller parameters and the set-point value, the computer will take consequent control action and the signal is sent to the I/P converter. Then the signal is converted to a pressure signal using the I/P converter. The pressure signal acts on a control valve which controls the inlet flow of water into the tank. A capacitive type level sensor is used to sense the level from the process and converts it into an electrical signal. Then the electrical signal is fed to the I/V converter, which in turn produces a corresponding voltage signal to the computer. The actual water level storage tank sensed by the level transmitter is fed back to the level controller, and then compared with a desired level to produce the required control action that will position the level control as needed to maintain the desired level. Now the controller decides the control action. It is first given to the V/I converter and then to I/P converter. The final control element (pneumatic control valve) is now controlled by the resulting air pressure. This in turn controls the inflow to the conical tank and the level is maintained.

The system specifications (Rajesh et al., 2014) of the tank are as follows:

- Conical tank – Stainless steel body, height 70 cm, top diameter 35 cm, bottom diameter 2.5 cm
- Pump – Centrifugal 800 LPH
- Valve coefficient – K = 2
- Control Valve – Size ¼ pneumatic actuated type: Air to open, input 3–15 psi
- Rota meter range – 0–600 LPH.

2.1 *Mathematical modeling*

A mathematical model is a description of a system using mathematical concepts and language. Generally modeling of linear systems involves direct derivations whereas nonlinear systems require certain approximations to arrive at the solution (Vijayan & Avinashe, 2015). The Taylor's series method is simple and accurate over certain ranges near the steady state point. The dynamic behavior of the liquid level **h** in the conical storage tank system is shown in Figure 2.

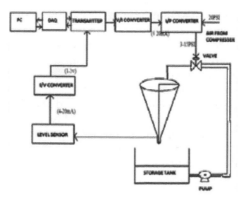

Figure 1. Level control of conical tank system.

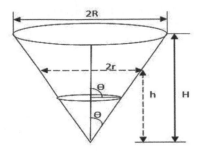

Figure 2. Tank cross sections.

$$\tan\theta = \frac{r}{h} \text{ and also } \tan\theta = \frac{R}{H} \qquad (1)$$

where R = Top radius of the tank
   H = Total height of the tank
   r = Radius at the liquid level (h)
   h = Level of the liquid (variable)

By Newton's law:

$$F_{in} - F_{out} = \frac{1}{3} * \frac{dh}{dt}\left\{ A + 2\pi \left(\frac{R}{H}\right)^2 * h^2 \right\} \qquad (2)$$

$$\text{Output flow rate, } F_{out} = K\sqrt{h} \qquad (3)$$

$$\frac{dh}{dt} = \alpha F_{in} h^{-2} - \beta h^{-3/2} \qquad (4)$$

where (Vijula et al., 2014).

$$\alpha = \frac{1}{\pi\left(\dfrac{R}{H}\right)^2} \qquad (5)$$

$$\beta = K\alpha \qquad (6)$$

$$\text{At steady state, } \frac{dh_s}{dt} = \alpha F_{is} h_s^{-2} - \beta h_s^{-3/2} = 0 \qquad (7)$$

$$\text{Let } y = (h - h_s) \text{ and } U = F - F_{is} \qquad (8)$$

$$\frac{dy}{dt} = -\left(\frac{1}{2}\right)\beta h_s^{-5/2} y + \alpha h_s^{-2} U \qquad (9)$$

$$\left(\frac{2}{\beta}\right) h_s^{-5/2}\left(\frac{dy}{dt}\right) + y = \left(\frac{2\alpha}{\beta}\right) h_s^{-1/2} U \qquad (10)$$

$$\tau\left(\frac{dy}{dt}\right) + y = CU \qquad (11)$$

Taking the Laplace transform:

$$\frac{Y(s)}{U(s)} = \frac{C}{\tau s + 1} \qquad (12)$$

577

Table 1.   Transfer function for the different height of the tank. [5].

Model	Height	Transfer function
1	10	$\dfrac{y(s)}{U(s)} = \dfrac{3.16}{62.08s+1}$
2	15	$\dfrac{y(s)}{U(s)} = \dfrac{3.87}{170.66s+1}$

Table 2.   Process transfer function.

Model	Height	Transfer function
1	10	$\dfrac{y(s)}{U(s)} = \dfrac{0.03097}{9.312s^2 + 28.15s + .451}$
2	15	$\dfrac{y(s)}{U(s)} = \dfrac{0.03793}{36.942s^2 + 12.16s + 0.451}$

where

$$\tau = \left(\frac{2}{\beta}\right) h_s^{-5/2} \tag{13}$$

$$C = \frac{2\alpha}{\beta} h_s^{-1/2} \tag{14}$$

The transfer function of the sensor is:

$$G_v(s) = \frac{0.0098}{0.15s + 0.451} \tag{15}$$

The modeling of transfer function for the conical tank with height 70 cm, top diameter 35 cm and height of liquid level at 15 cm is obtained as:

$$\frac{y(s)}{U(s)} = \frac{0.038}{36.942s^2 + 12.16s + 0.451} \tag{16}$$

3   CONTROLLER TUNING METHODS

3.1   *PID Controller*

The PID control is simple in principle, easy to tune, robust and a successful realistic application, which is still widely used in industrial process control. The PID controller is a fundamental part of the control loop in the process industry. Even though many advanced control modes are based on a PID control algorithm, the conventional PID control algorithm cannot achieve ideal control effect in any practical production process with nonlinear and time varying uncertainty.

$$u(t) = K_p e(t) + K_d \frac{de(t)}{dt} K_i \int_0^t e(\tau)d\tau \tag{17}$$

3.2 *IMC-BASED pid controller*

The Internal Model Control (IMC) is commonly used to provide a transparent mode for the design and tuning of various types of control. The IMC-based PID tuning method is a clear trade-off between closed-loop performance and robustness to model inaccuracies, and the tuning is achieved with a single tuning parameter. Also, it allows good set-point tracking but sulky disturbance response, especially for the process with a small time-delay/time-constant ratio. However, for many process control applications, disturbance rejection for the unstable processes is much more important than set-point tracking. Hence, controller design that emphasizes disturbance rejection rather than set-point tracking is an important design problem that must be taken into consideration. In process control applications, model-based control systems are often used to track set-points and reject low disturbances. The IMC design procedure is the same as the open loop control design procedure. Unlike open loop control, the IMC structure compensates for disturbances and model uncertainties. The IMC filter tuning parameter 'λ' is used to avoid the effect of model uncertainty.

The various steps in the IMC system design procedure (Vijayan & Avinashe, 2015). are:

Step 1: Factorization: It includes factorizing the transfer function into invertible and non-invertible parts.

Step 2: Form the idealized IMC controller. The ideal internal model controller is the inverse of the invertible portion of the process model.

Step 3: Adding a filter. Now a filter is added to make the controller stable.

IMC (Fathima *et al.*, 2015) is compared with conventional PID, giving:

$$K_c = \frac{\tau + \frac{\theta}{2}}{K\left(\lambda + \frac{\theta}{2}\right)} \qquad T_i = \frac{\theta}{2} + \tau \qquad T_d = \frac{\frac{\theta}{2} + \tau}{2\left(\frac{\theta}{2} + \tau\right)} \qquad (18)$$

IMC filter tuning parameter, λ = 1.

3.3 *Fuzzy logic controller*

The ideas of fuzzy set and fuzzy control were introduced by Zadeh (1996) to control systems that are structurally difficult to model. Mamdani was the first person to use fuzzy logic for control purposes (Lee 1990). Fuzzy systems can transform vague information and expert knowledge into computable numerical data. It incorporates heuristics, developed by experts and

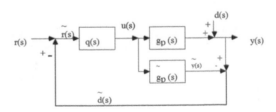

Figure 3. The IMC structure.

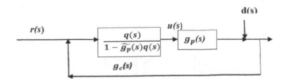

Figure 4. The equivalent feedback form to IMC.

Table 3.   Comparison of the PID and IMC-based PID parameter values.

	K_p	T_i	T_d
PID	57.16	10.78	22.26
IMC-based PID	50.21	1.16	0.36

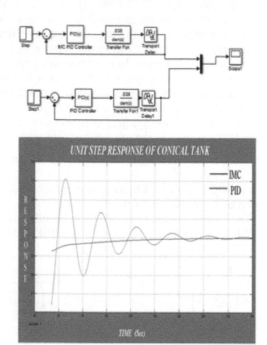

Figure 5.   Simulink diagram and response of an IMC-based PID and a PID.

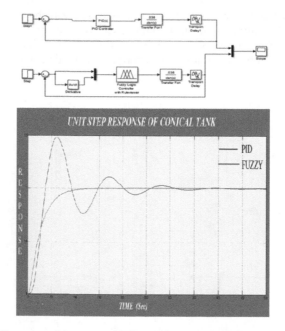

Figure 6.   Simulink diagram and response of PID and fuzzy controller.

Table 4. Performance analysis.

	Rise time (sec)		Setting time (sec)
IMC based PID	5.4766		65.9554
PID	9.8523		83.2704
	ISE	IAE	ITAE
PID	26.57	55.3	4000
IMC based PID	16.81	41.99	2763
Fuzzy	14.93	27.56	967.9

operators into automatic control. MATLAB/Simulink provides tools to create and edit fuzzy inference systems within the framework. It is also possible to integrate the fuzzy systems into simulations with Simulink. These are: fuzzy linguistic variable representing the level error (e); change of level error (de); and, the output control effort (u), respectively. A Fuzzy Logic System (FLS) can be defined as the nonlinear mapping of an input data set to a scalar output data.

Firstly, a crisp set of input data is gathered and converted to a fuzzy set using fuzzy linguistic variables, fuzzy linguistic terms and membership functions. This step is known as fuzzification. Afterwards, an inference is made based on a set of rules. Lastly, the resulting fuzzy output is mapped to a crisp output using the membership functions, in the defuzzification step. Mainly Fuzzy Logic Controllers (FLC) are implemented on nonlinear systems which yield better results. In designing the controller, the number of parameters (Ilyas et al., 2013) needs to be selected, and then the membership function and rules are selected based on heuristic knowledge.

4 RESULTS AND DISCUSSION

The response of IMC and fuzzy controllers are compared with convectional a PID controller.

The performance is studied by evaluating Rise Time, Settling Time, Integral Square Error, Integral Absolute Error and Integral Time Absolute Error.

5 CONCLUSION

The controlling of nonlinear process is a challenging task and nonlinearity of the conical tank is analyzed. Modeling of transfer function of the system is done by using system identification. Various controllers are simulated in MATLAB/Simulink. An open loop step test method is used to find the proportional gain, delay time and dead time. Here Taylor series approximation is used for the nonlinear approximation, because of its accuracy compared to other nonlinear approximation techniques. The simulation results show that the IMC-based PID controllers have minimum settling time and rise time in order to reach steady state value, as compared to the conventional controller. After analyzing simulated response of models, the fuzzy controller is found to be the more excellent controller than the IMC-based PID and PID controllers.

REFERENCES

Fathima, M.S., Banu, A.N. Nisha, A. & Ramachandran, S. (2015). Comparison of controllers for a flow process in a conical tank. *International Journal*, *1*, 145–148.
Ilyas, A., Jahan, S. & Ayyub, M. (2013). Tuning of conventional PID and fuzzy logic controller using different defuzzification techniques. *International Journal of Scientific & Technology Research*, *2*(1), 138–142.

Lee, C.C. (1990). Fuzzy logic in control systems: fuzzy logic controller. I. *IEEE Transactions on systems, man, and cybernetics*, *20*(2), 404–418.

Rajesh, T., Arun., S. & Siddharth, S.G. (2014). Design and implementation of IMC based PID controller for conical tank level control process. *International Journal of Innovative Research in Electrical, Electronics, Instrumentation and Control Engineering*, *2*(9), 2041–2045.

Sharma, A. & Venkatesan, N. (2013). Comparing PI controller performance for non-linear process model. *International Journal of Engineering Trends and Technology*, *4*(3), 242–245.

Vijayan, S. & Avinashe, K.K. (2015). *IMC based PID and fuzzy controller for nonlinear conical tank level process*. IJEECS ISSN 2348–117X Volume 4, Special Issue September 2015.

Vijula, D.A., Vivetha, K., Gandhimathi, K. & Praveena, T. (2014). Model based controller design for conical tank system. *International Journal of Computer Applications*, *85*(12), 8–11.

Zadeh, L.A. (1996). Fuzzy logic = computing with words. *IEEE transactions on fuzzy systems*, *4*(2), 103–111.

Emerging Trends in Engineering, Science and Technology for Society,
Energy and Environment – Vanchipura & Jiji (Eds)
© *2018 Taylor & Francis Group, London, ISBN 978-0-8153-5760-5*

Study on multi walled carbon nanotubes synthesis, modeling and applications

K. Krishnarchana & N. Manoj
Department of Chemical Engineering, Government Engineering College, Thrissur, Kerala, India

Sushree Sangita Dash
Propellant Fuel Complex, VSSC, ISRO, Thiruvananthapuram, India

ABSTRACT: This study investigates the synthesis of multi walled nanotubes by the decomposition of acetylene over misch metal catalyst in a chemical vapour deposition reactor. The synthesized CNTs were analyzed by different spectroscopic techniques. Also, a kinetic model for MWNT growth is proposed to investigate the dependence of the flow rate of precursor on CNT production rate. The model is validated by comparing its predictions with a set of experimental measurements and is simulated in MATLAB software. The experimental results were found to agree well with the theoretical predictions obtained from the model. In addition to the synthesis and modeling of CNTs, this work also embodies a technique for conductive coating using multi walled carbon nanotubes.

1 INTRODUCTION

Carbon nanotubes attracted a lot of researchers from academia to industry because of their remarkable mechanical and electronic properties, viz. high thermal and electrical conductivity, high aspect ratio, high tensile strength and low density, when compared with conventional materials. And also finds promising applications in many fields such as field and light emission, biomedical systems, nanoelectronic devices, nanoprobes, nanosensors, Conductive composites and energy storage. The CNTs are either single walled carbon nanotubes (SWNTs) or multi walled carbon nanotubes (MWNTs). The biggest challenge in developing potential applications for CNTs lies in the production of pure CNTs at affordable prices. The commonly used CNT synthesis techniques are Arc discharge, Laser ablation, Chemical vapor deposition (CVD), Electrolysis, Flame synthesis, etc. Among these methods, Chemical vapor deposition is considered as cheap, simple and most promising way for large-scale synthesis of CNTs [1–3].

The CNT deposition profiles inside a CVD reactor strongly depend on various parameters such as reaction temperatures, feed gas flow rates, carrier gas flow rates, catalyst type, etc. These reaction conditions can vary throughout the reactor, affecting the yield as well as rate of the reaction. Therefore it is very important to develop a model of the system as an important aid in studying the CNT growth process since it can envisage the yield without undertaking expensive experimental studies. Also helps in optimizing the process thereby enhancing CNT scale up process. This work is solely dedicated to the formation of MWNTs via the decomposition of acetylene over misch metal catalyst and its characterization. Also, aims to the modeling of CNT synthesis process and finally the development of conductive coating using MWNTs.

2 SYNTHESIS AND CHARACTERIZATION OF MWNTS

Multi walled nanotubes are synthesized by the decomposition of acetylene over an alloy of misch metal catalyst powder in a CVD reactor. The CVD reactor consist of a tubular furnace with a quartz tube (55 mm diameter and 1.5 m long for small scale production) and feed inlets for C_2H_2, CH_4, H_2 and Ar. The quartz tube has a heated zone, which acts as the reaction chamber. The furnace temperature was maintained at 750°C. The flow rates of the various gases were monitored and controlled using flow meters and solenoid valves. The process is carried out in presence of catalyst and the carrier gas. The quartz boat was cleaned with acetone, dried before the process and catalyst as taken in the quartz boat for the synthesis of CNTs. The synthesis of CNTs in CVD was done by using LABVIEW software.

A particular amount of the catalyst is taken in a quartz boat and is inserted in the center of a quartz tube. Then the furnace is heated to 500°C in argon atmosphere. The flow rate of argon is maintained at 160 sccm for 40 minutes. When the temperature reaches 500°C, hydrogen is introduced into the quartz tube for about 30 minutes, in order to remove any oxygen on the surface of the alloy hydride catalyst. The hydrogen flow is then stopped and the furnace is heated up to the desired reaction temperature of 750°C and this is followed by the introduction of acetylene at a flow rate of 70 sccm. The reaction is carried out for 90 minutes and thereafter the furnace is cooled to room temperature. Argon flow is maintained throughout the experiment. The weight of as grown CNT was taken. The procedure is repeated for various flow rate of acetylene.

The synthesized CNTs may contain impurities such as graphite nanoparticles, amorphous carbon, smaller fullerenes, and metal catalyst particles. These impurities were separated from the carbon nanotubes by following purification steps; Air Oxidation (Removes carbonaceous impurities such as amorphous carbon and helps expose the catalytic metal surface enclosed in the carbon nanotube for further purification techniques), Acid Treatment (Removes metal catalyst and small fullerene isomers from CNTs and, also helps in its functionalization), Washing (Based on size or particle separation that separates CNTs from metal nanoparticles, polyaromatic carbons and fullerenes) and Drying (Remove any volatile products, if present). After purification, the CNTs are characterized by Raman spectroscopy, HRTEM, SEM, TGA and FTIR.

3 KINETIC MODELING

Catalytic graphitisation of carbon was used to explain the synthesis of multi walled carbon nanotubes from acetylene using the catalytic chemical vapour deposition method.

Catalytic graphitisation involves carbon dissolution, adsorption and reaction to produce CNTs. Equation (1) presents the mechanisms of catalytic graphitisation of acetylene to CNTs using the CVD technique, where acetylene and possibly the cracked fractions under heat are dissociated into carbon atoms. The carbon atoms are deposited and adsorbed on the catalyst surface, which in turn reacted with each other to form C-C bonds to produce the carbon nanotubes.

$$2C_2H_2 \xrightarrow[750^0C]{Mm} 4C + 2H_2$$

$$nC \leftrightarrow C_n - Mm \rightarrow m[CNT_s] + Mm \tag{1}$$

The rate of catalytic graphitization,

$$r_{cc} = fn \text{ (Dissolution, Adsorption, chemical reaction)} \tag{2}$$

At temperature less than 750°C, no chemical reaction or no CNTs production was observed, but as the temperature reaches 750°C, production of CNTs was occur which indicating the complete decomposition of acetylene. Therefore, at this temperature, dissolution process was assumed to become non rate limiting, hence the rate of catalytic graphitisation formation and acetylene consumption becomes equal.

$$r_{cc} = -r_{c_{2}H_{2}} = fn \text{ (Adsorption, chemical reaction)} \tag{3}$$

The catalytic graphitization of acetylene to CNT is normally the case with solid catalyzed reaction that can be expressed by the rates of reaction catalyzed by solid surfaces per unit mass as,

$$-r_{c_{2}H_{2}} = \frac{1}{W_{Mm}} \frac{dN_C}{dt} = k\theta_c n \tag{4}$$

Here, Langmuir Hinshelwood mechanism is adopted to obtain the reaction rate and equilibrium constants.

$$-r_{c_{2}H_{2}} = \frac{kKC^n}{1+KC^n} \tag{5}$$

$$\frac{1}{r_{c_{2}H_{2}}} = \frac{1}{kKC^n} + \frac{1}{k} \tag{6}$$

Equation (6) is used to determine the kinetics parameters used in computing the model. The model which represents the production CNT by CVD is obtained as,

$$r = \frac{kKC_A^n}{1+K} \frac{\exp(1-\theta)}{k_1 t} C \tag{7}$$

4   DEVELOPMENT OF CONDUCTIVE COATING

The high electrical conductivity and low density of CNTs makes them a suitable material for coating applications. Here, the coating is to be developed particularly for cryogenic tank exteriors. Currently a PU base gray conductive coating is used for the purpose which is of higher density. CNTs, because of its innate low density are expected to perform much better than this conventional strategy.

The experimental procedure for the development of conductive coating involves two steps; Substrate preparation and CNT dispersion in solvents. In Substrate preparation, PU foam of $10 \times 10 \times 2$ cm was taken. Two coatings were applied over the foam; first one being the PU coating as VBC to prevent any moisture permeation into the foam from the outside atmosphere and second one, the CNT conductive coating. Electrically conductive coatings were prepared by dispersing CNT in solvent. Before applying the second coating, the CNTs were totally dispersed in either acetone or toluene by sonication for 3 hours. The dispersed solution is then applied on the substrate, weighed and dried for 15 minutes to obtain a uniformly coated conductive layer on the PU substrate. The conductivity of the coating was measured by means of surface resistivity meter. The procedure was repeated until the expected conductivity (Conductivity of MWNT, 10^{-3} S/m) was obtained.

5   RESULTS AND DISCUSSION

5.1   *Synthesis and characterization*

The MWNT yielded by the chemical vapor deposition of acetylene gas over misch metal catalyst was 87.06%. Fig. 1 shows the SEM image of the surface morphology of MWNTs. The SEM image discloses that the CNTs are less straight and length in several micrometers. The white region indicates the aggregated CNTs and the regions are magnified in order to study the morphology.

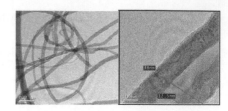

Figure 2.   HRTEM image of MWNT.

Figure 1.   SEM image of MWNTs.

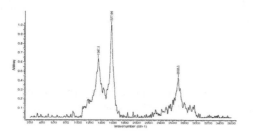

Figure 3.   Raman spectra of MWNT.

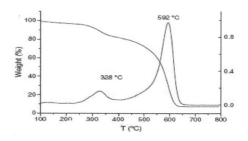

Figure 4.   TGA of MWNT.

Figure 2 is the HRTEM image of MWNTs which shows that MWNTs have a hollow structure. And image reveals the diameter of the CNTs. It shows that MWNT has an inner diameter of 12.5 nm and outer diameter of 38 nm.

Fig. 3 shows the Raman spectra of MWNTs, from which Multiwall structure of CNT was identified. The peaks present at 1577.96 cm^{-1}, 1347.1 cm^{-1} and 2693.3 cm^{-1} represents the G, D and G' modes in Raman spectra. The presence of CNT is identified by a G line at 1577.96 cm^{-1}. The RBM mode was not there. Hence, the multi walled structure of CNT was confirmed. The D band corresponds to 1347 cm^{-1} is related defects of graphitic sheets or carbonaceous particle at the surface of the tube.

Figure 4 is a plot for the weight loss in % vs. the oxidation temperature, measured by heating up the MWNTs in a TGA. The weight loss curve between 100 and 800°C was plotted by adjusting about 100% for the weight loss at 800°C, in which the actual weight was presumably the weight of catalyst (usually 10% of total weight). Weight losses below 200°C and From the TGA plot, it can be seen that there was no significant weight loss up to 300°C. After this temperature, a slight decrease in the weight can see due to the burning of amorphous carbon until the temperature reaches 500°C. While reaching 500°C, a sharp decrease in the weight can be seen due to the burning of MWNTs. On reaching 650°C, all the MWNTs have burned. There was no residual weight percentage at 650°C which implies that the MWNTs produced at 750°C were 100% pure.

Fig. 4 shows the FTIR image of MWNTs synthesized over misch metal catalyst at 750°C. The wave number 3410 cm^{-1}, 1726 cm^{-1} and 1594 cm^{-1} represents the O-H stretches of the terminal carboxyl group, the carboxyl C = O groups and the C = C stretching respectively. From this it was clear that by the acid treatment of MWNTs improves its interfacial interaction. Hence it can be used to make matrix structures.

5.2   Simulation results

The effects of varying acetylene concentration on CNTs production rate were investigated at the range of 50 ml/min – 90 ml/min at 750°C, while 160 ml/min Ar flow was used as the car-

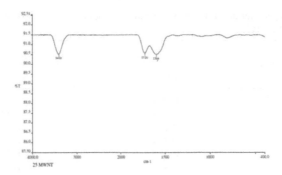

Figure 5.   FTIR of MWNT.

rier gas flow and the results obtained are presented in Figure 6(a). A kinetic model equation was developed to predict the production rate of MWNTs as

$$r = \frac{0.775 \times 2.206 \times 10^{-15}}{1 + 2.206 \times 10^{-15}} \frac{\exp(1-1)}{5400 \times 2.86 \times 10^{-15}} C_{C2H2}{}^4 \tag{8}$$

The modeled equation predicts the rate of production of CNTs at various acetylene concentrations. The results obtained were shown in Table 1.

It is evident that the plots are comparable, even though they are not exact similar. This may due to the assumption that the carbon atoms are occupied on the entire surface of the catalyst.

Hence, the value of fraction of surface area occupied by the carbon atom (θ) is taken as 1. It is expected that a more accurate value of θ may yield better results.

5.3   Conductive coating

Assessment of MWNTs as conductive coating material investigated through several experiments. The best results obtained were shown in Table 2.

Trials 1,2,3 and 4 showed promising results, which make CNT based conductive coating a suitable candidate for cryogenic tank insulation. Especially, the trial 15 evolved a light weight system which improves the ease of application with enhanced performance. The optimization of the process was also undertaken which enhances the repeatability and reliability of the technique.

6   CONCLUSION

Chemical vapor deposition based production of carbon nanotubes yields good quality, uniformly and well aligned nanostructures. In this work multi walled nanotubes (87.06%) are synthesized by chemical vapor deposition of acetylene over misch metal catalyst at 750°C. The CNTs were characterized by SEM, HRTEM, Raman spectroscopy, TGA and FTIR. From Raman spectra and HRTEM, the presence of CNTs was identified and confirmed. FTIR analysis revealed that the CNTs are functionalized during acid treatment. SEM image gives an idea about the morphology and structure of MWNTs. The TGA results show that the CNT synthesized at 7500C was almost 100% pure.

In the second section of the work, a kinetic model was developed, for the MWNT synthesis by CVD, to study the effect of flow rate of acetylene on the CNT production rate. The model equation was based on the experimental data. The theoretical prediction from the model equation and experimental data are comparable. Here the maximum yield obtained is 0.374 mg/sec at acetylene concentration of 4287.32 ppm.

Synthesized MWNTs are used to develop a CNT based conductive coating (conductivity: 10–4 S/m) which is a suitable coating for cryogenic tank insulation. The obtained coating

Table 1. Experimental and computed rate of production of CNTs at varying concentration of acetylene.

Concentration of acetylene, C_A (ppm)	CNT production rate, r (mg/sec)	
	Experimental value	Computed value
2836.48	0.098	0.071
3248.34	0.114	0.123
3439.72	0.135	0.155
3624.75	0.214	0.191
3968.74	0.351	0.274
4287.32	0.37	0.374

Table 2: Results of trials conducted on conductive coating.

No. of trials	Materials used	Sonication	Results
1	CNT: 1.5 g Acetone: 50 ml	2 hr	Conductive, 10^{-5} S/m (spray coating) But, CNTs are coming out
2	CNT: 0.03 g Acetone: 5 ml PU: 3.21 g	3 hrs	Conductive, 10^{-4} or 10^{-5} S/m
3	CNT: 0.03 g Acetone: 10 ml PU: 5.58 g	3 hrs	Conductive, 10^{-4} S/m
4	CNT: 0.05 g Acetone: 20 ml PU: 0.14 g	3 ½ hrs	Conductive, 10^{-4} S/m

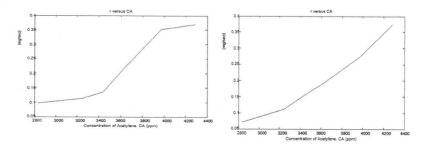

Figure 6. (a) Effect of flow rate of acetylene on CNTs production rate (b) Computed results of CNTs production rate.

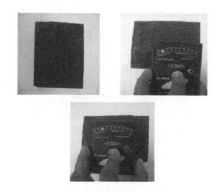

Figure 7. Conductive coating (Conductivity: 10–4 or 10–5 S/m, Area: 0.01 m²).

is comparatively light in weight which improves the easy of application with enhanced performance. The optimization of the process was also undertaken which enhances the repeatability and reliability of the technique.

ACKNOWLEDGEMENT

The satisfaction and euphoria on the successful completion of any task would be incomplete without mentioning the people who made it possible whose constant guidance and encouragement crowned out effort with success. I express my heartfelt thanks to Sushreesangita dash (External guide), Manoj N (Internal guide), V.O. Rejini (HOD), S.K. Manu (Dy. Manager, PFC, VSSC), Sriram P. (Engineer SC), all other staffs of VSSC, all the faculty members of the department of Chemical Engineering, my friends and my family.

NOTATIONS

K_a: Rate constants for adsorption
K_d: Rate constant for desorption
K: Rate constant for chemical reaction.
WMm: Weight of catalyst used
N_c: No. of moles of carbon
N: Order of reaction
Θ: Fraction of Mm surface occupied by carbon atoms
C_A: Concentration of reactant
K_1:   Rate constant, proportional to the diffusion coefficient of carbon
T: Time

REFERENCES

[1] Andrea Szabó, CaterinaPerri, Anita Csató, Girolamo Giordano, DaniloVuono and János B. Nagy, "Synthesis Methods of Carbon Nanotubes and Related Materials", Materials 2010, 3, 3092–3140.
[2] KalpanaAwasthi, Anchal Srivastava and O.N. Srivastava, "Synthesis of carbon nanotubes ", Physics Department, Banaras Hindu University, Varanasi-221 005, India.
[3] Adedeji E. AgboolaRalph W. Pike T.A. Hertwig Helen H. Lou, "Conceptual design of carbon nanotube processes", Clean Techn Environ Policy (2007) 9:289–311.
[4] KochandraRaji and Choondal B. Sobhan, "Simulation and modeling of carbon nanotube synthesis: current trends and investigations", Nanotechnolgy Rev 2013; 2(1): 73–105.
[5] Sunny EsayegbemuIyuke, SakaAmbaliAbdulkareem, Samuel Ayo Afolabi, and Christo H. vZPiennar, "Catalytic Production of Carbon Nanotubes in a Swirled Fluid Chemical Vapour Deposition Reactor", International Journal of Chemical Reactor Engineering, Volume 5, 2007, Note S5.
[6] K. Raji, Shijo Thomas, C.B. Sobhan, "A chemical kinetic model for chemical vapor deposition of carbon nanotubes", Applied Surface Science 257 (2011) 10562–10570.
[7] O. Levenspiel, "Chemical Reaction Engineering" third ed., Wiley India Pvt. Ltd., 2006.
[8] M.N. Masri, Z. M. Yunus, A.R.M Warikhand A.A Mohamad (2010), "Electrical conductivity and corrosion protection properties of conductive paint coatings", anticorrosion methods and materials, vol 57, issue 4, pp. 204–208 (2010).

Emerging Trends in Engineering, Science and Technology for Society,
Energy and Environment – Vanchipura & Jiji (Eds)
© 2018 Taylor & Francis Group, London, ISBN 978-0-8153-5760-5

Removal of nickel from aqueous solution using *sargassum tenerrimum* powder (brown algae) by biosorption: Equilibrium, kinetics and thermodynamic studies

M. Tukaram Bai, P. Venkateswarlu & Y.V. Anudeep
Department of Chemical Engineering, Andhra University, Visakhapatnam, India

ABSTRACT: In the present study the biosorption of nickel onto *sargassum tenerrimum* powder (brown algae) from an aqueous solution was studied. The equilibrium study was carried out for parameters: agitation time (1–210 min) (t), biosorbent size (45–300 µm) (d_p), biosorbent dosage (2–24 g/L) (w), pH of aqueous solution (1–8), initial concentration of nickel in aqueous solution (5–150 mg/L) (C_0), and temperature (283–323°K) of aqueous solution on biosorption of the metal (nickel) were studied. In the present investigation the equilibrium data was well explained by Langmuir, Temkin, and Redlich and Peterson with a correlation coefficient of 0.99, and followed by a Freundlich isotherm. The kinetic studies reveal that the biosorption system obeyed the pseudo second order kinetic model by considering the correlation coefficient value as 0.99. From the values of ΔS, ΔH and ΔG it is observed that the biosorption of nickel onto *sargassum tenerrimum* powder was irreversible, endothermic and spontaneous.

Keywords: Nickel, biosorption, *sargassum tenerrimum*, algae, isotherms, kinetics, thermodynamics

1 INTRODUCTION

All living organisms require heavy metals in low concentrations, but high concentrations of heavy metals are toxic and can cause cancer (Koedrith et al., 2013). Nowadays the environment is threatened by an increase in heavy metals. Therefore, in recent years the removal of heavy metals has become an important issue (Nourbakhsh et al., 2002). Methods for removing metal ions from aqueous solution mainly consist of physical, chemical and biological technologies. Conventional methods for the removal of heavy metal ions from wastewater, such as chemical precipitation, flocculation, membrane filtration, ion exchange, and electrodialysis electrolysis, are often costly or ineffective for the treatment of low concentrations of pollutants (Wang & Chen, 2009). Biological uptake is a promising approach that has been studied in the past decade. This process is a good candidate for replacing old methods (Pinto et al., 2011). High efficiency, removal of all metals even at low concentrations, being economical, and energy independence are the main advantages of biological uptake which present this process as being a viable new technology (Bai & Abraham, 2002). Biosorption is used to describe the passive non-metabolically mediated process of metal binding to living or dead biomass (Rangsayatorn et al., 2002). Water pollution by heavy metals is globally recognized as being an increasing environmental problem since the start of the Industrial Revolution in the 18th century (Dàvila-Guzmàn et al., 2011). Heavy metals may come from different sources such as electroplating, textile, smelting, mining, glass and ceramic industries as well as storage batteries, metal finishing, petroleum, fertilizer, pulp and paper industries. Nickel is one of the industrial pollutants, possibly entering into the ecosystem through soil, air, and water. Nickel is a toxic heavy metal found in the environment, as a result of various natural and industrial activities. The higher concentration of the nickel causes poisoning effects like headache, dizziness, nausea, tightness of the chest, dry cough and extreme weakness

(Krishna & Swamy, 2011). So, it is very much essential to remove nickel from wastewater. The objective of the present work is to explore the potential of a biosorption technique for the removal of nickel from aqueous solutions using cheap and abundantly available materials like *sargassum tenerrimum* powder. The experiments are undergone in batch process for the equilibrium studies, kinetics and thermodynamics of biosorption.

2 MATERIALS AND METHODS

2.1 *Preparation of the biosorbent*

Fresh samples of *sargassum tenerrimum,* a species of the brown algae, were collected from Rushikonda beach in Visakhapatnam, Andhra Pradesh. The brown algae were washed thoroughly with distilled water ten times to remove sand and dirt completely. The cleaned brown alga was dried in sunlight, until the moisture was completely removed. The dried biomass was ground to powder by using a mechanical grinder. The ground powder was then sieved and separated into various sizes (45, 53, 75, 150 and 300 μm) by using British Standard Sieves. The powders of these size fractions were then stored in separate plastic airtight containers at room temperature and used as biosorbent for subsequent analysis.

2.2 *Preparation of aqueous nickel stock solution*

The quantity of 4.1582 g of 97% pure of $NiCl_2.6H_2O$ was dissolved in 1 L of distilled water to prepare 1,000 mg/L of nickel solution. 100 mg/L nickel solution was prepared by diluting 100 mL of 1,000 mg/L nickel stock solution with distilled water in a 1,000 mL volumetric flask up to the mark. The pH of the aqueous solution was adjusted to the desired value by the addition of 0.1 N HCl or 0.1 N NaOH solutions.

2.3 *Procedure*

The procedures adopted to evaluate the effects of various parameters viz. agitation time (t), biosorbent size (d_p), biosorbent dosage (w), pH of aqueous solution, initial concentration of nickel in aqueous solution (C_0), and temperature of aqueous solution, on the biosorption of metal (nickel) are explained below.

50 mL of aqueous solution containing 20 mg/L of initial concentration of nickel was taken in a 250 mL conical flask. 10 g/L of 45 μm size biosorbent was added to the flask. The conical flask was then kept on an orbital shaker at room temperature (30°C) and was shaken for one min. Similarly, 21 more samples were prepared in conical flasks by adding 10 g/L of biosorbent and agitating for different time periods from 2 to 210 min. For the resulting agitation equilibrium time of 120 min the further experiments were repeated for varying biosorbent sizes viz. 75, 150 and 300 μm. The resulting optimum biosorbent size was 45 μm. The above procedure was repeated for different adsorbent dosages of 4, 6, 8, 10, 12, 14, 16, 18, 20, 22 and 24 g/L at equilibrium agitation time (120 min) and biosorbent size is optimum (45 μm). The equilibrium biosorbent dosage was found to be 18 g/L. To determine the effect of pH on nickel biosorption, 50 mL of aqueous solution was taken in each of 12 conical flasks. The pH values of aqueous solutions were adjusted to 1, 2, 3, 3.5, 4, 4.5, 5, 5.5, 6, 6.5, 7 and 8 in separate 250 mL conical flasks. 18 g/L of 45 μm size biosorbent was added to each of the conical flasks. The influence of initial concentration on biosorption of nickel was determined as follows: 50 mL of aqueous solutions, each of different nickel concentrations of 5, 10, 20, 30, 40, 50, 60, 75, 100, 125 and 150 mg/L were taken in 11 250 mL conical flasks. 18 g/L of 45 μm size biosorbent was added to each of the conical flasks. The flasks were agitated on an orbital shaker for equilibrium agitation time at room temperature. The samples were allowed to settle and then filtered separately. The samples thus obtained were analyzed in atomic absorption spectroscopy (AAS) for the final concentrations of nickel in aqueous solutions.

3 RESULTS AND DISCUSSIONS

3.1 *Effect of agitation time*

The effect of agitation time on percentage biosorption is shown in Figure 1. For 45 μm biosorbent size, the % biosorption is significant in the first 1 min up to 81.25%. The % biosorption is increased gradually from 81.25% to 93% between the time intervals of 1 to 120 min. After the agitation time of 120 min, % biosorption remains constant, indicating the attainment of equilibrium conditions. The same equilibrium agitation time of 120 min was reported for the biosorption of nickel by brown alga *fucus vesiculosus* (Mata et al., 2008) and immobilized algal cells (Al-Rub et al., 2004).

3.2 *Influence of biosorbent size*

The experiments are carried out for various biosorbent sizes – 45, 75, 150 and 300 μm—keeping other parameters constant. Figure 2 shows the graph drawn between % biosorption of nickel against biosorbent size. The biosorption of nickel decreases from 93% (1.86 mg/g) to

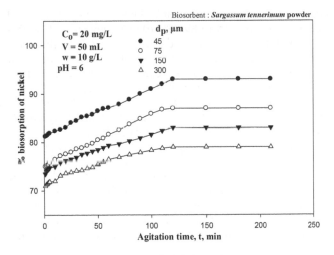

Figure 1. Effect of contact time % on removal of nickel metal.

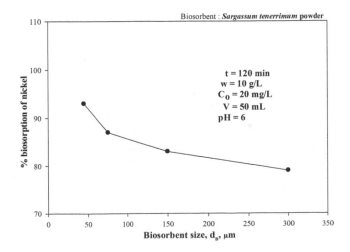

Figure 2. Effect of biosorbent size on % removal of nickel.

79% (1.58 mg/g) as the biosorbent size increases from 45 to 300 μm. With a decrease in bio-sorbent particle size, the surface area of the biosorbent increases and the number of active sites available on the biosorbent are better exposed to the biosorbate.

3.3 Effect of biosorbent dosage

To study the effect of biosorbent dosage on % biosorption of nickel, the biosorbent dosage is varied from 2 to 24 g/L (2, 4, 6, 8, 10, 12, 14, 16, 18, 20, 22 and 24 g/L). A plot is drawn between % biosorption of nickel and biosorbent dosage in Figure 3. The biosorption of nickel is increased from 90.5% (9.05 mg/g) to 96.35% (1.07 mg/g) with an increase in biosorbent dosage from 2 to 18 g/L. Such behavior is obvious because with an increase in biosorbent dosage, the number of active sites available for nickel biosorption would be more. The change in percentage biosorption of nickel is minimal, from 96.35% (1.07 mg/g) to 96.58% (0.804 mg/g) when 'w' is increased from 18 to 24 g/L. Hence, optimum biosorbent dosage of 18 g/L is considered to study all other parameters.

3.4 Effect of pH

The pH of aqueous solution is drawn against % biosorption of nickel in Figure 4. The % biosorption of nickel is increased from 89.65% (0.996 mg/g) to 97% (1.077 mg/g) as pH is increased from 1 to 4.5 and decreased beyond pH value of 4.5. In the case of lower pH values, the occupation of the negative sites of the biosorbent by H^+ ions leads to a reduction of vacancies for nickel ion and consequently causes a decrease in nickel ion biosorption. As the pH is raised, the ability of the nickel ions to compete with H^+ ions also increases. Although the sorption of nickel ions is raised by a growing pH, a further increment of pH causes a decline in biosorption due to the precipitation of nickel hydroxides. The predominant adsorbing forms of nickel are nickel and $NiOH^+$, which occur in the pH range of 4–6. The value of pH at 4.5 is considered as being optimum for the study of other parameters. The functional groups like aliphatic C-H, SO_3 stretching, C-O and C = O stretching, aromatic—CH stretching and amine groups of the biosorbent were responsible for nickel biosorption. Similar results were reported for biosorption of nickel by *waste pomace of olive oil factory* (Nuhoglu & Malkoc, 2009). The optimum pH ranging from 4 to 5 was reported by Aksu et al. (2006), Özer et al. (2008) and Congeevaram et al. (2007) for the biosorption of nickel by using various biosorbents like dried *chlorella vulgaris*, *enteromorpha prolifera* and *aspergillus* species respectively.

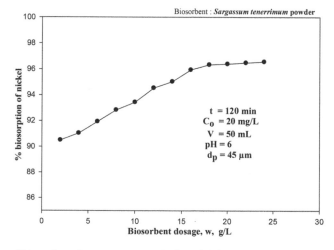

Figure 3. Effect of biosorbent dosage on removal of % nickel.

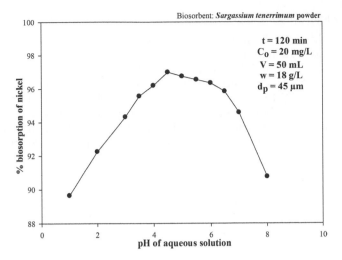

Figure 4. Effect of pH on removal of % nickel.

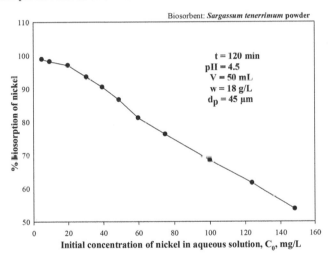

Figure 5. Effect of pH on removal of % nickel.

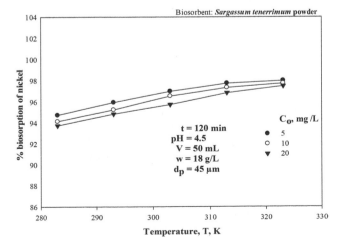

Figure 6. Effect of temperature on % removal of nickel.

3.5 Effect of initial concentration of nickel

Figure 5 shows the graph between initial concentration of aqueous solution (C_0) and % biosorption of nickel at w = 18 g/L and pH = 4.5. The effect of initial nickel ion concentration is investigated in the range of 5–150 mg/L. The percentage biosorption of nickel is decreased from 98.85% to 53.75% with an increase in C_0 from 5 to 150 mg/L while the uptake capacity is increased from 0.274 to 4.419 mg/g respectively. Such behavior can be attributed the increase in the amount of biosorbate to the unchanging number of available active sites on the biosorbent. Similar results were reported by Pahlavanzadeh et al. (2010) and Gupta et al. (2010), using *brown alga* and treated alga (*oedogonium hatei*) for the biosorption of nickel.

3.6 Effect of temperature

The effect of temperature on % biosorption of nickel is shown in Figure 6. The study was further extended at initial concentrations of 5, 10 and 20 mg/L. For the initial concentration (C_0) of 5 mg/L, the % biosorption is increased from 96.25% to 99% as the temperature increases from 283°K to 333°K.

4 ADSORPTION ISOTHERMS

4.1 Freundlich isotherm

Freundlich (1907) presented an empirical adsorption isotherm equation that can be applied in cases of low and intermediate concentration ranges.

Taking logarithms on both sides, we get:

$$\log qe = \log Kf + n \log Ce \tag{1}$$

A Freundlich isotherm is drawn between log q_e and log C_e, in Figure 7 for the present data. The equation obtained is:

$$\log qe = 0.381 \log Ce + 0159 \tag{2}$$

with a correlation coefficient of 0.97. The Freundlich constant (K_f) is found to be 1.054 and the n-value of 0.381 lies between 0 and 1, indicating the applicability of the Freundlich isotherm to the experimental data.

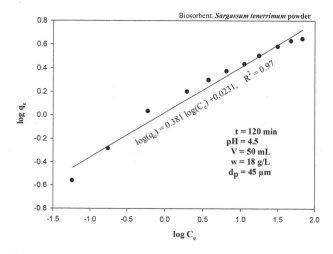

Figure 7. Freundlich isotherm for biosorption of nickel.

4.2 *Langmuir isotherm*

Irving Langmuir developed an isotherm named as the Langmuir isotherm (Langmuir, 1918). It is the most widely used simple two-parameter equation. The Langmuir relationship is hyperbolic and is:

$$\left(\frac{qe}{qm}\right) = bCe/(1+bCe) \tag{3}$$

where
    C_e is the equilibrium concentration (mg/L),
    Q_e is the amount of nickel adsorbed (mg/g).
    Equation 3 is rearranged as:

$$\left(\frac{Ce}{qe}\right) = \frac{1}{bqm} + (1/qm)Ce \tag{4}$$

The Langmuir isotherm (Figure 8) for the present data is represented as:

$$\left(\frac{Ce}{qe}\right) = 0219Ce + 0.920 \tag{5}$$

with a good linearity (correlation coefficient, $R^2 = 0.98$) indicating a strong binding of nickel ions to the surface of *sargassum tenerrimum* powder. The q_m and b values are 4.566 mg/g and 0.98 respectively. The value of separation factor is 0.508 and it indicates favorable biosorption ($0 < R_L < 1$) of nickel onto *sargassum tenerrimum* powder.

4.3 *Temkin isotherm*

The Temkin and Pyzhev isotherm (King et al., 2007) equation describes the behavior of many adsorption systems on the heterogeneous surface and it is based on the following Equation:

$$q_e = RT \ln(A_T C_e)/b_T \tag{6}$$

The linear form of the Temkin isotherm can be expressed as:

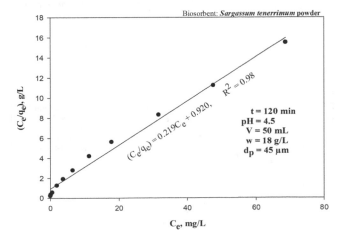

Figure 8.   Langmuir isotherm for biosorption of nickel.

597

$$q_e = (RT/\ b_T)\ \ln(A_T) + (RT/b_T)\ \ln(C_e) \tag{7}$$

where
  R = Universal gas constant (8.314 J/mol.K)
  T = Temperature of dye solution, K
  A_T, b_T = Temkin isotherm constants

$$A_T = \exp\ [b(0) \times b(1)\ /\ RT] \tag{8}$$

$b(1) = RT/\ b_T$ is the slope
$b(0) = (\ RT/b_T\ )\ \ln\ (A_T)$ is the intercept and $b = RT/b$ (1). The Temkin isotherm is applied to the present data and the linear plot is shown in Figure 9. The Equation obtained is:

$$qe = 0.606 \ln Ce + 1.154 \tag{9}$$

The resulting b_T and A_T values are 4,157, A and 2.16 respectively.

4.4 Redlich-Peterson isotherm

Redlich and Peterson (1959) proposed a three-parameter isotherm to incorporate features of both the Langmuir and Freundlich equations. It can be described as follows:

$$q_e = \frac{AC_e}{1 + BC_e{}^g} \tag{10}$$

where A (L/g) and B (L/ mg) are the Redlich-Peterson isotherm constants and 'g' is the Redlich-Peterson isotherm exponent, which lies between 0 and 1.
  The linear form of the equation is:

$$\ln\left(A\frac{C_e}{q_e} - 1 \right) = g\ln(C_e) + \ln B \tag{11}$$

Although a linear analysis is not possible for a three-parameter isotherm, the three iso-therm constants – A, B and g – can be evaluated from the pseudo linear plot using a trial and error optimization method. A general trial and error procedure is applied to determine the

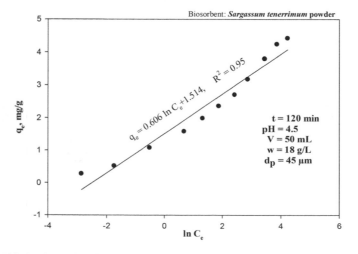

Figure 9. Temkin isotherm for biosorption of nickel.

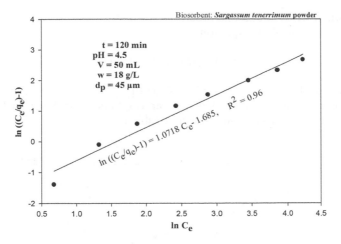

Figure 10.   Redlich-Peterson isotherm for biosorption of nickel.

coefficient of determination (R^2) for a series of values of 'A' for the linear regression of $\ln (C_e)$ on $\ln [A(C_e/q_e)-1]$ and to obtain the best value of 'A' with maximum 'R^2'. Figure 10 shows the Redlich-Peterson plot drawn between $\ln [A(C_e/q_e)-1]$ and $\ln C_e$. For the present experimental data, the equation obtained is:

$$\ln\left(\left(\frac{Ce}{qe}\right)-1\right)=1.0718\ln Ce-1.685 \qquad (12)$$

The Redlich-Peterson isotherm constants, B (L/mg) and the Redlich-Peterson isotherm exponent (g) are 0.185 L/mg and 1.0718 respectively. The correlation coefficient of 0.96 suggests that the Redlich-Peterson isotherm model is suited to describe the biosorption of nickel.

The biosorption data is well represented by the Langmuir ($R^2 = 0.98$), Freundlich ($R^2 = 0.97$), Redlich-Peterson ($R^2 = 0.96$) and Temkin ($R^2 = 0.95$) isotherms.

5   BIOSORPTION KINETICS

5.1   *First order and second order kinetics*

The pseudo first order rate equation of Lagergren is:

$$(dq_t/dt) = K_1 (q_e - q_t) \qquad (13)$$

The above equation can be presented as:

$$\int (dq_t/(q_e - q_t)) = \int K_1 \, dt \qquad (14)$$

Applying the initial condition $q_t = 0$ at $t = 0$, we get:

$$\log (q_e - q_t) = \log q_e - (K_1/2.303) \, t \qquad (14)$$

The plotting of $\log (q_e - q_t)$ versus 't' gives a straight line for first order kinetics, facilitating the computation of adsorption rate constant (K_1). If the experimental results do not follow the above Equation, in such cases the pseudo second order kinetic equation:

$$(dq_t/dt) = K_2 (q_e - q_t)^2 \qquad (15)$$

599

is applicable, where 'K_2' is the second order rate constant.
The other form of the above equation is:

$$(t/q_t) = (1/ K_2q_e^2) + (1/q_e)\, t. \tag{16}$$

In the present study, the Lagergren plot of $\log (q_e - q_t)$ vs. 't' is shown in Figure 11. The pseudo second order rate equation plot between (t/q_t) and 't' is drawn in Figure 12. The resulting equations and constants are shown in Table 1.

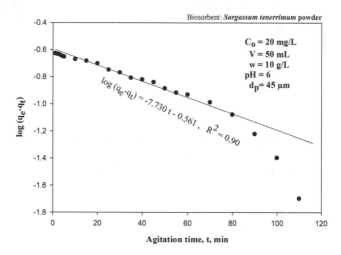

Figure 11.   First order kinetics for biosorption of nickel.

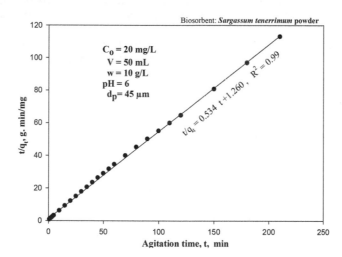

Figure 12.   Second order kinetics for biosorption of nickel.

Table 1.   Kinetic equations and rate constants.

Order	Kinetic equation	Rate constants	R^2
Pseudo first order	$\log(q_e - q_t) = -7.730t - 0.561$	$K_1 = 17.802$ min^{-1}	0.90
Pseudo second order	$t/q_t = 0.534\, t + 1.260$	$K_2 = 0.226$ g/mg-min	0.99

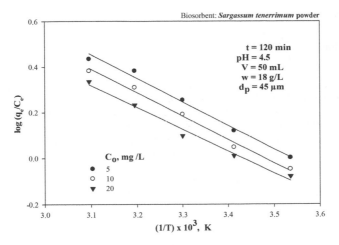

Figure 13.   Van't Hoff plot for biosorption of nickel.

Table 2.   Thermodynamic parameters of biosorption for nickel.

C_0, mg/L	ΔS, J/(mol–°K)	ΔH, J/mol	$-(\Delta G)$, kJ/mol				
			283 K	293 K	303 K	313 K	323 K
5	70.23	19.85	19.85	20.55	21.26	21.96	22.66
10	68.58	19.70	19.38	20.07	20.76	21.44	22.13
20	62.99	18.34	17.80	18.43	19.06	19.69	20.32

As the correlation coefficient for the pseudo second order kinetics is 0.999, it describes the mechanism of nickel–*sargassum tenerrimum* powder interactions better than first order kinetics ($R^2 = 0.93$).

5.2   *Thermodynamics of biosorption*

The Van't Hoff's plot for the biosorption data are shown in Figure 13.

The more negative values reflect a more energetically favorable biosorption [15]. The negative value of ΔG indicates the feasibility of the process and the spontaneous nature of sorption with a high affinity for nickel biosorption onto *sargassum tenerrimum* powder. The positive value of enthalpy change has confirmed that the biosorption tends to be endothermic in nature. The values of ΔS, ΔH and ΔG obtained in the present investigation for different initial concentrations of nickel are shown in Table 2.

6   CONCLUSION

The equilibrium agitation time for biosorption of nickel is 120 min. The optimum dosage is 18 g/L. % biosorption is increased up to pH = 4.5. The experimental data is well represented by the Langmuir isotherm with a higher correlation coefficient ($R^2 = 0.98$). The biosorption of nickel is better described by pseudo second order kinetics ($K_2 = 0.226$ g/(mg-min)). The biosorption is endothermic as ΔH is positive, irreversible as ΔS is positive, and spontaneous as ΔG is negative.

REFERENCES

Aksu, Z. & Donmez, G. (2006). Binary biosorption of cadmium (II) and nickel (II) onto dried Chlorella vulgaris: Co-Ion effect on mono-component isotherm parameters. *Process Biochemistry*, *41*(4), 860–868.

Al-Rub, F.A., El-Naas, M.H., Benyahia, F. & Ashour, I. (2004). Biosorption of nickel on blank alginate beads, free and immobilized algal cells. *Process Biochemistry*, *39*(11), 1767–1773.

Bai, R.S. & Abraham, T.E. (2002). Studies on enhancement of Cr (VI) biosorption by chemically modified biomass of Rhizopus nigricans. *Water Research*, *36*(5), 1224–1236.

Congeevaram, S., Dhanarani, S., Park, J., Dexilin, M. & Thamaraiselvi, K. (2007). Biosorption of chromium and nickel by heavy metal resistant fungal and bacterial isolates. *Journal of Hazardous Materials*, *146*(1–2), 270–277.

Dàvila-Guzmàn, N., Cerino-Cordova, F., Rangel-Méndez, J. & Diaz-Flores, P. (2011). Biosorption of lead by spent coffee ground: Kinetic and isotherm studies. In *AIChE Annual Meeting, Conference Proceedings*, 1–9.

Freundlich, H. (1907). Über die Adsorption in Lösungen (Adsorption in solutions). *Z. Physiol. Chem.*, *57*(1), 384–470.

Gupta, V.K., Rastogi, A. & Nayak, A. (2010). Biosorption of nickel onto treated alga (Oedogonium hatei): Application of isotherm and kinetic models. *Journal of Colloid and Interface Science*, *342*(2), 533–539.

King, P., Rakesh, N., Beenalahari, S., Kumar, Y.P. & Prasad, V.S.R.K. (2007). Removal of lead from aqueous solution using Syzygium cumini L.: Equilibrium and kinetic studies. *Journal of Hazardous Materials*, *142*(1–2), 340–347.

Koedrith, P., Kim, H., Weon, J.I. & Seo, Y.R. (2013). Toxicogenomic approaches for understanding molecular mechanisms of heavy metal mutagenicity and carcinogenicity. *Int J Hyg Environ Health*, *216*(5), 587–598.

Krishna, R.H. & Swamy, A.V.V.S. (2011). Studies on the removal of Ni (II) from the aqueous solutions using powder of mosambi fruit peelings as a low cost adsorbent. *Chemical Sciences Journal*, *2011*, 1–13.

Langmuir, I. (1918). The adsorption of gases on plane surfaces of glass, mica and platinum. *J. Am. Chem. Soc.*, *40*(9), 1361–1403.

Mata, Y.N., Blazquez, M.L., Ballester, A., Gonzalez, F. & Munoz, J.A. (2008). Characterization of the biosorption of cadmium, lead and copper with the brown alga Fucus vesiculosus, *Journal of Hazardous Materials*, *158*(2–3), 316–323.

Nourbakhsh, M.N., Kiliçarslan, S., Ilhan, S. & Ozdag, H. (2002). Biosorption of Cr6+, Pb2+ and Cu2+ ions in industrial waste water on Bacillus sp. *Chem Eng J*, *85*(2–3), 351–355.

Nuhoglu, Y. & Malkoc, E. (2009). Thermodynamic and kinetic studies for environmentally friendly Ni (II) biosorption using waste pomace of olive oil factory. *Bioresource Technology*, *100*(8), 2375–2380.

Özer, A., Gürbüz, G., Çalimli, A. & Körbahti, B.K. (2008). Investigation of nickel (II) biosorption on Enteromorpha prolifera: Optimization using response surface analysis. *Journal of Hazardous Materials*, *152*(2), 778–788.

Pahlavanzadeh, H., Keshtkar, A.R., Safdari, J. & Abadi, Z. (2010). Biosorption of nickel (II) from aqueous solution by brown algae: Equilibrium, dynamic and thermodynamic studies. *Journal of Hazardous Materials*, *175*(1–3), 304–310.

Pinto, P.X., Al-Abed, S.R. & Reisman, D.J. (2011). Biosorption of heavy metals from mining influenced water onto chitin products. *Chem Eng J*, *166*(3), 1002–1009.

Rangsayatorn, N., Upatham, E.S., Kruatrachue, M., Pokethitiyook, P., & Lanza, G.R. (2002). Phytoremediation potential of Spirulina (Arthrospira) platensis: Biosorption and toxicity studies of cadmium. *Environmental Pollution*, *119*(1), 45–53.

Redlich, O.J.D.L. & Peterson, D.L. (1959). A useful adsorption isotherm. *J Phys Chem*, *63*(6), 1024.

Wang, J. & Chen, C. (2009). Biosorbents for heavy metals removal and their future. *Biotechnol. Adv.*, *27*(2), 195–226.

Emerging Trends in Engineering, Science and Technology for Society,
Energy and Environment – Vanchipura & Jiji (Eds)
© 2018 Taylor & Francis Group, London, ISBN 978-0-8153-5760-5

Automation of IGCC power plant using Yokogawa DCS

K.J. Jyothir Rose & R. Anjana
Department of Chemical Engineering, Government Engineering College, Thrissur, Kerala, India

Lydia Jenifer
Yokogawa India Ltd., Bangalore, India

ABSTRACT: Integrated Gasification Combined Cycle (IGCC) power plants which are working on high-efficiency coal gasification technologies, are operated commercially or semi commercially worldwide. Various coal gasification technologies are embodied in these plants including different coal feed systems (dry or slurry), fireproof interiors walls (fire brick or water-cooled tubes), oxidants (oxygen or air), and other factors. These designs which are several decades old, but using new systems and cycles are emerging to further improve the efficiency of the coal gasification process. The development of Distributed Control System (DCS) for automated operation, monitoring and control functions which combines Human Machine Interface (HMI), interlocks, logic solvers, historian, common database, report generation, alarm management and a common engineering suite into a single automated system. The implementation is done using automatic methods of distribution which guarantee the preservation of behavior of the whole system. The purpose of the project is to develop a suitable control strategy using DCS for the futuristic power plant for better monitoring, operation and availability.

1 INTRODUCTION

Integrated Gasification Combined Cycle (IGCC) is a power plant in which a gasification process provides syngas to a combined cycle under an integrated control system. It has potentially many advantages including high thermal efficiency, good environmental characteristics, reduced water consumption etc. Hence Gasification based Power plants are future of power production from coal or biomass. The aim of the project is to develop a Distributed Control System (DCS), which are dedicated systems used in manufacturing processes that are continuous or batch-oriented which is used for monitoring and controlling of an IGCC power plant which will definitely play a significant role in future.

Studying different controlling methods for plant instrumentation and control and for this project implementation DCS have been opted considering the advantages of DCS over other systems. Proprietary interconnections and communications protocol is used in DCS which uses custom designed processors as controllers. Input and output modules form component parts of the DCS. The input modules receive information from input instruments in the process (or field) and transmit instructions to the output instruments in the field. Computer buses or electrical buses are used to connect the processor and modules through multiplexer or de multiplexer. Buses also connect the distributed controllers with the central controller and finally to the Human machine interface (HMI) or control consoles. An HMI is a software application that presents information to an operator or user about the state of a process, and to accept and implement the operators control instructions.

2 IGCC TECHNOLOGY

IGCC has potentially many advantages including high thermal efficiency, good environmental characteristics, reduced water consumption etc. Hence Gasification based Power plants are future of power production from coal or biomass. The aim of the project is to develop a Distributed Control System for monitoring and controlling of an IGCC power plant which will definitely play a significant role in future. The fuel Syngas is generated using a high efficiency power generation technology which gasifies coal and combustion takes place in high efficiency gas turbine (GT) in IGCC power plant. Compared with conventional pulverised coal (PC) fired power plants IGCC has potentially many advantages including:

High thermal efficiency: The Shell gasifier efficiency of IGCC generation is estimated to be 46–47% net, low heating value (LHV) basis (44–45% net, high heating value (HHV) basis), for FB class gas turbine using bituminous coal. The highest reported efficiency for an IGCC is Efficiency of 41.8% HHV basis. Good environmental characteristics that match or exceed the latest PC plants. The plant's high thermal efficiency means that emissions of CO_2 are low per unit of generated power. In addition, emissions of SOx and particulates are reduced by the requirement to deep clean the syngas before firing in the gas turbine.

Reduced water consumption: IGCC uses less water. Since 60% of its power is derived from an air based Bray ton cycle reducing the heat load on the steam turbine condenser to only 40% of that of an equivalent rated pulverised coal fired plant. Additionally, through the direct de-sulfurization of the gas, IGCC does not require a large flue gas de-sulfurization unit which consumes large amounts of water, thereby reducing water consumption in comparison with a conventional pulverised coal fired power plant. Further gains in reducing water use can be achieved when CCS is incorporated into the plant.

Table 1.   Evolution of Yokogawa DCS.

Systems	Year	Platform
Centum/Centum V	1975	MS-DOS based
Centum-XL/Micro-Xl	1988	MS-DOS/UNIX based
Centum CS	1993	UNIX based
CS 3000	1998	Windows based
CENTUM VP	2008	Windows based

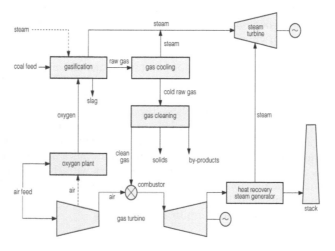

Figure 1.   Basic IGCC diagram without CO_2 Capture.

An IGCC system is composed of a gasification unit, an air separation unit (ASU), a syngas purification unit, and a combined cycle involving a gas turbine and a steam turbine. A simplified version of a coal-fuelled IGCC cycle is shown in Figure 1.

2.1 *Yokogawa CENTUM VP*

A distributed control system (DCS) is a control system used in various industries as manufacturing system, process or any kind of dynamic system, in which the there is no central control location for controller elements but are distributed throughout the system for one or more controllers for each component sub-system controlled. The DCS used in the project is YOKOGAWA DCS.

3 P&ID DESIGN AND DEVELOPMENT

This approach requires a great deal of input from the various design engineers before all of the details have been worked out. It recommends the members of the design team to consider all of the problems involving successful instrumentation operation. The limitation to the detailed P&IDs design before detailed layout is complete is that the P&IDs must synchronise with requirements in the electrical, process, instrumentation and piping as closely as possible. If there are major modifications to be implemented to the project during detailed process and instrumentation diagram (P&ID) development, the P&IDs must be modified. If the P&IDs does not sync with the work of the various departments, design team members may use incorrect information.

The second approach is to allows the P&IDs to show the instrumentation connections only. Instrumentation designers and engineers (and possibly the electrical engineers to double check instrumentation wiring requirements) use this approach when P&IDs are only used among them. These diagrams do not detail the same level of information as the first approach. The intention is to show how instrumentation and the process are related which possibly show the electrical requirements.

4 PROCESS EXPLANATION AND METHODOLGY

One of the most popular air separation process used is cryogenic air separation, frequently in medium to large scale plants. This technology is mostly preferred for producing nitrogen, oxygen, and argon as gases and/or liquid products and supposed to be the most cost effective for high production rate plants. The process of cryogenic air separation is studied and P&ID is developed as shown in Figure 2.

Similarly the IGCC is divided into several sub sections based on operations and P&ID for each section is developed separately. The control loops are identified and developed are shown in Figure 3.

The Coal Grinding System provides a means to prepare the coal as a slurry feed for the gasifier. Coal is continuously fed to the Coal Weigh Feeder, which regulates and weighs the coal fed to the Grinding Mill. The unloading of coal, its crushing, storage and filling of boiler bunkers in a thermal power station is covered by The coal handling plant (CHP). The main function of coal handling unit is crushing of coal into very fine particles for gasification and regulate the ratio of the coal and lime mix with the requirements. The lime ratio is increased to decrease the SOx level in the syngas produced from gasification chamber.

The process of gasification to produce combustible gas also known as syngas or producer gas from organic feeds is used. The gasification of Biomass is a thermo-chemical process that produces relatively clean and combustible gas through pyrolytic and reforming reactions. The product of gasification is a combustible synthesis gas, or syngas. Because gasification involves the partial, rather than complete, oxidization of the feed, gasification processes operate in an

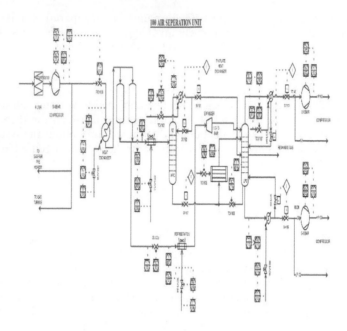

Figure 2.　P&ID developed for air seperation unit.

Figure 3.　P&ID developed for air separation unit.

oxygen-lean environment. The ratio of the combustible hydrogen (H_2), methane (CH_4), carbon monoxide (CO) and moisture determines the heating value of the obtained fuel.

The systems used to remove some particulates and/or gases from industrial exhaust streams are known as Scrubber system which is a group of air pollution control devices. Gas is contaminated and has high temperature (500–800°C) as it leaves atmospheric fluid gasifier which focuses mostly on tar elimination and dust removal. Water or specific organic liquid or both may be applied in tar elimination from gas. Boiling point (volatility), availability, and price of organic liquid are major criteria for selection of a proper material. From gasifier, syngas is produced and is fed into the scrubber to remove particulates and ash formed from the gasification process.

Gas separation allows H_2 to be isolated from syngas and used as a clean fuel or feedstock in chemical production. The CO_2 extracted from the process can be captured and used in chemical production or sequestered, rather than released into the atmosphere. Conventional options for CO_2 removal, such as solvent-based absorption, are energy intensive and require cooling of syngas. The newly purified H_2 must be re-pressurized and/or reheated, imposing additional energy penalties for many applications. The exothermic shift reaction (or water gas shift reaction) transfers the fuel heating value from CO to H_2 and transfers the carbon from CO to CO_2.

Depending on the gasification process used, the temperature of syngas leaving a gasifier can be as high as 1600°C, recovery of heat from the high temperature syngas is essential for attaining high process efficiency. Using heat recovery systems, depending on the technology used a significant portion (5–25%) of the energy in the feed can be utilised. The syngas cooler is one of the most crucial and highly loaded components in gasification plants. It operates with gas inlet temperatures ranging from 1600°C to 400°C and gas-side pressures up to 8 MPa.

Removing Hg from the syngas prior to combustion is more effective in IGCC systems. This may already occur, to some extent, via the acid gas scrubbing system, but more data are required to verify this. The process of Syngas removal has the advantages like higher mercury concentration, lower mass flow rates, and higher pressure than the stack gas. HgSIV, is a molecular sieve (MS) that removes very low levels of elemental mercury from natural gas or syngas via a regenerable adsorption process. It uses a 2-bed thermal-swing MS adsorption system.

Sulfur is naturally present as an impurity in fossil fuels. When the fuels are burned, the sulfur is released as sulfur dioxide—an air pollutant responsible for respiratory problems and acid rain. The fuel processors have to remove the sulfur from both fuels and exhaust gases as Environmental regulations have increasingly restricted sulfur dioxide emissions. This multistep process has low operating costs but high capital costs too expensive for plants recovering less than about 20 tons of sulfur per day. These plants use liquid-phase reduction-oxidation (redox) processes to remove sulfur content from the syngas produced.

The separation and capture process for producing CO_2 stream comprehensively includes all the operations that take place at the power plant site, including compression. For ease of transport, CO_2 is generally compressed to the order of 100 atm. The NOx and SO_2 should be cleaned up prior to CO_2 separation as required by flue gas approach in use today. Integrated coal gasification combined cycle (IGCC) plants are an example of the hydrogen route.

The Heat Recovery Steam Generation (HRSG) receives the exhaust gases from the GT discharge. The exhaust gas, flowing in counter flow with respect to the steam/water coils, cools down by transferring heat to steam/water. The flue gas temperature at the stack is about 230°F (110°C), even though lower temperatures [200°F (93°C)] can be used if the fuel gas is very clean and Sulfur-free. The HRSG is, therefore, similar to a heat exchanger in which the shell side carries the flue gas and the various sections of the tube side carry steam or water. It has also the characteristics of a boiler because there are one or more steam drums, where the generated steam is separated from boiling water before entering the super heaters.

5 RESEARCH FINDINGS AND EXPLORATION

The objective of this paper is to develop a DCS for IGCC Power Plant. Initially the complete processes of Integrated Gasification Combined Cycle (IGCC) Power plant is studied and found out the requirements for controlling the processes and the hardware used for implementing the control system for different parameters like pressure, temperature, flow, level etc. The whole power plant consists of several sections of which each section is studied and P&ID for the process is developed. The P&ID is a specialized document shown on a side view representation of all equipment. Integration and properly designed interfacing between the DCS and other digital control packages is essential. The serial links should be made redundant to ensure the maximum operating continuity. The system bus and the input/output (I/O) buses are also implemented to be redundant, for guaranteeing the maximum uptime. The sequence of events

control function carried out in the DCS, the appropriate I/O cards have to be correctly anticipated or a dedicated system must be connected to the DCS via a serial link. The next step is to
identify the control loops and function blocks required to realise the control strategies required
for the project. After this two procedures the whole plant has to be displayed and controlled
using Human Machine Interface (HMI). Next Step is to develop the Graphics section to represent the whole plant and to view different operations and to control each controllers. In graphics whole process of the plant can be visualized. The Alarm Processing is a message processing
function performed by the HMI when alarm occurs during plant operation. The issued alarm
can be used by the operator to determine what action is to be taken for the abnormality and
then process the alarm. The operator can perform an acknowledgment action for the alarm
output. These alarms are identified and implemented in the project.

The graphical user interface to interact with the functioning and overall processes of the
plant. From the graphical interface the user can know the present position or value of the
controller and the user can take necessary actions to change the values such as set value or
put a motor to on/off position depending on the situations arising. The user can monitor the

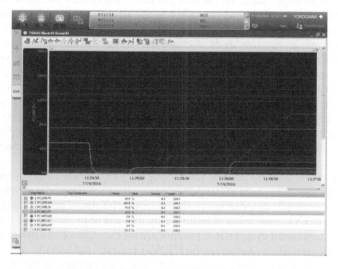

Figure 4.   Graphics developed for the ASU of power plant.

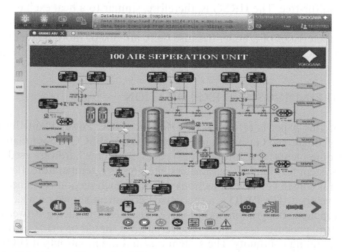

Figure 5.   Controller action with varying PV.

functions of the plant from the interface. The representations in the interface can be configured during the designing of the graphics development. The graphics developed for each section is shown in Figure 4.

The controller action for different values of Process Variable (PV) using field instruments is shown in Figure 5.

6 CONCLUSION

The proposed control system using Yokogawa CENTUM VP and procedure presents an effective approach for monitoring and control. In real time the implementation of this method is very effective and system is available almost 99% time. The design of P&ID of the system developed by carefully studying the operations and plant processes of IGCC Power Plant.

The present study has demonstrated the design of DCS by I/O station inputs from the different equipments. The quantities are measured and controlled, and control valves of the processes are manipulated in real time to implement temperature, pressure, level and flow rate control, breakdowns are detected and the system is maintained. Live measured values and status indications reveal the current situation. Process operators monitor and control the long-distance processes from the console.

ACKNOWLEDGMENT

This work is mainly carried out at Yokogawa India Ltd. Bangalore, India. I would like to thank all co-authors for the important discussions about the work.

REFERENCES

[1] Sher shah Amarkhail, 'Air Separation', Development of human resource capacity of Kabul polytechnic university".
[2] Monika Kurková-Zdeněk Klika-Petr Martinec-Jaroslava Pěgřimočová, 'Composition of bituminous coal in dependence on environment and temperature of alteration', Kurkova, M. et al. 2003.
[3] Anil Bose, 'Classification of Coal in India', Indian Geography.
[4] Jeffery Phillips, 'Different types of gasification and their integration with gas turbines', EPRI/ Advanced Coal Generation.
[5] Younes Chhiti, Mohammed Kemiha, 'Thermal Conversion of Biomass, Pyrolysis and Gasification', The IJES.
[6] Yongseung Yun, Seung Jong Lee and Seok Woo Chung, 'Considerations for the Design and Operation of Pilot-Scale Coal Gasifiers', Institute for Advanced Engineering, Suwon, Republic of Korea.
[7] Marek Balas, Martin Lisy, Zdenek Skala, Jiri Pospisil, 'Wet scrubber for cleaning of syngas from biomass gasification', Indiana Council of Administrators of Special Education.
[8] Andrej Lotrič, – Mihael Sekavčnik – Christian Kunze – Hartmut Spliethoff, 'Simulation of Water-Gas Shift Membrane Reactor for Integrated Gasification Combined Cycle Plant with CO_2 Capture', Strojniški vestnik – Journal of Mechanical Engineering.
[9] J.E. Jamison, B.G. Lipták, A. Rohr, 'Power Plant Controls: Cogeneration and Combined Cycle'.
[10] Hyungwoong Ahn, Zoe Kapetaki, Pietro Brandani, Stefano Brandani, 'Process simulation of a dual-stage Selexol unit for pre-combustion carbon capture at an IGCC power plant'.
[11] Howard Herzog, 'An Introduction to CO_2 Separation and Capture Technologies', MIT Energy Laboratory.
[12] Alan Darby, "Hydrocarbon upgrading gasification program', Alberta Energy Research Institute Grant.
[13] Steve Fusselman, Alan Darby and Fred Widman, "Advanced gasifier pilot plant concept definition", September 2006.
[14] Yongseung Yun, Seung Jong Lee and Seok Woo Chung, 'Considerations for the Design and Operation of Pilot-Scale Coal Gasifiers'.

[15] Kosan Roh and Jay H. Lee, 'Selection of Control Structure of Elevated Pressure Air Separation Unit in an IGCC'.

[16] Makarand M. Joshi, 'Development of Condition Based Maintenance for Coal Handling Plant of Thermal Power Stations', Plant Maintenance Resource Center.

[17] Taylor and Francis Group, 'Combined Cycle Power Plants'.

[18] Everret B Woodruff, Herberrt B Lammers, Thomas F Lammers, 'Steam Plant Operations'.

[19] Van Dijka, K. Damenb, M. Makkeec, C. Trappd, 'Water-gas shift (WGS) operation of pre-combustion CO_2 capture pilot plant at the Buggenum IGCC', H.A.J.

[20] Qian Zhu, 'High Temperature Syngas coolers', IEA Clean Coal Centre.

[21] John Markovs, 'Optimized Mercury Removal In Gas Plants', Adsorption Solutions LLC Cross Junction, Virginia, U.S.

Emerging Trends in Engineering, Science and Technology for Society,
Energy and Environment – Vanchipura & Jiji (Eds)
© 2018 Taylor & Francis Group, London, ISBN 978-0-8153-5760-5

Deprotonation studies on polyaniline polymethyl methacrylate blends processed from formic acid

V.O. Rejini, A.H. Divya & Rakesh S. Nair

Department of Chemical Engineering, Government Engineering College, Thrissur, Kerala, India

ABSTRACT: Polyaniline (PANI) and Polymethylmethacrylate (PMMA) blend films were prepared from homogeneous solutions in formic acid, which serves both as a dopant as well as a solvent. The absorption spectra revealed that formic acid processed PANI and its blends with PMMA to be deprotonated in atmospheric conditions. Stability studies were done in three ways: with and without an additional protonating agent, Camphoursulphonic Acid (CSA), and with an addition of a plasticizer to the host matrix. The results showed that the presence of CSA imparts stability or resistance against deprotonation.

Keywords: Polyaniline; PMMA; solution blending; blends; plasticizer; deprotonation

1   INTRODUCTION

Solution processing has been reported to be a quite efficient method when both polyaniline and the polymer matrix are 'compatible' with each other and 'soluble' in a common solvent (Barra et al., 2002). The conductivity of solution-cast Polyaniline (PANI)/polymer blends depends upon the ability of the solvent to finely disperse the conducting polymer, and on the flocculation of the dispersed PANI in the blend (Paul & Pillai, 2002). The deprotonation nature or the stability of the dopant also play a vital role on the overall conductivity. Most polyaniline blends described in the literature were processed from m-cresol, a solvent which is both acidic and high boiling point. If the emeraldine base and the host polymer are co-soluble in an acid dopant, the blend can be obtained as a film in the conducting form by solution casting.

The conductive PANI/polyamide-11 blend fibers were prepared by wet-spinning technology from concentrated sulfuric acid with relatively high electrical conductivity (Zhang et al., 2001). High strength and high modulus electrically conducting PANI composite fibers were also reported (Hsu et al., 1999) from PANI/PPD-T (poly (para-phenylenediamine) terephthalic acid) sulfuric acid solutions. Due to the ease of handling and solvent removal, it is more convenient to use liquid organic acids than sulfuric acid as solvents. Abraham et al., (1996) used formic acid as the solvent as well as dopant for a polyaniline–nylon 6 blend system. The chemical modification or blending of polyaniline with nylon 6 does not affect the crystal structure of either polyaniline or nylon 6. This was confirmed by X-ray diffraction. The maximum conductivity of the films was about 0.2S/cm, corresponding to a weight ratio of 0.5 (w/w) for PANI and Nylon 6. Formic acid was also used by Anand et al. (2000) for the preparation of blends of PANI derivatives (Poly (O-Toluidine) (POT), Poly (M-Toluidine) (PMT)) with Polymethylmethacrylate (PMMA). The blend was precipitated by the addition of the formic acid solution to water (non-solvent). The thermal stability of the blends was reported to be greater than that of their respective salts. Zagórska et al. (1999a) studied the stability against deprotonation of polyaniline/polyamide 6 blends processed from formic acid. They prepared polyaniline-polyamide 6 blends in two different ways: one with an additional protonating agent and the other without an additional protonating agent. The blends of polyaniline and polyamide 6 processed from formic acid were prepared without an additional protonating agent. Performed stability studies found them to be unstable at ambient laboratory conditions and to have a tendency to deprotonate and gradually lose their conductivity, which is in contradiction to the observation made by Abraham et al. (1996). UV-VIS spectra of polyaniline/polyamide 6 blend films were collected for increasing exposure time to laboratory atmosphere. They verified the deprotonation behavior by instantaneous increase of the peak 630 nm, which is the characteristic of non-protonated PANI. An addition of a supporting protonating agent to the polyaniline/polyamide 6 systems improved their environmental stability with respect to the deprotonation process. Anand et al. (2000) also reported UV-visible spectra of poly (o- and m-toluidine) polymethylmethacrylate blends, which showed only two peaks at 315 and 610 nm for all blend compositions; this is the characteristic of a nonprotonated state. The absorption spectrum of formic acid processed POT-PMMA blends in Dimethyl Sulphoxide (DMSO) exhibited two bands around 315 and 610 nm. The indication of insolubility of POT salt in DMSO and the presence of the base in the salt solution were indicated. There was no mention of the environmental instability of formic acid processed systems. Juvin et al. (1999) studied conductive blends of polyaniline with plasticized polymethylmethacrylate and reported that plasticizer dibutyl phthalate increases interaction between the blend components and thereby improves the conductivity of the blend. Also, these blends show good resistance to deprotonation against basic media for all the dopants such as Camphorsulphonic Acid (CSA) and three different polyalkylene phosphates studied. Solution-cast blends of PANI/PMMA were discussed in the literature for their electrical properties with respect to their interaction and compatibility (Yang et al., 1993).

In the present study we used formic acid as a co-solvent for emeraldine base and PMMA. For deprotonation studies, three types of PANI/PMMA blend films were prepared: (1) with additional protonating agent; (2) without an additional protonating agent; and (3) with plasticized PMMA. Four supporting protonating agents were reported in the literature: phenylphosphonic acid, camphorsulphonic acid, 2-acrylamide-2-methyl-1-propane sulphonic acid and dibutyl phosphate. This do not promote phase separation. We used camphorsulphonic acid as an additional protonating agent. Formic acid is known to be a very good solvent for CSA doped polyaniline (Cao et al., 1992). The additional/supporting protonating agent would exchange with formic acid as a polyaniline dopant at the final stage of the composite casting.

2 EXPERIMENTAL

2.1 *Materials*

The monomer aniline was double-distilled under reduced pressure prior to use. Ammonium persulphate, ammonium hydroxide, formic acid and hydrochloric acid are analytical grade reagents and used without purification. PMMA was supplied by SUMIPEX, Korea.

2.2 *Synthesis of polyaniline*

PANI powder in emeraldine form was synthesized by chemical oxidation of aniline with ammonium persulphate $((NH_4)_2S_2O_8)$ as the oxidant in 1M aqueous hydrochloric acid solution (Angelopoulos et al., 1988). The mixture was constantly stirred in an ice bath for 6 h. The precipitate collected was washed, first with aqueous hydrochloric acid and then with acetone. PANI salt was dedoped using 0.2 M NH_4OH.

2.3 *Preparation of PANI/PMMA blends*

The approximate solubility of emeraldine base in formic acid was evaluated using the procedure reported in the literature (Zagórska et al., 1999a). The emeraldine base and PMMA (10% wt/vol.) were dissolved separately in formic acid. The blend mixtures were prepared by varying the weight ratios of emeraldine base to PMMA. A specific composition of 10/90 wt% of polyaniline and PMMA respectively were prepared. The mixtures were sonicated for ten min. and then magnetically stirred for 6h at room temperature. The resulting homogeneous solution was made into films by casting. The blend films obtained were dried for 48 h under vacuum. The blends are denoted by using the initial weight percentage of emeraldine base and PMMA.

2.4 *Preparation of PANI/PMMA blends with an additional protonating agent*

A specific composition of polyaniline–polymethylmethacrylate (10/90) blends were prepared by co-solvation method with an additional protonating agent. CSA as an additional protonating agent was added in the molar ratio of 0.5 per PANI repeat unit, involving one ring and one nitrogen atom (Zagórska et al., 1999b). Blend films were prepared by solution casting.

2.5 *Preparation of PANI with plasticized PMMA blends*

Polymethyl methacrylate (10% wt/vol.) in formic acid was plasticized using dimethyl phthalate (35 wt%). Plasticized PMMA solution was mixed with polyaniline solution in formic acid for 10/90 wt% (PANI/PMMA) composition and films were prepared by solution casting.

2.6 *Deprotonation studies*

For deprotonation studies, blend films of a particular composition (10/90) were exposed to laboratory atmosphere (29°C) for seven days. UV spectra were measured every day at a particular time for seven days. The effect of additional protonating agent, CSA, on PANI/PMMA blend system was determined and also any influence was found on using plasticized (dimethyl phthalate) PMMA on the deprotonation behavior of the formic acid doped polyaniline system.

3 CHARACTERIZATION

The UV-VIS absorption spectra of the blend films were recorded using a Varian Cary 5E model UV-VIS near–IR spectrophotometer.

4 RESULTS AND DISCUSSION

The base form shows two major peaks at 630 and 330 nm, which is the characteristic absorption spectrum of the base form of the PANI. The peak at 330 nm regions is assigned to the

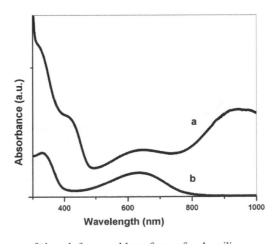

Figure 1.   UV-VIS spectra of the salt form and base form of polyaniline.

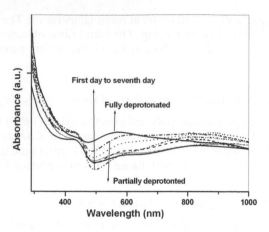

Figure 2. UV-VIS spectra collected for PANI/PMMA blend film, exposed to the laboratory atmosphere for seven days.

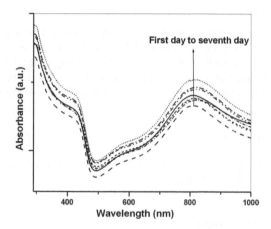

Figure 3. UV-VIS spectra of PANI/PMMA blends with camphoursulphonic acid.

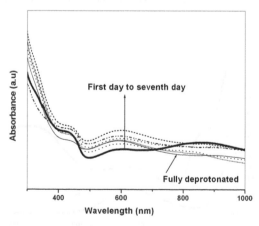

Figure 4. UV-VIS spectra of formic acid doped polyaniline with plasticized polymethylmethacrylate blend film, exposed to atmosphere for seven days.

Π-Π* transition, related to the extent of conjugation between adjacent phenyl rings in the polymer chain (Zhang et al., 2001). The peak at 630 nm is a measure of extended conjugation, corresponding to the excitation-like transition from the highest occupied benzenoid ring to the lowest unoccupied quinoid ring, caused by interchain or intrachain charge transfer. For fully protonated PANI upon UV irradiation, the intensity of the 320 nm band decreases and a new absorption band appears at 440 nm, assigned to the polaron transition. An absorption tail extending into the near-IR region is also observed. These features are characteristic of protonated PANI. The absorption value of peaks at 950 nm increased with an increase of protonation state. The peak at 630 nm indicates that the quinoid segment on the polyemer-aldine chain is present in partially protonated state, or that formic acid doped polyaniline undergoes slow deprotonation.

The absorption spectra of the blend film indicate that formic acid polyaniline undergoes deprotonation as time proceeds. Within seven days, the blend film was completely deproto-nated, the peaks at 440 nm and 900 nm disappear and a specific peak at around 600 nm appears.

In the presence of an additional protonating agent, CSA, the system retains its conductiv-ity or stability against deprotonation, even after being exposed to the laboratory atmosphere for seven days. Absorbance spectra show energetic peaks at 430 nm (related to the Π–Π* transition on the polymer chain) and around 800 nm, attributed to polaron transitions, indi-cating the presence of polyaniline in the salt form.

Absorbance spectra indicate that the presence of the plasticizer does not make any change to the deprotonating nature of the formic acid processed system. The blend film was deprotonated within seven days. The PANI-CSA-plasticized PMMA system was reported for its improved conductivity and resistance against deprotonation rate (Barra et al., 2002).

5 CONCLUSIONS

Deprotonation studies show that PANI/PMMA blends and PANI-plasticized PMMA blends processed from formic acid are not environmentally stable. The interaction between the blend components and compatibility do not impart any resistance to deprotonation. As per the literature, formic acid doped PANI with nylon 6 blends are immiscible and PANI/PMMA blend systems are highly compatible. From the results obtained, we could conclude that the nature of host matrix and miscibility or immiscibility of the blend components do not make any change in the deprotonating nature of formic acid processed polyaniline blend systems. PANI/PMMA formic acid system with an additional protonating agent, camphoursulphonic acid, showed good stability against deprotonation.

REFERENCES

Abraham, D., Bharathi, A. & Subramanyam, S.V. (1996). Highly conducting polymer blend films of polyaniline and nylon 6 by co-solvation in an organic acid. *Polymer*, *37*(23), 5295–5299.
Anand, J., Palaniappan, S. & Sathyanarayana, D.N. (2000). Solution blending of poly (o- and m-toluidine) with PMMA in formic acid medium: Spectroscopic, thermal and electrical behaviour. *Eur Polym J*, *36*(1), 157–163.
Angelopoulos, M., Asturias, G.E., Ermer, S.P., Rey, A., Scherr, E.M., Macdiarmid, G., & Epstein, A.J. (1988). Polyaniline: Solutions, films and oxidation state. *Molecular Crystals and Liquid Crystals*, *160*(1), 151–163.
Barra, G.M., Levya, M.E., Soares, B.G. & Sens, M. (2002). Solution-cast blends of polyaniline-DBSA with EVA copolymers. *Synth Met*, *130*(3), 239–245.
Cao, Y., Smith, P. & Heeger, A.J. (1992). Counter-ion induced processibility of conducting polyaniline and of conducting polyblends of polyaniline in bulk polymers. *Synth Met*, *48*(1), 91–97.
Cao, Y. & Smith, P. (1993). Liquid-Crystalline solutions of electrically conducting polyaniline. *Polymer*, *34*(15), 3139–3143.
Dan, A. & Sengupta, P.K. (2004). Synthesis and characterization of polyaniline prepared in formic acid medium. *J Appl Polym Sci*, *91*(2), 991–999.

Hsu, C.H., Shih, H., Subramoney, S. & Epstein, A.J. (1999). High tenacity, high modulus conducting polyaniline composite fibers. *Synth Met, 101*(1–3), 677–680.

Juvin, P., Hasik, M., Fraysse, J., Planès, J., Pron, A. & Kulszewicz-Bajer, I. (1999). Conductive blends of polyaniline with plasticized poly (methyl methacrylate). *Appl Polym Sci, 74*(3), 471–479.

Paul, R.K. & Pillai, C.K.S. (2002). Melt/solution processable conducting polyaniline: Elastomeric blends with EVA. *J Appl Polym Sci, 84*(7), 1438–1447.

Terlemezyan, L., Mihailov, M. & Ivanova, B. (1992). Electrically conductive polymer blends comprising polyaniline. *Polymer Bulletin, 29*(3–4), 283–287.

Yang, C.Y., Cao, Y., Smith, P. & Heeger, A.J. (1993). Morphology of conductive, solution-processed blends of polyaniline and poly (methyl methacrylate). *Synth Met, 53*(3), 293–301.

Zagórska, M., Harmasz, E., Kulsewicz-Bajer, I. Proń, A. & Niziol, J. (1999a). Blends of polyaniline with polyamide 6 processed from formic acid. *Synth Met, 102*(1–3), 1240.

Zagórska, M., Taler. E, Kulsewicz-Bajer, I., Proń, A. & Niziol, J. (1999b). Conductive polyaniline-polyamide 6 blends processed from formic acid with improved stability against deprotonation. *J Appl Polym Sci, 73*(8), 1423–1426.

Zhang, Q., Jin, H., Wang, X. & Jing, X. (2001). Morphology of conductive blend fibers of polyaniline and polyamide-11. *Synth Met, 123*(3), 481–485.

Emerging Trends in Engineering, Science and Technology for Society,
Energy and Environment – Vanchipura & Jiji (Eds)
© 2018 Taylor & Francis Group, London, ISBN 978-0-8153-5760-5

Characterization of brushless DC motor for control valve actuation in rocket propulsion systems

K.N. Ajeesh & Suhana Salim
Department of Chemical Engineering, Government Engineering College, Thrissur, Kerala, India

R. Sujith Kumar
VSSC, Trivandrum, Kerala, India

ABSTRACT: Control valves are usually actuated by pneumatic signals. Electrical actuators can also be used, which are more accurate and cheap. Laboratory research on a flow control valve actuated by brushless DC (BLDC) motor has been carried out. The characterization of a control valve driven by a DC motor has been obtained to determine whether the outlet flow depends on motor parameters such as the motor speed, stepping angle etc. Laboratory test for the characterization work is done by using nitrogen as the test fluid. LabVIEW is used for the data acquisition. Different tests were carried out with varying speeds and from the time constants obtained, first order systems were designed. The theoretical response curve for first order systems were generated using MATLAB software and compared with responses of controllers with conventional pneumatic actuator. The comparison showed electrical actuators were much faster than pneumatic actuators.

1 INTRODUCTION

Control valves are essential components of any piping system. When we use the term valve, it is manually operated whereas a control valve is one with an actuator that automatically opens or closes the valve fully or partially to a position dictated by signals transmitted from the controlling instruments. Based on actuation, control valves can be mainly classified into quarter turn valves, multi-turn valves and check valves. Quarter turn valve allows only 90° rotation and it includes ball valves, butterfly valves spherical valves and plug valves. In this work a ball valve is used for characterization study. A multi-turn valve allows 360° rotation and it requires 4–5 turns to completely open or close the valve. Gate valves, globe valves and pinch valves are all multi-turn valves.

Apart from process industries, control valves find lot of other applications like power stations, rockets & space crafts, automobile systems etc. For such applications, sometimes it is required to develop non-conventional type of technologies. Control valves are actuated by using pneumatic, hydraulic and electrical actuators. Most of the industries are using conventional pneumatic signals for valve actuation. Control valve can also be actuated by using DC motor.

The objective of this work was to find the response time for a ball valve coupled with a BLDC motor at various speeds so as to characterize the system. Knowledge about BLDC motor and its drive is essential for the characterization study. The design and implementation of BLDC motor drive for automotive applications was reported to give reliable results (Park et al. 2012). Most of the rocket systems use light weight propulsion systems. One of the most desired technological requirements for an efficient aerospace launch vehicle is the use of light weight propulsion system. Brushless DC motors whose efficiency can be greater than 90% are useful for this purpose in this sense (Van Neikerk 2015).

2 TEST SETUP

2.1 *System description*

A rectangular aluminium sheet is taken and it is bent into an L shaped structure. The power supply circuit is mounted on its base. The control circuit (driver) of the motor is also mounted near to it. Five switches and a 20 K POT are placed on the uppermost part. So for developing the system, the hardware components required are a power supply circuit, BLDC motor and its control unit, ball valve and two pressure transmitters. The motor requires 24V DC supply for its operation. Control circuit for opening and closing of the valve and the BLDC motor is shown in the Figure 1.

The valve and the actuator portion are placed on field and the DAQ and PC is placed on the control room. Figure 2 shows the arrangement of switches along with motor-ball valve interconnection.

The placement of upstream and downstream pressure transmitters along with valve-motor interconnection is shown in Figure 3. This developed module was placed on the test bay. It only possesses the valve and actuator portion.

2.2 *Instrumentation scheme*

The instrumentation scheme provides for measurement of pressure and flow. The details of the sensors, sensor gauges and sensor location are given below.

a. Two strain gauge type pressure sensors with range 0–10 kg/cm^2 for upstream and downstream pressure measurement.
b. 24V DC power supply unit for motor triggering.

Figure 1.   BLDC motor and its control unit.

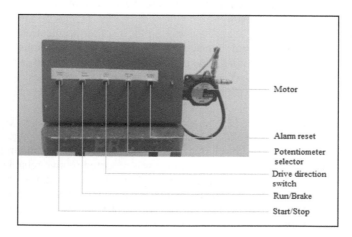

Figure 2.   System configuration.

Figure 3.   Valve-Motor interconnection along with pressure transmitters.

c. Signal conditioning module.
d. USB Data Acquisition module.
e. 10 volts power supply for sensor excitation.
f. Digital panel meters for continuous monitoring of the pressure lines.

The data acquisition is carried out by using a low cost USB data acquisition module and a laptop with USB interface.

A LabVIEW based software and real time display panel was developed for parametric display and acquisition. The software was also provided with a provision for offline analysis of acquired data.

3   EXPERIMENT

Stored nitrogen gas at a pressure of 150 bar is regulated to 10 bar by using a spring loaded pressure regulator. It is then passed through a solenoid valve. A pressure transducer is provided next to the solenoid valve in the flow line to measure the incoming pressure. The gas then passes through the motorized ball valve. The upstream pressure can be measured by using a strain gauge type pressure transducer. By using pressure transmitter, the measured pressure is transmitted to the DAQ system. For DAQ purpose, NI data acquisition unit and LABVIEW software is used. The details of the methodology adopted for the test and test sequence are explained below.

3.1   *Test methodology*

The methodology adopted for the test includes the following steps:

a. Turn on the DAQ system.
b. Turn on the regulator. Set the regulator pressure as 10 bar.
c. Turn ON the power supply of the motor.
d. Switch on the START button.
e. Switch ON the RUN button.
f. After specified time delay switch OFF the RUN button.
g. Press STOP.
h. Turn OFF power supply of motor.-driver.
i. Turn off regulator.
j. Turn off the DAQ system.

Table 1. Test sequence.

Time (s)	Events
T-15	Start count down
T-10	DAQ ON
T-6	GN2 supply ON
T-3	Switch on the power supply of motor
T-2	Switch on START/STOP
0	Switch on RUN/BRAKE
T+3	Switch off START/STOP
T+4	Switch off RUN/BRAKE
T+5	Switch off the power supply of motor
T+6	GN2 supply off
T+10	DAQ off

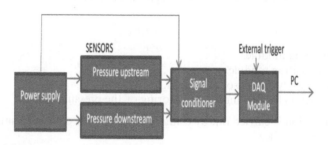

Figure 4. Block diagram of measurement system.

3.2 *Test sequence*

The sequence of operation adopted for the test is shown in Table 1. Six tests with different speeds were carried out. All the tests were conducted in the test bay equipped with nitrogen cylinders, pressure regulators, motorized valves, pressure transducers etc.

Figure 4 shows the block diagram of the measurement scheme. 10V DC Power supply is used for sensor excitation. The pressure measurement is done by using pressure sensors.

4 RESULTS

The obtained experiment results were analysed and graphs were plotted. Time constants obtained and first order systems were designed. Figure 5 shows first order system with different time constants.

4.1 *MATLAB responses*

For obtaining the transient characteristics such as rise time, peak time and settling time of the response curves, MATLAB software is used. The MATLAB responses obtained for one trial test (time constant = 0.0215 seconds) is shown in Figure 6.

The time constant, rise time, peak time and settling time obtained for six different trial tests are given in Table 2.

Comparative study

A comparison of the results obtained in this work with a previous work using pneumatically actuated controller (Sanoj et al. 2013) is tabulated as shown in Table 3.

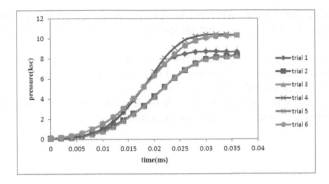

Figure 5. First order system with different time constants.

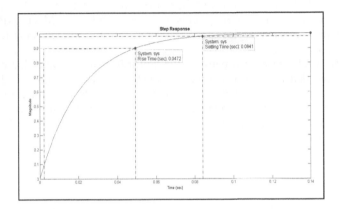

Figure 6. MATLAB response for trial 1.

Table 2. Transient characteristics of different test trials.

	Time constant (s)	Rise time (s)	Settling time (s)	Peak time (s)
Trial 1	0.0215	0.0473	0.0841	>0.14
Trial 2	0.0266	0.0585	0.1040	>0.16
Trial 3	0.0272	0.0598	0.1060	>0.16
Trial 4	0.0199	0.0437	0.0779	>0.12
Trial 5	0.0201	0.0442	0.0786	>0.12
Trial 6	0.0303	0.0666	0.1190	>0.18

Table 3. Comparison of pneumatic and electrical actuators.

		Rise time (s)	Peak time (s)	Settling time (s)
Electrical actuator	Trial 4	0.044	>0.12	0.078
	Trial 2	0.059	>0.16	0.104
	Trial 6	0.066	>0.18	0.119
Pneumatic actuator*	Ziegler-Nichols	0.564	1.68	5.346
	Tyreus-Leyben	0.595	–	17.69
	Fuzzy	1.124	2.64	1.889
	Fuzzy PID	0.52	2.11	1.691

*(Sanoj et al. 2013).

For comparison, three trials were taken for the study. Considering the rise time, the response characteristics of electrical actuators are about ten times faster than that of pneumatic actuators.

While, in the case of pneumatic actuators, even with the use of Fuzzy PID controller, the peak time obtained was 2.10 seconds whereas in case of electrical actuator, the peak time is in the range of 0.12 to 0.18 seconds which indicates a much better response.

Thus it is clear that valve actuation using brushless DC servo motors is much faster and advantageous than the conventional pneumatic actuators.

5 CONCLUSION

The characterization of a ball type control valve actuated by brushless DC motor has been carried out. Different tests were carried out with varying speeds by using an external potentiometer and from the time constants obtained, first order systems were designed. The theoretical response curve for first order systems were generated using MATLAB software and response characteristics such as rise time, peak time, settling time etc were found out from this output response curve.

A comparative study performed between the motor driven valve and that of a pneumatic valve has indicated that the response time are much lower for the motorised valve making the valve responses much faster. Hence a motorised valve can be used for control application that warrants faster response. Rocket engines used in propulsion systems require much faster response, so motorized valves are found to be a very good choice for control valve actuation in piping systems such as those used in the aerospace industries.

REFERENCES

Avila, A., Carvajal, C. & Carlos Cotrino, C. 2014. Characterization of a Butterfly-type Valve, *IEEE transactions, 16(1): 213–218.*

Fisher Controls International. 2005. *Control Valve Handbook.*

Ireneusz, D. & Stanislaw, F. 2014. Characteristics of flow control valve with MSMA actuator, *International Carpethian Control Conference, Krakow, Poland.*

Li, T. Huang, J. Bai, Y. Quan, L. & Wang, S. 2015. Characteristics of a Piloted Digital Flow Valve Based on Flow Amplifier, *International Conference on Fluid Power and Mechatronics.*

Miller, R.W. 1996. *Flow Measurement Engineering Handbook.* New York: McGraw Hill.

Oriental Motors, *Brushless DC Motor and Driver Package,* BLH Series Operating Manual.

Park, J.S. Bon-GwanGu, Kim, Jin-Hong, Choi, Jun-Hyuk & Jung, In-Soung. 2012. Development of BLDC Motor Drive for Automotive Applications, *Electrical Systems for Aircraft, Railway and Ship Propulsion (ESARS).*

Sanoj, K.P., Ajeesh, K.N., Ganesh, P. & Sujithkumar, R. 2013. Characterization of Mass Flow Control System for Liquid Rocket Engine Application, *'Proceedings of National Conference on Advanced Trends in Chemical Engineering', Govt. Engineering College, Thrissur, India.*

Van Neikerk, D. 2015. Brushless Direct Current Motor Efficiency Characterization, *Electrical Machines & Power Electronics (ACEMP).*

Emerging Trends in Engineering, Science and Technology for Society,
Energy and Environment – Vanchipura & Jiji (Eds)
© *2018 Taylor & Francis Group, London, ISBN 978-0-8153-5760-5*

Sequestration of hydrogen sulfide and carbon dioxide from biogas

Antony P. Pallan & S. Antony Raja
Department of Mechanical Engineering, Karunya University, Tamil Nadu, India

B. Sajeenabeevi
Department of Chemical Engineering, Government Engineering College, Kerala, India

C.G. Varma
Department of Live Stock Production Management, Kerala Veterinary and Animal Sciences University, Kerala, India

ABSTRACT: Anaerobic Digestion (AD) of wastes is one of the best treatment methods in the arena of waste management. Biogas, the end product of AD, comprises of 40%–75% CH_4 and 25%–55% CO_2, with other minor components such as H_2S and SO_2. Increased concentration of minor gases will cause the corrosion of the pipe lines which are being used in bio-energy generation. Hence, the present study is undertaken to develop a low-cost scrubbing mechanism for toxic gas removal. A biogas purification system with multi scrubbing apparatus (Phase I and Phase II) was designed and utilized for the study. The principle of chemical absorption was employed and the efficiency of the different caustic solutions at saturated concentration was investigated. Scrubbing at Phase I reported a 5% increase in methane and a 5.8% removal of CO_2. The removal rate of 43% and 37% was observed for H_2S and NH_3 respectively. Carbon dioxide was removed at a rate of 34,9% for KOH, followed by NaOH and $Ca(OH)_2$ at a rate of 34.1% and 33.9% respectively, in a time duration of three minutes. It has been found that the absorption capacity of caustic solution was dropping 'within' a short time period. Hence, it is necessary to replace the caustic solution in order to uphold the chemical at saturation point.

Keywords: anaerobic digestion, scrubbing, chemical absorption

1 INTRODUCTION

Anaerobic Digestion (AD), popularly known as biogas technology, has gained a lot of momentum in the arena of waste management because of its 'dual role', that is, the conversion of waste to energy and the mitigation of Green House Gas emission from the disposal of waste. The end product of AD is mainly biogas, which is principally comprised of 40%–75% CH_4 and 25%–55% CO_2 with other minor components such as H_2S and SO_2 [Kadam & Panwar, 2017]. In India, biogas technology is chiefly employed in the management of manure and farm waste, produced by the activities of agriculture and its allied sectors. These wastes are rich in carbon and nitrogen, which is a prerequisite for the bacteria involved in AD. The C:N ratio of 25–30 is ideal for AD [Sanaei-Moghadam et al., 2014], but due to the practices adopted there is an alteration in the C:N ratio to be either too low or too high. This in turn will have an effect on the metabolism of anaerobic bacteria, finally affecting the composition of the biogas with an increased concentration of undesirable gases such as H_2S, CO_2 and NH_3 [Scano et al. 2014].

An increased concentration of these gases is not suitable for the combustion systems. Carbon dioxide will decrease the calorific value of the biogas because of its non-combustible nature. It is non-toxic, but other gases present such as H_2S are toxic when inhaled [Tippayawong & Thanompongchart, 2010]. Thus, the condition of the biogas plant will go down, affecting the economic returns of the farm and the health of the farmer. Earlier, studies have also revealed that the direct utilization of the pure biogas had decreased the efficiency of the machinery employed [Surata et al, 2014]. Hence, due to this current situation many experimental methods were developed for the purification of biogas, which is broadly known as a scrubbing mechanism. Each method works on a unique principle that targets the removal of gases [Kapdi et al, 2005]. There were many models of scrubbers developed by various researchers, but this present study was conducted to develop a holistic scrubbing mechanism for harvesting purified biogas. The purified biogas, being used for cooking purposes, electricity generation and vehicle fuel, helps to improve the world's energy security. The purified biogas provides the added advantage of an increased efficiency.

2 MATERIALS AND METHODS

2.1 Theory

Chemical absorption is an adequate technology for the removal of CO_2, NH_3 and H_2S from the biogas. In a scrubber system, effluvium from biogas is transferred from gas to liquid as a part of the reaction [Privalova et al., 2013]. Amines, caustic solvent and amino acid salt solutions are the various chemicals which are used in biogas-purification [Abdeen et al., 2016], but caustic solvent is mostly chosen for the purpose of cost efficiency. All aqueous solution for chemical scrubbing was prepared at St Thomas College chemical lab, Thrissur, Kerala. The dissolved caustic salt reacts with CO_2 as part of the purification process [Üresin et al., 2015].

2.2 Experimental setup

A 0.75 m^3 capacity bio-digester (floating drum type) was utilized for this study, located at Eco Farm, Kerala Veterinary and Animal Sciences University (KVASU), Mannuthy, Kerala. Outlet from the digester was connected to a total biogas purification system with multi scrubbing apparatus (Phase I and Phase II), as shown in Figure 1. Both the phases were fabricated with PVC material, 100 cm in height and 20 cm in diameter, with an overall volume ratio of 0.031 m^3. Phase II contains a cup structured reducer (Figure 2) for efficient purification, to delay biogas from bubbling inside the apparatus. The Phase I scrubbing unit is filled with a mixture of limestone with metallic waste, to remove the majority of the NH_3 and H_2S. CO_2 was also filtered out to a certain extent in Phase I, reducing the cost of chemical absorbent in Phase II. Caustic solution in Phase II exchanges CO_2 and forms a precipitate at the bottom, finally yielding purified biogas. Biogas was circulated inside the caustic solution using an external compressor. The raw and purified biogas concentrations at the inlet and outlet from the scrubber units were monitored using an Infra Red Gas analyzer. Pressure and temperature inside the scrubber was

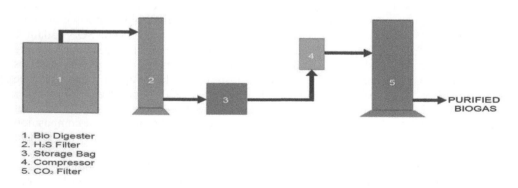

1. Bio Digester
2. H₂S Filter
3. Storage Bag
4. Compressor
5. CO₂ Filter

Figure 1. Basic layout of experimental setup for electricity generation from biogas.

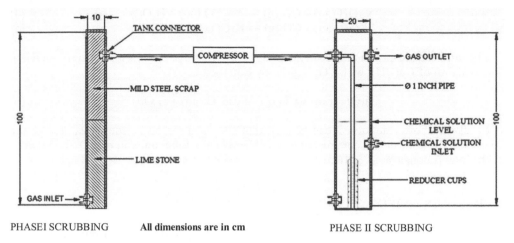

PHASE I SCRUBBING   **All dimensions are in cm**   PHASE II SCRUBBING

Figure 2.   Biogas purification systems.

above atmospheric condition. Sedimentation contents from the caustic scrubbing deposition had accumulated at the bottom side and regular pH measurements were made using a digital pH meter. All chemical reactions were carried out multiple times for precision.

A pilot study was carried out with various combinations of biodegradable waste that had been generated at the university canteen, KVASU, Mannuthy, Kerala. Co-digestion of the veg-etable leftovers with cow dung, maintained at a constant temperature (37°C), gave the maximum biogas yield. Hence, this combination was used when evaluating the efficiency of scrubbing units.

2.3   Chemical reactions

Chemical reactions taking place in the scrubbers are responsible for the removal of unwanted gases from the biogas. The reactions taking place in both the scrubbers are as follows.

2.4   Chemical reactions in Phase I

Phase I is filled with limestone and mild steel scraps. Iron oxide in the mild steel removes H_2S, whereas calcium oxide in limestone removes CO_2 and NH_3, as shown in Equations 1, 2 and 3.

$$H_2S \text{ removal: } 2Fe_2O_3 + 6H_2S \rightarrow 2Fe_2S_3 + 6H_2O \tag{1}$$
$$2Fe_2S_3 + 3O_2 \rightarrow 2Fe_2O_3 + 6S \tag{2}$$
$$CO_2 \text{ removal: } CaO + CO_2 \rightarrow CaCO_3 \tag{3}$$

The calcium carbonate, iron sulfide, calcium and nitrogen formed during these reactions do not have much effect on the performance of the biogas and can be removed along with scrubbing material at regular intervals. Thus, the biogas coming out of scrubber 1 has a lesser amount of CO_2, H_2S and NH_3 than it does at the inlet [Katare et al. 2016].

2.5   Chemical reactions in Phase II

Chemicals like NaOH, KOH and $Ca(OH)_2$ were used for the removal of CO_2. Sodium hydroxide is also a caustic solvent, which his categorized as being a strong alkaline. NaOH ionizes into Na^+ and OH^- in water. CO_2 will be chemically absorbed by NaOH, as shown in Equation 4. Potassium hydroxide (KOH) is an abundant caustic solvent, used for up-gradation of biogas [Cebula, et al. 2009]. KOH reacts with CO_2, as shown in Equation 5. There was no difference between the chemical absorption of CO_2 using KOH and NaOH. KOH is more expensive than NaOH, but it is advantageous due to the K_2CO_3 formed, which has several industrial applications [Ghatak et al 2016].

$$2NaOH\ (aq) + CO_2(g) \rightarrow Na_2CO_3(aq) + H_2O(l) \qquad (4)$$
$$2KOH\ (aq) + CO_2(g) \rightarrow K_2CO_3(aq) + H_2O(l) \qquad (5)$$

The third caustic solvent utilized in the study to absorb CO_2 is calcium hydroxide ($Ca(OH)_2$). The reaction of $Ca(OH)_2$ with CO_2 is given by Equation 6.

$$Ca(OH)_2(aq) + CO_2(g) \rightarrow CaCO_3(aq) + H_2O(l) \qquad (6)$$

In all the above reactions, it can be seen that CO_2 is absorbed by chemicals used in the scrubber. Thus, the biogas coming out of scrubber II has a lesser amount of CO_2 than it does at the inlet [Leonzio et al. 2016].

3 RESULTS AND DISCUSSION

The raw biogas is allowed to pass through the scrubbing system. Initial purification was done using a single column packed unit for the removal of CO_2, H_2S and NH_3, as shown in [Fig. 2]. The composition of raw biogas (i.e. at the inlet and outlet of the Phase I scrubber) is as shown in [Table 1] and [Fig. 3].

Scrubbing at Phase I reported a 5% increase in methane and a CO_2 removal of 5.8%. A removal rate of 43% and 37% was observed for H_2S and NH_3 respectively. Experimental data from Phase I reported better H_2S removal, as compared with [Katare, et al. 2016]. A further purification process was carried out using saturated caustic solution of NaOH, KOH and $Ca(OH)_2$. The results for CO_2 removal are shown in Table 2 and Figure 3. The caustic absorption process varied with time duration (Fig. 4).

It was found that different caustic solutions gave dissimilar CO_2 levels at the scrubber outlet. A nonlinear graph plotted clearly indicated a variation in the CO_2 removal rate, along with the type of caustic solution, and was dependent on time. Carbon dioxide sequestration

Table 1.   Biogas composition at inlet and outlet of Phase 1 scrubber.

Gas	Inlet	Outlet
CH_4	53.1%	58.1%
CO_2	45.8%	40.3%
H_2S	721 ppm	315 ppm
NH_3	75 ppm	28 ppm

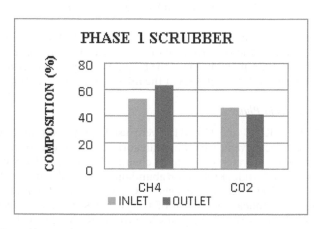

Figure 3.   Phase I scrubber performance.

using different caustics is represented in the Fig. 4. The unique design to delay biogas from bubbling inside the apparatus is the key object for efficient purification. Therefore, the final results reported were better than those of [Ghatak, et al. 2016].

Carbon dioxide was removed at a rate of 34.9% for KOH, followed by NaOH and $Ca(OH)_2$ at a rate of 34.1% and 33.9% respectively, in a time duration of three minutes. All the data in [Table 2] was greater than reported by [Cebula, J. et al. 2009]. It has undoubtedly been indicated that the quality of biogas increased with the use of caustic scrubbing solution. Among all caustic chemicals, KOH reported the maximum efficiency of CO_2 removal rate (Fig. 5).

All caustic solution stability used to purify biogas is clearly plotted in Figure 6 with respect to time, and it was found that KOH had maintained pH stability for a longer time period. $Ca(OH)_2$ solution is very unstable, and the pH dropped rapidly in a time period of 15 minutes.

Table 2. CO_2 removal with caustic solutions for 3 minutes.

Biogas content	$Ca(OH)_2$	NaOH	KOH
CH_4 (%)	94.3	94.6	96.1
CO_2 (%)	6.4	6.2	5.4
CO_2 removal rate (%)	33.9	34.1	34.9

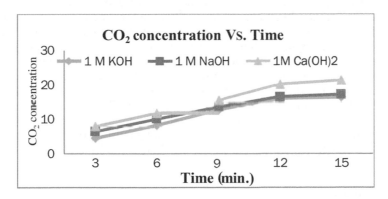

Figure 4. CO_2 removal rate with time.

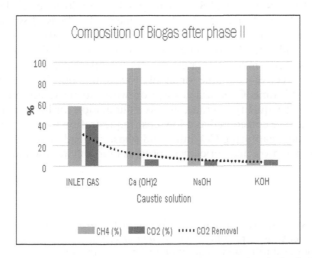

Figure 5.   Variation of composition of CO_2 and CH_4 with different chemical solution.

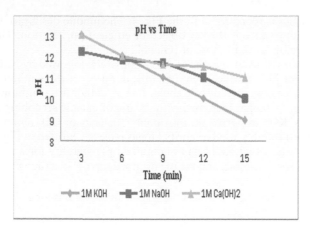

Figure 6.   Variation of pH with time.

It can be seen from the findings that saturated caustic solutions are highly suitable for the continuous purification process. Final results were in accordance with the reported literature. Results obtained were inferior to those from [Tippayawong & Thanompongchart, 2010]. The maximum H_2S and CO_2 removal rate was obtained when compared to the caustic scrubbing system developed by [Ürsein et al. 2015]. Enhanced H_2S absorption was observed, compared to that stated by [Miltner, M., Makaruk et al. 2012]. Efforts were made to create purified methane enriched gas for a sustained period. It has been found that absorption capacity of caustic solution was dropping within a short time period. Hence, it is necessary to replace the caustic solution is necessary in order to uphold the chemicals at saturation point. Regular replacement results in concentration instability and the major drawback in consuming caustic solvents was that they are very problematic to recycle. Although they are comparatively cheaper, a huge quantity of chemicals is essential to fulfill the purification process and to overcome the drop in engine efficiency accounted for by the presence of CO_2 in biogas. Amount utilized in this present study is high for making purified biogas. So, future work is mandatory to reduce the capital cost in biogas enrichment and its applications.

4   CONCLUSION

Sequestration of CO_2, NH_3 and H_2S from biogas by a multistage scrubber was studied. Enhanced H_2S scrubbing effectiveness has been gained in the Phase 1 scrubber. Meanwhile, the effectiveness was found to be dropping with time. NaOH, KOH and $Ca(OH)_2$ were used in the current study and their absorption behaviors were observed. The absorption characteristics of all caustic solvents indicated similar results. Chemical absorption by caustic solution was found to be an effective technique for small process time but their absorption capability weakened quickly with time. Chemical absorption with caustic solution is not advisable as an alternative for biogas quality upgrade, due to its limitation for being reused. Still, caustic scrubbing techniques are considered as a low-cost sequestration method for sub-continental conditions. In addition, capturing CO_2 into solid phase, instead of it being released into the atmosphere, makes the projected enrichment process more environmentally friendly.

REFERENCES

Abdeen, F.R., Mel, M., Jami, M.S., Ihsan, S.I. & Ismail, A.F. (2016). A review of chemical absorption of carbon dioxide for biogas upgrading. *Chinese Journal of Chemical Engineering*, 24(6), 693–702.
Cebula, J. (2009). Biogas purification by sorption techniques. *Architecture Civil Engineering Environment Journal*, 2, 95–103.

Ghatak, M.D. & Mahanta, P. (n.d.) Biogas Purification using Chemical Absorption (2016).

Kadam, R. & Panwar, N.L. (2017). Recent advancement in biogas enrichment and its applications. *Renewable and Sustainable Energy Reviews, 73*, 892–903.

Kapdi, S.S., Vijay, V.K., Rajesh, S.K. & Prasad, R. (2005). Biogas scrubbing, compression and storage: Perspective and prospectus in Indian context. *Renewable Energy, 30*(8), 1195–1202.

Katare, S. (2016). Biogas purification & enrichment through stepped scrubbing. *International Journal of Engineering Sciences & Research Technology, 1*(5), 795–799.

Leonzio, G. (2016). Upgrading of biogas to bio-methane with chemical absorption process: Simulation and environmental impact. *Journal of Cleaner Production, 131*, 364–375.

Miltner, M., Makaruk, A., Krischan, J. & Harasek, M. (2012). Chemical-Oxidative scrubbing for the removal of hydrogen sulphide from raw biogas: Potentials and economics. *Water Science and Technology, 66*(6), 1354–1360.

Osorio, F. & Torres, J.C. (2009). Biogas purification from anaerobic digestion in a wastewater treatment plant for biofuel production. *Renewable Energy, 34*(10), 2164–2171.

Privalova, E., Rasi, S., Mäki-Arvela, P., Eränen, K., Rintala, J., Murzin, D.Y. & Mikkola, J.P. (2013). CO_2 capture from biogas: Absorbent selection. *RSC Advances, 3*(9), 2979–2994.

Sanaei-Moghadam, A., Abbaspour-Fard, M.H., Aghel, H., Aghkhani, M.H. & Abedini-Torghabeh, J. (2014). Enhancement of biogas production by co-digestion of potato pulp with cow manure in a CSTR system. *Applied Biochemistry and Biotechnology, 173*(7), 1858–1869.

Scano, E.A., Asquer, C., Pistis, A., Ortu, L., Demontis, V. & Cocco, D. (2014). Biogas from anaerobic digestion of fruit and vegetable wastes: Experimental results on pilot-scale and preliminary performance evaluation of a full scale power plant. *Energy Conversion and Management, 77*, 22–30.

Surata, I.W., Nindhia, T.G.T., Atmika, I.K.A., Negara, D.N.K.P. & Putra, I.W.E.P. (2014). Simple Conversion method from gasoline to biogas fueled small engine to powered electric generator. *Energy Procedia, 52*, 626–632.

Tippayawong, N. & Thanompongchart, P. (2010). Biogas quality upgrade by simultaneous removal of CO_2 and H_2S in a packed column reactor. *Energy, 35*(12), 4531–4535.

Üresin, E., Saraç, H.İ., Sarıoğlan, A., Ay, Ş. & Akgün, F. (2015). An experimental study for H_2S and CO_2 removal via caustic scrubbing system. *Process Safety and Environmental Protection, 94*, 196–202.

Emerging Trends in Engineering, Science and Technology for Society,
Energy and Environment – Vanchipura & Jiji (Eds)
© 2018 Taylor & Francis Group, London, ISBN 978-0-8153-5760-5

Development of ceramic membrane for microfiltration application in biotechnology field

C. Mohit Kumar & D. Vasanth
Department of Biotechnology, National Institute of Technology, Raipur, India

ABSTRACT: A new composition of precursors was identified using china clay, quartz and calcium carbonate to fabricate the microfiltration membrane. The membrane was fabricated by pressing method and sintered at 1000°C. Various characteristics of membrane such as porosity, average pore size, water permeability and chemical resistance were evaluated. Energy Dispersive X-ray analysis (EDX) was conducted to identify the elements present in the membrane. The porosity, water permeability and pore size of membrane are found to be 37%, 2.88×10^{-3} L/m^2.h.Pa and 555 nm respectively. Corrosion resistance test indicates that the membrane can be subjected to acid and alkali based cleaning procedure.

1 INTRODUCTION

In recent years, the preparation of clay based inexpensive ceramic membrane is getting significant attention due to its cost benefits. The ceramic membrane could be deployed to highly corrosive medium and high pressure applications. Numerous articles have been published for the preparation of clay based ceramic membranes. Nandi et al. (2009) formulated new composition of raw materials using kaolin, quartz, sodium carbonate, calcium carbonate, boric acid and sodium metasilicate to synthesize circular membrane. The prepared membrane was deployed for the purification of oil-water emulsions. Abbasi et al. (2010) utilized kaolin, clay and α-alumina to synthesize mullite and mullite–alumina based ceramic membranes for the separation of oil emulsions. Similarly, the ceramic membrane was manufactured using mixture of kaolin, pyrophyllite, feldspar, quartz, calcium carbonate, ball clay and titanium dioxide by uniaxial compaction method. The membrane was performed well for the treatment of oil-water emulsions (Monash & Pugazhenthi 2011). Using perlite materials, Al-harbi et al. (2016) prepared the super-hydrophilic membrane (mean pore size of 16 µm) for the wastewater treatment applications. In another work, dairy wastewater was treated using novel tubular ceramic membrane. The membrane was prepared using mixture of naturally available clays (Kumar et al. 2016). Jeong et al. (2017) used the pyrophyllite and alumina to prepare composite ceramic membrane (pore size of 0.15 µm). The membrane performance was investigated for the treatment of low-strength domestic wastewater.

The detailed investigation of above literatures indicates that the clay based membranes were mainly deployed for wastewater treatment applications. To best of our knowledge, the applicability of clay based ceramic membrane in biotechnological field is less studied. In this context, the applicability of clay based ceramic membrane for biotechnological field needs to be investigated. Such research would be useful to understand upon the suitability of ceramic membrane for biotechnological applications.

This article addresses the preparation of ceramic membrane using china clay, quartz and calcium carbonate. Primary characteristics such as porosity, average pore size, water permeability was evaluated. Corrosion resistance test was conducted to identify the suitable cleaning procedures and its stability in highly corrosive medium. Eventually, the production cost of membrane was determined to compare the cost of other reported membrane in literatures.

2 EXPERIMENTAL

2.1 *Precursors*

The raw materials namely china clay, quartz and calcium carbonate were used to develop ceramic membrane and its composition is presented in Table 1. China clay and quartz were purchased from Royalty minerals, Mumbai, India. The calcium carbonate was procured from Loba Chemie, Ltd. The materials were used without any pretreatment.

2.2 *Synthesis of membrane*

Using uni-axial compaction method, the ceramic membrane was prepared with newly identi-fied precursor composition. Different sequential steps involve in the preparation of mem-branes is illustrated in Fig. 1. Initially, the raw materials were uniformly mixed with 4 ml of 2% polyvinyl alcohol (binder) using mortar and pestle. The necessary amount of powder was then poured in SS mould and pressed at higher pressure (50 MPa) using hydraulic press (Guru Ramdas Machine Works, Raipur, India) to cast the membrane (52 mm diameter and 5 mm thickness). The obtained membrane was dried at 100°C to remove moisture present in the membrane using hot air oven (Pooja Scientific Instruments, India). Subsequently, the membrane was sintered at 1000°C for 5 h in a muffle furnace (Nanotec, Chennai, India). After that, the membrane was polished at both sides using abrasive sheet (No. 220) and washed in ultrasonicator (PCI Analytics, India). Eventually, the membrane was dried at 100°C for 12 and subjected to the various characterizations. The photograph of prepared membrane is shown in Fig. 2.

Table 1.   Composition of raw materials.

Materials	Quantity (gm)
China clay	50
Quartz	25
Calcium carbonate	25

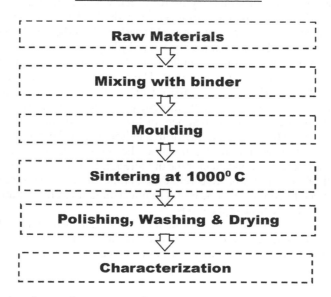

Figure 1.   Flow chart for membrane preparation.

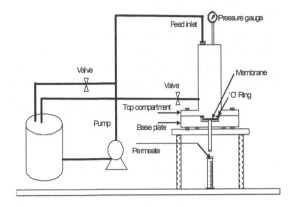

Figure 2.   Image of prepared membrane.   Figure 3.   Experimental setup.

2.3   *Characterization*

The primary characteristics of membrane such as porosity, water permeability, average pore size, chemical resistance were evaluated. The porosity of membrane was evaluated by Archemede's principle using water as wetting liquid (Nandi et al. 2009). The water flux (J, L/m^2h) of membrane was measured using indigenous continuous dead-end filtration setup (Fig. 3). The flux was measured at different applied pressure (69–345 kPa) at room temperature (25°C). This involves the measurement of permeate volume at an interval of 5 min during the total run time of 25 min. The flux was calculated using the relation.

$$J = Q/A \times t. \qquad (1)$$

where, Q is the volume of permeate collected, t is the time and A is effective membrane area (m^2) for permeation.

The water permeability (L_h) and average pore size (r_l) of the membrane was determined from water flux data according to the following expression

$$J_v = \frac{\varepsilon r^2 \Delta P}{8\mu l} = L_h \Delta P \qquad (2)$$

where J_v (L/m^2h) is the water flux through the membrane, ΔP (kPa) is the trans-membrane pressure drop across the membrane, μ is the viscosity of water, l is pore length, ε is the porosity of the membrane.

The corrosion resistance of membrane was tested in acid and alkali solutions individually at different pH levels (1–14) using HCl and NaOH. To do so, the membrane was kept in contact with acid and alkali solutions for seven consecutive days at room temperature. After that, the weight loss of membrane was measured that characterizes the corrosion resistance of membrane.

In addition, EDX was performed to confirm the elements present in the membranes.

3   RESULTS AND DISCUSSION

3.1   *Porosity*

The porosity was determined by Archimedes principle using water as wetting liquid. Generally, the pore size of membrane is depending upon the porosity. In this work, the porosity of membrane is found to be 37%. In this context, it can be pointed out that the obtained porosity (37%) is comparable even higher than the porosity of cordierite (36%), mullitte (32%) and kaolin (36%) (Dong et al. 2007, Abbasi et al. 2010, Monash & Pugazhenthi 2011). Thus, it is inferred that the formulated composition of raw material provides higher porosity.

3.2 *Water flux and permeability*

The water flux was measured at different trans-membrane pressure (69–345 kPa) for total run time of 25 min at time interval of 5 min. The time versus flux at different applied pressure is presented in Fig. 4. It can be observed that the flux is constant over entire studied time. It indicates that there is no any resistance for pure water flow. Further, it is noted that water flux increases linearly with increasing pressure (Fig. 5). Using the water flux data, water permeability of membrane is evaluated to be 2.88×10^{-3} (L/m²hPa).

3.3 *Average pore size*

Using the water flux data, the average pore diameter of membrane is evaluated to be 555 nm. In this context, it can be pointed out that the reported pore size (555 nm) is significantly lesser than the average pore size obtained with different precursors namely Moroccan clay (10.75 μm), cordierite (8.66 μm) and apatite (5 μm) (Saffaj et al. 2006, Dong et al. 2007, Masmoudi et al. 2007). The obtained membrane with lesser pore size can be applied for various biotechnology applications.

3.4 *Corrosion resistance*

Corrosion resistance test was performed to identify the suitable chemical cleaning procedure and its application in highly corrosive medium. In view of this, the membrane was subjected to treat in acid and alkali solutions at different pH range (1–14). It is observed that there is no significant weight loss in both acid and alkali. This result shows the membrane have excellent corrosive resistivity that can be used for any separation application whose pH range between 1 and 14. Further, the membrane can be subjected to acid and alkali based cleaning procedure.

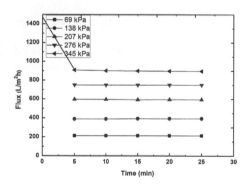

Figure 4.   Time vs Flux.

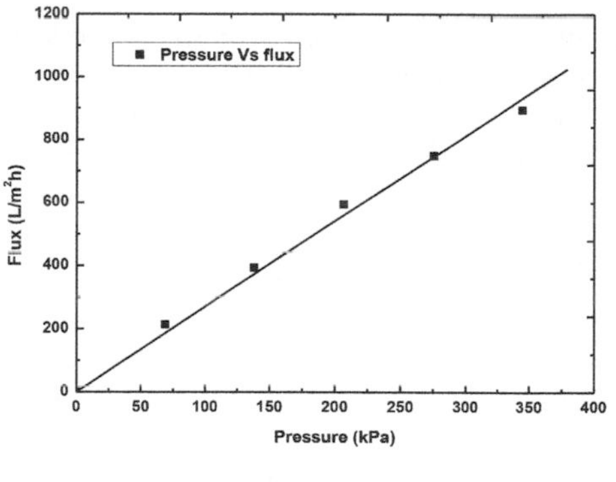

Figure 5.   Pressure vs Flux.

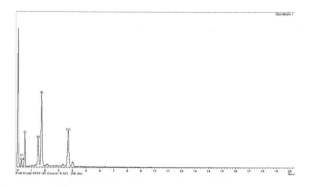

Figure 6.   EDX analysis of membrane.

Table 2. EDX analysis of membrane.

Elements	Weight (%)
Carbon	16.24
Oxygen	58.01
Aluminum	5.95
Silicon	14.85
Calcium	4.95

Table 3. Determination of cost.

Raw materials	Unit price/kg (Rs.)	Raw materials used for preparation of one membrane (gm)	Cost of one membrane (Rs)
China clay	11.44	10	0.11
Quartz	11.40	5	0.06
Calcium carbonate	297	5	1.48
Total			1.65

Table 4. Summary of characterization results.

Properties	Values
Porosity (%)	37
Water permeability (L/m^2 h Pa)	2.88×10^{-3}
Pore size (nm)	555

3.5 *EDX analysis*

Fig. 6 illustrates the EDX analysis of membrane. The EDX graph was obtained to identify the elemental composition of membrane. It is analyzed that the main elements that are
present in the membrane are carbon, oxygen, aluminum, silicon and calcium and their relevant compositions are present in Table 2.

3.6 *Cost estimation*

The research in the preparation of low cost ceramic membrane is intensified worldwide.
Generally, the cost of any membrane is depending upon the raw material cost and fabrica-
tion method. Therefore, the selection of inexpensive raw materials and simple fabrication
technique is vital. Based on the raw material cost, the cost of membrane is estimated to be
Rs. 1.65 (5 mm thickness and 52 mm diameter). The details of cost estimation are shown in
Table 3. Moreover, the overall characterization results is presented in Table 4.

4 CONCLUSIONS

A new raw material composition was identified for the preparation of ceramic microfiltration membrane. The porosity, water permeability and pore size of membrane are found to
be 37%, 2.88×10^{-3} L/m^2.h.Pa and 555 nm, respectively. Corrosion resistance test indicates
the membrane can be subjected to acid and alkali based cleaning procedure. Henceforth, it
is concluded that the fabricated ceramic membrane possesses very small pore size which is
suitable for various biotechnology applications.

ACKNOWLEDGEMENT

The authors wish to express their sincere thanks to DST-SERB for the financial support.

REFERENCES

Abbasi, M. Mirfendereski, M. Nikbakht, M. Golshenas, M. & Mohammadi, T. 2010. Performance study of mullite and mullite–alumina ceramic MF membranes for oily wastewaters treatment. *Desalination* 259: 169–178.

Al-Harbi, O.A. Mujtaba, K.M. & Ozgur, C. 2016. Designing of a low cost super-hydrophilic membrane for wastewater treatment. *Materials and Design* 96: 296–303.

Dong, Y. Feng, X. Dong, D. Wang, S. Yang, J. Gao, J. Liu X. & Meng, G. 2007. Elaboration and chemical corrosion resistance of tubular macro-porous cordierite ceramic membrane supports. *Journal of Membrane Science* 304: 65–75.

Jeong, Y. Lee, S. Hong, S. & Park, C. 2017. Preparation, characterization and application of low-cost pyrophyllite alumina composite ceramic membranes for treating low-strength domestic wastewater. *Journal of Membrane Science* 536: 108–115.

Kumar, R.V. Goswami, L. Pakshirajan, K. & Pugazhenthi, G. 2016. Dairy wastewater treatment using a novel low cost tubular ceramic membrane and membrane fouling mechanism using pore blocking models. *Journal of Water Process Engineering* 13: 168–175.

Masmoudi, S., Larbot, A., El Feki, H. & Amar, R.B. 2007. Elaboration and characterisation of apatite based mineral supports for microfiltration and ultrafiltration membranes. *Ceramics International* 33: 337–344.

Monash, P. & Pugazhenthi, G. 2011. Effect of TiO_2 addition on the fabrication of ceramic membrane supports: a study on the separation of oil droplets and bovine serum albumin (BSA) from its solution. *Desalination* 279: 104–114.

Nandi, B.K. Uppaluri, R. Purkait, M.K. 2009. Treatment of oily wastewater using low cost ceramic membrane: flux decline mechanism and economic feasibility. *Separation Science and Technology* 44: 2840–2869.

Saffaj, N. Persin, M. Younsi, S.A. Albizane, A. Cretin M. & Larbot, A. 2006. Elaboration and characterization of microfiltration and ultrafiltration membranes deposited on raw support prepared from natural moroccan clay: application to filtration of solution containing dyes and salts. *Applied Clay Science* 31: 110–119.

Emerging Trends in Engineering, Science and Technology for Society,
Energy and Environment – Vanchipura & Jiji (Eds)
© *2018 Taylor & Francis Group, London, ISBN 978-0-8153-5760-5*

Studies on carbon felt/phenolic composites as light weight advanced ablative TPS for launch vehicle programmes

C.Y. Lincy
Department of Chemical Engineering, GEC Thrissur, Kerala, India

V. Sekkar, Vijendra Kumar & Ancy Smitha Alex
Chemical Systems Group, PCM Entity, Vikram Sarabhai Space Centre, Thiruvananthapuram, Kerala, India

Saurabh Sahadev
Department of Chemical Engineering, GEC Thrissur, Kerala, India

ABSTRACT: Carbon felt based phenolic composites were prepared at various densities viz. 0.3, 0.4, 0.5, 0.6 and 0.7 g/cc. The mechanical and thermal characteristics of the composites were evaluated adopting standard test procedures. With the change in density from 0.3 to 0.7 g/cc, flexural strength enhances from 5.2 to 32 MPa, the compressive strength increases from 1.1 to 6.9 MPa, tensile strength is pushed up to 225 from 45 KSC while resilience improves from 1.5 to 1.8 kJ/m². Over the change of density, thermal conductivity of the composites marginally increases from 0.1 to 0.2 W/mK. Char yield for the composites as determined by thermogravimetric analysis was about 60% up to 900°C. Ablative characteristics were determined through plasma arc jet simulated test procedures at an energy flux of 50 W/cm² and time duration of 10 seconds; the heat of ablation was 2044 cal/g, and composite with 0.7 g/cc density had survived the erosion test with an erosion rate of 0.002 mm/s and a mass loss of 0.005 g.

1 INTRODUCTION

In a launch vehicle momentum for the rocket is gained through conversion of chemical energy to mechanical energy. In a typical propulsion system, fuel and oxidizer undergo combustion thereby generating low molecular weight gases at high temperatures (3500 K) and pressure (200 bar). When this hot high pressure gas is expanded through nozzle, thrust is created and the rocket experiences forward motion. For most of inner wall portions of the nozzle temperature would be in the range 700–1200 K. All known structural materials cannot survive under such severe erosive and thermal shock conditions. Thus the metals form nozzle needs protection from very high speed extremely hot gas streams. Traditionally, highly dense silica/phenolic and carbon/phenolic composites (1.8 g/cc) are employed as nozzle liners. Ablation lining thickness is about 8–12 mm. It is envisaged that the fully dense liner is replaced with a porous, low dense and light weight material with a thick anti erosion coating, a huge reduction in weight of the nozzle with improved performance as the porous material would have a much lower thermal conductivity than the fully dense counterparts. In view of the above, carbon felt impregnated phenolic composites are investigated for their role as light weight ablative liners. Carbon felt-based ablators have several advantages over classical fully dense ablative liner materials. Most importantly, they reduce the limited strain response of large rigid substrates. Carbon felt materials are known for their benign insulating property, uniform bulk density and better shape retention properties with macro and micro communication channels among the cells allowing efficient resin infiltration. Enabling manufacturing in larger sizes, felt based substrates reduce the number of independent parts mitigating the need of gap fillers. They also offer improved robustness in absorbing loads and deflections, and they allow shaping the substrate around complex geometries with the possibility of maintaining a uniform and low thermal conductivity in the system. In this study, the evaluation of mechanical as well as thermal performance of the composites are carried out as the preliminary screening exercise before such low density composites are considered for the intended applications. Because of the highly porous nature, it enhances efficient infiltration of the resin and accounts for its uniform properties.

2 MATERIALS AND METHODS

2.1 *Materials required*

The materials used for the preparation of carbon felt based phenolic composites are phenolic resin PF108, carbon felt and acetone as the solvent. Phenol formaldehyde resin (PF 108) was produced at PFC, VSSC, Trivandrum. Carbon felt was procured from Mersens India Limited, Bangalore. Acetone was procured from Merck Specialities India Limited, Bangalore.

2.2 *Preparation of carbon felt/phenolic composites*

The carbon felts were cut to the required size. Required amount of PF 108 resin was dissolved in acetone and the solution was completely absorbed over carbon felts. The solvent was driven

Table 1. Calculation of weight of PF108 for preparation of composites of each densities.

Composite name	Required density (g/cc)	Felt volume (cc)	Felt weight (g)	Weight of PF108 (g)
A	0.3	402	49.72	98.42
B	0.4	393	50.39	161.44
C	0.5	395	51.5	204.53
D	0.6	393	52.5	254.58
E	0.7	391	53.6	305.65

Figure 1. Carbon felt/phenolic composites after preparation.

off by drying the soaked felt at 50°C for 18 hrs. Curing was undertaken at 120°C for 1 hr and at 180°C for 4 hrs. In a similar way felt composites with densities ranging from 0.3 to 0.7 g/cc were prepared. The cured composites were characterized for their mechanical and thermal properties, such as resilience, tensile strength, compressive strength, flexural strength and thermal conductivity, adopting procedures conforming to relevant ASTM standards.

3   RESULTS AND DISCUSSIONS

3.1   *Mechanical properties*

Carbon felt based phenolic composites with densities varying between 0.3 and 0.7 g/cc were prepared and characterized for both mechanical and thermal characteristics relevant to their application as low density ablative nozzle liner applications. Phenolic Carbon felt composites may be better suited to intended applications chiefly due to spectacular specific strength capabilities and very low thermal conductivities. PF resin during the curing process looses as much as 25% by weight which is considered in calculating the resin intake. In addition to that, felt offers easy handling and structural tailor ability for the composites. Phenolic composites are best suited for ablation. Phenolic composites are not inflammable but undergo endothermic decomposition which effectively remove heat from the system and cool it. Further, carbon derived from ablation of phenolic composites belongs to the class of glassy carbons which are known for their extremely low thermal conductivity at fairly high temperatures.

Carbon felt/phenolic composites thus satisfy three important requirements for an ablative TPS, viz.: good mechanical characteristics, low thermal conductivity and low density. In particular, the density is characterized by a lower limit dictated by the necessity of a low recession rate, and an upper limit dictated by the light weight requirements of all the aerospace parts. Carbon felt based phenolic with low densities during ablative process contribute to the formation of a porous char skeleton where the gases can flow through as the degradation reaction evolves. Thus it helps in ablative cooling and limiting heat transfer and serves as good TPS material.

In general, all mechanical characteristics increase with density. Structural strength component comes from the phenolic matrix. However, the fibrous portion of the felt thoroughly and uniformly reinforce the phenolic matrix, which contributes surprisingly very high strength for the densities encountered in this study. Compressive strength, flexural strength and tensile strength increase linearly with density while impact strength has non linear dependence on density. Impact strength obtained in this study is fairly high for the densities of the composites. Materials at the similar densities without felt would be significantly brittle and would associate with much lower strength figures. It is evident that the resin is completely responsible.

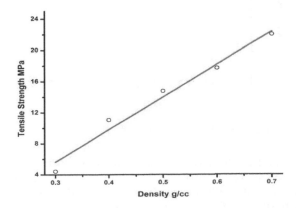

Figure 2.   Variation of tensile strength with density.

639

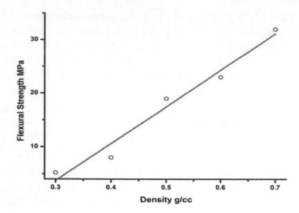

Figure 3. Variation of flexural strength with density.

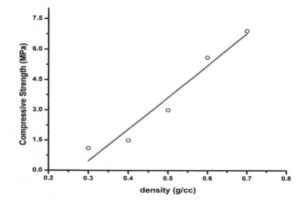

Figure 4. Variation of compressive strength with density.

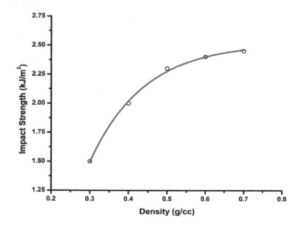

Figure 5. Variation of impact strength with density.

With a density variation from 0.3 g/cc to 0.7 g/cc, the flexural strength changed from 5.2 to 32 MPa, the resilience changed from 1.5 to 2.5 kJ/m^2, compressive strength changed from 1.1 to 6.9 MPa and tensile strength changed from 4 to 22 MPa. The compressive and tensile strength of the composite with 0.7 g/cc density is highly remarkable when compared to the conventional TPS materials for the given density.

640

3.2 Thermal conductivity

Thermal conductivity does not get altered much during the density range. All the composites have thermal conductivity between 0.1 and 0.2 W/mK. This is a very favorable aspect as the felt composites have a wonderful combination of remarkable mechanical strength, low density and very low thermal conductivity. Significantly larger pore spaces are reason for extraordinary lower thermal conductivity. The thermal conductivity values are much lower than the typical high density counterparts. Thus, combination of low thermal conductivity with fairly higher mechanical strength parameters makes them ideal candidates for ablative TPS applications.

3.3 Thermo gravimetric analysis

The thermo gravimetric analysis conducted on the sample of density 0.7 g/cc and weight 5.92 mg showed that the decomposition of the sample occurs in the temperature range between 300°C to 600°C.

The TGA curve is showed in Figure 7. It is evident from the analysis that the residue remained at 900°C is 59.64%. From the different analyses conducted on the specimen, it is obvious that the carbon felt/phenolic composites show superior performance. In order to investigate more on its ablation characteristics, a plasma arc jet test was also conducted which is explained in the following section.

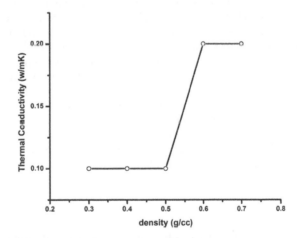

Figure 6.   Variation of thermal conductivity with density.

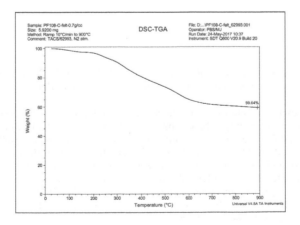

Figure 7.   TGA curve for carbon felt/phenolic composites.

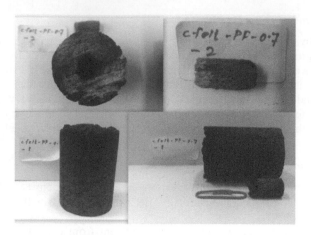

Figure 8. Specimen after ablation test.

3.4 *Ablation test*

In order to screen a material for its aptitude for ablation application, it needs to be subjected to plasma arc jet test, by which important ablation characteristics, such as heat of ablation, erosion rate and mass loss are evaluated. Plasma arc jet test was performed adopting the procedure conforming to test standard ASTM E258. The samples are tested at a heat flux of 50 W/cm^2 for 10 seconds; the test was conducted in environment of argon Ablative properties were evaluated for carbon felt reinforced phenolic with varying densities viz. 0.3 and 0.7 g/cc. However, the composite with 0.3 g/cc could not tolerate the thermo chemical and mechanical stresses and it failed during the ablative evaluation. Composite at 0.7 g/cc survived the erosion test with an erosion rate of 0.002 mm/s and a mass loss of 0.005 g. For a heat flux of 50 W/cm^2, the heat of ablation was found to be 2044 cal/g. The values obtained are very promising qualifies the composite as a possible TPS material. Further investigations at higher heat flux are necessary for the complete analysis of its ablation property.

4 CONCLUSION

The carbon felt phenolic composites exhibit superior mechanical properties. Thermal conductivity values are very small. Low density felt composites are thus attractive materials for nozzle liner end uses.

REFERENCES

de Almeida, L., Cunha, F., Batista N., Roccol, J., Iha K. & Botelho E. 2014. Processing and characterization of ablative composites used in rocket motors. *Reinforced Plastics and Composites* 33(16): 1474–1484.
Francesco Panerai, Joseph C. Ferguson Jean Lachaud, Alexandre Martin, Matthew J. Gasch & Nagi N. Mansour. 2016. Micro-tomography based analysis of thermal conductivity, diffusivity and oxidation behavior of rigid and flexible fibrous insulators. *Heat and Mass Transfer.* 108(2017) 801–811.
Ganeshram, V. & Achudhan, M. 2013. Synthesis and characterization of phenol formaldehyde resin as a binder used for coated abrasives. *Science and Technology.* (6S), 0974-6846.
Varadarajulu & Rama Dev, R. 2008. Flexural properties of ridge gourd/phenolic composites and glass/ridge gourd/phenolic hybrid composites. *Composite materials.* 42, 6/2008.
Vineta, S. & Gordana B. 2009. Composite material based on an ablative phenolic resin and carbon fibers. *Serbian chemical society.* 74(4): 441–453.

Emerging Trends in Engineering, Science and Technology for Society,
Energy and Environment – Vanchipura & Jiji (Eds)
© 2018 Taylor & Francis Group, London, ISBN 978-0-8153-5760-5

Desalination studies on two single-sloped solar stills with heat-absorbing materials as coating material on aluminum basin

Pativada Suman, Deepa Meghavathu & Sridevi Veluru
Department of Chemical Engineering, Andhra University College of Engineering,
Visakhapatnam, Andhra Pradesh, India

Zakir Hussain
Department of Chemical Engineering, Rajiv Gandhi Institute of Petroleum Technology, Jais,
Amethi, Uttar Pradesh, India

ABSTRACT: Fresh water is the component required for the survival of many living organisms. Though 71% of the earth's surface is covered by water, only 3.5% is available for human needs, with the remaining 96.5% in the form of oceans. Large quantities of water are consumed by industries due to rapid industrialization, thus polluting the fresh water and further causing its scarcity. To meet the demand, solar energy can be used to convert seawater into fresh water in a solar distillation unit using a solar still. The solar still should possess high absorptivity so that it captures the maximum solar energy and converts the seawater to fresh water. In this paper, the effect of different materials such as black paint, paraffin wax, coal bed, sand and ceramic packings, as coating materials in various combinations to the base aluminum basin, was studied in a detailed way. Energy balance and heat transfer coefficients were estimated for all the cases. Further, the overall efficiency was also estimated for all the cases and compared. The black paint coated basin (base one) has shown the least efficiency (42.19%) among all considered cases, and the basin coated with the combination of black paint, sand and ceramics has shown the highest efficiency (65.06%).

Keywords: desalination, solar stills, heat-absorbing, energy balance, heat transfer coefficient

1 INTRODUCTION

Water is the main source of survival for many living organisms, and humans especially rely on fresh water for consumption and household purposes. Around 71% of the earth's surface is covered with water but only 3.5% is available in the form of fresh water, while 1.7% is in the form of glaciers, 1.7% is in the form of groundwater and 0.1% is in the form of rivers and lakes (Sethi & Dwivedi, 2013). Not only humans but for all the life forms, water is an essential commodity for their survival. The fresh water availability is less but the demand for it is increasing day by day at a rapid rate, due to the increase in the population and also due to rapid industrialization. Industrial wastes are disposed directly into the available fresh water sources, thus polluting them and creating a rapid decline in the availability of the fresh water (Panchal, 2015). According to the WHO (World Health Organization, 2017), around 25% of the human population does not have the provision of safe drinking water. With the population growing by 82 million every year, the need for safe drinking water is increasing day by day and as many as a third of humans will face a shortage of water by 2025 (Omara & Kabeel, 2014). To mitigate and overcome this fresh water problem, many water purification techniques are available for the production of clean water. Solar distillation is one such technique that converts seawater into potable water. Solar energy is abundantly available in nature and can be successfully utilized in treating the saline water, thus creating a demand for solar distillation that is increasing every year. The basic purpose of solar distillation is

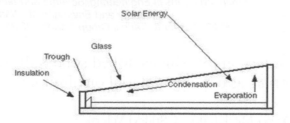

Figure 1.   Schematic diagram of solar desalination unit.

to provide fresh drinking water from seawater in order to meet the demand for fresh water (Somanchi et al., 2015). The process of solar distillation occurs in two ways based on the energy consumption, namely thermal and non-thermal processes. In the solar distillation unit, there are two types of solar stills that are used based on the energy utilization, namely active and passive solar stills. Among these, active solar stills need an energy input in the form of pumps for the input of feed into the system, whereas passive systems do not require energy input. Hence, passive solar stills are preferred over active as they are more cost-effective. There are many designs for solar stills, such as a single-slope solar still (Krishna et al., 2016), double slope solar still (Murugavel et al., 2008), wick-type still (Suneesh et al., 2016), spherical still (Dhirman, 1988), vertical still (Boukar, 2004), and multi-effect still (Tanaka et al., 2009). The single-slope solar still provides a better productivity in the winter whereas the double slope provides better results in the summer (Yadav & Tiwari, 1987). The design and fabrication of a single-slope solar still is much easier when compared to that of the other designs as this can be made from locally available materials like wood and polyurethane. Further, the design involves a low maintenance cost and skilled labor is not required (Gugulothu et al., 2015). A single-slope solar still is considered in the present work. The inclination of the single-slope solar still must be equal to the latitude of the experimental location, in order for the maximum solar irradiance to fall on the still to get the maximum yield (Malaeb et al., 2014). The addition of a sand bed to the conventional solar still gives a higher productivity than with the normal conventional solar still; the efficiency is increased from 35% in a conventional solar still, to 49% in a sand-bed solar still with a sand bed of height 0.01 m above the normal conventional solar still basin (Omara & Kabeel, 2014). The operating principle of the solar still involves the evaporation of the pure water molecules from the saline water due to the impingement of the solar energy onto the saline water, evaporating out the water molecules and leaving behind the dissolved salts and impurities at the bottom of the solar still. The process of impingement of solar energy occurs by the incidence of the sun's radiation on the glazing material of the still. This allows the heat-absorbing material to absorb the radiation and heats up the seawater present, with the evaporated water condensing and being collected through a collecting channel (Aburideh et al., 2012). A schematic diagram showing the principle is shown in Figure 1.

2   MATERIALS AND METHODS

A single-slope distillation unit is made up of wood for the external body of the box with dimensions of 0.715 m in length, 0.415 m in breadth, the height of the smaller edge is 0.125 m, the height of the longer edge is 0.36 m, and with a thickness of 0.01 m. Inside the box, there is 0.02 m thick insulation that is made up of thermocol sheets. The basin that holds the seawater is made up of aluminum (length 0.64 m, breadth 0.33 m, and height of 0.085 m) and is coated with black paint. The glazing material, which is the glass cover that covers the top of the unit, is 0.004 m thick, with an inclination of 18° that results in a heat-absorbing area of 0.3154 m². A collector is attached on the glass cover at the lower end to collect the condensate. To avoid the glass slipping down and to reduce the vapor losses, a rubber seal is provided between the wooden box and the glass cover. The base experimental setup is shown in Figure 2.

Figure 2.   Solar distillation unit with black paint coated basin (base case).

2.1   *Different parts of the solar distillation unit*

2.1.1   *Still basin*
This is an important part of the solar distillation unit, which is used to hold the water to be distilled. Materials with high absorptivity are preferred as basin materials, so that the maximum heat energy can be absorbed. Various materials can be used as basin materials, such as leather sheet, GE silicon, aluminum, copper, paraffin wax, sand, coal and black paint. The present study is carried out using a base case consisting of an aluminum basin coated with black paint. Further, different materials such as coal bed, sand, paraffin wax and ceramic packing are used as extra layers in different combinations.

2.1.2   *Side walls*
The side wall material for the construction of a box to house the solar still should possess enough rigidity to hold the still, without any disturbance due to external forces such as wind. Further, the side wall acts as an insulation for effective heat transfer that takes place from the system to the surroundings. The material of the side wall should possess very low thermal conductivity in order for it to act as a good insulation agent. Different materials that can be utilized as side wall material are wood, RPF (Reinforced Plastic with Fiber), thermocol, and concrete. In the present study, wood is considered as the side wall material and for better insulation purposes, thermocol sheets (0.02 m thickness) are used inside the wooden box.

2.1.3   *Top cover*
The passage where irradiation occurs on the surface of the basin is the top cover. Also, it is the surface where condensate collects. A suitable material is selected as a top cover, based on a few features such as its transparency to solar radiation, its non-absorbency of water, and having a clean and smooth surface. The materials that can be used are glass and polyethylene. In the present study, the glass material of 4 mm thickness is selected as the top cover (size: 76×41.5 cm).

2.1.4   *Channel*
The channel is used to collect the condensate that has formed on the surface of the top glass cover. The materials that can be used are PVC, galvanized steel, and RPF. In the present study, PVC material of 1 inch diameter is used.

2.2   *Experimental procedure*

The single-slope solar distillation unit with an aluminum basin coated with black paint was kept in the sunlight from 9:30 a.m. to 12:30 p.m. in the easterly direction, and from 12:30 p.m.

to 5:30 p.m. in the westerly direction. For paraffin wax and coal bed materials, experimentation continued until 7:30 p.m. For every hour, the solar radiation was measured by a pyranometer. The temperatures were measured at different locations: the ambient temperature (T_a), the top glass cover outside temperature (Tg_o), the top glass cover inside temperature (Tg_{in}), the moist temperature (T_m), the basin temperature (T_b) and the distillate temperature (T_d), with the help of thermometers. Further, the produced distillate was collected for every hour using a measuring cylinder. The same experimental procedure is carried out for treating different capacities of seawater (2, 2.5, 3 liters) and for different materials used as coatings on the aluminum basin in different combinations.

2.3 Data analysis

The efficiency of a still (η) is calculated by using the following Equation:

$$\eta = \frac{\text{m*latent heat of vaporization of water}}{\text{rg*Is}} * 100 \tag{1}$$

2.3.1 Energy balance

The energy balance for the distillation unit is given as:

$$
\begin{aligned}
I_s A_g/A_b = I_s r_g A_g/A_b + q_{g,s} A_g/A_b + q_{h,g} A_g/A_b + q_{k,air} A_{k,air}/A_b \\
+ q_{k,l} A_{k,l}/A_b + q_{k,b} A_b/A_b + (m_{cw} h_{sat,g}) A_b
\end{aligned}
\tag{2}
$$

where

I_s	Solar radiation intensity, W/m²;
A_g	Area of glass surface, m²;
A_b	Area of basin surface, m²;
r_g	Reflectivity of the glass cover for visible light;
$A_{k,air}$	Circumferential area of solar still covered by inside moist air, m²;
$A_{k,l}$	Circumferential area of solar still covered by seawater, m²;
m_{cw}	Mass velocity of condensed water, kg/m².sec;
$h_{sat,g}$	Enthalpy of water at saturation temperature, kJ/kg.

Considering the heat transfer from the cover to the atmosphere by convection:

$$q_{h,g} = h_g(T_g - T_a), \quad \text{W/m}^2 \tag{3}$$

where T_g is the glass temperature, T_a is the ambient temperature, and h_g is the convective heat transfer coefficient given by the following formula:

$$h_g = 5.7 + 3.8\ v, \quad \text{W/m}^2 \tag{4}$$

where the forced convection coefficient dependent on the wind velocity, v (m/s) = 3.5 m/sec.

The radiative heat transfer from the glass cover to the atmospheric air is given by the formula:

$$q_{g,s} = \varepsilon_g\ C_s\ [(T_g/100)^4 - (T_{sky}/100)^4], \quad \text{W/m}^2 \tag{5}$$

where

Glass emissivity	0.88;
Constant C_s	5.667 W/m² K⁴;
T_{sky}	Sky temperature (Ta-20°C).

The conductive heat transfer from the bottom to the atmosphere may be formulated as:

$$q_{k,b} = k_b\ (T_b - T_a) \quad \text{W/m}^2 \tag{6}$$

where

$$(1/k_b) = (1/h_{in}) + (\Sigma(\delta_i/\lambda_i)) + (1/h_a) \text{ m}^2 \text{ K/W} \tag{7}$$

h_{in} = Convective heat transfer coefficient at seawater interface, W/m².

$$\Sigma(\delta_i/\lambda_i) = (\delta_g/\lambda_g) + (\delta_b/\lambda_b) + (\delta_w/\lambda_w) + (\delta_{th}/\lambda_{th}) \text{ m}^2 \text{ K/W} \tag{8}$$

where
δ_g Thickness of glass;
λ_g Thermal conductivity of glass (0.96 W/m.K);
δ_b Thickness of basin;
λ_b Thermal conductivity of basin (205 W/m-K for aluminum, 1.6 W/m-K for black paint, 2.05 W/m-K for sand, 0.25 W/m-K for paraffin wax, and 0.33 W/m-K for coal);
δ_w Thickness of wood;
λ_w Thermal conductivity of wood (0.17 W/m-K);
δ_{th} Thickness of thermocol (20 mm);
λ_{th} Thermal conductivity of thermocol (0.036 W/m-K);
h_a Convective heat transfer coefficient at ambient temperature, W/m².K;
T_b Temperature of the basin, K;
T_a Ambient temperature, K.

Considering the heat transfer from the circumferential area of the still by conduction. From inside moist air to the atmosphere:

$$q_{k,air} = k_m (T_r - T_a), \quad \text{W/m}^2 \tag{9}$$

where

$$(1/k_m) = (1/h_r) + (\Sigma (\delta_i/\lambda_i)) + (1/h_a) \tag{10}$$

h_r Convective heat transfer coefficient of moist air, W/m²
   From liquid to the atmosphere,

$$q_{k,l} = k_1 (T_b - T_a) \quad \text{W/m}^2 \tag{11}$$

where

$$(1/k_1) = (1/\infty) + (\Sigma(\delta_i/\lambda_i)) + (1/h_a) \text{m}^2.\text{K/W} \tag{12}$$

2.3.2   *Heat transfer coefficients (W/m²-K)*

$$h_{wc} = 0.884 \left[(T_w - T_{gi}) + \left(\frac{(p_w - p_{gi})(T_w + 273)}{268.9*10^3 - p_w} \right)^{\frac{1}{3}} \right] \tag{13}$$

$$Pw = \exp\left[25.317 - \frac{5144}{Tw + 273} \right], \text{J/K} \tag{14}$$

$$Pgi = \exp\left[25.317 - \frac{5144}{Tgi + 273} \right], \text{J/K} \tag{15}$$

$$he = 16.273*10^{-3} * hwc * \frac{Pw - Pgi}{Tw - Tgi} \tag{16}$$

3 RESULTS AND DISCUSSION

The whole experiment was carried out during the months of March and April 2017. The effects of various materials as coating for the aluminum basin in different combinations were studied. Further, the effect of treating capacity of the solar still on the overall efficiency, energy balance and heat transfer coefficients were studied by varying the treating volumes of the seawater by 2, 2.5, and 3 liters respectively. For all the cases, the overall efficiency, energy utilized (%) and heat transfer coefficients were estimated.

The effect of sand, ceramic, paraffin wax and coal bed materials, along with black paint, for varying treating capacities of water, is shown in Figures 3, 4, 5, 6, 7, 8 and 9.

As shown in Figures 3 to 7, the optimum exposed time was found to be 1:30 p.m. for the materials, black paint (with and without ceramics), black paint (with and without sand), black paint (with sand + ceramics), and black paint + paraffin wax (with and without sand), as the maximum incidence of solar irradiance was observed at that time.

As shown in Figures 8 and 9, the optimum exposed time was found to be 12:30 p.m. in the case of black paint + coal bed (with and without ceramics). The optimum exposed time was

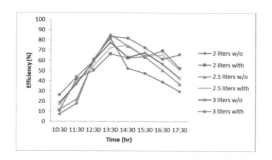

Figure 3. Effect of ceramics on base case.

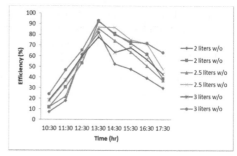

Figure 4. Effect on sand on base case.

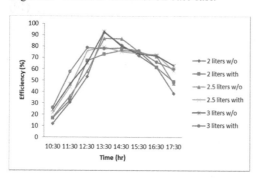

Figure 5. Effect of (sand + ceramics) on base case.

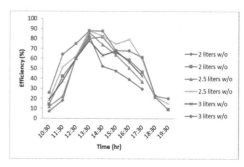

Figure 6. Effect of paraffin wax on base case.

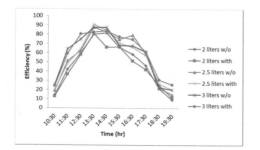

Figure 7. Effect of sand on paraffin wax.

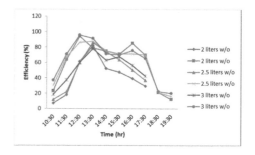

Figure 8. Effect of coal bed on base case.

observed earlier when compared to other combination of materials as the coal is considered to be a heat-absorbing material and black in color.

As shown in Figures 6 to 9, it was observed that the distillate production is continued till 7:30 p.m., even in the absence of the sunlight.

The overall efficiency was estimated for all the considered cases, as shown in Figure 10. The highest efficiency of 65.06% at 3 liters of treating volume capacity of seawater was exhibited by the solar still coated with (sand + ceramics), with the lowest (53.01%) being the base case.

The energy utilized during the process was calculated for all the considered case with 3 liters treating volume capacity and the same is shown in Figure 11. The energy utilization was found to be maximum in the case of solar still basin coated with paraffin wax + sand. The combination of paraffin wax along with sand bed was proved to have an improved heat-absorbing capacity, with a minimum amount of energy losses compared to the other combinations.

The evaporative heat transfer coefficient of water to the glass surface was calculated for all the cases and compared. Figure 12 shows the comparison of the heat transfer coefficients for the treating volume of 3 liters.

The evaporative heat transfer coefficient was observed to be the highest (2964.83 W/m² K) in the case of a solar still basin coated with coal bed. The lowest heat transfer coefficient of 1779.38 W/m² K was observed in the case of a sand on base case basin.

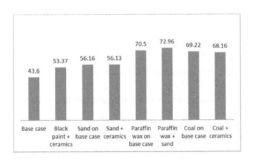

Figure 9.   Effect of ceramics on coal bed.

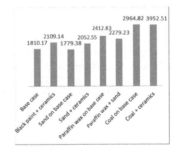

Figure 10.   Overall efficiency for 3 liters treating volume.

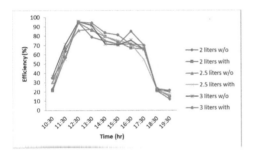

Figure 11.   Energy utilized for 3 liters treating volume.

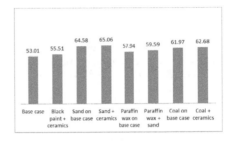

Figure 12.   Heat transfer coefficient for 3 liters treating volume.

Table 1.   Analysis of water before and after solar distillation.

Property	Seawater	Distillate
pH	8.2	7.6
TDS, ppm	35,430	3
TSS, ppm	30,620	8
COD, ppm	750	6
BOD, ppm	300	2
Chlorides, ppm	17,600	50
Hardness, ppm	6,620	22

3.1 Analysis of water

The complete analysis of both the seawater and the distillate obtained from the present study was carried out and is shown in Table 1.

For the distillate, all the properties such as pH, TDS, TSS, COD, BOD, Chlorides and hardness were reduced drastically, and almost distilled water was obtained.

4 CONCLUSIONS

Desalination of seawater by using the solar distillation unit is the most economical efficient process and is also the low external energy requirement process. All the materials used for coating were proved to be suitable materials, with improved efficiency when compared to the base case (black paint coated solar still). The combination of base solar still basin coated with sand and ceramics proved to be best in view of its highest efficiency (65.06%). The least efficiency was given by black paint coated basin (42.19%). The use of paraffin wax (which is a phase change material) and coal bed proved to be fruitful as the production of distillate occurred even in the absence of sun until 7:30 p.m.

REFERENCES

Aburideh, H., Deliou, A., Abbad, B., Alaoui, F., Tassalit, D. & Tigrine, Z. (2012). An experimental study of a solar still: Application on the sea water desalination of Fouka. *Procedia Engineering*, *33*, 475–484.
Boukar, M. & Harmim, A. (2004). Parametric study of a vertical solar still under desert climatic conditions. *Desalination*, *168*, 21–28.
Dhirman, N.K. (1988). Transient analysis of a spherical solar still. *Desalination*, *69*(1), 47–55.
Gugulothu, R., Somanchi, N.S., Devi, R.S.R. & Banoth, H. (2015). Experimental investigations on performance evaluation of a single basin solar still using different energy absorbing materials. *Aquatic Procedia*, *4*, 1483–1491.
Krishna, P.V., Sridevi, V. & Priya, B.S.H. (2016). Comparative studies on a single slope solar distillation unit with and without copper electroplating on aluminium basin. *International Journal of Advanced Research*, *4*(9), 1028–1039.
Malaeb, L., Ayoub, M.G. & Al-Hindi, M. (2014). The experimental investigation of a solar still coupled with an evacuated tube collector. *Energy Procedia*, *50*, 406–413.
Murugavel, K.K., Chockalingam, K.K. & Srithar, K. (2008). Progresses in improving the effectiveness of the single basin passive solar still. *Desalination*, *220*(1–3), 677–686.
Omara, Z.M. & Kabeel, A.E. (2014). The performance of different sand beds solar stills. *International Journal of Green Energy*, *11*(3), 240–254.
Panchal, H.N. (2015). Enhancement of distillate output of double basin solar still with vacuum tubes. *Journal of King Saud University-Engineering Sciences*, *27*(2), 170–175.
Sethi, A.K. & Dwivedi, V.K. (2013). Exergy analysis of double slope active solar still under forced circulation mode. Science Direct. *51*(40–42), 7394–7400.
Somanchi, N.S., Sagi, S.L.S., Kumar, T.A., Kakarlamudi, S.P.D. & Parik, A. (2015). Modelling and analysis of single slope solar still at different water depth. Science Direct. *4*, 1477–1482.
Suneesh, P.U., Paul, J., Jayaprakash, R., Kumar, S. & Denkenberger, D. (2016). Augmentation of distillate yield in "V"-type inclined wick solar still with cotton gauze cooling under regenerative effect. *Cogent Engineering*, *3*(1), 1–10.
Tanaka, H. & Nakatake, Y. (2009). One step azimuth tracking tilted-wick solar still with a vertical flat plate reflector. *Desalination*, *235*(1–3), 1–8.
World health organization (2017) Progress on drinking-water, sanitation and hygiene, 2017.
Yadav, Y.P. & Tiwari, G. (1987). Monthly comparative performance of solar stills of various designs. Desalination, *67*, 565–578.

Emerging Trends in Engineering, Science and Technology for Society,
Energy and Environment – Vanchipura & Jiji (Eds)
© 2018 Taylor & Francis Group, London, ISBN 978-0-8153-5760-5

Towards current induced magnetization reversal in magnetic nanowires

Aicy Ann Antony
Department of Chemical Engineering, Government Engineering College, Thrissur, Kerala, India

P.S. Anilkumar
Department of Physics, Indian Institute of Science, Bangalore, Karnataka, India

V.O. Rejini
Department of Chemical Engineering, Government Engineering College, Thrissur, Kerala, India

ABSTRACT: The effect of pulsed currents by passing through single ferromagnetic nanowires of thickness 20 nm and width 450 nm were studied here. A pulsed current of about 10^{12} A/m^2 was injected at different values of the applied field. During this work, the automation of the measurement system is carried out with a Pico pulse generator and the output is displayed on a DPO (Digital Phosphor Oscilloscope) to get the shape of the waveform. With this setup, transport and optimization using different pulse widths of electrical pulses in the range 10 ns to 80 ns were carried out. The statistical analysis of these results give information about the magnetization processes in nanowires.

1 INTRODUCTION

Spintronics or magneto-electronics is an area of active research because of the tremendous potential both in terms of fundamental physics and technology. Here, one exploits the spin degree of freedom of the electrons along with its charge. So we can expect new generation of devices with completely different functionality. The advantages of these magnetic devices would be non-volatility, increased data processing speed, decreased electric power consumption and increased integration densities compared to semiconductor devices. This has led to a revolution in magnetic data storage technology. Present day data storage heavily relies on magnetic hard disc drives in the form of magnetic thin film media and one of the key concerns is the magnetization reversal of the magnetic data bits by applying local magnetic fields (1). The replacement of magnetic field assisted reversal by more sophisticated current induced magnetic reversal mechanism is emerging as a viable and attractive option for next generation technology.

This mechanism requires source of spin polarized electrons from a ferromagnet to switch the magnetization. The current density required to switch the magnetization is 10^{12} A/m^2. So magnetic material is patterned into nanowires and hence the current requirement is reduced. However, Joule heating will destroy these wires. Hence we need to use pulsed current of the order of nanoseconds to prevent this damage. In this work, an experiment has been set up to pass 10–80 ns current pulses through a magnetic nanowire ($Ni_{81}Fe_{19}$) and achieve current induced magnetization reversal. A Pico-pulse generator will be used for this. A high frequency oscilloscope will be used to understand the pulse shape and its distortion while passing through the nanowire. During this work the system was automated and various parameters were evaluated. The important instruments included in this work are the Pulse Generator and the DPO (Digital Phosphor Oscilloscope). The programming language which interfaces these Pulse Generator and DPO with the controller is LabVIEW (Laboratory Virtual Instrument Engineering Workbench).

The Pulse Generator used here is PSPL10300B and is manufactured by Tektronix Company. It provides high amplitude positive or negative pulses and its output is designed for 50 Ω impedance. Its other features include 300 ps rise time and 750 ps fall time, adjustable amplitude from −4 mV to 50 V in 1 db steps, adjustable duration from 1 ns to 100 ns etc: The DPOs captures, stores, displays and analyzes in real time the three dimensions of signal information such as Amplitude, Time and distribution of Amplitude over Time. DPO uses a separate parallel processor. So the speed of the display is not limited. TDS794D model DPO manufactured by Tektronix company is used here. Also its features include 2 GHz bandwidth, 2 or 4 channels, advanced triggering, 1 mV to 10 V/div Sensitivity etc:

The programming language which interfaces the pulse generator and DPO with the controller is the LabVIEW. It is interfaced to the instruments by using GPIB (General Purpose Interface Bus) cable. LabVIEW is a system-design platform and developing environment for a visual programming language from National Instruments. It is a graphical programming language. LabVIEW programs are called virtual instruments. It is because their appearance and operation imitate physical instruments. There is a user interface or front panel with controls and indicators. A block diagram is there which contain codes to control the front panel objects. Here we are passing 10 ns to 80 ns current pulses which are of very high frequency from a Pulse generator to a nanowire and the magnetic reversal is studied and the result is displayed on a digital phosphor oscilloscope.

2 METHODOLOGY

In this work the main parameters are the Pulse generator, Oscilloscope and the interfacing language LabVIEW. The Pulse generator used here is a Pico pulse generator with 300 ps

Figure 1.   Pulse generator.

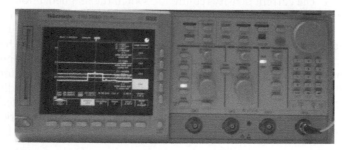

Figure 2.   Digital phosphor oscilloscope.

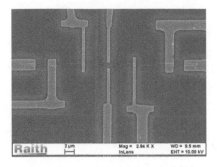

Figure 3.   SEM image of ring device with width 450 nm and inner diameter 1.86 μm.

rise time and 750 ps fall time. This study will focus on the magnetization reversal in magnetic nanowires through electrical transport. The automation of the measurement system is carried out using Pico pulse generator. With this set up the transport and optimization using different pulse widths of electrical pulses is carried out. The statistical analysis of these results are expected to give information about the magnetization processes in nanowires.

Today, data storage heavily relies on the magnetic hard disc drives in the form of thin film. The speed of the hard disc depends on the speed of rotation of the disc so that sensor can read data fastly. The interchanging of north and south poles cannot be done manually. It is done by Magnetization switching. Magnetic field is required for magnetization switching to take place.

The replacement of magnetic field assisted reversal or switching by more sophisticated current induced magnetic reversal is emerging as a viable and attractive option for next generation technology. This mechanism requires spin polarized electrons from a ferromagnet to switch the magnetization. The current density required for magnetic switching is 10^{12} A/m^2. So, for reducing the huge current, magnetic material is patterned into nanowires. But Joule heating will destroy these wires. Hence we are using pulsed current of the order of nanoseconds.

The above figure is the SEM image of the nanowire used to pass current pulses which is having width 450 nm, thickness 20 nm and inner diameter 1.86 μm.

3   RESULTS AND DISCUSSION

Experiments were conducted by connecting PSPL5333, which is a power divider, between the input and output and the reflections and the current passed were studied from a high frequency DPO (Digital Phosphor Oscilloscope). At first when trial and error has done with different lengths of a wire it was found that as the length of the wire is reduced the attenuation or reflections gets reduced and it can be understood from the voltage variations on the DPO. At high frequency, obtained by shorter durations or pulse widths the reflection is less. After trial and error I moved on with the sample which is $Ni_{81}Fe_{19}$. When BNC cables were used 100% transmission was obtained. But when devices consisting of a nanoring, that is, nanowires are connected, it is found that only 10% of the signal gets transmitted in the pulse width range 10 ns to 80 ns.

It has been realized that this transmission is sufficient to achieve current induced magnetization reversal in these structures. The observations done with the sample $Ni_{81}Fe_{19}$ for different pulse widths is plotted in Origin Software and is given below.

From the graph we can understand that for a pulse width of 20 ns the input signal has a voltage of 200 mV, but after passing through the structure the signal is attenuated to 20 mV. However the width of the pulse remain unaltered.

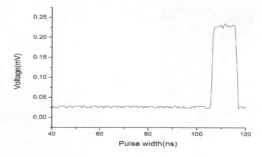

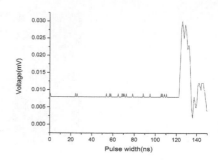

Figure 4. Input displayed on DPO by giving 10 ns pulse width through nanowire.

Figure 5. Output obtained from DPO when 10 ns pulse width applied through nanowire.

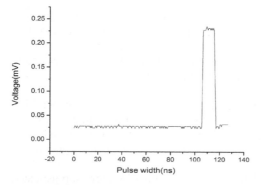

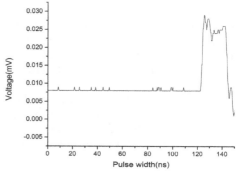

Figure 6. Input displayed on DPO when 20 ns pulse width applied through nanowire.

Figure 7. Output obtained from DPO when 20 ns pulse width applied through nanowire.

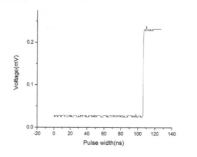

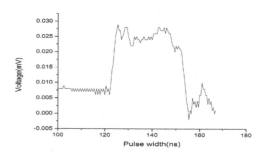

Figure 8. Input displayed on DPO when 30 ns pulse width applied through nanowire.

Figure 9. Output obtained from DPO when 30 ns pulse width applied through nanowire.

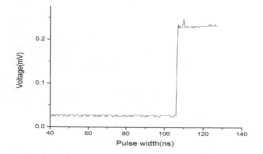

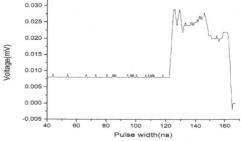

Figure 10. Input displayed on DPO when 40 ns pulse width applied through nanowire.

Figure 11. Output obtained from DPO when 40 ns pulse width applied through nanowire.

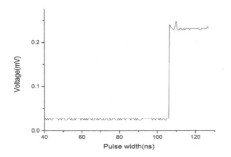

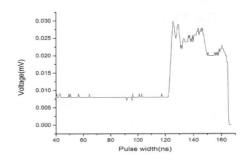

Figure 12. Input displayed on DPO when 50 ns pulse width applied through nanowire.

Figure 13. Output obtained from DPO when 50 ns pulse width applied through nanowire.

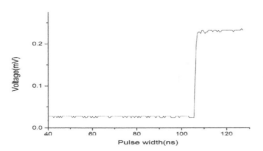

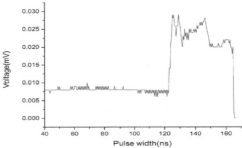

Figure 14. Input displayed on DPO when 60 ns pulse width applied through nanowire.

Figure 15. Output obtained from DPO when 60 ns pulse width applied through nanowire.

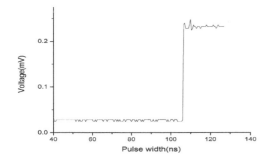

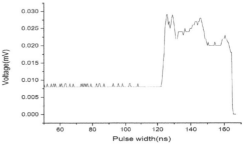

Figure 16. Input displayed on DPO when 70 ns pulse width applied through nanowire.

Figure 17. Output obtained from DPO when 70 ns pulse width applied through nanowire.

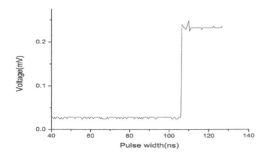

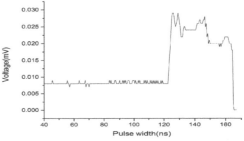

Figure 18. Input displayed on DPO when 80 ns pulse width applied through nanowire.

Figure 19. Output obtained from DPO when 80 ns pulse width applied through nanowire.

4 CONCLUSION

In this work, a Pico pulse generator was studied, programmed and interfaced with the controller. Also the rise time and fall time of the pulse generator was verified by passing current pulses with 10 ns to 80 ns pulse width range. The pulse output shape and distortion were understood from a DPO (digital phosphor oscilloscope) which is interfaced with the controller. DPO was also programmed for this study. At first, when BNC cables were used for the transmission of the current pulses, nearly 100% transmission was obtained. But when devices consisting of a nanoring, that is, a nanowire was connected it was found that only 10% of the signal gets transmitted in the pulse width range from 10 ns to 80 ns. So it has been realized that this transmission is sufficient to achieve current induced magnetization reversal in nanowire structures. For future scope we can say that under these conditions, we will be able to image magnetic domain and the velocity of the magnetic domain walls at higher current density.

REFERENCES

[1] Masamitsu Hayashi, Luc Thomas, Rai Moriya, Charles Rettner and Stuart S.P Parkin (2008), Current-Controlled Magnetic Domain-Wall Nanowire Shift Register, *Science*, Vol. 320, 209–211.
[2] J.E. Wegrowe, D. Kelly, Y. Jaccard, Guittenne and Ansermet (1998), Current Induced Magnetic reversal in magnetic nanowires, *Europhysics Letters*, 45, 626–631.
[3] Tao Yang, Takashi Kimura and Yoshichika Ottani (2008), Giant spin accumulation signal and pure spin-current induced reversible magnetization switching, *Nature Physics*, Vol. 4, 851–854.
[4] E.A. Rando and S. Allende (2015), Magnetic reversal modes in multisegmented nanowire arrays with long aspect ratio, Vol. 118, 013905-1 to 013905-8.
[5] A. Thiaville, Y. Nakatani, J. Miltat, Y. Suzuki (2005), Micromagnetic understanding of current-driven domain wall motion in patterned nanowires, *Europhysics Letters*, Vol. 69, 990–996.

Emerging Trends in Engineering, Science and Technology for Society,
Energy and Environment – Vanchipura & Jiji (Eds)
© 2018 Taylor & Francis Group, London, ISBN 978-0-8153-5760-5

Synthesis and surfactant size regulation of nanoparticles of maghemite (γ-Fe$_2$O$_3$)

Mushtaq Ahmad Rather

Chemical Engineering Department, National Institute of Technology, Hazratbal, Srinagar Kashmir, India

ABSTRACT: In this study, nanoparticles of maghemite (γ-Fe$_2$O$_3$) were synthesized and their size regulated by utilizing different amounts of surfactant, Polyethylene Glycol (PEG)-4000. The amount of surfactant was varied in order to analyze its effect upon the particle size. The final synthesized particles had a mean size of 40, 23, 15, and 11 nm for various surfactant quantities. X-ray diffraction results confirmed that the as-synthesized iron oxide nanoparticles were γ-Fe$_2$O$_3$. Superconducting Quantum Interference Device (SQUID) measurements showed the highest saturation magnetization value of 68 emu/gm at room temperature when a surfactant quantity of 0.5 g was used under the given conditions. The particle shape was almost a sphere, as confirmed by transmission electron microscopy.

Keywords: Nanoparticles, Iron oxide, Surfactant, Paramagnetism

1 INTRODUCTION

Iron oxides are common compounds, which are widespread in nature and can be readily synthesized in the laboratory. Iron oxide nanoparticles find applications in the biomedical sector (in cellular labeling, cell separation, detoxification of biological fluids, tissue repair, drug delivery, Magnetic Resonance Imaging (MRI), hyperthermia and magneto faction) (Gupta & Gupta, 2005), electrode materials (Kijima et al., 2011), fabrication of pigments, sorbents, gas sensors (Afkhami & Moosavi, 2010), ferro fluids and wastewater purification (Shen, 2009a).

Eight iron oxides are known. Among these, hematite (α-Fe$_2$O$_3$), magnetite (Fe$_3$O$_4$) and maghemite (γ-Fe$_2$O$_3$) are very promising and popular candidates due to their polymor phism involving temperature-induced phase transition. These three iron oxides have unique properties (such as biochemical, magnetic, and catalytic), which make them suitable for specific technical and biomedical applications (Cornell & Schwertmann, 2003). Maghemite is ferrimagnetic at room temperature but its nanoparticles smaller than 10 nm are superparamagnetic. Maghemite is unstable at high temperatures and loses it susceptibility with time (Ray et al., 2008; Neuberger et al., 2005). The structure of γ-Fe$_2$O$_3$ is cubic. Oxygen anions give rise to a cubic close-packed array while ferric ions are distributed over tetrahedral sites (eight Fe ions per unit cell) and octahedral sites (the remaining Fe ions and vacancies). Therefore, the maghemite can be considered as being fully oxidized magnetite, and it is an n-type semiconductor with a band gap of 2.0 eV (Wu et al., 2015).

Particle agglomeration forms large clusters, resulting in an overall increase in particle size (Hamley, 2003). When two large-particle clusters approach one another, each of them comes under the influence of the magnetic field of its neighbor. Besides the arousal of attractive forces between the particles, each particle is within the magnetic field of the neighbor and becomes further magnetized (Tepper et al., 2003). The adherence of remnant magnetic particles causes a mutual magnetization, resulting in increased aggregation properties. Since particles are attracted magnetically in addition to the usual flocculation due to Van der Waals forces, surface modification is often indispensable. For effective stabilization of iron oxide nanoparticles, often a very high requirement of density for coating is desirable. Some stabilizer such as a surfactant or a polymer is usually added at the time of preparation to prevent aggregation of the nanoscale particulates. Most of these polymers adhere to surfaces in a substrate-specific manner (Mendenhall et al., 1996). Nanoparticle coatings may comprise of several materials, including both inorganic and polymeric materials. Polymeric coating materials can be classified into synthetic and natural types. Polymers based on, for instance, poly (ethylene-co-vinyl acetate), Poly (Vinyl Pyrrolidone) (PVP), Poly (Lactic Co-Glycolic Acid) (PLGA), Poly (Ethylene Glycol) (PEG), and Poly (Vinyl Alcohol) (PVA), are typical examples of synthetic polymeric systems (Miller et al., 1983; Ruiz & Benoit, 1991). Natural polymer systems include the use of substances such as gelatin, dextran, chitosan, and pullulan (Li et al., 1997; Massa et al., 2000).

As we know the addition of surfactant time and the amount that would regulate the growth of particles, in the present study PEG-4000 was used as a surfactant to stabilize the particles in aqueous medium. Different amounts of this surfactant were added to obtain Nanoparticles (NPs) of various sizes and to analyze its effect upon the particle size.

2 EXPERIMENTAL

Synthesis of nanoparticles was carried out using a coprecipitation process based on work by Yunabi et al. (2008), Shen et al. (2009b), and Maity and Agrawal (2007). The modified procedure adopted is briefly stated as follows.

A solution of 50 ml of ferrous chloride tetrahydrate was added to 100 ml of ferric chloride hexahydrate solution under constant agitation. The contents were heated to 50°C in a Borosil glass round-bottom reaction vessel. The 500 ml capacity reactor was specially designed for the purpose. Ammonium hydroxide solution was added drop-wise at the specified temperature. The initially black-brown colored solution formed changed into a completely black mixture after about half an hour at a pH of 9. The magnetic particles so formed were kept dispersed by an ultra-sonicator for about a further 15 minutes and then collected by settling under a ferro magnet. The particles were then transferred into a separate beaker and washed with water until the pH reached 7. Then the particles were washed twice with ethanol. The collected nanoparticles were dried in a vacuum oven at 60°C for about six hours.

PEG-4000 was used as surfactant to stabilize the particles in aqueous medium. In order to observe the effect of adding different amounts of PEG-4000 upon the final particle size, different 0.5 g, 1.0 g, 1.5 g, and 2.0 g amounts of it were used. It is pertinent to mention the fact that an increase of surfactant time and amount would regulate the growth of the particles. Four of the γ-maghemite NPs (S1, S2, S3, and S4) were obtained by adding PEG-4000 surfactant amounts as 0.5 g, 1 g, 1.5 g, and 2.0 g respectively. Crystallite size, magnetic properties and surface area of these NPs were determined.

The size of the particles and their shape were observed by Transmission Electron Microscopy (TEM). The type of crystal was verified by X-Ray Diffraction (XRD) study after comparison of the same with the JCPDS file. Standard patterns for bulk magnetite, maghemite, and hematite are respectively given by JCPDS file numbers. 19-0629, 39-1346 and 83-0664.

The crystallite size was calculated by using the Scherrer equation (Paterson, 1939). The magnetic properties of the nanoparticles were measured by Superconducting Quantum Interference Device (SQUID) measurements (Tepper *et al.*, 2003). A Malvern Dynamic Light Scattering (DLS) nanosizer was used to analyze the size of the particles in solution.

3 RESULTS AND DISCUSSION

3.1 *Transmission Electron Microscopic (TEM) results*

Interpretation of TEM pictures reveals the average size of particles to be 40, 23, 15, and 11 nm respectively for S1, S2, S3, and S4. Hence, as surfactant quantity increases from 0.5 g to 2.0 g, the particle size decreases from 40 to 11 nm. The photographs are shown in Figure 1 below.

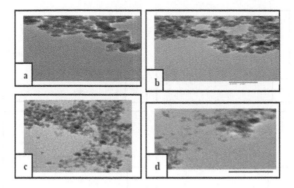

Figure 1.   TEM images: (a) S1; (b) S2; (c) S3; (d) S4.

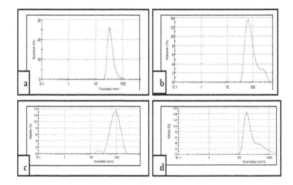

Figure 2.   DLS images: (a) S1; (b) S2; (c) S3; (d) S4.

3.2   *Dynamic Light Scattering (DLS) analysis*

The DLS images reveal the average size of particles to be 90, 60, 38, and 24 nm respectively for S1, S2, S3, and S4. Magnetic particles have a very high tendency of agglomeration due to the presence of additional magnetic interactions, in addition to the usual Van der Waals forces of attraction. Their size may not exactly correspond to that indicated by the TEM due to this agglomeration of NPs. The images are shown in the following Figure 2.

3.3   *X-Ray Diffraction (XRD) analysis*

In all the cases of S1, S2, S3, and S4, the bulk phase identified is maghemite (i.e. γ-Fe_2O_3). The majority of peaks in the XRD pictures and the peak of 100% intensity correspond to the γ-maghemite. Some of the peaks also are nearer to hematite, so it may also be present as an oxidation product together with γ-maghemite. The XRD pictures are shown in Figure 3, giving 2θ versus intensity data for the above γ-maghemite NPs S1, S2, S3, and S4. The crystallite size computed by the Scherrer equation was 35, 20, 19, and 10 nm respectively.

Peak-wise analysis for the identification of type of crystallite in S1, S2, S3, and S4, showed the predominant presence of maghemite (γ-Fe_2O_3) in each case. The above analysis clearly showed the presence of the γ-maghemite phase of iron oxide NPs. The presence of other phases was not ruled out as the 2θ versus intensity data for different iron oxide phases are very close, and particularly those for magnetite and maghemite.

3.4   *SQUID measurements*

SQUID measurements of samples S1, S2, S3, and S4 show respectively saturation magnetization values of 68, 56.5, 54.87, and 30.18 emu/gm at room temperature. This indicates that as

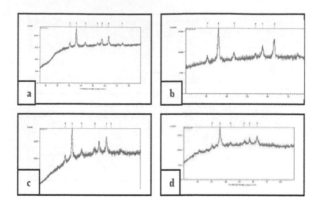

Figure 3.   XRD: (a) S1; (b) S2; (c) S3; (d) S4.

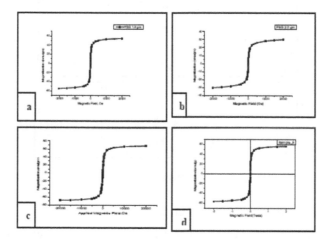

Figure 4.   Hysteresis curves: (a) S1; (b) S2; (c) S3; (d) S4.

the PEG-4000 surfactant amount increases, the size of the NPs and saturation magnetization value decrease. The saturation magnetization value of bulk maghemite is about 73 emu/gm (Maity & Agrawal, 2007), and the above values compare well with it. Figure 4 below shows the hysteresis curves of S1, S2, S3, and S4. All the magnetic hysteresis loops have passed the grid origin, which indicates good super paramagnetism.

3.5   *Brunauer, Emmett and Teller (BET) surface area studies*

The values for different NPs of γ- maghemite were 54.31, 61.8, 66.4, and 69.2 m²/gm respectively for S1, S2, S3, and S4.

4   CONCLUSION

In the given technique, four sizes of γ-maghemite NPs, viz. 40, 23, 15, and 11 nm, were obtained by using PEG-4000 surfactant amounts of 0.5 g, 1 g, 1.5 g, and 2.0 g respectively under the given conditions. The crystallite size, magnetic properties and surface area of these NPs were determined. PEG-4000 surfactant was successful in avoiding the huge agglomeration of NPs in solution. SQUID measurements showed the highest saturation magnetization value to be 68 emu/gm at room temperature when the surfactant quantity was 0.5 g. As surfactant amount was increased, the size of the NPs and the saturation magnetization value decreased. The synthesized NPs will be useful in various practical applications.

ACKNOWLEDGMENT

The author acknowledges the support received from National Institute of Technology, Srinagar, Kashmir for facilitating the work.

REFERENCES

Afkhami, A. & Moosavi, R. (2010). Adsorptive removal of Congo red, a carcinogenic textile dye, from aqueous solutions by maghemite nanoparticles. *J. Hazard. Mater.*, *174*(1–3), 398–403.

Cornell, R.M. & Schwertmann, U. (2003). *The iron oxides: Structure, properties, reactions, occurrences and uses* (2nd ed.). Weinheim: Wiley.

Gupta, A.K. & Gupta, M. (2005). Synthesis and surface engineering of iron oxide nanoparticles for biomedical applications. *Biomaterials*, *26*(18), 3995–4021.

Hamley, I.W. (2003). Nanotechnology with soft materials. *Angew Chem. Int. Ed. Engl*, *42*(15), 1692–1712.

Kijima, N., Yoshinaga, M., Awaka, J. & Akimoto, J. (2011). Microwave synthesis, characterization, and electrochemical properties of α-Fe_2O_3 nanoparticles. *Solid State Ionics*, *192*(1), 293–297.

Li, J.K., Wang, N. & Wu, X.S. (1997). A novel biodegradable system based on gelatin nanoparticles and poly (lactic-co-glycolic acid) microspheres for protein and peptide drug delivery. *Journal of Pharmaceutical Sciences*, *86*(8), 891–895.

Massia, S.P., Stark, J. & Letbetter, D.S. (2000). Surface-immobilized dextran limits cell adhesion and spreading. *Biomaterials*, *21*(22), 2253–2261.

Maity, D. & Agrawal, D. (2007). Synthesis of iron oxide nanoparticles under oxidizing environment and their stabilization in aqueous and non-aqueous media. *Journal of Magnetism and Magnetic materials*, *308*(1), 46–55.

Mendenhall, G.D., Geng, Y. & Hwang, J. (1996). Mendenhall, G.D., Geng, Y., & Hwang, J. (1996). Optimization of long-term stability of magnetic fluids from magnetite and synthetic polyelectrolytes. *Journal of Colloid and Interface Science*, *184*(2), 519–526.

Miller, E.S., Peppas, N.A. & Winslow, D.N. (1983). Morphological changes of ethylene/vinyl acetate-based controlled delivery systems during release of water-soluble solutes. *Journal of Membrane Science*, *14*(1), 79–92.

Neuberger, T., Schöpf, B., Hofmann, H., Hofmann, M. & Von Rechenberg, B. (2005). Superparamagnetic nanoparticles for biomedical applications: Possibilities and limitations of a new drug delivery system. *Journal of Magnetism and Magnetic materials*, *293*(1), 483–496.

Paterson, A.L. (1939). The Scherrer formula for X-ray particle size determination. *Physical Review*, *56*(10), 978–982.

Ray, I., Chakraborty, S., Chowdhury, A., Majumdar, S., Prakash, A., Pyare, R. & Sen, A. (2008). Room temperature synthesis of γ-Fe_2O_3 by sonochemical route and its response towards butane. *Sens Actuators B Chem*, *130*(2), 882–888.

Ruiz, J.M. & Benoit, J.P. (1991). In vivo peptide release from poly (DL-lactic acid-co-glycolic acid) copolymer 5050 microspheres. *J Control Release*, *16*(1–2), 177–185.

Shen, Y.F., Tang, J., Nie, Z.H., Wang, Y.D., Ren, Y. & Zuo, L. (2009a). Preparation and application of magnetic Fe_3O_4 nanoparticles for wastewater purification. *Separation and Purification Technology*, *68*(3), 312–319.

Shen, Y.F., Tang, J., Nie, Z.H., Wang, Y.D., Ren, Y. & Zuo, L. (2009b). Tailoring size and structural distortion of Fe_3O_4 nanoparticles for the purification of contaminated water. *Bioresource Technology*, *100*(18), 4139–4146.

Tepper, T., Ilievski, F., Ross, C.A., Zaman, T.R., Ram, R.J., Sung, S.Y. & Stadler, B.J.H. (2003). Faraday activity in flexible maghemite/polymer matrix composites. *J Appl Phys*, *93*(10), 6948–6950.

Wu, W., Wu, Z., Yu, T., Jiang, C. & Kim, W.S. (2015). Recent progress on magnetic iron oxide nanoparticles: Synthesis, surface functional strategies and biomedical applications. *Sci Technol Adv Mater*, *16*(2), 023501.

Yunabi, Z., Zumin, Q. & Huang, J. (2008). Preparation and analysis of Fe_3O_4 magnetic nanoparticles used as targeted drug carriers. *Chinese Journal of Chemical Engineering*, *16*(3), 451–455.

Emerging Trends in Engineering, Science and Technology for Society,
Energy and Environment – Vanchipura & Jiji (Eds)
© 2018 Taylor & Francis Group, London, ISBN 978-0-8153-5760-5

Hydrothermal carbonization: A promising transformation process of biomass into various product materials

Mushtaq Ahmad Rather
Chemical Engineering Department, National Institute of Technology, Hazratbal, Srinagar, Kashmir,
India

ABSTRACT: Hydrothermal Carbonization (HTC) conversion involves the reaction of biomass or other waste organic materials in water at high temperature and pressure, near the thermodynamic critical point of water (Tc = 373.95°C, Pc = 22.064 MPa) in a specially designed high-pressure autoclave reactor. HTC conversion leads to the formation of solid Hydrochar (HC) as a predominant product. This product is a versatile material with potential uses such as a fuel, adsorbent, or anode in Li-ion batteries. The present paper summarizes some of the important aspects of this recently emerging field in manufacturing and technology.

Keywords: Hydrothermal Carbonization, Hydrochar

1 INTRODUCTION

Hydrothermal carbonization (HTC) is a 100-year-old technique in the synthesis of materials, with increasing interest originating from the formation of charcoal. The Bergius process, firstly developed by German chemist Friedrich Bergius in 1913, is a method of production of liquid hydrocarbons for use as synthetic fuel by hydrogenation of high-volatile bituminous coal at high temperature and pressure (Hu et al., 2010). He was also awarded the Nobel Prize in Chemistry in 1931 for his development of high-pressure chemistry (Bergius, 1996). The pyrolysis in the presence of subcritical liquid water is called hydrothermal carbonization (Tekin et al., 2014). The process generates both solid and liquid products. The solid fuel generated in HTC conversion is referred to as Hydrochar (HC) (biochar). HTC of lignocellulosic biomass has received extensive research over the last two decades for both the production of solid and liquid fuels (subcritical conditions), and for gasification (supercritical conditions).

Hot compressed liquid water near its thermodynamic critical point (T_c = 373.95°C, P_c = 22.064 MPa) behaves very differently from liquid water at room temperature. As water is heated along its vapor–liquid saturation curve, its dielectric constant decreases due to the hydrogen bonds between water molecules being fewer and less persistent. The reduced dielectric constant enables hot compressed water to solvate small organic molecules, allowing organic reactions to occur in a single fluid phase. Additionally, the ion product of water increases with temperature up to about 280°C, but then decreases as the critical point is approached. This higher ion product leads to higher natural levels of hydronium ions in hot compressed water, which can accelerate the rates of acid-catalyzed hydrolytic decomposition reactions (Yeh et al., 2013). Hydrothermal carbonization processing wet biomass can produce a hydrochar that retains a large proportion of the chemical energy and lipids in the original biomass. The hydrothermal carbonization environment promotes the hydrolytic cleavage of ester linkages in lipids, peptide linkages in proteins, and glycosidic ether linkages in carbohydrates. These cleavage reactions can be accelerated by catalysts (Yeh et al., 2013).

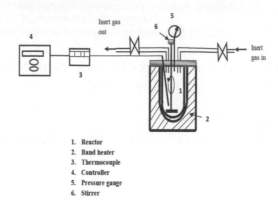

1. Reactor
2. Band heater
3. Thermocouple
4. Controller
5. Pressure gauge
6. Stirrer

Figure 1.   Schematic diagram of hydrothermal carbonization reactor setup.

2   EXPERIMENTAL

The HTC of biomass material to fine-powdered form is carried out in an autoclave reactor, usually made up of alloy steel. Controlled heating by a temperature controller is carried out by means of a heater of suitable power rating. The reactor is properly sealed and head-tightened for leak-proof operation. Under inert atmosphere, heating is started, and the autogenic pressure builds up. At the desired temperature, the carbonization reaction is allowed to proceed for the holding time for a suitable duration. The heating is then stopped and the reactor allowed to cool to room temperature. The carbonized solids are then separated from the liquid phase by means of vacuum filtration. The solid-product hydrochar is dried in an oven to remove any residual moisture. Dried HC is weighed and processed for further characterization. The schematic diagram of the experimental setup for HTC conversion is illustrated in Figure 1.

3   PROGRESS IN HTC PROCESS

Ozcimen et al. (2008) performed carbonization experiments on grape seed and chestnut shell samples, having the average particle size of 0.657 mm and 0.377 mm respectively, to determine the effect of temperature, sweep gas flow rate and heating rate on the biochar yield. It was found that the temperature had the dominant effect on the biochar yields, as compared to the effects of nitrogen gas flow rate and heating rate.

Sevilla and Fuertes (2009) produced highly functionalized carbonaceous materials by means of the HTC of cellulose at temperatures in the range of 220–250°C. They observed that the materials so formed were composed of agglomerates of carbonaceous microspheres (size 2–5 µm) as evidenced by scanning electron microscopy.

Funke and Ziegler (2010) elaborated the reaction mechanisms of hydrolysis, dehydration, decarboxylation, aromatization, and condensation polymerization during HTC. The mechanisms were important in studying the role of different operational parameters qualitatively for cellulose, peatbog and wood. The results were used to derive fundamental process design improvements for HTC.

Anastasakis and Ross (2011) subjected the brown macro-alga *Laminaria saccharina* to hydrothermal carbonization conversion for the generation of solid and liquid biofuels. Experiments were performed in a batch bomb-type stainless steel reactor (75 ml). The heating rate of the reactor was 25 K/min. The reactor was charged with the appropriate amounts of seaweed biomass and water. In the catalytic runs, an appropriate amount of KOH was added to the reactants. The influence of reactor loading, residence time, temperature and catalyst (KOH) loading were assessed. The experimental conditions were found to have a profound effect on the resulting char composition. The carbon content in the chars ranged from 14.9–52.6 wt.%.

Hoekman et al. (2011) in their experimental effort used a 2 L pressure vessel to apply the HTC process to a mixed wood feedstock. The effects of the reaction conditions on product compositions and their yields were examined by varying temperature over the range of 215–295°C and the reaction hold time over the range of 5–60 min. With increasing temperature and time, the amounts of gaseous products and produced water increased, while the amount of HTC char decreased.

Liu et al. (2012) studied HTC to upgrade waste biomass (coconut fiber and eucalyptus leaves) at temperatures ranging from 150–375°C and a residence time of 30 min. The kinetics of HTC was investigated and the fuel properties of the produced hydrochars were evaluated.

Roman et al. (2012) studied the HTC processes under different conditions using two different biomass materials: walnut shell and sunflower stem. Walnut shell and sunflower stem were subjected to HTC in order to increase their energy content, to provide a greater amount of energy per unit of mass. The hydrothermal carbonization processes were carried out using 5 g of raw material dispersed in deionized water (100–150 mL) in a stainless steel autoclave, which was heated up in an electric furnace at selected temperatures (190–230°C), during time intervals of 20–45 h. It was found that the temperature and water/biomass ratio were more important variables than the residence time for HTC.

Reza et al. (2013) used HTC as a pretreatment process to convert loblolly pine as feedstock to homogeneous energy-dense solid fuels. Experiments were conducted in a two-chamber reactor, maintaining isothermal conditions for 15 s to 30 min reaction times. Loblolly pine was treated at 200, 230, and 260°C. In the first few minutes of reaction, the solid-product mass yield decreased rapidly while the calorific value increased rapidly.

Wiedner et al. (2013) focused on chemical modification of wheat straw, poplar wood and olive residues through HTC at different temperatures (180°C, 210°C, and 230°C). Besides general properties such as pH, Electrical Conductivity (EC), ash content, elemental composition and yield, they evaluated bulk chemical composition and the contribution of specific compounds (lignin and black carbon). Hydrochar yields and carbon recovery decreased with increasing temperature to about 50% and 75%, respectively for all feedstocks at 230°C.

Eibisch et al. (2013) carbonized biomass via HTC to yield a Carbon (C) rich hydrochar. They investigated whether easily mineralizable organic components adsorbed on the hydrochar surface influenced the degradability of the hydrochars and so could be removed by repetitive washing.

Reza et al. (2014) in their study on HTC of cellulose, wheat straw, and poplar, evaluated the effects of reaction temperature and reaction time on both solid hydrochar and process liquid. The objective was to design a high pressure, high temperature slurry sampling system in an 18.6 L reactor. Several different reaction conditions by changing biomass feedstock, reaction time, and reaction temperature was studied. Hydrochar composition, as well as chemical components in HTC, varied with temperature and time.

Using HTC conversion, Zhao et al. (2014) produced nitrogen- and chlorine-free solid biofuel from high moisture and nitrogen content bio-wastes, such as Municipal Solid Waste (MSW), mycelia waste, sewage sludge and paper sludge. The work focused on energy recycling, and on optimizing the operating parameters and evaluating the energy efficiency of this fuel production process. The effect of the temperature and holding time on the biofuel recovering ratio, calorific value and energy recovery rate were investigated.

Guiotoku et al. (2014), in their work on hydrothermal carbonization, carbonized cellulose-based renewable raw materials in a microwave oven at 200°C for 60, 120, and 240 min. The values of fixed carbon were between 38 and 52%.

Danso-Boateng et al. (2015a) carried out HTC of primary sewage sludge using a batch reactor to convert wet biomass (sewage sludge) to lignite-like renewable solid fuel of high calorific value. The effect of temperature and reaction time were investigated on the characteristics of solid hydrochar, liquid and gas products, and the conditions leading to optimal hydrochar characteristics. The amount of carbon retained in hydrochars decreased as temperature and time increased, with carbon retentions of 64–77% at 140 and 160°C, and 50–62% at 180 and 200°C.

Danso-Boateng et al. (2015b), in their work on sewage sludge, carried out the mass and energy balances of semi-continuous HTC of fecal waste at 200°C and at a reaction time of 30 min.

Lin et al. (2015) used different temperatures in the range of 180–300°C to evaluate the effect of HTC temperature on hydrochar fuel characteristics and their thermal behavior. The hydrochar produced at 210°C had the maximum heating value (9,763 kJ/kg) with the highest energetic recovery efficiency (90.12%). Therefore, 210°C was the optimum temperature for HTC of paper sludge.

Yin et al. (2015) conducted batch hydrothermal carbonization tests for hydrothermal carbonization decomposition of sewage sludge using a tubular reactor (316 L stainless steel) with 5 mm internal diameter and 8.2 ml volume.

Benavente (2015) focused in the application of the HTC technology as a possible waste management treatment of moist agro-industrial waste. Through this technique, Olive Mill, Canned Artichoke and Orange Wastes (OMW, CAW and OW, respectively) were carbonized in a lab-scale high-pressure reactor at different temperatures (200–250°C) and durations (2, 4, 8, and 24 h) in order to obtain useful bioenergy feedstocks.

Pruksakit and Patumsawad (2016) investigated the HTC of sugar cane at different temperatures, (189, 220, and 250°C), and for different retention times (0.1, 1, and 2 h), to study the effect of operating conditions upon product yield and some other parameters. With increasing temperature, hydrochar yield, hydrogen and oxygen decreased. However, the carbon content of hydrochar increased.

Mau et al. (2016) studied the HTC of poultry litter under a range of process parameters (temperature, reaction time, and solids concentration). Results showed the production of hydrochar with caloric value of 24.4 MJ/kg, similar to sub-bituminous coal. Temperature had the most significant effect on processes and product formation. Solids concentration was not a significant factor once dilution effects were considered.

Wikberg et al. (2016) studied the progress of the conversion, the yield, the structure and the morphology of the produced carbonaceous materials as a function of time. Carbonaceous particles of different shape and size were produced with yields between 23% and 73% after 4 h, with the yields being higher for lignin than for carbohydrates. According to the results, potential pulp mill streams represent lignocellulosic resources for the generation of carbonaceous materials.

Lin et al. (2016) studied the resource utilization of MSW at several temperatures (210, 230, 250, and 280°C) and residence times (30, 60, and 90 min) to investigate the effects on the characteristics of HTC solid fuel from MSW. The results of thermogravimetric analysis illustrated that HTC did remarkably influence the thermal behavior and kinetics of MSW. In most cases, the HTC temperature had a more obvious effect than residence time.

Nizamuddin (2016) investigated the possible optimum conditions for maximum yield of hydrochar through HTC of palm shell. The hydrochar and the palm shell were characterized, and the chemical, dielectric, and structural properties of optimized hydrochar were examined. The effects of the reaction temperature, reaction time and biomass-to-water ratio were analyzed and optimized using the central composite design of response surface methodology. The optimized conditions for hydrochar production were found to be 180°C, 30 min, and 1.60 wt.%, temperature, time, and biomass-to-water ratio, respectively.

4 CONCLUSION

HTC in coming days is to play a great role in the transformation of waste biomass into versatile products that may for instance be in the form of fuels, adsorbents, anode in Li-ion batteries, and soil enrichment agents. We need to carry out an exhaustive study of biomass and its types that are available in India so that if the need arises, the hydrothermal carbonization may be used for transformation into various useful products.

ACKNOWLEDGMENT

The author acknowledges the support received from National Institute of Technology, Srinagar, Kashmir for facilitating the work.

REFERENCES

Anastasakis, K., & Ross, A.B. (2011). Hydrothermal liquefaction of the brown macro-alga Laminaria saccharina: effect of reaction conditions on product distribution and composition. *Bioresource technology*, *102*(7), 4876–4883.

Benavente, V., Calabuig, E., & Fullana, A. (2015). Upgrading of moist agro-industrial wastes by hydrothermal carbonization. *Journal of Analytical and Applied Pyrolysis*, *113*, 89–98.

Bergius, F. (1996). Chemical reactions under high pressure. *Nobel Lectures, Chemistry 1922–1941*, 244–276.

Danso-Boateng, E., Shama, G., Wheatley, A.D., Martin, S.J., & Holdich, R.G. (2015). Hydrothermal carbonisation of sewage sludge: Effect of process conditions on product characteristics and methane production. *Bioresource technology*, *177*, 318–327.

Danso-Boateng, E., Holdich, R.G., Martin, S.J., Shama, G., & Wheatley, A.D. (2015). Process energetics for the hydrothermal carbonisation of human faecal wastes. *Energy Conversion and Management*, *105*, 1115–1124.

Eibisch, N., Helfrich, M., Don, A., Mikutta, R., Kruse, A., Ellerbrock, R., & Flessa, H. (2013). Properties and degradability of hydrothermal carbonization products. *Journal of environmental quality*, *42*(5), 1565–1573.

Funke, A., & Ziegler, F. (2010). Hydrothermal carbonization of biomass: a summary and discussion of chemical mechanisms for process engineering. *Biofuels, Bioproducts and Biorefining*, *4*(2), 160–177.

Guiotoku, M., Rambo, C.R., & Hotza, D. (2014). Charcoal produced from cellulosic raw materials by microwave-assisted hydrothermal carbonization. *Journal of Thermal Analysis and Calorimetry*, *117*(1), 269–275.

Hoekman, S.K., Broch, A., & Robbins, C. (2011). Hydrothermal carbonization (HTC) of lignocellulosic biomass. *Energy & Fuels*, *25*(4), 1802–1810.

Hu, B.B., Wang, K., Wu, L., Yu, S.H., Antonietti, M. & Titirici, M.M. (2010). Engineering carbon materials from the hydrothermal carbonization process of biomass. *Advanced Materials*, *22*(7), 813–828.

Lin, Y., Ma, X., Peng, X., Hu, S., Yu, Z., & Fang, S. (2015). Effect of hydrothermal carbonization temperature on combustion behavior of hydrochar fuel from paper sludge. *Applied Thermal Engineering*, *91*, 574–582.

Lin, Y., Ma, X., Peng, X., Yu, Z., Fang, S., Lin, Y., & Fan, Y. (2016). Combustion, pyrolysis and char CO_2-gasification characteristics of hydrothermal carbonization solid fuel from municipal solid wastes. *Fuel*, *181*, 905–915.

Liu, Z., & Balasubramanian, R. (2012). Hydrothermal carbonization of waste biomass for energy generation. *Procedia Environmental Sciences*, *16*, 159–166.

Mau, V., Quance, J., Posmanik, R., & Gross, A. (2016). Phases' characteristics of poultry litter hydrothermal carbonization under a range of process parameters. *Bioresource technology*, *219*, 632–642.

Nizamuddin, S., Mubarak, N.M., Tiripathi, M., Jayakumar, N.S., Sahu, J.N., & Ganesan, P. (2016). Chemical, dielectric and structural characterization of optimized hydrochar produced from hydrothermal carbonization of palm shell. *Fuel*, *163*, 88–97.

Özçimen, D., & Ersoy-Meriçboyu, A. (2008). A study on the carbonization of grapeseed and chestnut shell. *Fuel Processing Technology*, *89*(11), 1041–1046.

Pruksakit, W., & Patumsawad, S. (2016). Hydrothermal Carbonization (HTC) of Sugarcane Stranded: Effect of Operation Condition to Hydrochar Production. *Energy Procedia*, *100*, 223–226.

Reza, M.T., Wirth, B., Lüder, U., & Werner, M. (2014). Behavior of selected hydrolyzed and dehydrated products during hydrothermal carbonization of biomass. *Bioresource technology*, *169*, 352–361.

Reza, M.T., Yan, W., Uddin, M.H., Lynam, J.G., Hoekman, S.K., Coronella, C.J., & Vásquez, V.R. (2013). Reaction kinetics of hydrothermal carbonization of loblolly pine. *Bioresource technology*, *139*, 161–169.

Román, S., Nabais, J.M.V., Laginhas, C., Ledesma, B., & González, J.F. (2012). Hydrothermal carbonization as an effective way of densifying the energy content of biomass. *Fuel Processing Technology*, *103*, 78–83.

Sevilla, M., & Fuertes, A.B. (2009). The production of carbon materials by hydrothermal carbonization of cellulose. *Carbon*, *47*(9), 2281–2289.

Tekin, K., Karagöz, S. & Bektaş, S. (2014). A review of hydrothermal biomass processing. *Renewable and Sustainable Energy Reviews*, *40*, 673–687.

Wiedner, K., Naisse, C., Rumpel, C., Pozzi, A., Wieczorek, P., & Glaser, B. (2013). Chemical modification of biomass residues during hydrothermal carbonization–What makes the difference, temperature or feedstock? *Organic Geochemistry*, *54*, 91–100.

Wikberg, H., Ohra-aho, T., Honkanen, M., Kanerva, H., Harlin, A., Vippola, M., & Laine, C. (2016). Hydrothermal carbonization of pulp mill streams. *Bioresource technology*, *212*, 236–244.

Yeh, T.M., Dickinson, J.G., Franck, A., Linic, S., Thompson Jr, J.G. & Savage, P.E. (2013). Hydrothermal catalytic production of fuels and chemicals from aquatic biomass. *Journal of Chemical Technology and Biotechnology*, *88*(1), 13–24.

Yin, F., Chen, H., Xu, G., Wang, G., & Xu, Y. (2015). A detailed kinetic model for the hydrothermal decomposition process of sewage sludge. *Bioresource technology*, *198*, 351–357.

Zhao, P., Shen, Y., Ge, S., & Yoshikawa, K. (2014). Energy recycling from sewage sludge by producing solid biofuel with hydrothermal carbonization. *Energy conversion and management*, *78*, 815–821.

Electronics, Signal Processing and Communication Engineering (E-SPACE)

Emerging Trends in Engineering, Science and Technology for Society,
Energy and Environment – Vanchipura & Jiji (Eds)
© *2018 Taylor & Francis Group, London, ISBN 978-0-8153-5760-5*

Denoising of musical signals using wavelets specific for musical instruments

P.V. Sreelakshmi & A. Gayathri
Department of Electronics and Communication Engineering, Government Engineering College,
Thrissur, Kerala, India

M.S. Sinith
Rajiv Gandhi Institute of Technology, Kottayam, Kerala, India

ABSTRACT: For musical signals, a waveform of a single note has a repeating element, as it contains fundamental frequency and its harmonics. A wavelet designed specifically for a musical instrument by taking this waveform as the scaling function can be used to analyze these musical signals. Since the waveform of a single note which is used as a scaling function does not satisfy orthogonality property, they can be designed as biorthogonal wavelets. In this paper, the filter bank coefficients corresponding to this wavelet are derived from the available analysis low-pass coefficients using the properties satisfied by biorthogonal wavelet. The musical signals can be decomposed and reconstructed using this set of filter bank coefficients. The coefficients thus obtained are modified using lifting technology for better performance. The lifting scheme is an approach to construct so-called second generation wavelets, which are not necessarily transalates and dilations of one function. Signal being corrupted with noise is found to be a major problem in signal processing. The musical signals are denoised using classical wavelets and two sets of filter bank coefficients obtained using the two methods. The denoising is performed by adopting a proper thresholding method. For the performance comparison and measurement of quality of denoising, the Signal to Noise Ratio (SNR) is calculated between original musical signal and the denoised signal. It is found that coefficients give better performance once modified using lifting technology.

Keywords: Wavelets, Biorthogonal Wavelets, Filter Bank, Thresholding, Denoising

1 INTRODUCTION

Transmitted signals are mostly corrupted by noise. Once corrupted by noise, a signal loses its pure signal characteristics. Recovering these characteristics from the corrupted signal is a major challenge in the signal processing area. The wavelet transform technique is a widely used method to denoise signal since it gives better results. Wavelet transform replaces Fourier transform in analyzing non-stationary signals. It analyzes a signal by truncating the signal using a window which has variable time frequency resolution called a wavelet. Daubechies introduced the wavelet transform as a tool that cuts up data or functions or operators into different frequency components, and then studied each component with a resolution matched to its scale [1]. Eventhough wavelet analysis replaces Fourier analysis, it is a natural extension of it. Wavelets have been called a mathematical microscope; compressing wavelets increases the magnification of this microscope, enabling us to take a closer look at small details in the signal [2]. The theory of wavelet analysis and design of the filter bank coefficients are given in [3]. The signals produced by musical instruments are found to be non stationary signals where small duration signals or small band-width musical pieces are placed at an effective temporal position to give special effects. Wavelet transform serves as a good technique to analyse those signals. In the existing wavelets, Coiflet5 is found to be the most suitable wavelet for analyzing musical signals. For the best analysis of a signal of interest, it is desirable to design a wavelet that matches the signal. Algorithms for designing wavelets that match the signal of interest are given in [4]. The scaling function and wavelet function of the wavelets present in the different signals are different. In the case of musical signals, the waveform of a single note played by a musical instrument has a repeating element. From this repeating element, the scaling and wavelet functions for that particular instrument can be found. This repeating element is taken as the scaling function and hence it should satisfy the necessary and sufficient conditions for a scaling function. There are different algorithms to derive the the filter coefficients corresponding to a scaling function. They are Adaptive filter algorithms like LMS [5], NLMS [6] or Recursive Least Square (RLS). Inorder to enhance the accuracy of results, there are several other variants of this algorithms [7]–[11]. Sinith et al. used the LMS algorithm to find out the filter coefficients. But the algorithm is very sensitive to step size [12]. NLMS and RLS algorithm are not sensitive to step size and worked well for standard wavelets. Eventhough adaptive filter algorithms work well for known wavelet, they fail to give satisfactory results for musical signals whose equations are unknown. This leads to adopting a method known as Particle Swarm Optimisation technique [13]–[14]. Hence the filter coefficients corresponding to the scaling function can be found accurately using PSO (Particle swarm optimization) algorithm. The algorithm is modified For better results. Thus using modified PSO algorithm, optimum values for filter coefficients, $h(n)$ are obtained that corresponds to a scaling function and this results in a new wavelet for analyzing musical signals known as SSM (Sinith-Shikha-Murthy)Wavelet [15].

Once the filter bank coefficient, $h(n)$, is obtained, the scaling function, $\phi(t)$ and the wavelet function, $\psi(t)$ can be obtained using the relation given below.

$$\phi(t) = \sum_{n=-\infty}^{\infty} h(n)\phi(2t-n) \tag{1}$$

$$\psi(t) = \sum_{n=-\infty}^{\infty} h(n)\phi(2t-n) \tag{2}$$

where

$$g(n) = (-1)^n h(1-n) \tag{3}$$

The method of generation of $\phi(t)$ is as shown in Figure 1. The input is an impulse function and after a few iterations the output obtained will be the scaling function, $\phi(t)$, as per the Equations given above.

In the case of classical wavelets, the analysis and synthesis filter bank coefficients are the same, since at a given scale the shifted versions of the scaling function and wavelet function are orthogonal to each other. The conditions are true for the case of standard wavelets like Daubechies and Morlet. However in the case of biorthogonal wavelets, analysis and synthesis

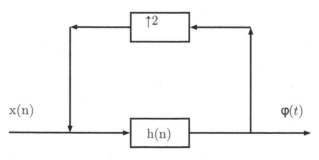

Figure 1.   Method for finding $\phi(t)$ iteratively.

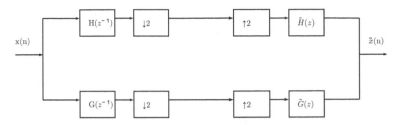

Figure 2.   Analysis and synthesis filter bank.

filter bank coefficients are not the same. Since the waveform of the single note of the musical signal which is used as the scaling function does not satisfy the orthogonality property they are designed to be biorthogonal wavelets. Hence the remaining analysis and synthesis filter bank coefficients can be found out using the properties of biorthogonal wavelet. Using the biorthogonal wavelet and scaling function coefficients, faithful reconstruction can be obtained. Inorder to get more accurate results the the filter bank coefficients can be modified using lifting technique [16]. Lifting results in a new wavelet with enough vanishing moments.

Denoising of noisy signals can be seen as an important application of any wavelet designed specifically for that signal. Wavelet based denoising is done by the thresholding of wavelet coefficients. In wavelet analysis high amplitude coefficients mainly represent signal and low amplitude coefficients with randomness represent noise. If a signal has its energy concentrated in a small number of wavelet coefficients, its coefficients will be large compared to any other signal or noise that has its energy spread over a large number of coefficients. Denoising is achieved by selecting an appropriate threshold for such high amplitude coefficients [17]–[18].

The rest of the paper is organised as follows. Section II briefly explains filter bank theory. Section III explains the need for biorthogonal wavelets and their design methodology. The method of modifying the wavelets using lifting technology is given in section IV. Section V explains the denoising of musical signals. The simulation results are given in Section VI. The paper is concluded in Section VII.

2   FILTER BANK THEORY

Wavelet decomposition and reconstruction of a musical signal based on the multi resolution theory can be obtained using digital FIR. Figure 2 shows the filter bank implementation of wavelets. The filters $h(n)$ and $g(n)$ are analysis (decomposition) filters which are low pass and high pass respectively. They are downsampled by two. The high frequency coefficients are detailed coefficients and low frequency coefficients are approximation coefficients. Since musical signals need biorthogonal wavelet, the synthesis reconstruction filters are dual of the analysis filters. They are denoted as $\tilde{h}(n)$ and $\tilde{g}(n)$ which are preceded by upsampling by two to give the reconstructed signal.

3   DESIGN OF BIORTHOGONAL WAVELETS

In the case of musical signals, the repeating elements are such that the scaling function obtained using them does not satisfy the orthogonal condition. They therefore need to be designed as biorthogonal. The analysis and synthesis filter coefficients are different in the biorthogonal case. In biorthogonal wavelets, there is a dual scaling function in addition to the scaling function generated by h(n). It is denoted by $\tilde{\phi}(t)$. In the case of orthogonal wavelets $\phi(t)$ is orthogonal to its own translates whereas in the case of biorthogonal wavelets, $\phi(t)$ is orthogonal to the translates of $\tilde{\phi}(t)$. Similarly, $\phi(t)$ is orthogonal to $\psi(t)$ for ordinary wavelets. But $\phi(t)$ is orthogonal to $\tilde{\psi}(t)$ for biorthogonal wavelets.

Mathematically biorthogonal wavelets imply,

$$\phi(t) = \sum_n h(n)\phi(2t - n) \tag{4}$$

$$\tilde{\phi}(t) = \sum_n \tilde{h}(n)\tilde{\phi}(2t - n) \tag{5}$$

$$\psi(t) = \sum_n g(n)\phi(2t - n) \tag{6}$$

$$\tilde{\psi}(t) = \sum_n \tilde{g}(n)\tilde{\phi}(2t - n) \tag{7}$$

where

$$g(n) = (-1)^n \tilde{h}(1 - n) \tag{8}$$

and

$$\tilde{g}(n) = (-1)^n h(1 - n) \tag{9}$$

The filter coefficients $h(n)$, are obtained using MPSO algorithm. The coefficients for the dual scaling function, $\tilde{h}(n)$, are designed from $h(n)$ so that they satisfy the following conditions. Normality of the dual scaling function:-

$$\sum_k \tilde{h}(k) = 2 \; \forall \; k \in Z \tag{10}$$

$\phi(t)$ should be orthogonal to the translates of $\tilde{\phi}(t)$:-

$$\sum_k h(k)\tilde{h}(k - 2n) = \delta_{n,0} \; \forall \; k \in Z \tag{11}$$

First vanishing moment of $\psi(t)$:-

$$\int_{-\infty}^{\infty} \psi(t)dt = 0 \Rightarrow \sum_k (-1)^n \tilde{h}(N - k - 1) = 0 \tag{12}$$

$\tilde{h}(n)$ is obtained by solving the Equation (10), Equation (11) and Equation (12). The obtained $h(n)$ and $\tilde{h}(n)$ values are used to design g(n) and $\tilde{g}(n)$ as shown in Equation (8) and Equation (9). These values are substituted in Equation (6) and Equation (7) to get the biorthogonal wavelet functions.

4 LIFTING TECHNOLOGY

The lifting scheme is an approach to construct so called second generation wavelets, which are not necessarily transalates and dilations of one function [16]. In the present work, lifting scheme in the Z domain is used for making new set of filter coefficients from the existing wavelet. It works in spatial domain. Using the lifting technology, a set of filter coefficients can be modified into a new set of filters without affecting the perfect reconstruction property. Figure 3 shows the basic idea of lifting. Filter coefficients $u(n)$ in the left part of the diagram modifies the high pass filtered signal by adding it a weighted sum of low pass filtered signal coefficient. On the right part of the diagram $u(n)$ nullifies this change by subtracting the same quantity. In the left part of the diagram $u(n)$ modifies the detail coefficients there by modifying high-pass analysis filter coefficients. In the right side $u(n)$ performs the undo operation giving the highpass filtered signal back. This results in a set of new analysis low-pass filter coefficients and synthesis highpass filter coefficients. Lifting of SSM wavelets is described as follows.

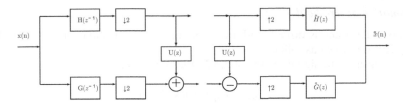

Figure 3. Analysis and synthesis filter bank using lifting.

Initially $h(n)$ and $\tilde{g}(n)$ are taken as such. $g(n)$ and $\tilde{h}(n)$ are taken as Haar wavelet coefficients. ie, $g(n) = [1\ -1]$ and $\tilde{h}(n) = [1]$. In the Z domain, the four filter coefficients can be written as

$$H(z) = h_1(n) + h_2(n)z^{-1} + \ldots + h_K(n)z^{-K} \tag{13}$$

$$G(z) = 1 - z^{-1} \tag{14}$$

$$\tilde{H}(z) = 1 + z^{-1} \tag{15}$$

$$\tilde{G}(z) = \tilde{g}_1(n) + \tilde{g}_2(n)z^{-1} + \ldots + \tilde{g}_K(n)z^{-K} \tag{16}$$

Suppose new wavelet function at the analysis side is

$$\psi^{new}(t) = \psi(t) + \alpha_1 \phi(t) + \alpha_2 \phi(t-1) \tag{17}$$

Applying the moment conditions to above equations, the constants α_1 and α_2 can be found out and the new wavelet function is obtained. From this modified analysis highpass filter, $g^{new}(n)$ can be obtained. The equation connecting $g^{new}(n)$ and $u(n)$ in Z domain is as follows:

$$G^{new}(z^{-1}) = G(z^{-1}) + U(z^2)H(z^{-1}) \tag{18}$$

$u(n)$ can be found out from above equation so that modified $\tilde{h}^{new}(n)$ can be obtained as follows:

$$\tilde{H}^{new}(z) = \tilde{H}(z) - U(z^2)\tilde{G}(z) \tag{19}$$

5 DENOISING USING WAVELETS

Consider a clean musical signal $x(n)$ of length N is corrupted by a additive white gaussian noise denoted by $w(n)$. Then the noisy signal v(n) is given by

$$v(n) = x(n) + \omega(n) \tag{20}$$

Here the idea is to recover back the signal x(n) from this noisy signal. For a particular musical instrument, the most suited wavelet will have maximum energy concentrated in the approximation coefficients rather than in the detailed coefficients. Hence only detailed coefficients are denoised. The denoising is done by proper thresholding of detailed coefficients obtained after decomposition of the noisy signal. Hence denoising can be viewed as three steps.

1. Decomposition of the noisy signal
2. Thresholding of the Detailed Coefficients
3. Reconstruction of the original signal

5.1 Estimation of threshold

In this paper denoising is done by adopting the universal threshold proposed by Donoho et al. [17]. It is described as follows:
If N is the length of the signal the threshold λ at level j is given by

$$\lambda_j = \sigma_j \sqrt{2\log(N)/N} \qquad (21)$$

where the scale estimate σ_j is given by

$$\sigma_j = MAD/0.6745 \qquad (22)$$

where MAD is the median absolute value of the wavelet coefficients at level j.

5.2 Thresholding of detail coefficients

If the noise energy is less than signal energy, the corresponding noise wavelet coefficients will be obviously less than the signal wavelet coefficients. By soft thresholding, the detail coeffficients can be thresholded. If $v_{j,k}$ is the wavelet coefficients at level j, soft thresholding is done by,

$$v_{j,k} = \begin{cases} v_{j,k} - \lambda, & \text{if } |v_{j,k}| \geq \lambda \\ 0, & \text{otherwise} \end{cases} \qquad (23)$$

6 SIMULATION RESULTS

6.1 Biorthogonal SSM filter coefficients

Once $h(n)$ is obtained using modified MPSO algorithm, $\tilde{h}(n)$ is obtained by solving equations formed by substituting different integer values for k in Equations (10)–(12). The coefficients, $\tilde{h}(n)$, are symmetrical and also the length is more than that of $h(n)$. The length of $\tilde{h}(n)$ for flute signal is 17 while that of $h(n)$ is 15. The coefficients obtained are as shown in the Table 1. It can be seen that $h(n)$ and $\tilde{h}(n)$ satisfy all the Eqns. (10)–(12). From these values, $g(n)$ and $\tilde{g}(n)$ are obtained using Equation (8) and Equation (9). The original signal and the reconstructed signal using biorthogonal SSM wavelet is shown in the Figure 4.

Table 1. Filter bank coefficients for flute signal.

h(n)	0.2290	0.1796	0.0235	0.2084	0.2928
	0.1962	−0.0004	−0.0201	0.1703	0.2593
	−0.0139	−0.1915	0.0359	0.0728	−0.0244
g(n)	−2.1465	−2.9536	−0.5659	−1.4113	0.4386
	−2.9206	0.6092	6.9320	4.0364	6.9320
	0.6092	−2.9206	0.4386	−1.4113	−0.5659
			−2.9536	−2.1465	
$\tilde{h}(n)$	−2.1465	2.9536	−0.5659	1.4113	0.4386
	2.9206	0.6092	−6.9320	4.0364	−6.9320
	0.6092	2.9206	0.4386	1.4113	−0.5659
		2.9536		−2.1465	
$\tilde{g}(n)$	−0.0244	−0.0728	0.0359	0.1915	−0.0139
	−0.2593	0.1703	0.0201	−0.004	−0.1962
	0.2928	−0.2084	0.0235	−0.1796	0.2290

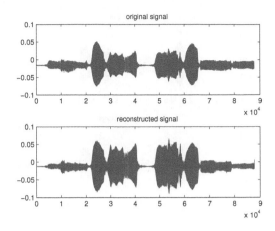

Figure 4.   Reconstructed signal using biorthogonal SSM wavelet.

Table 2.   Filter bank coefficients modified using lifting.

$h(n)$	0.2290	0.1796	0.0235	0.2084	0.2928
	0.1962	−0.0004	−0.0201	0.1703	0.2593
	−0.0139	−0.1915	0.0359	0.0728	−0.0244
$g(n)$	0.9531	−1.036	0.042	−0.00058	−0.055
	0.00025	0.06	0.04427	−0.35	−0.0571
	0.0377	0.0922	−0.00045	−0.0541	0.0123
			0.0728	−0.0244	
$\tilde{h}(n)$	1.0236	1.7047	−0.0103	−0.1122	−0.02245
	0.0597	−0.1509	0.2399	−0.1702	0.1698
	−0.283	0.3979	−0.3155	0.3822	−0.2451
			0.1796	−0.229	
$\tilde{g}(n)$	−0.0244	−0.0728	0.0359	0.1915	−0.0139
	−0.2593	0.1703	0.0201	−0.004	−0.1962
	0.2928	−0.2084	0.0235	−0.1796	0.2290

6.2   *Modified SSM filter coefficients after lifting*

The biorthogonal filter coefficients obtained above are modified using lifting technology as described before. The modified filter coefficients are as shown in the Table 2. The length of the filter coefficients are same as before. The reconstructed signal using modified SSM wavelets after lifting is shown in the Fig. 5. The signal is more similar to the original signal compared to Fig. 4.

6.3   *Denoising of flute signals*

1. *Using standard wavelets*: The set of flute signals are denoised using standard wavelets. Three wavelets families, Symlets 2 to 8, Daubechies 2 to 10 and Coiflet 1 to 5 are taken. The result shows that Coiflet5 gives best denoising performance.
2. *Using biorthogonal SSM wavelets*: The set of flute signals are denoised using biorthogonal SSM wavelets. The flute signal corrupted with noise is first decomposed to approximation and detail coefficients. The threshold is calculated as per Eqn. (21) and Equation (22). The detail coefficients are thresholded by using equation (23). Finally the signal is reconstructed back. Figure 6 shows the result.

677

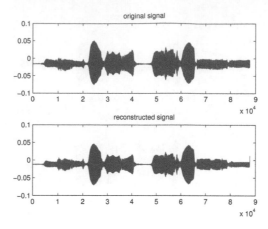

Figure 5. Reconstructed signal using biorthogonal SSM wavelet after lifting.

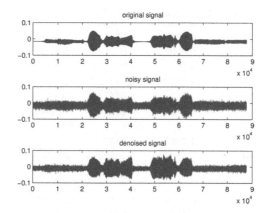

Figure 6. Denoised signal using biorthogonal SSM wavelet.

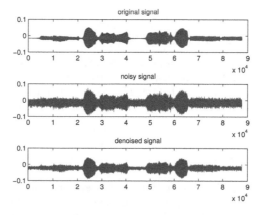

Figure 7. Denoised signal using modified biorthogonal SSM wavelet.

3. *Using modified SSM wavelets after lifting*: The set of flute signals are denoised using modified SSM wavelets using the same method described above. Figure 7 shows the result. The denoised signal is more similar to the original signal compared to the result in Figure 6.

Table 3. SNR comparison in DB.

Wavelet	Input SNR	Output SNR
Coiflet5	20.60	26.93
Coiflet1	20.60	24.48
Haar	20.60	23.97
daubechies10	20.60	26.52
Symlet8	20.60	26.46
Symlet1	20.60	23.92
Biorthogonal SSM	20.60	24.52
Modified SSM	20.60	29.03

7 COMPARISON OF DENOISING PERFORMANCE

In order to compare the quality and performance of the denoised signal using two set of filter coefficients, the signal to noise ratio of the signal is determined. SNR is calculated as follows.

$$SNR = 10\log\left[\frac{\sum_{k=0}^{N-1} x_k^2}{\sum_{k=0}^{N-1} (x_k - \tilde{x}_k)^2}\right] \tag{24}$$

where x is the original signal and $\tilde{x}$ is the denoised signal. The value of SNR obtained are as shown in the Table 3. Modified SSM wavelet gives the highest SNR.

8 CONCLUSION

Musical signals are analyzed using a wavelet specifically designed for these signals. Using the already available analysis low-pass coefficients of this wavelet, biorthogonal SSM analysis and synthesis filter coefficients are found out by adopting the properties satisfied by the biorthogonal wavelets. Using these coefficients, the musical signal is decomposed and reconstructed. The coefficients are modified using lifting technology with an aim to improve the reconstruction performance. The resulting reconstructed signal shows more similarity to the original signal than the previous one. Denoising of musical signals is performed using these filter bank coefficients. An appropriate threshold for removing the detail coefficients corresponding to noise is calculated. The noisy signal is denoised using standard wavelets. Among the standard wavelets, Coiflet5 is found to be the suitable wavelet for denoising a musical signal. It is able to reduce the noise in the noisy musical signal when it is denoised using the two sets of SSM coefficients obtained. The result shows that applying lifting improves the denoising performance.

REFERENCES

[1] I. Daubechies. Ten Lectures on Wavelets. SIAM, Philadelphia, PA, 1992.
[2] Hubbard, Barbara Burke, *"The World According to Wavelets The Story of a Mathematical Technique in the Making"*, Universities Press, 1998.
[3] Martin Vetterli and Cormac Harley, *"Wavelets and filter banks: Theory and design"*, IEEE Transactions on Signal Processing, Vol. 40, No. 9, September 1992, pp. 2207–2232.
[4] Joseph.O. Chapa and Raghuveer M.Rao, *"Algorithms for designing wavelets to match a specified signal"*, IEEE Transactions on Signal Processing, Vol. 48, No. 12, December 2000, pp. 3395–3406.

[5] J. Nagumo and A. Noda *"A learning method for system identification"*, IEEE Transactions on Automatic Control, Vol. AC-12, June 1967, pp. 283–287.

[6] Markus Rupp, *The Behaviour of LMS and NLMS Algorithms in the Presence of Spherically Invariant Process*, IEEE Transactions on Signal Processing. Vol. 41, No. 3, pp. 1149–1160, March 1993.

[7] Raghavendra Sharma and V Prem Pyara, *"A Comparative Analysis Of Mean Square Error Adaptive Filter Algorithms For Generation Of Modified Scaling And Wavelet Function"*, International Journal Of Engineering Science And Technology (IJEST), Vol. 4, No. 4, April 2012, pp. 1402–1407.

[8] Jyoti Dhiman, Shadab Ahmad and Kuldeep Gulia, *"Comparison Between Adaptive Filter Algorithms (LMS, NLMS and RLS)"*, International Journal Of Science, Engineering And Technology Research (IJSETR), Volume 2, Issue 5, May 2013, pp. 1100–1103.

[9] Thamer M. Jamel, *"Performance Enhancement of Adaptive Acoustic Echo Canceller Using a New Time Varying Step Size LMS Algorithm (NVSSLMS)"*, International Journal of Advancements in Computing Technology(IJACT), Korea, Vol. 3, No. 1, January 2013.

[10] R.H. Kwong and E.W. Johnston, *"A variable step size LMS algorithm"*, IEEE Transactions on Signal Processing, Vol. 40, July 1992, pp. 1633–1642.

[11] Junghsi Lee, Jia-Wei Chen, and Hsu-Chang Huang, *"Performance Comparison of Variable Step-Size NLMS Algorithms"*, Proceedings of the World Congress on Engineering and Computer Science 2009, Vol. 1, October 2009.

[12] M.S. Sinith, Madhavi N. Nair, Niveditha P. Nair and Parvathy S. *Identification of Wavelets and Filter Bank Coefficients in Musical Instruments*, International Conference in Audio Language and Image Processing (ICALIP 2010), Shanghai, China pp. 727–731, Nov. 2010.

[13] J Kennedy and R Eberhart, *Particle swarm optimization*, Proceedings of IEEE International Conference on Neural Networks, vol. 4, pp. 1942–1948, 1995.

[14] R Poli, J Kennedy and T Blackwell, *Particle swarm optimization*, International Journal of Swarm Intelligence, Volume 1, Issue 1, pp. 33–57, June 2007.

[15] M.S. Sinith, Shikha Tripathi, K.V.V. Murthy, *"SSM wavelets for analysis of music signals using particle swarm optimization"*, IEEE International Conference on Signal Processing and Communication (ICSC), Noida, India, pp. 247–251, Dec. 2013.

[16] K.P. Soman and K.I. Ramachandran, *"Insight into Wavelets:From Theory to Practise"*, Prentice Hall.

[17] Donoho D.L., Johnston I.M., *"De-noising by soft-thresholding"* IEEE transactions on information theory, vol. 41, No. 3, 1995, pp. 613–627.

[18] Donoho D.L., Johnston I.M., "Ideal spatial adaptation by wavelet shrinkage." Biometrika, Vol. 81, No. 3, 1994, pp. 425–455.

Emerging Trends in Engineering, Science and Technology for Society,
Energy and Environment – Vanchipura & Jiji (Eds)
© *2018 Taylor & Francis Group, London, ISBN 978-0-8153-5760-5*

Low complexity encoding of M-ary QAM constellation for linear index codes

Anjaly Shaju & Senthilkumar Dhanasekaran
Department of Electronics and Communication, Ramaiah Institute of Technology, Bangalore, India

ABSTRACT: An index coding problem consists of a single source S with K messages transmitted across a Gaussian broadcast channel, where each receiver demands a set of information from the source while having a subset of prior information in its cache, known as 'side information'. The power of the receiver which efficiently exploits receivers' side information is called the 'side information coding gain'. The known index codes have the heaviest quantity of side information gain, but finding the encoding matrix is a tedious process. This presented work aims to find the encoding matrix to construct the multidimensional Quadrature Amplitude Modulation (QAM) constellation proposed by Natarajan et al. (2015a), with a fewer number of computer searches. Furthermore, the side information gain achieved is compared with the existing method.

1 INTRODUCTION

Network coding is a promising technology which offers advantages to the communication networks, such as throughput and performance of the network (Koetter & Médard 2003). In a multicast communication network, source nodes broadcast data to the intermediate nodes. The intermediate nodes combines the data from the source nodes and broadcast to the destination nodes. Hence, it helps in minimizing the bandwidth and the delay.

An equivalence between the network coding and index coding was proved by Effros et al. (2015). Any network coding problem can be converted into the index coding problem, the solution obtained to the index coding problem, and then the solution returned back into the original network coding problem. Index coding is the reduced form of the network coding. The equivalence between these codes holds good for both the linear and nonlinear codes.

Index coding was first proposed (Birk & Kol 2006) for satellite communication, where a single source S with a set of receivers $(R_1, R_2, ..., R_n)$ are considered and each receiver has some subset of information as *a priori*. Information Source Coding on Demand (ISCOD) entails exploiting full knowledge of client side information to fulfill the receivers' demand. In satellite communication, during the main transmission, all the clients may not satisfy its demand due to insufficient storage capacity in the client's cache or due to any other interferences. There is no direct communication between the clients, but a feedback channel is used to send the requests back to the server. In this case, index coding is used in wireless networks to satisfy the receivers' demand with a fewer number of binary transmissions. Optimal index codes can be used to reduce the number of binary transmissions by using the min-rank (Bar-Yossef et al. 2011, Mahesh and Rajan 2016). The optimal index codes is constructed by using the encoding matrix, which consists of receiver-side information and its demand.

A system model for lattice index codes (Natarajan et al. 2015b) consists of a single source S and a set of K messages, where each receiver demands some subset of messages from the source while having some information in its cache as *a priori*. Lattice index codes efficiently exploit the receiver-side information and convert it into apparent coding gain. The lattice index codes for the Gaussian broadcast channel, in which the K messages are individually mapped into the K modulo lattice constellation and the transmitting symbols are generated by taking the sum of the individual symbols. Natarajan et al. (2015a) proposed index codes based on a multidimensional Quadrature Amplitude Modulation (QAM) constellation for the Gaussian broadcast channel, where every receiver demands all the messages from the source. Side information coding gain is obtained by efficiently exploiting the receiver-side information present in its cache, over a non-fading broadcast channel with K independent messages. The known index codes have larger complexity in the construction of the encoding matrix, which is uniquely decodable.

The noisy index coding (Natarajan et al. 2015b, Natarajan et al. 2015a) is a specific case of index coding where all the receivers demands K independent messages from the source, indicating $(w_1, w_2, ..., w_k)$ that assumes the value from $(\mathcal{W}_1, \mathcal{W}_2, ..., \mathcal{W}_k)$. The lattice codes (Natarajan et al., 2015b) are constructed by using the Chinese remainder theorem with larger side information gain. The K messages are individually mapped into the K modulo lattice constellation and the transmitting symbols are generated by taking the sum of the individual symbols. The source is assumed to be operated in average power constraints and the receiver experiences Additive White Gaussian Noise (AWGN). Each receiver is denoted by (SNR, S), where SNR denotes the signal-to-noise ratio and $S = 1, ..., K$ where S represents the side information present in each receiver.

The terminology $S = \phi$ indicates that there is no side information present in the receivers. Let $(R_1, R_2, ..., R_K)$ be the rate of the each messages in bits per dimension. Suppose the source entropy is $R = (R_1, R_2, ..., R_K)$, and the side information rate is represented by $R_s = \sum_{K \in S} R_K$ (Natarajan et al. 2015b). The index code presented by Natarajan et al. (2015b) offers larger receiver side information coding gains, but there are two practical limitations: i) It does not encode all the messages at equal rates; and ii) It does not consider the messages that are a power of two. Recently, Natarajan et al. (2015a) have proposed multidimensional QAM constellation mapping for index codes over the Gaussian broadcast channel, where all the receivers demand an equal number of messages from the source. With the help of computer searches, an encoding matrix is constructed with the symbols in the set Z_M, with the determinant as the odd integer in the set, and up to five messages with message size of $2^m \le 6$. The complexity is exponentially increased in the construction of the encoding matrix when a larger number of messages are considered.

In this paper, we proposed an algorithm which reduces the computer searches, in turn, minimizing the complexity in the construction of the encoding matrix to map the symbols on the multidimensional QAM constellation. Furthermore, we have shown that the side information coding gain achieved by the code for a different number of side information is known at the receiver. The obtained results are compared with the recently proposed algorithm, which clearly shows that the obtained encoding matrix offers better side information gain in some cases but inferior in some other scenarios.

2 CONSTRUCTION OF ENCODING MATRIX FOR QAM CONSTELLATION

In this section, we discuss an algorithm to construct the encoding matrix which linearly transforms the K-tuple symbols from the Z_M^K message symbol space into an Z_M^K coded symbol space. This transformation includes all the possible K-tuple symbols in the construction of the transformation matrix (i.e., circulant encoding matrix). For example, K-tuple symbols with the symbol set size of M, there are M^K possible encoding matrices. Hence, this will exponentially increase the complexity with K, in the construction of an optimum encoding matrix to map the message symbols onto the coded symbols. The simplest way to find K linearly independent code vectors is to construct the circulant matrices using all possible symbols, which significantly reduces the complexity of the computer searches (Natarajan et al. 2015a).

In order to span the K-tuple message symbol space over modulo M, we consider a symbol set, $Z_M \left\{ \frac{-M}{2}, \frac{-(M-2)}{2}, ..., 0, ..., \frac{(M-2)}{2} \right\}$ for even values of M, and for odd values of M, $Z_M \left\{ \frac{-(M-1)}{2}, \frac{-(M-3)}{2}, ..., 0, ..., \frac{(M-1)}{2} \right\}$. The Z_M has the structure of the commutative ring with addition and multiplication performed over an integer modulo M. A unit, $U(M)$, of the defined set Z_M, is the odd integers in the set Z_M when M is even, and even integers when M is odd. A linear index code with K messages consists of a set of K generators where the linear encoder $X = \rho(w_1, w_2, \ldots w_K) = \sum_{k=1}^{K} w_k c_k \mod M$ is injective.

The injectivity of the linear encoder X gives the unique decodability at the receiver side. A linear index code is completely characterized by the matrix C whose rows are K generators $c_1, c_2, \ldots, c_K$. The encoding matrix C defines a linear transformation in which the matrix multiplies with message symbols to form the codewords. Thus the encoder mapping is injective if and only if C is invertible. The matrix is constructed with the determinant as the unit of the symbol set Z_M.

3 SIMPLIFIED ALGORITHM TO CONSTRUCT ENCODING MATRIX

3.1 Proposed algorithm

In this section, we propose an algorithm as shown in Figure 1, which minimizes the complexity in the computer searches significantly, to construct the encoding matrix. We consider the following issues to frame a simplified algorithm: i) Exchanging of the rows in the circulant matrices will provide the same coding gain; and ii) An odd number of odd integers in the first row of the circulant matrix will provide the determinant value of an odd integer.

Remark 1 Circulant matrix which has an odd number of odd integers in the first row, will offer the determinant value of an odd integer.

Example 1. For a 2×2 matrix, the odd number of odd integers in the first row of the circulant encoding matrix gives an odd integer as the determinant.

$$\begin{vmatrix} o_1 & e_1 \\ e_1 & o_1 \end{vmatrix} = o_1.o_1 - e_1.e_1 = o - e = \text{odd number} \qquad (1)$$

Similarly, for a 3×3 matrix:

$$\begin{vmatrix} o_1 & e_1 & e_2 \\ e_2 & o_1 & e_1 \\ e_1 & e_2 & o_1 \end{vmatrix} = \text{odd number} \qquad (2)$$

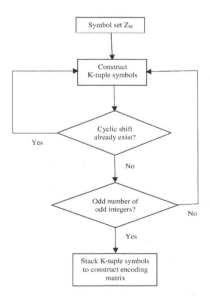

Figure 1.   Flowchart for simplified algorithm to construct the encoding matrix.

Algorithm

Step 1: Construct the symbol set Z_M
Step 2: Construct all possible K-tuple symbols
Step 3: Construct the circulant matrix
Step 4: If cyclic shift of that matrix already exists, repeat step 2
Step 5: If even number of odd integers in the K-tuple symbols, repeat step 2
Step 6: Stack the symbols to construct the encoding matrix

Example 2. Let $K = 2$ and $M = 8$. The symbol sets are mapped into the set Z_M to form a commutative ring. For $K = 2$, we construct the 2×2 encoding matrix, which is circulant, and the symbols with an odd number of odd integers are chosen as the first row to construct the circulant matrix. The encoding matrix should satisfy the conditions such that the determinant should be in the unit of the ring.

For $K = 2$ and $M = 8$ the two generators are $c_1 = (-2, -3)$ and $c_2 = (-3, -2)$. The encoder $x = (-2w_1 \ -3w_2, \ -3w_2 \ -2w_1)$ mod 8, is calculated from the rows of the circulant encoding matrix. The circulant encoding matrix for $K = 2$ and $M = 8$ is given by:

$$C = \begin{bmatrix} -2 & -3 \\ -3 & -2 \end{bmatrix}$$

The determinant for the encoding matrix is, $-5 \bmod 8 = -1$, which is in the unit of the set Z_M^K. Therefore, this linear encoder is injective. The labelling scheme for a two dimensional 64- QAM constellation is shown in Figure 2.

3.2 *Complexity*

In an existing algorithm (Natarajan et al. 2015a), in order to reduce the complexity in the exhaustive search space, the authors have chosen the circulant matrices whose determinant is in unit of a set Z_M as the encoding matrices, which provides the largest minimum Euclidean distance between the coded symbols. As discussed in Section 3.1, the proposed algorithm minimizes the complexity by reducing the computer searches significantly. For example, for $K = 2$ and $M = 32$, there are $32^2 = 1024$ possible symbols (i.e., encoding matrices). The number of searches, computed from the simulation, in the proposed algorithm is 257, whereas in (Natarajan et al. 2015a) it is, $K \times 257 = 2 \times 257 = 514$.

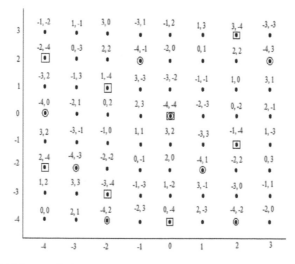

Figure 2. The 64-QAM constellation. The eight points forming the subcode corresponding to the side information when $w_1 = -4$ are highlighted with circles and the subcode for $w_2 = -4$ is marked with squares.

3.3 Side information gain

The ability of the receiver to efficiently exploit the receiver-side information is called 'side information coding gain'. It is assumed that all K messages have equal transmission rate. The transmission rate is given by:

$$R_k = \frac{1}{K}\log_2 M \quad \text{b/dim} \tag{3}$$

The side information rate at each receiver is given by:

$$R_S = \sum_{k \in S} R_k = \frac{|S|}{K}\log_2 M \quad \text{b/dim} \tag{4}$$

The coding gain achieved with side information available at the receiver, is given by:

$$\Gamma \triangleq \frac{10\log_{10} d_s^2 / d_0^2}{R_S} \quad \text{dB/b/dim} \tag{5}$$

Table 1. Comparison between the existing algorithm and the proposed algorithm (one side information is known) with first row of the circulant encoding matrix and the gain.

	Algorithm in (Natarajan et al. 2015a)			Proposed Algorithm		
M	$K=2$	$K=3$	$K=4$	$K=2$	$K=3$	$K=4$
4	(1,–2) 6.02	(1,–2,–2) 4.515	(1,1,–1,0) 6.02	(–1,–2) 6.02	(–2,1,–2) 4.515	(–2,–2,–2,–1) 6.02
8	(1,2) 4.65	(1,2,0) 6.02	(1,0,3,3) 4.01	(–2,–3) 4.65	(3,–2,0) 6.02	(–4,–1,2,2) 4.01
16	(1,–4) 6.02	(1,2,–6) 5.24	(1,4,–6,–8) 6.02	(–3,–4) 6.02	(2,–6,7) 5.24	(–8,–5,–4,2) 6.02
32	(1,6) 5.85	(1,–10,14) 5.73	(1,10,14,2) 6.22	(–10,–9) 5.85	(–15,–10,14) 5.73	(–15,10,14,2) 6.22
64	(1,–28) 6.04	(1,–26,–4) 5.73	(1,–26,20,30) 6.36	(–11,–12) 6.04	(–20,8,17) 5.57	(–31,–30,20,26) 6.36

Table 2. Comparison between the existing algorithm and the proposed algorithm (two side information are known) with first row of the circulant encoding matrix and the gain.

	Algorithm in (Natarajan et al. 2015a)		Proposed Algorithm	
M	$K=3$	$K=4$	$K=3$	$K=4$
4	(1,–2,–2) 4.515	(1,1,–1,0) 3.01	(–2,1,–2) 4.515	(–2,–2,–2,–1) 3.01
8	(1,2,0) 3.49	(1,0,3,3) 5.18	(3,–2,0) 3.49	(–4,–1,2,2) 4.65
16	(1,2,–6) 5.83	(1,4,–6,–8) 5.57	(2,–6,7) 5.83	(–8,–5,–4,2) 5.57
32	(1,–10,14) 5.84	(1,10,14,2) 6.02	(–15,–10,14) 5.84	(–15,10,14,2) 6.02
64	(1,–26,–4) 5.815	(1,–26,20,30) 5.85	(–20,8,17) 6.02	(–31,–30,20,26) 5.85

Table 3. Comparison between the existing algorithm and the proposed algorithm (three side information are known) with first row of the circulant encoding matrix and the gain.

M	Algorithm in (Natarajan et al. 2015a) K = 4	Proposed Algorithm K = 4
4	(1,1,–1,0) 3.18	(–2,–2,–2,–1) 4.01
8	(1,0,3,3) 4.62	(–4,–1,2,2) 5.35
16	(1,4,–6,–8) 6.02	(–8,–5,–4,2) 6.02
32	(1,10,14,2) 5.80	(–15,10,14,2) 5.80
64	(1,–26,20,30) 6.08	(–31,–30,20,26) 6.08

where d_S is the distance between the constellation points when either of the side information is known to the receiver and d_0 is the distance between any two adjacent points in the constellation.

4 RESULTS AND DISCUSSION

In this section, we present the results for $K = 2, 3, 4$ and $M = 4, 8, 16, 32, 64$. With the help of a computer search we find the best linear index codes which provide maximum coding gain. The coding gains achieved by the code for one, two, and three side information are known, and respectively, detailed in Tables 1, 2, and 3. The results presented in Tables substantiate that the encoding matrix suggested by the proposed algorithm offers the comparable side information gain compared to the existing work, with low computational complexity.

5 CONCLUSION

The construction of an encoding matrix for multidimensional QAM constellation mapping of linear index codes is presented. Moreover, it is substantiated that the proposed simplified algorithm significantly reduces the computer searches, and in turn, the complexity in the construction of the encoding matrix for index codes is significantly less, compared to the existing algorithm (Natarajan et al. 2015a). Simulation results presented in Tables have shown that the proposed algorithm performs similar to its counterpart (Natarajan et al. 2015a), with low complexity. It is shown that the complexity in the proposed algorithm is reduced by a factor of K.

REFERENCES

Bar-Yossef, Z., Y. Birk, T. Jayram, & T. Kol (2011). Index coding with side information. *IEEE Transactions on Information Theory 57*(3), 1479–1494.

Birk, Y. & T. Kol (2006). Coding on demand by an informed source (ISCOD) for efficient broadcast of different supplemental data to caching clients. *IEEE/ACM Transactions on Networking 14(5)*, 2825–2830.

Effros, M., S. El Rouayheb, & M. Langberg (2015). An equivalence between network coding and index coding. *IEEE Transactions on Information Theory 61*(5), 2478–2487.

Koetter, R. & M. Médard (2003). An algebraic approach to network coding. *IEEE/ACM Transactions on Networking 11*(5), 782–795.

Mahesh, A.A. & B.S. Rajan (2016). Noisy index coding with PSK and QAM. *arXiv preprint arXiv:1603.03152*.

Natarajan, L., Y. Hong, & E. Viterbo (2015a). Index codes for the gaussian broadcast channel using quadrature amplitude modulation. *IEEE Communications Letters 19*(8), 1291–1294.

Natarajan, L., Y. Hong, & E. Viterbo (2015b). Lattice index coding. *IEEE Transactions on Information Theory 61*(12), 6505–6525.

Emerging Trends in Engineering, Science and Technology for Society,
Energy and Environment – Vanchipura & Jiji (Eds)
© *2018 Taylor & Francis Group, London, ISBN 978-0-8153-5760-5*

Improving myoelectric grasp recognition using empirical mode decomposition and differential evolution based approach

C.K. Anusha & K. AjalBabu
Department of Electrical and Electronics Engineering, TKM College of Engineering, Kollam, India

Nissan Kunju
Department of Electronics and Communication Engineering, TKM College of Engineering, Kollam, India

ABSTRACT: This paper presents a surface Electromyographic (sEMG) signal based hand grasp recognition technique utilizing Empirical Mode Decomposition (EMD) and Differential Evolution based Feature Selection (DEFS). A series of features were derived from both the raw signal and its corresponding Intrinsic Mode Functions (IMF's), obtained by performing EMD. Differential Evolution (DE) is a relatively new soft computing technique with wide range of applications. Being a promising stochastic population based optimization method, a feature selection framework using DE is utilized in this research to identify the optimum feature subset. sEMG signals recorded from eleven healthy subjects are used for this study. The proposed method is further validated using other popular feature extraction techniques and different pattern recognition algorithms. Outcome of our research shows that the methodology of using EMD with DEFS can achieve significant improvement in the overall recognition rate.

1 INTRODUCTION

Assistive devices for neurological rehabilitation, for example active prostheses, are controlled by man machine interfacing. Nowadays myoelectric control is evolved as the most promising approach to control devices utilized in clinical and commercial applications (Jiang et al.2012, Fougner et al. 2012, Scheme & Englehart 2009). In spite of the fact that nerve and brain recording are exceptionally encouraging for a direct neural interfacing, they often require invasive methods for electrode placement that limits their practical applicability to laboratory research or small-scale clinical testing (Micerra & Navarro 2009). Although Industrial developers like Otto Bock (Germany) and Touch Bionics (USA) have introduced surface EMG based artificial limbs in the market, EMG based control is still in a premature state being limited to few hand postures and higher EMG-channels required for effective control.

In this paper a scheme for classification of human hand grasps from surface EMG signals is presented. The novelty of our approach dwelled on the use of EMD for feature extraction combined with Differential Evolution Based Feature Selection. The feature selection framework that have been utilized in this study also gives a versatile approach to improve the developed models comprehensibility by selecting the optimum feature subset adaptively (Khushaba R. et al. 2008, Storn R. 2008, Ahmed Al-Ani et al. 2013). Our results prove that the methodology of using EMD with DEFS can achieve significantly good results.

2 PROPOSED METHODOLOGY

The work presented in this paper stems from the desire to design a self-contained prosthetic system. For the laboratory stage of the work a standard PC installed with windows

OS is used. EMG recordings were made from a generic Data Acquisition System-CMCdaq (Nissan Kunju et al. 2013). EMG recordings were processed offline using Matlab 2014(a) software. The extracted signal is bandpass filtered and non-contracting portions at the beginning of the recording are effectively removed by carrying out signal thresholding. The sEMG signal after preprocessing is segmented using an overlapping windowing approach and various features were extracted both from the raw signal and its corresponding intrinsic mode functions (IMF's), derived by performing EMD. A feature selection stage using Differential Evolution Algorithm is employed after feature extraction to avoid the effects of dimensionality. Pattern recognition phase uses the extracted features and classify each segment to one of the six hand grasps.

3 ELECTROMYOGRAM ACQUISITION

The EMG data is recorded from eleven healthy subjects (aged between 20–30 years). Before the start of the experiment, subjects were thoroughly familiarized with the experimental protocol and the EMG equipment. A four channel generic EMG data acquisition system CMCdaq is used to acquire the data at a sampling rate of 1000 Hz. The Ag/AgCl electrodes were attached over the muscle belly in line with the muscle fibres in accordance with the standard procedure in literature (Shrirao N.A. et al. 2009). The four surface EMG electrodes were placed on forearm muscles Flexor Capri Ulnaris, Extensor Capri Radialis, Extensor Digitorum, Flexor Digitorum Superficialis and ground electrode was placed onthe contralateral upper limb. This follows the electrode placement discussed in (Frank F.H. 1989). Three trials of the six different grasps, each with a duration of six second was performed and the speed and force was intentionally left to the subject's will. The Maximum Voluntary Isometric Contraction (MVIC) test was also executed by having the subject to flex and extend his/her hand at the wrist joint by exerting maximum possible force to the maximum possible inclination and sustaining it up to six seconds. The six basic hand grasps (Schlesinger G. 1919) and the experimental setup is shown in Fig. 1 and Fig. 2 respectively.

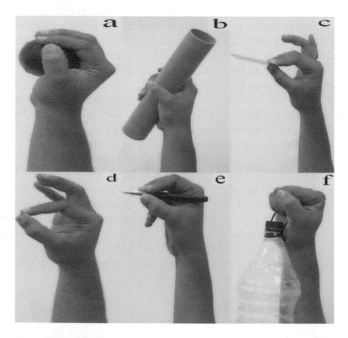

Figure 1.   Illustration of basic hand grasps.

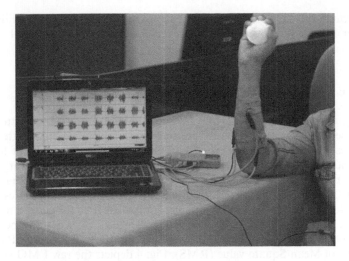

Figure 2.   Experimental setup.

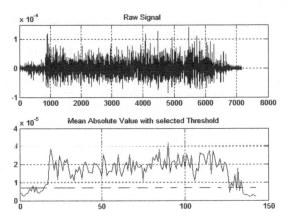

Figure 3.   Illustration of evolution of the MAV values of EMG signal starting from rest along with the threshold value (shown as dotted line) for spherical grasp.

4   SIGNAL PRE PROCESSING

Surface Electromyographic (sEMG) signals are usually affected by noise and in practice, the acquired signal may be corrupted hence the raw signal need to be preprocessed before further processing. The usable frequency range of EMG signal is considered to be in the range of 15–500 Hz as most of the energy is concentrated at this specified frequency range (De Luca C.J 1998). The acquired signal is filtered using a 4th order butterworth bandpass filter with high pass cut off frequency at 15 Hz and low pass cut off frequency at 500 Hz. After filtering each channel is normalized using the MVIC obtained for each muscle.

For each grasp the level of involvement of each muscle will be different and a dynamic threshold selection approach based on local characteristics of the signal is implemented for offline analysis (Fig. 3). Sliding window approach is used to focus only on segments where muscle is contracted. Keeping sliding window size as 50 ms, Mean Absolute Value (MAV) is calculated for each window and once that value exceeds a threshold the muscle is no longer considered to be in resting phase. In order to preserve the temporal information, recordings were taken for processing on due activation of any one of the four channels. Even though this thresholding cannot be used directly in online validation since it requires prior knowledge about the signal, this scheme worked out well in our offline analysis.

5   FEATURE EXTRACTION AND SELECTION

Each grasping operation is characterized by a unique motor unit firing pattern (Kilbreath et al. 2002, Boser B.E. et al. 1992), identifying and describing patterns that better discriminate different classes of grasp over different trials. This is the core philosophy of feature extraction. The adaptive nature of decomposition and its ability to preserve the varying frequency in time makes EMD a powerful choice in the analysis of EMG signals (Huang et al. 1998, Andrade A.O. 2004, Flandrin P. 2004). In this study 10 most popular features (Ericka Janet & Housheng Hu 2011) from time and frequency domain are used for pattern recognition and these features are Mean Absolute Value, Variance, Kurtosis, Skewness, Slope Sign Change, Waveform length, Zero Crossing, Mean Power Spectrum Density, Median Power spectrum Density and Root Mean Square value (RMS). Fig. 4 depicts the raw EMG signal with corresponding IMF's from flexor digitorum muscle during lateral grasp. The feature selection framework (Khushaba R. et al. 2008, Storn R. 2008, Ahmed Al-Ani et al. 2013) employed in this work is depicted in Fig. 5.

According to Christos Sapsanis et al. (Christos Sapsanis et al. 2013) incorporation of features derived from first three IMF's improves the overall recognition rate. However in our study it is found that no significant contributions in terms of classification accuracy is received from feature set derived from IMF's beyond second decomposition level. Sometimes feature sets derived from higher order IMF's seems to deteriorate the overall performance hence ensemble of aforementioned features from raw signal and from the first two IMF's taken, this feature set is denoted as TDEMD. The proposed TDEMD method is further validated using Discrete Wavelet Packet Transform (DWPT) utilizing energy of wavelet coefficient at each node using Daubechies family of wavelets at 4 level of decomposition, ensemble of features extracted from statistical and auto regressive modeling (TA) and Discrete Wavelet Transform (DWT) utilizing Standard Deviation, Entropy, Waveform Length and Energy of wavelet coefficients using Daubechies family of wavelets at 4 level of decomposition.

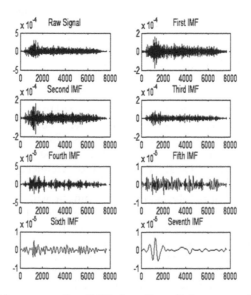

Figure 4. EMG signal with corresponding IMF's from flexor digitorum muscle during lateral grasp.

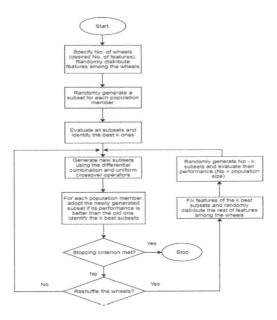

Figure 5.   Feature selection algorithm.

6   PATTERN RECOGNITION

There are several schemes based on artificial intelligence, statistical methods available for pattern recognition. All these schemes are tested with mixed success. In order to assess the effects of different pattern recognition algorithms on the performance of the proposed TDEMD approach with different pattern recognition algorithms were tested in our study. Owing to the wide acceptance in EMG based applications classifiers including Quadratic Discriminant Analysis (QDA) (Oskoei M.A. & Hu H. 2007), Support Vector Machine (LIBSVM) (Oskoei M.A. & Hu H. 2008), K-Nearest Neighbor (KNN, k = 1) (Cover T.M. & Hart P.E. 1967), Extreme Learning Machines (ELM) (Huang et al. 2012) were utilised in this study. The parameters of SVM are estimated by conducting grid search with cross validation.

7   RESULTS AND DISCUSSIONS

Even though the feature selection algorithm is designed to select the optimum number of features an 'inner' and 'outer' loop validation scheme is used to create a generalized result since the algorithm used for feature selection is of wrapper feature selection which has dependency on the training and testing samples. The outer loop performs the feature selection with randomly partitioning the feature set into training and testing samples, 70% samples from feature set is chosen as training samples and remaining as testing samples. This is repeated ten times and during each iteration of the outer loop, the inner loop performs classification with selected feature subset with 10-fold cross validation. The average results obtained are taken as the final result. This inner and outer loop validation also helps to reduce the effect of initial population on the final result.

7.1   *Comparison of different classifiers*

Fig. 6 represents the average error rate when the classifiers are trained using TDEMD based approach. The aggregated confusion matrix depicting average recognition rate of each grasp are shown in Appendix. The misclassification rate of each grasp (average of eleven subjects) using KNN, ELM, LIBSVM and LDA were found to be 4.79%, 2.89%, 2.08%, 5.07% respectively. However, there are different grasps that are difficult to differentiate between each other, for example confusions arise between S-C-H and P-L-T using KNN/LDA. But these confusions

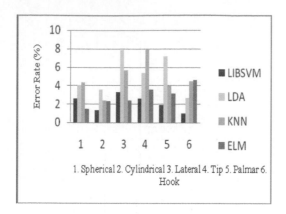

Figure 6.   Performance of TDEMD approach with different classifiers.

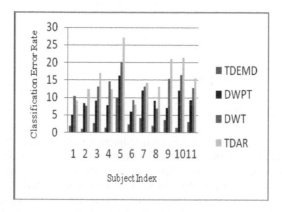

Figure 7.   Performance of different feature extraction methods.

are considerably reduced using **LIBSVM/ELM** classifier. Among the four classifiers **LIBSVM** is chosen to compare the performance of proposed **TDEMD** approach with other feature extraction techniques. Though the overall results are promising it is noted that there is a subject dependency in the recognition of each grasp which has to be further evaluated.

7.2   *Comparison with other feature extraction methods*

The classification error rate across all subjects using TDEMD, DWT, TA and DWPT are shown in Fig. 7. The aggregated classification errors were found to be 2.08%, 10.1%, 14.6%, and 19.8% for each of the TDEMD, DWPT, DWT and TA features respectively. From Fig 5, it is evident that in all subjects TDEMD approach outperformed all other feature extraction techniques. DWT, TA and DWPT provide mixed response but DWPT seems to perform better compared to DWT and TA methods. Relatively high error rate in contrast with other subjects exhibited by subject 5 is ascribed to issues during data acquisition.

8   CONCLUSIONS

This paper presents a preliminary report made on behalf of an ongoing research to develop a dexterous and natural control of powered upper limbs using EMG signals. The outcome of this stage of our research proved that the methodology of using EMD with DEFS can achieve significantly good results. The variation of grasp recognition rate among different subjects uncovers the requirement of fine tuning of the algorithm. Detailed analysis will be further carried out in future with a database created by involving more subjects including amputees.

REFERENCES

Al-Ani, Ahmed. Alsukke, Akram., Khushaba, R. 2013. Feature subset selection using differential evolution and a wheel based search strategy Swarm and Evolutionary Computation Volume 9, Pages 15–26.

Andrade, A.O., Kyberd, P.J., Nasuto, S. 2004. Time–frequency analysis of surface electromyographic signals via Hilbert spectrum, in: S.H. Roy, P. Bonato, J. Meyer (Eds.), XVth ISEK Congress—An Invitation to Innovation, Boston, MA, USA.

Boser, B.E., Guyon, I.M., and Vapnik, V.N. 1992. A training algorithm for optimal margin classifiers.5th Annual ACM Workshop on COLT, Pittsburgh.

Cover, T.M., and Hart, P.E. Hart, 1967. Nearest neighbor pattern classification. *IEEE Trans. Inform. Theory*, vol. IT-13, pp. 21–27.

De Luca, C.J. May 1998. The Use of Surface Electromyography in Biomechanics *Journal of Applied Biomechanics*, Volume 13 Issue 2.

Ericka Janet Rechy-Ramirez and Huosheng Hu, Stages for Developing Control Systems using EMG and EEG Signals: A survey, *TECHNICAL REPORT: CES-513, ISSN* 1744-8050.

Flandrin, P., Rilling, G. and Goncalv, P. 2004. Empirical mode decomposition as a filter bank. *IEEE Signal Process*.Lett.vol. 11, no. 2, pp. 112–114.

Fougner, A., Stavdahl, O., Kyberd, P.J., Losier, Y.G., and Parker, P.A. Sep.2012. Control of upper limb prostheses: Terminology and proportional myoelectric control—A review. *IEEE Trans. Neural Syst. Rehabil. Eng.*, vol. 20, no. 5, pp. 663–677.

Frank, F.H., 1989. *Atlas of Orthopedic Anatomy.* Ciba—Geigy, Switzerland.

Huang, G.-B., Zhou, H., Ding, X., & Zhang, R. 2012. Extreme learning machine for regression and multiclass classification.*IEEE Transactions on Systems, Man, and Cybernetics*, Part B: Cybernetics, 42(2), 513–529.

Huang, N.E., Shen, Z., Long, S.R., Wu, M.C., Shih et al.; March 1998. The Empirical Mode Decomposition and the Hilbert Spectrum for Nonlinear and Non stationary Time Series Analysis. *Royal Society Proceedings on Math, Physical, and Engineering Sciences,* Vol. 454, No. 1971 pp. 903–995.

Jiang, N., Dosen, Muller, K.R. and Farina, D. Sep. 2012. Myoelectric control of artificial limbs—Is there a need to change focus?[In the spotlight] *IEEE Signal Processing.* vol. 29, no. 5, pp. 150–152.

Khushaba, R., Al-Ani, A., Al-Jumaily, 2008. A. Differential evolution based feature subset selection. *Proceedings of the International Conference Pattern Recognition (ICPR'08).*

Kilbreath, S.L., Gorman, R.B., Raymond, J. and Gandevia, S.J. Distribution of the forces produced by motor unit activity in the human flexor digitoriumprofundus 2002. *Journal of Physiology*, vol. 543, no. 1, pp. 289–296.

Kunju, Nissan., Ojha, Rajdeep. R, Suresh. Devasahayam. 2013. A palmar pressure sensor for measurement of upper limb weight bearing by the hands during transfers by paraplegics. *Journal of Medical Engineering and Technology*, Vol. 37, No. 7, Pages 424–428.

Kunju, Nissan. Tharion, George. R, Suresh. Devasahayam, M. Manivannan. 2013. Muscle Activation Pattern and Weight Bearing of Limbs during Wheelchair Transfers in Normal Individuals—a step towards Lower Limb FES Assisted Transfer for Paraplegics. *Converging Clinical and Engineering Research on NeuroRehabilitation*, pp. 197–201. Biosystems and Biorobotics Series Springer (doi:10.1007/978-3-642-34546-3_31).

Micera, S.and Navarro, X. Jan. 2009. Bidirectional interfaces with the peripheral nervous system. *Int. Review of Neurobiology.* vol. 86, pp. 23–38.

Oskoei, M. A., & Hu, H. 2007. Myoelectric control systems—a survey. *Biomedical Signal Processing and Control*, 2(4), 275–294.

Oskoei, M. A., & Hu, H. 2008. Support vector machine-based classification scheme for myoelectric control applied to upper limb. *IEEE Transactions on Biomedical Engineering,* 55(8).

Sapsanis, Christos, Georgoulas, Georgoulas and Tzes. 2013. Anthony Tzes, EMG based classification of basic hand movements based on time-frequency features. *21 st Mediterranean Conference on Control and Automation (MED).*

Scheme, E. and Englehart, K.. 2011. Electromyogram pattern recognition for control of powered upper-limb prostheses: State of the art and challenges for clinical use. *Journal of Rehabilitation Research Development*.vol. 48, no. 6, p. 643.

Schlesinger, G. 1919. The mechanical construction of the artificial limb *Verlag von Julius Springer*, pp. 321–661.

Shrirao, N.A., Reddy, N.P. and Kosuri, D.R. 2009. Neural network committees for finger joint angle estimation from surface Emg signals. *Journal of Bio Medical OnLine*, vol. 15, no. 2, pp. 529–535.

Storn, R. 2008. Differential evolution research—trends and open questions, in: U.K. Chakraborty (Ed.), *Advances in Differential Evolution SCI*, vol. 143, Springer-Verlag, Berlin, Heidelberg, pp. 1–31.

APPENDIX

The aggregated confusion matrix (Average of eleven subjects) illustrating the performance of TDEMD method with different classifiers are attached below.

Table 1.   Aggregated confusion matrix (KNN classifier).

		PREDICTED					
		S	C	L	T	P	H
T	S	5728	68	32	0	22	56
R	C	48	5508	36	0	33	22
U	L	22	40	5570	106	76	44
E	T	28	24	84	5696	56	18
	P	2	12	64	96	5578	28
	H	88	48	32	14	44	5792

Table 2.   Aggregated confusion matrix (LIBSVM classifier).

		PREDICTED					
		S	C	L	T	P	H
T	S	5432	30	0	0	7	19
R	C	40	5419	0	0	24	9
U	L	2	0	5586	52	46	13
E	T	16	0	86	5715	31	37
	P	2	8	14	26	5644	4
	H	0	4	22	0	0	5526

Table 3.   Aggregated confusion matrix (ELM classifier).

		PREDICTED					
		S	C	L	T	P	H
T	S	7944	45	0	0	9	204
R	C	36	8292	0	24	24	102
U	L	9	0	7992	48	108	78
E	T	0	9	48	7572	96	15
	P	0	12	24	69	7188	36
	H	63	33	48	60	0	7464

Table 4.   Aggregated confusion matrix (LDA classifier).

		PREDICTED					
		S	C	L	T	P	H
T	S	5236	176	18	0	44	182
R	C	126	5576	0	4	12	84
U	L	62	14	5432	196	224	12
E	T	16	0	156	5224	224	6
	P	32	6	52	40	5684	2
	H	36	10	2	2	8	5620

Emerging Trends in Engineering, Science and Technology for Society,
Energy and Environment – Vanchipura & Jiji (Eds)
© *2018 Taylor & Francis Group, London, ISBN 978-0-8153-5760-5*

Automated aquaponics system

B. Sreelekshmi & K.N. Madhusoodanan
Department of Instrumentation, Cochin University of Science and Technology, Cochin, India

ABSTRACT: Agriculture has proven to be an arena in which technology has crucial roles to play. Presently, agriculture automation has a wider scope with emerging trends such as Controlled Environment Agriculture, Precision Farming etc. This paper describes the design and implementation of an automated aquaponics system in the framework of Internet of Things. Aquaponics is the integration of recirculating aquaculture with hydroponics. It is a sensitive system in which several parameters are to be maintained at certain optimum values inorder to ensure proper functioning of the system. The implemented automated aquaponics system enables web-based remote monitoring of the important parameters via the ThingSpeak IoT platform using Arduino Uno, ESP8266-01 and several sensors. Arduino-based control of certain parameters is also possible in case of deviation from setpoint. In addition, there is an SD-card data-logger used to store the data for future analysis.

1 INTRODUCTION

Integration of technology with agriculture has remarkably increased the ease and efficiency of agriculture. Presently, technology-related agriculture is adopting newer dimensions such as Controlled Environment Agriculture, Precision Farming (Mondal & Basu 2009) etc. These emerging agricultural trends make use of various technological aspects such as WSN (Ojha et al. 2015), IoT, Artificial Intelligence (Hashimoto et al. 2001, Lee 2000), control systems (De Baerdemaeker et al. 2001) and so on with an aim to improve factors such as sustainability and food security.

Aquaponics is the integration of recirculating aquaculture with hydroponics in a single system (Diver 2000). It is a form of Controlled Environment Agriculture and is thus a technology-related agricultural practice. Aquaculture is the rearing of fish in controlled conditions whereas hydroponics involves soilless growth of plants. Thus, the combination of both the techniques into a single system enables the organisms to benefit mutually wherein the plants absorb the required nutrients and the fishes are provided with purified water. The aquaculture effluent consists of ammonia which is toxic to fish. The water from the aquaculture tank is pumped to a grow-bed (which serves as the substrate for plant growth) and recirculated back into the aquaculture system with the help of a siphon. During the circulation, the water is subjected to a two-step nitrification process (Somerville et al. 2014) in which *Nitrosomonas* bacteria converts ammonia into nitrite which is then converted into nitrate, an absorbable nutrient for plants, by *Nitrobacter* bacteria (Klinger & Naylor 2012). The nitrate is absorbed by the plants (Buzby & Lin. 2014) and the filtered water is recirculated back into the aquaculture tank (Graber & Junge 2009, van Rijn 2013) with the help of a flood and drain mechanism operated by a siphon.

For the aquaponics system to be properly balanced, several parameters should be maintained at certain optimum values. The important parameters include temperature, humidity, light intensity, water level in the aquaculture tank, pH, nitrate and ammonia content, dissolved oxygen level etc. Regular manual monitoring and control of such parameters is a difficult task for the farmers (Goddek et al. 2015). The requirement of an automated aquaponics system (Saaid et al. 2013) lies in this aspect.

This paper describes the design and implementation of an automated aquaponics system in the framework of Internet of Things. The implemented system enables remote monitoring of the important parameters via the ThingSpeak IoT platform. Two wireless sensor nodes are deployed in the aquaponics system (one in the aquaculture tank and other in the growbed). Each sensor node consists of an Arduino Uno, several sensors and an ESP8266-01 Wi-Fi transceiver. The Wireless Sensor Network reports the sensor readings to the ThingSpeak IoT platform via the Wi-Fi for real-time monitoring purpose. An arduino-based control of water level in the aquaculture tank is also provided to maintain optimum water level. In addition, there is an SD-card data-logger for storing the sensor data into a microSD card for future analysis. The farmers can thus perform an analysis of the important parameters at different stages so as to seek better farming strategies.

2 OVERALL SYSTEM ARCHITECTURE

2.1 Small-scale IBC aquaponics system

A conventional small-scale IBC (Intermediate Bulk Container) type aquaponics system was designed and implemented inorder to study the various aspects of the system. The dimensions (shown in Figure 1) and design features of the implemented system were chosen as described by Somerville et al. (2014).

The design technique adopted was Media-Bed technique which is the most popular aquaponic design technique (Lennard & Leonard 2006). A medium like clay, metal etc. serves as substrate for plant growth. The fish species used was GIF (Genetically Improved Farming) Tilapia. The plants used *Solanum lycopersicum* (tomato), *Phaseolus vulgaris* (beans) and *Abelmoschus esculentus* (Ladies finger). A bell siphon was used for recirculating the water from the media-bed into the aquaculture tank through a flood and drain mechanism.

2.2 Architecture of automated system

The implemented automated system includes three subsystems deployed in the IBC Aquaponics farm: an IoT-based Aquaponics Monitoring System, Arduinobased level contol and an SD-card data-logger.

2.2.1 IoT-based aquaponics monitoring system

This part enables real-time monitoring of the important parameters via the ThingSpeak IoT platform. Two sensor nodes are deployed in the IBC aquaponics system (one in the grow bed and other in the aquaculture tank). Each sensor node consists of an Arduino Uno (microcontroller), several sensors and ESP8266-01 Wi-Fi transceiver. The Wireless Sensor Network is responsible for reporting the sensor readings to the ThingSpeak IoT platform via the Wi-Fi for realtime monitoring purpose. The parameters monitored include ambient light intensity, ambient temperature, relative humidity, grow-bed moisture, level and temperature of water in the aquaculture tank.

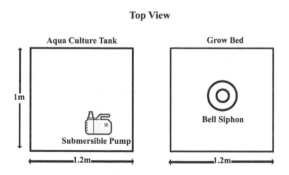

Figure 1. Dimensions of IBC aquaponics system (Top view).

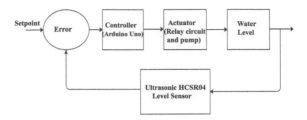

Figure 2.   Arduino-based level control.

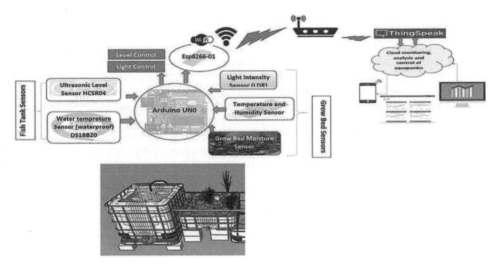

Figure 3.   Overall system architecture of automated aquaponics system: IoT-based monitoring system and Arduino-based control.

2.2.2   *Arduino-based level control*

The level of water in the aquaculture tank is a crucial parameter while considering an aquaponics system. An outdoor aquaponics system which is prone to variant environmental conditions demands an accurate level control mechanism inorder to maintain optimum water level. The proposed level control system (as shown in Figure 2) employs Arduino Uno as the microcontroller and a two channel 5 V relay module is used to give the actuating signal to a pump depending upon the deviation of water level from the set-point value (either increase or decrease).

2.2.3   *SD-card data-logger*

The sensor readings can be stored in a micro-SD card by deploying an SD card data-logger in the aquaponics system. An SD-card module (which holds the micro-SD card) is interfaced with Arduino Uno. The sensor data stored in micro-SD card can later be imported as excel files for future analysis.

The overall conceptual diagram of the implemented automated aquaponics system is shown in Figure 3.

3   SYSTEM DESIGN

The hardware and software design of the implemented system is briefly described in this section.

3.1 Hardware design

The hardware requirements for the real-time monitoring purpose include Arduino Uno, sensors and ESP8266-01 Wi-Fi transceiver module. The Arduinobased level controller requires a two channel 5 V relay module as the actuator. In addition, an SD-card module is required for the SD-card data-logger.

3.1.1 Arduino Uno
Arduino Uno is a microcontroller board based on ATmega328P. It consists of 16 MHz quartz crystal, USB connection, powerjack, an ICSP header and reset button. It has 14 digital input/output pins (out of which six can be used as PWM outputs) and six analog input pins. Operating voltage is 5 V.

3.1.2 Sensors
Six parameters of the aquaponics system are monitored, namely ambient light intensity, ambient temperature, relative humidity, moisture content in the grow-bed, level and temperature of water in the aquaculture tank. Five different sensors are used for this purpose. LDR (Light Dependent Resistor) is used to measure the ambient light intensity. DHT11 is the sensor used for measuring the ambient temperature (in degree celsius) and relative humidity (in percentage). The grow-bed moisture sensor consists of two probes which act as a variable resistor depending upon the moisture content of the grow-bed media. Ultrasonic level sensor HCSR04 is used for measuring the level of water in the aquaculture tank. The temperature of water in the aquaculture tank is measured using DS18B20 waterproof temperature sensor.

3.1.3 ESP8266-01 Wi-Fi module
ESP8266-01 is a Wi-Fi transceiver module widely used for IoT applications. This module enables Wi-Fi access to microcontrollers. The operating voltage is 3.3 V. In the proposed system, ESP8266-01 is used to send sensor data into ThingSpeak IoT platform.

3.1.4 Two channel 5 V relay module
The 5 V relay module can be controlled directly by the Arduino microcontroller. For a relay to be switched ON, the digital output given from arduino should be LOW. A HIGH digital output from arduino will switch the relay module to OFF position.

3.1.5 SD card module
The SD card module is interfaced to Arduino Uno to store the sensor data. A micro-SD card is mounted on the SD card module for storing data. The operating voltage is 5 V.

3.2 Software requirements

The software requirements of the system include Arduino IDE and ThingSpeak.

3.2.1 Arduino IDE
Arduino Integrated Development Environment (IDE) is an open-source software used to upload programs into the arduino hardware and communicate with them. The programs written in Arduino IDE are called sketches. These sketches are written in the text editor and are saved with the file extension '.ino'. The output is displayed in the serial monitor.

3.2.2 ThingSpeak
ThingSpeak is an open-source IoT platform used for real-time monitoring purposes. ThingSpeak channels can be created by logging into ThingSpeak using MathWorks account. The data from the sensors get stored in ThingSpeak channels in various fields (Pasha 2016, Rao & Ome 2016). The data is displayed in the form of charts. Certain details are required to be entered in the Arduino IDE sketch inorder to send sensor data from Arduino to ThingSpeak using ESP8266-01 Wi-Fi module. These include the write API key of ThingSpeak channel, ThingSpeak IP, the SSID and password of the Wi-Fi network to be accessed and HTTP GET request.

4 RESULTS AND DISCUSSION

Real-time monitoring of the parameters of the automated system is possible from any remote location by logging into the ThingSpeak channel created using Mathworks account. The link to the ThingSpeak login page is https://thingspeak.com/login. ThingSpeak channel consists of fields corresponding to each parameter. The measured readings are displayed in the form of charts.

Figure 4 shows the real-time graphical display of measured parameters as obtained from the ThingSpeak channels. The y-axis of all the charts are labelled with the parameter concerned and the x-axis is labelled as date with the time at which reading is taken.

The status of actuation of the arduino-based level controller can also be monitored using ThingSpeak channel. Two fields are created corresponding to each relay of the two channel relay module. When a relay turns ON, the chart corresponding to the particular relay displays 1. When the relay turns OFF, the status displayed in the chart changes to zero.

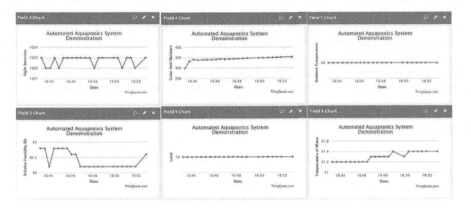

Figure 4. Fields of ThingSpeak channel showing the real-time values of measured parameters against corresponding time.

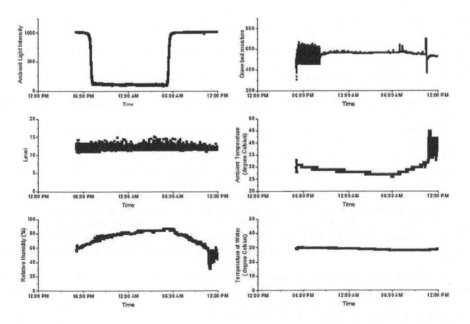

Figure 5. Plot of measured parameters aginst time from data stored in microSD card using SD-card datalogger.

699

The data stored in the micro-SD card using the SD card data-logger is imported and saved as excel files for future analysis purpose. A plot of the measured parameters against time for a period of twenty-four hours is shown in the Figure 5.

5 FUTURE DIRECTIONS AND CONCLUSIONS

The proposed automated system can be modified by including other sensors such as dissolved oxygen, pH and ammonia sensors. The same methodology can be adopted to other Controlled Environment Agriculture practices such as Greenhouse, Hydroponics and Aquaculture. Automation can also be carried out using Wireless Sensor Networks employing Raspberry Pi and zigbee protocol (Ferdoushi & Li 2014) wherein image processing techniques can be applied for analysis of effects of various environmental factors on plant growth (Liao et al. 2017). Image processing techniques can also be applied for weed detection. Intelligent controllers utilizing Artificial Neural Networks and Expert Systems can also improve the performance of Controlled Environment Agriculture systems to a greater extent (Hashimoto et al. 2001, Lee 2000).

The scope of technology related agricultural practices is increasing due to their sustainable nature and efficient utilization of resources such as land, water etc. Application of technological aspects enable the farmers to adopt better farming strategies and thereby increase the food productivity. These methods are thereby capable of ensuring sustainability as well as food security.

REFERENCES

Buzby, K. & L.-S. Lin. (2014). Scaling aquaponic systems: Balancing plant uptake with fish output. *Aquacultural Engineering 63*, 39–44.
De Baerdemaeker, J., A. Munack, H. Ramon, & H. Speckmann (2001). Mechatronic systems, communication, and control in precision agriculture. *IEEE Control Systems Magazine*, 48–70.
Diver, S. (2000). Aquaponics integration of hydroponics with aquaculture. Technical report, ATTRA, NCAT.
Ferdoushi, X. & X. Li (2014). Wireless sensor network system design using raspberry pi and arduino for environmental monitoring applications. *Procedia Computer Science 34*, 103–110.
Goddek, S., B. Delaide, U. Mankasingh, K. Ragnarsdottir, H. Jijakli, & R. Thorarinsdottir (2015). Challenges of sustainable and commercial aquaponics. *Sustainability 7*, 4199–4224.
Graber, A. & R. Junge (2009). Aquaponic systems: Nutrient recycling from fish wastewater by vegetable production. *Desalination 246*, 147–156.
Hashimoto, Y., H. Murase, T. Morimoto, & T. Torii (2001). Intelligent systems for agriculture in japan. *IEEE Control Systems Magazine*, 71–85.
Klinger, D. & R. Naylor (2012). Searching for solutions in aquaculture: Charting a sustainable course. *Annual Review of Environment and Resources 37*, 247–276.
Lee, P.G. (2000). Process control and artificial intelligence software for aquaculture. *Aquacultural Engineering 23*, 13–36.
Lennard, W. & B. Leonard (2006). A comparison of three different hydroponic sub-systems (gravel bed, floating and nutrient film technique) in an aquaponic test system. *Aquacult. Int. 14*, 539–550.
Liao, M., S. Chen, C. Chou, H. Chen, S. Yeh, Y. Chang, & J. Jiang (2017). On precisely relating the growth of phalaenopsis leaves to greenhouse environmental factors by using an iot-based monitoring system. *Computers and Electronics in Agriculture 136*, 125–139.
Mondal, P. & M. Basu (2009). Adoption of precision agriculture technologies in India and in some developing countries: Scope, present status and strategies. *Progress in Natural Science 19*, 659–666.
Ojha, T., S. Misra, & N.S. Raghuwanshi (2015). Wireless sensor networks for agriculture: The state-of-the-art in practice and future challenges. *Computers and Electronics in Agriculture 118*, 66–84.
Pasha, S. (2016). Thingspeak based sensing and monitoring system for iot with matlab analysis. *IJNTR 2*, 19–23.
Rao, S. & N. Ome (2016). Internet of things based weather monitoring system. *IJARCCE 5*, 312–319.
Saaid, M., N. Fadhil, M. Ali, & M. Noor (2013). Automated indoor aquaponic cultivation technique. *Proc. of 2013 IEEE 3rd International Conference on System Engineering and Technology.*, 285–289.
Somerville, C., M. Cohen, E. Pantanella, A. Stankus, & A. Lovatelli (2014). Small-scale aquaponics food production. Technical report, Food and Agricultural Organization of the United Nations, Rome.
van Rijn, J. (2013). Waste treatment in recirculating aquaculture systems. *Aquacultural Engineering 53*, 49–56.

Emerging Trends in Engineering, Science and Technology for Society,
Energy and Environment – Vanchipura & Jiji (Eds)
© 2018 Taylor & Francis Group, London, ISBN 978-0-8153-5760-5

A review of the methods for despeckling in optical coherence tomography

K. Athira, K. Brijmohan, Varun P. Gopi, K.K. Riyas, Garnet Wilson & T. Swetha
Department of Electronics and Communication Engineering, Government Engineering College, Wayanad, India

ABSTRACT: In medical image processing, the Optical Coherence Tomography (OCT) imaging technique is widely used for disease detection. It is a non-invasive imaging technique that provides images of tissue structures with high resolution. Speckle noise is a granular noise. Speckle noise is introduced in OCT images due to the constructive and destructive interference of optical waves that undergo multiple scattering in different directions, when waves propagate through the tissue. The quality of OCT images degrades due to speckle noise. This paper is a review of different methods for speckle noise reduction in OCT images.

Keywords: Speckle Noise; Optical coherence Tomography; Denoising Filters

1 INTRODUCTION

Optical coherence tomography (OCT) is a non-invasive imaging technique that provides high resolution images of tissue structures and cross-sectional imaging of many biological systems (Drexler & Fujimoto, 2008). The OCT imaging technique is widely used by ophthalmologists in the diagnosis of eye disease such as Glaucoma, Macular Edema and Diabetic Retinopathy. Now-a-days this imaging technique is also used in the detection of skin disorders.

The working of the OCT imaging technique is based on the Michelson Interferometer, using Low Coherence Interferometry (Schmitt et al., 1999). Typically, near-infrared laser light is used as a light source to penetrate into the scattering medium, before capturing the backscattered optical waves. Due to heat produced by the image sensors, or due to the physical properties of light photons the image is corrupted during the image acquisition process. The light reflected from the micro-structural tissue contains the features of the image. The combination of various crests and troughs of backscattered light waves from the tissue produces granular structures in an image. This grainy representation is known as speckle noise. It can change the important details in an image used to diagnose disease. Therefore leads to image quality degradation. This degradation makes it difficult for humans to differentiate pathological tissues from the normal tissues. Speckle noise is a multiplicative noise which contains information about the image. Therefore it is difficult to remove speckle noise without any change in important features in an image. The primary aim of the OCT research is to denoise the speckle noise and preserve the edges clearly. Several filtering methods are proposed for reducing speckle noise. The limitation is that filtering techniques remove some parts of the information in an image along with speckle noise. This paper presents a comparative study on the performance of different filters.

2 NOISE MODEL

Speckle noise can be modeled as multiplicative noise. It is known to have a Gamma distribution. It is a granular type noise which appears in the lighter regions of the image as bright specks. Speckle noise can be modeled as:

$$Y(x, y) = S(x, y).N(x, y)$$

where Y, S and N represent the noisy image, signal and speckle noise, respectively. A logarithmic transformation is applied to the image data to change the multiplicative nature. Then the model can be rewritten as:

$$f(x,y) = s(x,y) + e(x,y)$$

where f, s and e represent the logarithm of the noisy image, signal and noise respectively.

3 RELATED WORKS

3.1 *Nonlocal means denoising filter with double Gaussian anisotropic kernels*

The Non-local means (NLM) filter is one of the important denoising filters (Aum et al., 2015). It is a denoising algorithm which utilizes the presence of similar features in an image and then takes the average of those features to remove speckle noise in an OCT image. This method provides a low signal-to-noise ratio due to the low performance of noise reduction around the edges of an image. To overcome this limitation, the conventional NLM filter converts into an NLM filter with double Gaussian anisotropic kernels. The conventional NLM filter contains a Gaussian kernel to measure the similarities in an image. It may be able to measure the distinct similar features from the image. Since the same Gaussian kernel is used on every pixel, the speckle noise corrupted edges cannot be denoised correctly. Therefore, the new algorithm proposes new kernels and their shapes are adaptively varied. Various kernels were used for calculating the similarity between the local neighborhoods from the pixel positions. The modified NLM method produced a PSNR of 31.01db. Figures 1(a) and 1(b) show the denoised images, using the conventional NLM filter and the modified NLM filter, respectively.

3.2 *Noise adaptive wavelet thresholding*

The working of optical the OCT imaging technique is based on coherence detection of interferometric signals. Speckle noise is introduced into the OCT image, due to the constructive and destructive interference of optical waves that undergo multiple scattering in different directions. Wavelet domain thresholding provides much noise suppression by preserving image sharpness whilst removing speckle noise in the OCT image. Speckle noise has different characteristics, but it has not been considered in conventional wavelet thresholding algorithms. A Noise Adaptive Wavelet Thresholding (NAWT) algorithm is introduced in this paper (Adler et al., 2004). Graphical representation of the optimized adaptive wavelet thresholding algorithm is shown in the Figure below.

The algorithm uses two types of images:- the reference image and the original OCT image. The speckle noises can be characterized by using the reference image received from a uniform scattering sample, since the variation in magnitude for such an image attributes to random noise. Wavelet transform is applied to both the original image and the reference image. Then the signal variance and noise variance are calculated in all subbands:-

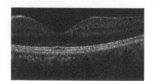
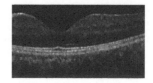

(a)   NLM Image   (b) Modified NLM filter

Figure 1.   OCT images obtained from a human index fingertip: (a) image processed with the conventional NLM: - (b) image processed with the modified NLM.

702

$$T = \sigma_w^2 / \sigma_x$$

where T is the thresholding value. Thresholding is applied on an OCT image and performing inverse wavelet transform.

3.3 *Wavelet denoising of multiframe OCT image*

In this method, the wavelet technique is used for denoising of multiframe OCT data (Mayer et al., 2012). This algorithm makes use of wavelet decompositions in single frames for noise estimation, instead of taking the average of multiple image frames. The flow diagram of a multiframe denoising algorithm is shown below.

Speckle noise is a multiplicative noise. The preprocessing stage is logarithmic transformation, which is applied to the input image frames. The multiplicative noise is therefore converted into additive noise. The single frames are decomposed by wavelet transformations. Two different wavelet transformations are used and compared with each other such as with discrete stationary wavelet transformation and dual tree complex wavelet transformation. A significance weight and a correlation weight are proposed in this algorithm. The significance weight provides local noise estimation and correlation weight gives details of a structure. Finally, the despeckled image is attained by taking the inverse wavelet transform.

3.4 *Combination of wiener filter and wavelet transform*

This method proposes an effective algorithm based on the combination of spatial and frequency domain techniques for despeckling in OCT image (Rajesh et al., 2016). The wiener filtering method is applied to the speckled image as a preprocessing stage. This filter works efficiently for the suppression of additive and multiplicative noise in an image. The speckle

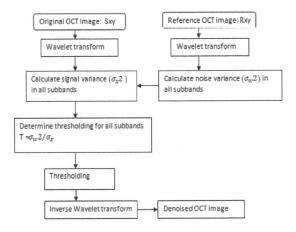

Figure. 2. Flow chart of the adaptive thresholding algorithm.

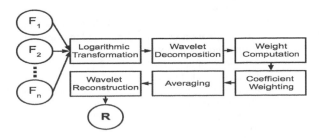

Figure 3. Graphical representation of the wavelet multiframe denoising algorithm.

703

noise model is multiplicative in nature. The resultant image of the wiener filtering is transformed into an additive one. Then discrete wavelet transform is applied to this output image. Thresholds are computed for each sub band except for the lowest level. Then thresholding is applied to all sub bands and inverse discrete wavelet transform is applied. The exponent is taken to get the despeckled image.

3.5 *Adaptive bilateral filtering technique*

The bilateral filter is proposed by Tomasi and Manduchi [7]. This non-linear filtering technique smoothes the edges as well as preserving the edge structures. The kernel of the conventional bilateral (CB) filter contains two components, such as a range filter kernel and a domain filter kernel. CB filter response at a given pixel location is given as:

$$\hat{I}(m) = \frac{1}{Z} \sum_{n \in N(m)} w_D(m,n) w_R(m,n) I(n),$$

The weight function WD is related to the domain filter which provides larger weight to pixels that are spatially close to the center pixel. Similarly, the other weight function WR

Table 1. Comparison of despeckling algorithm.

Algorithm	Properties	Remarks
Non-local mean	Estimates pixel intensity based on the presence of similar patterns and features of an image.	➤ It is not able to suppress speckle noise for non—repetitive neighborhoods. ➤ It does not preserve edge details.
Modified Non-Local mean	Using non-local means denoising filter with double Gaussian anisotropic kernels. The shape of Gaussian kernel varies adaptively.	➤ It is able to suppress speckle noise at the edges of an image.
Noise adaptive wavelet Thresholding	Based on variance of reference image and OCT image the thresholding value is estimated in all subbands.	➤ It eliminates a significant amount of noise. ➤ It does not preserve edge details completely. ➤ Signal-to-Noise ratio is less.
Wavelet denoising of multiframe OCT	This algorithm makes use of wavelet decompositions in single frames for noise estimation instead of taking the average of multiple image frames.	➤ It removes the noise but accuracy depends on number of input frames.
Combination of wiener filter and wavelet transform	Algorithm based on the combination of spatial and frequency domain techniques for reducing speckle noise	➤ It is able to suppress both additive and multiplicative noise in an image.
Adaptive Bilateral Filtering Technique	Combination of two filters based on spatial distance and intensity difference	➤ Eliminates significant amount of noise. ➤ Performs better in preserving sharp edges and fine details.

is linked to range filtering and the range lowpass Gaussian filter provides larger weight to pixels. These pixels are the same as the center pixel in gray value. These weight functions can be defined as:

$$w_D(m,n) = \exp\left(-\frac{|m-n|^2}{2\sigma_d^2}\right)$$

$$w_R(m,n) = \exp\left(-\frac{|I_m-I_n|^2}{2\sigma_r^2}\right)$$

where Im and In are the intensities at m and n, respectively. The σd is the geometrical spread in the domain which calculates the blurring effect in an image. Then the value of σd is independent to noise. Similarly, the σr is the photometric spread in the image range and its optimal value is linearly proportional to the noise standard deviation σ. The photometric spread in the range filter is a constant value. This is one of the main disadvantages of the bilateral filter, as fixing the optimal value is difficult. In order to enhance the sharpness of an image, the bilateral filter needs some modifications. It contains two modifications:- an offset is introduced to the range filter, and the width of the range filter varies in an adaptive manner.

4 CONCLUSION

The speckle noise reduction algorithms are described in this paper. Among filters used for despeckling in optical coherence tomography images, the bilateral filtering technique is efficient. This technique eliminates a significant amount of noise and preserves the edges of the denoised image. Adaptive filter methods are more efficient than filtering as they preserve the fine details of edges.

REFERENCES

[1] W. Drexler, J.G. Fujimoto, "Optical Coherence Tomography: Technology and Applications", Springer International Publishing, Switzerland, 2015.
[2] J.M. Schmitt, S.H. Xiang, and K.M. Yung, "Speckle in Optical coherence Tomography: An Overview" J. Biomed. Opt. 4 (1) (1999) 95–105.
[3] Jaehong aum, Ji-hyun kim, and Jichai jeong, "Effective speckle noise suppression in optical coherence tomography images using nonlocal means denoising filter with double Gaussian anisotropic kernels" published 3 April.
[4] Desmond C. Adler, Tony H. Ko, and James G. Fujimoto "Speckle reduction in optical coherence tomography images by use of a spatially adaptive wavelet filter", December 15, 2004/Vol. 29, No. 24/OPTICS LETTERS.
[5] Markus A. Mayer, Anja Borsdorf, Martin Wagner, Joachim Hornegger, Christian Y. Mardin, and Ralf P. Tornow, "Wavelet denoising of multiframe optical coherence tomography data", 1 March 2012/Vol. 3, No. 3/BIOMEDICAL OPTICS EXPRESS.
[6] Rajesh Mohan R., S. Mridula, P. Mohanan, "Speckle Noise Reduction in Images using Wiener Filtering and Adaptive Wavelet Thresholding", 978-1-5090-2597-8/16/$31.00_c 2016 IEEE.
[7] Ch. Ravi Kumar, *Member, IACSIT* and S.K. Srivatsa, "Enhancement of Image Sharpness with Bilateral and Adaptive Filter", International Journal of Information and Education Technology, Vol. 6, No. 1, January 2016.
[8] A. Ozcan, A. Bilenca, A.E. Desjardins, B.E. Bouma, G.J. Tearney, "Speckle reduction in optical coherence tomography images using digital filtering," J. Opt. Soc. Am. A 24 (7) (2007) 1901–1910.
[9] V. Frost, J. Stiles, K. Shanmugan, J. Holtzman, "A model for radar images and its application to adaptive digital filtering of multiplicative noise," IEEE Trans. Pattern Anal. Mach. Intell. (PAMI-4) (2) (1982) 157–166.
[10] T. Loupas, W. McDicken, P. Allan, "An adaptive weighted median filter For speckle suppression in medical ultrasonic images," IEEE Trans. Circuits Syst. 36 (1) (1989) 129–135.

Emerging Trends in Engineering, Science and Technology for Society,
Energy and Environment – Vanchipura & Jiji (Eds)
© 2018 Taylor & Francis Group, London, ISBN 978-0-8153-5760-5

Generic object detection in image using SIFT, GIST and SURF descriptors

Dona P. Joy, K.S. Shanthini & K.V. Priyaja
Department of Applied Electronics and Instrumentation Engineering, Government Engineering College,
Calicut, Kerala

ABSTRACT: This paper compares generic object detection method using three different feature extraction schemes. The query image could be of different types such as a real image or a hand-drawn sketch. The method operates using a single example of the target object. The feature descriptors emphasizes the edge parts and their distribution structures, so it is very robust and can deal with virtual images or hand-drawn sketches. The approach is extended to account for large variations in rotation. Good performance is demonstrated on several data sets, indicating that the object was successfully detected under different imaging conditions.

Keywords: DSIFT, keypoint, SURF

1 INTRODUCTION

In image processing, it is very important to analyze the visual objects in the image. Object detection means, to locate object of a certain class in a test image. This method uses only one query image as the template to detect the object without any training procedures. Such systems are applicable in different areas such as surveillance, video forensics and medical image analysis and so on.

The training free object detection with one query image has many applications such as automatic passport control at airports, where a single photo in the passport is the only example available. Another application is the image retrieval from the Web. In this case, only a single sample of the target is provided by the user and every database is compared with this single sample. Another application is for the classification of an unknown set of images into one of the training classes.

2 LITERATURE REVIEW

There are different methods for object detection. Most of them are based on training process. But the training-based methods are subject to sample restrictions. In cases such as frontal face detection, samples can accurately get aligned. But, in many cases, the collecting and aligning of samples is not possible which badly affects the performance of training-based detection methods. And the training method is not suitable for immediate task, because the collection of samples and training the model should be completed in advance. So when the target class changes, they must be redone.

2.1 *Keypoint-based matching*

Most methods use several key points in the image which are relatively stable and calculate the local invariant descriptor of the patch near the key point. Under these circumstances, the object image is converted into a set of local descriptors on the key points. Mikolajczyk and Schmid [2] had compared these methods, with SIFT algorithm shown to outperform other keypoint-based descriptors. And keypoints provide greater invariance and more compact coding, but they are not designed to select the most informative area for detection [3]. Another drawback is the object detection may have multiple objects or no objects in the given test image. So it complicates locating each object in the test image from the detected keypoints in the case of keypoint-based methods.

2.2 Densely computed descriptors

The local self-similarity descriptor [4] can be used as a descriptor for intensive calculations. It computes a simple sum of the squared differences between a center image patch and peripheral image patches. Some methods used a local regression kernel [4] as the descriptor. Rather than a point based representation, these densely computed descriptors give a global representation of the object. But these methods are not suitable to handle with hand-drawn images, which may differ from the real image in appearance.

2.3 Contour segment network

Contour-based methods such as the Contour Segment Network [6] can be used to detect object using a single hand drawn sketch. In these methods the original image is converted into a group of contour segments similar to the hand-drawn sketch. The object detection problem is developed as finding paths through the network resembling outlines of the model. The method allows the scale of the image to change and is computationally efficient. The disadvantage is that the contour detection is usually not very stable with complex background. Because, for complex background the generation of contour segments becomes very difficult.

2.4 Other descriptors

Zhang [10] proposed an efficient image matching technique based on SURF descriptor. They compared SURF to SIFT and found that SURF was more robust compared with SIFT and has a fast matching speed.

Most commonly developing object detection methods are based on local and global features. Global features are characteristics of regions in images such as area, perimeter, Fourier descriptor and moments. The local features are usually based on the boundary of an image or represent distinguishable small area. GIST features are a type of global feature. The GIST was first introduced by Murphy et al. [11]. They took global feature of an image which has been called as gist of a scene to overcome the ambiguity caused by only taking the local fragments of an image and interest point detector around it for object detection. In Ivan et al. [12], they described the effectiveness of GIST features in image classification. GIST describes the shape of the scene using low dimensional feature vectors, and it performs well in classification problems.

3   OVERVIEW

Object detection means, to locate any object of a particular class in a test image. This method uses only one query image as the template to detect the object without any training procedures. As shown in Figure 1, the query image should be a typical sample of the target class, containing only one object and as little background as possible. It can be a real image, a virtual image from a simulation model or even a hand-drawn sketch which only exhibits a rough profile of the object. The detection task is very similar to the template matching process. The query image is used as a standard template and the test images are matched to this template to recognize the objects.

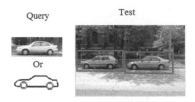

Figure 1. Object detection example using a single query image.

The test image T, is divided into overlapping patches, T_i, which have the same size as Q. Then the features of query image and the test image patches are extracted. These query image is compared with each of the test image patches and the most silimar patch is decided.

4 THEORY

The main steps of the object detection method are feature extraction, dimensionality reduction, similarity measurement and decision making, as shown in Figure 2. Feature extraction is the important step in the object detection process. The proposed method use three different feature extraction methods-Dense Scale Invariant Feature Transform (DSIFT), GIST and Speed UP Robust Features (SURF). Principal Component Analysis (PCA) is used to reduce the dimensionality of the features. Euclidean distance and Matrix Cosine Similarity (MCS) are used for similarity measurement. Decision process is based on minimum Euclidean distance or maximum MCS value.

4.1 *Feature extraction*

Feature extraction is an important step in object detection. Features should be easy to extract from images and unique to particular object. For the test image data, the features are extracted and compared with the query image features. In the proposed method, SIFT, GIST and SURF features are used for this purpose. The test image is divided into overlapping patches, T_i. Then features of the query image and test image patch are compared to find the similarity.

1. *DSIFT*: For the given image SIFT finds all the keypoints in the image with respect to the gradient feature of each pixel. Every keypoint contains the information of its location, local scale and orientation. Then, based on each keypoint, SIFT computes a local image descriptor which shows the gradient feature in the local region around the keypoint. Combining all the local descriptors, we get the complete features from the image. Densely computed SIFT (DSIFT) descriptors are very accurate and fast than SIFT. And for dense SIFT, the location of each keypoint is not from the gradient feature of the pixel, but from a pre-designed location. The DSIFT can grasp the most general characteristics of the object. The SIFT descriptors are calculated on every small patch of the image as shown in Figure 3, centered at dense sampling points [1]. Then, these densely computed SIFT features are arranged together to form the DSIFT descriptor.
2. *GIST*: GIST features are computational model of the recognition of scene categories that bypasses the segmentation and the processing of objects. The GIST descriptor focuses on the shape of scene itself, on the relationship between the outlines of the surfaces and their properties, and ignores the local objects in the scene and their relationships. The representation of the structure of the scene, termed spatial envelope is defined, as well as its five perceptual properties: naturalness, openness, roughness, expansion and ruggedness, which are meaningful to human observers. The input image is first pre-processed by converting it to grayscale, normalizing the intensities and locally scaling the contrast. The resulting

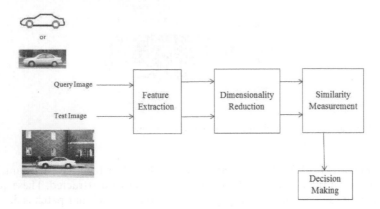

Figure 2.   Block diagram of the prposed system.

image is then split into a grid on several scales, and the response of each cell is computed using a series of Gabor filters. All of the cell responses are concatenated to form the feature vector.

3. *SURF*: SURF descriptors can be used to locate and recognize objects, people or faces, to reconstruct 3D scenes, to track objects and to extract points of interest. SURF is a fast feature extraction method based on the integral image and Hessian matrix and it is partially inspired from SIFT. To detect interest points, SURF uses an integer approximation of the determinant of Hessian blob detector. Its feature descriptor is based on the sum of the Haar wavelet response around the interested points.

SURF uses multi-resolution pyramid technology to convert images into coordinates to copy the original image with pyramid-shaped Gaussian or Laplacian pyramid shapes to obtain an image with the same size but reduced bandwidth. Thus achieves a special blurring effect on the original image, called Scale-Space and ensures that the interested points are invariant to scale.

4.2   *Dimensionality reduction*

Principal Component Analysis (PCA) is one of famous techniques for dimension reduction, feature extraction, and data visualization. In general, PCA is defined by a transformation of a high dimensional vector space into a low dimensional space. PCA provides an efficient way to reduce the dimensionality, so it is much easier to visualize the shape of data distribution.

4.3   *Similarity measurement*

The next step in the proposed system is a decision rule based on the measurement of distance between the computed features. The method used Matrix Cosine Similarity (MCS) and Euclidean distance to find the similarity between the features.

4.4   *Decision process*

1. *Matrix Cosine Similarity*: Cosine similarity is a measure of an inner product space that measures the cosine of the angle between them. The popularity of cosine similarity due to the reason that it is very efficient to evaluate, especially for sparse vectors, as only the non-zero dimensions need to be considered. Cosine similarity(ρ) between two feature vectors F_Q and F_T is given by:

$$\rho = \rho\left(F_Q, F_{T_i}\right) = \left\langle \frac{F_Q}{\|F_Q\|}, \frac{F_{T_i}}{\|F_{T_i}\|} \right\rangle \tag{1}$$

where F_Q is the feature descriptor for the query image and F_T is the descriptor for the test image patch.

The detection of the object is given by finding the most similar patch to the query image which has the similarity equal to 1.

2. *Euclidean distance*: The Euclidean distance is the straight line distance between two points in Euclidean space.

$$d_i = \sqrt{\| F_Q - F_{T_i} \|^2} \tag{2}$$

The patch with minimum Euclidean distance will be detected as the result.

5   RESULTS AND DISCUSSION

In order to demonstrate the identification performance, the system was tested for car detection, face detection, generic object detection and rotated object detection. For rotated object detection, generic objects were implanted in the background images.

5.1   *Car detection*

For the car detection tests, the UIUC car data set [13] was used. The data set includes 309 images of 100×40 pixels. Some detection examples are shown in Figure 3 with a real image or a hand-drawn sketch as the query image. The results show that the method is able to detect cars with a single query image and is robust to object variations, cluttered background, and partial occlusions.

5.2   *Face detection*

The system was also tested for face detection. The data is from the CMU data set [14]. The results are shown in Figure 4. There were images with one face or more than one faces. Hand-drawn sketch gives better results for the detection process than the real image.

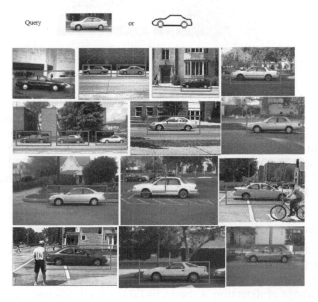

Figure 3.   Car detection results.

Figure 4. Face detection results.

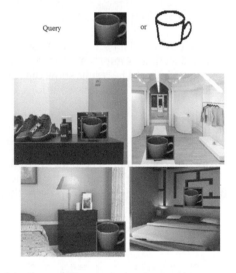

Figure 5. Generic object detection results.

5.3 *Generic object detection*

The proposed method can also be used for more generic object detection tasks. Coil-20P dataset [15] contains 20 object classes. Each class contains 72 gray scale images of a specific object in different positioning. All the images have pixel value 128×128, in PNG format. These object images are implanted in the background images. Coil-20P is obtained from Columbia Object Image Library. Figure 5 shows some examples of detecting the object(cup) with different background scenes.

5.4 *Rotated object detection*

For rotated object detection the Coil-20P objects with rotation are implanted on different backgrounds. Each of the objects has 72 images with 5° of rotation. The detected results are shown in Figure 6. When the object is rotated the detection is possible only upto some degree of rotation. This range of detection is shown in Table 1 for 4 different objects using MCS as the similarity measurement.

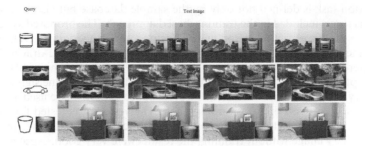

Figure 6. Rotated object detection results.

Table 1. Result for rotated object.

Object	Detected range (in degree)		
	DSIFT	GIST	SURF
Glass	0–360	0–360	0–360
Car	0–40	0–40	0–60
	150–215	150–215	100–210
	325–360	325–360	320–360
Cup	0–55	0–65	0–90
	165–185	150–215	110–215
	335–360	325–360	320–360
Vaseline	0–30	0–30	0–40
	145–205	140–200	145–300
	340–360	315–360	300–360

Table 2. Classification accuracy.

Feature extraction method	Car	Face	Generic object
DSIFT	88.7	86.6	88.1
GIST	92.3	90.3	94.2
SURF	90.1	87.3	95.7

The overall performance analysis of the system is shown in Table 2. In this analysis hand-drawn sketch is used as the query image and Cosine Similarity is used for distance measurement.

The method was tested using different objects. The query image can be a real image or a hand-drawn image of the object with a little background as possible. The results show that the approach is quite stable and is not affected by the choice of the query images. Even the performance with the hand-drawn sketch is quite good. Out of the three feature extraction methods GIST gives better detection rate. In the case of rotated object SURF gives good performance than others.

6 CONCLUSION

Object detection refers to find the position of a particular object in a given image. There are many object detection methods, mostly based on the training process. The target object for object detection task is defined not only by the sample database but also by human experience, which is rarely included in the training-based approach. The proposed system describes a training-free generic object detection method using a single query image. The query image can be become a typical real image, a virtual image, or even the hand-drawn sketch of the object of interest. The detection process is similar to template matching. SIFT, GIST and SURF descriptors are used for feature extraction. In order to determine significant features, dimensionality reduction is used. The Euclidean distance and MCS are used as the similarity measurement with a two-step decision approach. The system was tested for car detection, face detection, generic object detection and rotated object detection. Even if the system uses only a single query image without training, the results were very good, especially with hand-drawn sketches as the query images. But for rotated objects, the detection is possible only upto a certain degree of rotation.

REFERENCES

[1] Bin Xiong and Xiaoqing Ding, *"A Generic Object Detection Using a Single Query Image Without Training"*, Tsinghua Science and Technology, April 2012, 17(2): 194–201.

[2] Mikolajczyk K. and Schmid C., *"A performance evaluation of local descriptors"*, IEEE Trans. Pattern Analysis and Machine Intelligence, 2005, 27(10): 1615–1630.

[3] Jurie F. and Triggs B., *"Creating efficient codebooks for visual recognition"*, In: Proceedings of IEEE International Conference on Computer Vision. Beijing, China, 2005.

[4] Seo H, Milanfar P., *"Training-free, generic object detection using locally adaptive regression kernels"*, IEEE Trans. Pattern Analysis and Machine Intelligence, 2010, 32(9).

[5] Shechtman E. and Irani M., *"Matching local self-similarities across images and videos"*, In: Proceedings of IEEE Conference on Computer Vision and Pattern Recognition, 2007.

[6] Ferrari V., Tuytelaars T. and Van Gool L., *"Object detection by contour segment networks"*, In: Proceedings of European Conference on Computer Vision. Graz, Austria, 2006.

[7] Aude Oliva and Antonio Torralba, *"Building the gist of a scene: the role of global image features in recognition"*, Progress in Brain Research, 2006, Vol. 155, ISSN 0079-6123.

[8] Ivan Sikiric, Karla Brkic and Sinisa Segvic, *"Classifying traffic scenes using the GIST image descriptor"*, Proceedings of the Croatian Computer Vision Workshop, 2013.

[9] Bay H., Tuytelaars T. and Van Gool L., *"Surf: Speeded up robust features"*, Computer Vision, Springer Berlin Heidelberg, pp. 404 417, 2006.

[10] Baofeng Zhang, Yingkui Jiao, Zhijun Ma, Yongchen Li and Junchao Zhu, *An Efficient Image Matching Method Using Speed Up Robust Features*, Proceedings of 2014 IEEE International Conference on Mechatronics and Automation August 3–6, Tianjin, China.

[11] K. Murphy, A. Torralba, D. Eaton and W. Freeman, *Object detection and localization using local and global features*, Towards Category-Level Object Recognition, 2005, Vol. 4170: 382400.

[12] Ivan Sikiric, Karla Brkic, Sinisa Segvic, *Classifying traffic scenes using the GIST image descriptor*, Proceedings of the Croatian Computer Vision Workshop, 2013.

[13] Agarwal S, Awan A, Roth D, *Learning to detect objects in images via a sparse, part-based representation*, IEEE Trans. Pattern Analysis and Machine Intelligence, 2004, 26(2): 1475–1490.

[14] Rowley H, Baluja S, Kanade T., *Neural network-based face detection*, IEEE Trans. Pattern Analysis and Machine Intelligence, 1998, 20(1): 22–38.

Emerging Trends in Engineering, Science and Technology for Society,
Energy and Environment – Vanchipura & Jiji (Eds)
© 2018 Taylor & Francis Group, London, ISBN 978-0-8153-5760-5

Non-destructive classification of watermelon ripeness using acoustic cues

Rajeev Rajan & R.S. Reshma

Department of Electronics and Communication Engineering, Rajiv Gandhi Institute of Technology, Kottayam, India

ABSTRACT: A major issue in the post-harvest phase of the fruit production sector is the systematic determination of the maturity level of fruits, such as the ripeness of watermelons. Maturity assessment plays an important role while sorting in packing houses during the export. This paper proposes a support vector machine-based method for the automated non-destructive classification of watermelon ripeness by acoustic analysis. Acoustic samples are collected from ripe and unripe watermelons in a studio environment by thumping on the surface of watermelons. Sound samples are pre-processed to remove silence regions by fixing an energy threshold. Pre-processed sound signals are segmented into equal-length frames sized 200 ms, and Teager Energy Operator (TEO)—based features are extracted. The entire set of audio samples are divided into a training set with 60% of the total audio samples and the remaining 40% for testing. A support vector machine—based classifier is trained with features extracted from the training set. Twenty dimensional feature vectors are computed in the feature extraction phase and fed into the classification phase. The results show that the proposed TEO-based method was able to discriminate between ripe and unripe watermelons with overall accuracy of 83.35%.

1 INTRODUCTION

During the past decade, the inspection of quality and maturity of fruits and vegetables in harvest and postharvest conditions is highly demanding in the fruit production industry. Systematic determination of maturity assessment plays an important role in sorting. The need for automated non-contact techniques in sorting and grading is in high demand by the industry. Several techniques to measure firmness and quality have been listed in Abbott et al. (1968); Chen et al. (1993). Usually farmers identify the maturity and quality levels of fruits using certain indices, such as the number of days after full bloom, flesh color, thumping sound, fruit shape, fruit size and skin color. Traditional methods may have their own limitations. For example, judging watermelon ripeness, using its apparent properties such as size or skin color, is very difficult due to its thick skin. The most common way by which people used to determine the watermelon ripeness was by tapping the skin of the melon and then judging the ripeness using the reflected sound. If the sound is dense, then the watermelon is under-ripe, while if the sound is hollow, then the watermelon is ripe. Examples of a ripe and unripe watermelon are shown in the Figure 1. The use of automated inspection of fruits and vegetables has increased in recent decades to achieve higher quality sorting before packaging.

Acoustic features are widely used in many applications in day-to-day life Ayadi et al. (1995); Piyush et al. (2016). In the study of Miller and Delwiche Miller and Delwich (1989), spectral information and machine vision were used for bruise detection on peaches and apricots. Hyper-spectral imaging for detecting apple bruises was investigated by Xing and De Baerdemaeker Xing and De Baerdemaeker (2005). In Abbaszadeh et al. (2011), a non-destructive method for quality test using Laser Doppler Vibrometery (LDV) technology is presented. By means of a Fast Fourier Transform (FFT) algorithm and by considering the

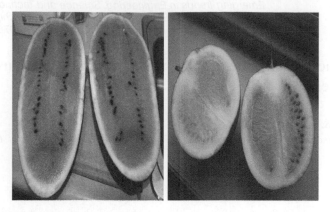

Figure 1.   Ripe and Unripe watermelon.

response signal to excitation signal ratio, vibration spectra of fruit are analyzed to classify ripeness. In this paper, we propose a method based on the Teager Energy Operator (TEO) and support vector machines. In the proposed work the Teager energy-based features give results par with other nondestructive techniques Baki et al. (2010); Diezma-Iglesias et al. (2004). The rest of the paper is organized as follows. Section 2 discusses the theory of the Teager energy operator and support vector machines in detail. Section 3 discusses the experimental description of the proposed method along with the description of the dataset. Section 4 presents results with analysis, followed by conclusion in Section 5.

2   THEORETICAL BACKGROUND

2.1   *Teager Energy Operator (TEO)*

The TEO approach was first proposed by Teager and further investigated by Kaiser H.M. Teager and S.M. (1990). The applications of TEO include speech analysis Teager and Teager (1983), speech emotion recognition Ayadi et al. (1995), image texture analysis and mechanical fault detection Liu et al. (2013).

The Teager energy operator is defined as:

$$\psi\{x(t)\} = \dot{x}^2(t) - x(t)\ddot{x}(t) \tag{1}$$

in the continuous case (where $\dot{x}$ means the first derivative of x, and $\ddot{x}$ means the second derivative). In the proposed work, a non-speech application of the TEO feature is demonstrated. The steps to compute a TEO-based feature are shown in Figure 2. Acoustic samples are collected by thumping on the surface of watermelon, as shown in Figure 3. Hamming windowed audio frame is transformed to the frequency domain using an FFT algorithm, and power spectrum $S(i)$ is computed, followed by a TEO transform, resulting in:

$$\psi[S(i)] = S^2(i) - S(i+1)(i-1) \tag{2}$$

A Mel-scale filter bank is used to filter the spectrum obtained from the TEO processing. Each filter in the filter bank is a triangle bandpass filter, H_m, which tries to imitate the frequency resolution of the human auditory system Hui et al. (2008).

The outputs of the filter bank are obtained by,

$$P_m = \psi[S(i) \cdot H_m(i) \, m = 1,2,...M] \tag{3}$$

716

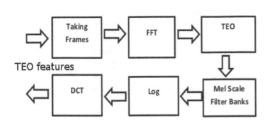

TEO features

Figure 2. Steps to compute Teager energy operator based feature.

Figure 3. Collecting acoustic samples by thumping watermelon.

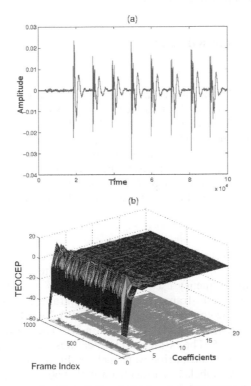

Figure 4. (a) Thumping sound collected from watermelon; (b) TEO features extracted for one acoustic sample.

where M is the number of filter banks. Then log compression is applied to the filter bank output. Finally Discrete Cosine Transform (DCT) applied in order to compress the spectral information into the low-order coefficients. The feature vector obtained is expressed by:

$$TEO_{CEP}(k) = \sum_{i=1}^{M} \cos\left[\frac{k(i-0.5)}{M} \cdot \pi\right] \cdot \log(P_i). \tag{4}$$

Thumping sound collected from a riped watermelon and its TEO feature are shown in Figures 4(a) and (b) respectively.

717

Table 1. Parameters for LIBSVM.

Sl. No.	Parameter	Libsvm$_{parameter}$	Value
1	Optimal Objective value (Q)	*obj*	−16.33
2	Bias term (b)	ρ	0.83
3	No. of SV (N)	*nsv*	21
4	Regularization parameter (C)	ν	0.9523

2.2 *SVM classifier*

Support Vector Machine (SVM) is a widely used classifier that computes an optimal hyperplane and maximizes the margin between two classes of data in the kernel-induced feature space Vapnik (1998). In the proposed experiment, we used a support vector machine-based classifier using LibSVM package. SVM classifier with a linear kernel is used in the experimental evaluation. The parameters used in the SVM classifier are listed in Table 1.

3 EXPERIMENTAL DESCRIPTION

3.1 *Data collection*

In the proposed system, 37 acoustic data samples were collected from 30 watermelons of a variety called *Carolina cross*, with different levels of ripeness. It has green skin, red flesh and commonly produces fruit weighing between 29 and 68 kg. It takes about 90 days from planting to harvest. Watermelons were labeled as ripe and unripe based on experience by the human expert. The maturity of the watermelons was determined for labeling purposes by manually thumping the watermelons and listening to the sound produced. Thumping sounds were recorded using a Audio-Technica AT2035 Large Diaphragm Studio Condenser microphone connected to a MAC with Pro Tools-9 digital audio workstation.

3.2 *Proposed method*

In the feature extraction phase, acoustic samples are pre-processed to remove silence region by fixing an energy threshold. Acoustic samples are segmented into equal-length frame size of 200 ms. The sampling frequency used is 44,100 Hz. 50% of successive frames are overlapped to smoothen the frame-to-frame transition. Each frame is multiplied with a hamming window in order to keep the continuity Baki et al. (2010). Twenty dimensional TEO-based feature vectors are extracted, as explained in Section 2.1. The various steps involved in the process are shown in Figure 5. The extracted TEO features are fed into the trained classifier to discriminate between ripe and unripe watermelons during the testing phase. The entire dataset was divided into two subclasses, in such a way that 60% of the acoustic samples could be used for the training phase and the rest for the testing phase.

4 RESULTS AND ANALYSIS

The confusion matrix of the proposed classification system is given in Table 2. The proposed method results in an overall classification accuracy of 83.35% with a 60: 40 (training: testing) pattern. The experimental results proved that the proposed TEO-based system outperforms the other acoustic-based methods cited in the literature. When the system identified all the riped watermelons correctly, 66.7% is the system accuracy for the unripe cases. The experimental results show that TEO_{CEP} features are very effective in capturing human perception sensitivity and energy distribution, which is more important for discriminating ripe and unripe watermelons.

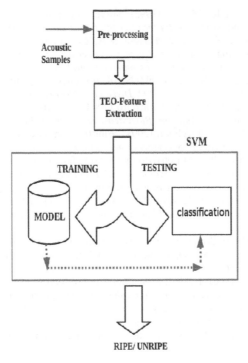

Figure 5.   Proposed method for classification of watermelon ripeness based on TEO feature.

Table 2.   Confusion matrix of RIPE/UNRIPE classification.

	Training set		Testing set	
	RIPE	UNRIPE	RIPE	UNRIPE
RIPE	10	0	8	0
UNRIPE	0	13	2	4

5   CONCLUSION

A potential method for non-destructive classification of watermelon ripeness using TEO based features and support vector machines is proposed in this paper. Teager energy operator based features have been extensively used for various speech processing applications such as speaker recognition and speech emotion recognition. In the proposed work, we effectively used TEO based features for a non-speech processing application, to discriminate ripe and unripe watermelons. In continuation of our research, we would like to focus on fusion of various features to improve classification accuracy of the proposed system.

REFERENCES

Abbaszadeh, R., A. Rajabipour, H. Ahmadi, M. Delshad, & M. Mahjoob (2011). Assessment of watermelon quality using vibration spectra. *Innovative Computing Technology Communications in Computer and Information Science 241*(2), 21–29.
Abbott, J., G.S. Bachman, R.F. Childers, J. Fitzgerald, & F.J. Matasik (1968). Sonic techniques for measuring texture of fruit and vegetables. *Food technology 5*(12), 101–112.

Ayadi, E., M.S. Kamel, & F.K. Moataz (1995). Survey on speech emotion recognition: Features, classification schemes, and databases. *Pattern Recognition, Elsavier 44*(3), 572–587.

Baki, S.R., M.A. Mohd, I.M. Yassin, H.A. Hassan, & A. Zabidi (2010). Non-destructive classification of watermelon ripeness using mel-frequency cepstrum coefficients and multilayer perceptrons. *Neural Networks (IJCNN), The International Joint Conference on*, 1–6.

Chen, P.,M.J. McCarthy, & R. Kauten (1993).NMR for internal quality evaluation of fruits and vegetables. *Austral. Postharv. Conf. C.S.I.R.O. Division of Horticulture, Paper No. 1119*, 355–358.

Diezma-Iglesias, B.,M. Ruiz-Altisent, & P. Barreiro (2004). Detection of internal quality in seedless watermelon by acoustic impulse response. *Biosystems Engineering, Elsevier 88*(2), 221–230.

Hui, G., C. Shanguang, & S. Guangchuan (2008). Emotion classification of infant voice based on features derived from teager energy operator. *Congress on Image and Signal Processing.*

Liu, H., J. Wangand, & C. Lu (2013, March). Rolling bearing fault detection based on the teager energy operator and elman neural network. *Mathematical Problems in Engineering 10.* Miller, B. & M. Delwich (1989). Color vision system for peach grading. *Trans. ASAE, 32*(4), 1484–1490.

Piyush, P., R. Rajan, M. Leena, & B. Koshy (2016, February). Vehicle detection and classification using audio visual cues. *in proceedings of Third international conference on signal processing and integrated networks(SPIN).*

Teager, H. & S. Teager (1983). Some observations on vocal tract operation froma fluid flow of view in: Vocal fold physiology: Biomechanics, acoustic, and phonatory control. *The Denver Center for performing arts 55*, 358–386.

Teager, H.M. & S. Teager (1990). Evidence for nonlinear production mechanisms in the vocal tract,. *Hardcastle W.J., Marchal A. (eds) Speech Production and Speech Modelling. NATO ASI Series (Series D: Behavioural and Social Sciences), Springer, Dordrecht 55*, 241–261.

Vapnik, V. (1998). Statistical learning theory. *New York Wiley.*

Xing, J. & J. De Baerdemaeker (2005). Bruise detection on jonagold apples using hyperspectral imaging. *Post harvest Biol. Technol. 37*(2), 152–162.

Emerging Trends in Engineering, Science and Technology for Society,
Energy and Environment – Vanchipura & Jiji (Eds)
© 2018 Taylor & Francis Group, London, ISBN 978-0-8153-5760-5

Predominant instrument recognition from polyphonic music using feature fusion

Roshni Ajayakumar & Rajeev Rajan
Department of Electronics and Communication Engineering, Rajiv Gandhi Institute of Technology, Kottayam, India

ABSTRACT: In this paper, the fusion of a modified group delay feature and a frame slope feature is effectively utilized to identify the most predominant instrument in polyphonic music. The experiment is performed on a subset of the Instrument Recognition in Musical Audio Signals (IRMAS) dataset. The dataset consists of polyphonic music, with predominant instruments such as acoustic guitar, electric guitar, organ, piano and violin. In the classification phase, a Gaussian Mixture Model (GMM)—based classifier makes the decision based on the log-likelihood score. The results show that a phase-based feature modified group delay works more effectively than magnitude spectrum based features, such as Mel-frequency cepstral coefficients and frame slope. The classification accuracy of 57.20% is reported from the fusion experiment. The proposed system demonstrates the potential of the fusion of features in recognizing the predominant instrument in polyphonic music.

1 INTRODUCTION

Music Information Retrieval (MIR) is a growing field of research with lots of real-world applications, and is applied well in categorizing, manipulating and synthesizing music. Music Information Retrieval (MIR) mainly focuses on the understanding and usefulness of music data through the research, development and application of computational approaches and tools. Automatic instrument recognition, one of the MIR tasks, has a wide range of applications ranging from source separation to melody extraction. In Computational Auditory Scene Analysis (CASA), musical instrument recognition and sound source recognition play a vital role. Predominant instrument recognition refers to the problem where the prominent instrument is identified from a mixture of instruments playing together. In the literature, we could see that the instrument recognition by source separation is widely studied in many music information retrieval applications. Since there are numerous approaches for source separation, such as polyphonic pitch estimation in Klapuri (2001) or the separation of concurrent harmonic sounds, it can act as a front-end for the successive monophonic recognition task.

2 RELATED WORK

In the literature, numerous attempts in instrument recognition have been reported for monophonic or polyphonic audio files. Features derived from a Root-Mean-Square (RMS) energy envelope via Principle Component Analysis (PCA) can be seen in Kaminsky and Materka (1995). In another approach Eronen and Klapuri (2000), cepstral coefficients combined with temporal features are used to classify 30 orchestral instruments with several articulation styles. The group delay-based feature has also been used for automatic instrument recognition in an isolated environment Diment et al. (2013). Line spectral frequencies-Gaussian Mixture Model (GMM) framework is used for instrument recognition in Yu and Yang (2014).

In a polyphonic environment, instrument recognition using a Non-Negative Matrix Factorization (NMF) based source-filter model with mel-filter bank cepstral coefficients and GMM is attempted in Heit tola et al. (2009). In Kashino and Nakadai (1998), a system transcribing random chords of clarinet, flute, piano, trumpet and violin was presented. In this paper, we propose a fusion of features for recognizing predominant instrument from polyphonic music.

The outline of the rest of the paper is as follows. Section 3 explains the overview of the system. The feature extraction task is discussed in Section 4. In Section 5, the performance evaluation is explained. Analysis of results is described in Section 6 and finally conclusions are drawn in Section 7.

3 OVERVIEW OF THE PROPOSED SYSTEM

The block diagram of the proposed system is shown in Figure 1. The experiment is conducted in four phases. In the first phase, a baseline system with MFCC features are used, followed by phases with frame slope, modified group delay feature and early fusion of these features. In all the phases, GMM—based classifiers are used. Sixty-four mixture GMMs are trained for all instrument models, using audio files in an isolated environment. For each instrument model, the likelihood score is computed for the test audio file and the model which reports maximum log-likelihood is declared as the decision. It can be formulated mathematically as finding the target – λ_i for which the following criteria is satisfied.

$$\arg\max_{1 \le i \le R} \sum_{m=0}^{M-1} \log\left[p\left(O_m / \lambda_i\right)\right] \tag{1}$$

where O_m, λ_i, M, R, represent the feature vectors, GMM model for an instrument, number of feature vectors, and the total number of instrument models, respectively.

4 FEATURE EXTRACTION

In the proposed experiments, three features, namely Mel-frequency cepstral coefficients, Modified Group Delay Feature (MODGDF), and frame slope features are frame-wise computed from the audio file. The steps to compute these features are discussed in subsequent sections.

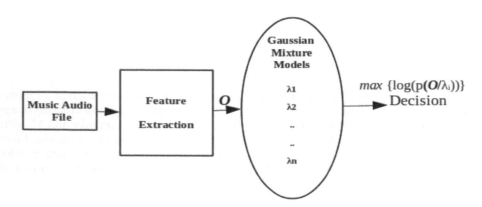

Figure 1.   The proposed system. λ_1 $\lambda 2$ … λ_n represent Gaussian mixture models for isolated instruments.

4.1 Mel-Frequency Cepstral Coefficients (MFCC)

In order to extract the MFCC feature, the Fourier analysis is performed on each speech frame, producing short-time discrete Fourier transform coefficients. The coefficients are then grouped together in critical bands and weighted by a triangular band pass filter called Mel-spaced filter banks. The MFCC can then be derived by taking the log of the band-passed frequency response and calculating the Discrete Cosine Transform (DCT).

4.2 Modified Group Delay Feature (MODGDF)

Group delay features have already been employed in numerous speech processing applications Murthy and Yegnanarayana (2011); Rajan and Murthy (2013a,b, 2017); Rajan et al. (2017). The group delay function, $\tau(e^{j\omega})$, is defined as the negative derivative of an unwrapped Fourier transform phase. The group delay function of minimum phase signals can be computed directly from the signal by Oppenheim and Schafer (1990):

$$\tau\left(e^{j\omega}\right) = \frac{X_R\left(e^{j\omega}\right)Y_R\left(e^{j\omega}\right) + Y_I\left(e^{j\omega}\right)X_I\left(e^{j\omega}\right)}{\left|X\left(e^{j\omega}\right)\right|^2} \qquad (2)$$

where the subscripts R and I denote the real and imaginary parts, respectively. $X(e^{j\omega})$ and $Y(e^{j\omega})$ are the Fourier transforms of $x[n]$ and $nx[n]$ respectively. The denominator is replaced by its spectral envelope to mask the spiky nature. The modified group delay function (MODGD) $\tau_m(e^{j\omega})$ is obtained as:

$$\tau_m\left(e^{j\omega}\right) = \frac{X_R\left(e^{j\omega}\right)Y_R\left(e^{j\omega}\right) + Y_I\left(e^{j\omega}\right)X_I\left(e^{j\omega}\right)}{\left|S\left(e^{j\omega}\right)\right|^{2\gamma}} \qquad (3)$$

where $S(e^{j\omega})$ is the cepstrally—smoothed version of $X(e^{j\omega})$. The group delay function and modified group delay function for the speech frame are shown in Figures 2(a) and (b), respectively. Modified group delay functions are converted to spectra using DCT, as shown in Hegde (2005). In the proposed experiment, 13-dimensional modified group delay features (MODGDF) are computed from the test and target audio files.

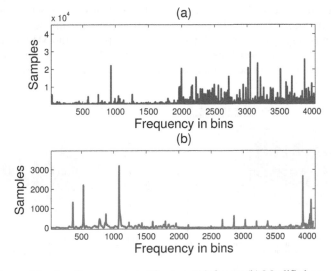

Figure 2.   (a) Group delay functions computed for a speech frame; (b) Modified group delay functions computed for the frame in (a).

723

4.3 *Frame slope feature*

The slope feature is determined by computing the slope of a linear fit to Mel filterbank energies. A Least Square Fit (LSF) is computed using the algorithm listed in Madikeri and Murthy (2011). The Mel filterbank slope feature emphasizes the system information better than do conventional MFCCs.

5 PERFORMANCE EVALUATION

5.1 *Dataset*

Instrument Recognition in Musical Audio Signals (IRMAS) dataset consists of more than 6,000 musical audio excerpts from various styles with annotations of the predominant instrument present. They are excerpts of three seconds for 11 pitched instruments. In the proposed experiment, initially Gaussian mixture models are built using monophonic instrument wave files with 1,000 files per instrument. We considered five classes, namely acoustic guitar, electric guitar, organ, piano and violin. In the testing phase, 250 files of five classes (50 each) are tested against these models. All audio files are stored in 16-bit stereo WAV format, sampled at 44.1 kHz.

5.2 *Experimental setup*

In the front-end, MFCC/MODGDF/frame slope feature sets are computed for every 10 ms with a window size of 30 ms. The MFCC system is used as a baseline system for comparative study. Initially GMM models are built for each class in the training phase. During the testing phase, feature vectors computed from the audio files are fed in to the model and the accumulated likelihood score is computed. Based on the likelihood score, the system ranks the matching from the best to worst. The experiment is conducted for five classes with 50 test files each. The performance is evaluated using classification accuracy in which, if the system output matches with ground truth, classification is treated as correct.

6 RESULTS AND ANALYSIS

The studies conducted by Diment et al. (2013) for monophonic instrument recognition, motivated to focus on system characteristics in the proposed polyphonic experiment. From the literature, it can be seen that conventional MFCC features are widely used for timbre analysis. But, as mentioned earlier, frame slope features have already been proven to be more effective in emphasizing system characteristics. The mapping of the MFCC and frame slope feature set into a two-dimensional feature space using PCA is shown in Figure 3. It is worth noting that classes are better separated in frame slope feature space than in the MFCC.

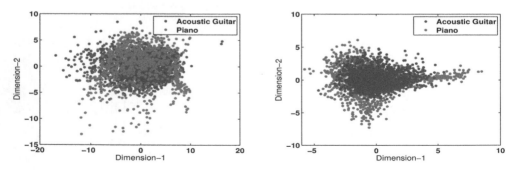

Figure 3. 2-dimensional mapping of (a) MFCC features; (b) Frame slope features.

Table 1.   Confusion matrix of MODGD experiment.

	A. Guitar	E. Guitar	Organ	Piano	Violin	%
A. Guitar	**35**	6	5	3	1	70.0
E. Guitar	8	**14**	15	1	12	28.0
Organ	24	7	**7**	9	3	14.0
Piano	8	0	1	**41**	0	82.0
Violin	0	0	0	6	**42**	84.0

Table 2.   Confusion matrix of frame slope experiment.

	A. Guitar	E. Guitar	Organ	Piano	Violin	%
A. Guitar	**32**	3	1	10	4	64.0
E. Guitar	2	**37**	1	6	4	74.0
Organ	4	15	**8**	13	10	16.0
Piano	5	3	0	**38**	4	76.0
Violin	28	2	0	0	**20**	40.0

Table 3.   Confusion matrix of MODGD + frame slope experiment.

	A. Guitar	E. Guitar	Organ	Piano	Violin	%
A. Guitar	**21**	26	1	0	2	42.0
E. Guitar	1	**44**	1	0	4	88.0
Organ	1	25	**15**	5	4	30.0
Piano	1	18	3	**28**	0	56.0
Violin	3	11	0	1	**35**	70.0

Table 4.   Overall classification accuracy.

Sl.No.	Feature Name	Accuracy (%)
1	MFCC	43.20
2	Frame Slope	54.00
3	MODGDF	55.60
4	**MODGDF + Frame Slope**	**57.20**

The confusion matrices for frame slope, MODGD and fusion experiments are shown in Tables 1–3. The overall accuracy is reported in Table 4. From the experiments, we observed that frame slope features and MODGD feature are giving complementary information. So, finally we combined the feature set and conducted the experiment.

The results show that the baseline system reports an overall accuracy of 43.2%. While the fusion of features improved the individual classification accuracy for electric guitar and organ, it deteriorated the performance for other classes. It is worth noting that overall results improved, when considering the experiment as a whole. The experimental results demonstrate the potential of the modified group delay feature in recognizing the predominant instrument in polyphonic music and related applications.

7 CONCLUSION

Predominant instrument recognition in polyphonic music is addressed in this paper. Three features namely MFCC, frame slope and modified group delay features, are computed in the experiment. A GMM Table based classifier is used in the classification phase. The performance is evaluated using a subset of the IRMAS dataset. Five classes are considered, namely acoustic guitar, electric guitar, organ, piano and violin. The results show that modified group delay features are very much more effective in predominant instrument recognition in polyphonic music as compared to conventional MFCC features. Moreover, the fusion of group delay feature and slope features is also promising in instrument recognition in a polyphonic context.

REFERENCES

Diment, A., P. Rajan, T. Heittola, & T. Virtanen (2013). Modified group delay feature for musical instrument recognition. *In Proceedings of 10th Int. Symp. Comput. Music Multidiscip. Res., Marseille, France, 2013,* 431–438.

Eronen, A. & A. Klapuri (2000). Musical instrument recognition using cepstral coefficients and temporal features. *In Acoustics, Speech, and Signal Processing, 2000. Icassp'00. Proceedings. 2000 IEEE International Conference on 2,* II753–II756.

Hegde (2005). *Fourier transform based features for speech recognition.* PhD dissertation, Indian Institute of Technology Madras, Department of Computer Science and Engg., Madras, India.

Heittola, T., A. Klapuri, & T. Virtanen (2009). Musical instrument recognition in polyphonic audio using source-filter model for sound separation. *In Proceedings of Int. Soc. Music Inf. Retrieval Conf.,* 327–332.

Kaminsky, I. & Materka (1995). Automatic source identification of monophonic musical instrument sounds. *in proceedings of the IEEE Int. Conf. on Neural Networks,* 185–194.

Kashino & T. Nakadai, Kinoshita (1998). Application of bayesian probability network to music scene analysis. *In Proceedings of the International Joint Conference on AI, CASA workshop,* 115–137.

Klapuri, A. (2001, May). Multipitch estimation and source separation by the spectral smoothness principle. *Acoustics, Speech and Signal Processing (ICASSP), 2001 IEEE International Conference on, 5,* 3381–3384.

Madikeri, S.R. & H.A. Murthy (2011). Mel filter bank energy based slope feature and its application to speaker recognition. *in proceedings of National Communication Conference(NCC),* 155–175.

Murthy, H.A. & B. Yegnanarayana (2011). Group delay functions and its application to speech processing. *Sadhana 36*(5), 745–782.

Oppenheim, A. & R. Schafer (1990). *Discrete time signal processing.* New Jersey: Prentice Hall, Inc.

Rajan, R., M. Misra, & H.A. Murthy (2017). Melody extraction from music using group delay functions. *International Journal of Speech Technology 20*(1), 185–204.

Rajan, R. & H.A. Murthy (2013a). Group delay based melody monopitch extraction from music. *in proceedings of the IEEE Int.Conf. on Audio, Speech and Signal Processing,* 186–190.

Rajan, R. & H.A. Murthy (2013b). Melodic pitch extraction from music signals using modified group delay functions. *In proceedings of the Communications (NCC), 2013 National Conference on,* 1–5.

Rajan, R. & H.A. Murthy (2017). Music genre classification by fusion of modified group delay and melodic features. *In proceedings of the Communications (NCC), 2017 National Conference on,* 1–5.

Yu, L. & Y. Yang (2014). Sparse cepstral codes and power scale for instrument identification. *In Proceedings of 2014 IEEE Int. Conf. Acoust., Speech Signal Process.,* 7460–7464.

Emerging Trends in Engineering, Science and Technology for Society,
Energy and Environment – Vanchipura & Jiji (Eds)
© *2018 Taylor & Francis Group, London, ISBN 978-0-8153-5760-5*

A technique for countering the integer boundary spurs in fractional phase-locked loops with VCO

R. Vishnu, S.S. Anulal & Jenitha Ravi
Broadcast and Communication Group, CDAC Trivandrum, Kerala, India

ABSTRACT: Integer boundary spur is a mechanism of fractional spur creation, caused by interactions between the RF VCO frequency and the reference frequency or phase detector frequency. An algorithm was developed which identifies the integer boundary frequencies and configures the synthesizer in such a way that the generated spurs are offsetted far away from the loop bandwidth, so that they are attenuated by the loop filter. The algorithm runs as a sub-routine in the micro-controller which programs the synthesizer, to modify the multiplier value accordingly. This algorithm was developed and tested on a fractional frequency synthesizer operating from 700–1000 MHz, even though the method is independent of frequency of operation.

1 INTRODUCTION

Frequency synthesizers (PLL and VCO combinations) that are capable of operating only at integer multiples of the phase frequency detector frequency are known as integer-N PLLs and those which can synthesize finer steps or fractions of the phase frequency detector frequency are known as fractional synthesizers. Fractional synthesizers generates two types of spurious signals, namely fractional spurs and integer boundary spurs (IBS). Modern PLLs use higher order $\Sigma\Delta$ modulators to reduce fractional spurs. Integer boundary spurs are caused by interactions between the RF VCO frequency and the harmonics of reference or PFD frequency. When these frequencies are not integer related, spur sidebands may appear as sidebands on the VCO output spectrum at offset frequencies that is the fundamental and harmonics of the difference frequency between an integer multiple of the reference/PFD frequency and the VCO frequency. They are spurs generated inside a PLL, but the system designer can predict these spurs and hence can be avoided. If the difference frequency can be increased in such a way that it is made larger than the loop bandwidth, then they will be filtered off by the loop filter.

2 BLOCK DIAGRAM AND CIRCUIT DESCRIPTION

The block diagram of the frequency synthesizer is shown below:
   The major blocks of this synthesizer are:

1. Reference Oscillator
2. Reference Multiplier
3. R, N counters and fractional interpolator
4. Phase frequency detector
5. Voltage Controlled Oscillator
6. Output divider

   This is a fractional N PLL circuit with an additional multiplier in the reference path. There is an output divider outside the PLL loop. The delta sigma modulator in the N-counter reduces the fractional spurs. The other major spurious contribution is by the IBS.

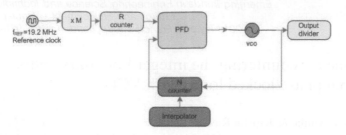

Figure 1. Block diagram.

The frequency range of operation for the synthesizer was from 700 MHz to 1000 MHz in 6.25 kHz steps. The output divider translates the VCO frequency range of 4300 MHz to 5300 MHz, to the required output frequency range. For each frequency, the optimum multiplier and R-counter settings are programmed to the synthesizer.

3 DESCRIPTION OF THE TECHNIQUE

3.1 Description

One of the widely used technique for avoiding the IBS is by changing the reference frequency. This technique will eliminate the IBS caused due to interactions of the VCO with reference as well as PFD. But it is more costly and space consuming. There must be an additional circuit for the clock synthesizer. This circuit is intended for handheld radio application, so there is space as well as cost constraints. The IBS generation due to interaction with reference cannot be avoided using this technique.

The algorithm works by computing the IBS offset frequency for each output frequency for various PFD frequencies. It starts with the highest PFD frequency and computes the IBS offset. A higher PFD frequency will result in a lower N counter and hence better phase noise performance. If the calculated offset is found to be greater than 2 MHz, the code exits the loop and program the PLL multiplier and R counter settings for that PFD frequency. The PFD frequency was constrained within 50 MHz to 120 MHz. The lower limit of 50 MHz is fixed based on a compromise between spurious performance and phase noise. The charge pump current must be re-programmed for a lower PFD frequency inorder to keep the loop dynamics constant. Charge pump current changes inversely with the PFD frequency, so when the PFD frequency increases, the charge pump current must decrease and vice versa.

3.2 Flowchart

1. Obtain the possible PFD values for multiplier M and R-counter values from 1 to 10.
2. Start with the maximum PFD frequency.
3. Assign value to the variable f_{PFD}. Calculate the value of f_{VCO} based on the present output divider value
4. For the desired VCO frequency (f_{VCO}), a formula check is done to evaluate whether this f_{VCO} will generate boundary spur.
   $\Delta f_1 = $ Absolute [{Roundup (f_{VCO}/f_{PFD})} $\times f_{PFD} - f_{VCO}$]
   $\Delta f_2 = $ Absolute [{Rounddown (f_{VCO}/f_{PFD})} $\times f_{PFD} - f_{VCO}$]
   Δf_1 and Δf_2 are the spur offset frequencies from the wanted frequency.
5. Check to see if these values are greater than 2 MHz. If this condition is not met, then go back and check the spur offsets for the next lower PFD frequency.
6. If the spur offset is greater than 2 MHz, then proceed to configure the PLL for that PFD frequency. The multiplier and R-counter settings must be programmed.

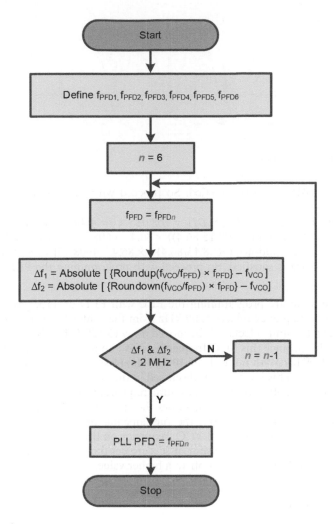

Figure 2.   Flowchart.

3.3   *Validation of the algorithm*

Consider an arbitrary frequency f_{OUT} = 768.025 MHz for spur analysis. The analysis can be done as per the flowchart sequence.

$$\text{Reference} = 19.2 \text{ MHz.}$$

There are 6 possible PFD values based on the various combinations of multiplier M and R-counter values from 1 to 10.

1. The possible values are 51.2, 67.2, 76.8, 86.4, 96, 115.2 MHz
2. Starting with the maximum value of 115.2 MHz
3. f_{PFD} = 115.2 MHz. Then $f_{VCO} = f_{OUT} \times DIV = 768.025 \times 6 = 4608.15$ MHz
4. Δf_1 = Absolute [{Roundup (4608.15/115.2)} × 115.2 − 4608.15] = 115.05 MHz
    Δf_2 = Absolute [{Rounddown (4608.15/115.2)} × 115.2 − 4608.15] = 0.15 MHz
5. Δf_1 & Δf_2 > 2 MHz is NOT satisfied. So proceed with the next lower value of PFD frequency.
6. f_{PFD} = 96 MHz. The $f_{VCO} = f_{OUT} \times DIV = 768.025 \times 6 = 4608.15$ MHz
7. Δf_1 = Absolute [{Roundup (4608.15/96)} × 96 − 4608.15] = 95.85 MHz
    Δf_2 = Absolute [{Rounddown (4608.15/96)} × 96 − 4608.15] = 0.15 MHz

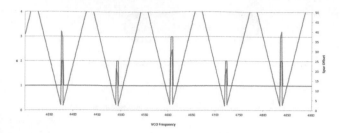

Figure 3.   Algorithm validation.

8. Δf_1 & $\Delta f_2 > 2$ MHz is NOT satisfied. So proceed with the next lower value of PFD frequency.
9. $f_{PFD} = 86.4$ MHz. The $f_{VCO} = f_{OUT} \times DIV = 768.025 \times 6 = 4608.15$ MHz
10. Δf_1 = Absolute [{Roundup (4608.15/86.4)} × 86.4 – 4608.15] = 57.45 MHz
    Δf_2 = Absolute [{Rounddown (4608.15/86.4) } × 86.4 – 4608.15] = 28.95 MHz
11. Δf_1 & $\Delta f_2 > 2$ MHz is satisfied. So proceed with programming the PLL with this PFD frequency.

So in this case if we are programming the PLL with PFD of 115.2 MHz or 96 MHz, there will be IBS at frequency offsets ±300 kHz from the center frequency. But if the PFD is made 86.4 MHz, the spur offsets will be moved away to >29 MHz. But this frequency has an additional problem, since this VCO frequency is also an integer multiple of the reference frequency of 19.2 MHz (240th multiple of 19.2 MHz) (4608.15 is not integer multiple, 4608 is tehe 240th multiple). This issue is later discussed in the evaluation results section.

The functioning of the algorithm over the entire frequency range of operation was validated in a spreadsheet software. The Figure 3, shows the spurious frequency offsets at different frequencies, with the spur offsets from the fundamental on the right y-axis, converging frequency number in the left y-axis and VCO frequency on the x-axis. It is evident that there are several frequency points where the spur offsets crosses the 2 MHz bound, where the PFD is changed and spurs offsets are moved off to a higher value.

4   EVALUATION RESULTS

Figure 4 shows the spectrum taken at the output frequency of 792.025 MHz. Using an output divider of 6, the VCO frequency will be 4752.15 MHz. 4752 MHz is an integer multiple of 86.4 MHz (55th multiple). 86.4 MHz is one of the choice of PFD frequency for a reference of 19.2 MHz.

Now applying the algorithm,

$\Delta f1$ = Absolute [{Roundup (4752.15/86.4)} × 86.4 – 4752.15] = 86.25 MHz
$\Delta f2$ = Absolute [{Rounddown (4752.15/86.4)} × 86.4 – 4752.15] = 0.15 MHz

So IBS can be observed as sidebands at 150 kHz offsets. This is seen in the YELLOW trace.

Now if we re-calculate with 115.2 MHz as PFD frequency,

$\Delta f1$ = Absolute [{Roundup (4752.15/115.2)} × 115.2 – 4752.15] = 86.25 MHz
$\Delta f2$ = Absolute [{Rounddown (4752.15/115.2)} × 115.2 – 4752.15] = 28.95 MHz

Hence the offset frequencies are increased to greater than 28 MHz so that it is easily filtered OFF by the loop filter. RED trace shows the same frequency at 115.2 MHz PFD frequency.

As mentioned earlier the IBS generation due to interaction of the VCO with the reference cannot be avoided using this technique. To demonstrate this phenomenon, an output

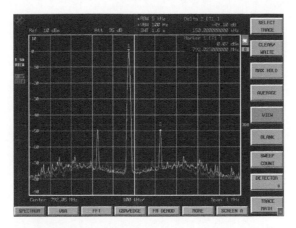

Figure 4.   Spectrum of 792.025 MHz.

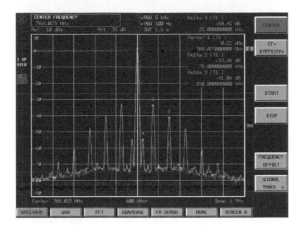

Figure 5.   Spectrum of 768.025 MHz.

frequency of 768.025 MHz maybe taken. The VCO frequency is $768.025 \times 6 = 4608.15$ MHz. The frequency 4608 MHz is an integer multiple of 19.2 MHz, 96 MHz as well as 115.2 MHz. But it is not an integer multiple of 86.4 MHz. The algorithm come up with the final PFD value of 86.4 MHz. But even then the boundary spurs are not fully suppressed due to the interaction of VCO with the reference frequency. This phenomenon cannot be evaded in this architecture, since the reference is not tunable.

This can be seen in the Figure 5 shown above. The yellow trace shows the output with the PFD frequency as 115.2 MHz and the red trace shows the output spectrum with the PFD frequency as 86.4 MHz. The reference frequency 19.2 MHz also interacts to produce the 150 kHz and their harmonics and divided down spurious.

REFERENCES

[1] "Wideband Synthesizer with Integrated VCO", ADF4350 Data Sheet Rev A04/2011, pp. 23–28.
[2] "An avoidance technique for mitigating the integer boundary spur problem in a DDS-PLL hybrid frequency synthesizer," 2015 International Conference on Communications and Signal Processing (ICCSP), Melmaruvathur, 2015, pp. 0443–0446. R. Vishnu and Anulal S. S.
[3] "Analyzing, Optimizing, and Eliminating Integer Boundary Spurs in Phase-Locked Loops with VCOs at up to 13.6 GHz" Analog Dialogue, Aug 2015, Vol. 49 by Robert Brennan.
[4] "Ultra-Low Noise PLLatinum Frequency Synthesizer With Integrated VCO datasheet (Rev. J)", LMX2541 Data Sheet Rev J.

Emerging Trends in Engineering, Science and Technology for Society,
Energy and Environment – Vanchipura & Jiji (Eds)
© 2018 Taylor & Francis Group, London, ISBN 978-0-8153-5760-5

Adaptive neuro-fuzzy based base station and relay station deployment for next generation wireless communication

Mariya Vincent

Department of Electronics and Communication Engineering, Rajagiri School of Engineering and Technology, Kakkanad, India

ABSTRACT: Improper deployment of Base Stations (BSs) and Relay Stations (RSs) at inappropriate locations can result in a decrease in power efficiency and throughput and an increase in deployment cost, transmission delay, transmission loss and power consumption. Hence, in order to exploit the advantages of BSs and RSs completely, an effective selection and deployment scheme for BS and RS, which attains the target Coverage Ratio (CR) and throughput, at an affordable deployment budget is required for next generation wireless communication. The superiority in performance of the proposed method over conventional clustering methods is illustrated through MATLAB simulations.

1 INTRODUCTION

Wireless communication have become a crucial part in our daily personnel and professional lives by exchanging information reliably from anywhere, any time. The enormous increase in the number of mobile subscribers has proportionally increased the data rate demand. The mobile subscribers also demand anytime-anywhere wireless broadband services with high quality of experience. One of the solutions to provide adequate signal-to-noise ratio (SNR), increase data throughput, overall coverage and system capacity is to decrease cell area and deploying more number of BSs which escalates the deployment cost as well as increase the inter-cell interference. Hence alternate intelligent solutions should be implemented in next generation wireless communication networks.

Multi-hop relay (MHR) network is one of the promising solutions recommended by the LTE-A standards to satisfy the above mentioned service requirements. Yang et al. (2009) in LTE-A and IEEE 802.16j standards, the concept of MHR network is introduced, where RSs are deployed along with BSs to improve the coverage and capacity. RSs are more suited solution in the locations where the backhaul connection is expensive or unavailable. The RSs also have other advantages like less carbon dioxide (CO_2) emission, less power consumption, easier and faster installation and low maintenance cost. Unlike the BS, which is connected to the backhaul by a wired connection, RS can be wirelessly connected to the BS. Moreover, MHR networks help to improve network throughput and at the same time covers more number of mobile users over a larger coverage area. Hence, MHR network is considered as one of the potential candidate to facilitate power efficient wireless communication. The use of RSs to extend battery was presented by Laneman and Womell (2000). Cho et al. (2009) proposed an RS deployment scheme to reduce the delay due to handovers by deploying the nodes at the boundaries of adjacent cell edges. A two stage joint BS and RS placement (JBRP) scheme is proposed by Lu and Liao (2009) where the authors used k-supplier concepts in the first stage to deploy BSs and greedy-heuristic concepts for the RS deployment in the second stage. Even though, the study considered joint BS and RS deployment problem, it ignores the trade-off between the CR and deployment cost.

The proposed scheme also suffers from unbalanced network load. Chang and Lin (2014), have proposed an uniform clustering based BSs and RSs placement scheme for MHR network.

In the recent years, the study on deployment strategies based on fuzzy logic has attracted active attention among academic and industry. Fuzzy logic based BS deployment scheme is proposed by Jau and Lin (2015). It has been shown that the fuzzy based deployment scheme is low complex and offers better coverage and throughput performance than the conventional uniform clustering based deployment scheme. Vincent et al. (2017) proposed a power-aware joint fuzzy based BS and RS deployment scheme for green radio communication. The authors used CR and traffic ratio (TR) as the input parameters for fuzzy inference engine. To account the greenness of the system, a novel parameter, power consumption effectiveness ratio (PCER) is given as the third input metric for the fuzzy inference engine. To maximize PCER, CR, TR, the fuzzy rules are formed and the corresponding selection factor (SF) is used to choose the BS and RS candidate locations for deployment. The fuzzy inference engine maps input characteristics to input membership functions and rules to obtained a single-valued SF. Such a system uses fixed membership functions and a rule structure that is essentially predetermined by the user's interpretation of the characteristics of the input variables. The shape of the membership functions depends on the parameters and scenario of the network, and changing these parameters change the shape of the membership function. So we need a system where the shapes of the membership functions and rules arbitrarily changes with the scenario. Moreover, by applying adaptive neuro fuzzy logic significant decisions can be taken which resemble those taken by human thinking and reasoning. In the present study, an effective BSs and RSs deployment based on adaptive-neuro fuzzy was proposed.

The paper is organized as follows: The MHR network model and background are presented in section 2. In section 3, Adaptive neuro fuzzy based BSs and RSs deployment and selection phase is described. Section 4 deals with the MATLAB based simulation of the proposed MHR network scenario and analyses the system performance. Section 5 concludes the paper.

2 MHR NETWORK MODEL AND BACKGROUND

MHR network model consist of BSs, RSs and MSs. The MSs are located within the geographical area are uniformly distributed in a large number. According to the geographic features of that particular area, there exist some feasible candidate positions where the RSs can be deployed. The RSs are deployed by the network operators at cell edges and coverage holes to improve the coverage as well as the capacity of the network. Fig. 1 shows the MHR network model.

In MHR network, the data can be transmitted directly by the BS to the MSs or be relayed through the RS, which transmits at relatively lesser power. The multi hop transmission process from the BS to the MSs through the RS results in a reduced hop distance between a pair of

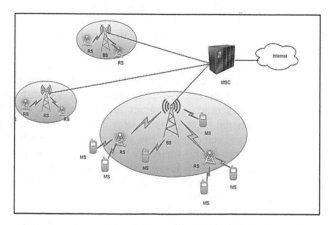

Figure 1.   MHR network model.

actively communicating devices at the same time extends the coverage area of the BS. Owing to the reduced transmission distance, the power consumption and path loss is reduced, which ultimately results in an improved data throughput. An RS transmit the signal at relatively low power than the BS and hence requires lower deployment cost. The distance between any two communication nodes varies with respect to the deployment scenario. As the distance increases, path loss increases and the effective signal to noise ratio (SNR) decreases and vice versa. The system opts for a higher order modulation if the distance between stations is less and will use lower order modulation if the distance is more. Therefore, the distance between the participating stations plays a vital role in determining the transmission mode and the data rate of the link. Hence the MHR networks uses an adaptive modulation coding (AMC) scheme in which the data rates are adaptively allocated according to the channel conditions between the communicating stations. Seven modulation coding schemes are considered for a geographical area which divided into seven non-overlapping regions. The seven distances from the center to the seven regions correspond to seven SNRs. According to the free space propagation model, the SNR of each link is given by

$$SNR = 10 \log_{10}\left(\frac{p_1}{p_n}\left(\frac{c}{4\pi fr}\right)^2\right) \tag{1}$$

where p_1, p_n, f, c and r represent the transmitted power, thermal noise power, center frequency, velocity of light and the distance between the transmission stations respectively. The modulation schemes, coding ratios and data rates for various distances and received SNRs are enumerated in Table 1.

Let D_{B_R} denotes the data rate between BS and RS and D_{R_M} indicates the data rate between RS and MS. The throughput for indirect transmission between BS and MS is given as,

$$D_{B\_R\_M} = \frac{D_{B\_R} \cdot D_{R\_M}}{D_{B\_R} + D_{R\_M}} \tag{2}$$

The transmission data rate of MS is decided based on throughput oriented scheme which is given as,

$$D = Max(D_{B\_R\_M}, D_{B\_M}) \tag{3}$$

The average system capacity is given as,

$$\bar{C} = \frac{\sum_{k=1}^{N} D(k)}{N} \tag{4}$$

Table 1. AMC transmission mode in Jau and Lin, (2015).

Mode	Modulation	Coding rate	Received SNR (dB)	Data rate (Mbps)	Distance (km)
1	BPSK	1/2	3.0	1.269	3.2
2	QPSK	1/2	6.0	2.538	2.7
3	QPSK	3/4	8.5	3.816	2.5
4	16-QAM	1/2	11.5	5.085	1.9
5	16-QAM	3/4	15.0	7.623	1.7
6	64-QAM	2/3	19.0	10.161	1.3
7	64-QAM	3/4	21.0	11.439	1.2

where $D(k)$ is the transmission data rate of the ith MS and N is the total number of MSs in the geographical area.

3 ADAPTIV NEURO FUZZY BASED BS AND RS DEPLOYMENT AND SELECTION PHASE

A two-phase effective BS and RS selection and deployment scheme is explored in the present study to increase the overall power efficiency and system performance. An adaptive neuro fuzzy based selection scheme makes an adaptive decision for the deployment of nodes from the candidate positions.

3.1 *BS and RS positions selection phase*

In the proposed method, we consider CR and TR as the adaptive neuro fuzzy input sets for each candidate positions of the BS and RS. Moreover fuzzy based selection strategies are utilized for the deployment of BS and RS in Vincent et al. (2017).

The CR of ith BS and RS candidate position can be calculated as,

$$CR_i = \frac{1}{N}(NC_i) \qquad (5)$$

where NC_i is the number of MSs covered under ith RS candidate position and N is the number of MSs in the geographical area.

The fuzzy sets for CR of ith RS candidate positions takes the linguistic variables like low, medium, high. The corresponding membership function plot is shown in Fig. 2.

TR_i of ith RS candidate position is given by,

$$TR_i = \frac{A_{d,i} - A_{t,i}}{\max\{A_{d,i}, A_{t,i}\}} \qquad (6)$$

where $A_{d,i}$ and $A_{t,i}$ are the average data transmission rate and average traffic demand of MSs covered by the ith RS candidate position respectively. CR and TR are the two fuzzy inputs which are considered in the conventional fuzzy based BS deployment schemes [Vincent et al. (2017)].

Similarly, the fuzzy sets for TR of ith RS candidate positions takes the linguistic variables like negative, center and positive. The corresponding membership function plot is shown in Fig. 3. The variations of the crisp values associated with the output with respect to inputs are shown in Fig. 4. It can be seen that the output is low in magnitude for smaller values of

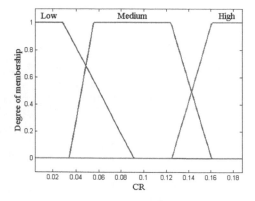

Figure 2. Membership diagram of CR.

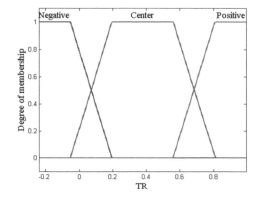

Figure 3. Membership diagram of TR.

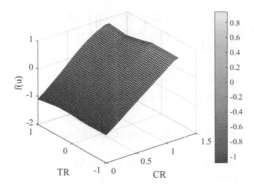

Figure 4. Surface plot of Output versus CR-TR combinations.

the combinations and increases with the increase in their values. The results obtained are in accordance with the fuzzy rules.

3.2 *BS an RS deployment phase*

The budget constraint for a total deployment budget (T_{DB}) is expressed as,

$$\sum_{i=1}^{u} DC_{BS} \cdot \beta_i + \sum_{i=1}^{v} DC_{RS} \cdot \gamma_i \leq T_{DB} \tag{7}$$

where DC_{BS} and DC_{RS} is the deployment cost of **BS** and **RS** respectively. u and v are the number of candidate positions of **BS** and **RS** respectively. β_i and γ_i are defined as follows:

$$\beta_i = \begin{cases} 1, & \text{if BS is deployed in the } i^{th} \text{ candidate position} \\ 0, & \text{otherwise} \end{cases} \tag{8}$$

$$\gamma_i = \begin{cases} 1, & \text{if RS is deployed in the } i^{th} \text{ candidate position} \\ 0, & \text{otherwise} \end{cases} \tag{9}$$

The coverage constraint is expressed as,

$$\bar{C} \geq E_{CR} \tag{10}$$

where E_{CR} is the expected **CR**.

The joint **BS** and **RS** deployment scheme based on the budget constraint is illustrated as follows:

Input: Candidate positions of BS, Candidate positions of RS, positions of the MSs, T_{DB}
Output: BS and RS deployment locations, CR of the MSs and average power consumed
Initialization: M = 0 (number of BS positions selected) and temp result = null.

Step 1: Load the training data from workspace to adaptive neuro fuzzy inference engine
Step 2: Set the membership numbers and patterns then test the trained data
Step 3: View the generated membership functions and rules generated.
Step 4: Output is calculated for all the candidate position of BSs.
Step 5: Find out the maximum number of BSs that can be deployed based on the budget constraint. Identify the candidate locations with highest output for BS deployment
Step 6: Deploy BS in the candidate positions with highest output. Output for all the candidate position of RSs in the cell edge of the deployed BSs is calculated.
Step 7: Select the RS candidate location with highest output and check for budget constraint in (7).

Step 8: If the condition in (7) is satisfied, go to step 6. Otherwise, go to step 4 and continue until either the budget constraint is satisfied or all the RSs within the coverage of selected BS are deployed.

Step 9: The temp result stores the final positions of BSs and RSs. In addition to that, the average CR, and average throughput per user of MSs are calculated.

The joint BS and RS deployment scheme based on coverage constraint is illustrated as follows:

Input: Candidate positions of BSs, Candidate positions of RSs, positions of the MSs, expected CR (E_{CR}).

Output: BS and RS deployment locations, CR of the MSs and average power consumed.

Initialization: M = 0 (number of BS positions to be selected) and temp result = null.

Step 1: Load the training data from workspace to adaptive neuro fuzzy inference engine

Step 2: Set the membership numbers and patterns then test the trained data

Step 3: View the generated membership functions and rules generated

Step 4: Output is calculated for all the candidate position of BSs.

Step 5: Select the BS with highest output crisp value. Output for all the RSs at the cell edge of selected BS is calculated.

Step 6: Select the RS with highest output and check for coverage constraint

Step 7: If the coverage constraint in (10) is satisfied, go to step 6. Otherwise, go to step 3 and continue until (10) gets satisfied or all the RSs within the coverage of selected BS are deployed.

Step 8: Increment M = M + 1 and go to step 2.

Step 9: The temp result stores the final positions of BSs and RSs. In addition to that the average CR and average throughput per user of MSs are calculated.

4   SIMULATION RESULTS AND ANALYSIS

The performance of the proposed method is analysed using MATLAB 2015a tool. The proposed fuzzy based BS and RS deployment scheme is compared with uniform clustering based deployment method. Chang, J. and Y. Lin (2014) has been proved that the coverage and throughput performance of uniform clustering scheme is far better than JBRP scheme. A practical wireless cellular network environment is simulated. The assumptions and simulation parameters taken are as follows:

- The geographic area is square of size 10 km × 10 km.
- The system consists of BSs, RSs and MSs.
- The candidate positions of BSs and RSs are randomly selected within the geographic area.
- The number of candidate position of BS is taken as 6 and the number of candidate position for RSs are varied from 20 to 60.
- MSs are uniformly distributed in the geographic area.
- The coverage radius of BS and RS are 3.2 km and 1.9 km respectively.
- The deployment cost of a BS and a RS is 9 and 3 units respectively.
- The traffic demand for each MS is uniformly selected between 5 and 15Mbps.
- The data rate between MS and BS or between MS and RS are calculated based on the Table 1

Fig. 5 and Fig. 6 shows the deployment result obtained for a sample simulation environment based on budget constraint of 50 units and coverage constraint of 80% respectively. From the simulation results, it can be inferred that the BSs and RSs are judiciously deployed such that maximum MSs are covered. In Fig 5, the proposed scheme achieves a coverage of more than 80% at a budget cost of 42 units, which is calculated using [7] and is less than the target budget. In Fig. 6, more number of BSs and RSs are deployed in order to achieve the expected coverage of 80% without any budget constraint.

Fig. 7 shows the average throughput comparison between the proposed fuzzy based scheme and uniform clustering schemes for the deployment budget of 50 units. It is observed

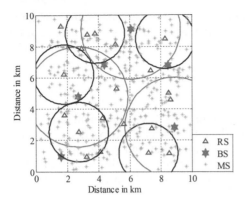

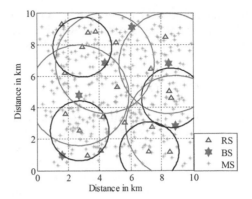

Figure 5. Network deployment scenario for a budget of 50 units.

Figure 6. Network deployment scenario for a expected coverage of 80%.

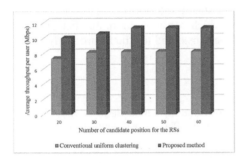

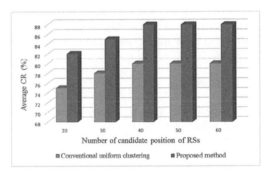

Figure 7. Average throughput per user vs. Number of candidate position for the RSs with TDB of 50 Units.

Figure 8. Average CR vs. Number of candidate position for the RSs with TDB of 50 Units.

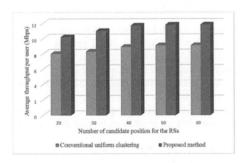

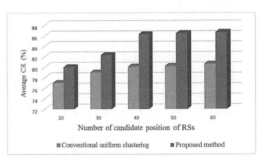

Figure 9. Average throughput per users vs. Number of candidate positions for the RSs with ECR of 80%.

Figure 10. Average CR vs. Number of candidate position for the RSs with ECR of 80%.

that the increase in the number of RSs will increase the throughput per user initially. But it is noticed that after certain number of RSs, the average system throughput remains constant due to the co-channel interference between RSs and between BSs and RSs. Fig. 8 shows the average CR for the deployment budget of 50 units.

The increase in the number of candidate positions of BS and the deployment of RSs at the cell edge will increase the CR. But there is no significant improvement in terms of CR, when the number of candidate locations of RS is above 50.

Fig. 9 and Fig. 10 shows the comparison results of average throughput per user (Mbps) and CR for a E_{CR} of 80%. Initially average throughput per user and CR will increase with the increase in the number of RSs. But, it maintains constant after certain number of RSs. The increase in the number of RSs, increases the interference, which also reduces signal to interference plus noise ratio (SINR). This makes the system to choose lower burst profiles.

5 CONCLUSION

In this paper, an adaptive neuro fuzzy based joint BS and RS deployment scheme is proposed for next generation wireless communication. The proposed deployment scheme is formulated for maximizing the network coverage, throughput and power efficiency of the system. The performance of the proposed scheme is compared with the conventional uniform clustering based scheme. Our proposed scheme satisfies both budget and coverage constraints. The proposed method shows improved performance than conventional scheme for all the considered combinations. The simulation results prove that the proposed scheme is computationally simple and sustainable for different channel and path loss conditions. Since the selection and deployment is carried out based on adaptive neuro fuzzy logic, the proposed method can be considered as a more suited solution for real-time channel conditions. An adaptive neuro fuzzy based scheme, which considers more number of input parameters like interference and CO_2 emission can be considered for the future study.

REFERENCES

[1] Akyildiz, I.F, D.M. Gutierrez-estevez, R. Balakrishnan and E. Chavarria-reyes (2014). LTE-Advanced and the evolution to Beyond 4G (B4G) systems, Physical Communication, Vol. 10, pp. 31–60.
[2] Chang, B.J., Y.H. Liang and S.S. Su (2015). Analyses of Relay Nodes Deployment in 4G Wireless Mobile Multihop Relay Networks, Wireless Personnel Communication, Vol. 83, No. 2, pp. 1159–1181.
[3] Chang, J. and Y. Lin (2014). A clustering deployment scheme for base stations and relay stations in multi-hop relay networks, Computers and Electrical Engineering, Vol. 40, pp. 407–420.
[4] Fettweis., G., and P. Rost, (2011). Green communications in cellular networks with fixed relay nodes, Cambridge University Press.
[5] Jau-Yang. C and Lin, Y.S. (2015). An Efficient Base Station and Relay Station Placement Scheme for Multi-hop Relay Networks. Wireless Personal Communications, Vol. 82, No. 3, pp. 1907–1929.
[6] Laneman, J.N. and G.K. Womell (2000). Energy-Efficient Antenna Sharing and Relaying for Wireless Networks, IEEE Wireless Communication Networking Conference, pp. 7–12.
[7] Lu, H. and W. Liao (2009). Joint Base Station and Relay Station Placement for IEEE 802.16 j Networks, IEEE Global Telecommunication Conference, pp. 1–5.
[8] Vincent, M., K.V. Babu, M. Arthi and P. Arulmozhivarman (2017). Corrigendum to "Power-aware fuzzy based joint base station and relay station deployment scheme for green radio communication" [J. Sustain. Comput.: Inform. Syst. 13 (2017) 1–14].
[9] Yang, Y., H. Hu, J. Xu and G. Mao (2009). Relay Technologies for WiMAX and LTE-Advanced Mobile Systems, IEEE Communication Magazine, Vol. 47, pp. 100–105.

International Conference on Emerging Trends in Computer Science (ICETICS)

Emerging Trends in Engineering, Science and Technology for Society,
Energy and Environment – Vanchipura & Jiji (Eds)
© 2018 Taylor & Francis Group, London, ISBN 978-0-8153-5760-5

A survey on sentence similarity based on multiple features

Riswana K. Fathima & C. Raseek
Computer Science and Engineering Department, Government Engineering College,
Palakkad, Kerala, India

ABSTRACT: Semantic Textual Similarity (STS) is to decide the level of semantic comparability between pairs of sentences. STS assumes an essential part in Natural Language Processing errands, which has drawn considerable attention in the field of research in recent years. This survey discusses multiple features including word alignment-based similarity, sentence vector-based similarity, and sentence constituent similarity to assess the correctness of sentence pairs.

1 INTRODUCTION

Semantic Textual Similarity (STS) is basically a measure used to compute the similarity between two textual snippets based on the likeliness of their meaning. Measuring semantic similarity is a difficult assignment since it is simple express a comparable idea in different ways. Therefore, STS is a deep natural language understanding problem. STS has been generally utilized as a part of natural language processing tasks such as machine translation (MT), summarization, generation, question answering (QA), short answer grading, semantic search, dialog and conversational frameworks and so on.

Previous researchers on semantic text similarity have been centered on records and paragraphs, while correlation protests in numerous NLP assignments are writings of sentence length, for example, Video descriptions, News headlines, and beliefs, etc. In this paper, we examine semantic similarity between two sentences. Given two input textual snippets, we have to consequently choose a score that demonstrates their semantic similarity. In general, the fundamental undertaking is to process semantic similarity for the given two English sentences in the range [0, 5], where the score increments with similarity (i.e., 0 indicates no similarity and 5 demonstrates indistinguishable). Similarity score with the explanation for some English sentences is shown in Figure 1. The assessment metric utilized is the Pearson correlation coefficient.

STS is also firmly identified with textual entailment (TE) (Dagan et al., 2006), paraphrase recognition (PARA) (Dolan et al., 2004) and semantic relatedness. All it differs from its tasks. Textual entailment recognition is the undertaking of choosing, given two text fragments, regardless of whether the significance of one text is entailed (can be gathered) from another text. On account of TE, the equivalence is directional. eg: an auto is a vehicle, however a vehicle isn't really an auto.

Paraphrase Recognition is a task expects to distinguish in the event that two sentences have a similar importance of using different words. Two important aspects of paraphrase is that: same meaning and different words. These two concepts are quite intuitive, but difficult to formalize. For example, consider the sentences below,

1. Hamilton Construction Company built the new extension.
2. The new extension was built by Hamilton Construction Company.

From the above example, could identify that (1) and the (2) are actually paraphrased. Since textual entailment and paraphrase detection catches degrees of meaning overlap rather than making binary classifications of particular relationships. Essentially, semantic relatedness communicates an evaluated semantic relationship. It is nonspecific about the possibility

5	*The two sentences are completely equivalent, as they mean the same thing.*
	The bird is bathing in the sink. Birdie is washing itself in the water basin.
4	*The two sentences are mostly equivalent, but some unimportant details differ.*
	Two boys on a couch are playing video games. Two boys are playing a video game.
3	*The two sentences are roughly equivalent, but some important information differs/missing.*
	John said he is considered a witness but not a suspect. "He is not a suspect anymore." John said.
2	*The two sentences are not equivalent, but share some details.*
	They flew out of the nest in groups. They flew into the nest together.
1	*The two sentences are not equivalent, but are on the same topic.*
	The woman is playing the violin. The young lady enjoys listening to the guitar.
0	*The two sentences are completely dissimilar.*
	The black dog is running through the snow. A race car driver is driving his car through the mud.

Figure 1.   Similarity scores with explanation.

of the relationship with contradicting material so far being a contender for a high score. For example, "night" and "day" are abundantly related yet not particularly comparable.

At the beginning of 2012, many efforts are undertaken to have STS over English sentence pairs. STS shared undertaking has been held yearly since 2012, which provides a platform to have new algorithms and models. During this period of time, diverse similarity method and datasets have been explored. One of such similarity signals emerged is a Sultan's alignment based method. Also found that deep learning is becoming an upcoming feature set for such an evaluation mechanism. All things considered, looks into are going ahead to discover the best performing feature set.

This paper is sorted out as takes after: Section 2 gives the methods used so far for STS. Section 3 discusses feature sets for STS and provides a comparison between them. Section 4 provides a discussion on datasets used. Section 5 discusses various future research directions and Section 6 gives a brief concluding comment.

2   METHOD

The methods used so far can be separated into three general classes: alignment approaches, vector space approaches, and machine learning approach (Cheng et al., 2016). Figure 2, shows the classification of methodologies based on approaches used.

Alignment approaches align words or phrases in a sentence pair and after that take similarity measure as the quality or scope of alignments. Vector space approaches speak to sentences as bag-of-words vectors and here the similarity measure will be the vector similarity. Machine learning approaches consolidate diverse similarity measures and feature utilizes supervised machine learning models.

This survey aims to consider the evidence from those three set of categories to measure semantic text similarity between two sentence pairs. Basically, from sentence pairs, we extract

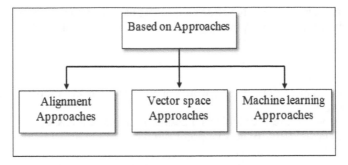

Figure 2.   Classification of methodologies based on the approach used.

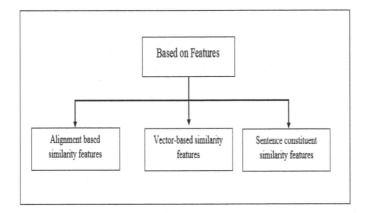

Figure 3.   Classification based on features.

the features called alignment-based similarity features, vector-based similarity features, and sentence constituent similarity features. From the study it found that the extracted feature set are then joined through a Support Vector Regression (SVR) model to convey the similarity score between the sentence sets. Figure 3, depicts the classification of feature sets used.

3   FEATURE EXTRACTION

In the following describe how to extract the different types of features.

3.1   *Alignment-based similarity feature*

Alignment approaches align words or phrases in a sentence pair and after that take similarity measure as the quality or coverage of alignments. Among this Sultan aligner is the one of the open source unsupervised word alignment tool (Sultan et al., 2014). The aligner adjusts related words in two sentences in light of two sets of properties of the words;

1. Whether they are semantically comparative
2. Whether they happen in comparative semantic contexts in the particular sentences

The previous one settles on choice in light of the data gave by the Paraphrase Database (PPDB) (Ganitkevitch et al., 2013). On account of the last mentioned, contextual similarity for a word match is computed as the aggregate of the word similarities for each combine of words with regards to them.

In order to have an overview of Sultans aligner let us denote the sentence being aligned as $S^{(1)}$ and $S^{(2)}$. Whereas here it just portrays content word alignment. It finds outs the word pair $w_i^{(1)} \in S^{(1)}$ and $w_j^{(2)} \in S^{(2)}$ such that:

1. $w_i^{(1)}$ and $w_j^{(2)}$ have non-zero semantic comparability, sim_{Wij}. The figuring of word similarity sim_{Wij} is as follows;
    i. On the off chance that the two words or their lemmas are indistinguishable, at that point $sim_W = 1$.
    ii. If the two words are accessible as a couple in the lexical XXXL corpus of the Paraphrase Database (PPDB), at that point sim_W.
    iii. For any other word pair, $sim_W = 0$.
2. Sometimes semantic context of word pair $w_i^{(1)}$ and $w_j^{(2)}$ have some similarity, denoted as sim_{Cij} implies the contextual similarity. $\left(w_i^{(1)}, w_j^{(2)}\right)$ is figured as the total of the word similarities for each combine of words with regards to $\left(w_i^{(1)}, w_j^{(2)}\right)$.
3. Finally, performs one-to-one word alignment.

This is how Sultan aligner aligns related words to STS score. We can summarize that STS score will be the extent of aligned content words in two input sentences. Sultan aligner found a STS framework that has effortlessness, high precision and speed and furthermore demonstrated that it can be sent with no supervision. Difficulty behind it is inability to demonstrate the semantics of units bigger than words.

The study revealed that extensive feature could be extracted from word alignment to have semantic textual similarity. One such plan is proposed by Hanig et al. in ExB Themis that enforced associate degree alignment rule supported named entities, temporal expressions, measuring expressions and dedicated negation handling. Like Sultan's aligner this also follows a strict sequential order;

1. Align **named entities** nearer to each other. It aligns all name pairs that share no less than one indistinguishable token. Eg: Michael in one content and both Michael Jackson and Michael Schumacher in the other, each carries Michael as the identity token and therefore it gets aligned.
2. Normalized **temporal expressions** are aligned only if it shares just on the off chance that it has a similar purpose of time or a similar time interval *(e.g. 13:05 and 1.05 pm)*.
3. **Measurement expressions** are adjusted iff they convey same outright esteem set *(e.g. $200 k and $200.000)*.
4. **Arbitrary token** sequence is then aligned. Which requires different advances and tedious as well. Make utilization of synonym-lookups and overlooks case information, accentuation characters and symbols. Matches expressions like *long term and long-term* and *US and United States*. Language assets like WordNet and ConceptNet to acquire data of synonymy (e.g. does and do), antonymy (e.g. doesn't and does), and hypernymy.
5. Aligns **negations** *(e.g. You are a Christian. versus In this way you are not a Christian)*.
6. Staying content words are aligned based on cosine similarity on word2vec vectors.

ExB Themis turned out to be the best multilingual system—Spanish and English. Effectively can be adapted to further languages. Likewise extensive feature extraction from word alignments is an extremely hearty approach. Though it couldn't advance improved phrase similarity computation.

3.2 *Vector-based similarity features*

The framework extracts two vector-based similarity features in light of the two-word vectors subsequently shaped. So for the given two sentence pairs first it generates two vector representation. At that point register the cosine similarity between those vectors to have their similarity score.

The primary word vector is learned by the Skip-Gram framework (Baroni et al. 2014). Skip-Gram model makes use of current words to predict the surrounding window of context words. Do the better job for infrequent words.

The second type of word vectors uses Latent Dirichlet Allocation (LDA) which is a vector space model (Liu Y et al., 2015). LDA is an example of the topic model. It explains why some part of data is similar. In LDA first words gathered into documents and each document is a mix of the humble number of topics. For example, a document wants to be classified as CAT-related and DOG-related. Then compare the probability of generating various words namely milk, meow and kitten then classified to CAT-related. Similarly if words like bone, bark and puppy then it classifies to DOG-related. These are determined using term co-occurrence. Gibbs algorithm (Griffiths & Steyvers, 2004) is the one of the main algorithm used to have topic related classifications.

3.3 *Sentence constituent similarity features*

Cheng et al., proposed to extract the accompanying arrangements of sentence constituent similarity features:

- **Subject similarity:** When it convey the subject similarity between two sentences are of the same then assign 1 for subject similarity else 0.
- **Predicate similarity:** Assign predicate similarity 1 if the predicate of two sentences are of same, otherwise 0.
- **Object similarity:** Assign object similarity 1 if the object of two sentences are of same, otherwise 0.
- **Complement similarity:** Assign complement similarity 1 if the complements of two sentences are of same, otherwise 0.
- **Named entity similarity:** First, it checks whether they are aligned time sets, area sets or individual matches between looked at sentences. If suppose area could be aligned between the given sentence pairs means assign 1 for the named entity similarity, otherwise 0.
- **Keyword similarity:** Assign 1 as keyword similarity view of the proportion of aligned keywords, otherwise 0.

4 DISCUSSION ON DATASETS

The data set used for the task is of a combination of different settings. The survey discusses three different settings. Five sets of the dataset in (Cheng et al., 2016) consist of dataset namely Questions and answers (Q&A) Answer-Answer data set, Headlines data set, Plagiarism Detection data set, Post Edited Machine Translations data set and Q&A Question-Question data set. These datasets were taken from the previous four years (2012–2015) of the SemEval English STS assignment.

Sultan et al. alignment mechanism in (Sultan et al., 2014) was based on six datasets in particular deft-forum, deft-news, headlines, images, OnWN and tweet-news. These mentioned datasets are from SemEval STS 2014. The sentences were gathered from a variety of sources like discussion forums, news articles, and news headlines and so on. Consist each dataset with a minimum of 750 sentence pairs.

ExB Themis (Hanig et al., 2015) is a multilingual STS system for English and Spanish languages. Therefore dataset on English and Spanish are utilized for the task. For English accessible dataset proposed by Agirre et al. in 2012–2014 are utilized. These datasets are taken from 2012–2014. The data set collection in (Agirre et al., 2012) are namely MSRpar, MSRvid, OnWN, SMTnews, and SMTeuroparl. Similarly, dataset in (Agirre et al., 2013) comprises of HDL, FNWN, OnWN, SMT and TYPED. Whereas in (Agirre et al., 2014) it is of namely HDL, OnWN, OnWN, Deft-news, Images and Tweet-news. ExB Themis formed a dataset from above mentioned as a new combination which comprises of the domain namely forum, students, belief, headlines and images. For Spanish has taken datasets from Wikipedia and Newswire. Table 1, shows the overview of the feature sets discussed so far and the datasets used by them in a nutshell.

The performance on each data set is evaluated using the metric called Pearson Correlation Coefficient. It found that (Cheng et al., 2016) proved a mean of 0.69996 Pearson correlation.

Table 1. Overview of the compared features and the datasets used.

| Types of features | Proposed approaches | |
	Model	Data set
Alignment based similarity features	F. Cheng et al.	Recent years (2012–2015) SemEval English STS task
	Sultan et al.	Dataset in SemEval STS 2014
	Hanig et al.	Available SemEval English dataset from 2012–2014
Vector-based similarity features	F. Cheng et al.	Recent years (2012–2015) SemEval English STS task
Sentence constituent similarity features	F. Cheng et al.	Recent years (2012–2015) SemEval English STS task

Sultan aligner (Sultan et al., 2014) found a weighted mean Pearson correlation between 0.7337 to 0.7610. In ExB Themis (Hanig et al., 2015) found a mean of 0.7942 Pearson correlation for English dataset and a mean of 0.6725 Pearson correlation for Spanish data.

5 FUTURE RESEARCH DIRECTIONS

Semantic Textual Similarity (STS) drawn a considerable attention in recent years. It reveals that STS can be improved further to a greater extent. So far discussed sentence similarity of monolingual ones. Therefore it shows an open idea to extend the STS in multilingual semantic similarity.

The existing study shows that it is difficult to assign semantics of units larger than words. In this way, it is an ideal opportunity to create calculations that can adjust to necessities looked by different information areas and application.

6 CONCLUSION

Semantic Textual Similarity measures the extent of similarity between two sentences. Different features including alignment-based similarity features, vector-based similarity features, and sentence constituent similarity features are used to have the semantic score between the sentences. Pearson correlation coefficient is used as the evaluation metric. Sentence similarity based on support vector regression shows the best performance among all other evaluation strategies with the feature sets.

REFERENCES

Agirre, E., Banea, C., Cardie, C., Cer, D.M., Diab, M.T., Gonzalez-Agirre, A., ... & Wiebe, J. (2014). SemEval-2014 Task 10: Multilingual Semantic Textual Similarity. *SemEval@ COLING*, 81–91.
Agirre, E., Banea, C., Cardie, C., Cer, D.M., Diab, M.T., Gonzalez-Agirre, A., ... & Rigau, G. (2015). SemEval-2015 Task 2: Semantic Textual Similarity, English, Spanish and Pilot on Interpretability. *SemEval@ NAACL-HLT*, 252–263.
Agirre, E., Cer, D., Diab, M., Gonzalez-Agirre, A., & Guo, W. (2013). sem 2013 shared task: Semantic textual similarity, including a pilot on typed-similarity. *In* SEM 2013: The Second Joint Conference on Lexical and Computational Semantics. Association for Computational Linguistics.*
Agirre, E., Diab, M., Cer, D., & Gonzalez-Agirre, A. (2012). Semeval-2012 task 6: A pilot on semantic textual similarity. In *Proceedings of the First Joint Conference on Lexical and Computational Semantics-Volume 1: Proceedings of the main conference and the shared task, and Volume 2: Proceedings of the Sixth International Workshop on Semantic Evaluation*, Association for Computational Linguistics, 385–393.

Bar Haim, R., Dagan, I., Dolan, B., Ferro, L., Giampiccolo, D., Magnini, B., & Szpektor, I. (2006). The second pascal recognising textual entailment challenge.

Bär, D., Biemann, C., Gurevych, I., & Zesch, T. (2012). Ukp: Computing semantic textual similarity by combining multiple content similarity measures. In *Proceedings of the First Joint Conference on Lexical and Computational Semantics-Volume 1: Proceedings of the main conference and the shared task, and Volume 2: Proceedings of the Sixth International Workshop on Semantic Evaluation*, 435–440.

Baroni, M., Dinu, G., & Kruszewski, G. (2014). Don't count, predict! A systematic comparison of context-counting vs. context-predicting semantic vectors. *In ACL (1)*, 238–247.

Blei, D.M., Ng, A.Y., & Jordan, M.I. (2003). Latent dirichlet allocation. *Journal of machine Learning research*, 3(Jan), 993–1022.

Corley, C., & Mihalcea, R. (2005, June). Measuring the semantic similarity of texts. In *Proceedings of the ACL workshop on empirical modeling of semantic equivalence and entailment*. Association for Computational Linguistics, 13–18.

Dagan, I., Dolan, B., Magnini, B., & Roth, D. (2009). Recognizing textual entailment: Rational, evaluation and approaches. *Natural Language Engineering*, 15(4), i-xvii.

Fu, C., An, B., Han, X., & Sun, L. (2016). ISCAS_NLP at SemEval-2016 Task 1: Sentence Similarity Based on Support Vector Regression using Multiple Features. *SemEval@ NAACL-HLT*, 645–649.

Ganitkevitch, J., Van Durme, B., & Callison-Burch, C. (2013, June). PPDB: The Paraphrase Database. *HLT-NAACL*, 758–764.

Han, L., Kashyap, A.L., Finin, T., Mayfield, J., & Weese, J. (2013). UMBC_EBIQUITY-CORE: Semantic Textual Similarity Systems. * *SEM@ NAACL-HLT*, 44–52.

Han, L., Martineau, J., Cheng, D., & Thomas, C. (2015). Samsung: Align-and-Differentiate Approach to Semantic Textual Similarity. *SemEval@ NAACL-HLT*, 172–177.

Hänig, C., Remus, R., & De La Puente, X. (2015). ExB Themis: Extensive Feature Extraction from Word Alignments for Semantic Textual Similarity. *SemEval@ NAACL-HLT*, 264–268.

Islam, A., & Inkpen, D. (2008). Semantic text similarity using corpus-based word similarity and string similarity. *ACM Transactions on Knowledge Discovery from Data (TKDD)*, 2(2), 10.

King, M., Gharbieh, W., Park, S., & Cook, P. (2016). UNBNLP at SemEval-2016 Task 1: Semantic Textual Similarity: A Unified Framework for Semantic Processing and Evaluation. *SemEval@ NAACL-HLT*, 732–735.

Li, Y., McLean, D., Bandar, Z.A., O'shea, J.D., & Crockett, K. (2006). Sentence similarity based on semantic nets and corpus statistics. *IEEE transactions on knowledge and data engineering*, 18(8), 1138–1150.

Liu, Y., Liu, Z., Chua, T.S., & Sun, M. (2015). Topical Word Embeddings. *AAAI*, 2418–2424.

Mikolov, T., Chen, K., Corrado, G., & Dean, J. (2013). Efficient estimation of word representations in vector space. ar*Xiv preprint* ar*Xiv*:1301.3781.

Pilehvar, M.T., Jurgens, D., & Navigli, R. (2013). Align, Disambiguate and Walk: A Unified Approach for Measuring Semantic Similarity. *IACL (1)*, 1341–1351.

Šarić, F., Glavaš, G., Karan, M., Šnajder, J., & Bašić, B.D. (2012). Takelab: Systems for measuring semantic text similarity. In *Proceedings of the First Joint Conference on Lexical and Computational Semantics-Volume 1: Proceedings of the main conference and the shared task, and Volume 2: Proceedings of the Sixth International Workshop on Semantic Evaluation*. Association for Computational Linguistics, 441–448.

Srivastava, S., & Govilkar, S. (2017). A Survey on Paraphrase Detection Techniques for Indian Regional Languages. *International Journal of Computer Applications*, 163(9).

Sultan, M.A., Bethard, S., & Sumner, T. (2014). DLS @ CU: Sentence Similarity from Word Alignment. In *SemEval@ COLING*, 241–246.

Sultan, M.A., Bethard, S., & Sumner, T. (2015). DLS @ CU: Sentence Similarity from Word Alignment and Semantic Vector Composition. In *SemEval@ NAACL-HLT*, 148–153.

Emerging Trends in Engineering, Science and Technology for Society,
Energy and Environment – Vanchipura & Jiji (Eds)
© *2018 Taylor & Francis Group, London, ISBN 978-0-8153-5760-5*

A review of extractive sentence compression

Shirin A. Fathima & C. Raseek
Computer Science and Engineering Department, Government Engineering College,
Palakkad, Kerala, India

ABSTRACT: Sentence compression makes a sentence shorter whilst preserving its meaning and grammar; it is applicable in different fields such as text summarization. Among the
various methods to shorten sentences, extractive sentence compression is still far from being
solved. Reasonable machine-generated sentence compressions can often be obtained by considering subsets of words from the original sentence. This paper presents a survey of the
different approaches to extractive sentence compression.

1   INTRODUCTION

Sentence compression is the process of shortening a text whilst keeping the idea of the original sentence. It is in such a way that the grammar and structure of the sentence is greatly
simplified, while the underlying meaning and information remains the same. It makes the
necessary to speak by eliminating unnecessary words or phrases. Sentence compression is
also known as text simplification or summarization.

Sentence compression is an important area of research because it is a backbone for different Natural Language Processing (NLP) applications. Normally, human languages contain
complex compound constructions which causes difficulties in automatic modification, classification or processing of human-readable text. Today, sentence compression techniques are
widely used in many industries, mainly as a part of data mining and machine learning. It has
been widely used for displaying on small screens (Corston-Oliver, 2001), such as in television
captions, automatic title generation (Vandeghinste & Pan, 2004), search engines, topic detection, summarization (Madnani et al., 2007), machine translation, paraphrasing, and so forth.

There are various techniques for sentence compression, including word or phrase removal,
using shorter paraphrases, and common-sense knowledge. There are, primarily, two types
of sentence compression: extractive and abstractive. In the extractive method, objects are
extracted from the source sentence without modifying the objects themselves. The main idea
here is to find the subset of important words that contain the information of the entire sentence. An example of this is key phrase extraction, where the goal is to select individual words
or phrases that are important (without modifying them) to create a short sentence; it should
preserve the meaning of the original sentence. Abstractive sentence compression considers
semantic representation of the sentence in order to make it simpler.

The task of extractive sentence compression is very complex. It is not simply shortening
a sentence; the properties of the original sentence should be preserved. The performance of
a technique depends upon the compression rate, the grammar and keeping the important
words from the original sentence. Various approaches for extractive sentence compression
include: generative noisy channel models (Knight & Marcu, 2002); tree transduction model
(Knight & Marcu, 2002; Cohn & Lapata, 2007, 2009; Yao et al., 2014); structured discriminative compression model; Long-Short Term Memory (LSTM) (Clarke & Lapata, 2008); ILP
(Yao & Wan, 2017; De Belder & Moens, 2010; Wang et al., 2013). In addition, the different
techniques make use of machine-learning algorithms such as the Maximum Entropy Model
and Support Vector Machines (SVMs).

This survey discusses different methodologies of extractive sentence compression and makes a competitive study by considering the advantages and disadvantages of each. This study is helpful because sentence compression has a wide range of applications and many of them favor extractive compression. This survey will assist in the selection of the right techniques for a particular application.

This paper is organized as follows. Section 2 discusses the machine-learning approach, the decision-tree model, the Integer Linear Programming (ILP) approach, and the LSTM approach for extractive sentence compression and provides. Section 3 conducts a comparative critical analysis of these approaches. Section 4 discusses the datasets and evaluation approaches and makes a comparative study between them. Section 5 gives a brief conclusion.

2 LITERATURE REVIEW

Different approaches are put forward for extracting the important words from a sentence in order to form a new compressed sentence. Figure 1 shows the approaches discussed here for extractive sentence compression.

2.1 Tree transduction method

One of the common approaches for extracting the important words from a sentence is the tree to tree transduction method, which typically makes use of Synchronous Context-Free Grammar (SCFG). Here, the compression is generated by pruning the dependency tree or constituency tree (Cohn & Lapata, 2009; Knight & Marcu, 2000; Berg-Kirkpatrick et al., 2011; Filippova & Altun, 2013). In Cohn and Lapata (2009), the source sentence and target sentence are in the form of trees. Synchronous grammar will license more than one compression for a source tree where the SCFG is also in the form of a tree. This rule, in the form of tree shown in Figure 2 (Cohn & Lapata, 2009), helps to eliminate different branches of the source tree in order to generate the target tree. Each grammar has a score which helps to calculate the score of the candidates.

The noisy channel model and the statistical model are implemented in the work of Knight et al. (2011). The statistical model makes use of Probabilistic Context-Free Grammar (PCFG) on the tree representation in order to extract the relevant phrase. Firstly, the input sentence is parsed as per Collins (1997). Subsequently, various small output trees are hypothesized and ranked using PCFG.

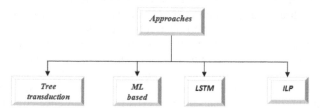

Figure 1.   Classification of methodologies.

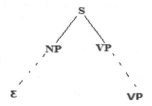

Figure 2.   Example of SCFG; dotted lines denote variable correspondences, and Ɛ denotes node deletion.

2.2 Approach based on machine learning

A two-stage approach is taken by Galanis and Androutsopoulos (2010) where each stage utilizes different machine-learning classification techniques. In the first stage, candidates are generated using the maximum entropy model. Candidates are ranked based on grammaticality and importance. Choosing the appropriate compressed sentence from the candidates is the second stage of the process and is done using the SVM regression model.

Galanis and Androutsopoulos (2010) also considered the tree structure of the input source sentence. The candidates are ranked based on the probability estimations made using different properties of the edges. Properties such as the Parts Of Speech (POS) tag denote whether to remove an edge or not, whether to remove a head or a modifier of the edge, and so on. SVM is applied to the ranked candidates in order to choose the best one. The model is trained using the training vector (x_i, y_i) where x_i is a feature vector representing a candidate compression, and y_i is a score indicating how good the compression is. 98 different features are used for this. Finally, the compression which is most similar to the gold compression is selected.

2.3 Modeling with LSTM recurrent neural network

Thus far, the discussed approaches have to wait until the end of the sentence in order to determine whether a particular word is important or not. In LSTM, decisions can be made at each time step. Sakti et al. (2015) consider the basic LSTM Recurrent Neural Network (LSTM-RNN) and perform a few additional modifications. In basic LSTM, the input layer represents the current word and the output layer gives the probability representation that determines whether the word is important or unimportant. The hidden layer gets activated by the weights associated with that layer, the current input, and any previous hidden activations related to other words.

In order to handle the sparseness of the input representation and make a better decision, modifications are performed on the basic structure. For this purpose, a continuous representation layer is introduced between the hidden layer and input layer, which has a connection to the output layer. Sakti et al. (2015) also propose a method for pretraining on large unsupervised data to learn sentence representation.

A sequence to sequence model is considered by Filippova et al. (2015). Here the sentences are considered as sequences of zeroes and ones. LSTM is used for the purpose of remembering long-distance dependencies from the input sentence. Probability estimation makes use of the model from Sutskever et al. (2014). The input sentence is parsed more than once. In the first pass, the network learns the sentence representation and in later passes it starts to make predictions on the importance of the words.

2.4 Modeling with ILP

Approaches based on ILP consider an integer-value representation of sentences. The sentences are represented as integers. Most methods use a binary representation where each word can have values of either 0 or 1. The work of Yao and Wan (2017) presents a model of deleting unimportant words by applying grammatical or language constraints on the binary representation of the input sentence. The methodology considers the input long sentence as $x = \{x_1,..., x_n\}$. The integer linear representation of x is $\delta = \{\delta_1,..., \delta_n\}$ where $\delta_i \pounds \{0,1\}$; 0 corresponds to the exclusion of that particular word and 1 represents the inclusion of that particular word in the sentence.

In Yao and Wan (2017), the procedure starts with an initial random bit vector, and the bits are flipped according to the constraints in Clarke and Lapata (2008), and Yao and Wan (2017). The constraints are checked at each node (where the sentence is considered as a dependency tree). Yao and Wan (2017) introduced a top-down, randomized, constrained, greedy flipping algorithm, which enabled a better scoring function (McDonald, 2006).

3 CRITICAL ANALYSIS

An analysis of the above discussed approaches for extractive sentence compression is conducted as follows by considering the particular contribution of each approach.

Word deletion based on tree transduction, Cohn and Lapata (2009) supports an ample rewrite operation. That is, it is not only deletion specific; we can rearrange the words to form a new sentence. The advantage of using STSG is that it can model non-isomorphic tree structures while having efficient inference algorithms. The method introduced can also be used for structural matching. The grammar rules introduced here are general and can be applied for more elaborate tree divergences. Moreover, by adopting a more expressive grammar formalism, we can naturally model syntactically complex compressions without having to specify additional rules. Also, synchronous grammar will provide a large number of compressions for a given source tree. The decoding algorithm used here can preserve the grammaticality of the sentence. The tree model provides greater flexibility.

Moving on to the disadvantages of Cohn and Lapata (2009); because this model uses a parallel corpus to learn, the accuracy of the grammar generated depends on the accuracy of the corpus. As tree transduction requires a tree-editing operation, designing the tree-editing operation is complex. There is no guarantee that the induced rules will have good coverage on unseen trees. Tree fragments containing previously unseen terminals or non-terminals, or even an unseen sequence of children for a parent non-terminal, cannot be matched by any grammar productions. In this case, the transduction algorithm fails as it has no way of covering the source tree, but it can be recovered by adding a new rule. For unseen productions there is the problem of under compression. There is also the possibility of spurious ambiguity; the situation in which more than one rule produces the same target tree.

As discussed in this survey, the approach of Galanis and Androutsopoulos (2010) is a two-stage process which enables the consideration of many compression candidates by considering different constraints. Different ranking models can be used which makes the system user friendly and the method can be chosen according to the purpose. This method compares favorably to a state-of-the-art extractive system. Unlike other recent approaches, our system uses no handcrafted rules. It also uses a trigram scoring function.

The ranking in Galanis and Androutsopoulos (2010) is performed on the basis of grammaticality and importance; it gives less importance to the compression rate. In addition, because it uses more than one machine-learning process, the system is more complex.

One of the main advantages of the LSTM-based approach of Sakti et al. (2015) is that it does not require a full parse of the input sentence. As well as considering the pretraining method, it handles a sparseness of data too. With regard to the pretraining method, the compression rate is near to the human annotation compression rate. As LSTM gives a more accurate prediction of the importance of the word, it can be applied in many sentence compression techniques.

However, one of the main disadvantages we could take from Sakti et al. (2015) is that of giving importance to grammar. Sakti et al. (2015) mainly focus on the compression rate and the importance of the word. A neural network-related method is quite difficult to implement and from which the basic LSTM gives a compression rate of close to one, that is, no compression took place.

The methodology of Yao and Wan (2017) is a combination of ILP and randomized greedy flipping techniques. This methodology proposes an effective and simpler approach for sentence compression. Comparing this model with that which uses only ILP, it does not have a large number of variables. Even though this Randomized C Greedy Flipping (RCGF) algorithm is iterative, it is effective and fast. A preorder traversal of the dependency tree is considered, which helps to make decisions during the tree traversal. As the random search procedure is independently proceeding, so the inference algorithm can be implemented by introducing a parallel implementation. Using this method, a global optimal solution for sentences can be obtained effectively and the number of local optimal points are small. This property is particularly significant for shorter sentences with simpler structures. Some of the local solutions might be close to the global optimum as well. In addition, this method uses a trigram scoring function that is more advanced. It can be applied with any kind of objective function.

Moving on to the limitations of the approach (Yao and Wan 2017); it requires multiple random restarts as a part of the iterations. Due to the complex nature of the problem structure, with different types of constraints, it is difficult to give a formal mathematical analysis for worst-case convergence bounds. It is difficult to reach an optimal solution for longer sentences of more than 15 words. Longer and complex sentences attain only sub-optimal solutions.

Table 1. Results comparison.

Model	Methodology	Written corpus		Spoken corpus	
		Compression rate	Accuracy	Compression rate	Accuracy
Cohn & Lapata (2009)	Tree transduction	70.4	58.8 (Grammar)	75.5	59.5 (Grammar)
Galanis & Androutsopoulos (2010)	ML-based	71.5	60.2 (Grammar)	71.7	59.2 (Grammar)
Sakti et al. (2015)	LSTM (pretraining)	81.9	68.8 (Importance)	x	72.0 (Importance)
Yao et al. (2017)	ILP	71.7	66.4 (Grammar)	72.7	66.2 (Grammar)

4 DATASETS AND RESULTS DISCUSSION

There are different datasets available for analyzing different sentence compression tasks. Most of the approaches discussed in this survey (Cohn & Lapata, 2009; Sakti et al., 2015; Galanis & Androutsopoulos, 2010; Yao & Wan, 2017) use the same corpora for evaluation that was annotated by human annotators.

These four papers consider both written corpora and spoken corpora. For written corpora, they consider sentences from the British National Corpus (BNC) and sentences from the American News Text Corpus. The spoken corpora were annotations from broadcast news. Mostly, compression rate and F1 score measures are considered for automatic evaluation. The grammatical relationship of generated compression against the gold standard compression is measured by F1 score. Whereas the LSTM-based method of Sakti et al. (2015) does not focus on grammar; instead, it considers the importance measure along with compression rate. Table 1 gives the comparison data.

5 CONCLUSION

Basically, longer sentences are simply a waste of memory and time. Therefore, sentence compression is a necessary task. Extractive sentence compression has a wide range of applications in different domains. It is one of the key activities of text processing in NLP. Google's Hand-Fed AI (artificial intelligence) is one of the best examples. It can also be applied in fields such as phrasal substitution, especially for figurative expressions (Liu and Hwa, 2016). Hence a survey on extractive sentence compression may be helpful in order to choose the best technique.

There is a wide range of methodologies for extracting the relevant words from a sentence in order to form a new compressed sentence. Different authors choose different techniques for extracting and for making a good sentence. We have discussed a few approaches to extractive sentence compression and made a comparative study of them.

Extractive sentence compression can be improved further. Most of the methods discussed here are suitable for smaller sentences but less suitable for larger ones. In future, we can train with data on a large scale. Thus, we can improve the compression accuracy on large sentences as well. Extractive compression can be modeled using a sampling-based method. It will be applicable in different areas such as ECG compression.

REFERENCES

Berg-Kirkpatrick, T., Gillick, D. & Klein, D. (2011). Jointly learning to extract and compress. In *Proceedings of the 49th Annual Meeting of the Association for Computational Linguistics: Human Language Technologies* (pp. 481–490). Stroudsburg, PA: Association for Computational Linguistics.

Clarke, J. & Lapata, M. (2008). Global inference for sentence compression: An integer linear programming approach. *Journal of Artificial Intelligence Research*, *31*, 399–429.

Cohn, T. & Lapata, M. (2007). Large margin synchronous generation and its application to sentence compression. In *Proceedings of the 2007 Joint Conference on Empirical Methods in Natural Language Processing and Computational Natural Language Learning, 28–30 June 2007, Prague, Czech Republic* (pp. 73–82).

Cohn, T.A. & Lapata, M. (2009). Sentence compression as tree transduction. *Journal of Artificial Intelligence Research*, *34*, 637–674.

Collins, M. (1997). Three generative, lexicalised models for statistical parsing. In *Proceedings of the Eighth Conference of the European Chapter of the Association for Computational Linguistics* (pp. 16–23). Gothenburg, Sweden: Association for Computational Linguistics.

Corston-Oliver, S. (2001). Text compaction for display on very small screens. In *Proceedings of the NAACL Workshop on Automatic Summarization* (pp. 89–98). Stroudsburg, PA: Association for Computational Linguistics.

De Belder, J. & Moens, M.F. (2010). Integer linear programming for Dutch sentence compression. In A. Gelbukh (Ed.), *Computational linguistics and intelligent text processing. CICLing 2010. Lecture notes in computer science* (Vol. 6008, pp. 711–723).

Filippova, K., Alfonseca, E., Colmenares, C.A., Kaiser, L. & Vinyals, O. (2015). Sentence compression by deletion with LSTMs. In *Proceedings of the 2015 Conference on Empirical Methods in Natural Language Processing, 17–21 September 2015, Lisbon,* Portugal (pp. 360–368). Stroudsburg, PA: Association for Computational Linguistics.

Filippova, K. & Altun, Y. (2013). Overcoming the lack of parallel data in sentence compression. In *Proceedings of the 2013 Conference on Empirical Methods in Natural Language Processing, 18–21 October 2013, Seattle, Washington, USA* (pp. 1481–1491). Stroudsburg, PA: Association for Computational Linguistics.

Galanis, D. & Androutsopoulos, I. (2010). An extractive supervised two-stage method for sentence compression. In *Proceedings of the 2010 Annual Conference of the North American Chapter of the Association for Computational Linguistics: Human Language Technologies* (pp. 885–893). Stroudsburg, PA: Association for Computational Linguistics.

Knight, K. & Marcu, D. (2000). Statistics-based summarization – step one: Sentence compression. In *Proceedings of the Seventeenth National Conference on Artificial Intelligence and Twelfth Conference on Innovative Applications of Artificial Intelligence, 30 July – 3 August 2000* (pp. 703–710). Palo Alto, CA: AAAI Press.

Knight, K. & Marcu, D. (2002). Summarization beyond sentence extraction: A probabilistic approach to sentence compression. *Artificial Intelligence*, *139*(1), 91–107.

Liu, C. & Hwa, R. (2016). Phrasal substitution of idiomatic expressions. In *Proceedings of NAACL-HLT 2016, 12–17 June 2016, San Diego, California* (pp. 363–373). Stroudsburg, PA: Association for Computational Linguistics.

Madnani, N., Zajic, D., Dorr, B., Ayan, N.F. & Lin, J. (2007). Multiple alternative sentence compressions for automatic text summarization. In *Proceedings of the 2007 Document Understanding Conference (DUC-2007) at NLT/NAACL 2007, April 2007, Rochester, New York.*

McDonald, R.T. (2006). Discriminative sentence compression with soft syntactic evidence. In *Proceedings of the 11th Conference of the European Chapter of the Association for Computational Linguistics, 3–7 April 2006, Trento, Italy* (pp. 297–304). Gothenburg, Sweden: Association for Computational Linguistics.

Sakti, S., Ilham, F., Neubig, G., Toda, T., Purwarianti, A. & Nakamura, S. (2015). Incremental sentence compression using LSTM recurrent networks. In *Proceedings of 2015 IEEE Workshop on Automatic Speech Recognition and Understanding (ASRU)* (pp. 252–258). New York, NY: IEEE.

Sutskever, I., Vinyals, O. & Le, Q.V. (2014). Sequence to sequence learning with neural networks. In Z. Ghahraman, M. Welling, C. Cortes, N.D. Lawrence & K.Q. Weinberger (Eds.), *Advances in Neural Information Processing Systems 27* (pp. 3104–3112).

Vandeghinste, V. & Pan, Y. (2004). Sentence compression for automated subtitling: A hybrid approach. In *Proceedings of the ACL Workshop on Text Summarization, Barcelona, July 2004* (pp. 89–95).

Wang, H., Zhang, Y. & Zhou, G. (2013). Sentence compression based on ILP decoding method. In G. Zhou, J. Li, D. Zhao & Y. Feng (Eds.), *Natural language processing and Chinese computing. Communications in computer and information science* (Vol. 400, pp. 19–29). Berlin, Germany: Springer.

Yao, J.G. & Wan, X. (2017). Greedy flipping for constrained word deletion. In *Proceedings of the Thirty-First AAAI Conference on Artificial Intelligence (AAAI-17)* (pp. 3518–3524). Menlo Park, CA: Association for the Advancement of Artificial Intelligence.

Yao, J.G., Wan, X. & Xiao, J. (2014). Joint decoding of tree transduction models for sentence compression. In *Proceedings of the 2014 Conference on Empirical Methods in Natural Language Processing (EMNLP), 25–29 October 2014, Doha, Qatar* (pp. 1828–1833). Stroudsburg, PA: Association for Computational Linguistics.

Emerging Trends in Engineering, Science and Technology for Society,
Energy and Environment – Vanchipura & Jiji (Eds)
© 2018 Taylor & Francis Group, London, ISBN 978-0-8153-5760-5

A survey of figurative language detection in social media

P.D. Manjusha & C. Raseek
Computer Science and Engineering Department, Government Engineering College,
Palakkad, Kerala, India

ABSTRACT: A figurative language often uses words with a meaning different from its literal meaning. Detection of figurative language helps in computational linguistics and social media sentiment analysis. The majority of social media comments make use of figurative language for better expression of user emotions. Figurative language such as simile, humor, sarcasm, and irony have widespread uses within social media. The purpose of this paper is to survey various methods for detection of figurative language based on their distinguishing features.

1 INTRODUCTION

Figurative language expresses the feeling of a writer rather than its literal interpretation. In the literal interpretation, the writer expresses things as such it is. Figurative language offers a better understanding and analysis of the text than its literal interpretation. In the current social media arena, people make use of various figurative language like simile, humor, irony and sarcasm to express their emotions. Identifying features that better distinguish figurative language is important for the detection process.

Most of the figurative language has an implicit meaning embedded within it. This provides difficulty in identifying actual sense meant by the figurative language. Each figurative language has specific features exclusively determined for them to better distinguish them, which can be different for different languages. Recognition of this figurative language, whilst also combining various features that better distinguish them, acts as 'trending' in social media.

There exists a wide range of figurative language and this survey focuses on simile, sarcasm, irony and humor because most micro-blogging platforms make use of these figures of speech mostly. A simile is a figure of speech that essentially compares two different things with the help of words such as: *like, as, than.* They can have implicit or explicit properties mentioned. For example: *Our room feels like Antarctica.* Sarcasm usually makes use of words to express meaning that is the opposite of its actual meaning. It is mainly used to criticize someone's feelings. For example: *I love the way my sweetheart cheats on me.* Irony includes words that are the opposite of the actual situation. For example: *Butter is as soft as a slab of marble.* Humor is also a figure of speech which is used to produce the effect of laughter and to make things funny. Humor provides the direct implication of the situation. For example: *He faces more problems than a math book has.* Identifying features that better distinguish these figures of speech act as a trending topic in social media.

Figurative language detection is necessary in different applications such as computational linguistics, social multimedia, and psychology. Text summarization, machine translation systems, advertisements, news articles, sentiment analysis and review processing systems make use of figurative language.

This survey mainly focuses on different methods that make use of various distinguishing features of different figurative language. It also helps to identify features that better distinguish each figurative language.

This paper is organized as follows. Section 2 gives a formal definition of the figurative language detection with an example; also identified are various distinguishing features for different figurative language term. Section 3 discusses various future research directions for figurative language detection. Section 4 gives brief concluding comments.

2 FIGURATIVE LANGUAGE DETECTION

A figurative language detection system mainly focuses on identification of various figurative language terms based on certain distinguishing features.

The main objective of the figurative language detection system is defined as: *Given an input sentence, the system aims to identify the figurative language class in which it belongs. The class can be simile, sarcasm, irony or humor because we are dealing with social media.* An important step in the figurative language detection is dataset collection. After gathering a collection of tweets, preprocessing of them is needed in order to filter out the actual structure of each figurative language. Then the features are extracted for each figure of speech for the purpose of better classification. Extracted features can then be used to train any classifier. Most of the existing work makes use of the classifier Support Vector Machine (SVM). So, when a new instance enters the classifier model, it will produce the figurative language class, in which it belongs as the predicted output. For example, given the sentence: *He looks completely like a zombie*, the system produces an output that it is a simile.

Features that can better distinguish simile, sarcasm, irony and humor includes lexical, syntactic and semantic, pragmatic and finally emotion and sentimental features.

2.1 *Lexical features*

Lexical features include the presence of N-grams. According to this feature words or phrases that occur frequently are considered. In Joshi et al. (2015), the proposed method is to detect sarcastic tweets and it makes use of a unigram feature. Frequently occurred unigrams were counted in the approach. In Thu and New (2017), the proposed method is to detect irony, sarcasm, simile and humor at the same time based on supervised classification method. The method uses N-gram as a feature for classification. It considers unigram, bigram and trigram that occur frequently.

Barbieri and Saggion (2014) proposed a method to automatically detect irony and humor. The method helps to study aspects of frequency imbalance in tweets based on the most frequent and rare words. Frequent and rare words are determined using the American National Corpus (ANC) frequency data corpora. The presence of frequent words and rare words in a sentence can cause unexpectedness. This unexpectedness and incongruity can be used to distinguish irony and humor. This method also considers the length of the sentence as a feature in which it checks whether the sentence is long or short.

Khokhlova et al. (2016) also consider N-grams for detection of irony and sarcasm automatically. Certain observations are identified based on these obtained N-grams. The method identifies that a larger number of constructions with negations are found in ironic texts than sarcastic texts. Sarcastic texts include a larger number of proper names than ironic texts.

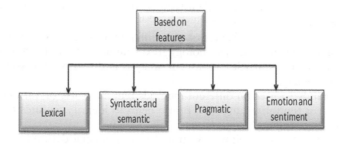

Figure 1.   Classification of methodologies for figurative language detection based on the features used.

A high frequency of verb 'love' is found in sarcasm. Sarcastic texts are more egocentric in that the writer makes use of the pronoun *I* mostly.

2.2 Syntactic and semantic features

Each figurative language term has a different syntactic structure. A simile consists of different components such as tenor, vehicle, event and comparator, Qadir et al. (2015). The tenor is the subject of comparison, the vehicle is the object of comparison, the event is an action or state and the comparator includes words such as: *like, as, than*. Qadir et al. (2016) applied rules to recognize the syntactic patterns of similes. The rules used include: *NP VERB like NP*; *ADJ as [a, an] Noun*; *ADJ like [a, an] Noun*.

Parts of speech are an important feature for detection of figurative language. Khokhlova et al. (2016) used parts of speech as a distinguishing feature for detection of irony and sarcasm. The method identified the ratio of parts of speech in both corpora. Ironic texts include the presence of a large number of nouns. It was also observed that other parts of speech are common in sarcasm. Sarcasm tends to be more emotional due to the presence of adjectives and adverbs. The hashtag is another feature used by Khokhlova et al. (2016), in which the hashtag of ironic texts, giving reference to other authors or objects, are more structured. Ironic texts are more structured than sarcastic texts.

Barbieri and Saggion (2014) consider synonyms and ambiguity as features for detecting irony and humor. Irony often conveys two messages at the same time and the choice of the term rather than its synonym is important. The method used identified synonyms of words in tweets using WordNet. They then calculated their ANC frequencies and sorted them in a decreasing rank list. This acts as a distinguishing feature for irony and humor. Another important feature for irony and humor is ambiguity. The method identifies that ironic text includes a greater arithmetical average of synsets than humor. Ironic tweets present words with more meaning, whereas humorous tweets present words with less meaning.

2.3 Pragmatic features

Pragmatic features are mainly used to understand other speakers' intended meaning. González-Ibánez et al. (2011) reported a method for identifying sarcasm in Twitter tweets. The authors investigated the impact of lexical and pragmatic features on the task. Pragmatic features include the use of positive emoticons such as smileys, and negative emoticons such as frowning faces. Another feature is ToUser, which indicates whether the tweet contains any marks that are a reply to another tweet. Sarcastic tweets include a larger number of these pragmatic features than non-sarcastic tweets. Emoticons are mostly found in large numbers in sarcastic texts. Barbieri and Saggion (2014) also used emoticons as a feature for detection of irony and humor. An emoticon feature is the number of :), :D, :(, and ;) in a tweet. Humorous text includes emoticons around four times more than ironic texts. Humorous texts are more explicit than ironic ones and there is a need to understand humorous texts without the use of explicit signs such as emoticons. Another feature used by Barbieri and Saggion (2014) indicates the presence of laughter, such as hahah, lol, rofl, imao.

Capitalization is another feature that can distinguish figurative language. Joshi et al. (2015) used capital letters as a feature for detection of sarcastic texts. Sarcastic texts include a larger proportion of capital letters than non-sarcastic texts. Khokhlova et al. (2016) added another feature, interjection. The presence of various interjections in the tweet is taken into consideration. The presence of punctuation acts as a better pragmatic feature. Punctuation includes the number of commas, exclamation and quotation marks, full stops, semicolons, ellipses and hyphens. These are included in the method proposed by Barbieri and Saggion (2014).

2.4 Emotion-based and sentimental features

Emotions and sentiment can be used as better features for figurative language detection. There are eight basic EmoLex (Emotion Lexicon) emotions: anger, anticipation, disgust, fear, joy, sadness, surprise, trust. Sentiment includes positive, negative and neutral polarities. Emotion-based features, along with polarity features, can improve the efficiency of a detection system.

Thu and New (2017) proposed a method to detect several figurative language terms at once from an emotional point of view. It is used for detection of figurative language such as sarcasm, irony, simile, and humor, while at the same time using multi-class supervised classification. Most of the previous work was based on either detecting figurative language separately or a combination of the two. The method uses a multi-class supervised classification problem for better detection of figurative language based on emotions. Ensemble Bagging Classifier and standard SVM are used for classification, thus performing a comparison. This method focuses on the impact analysis of different features used by the method. Features used in the approach include word-based, emotion, sentiment, Bag of Sorted Emotion (BOSE) and BOSE-TFIDF (Term Frequency–Inverse Document Frequency).

Emotion includes eight basic emotions excerpted from the lexicon EmoLex. Sentiment features includes positive, negative and neutral polarity for each word in the sentence. BOSE is created from eight basic emotional features and three sentimental features. First, individually sort the emotional features and sentiment features according to the value they obtained from each tweet. As a result, a BOSE is obtained that includes a bag of emotions and sentiment together. This could better act as a distinguishing feature. BOSE-TFIDF incorporates document frequency and inverse document frequency along with BOSE.

Khokhlova et al. (2016) proposed a method for automatic detection of irony and sarcasm. Their method considers eight emotions along with syntactic features for the detection of irony and sarcasm. EmoLex is used to assign emotion and polarity to the words of tweets. A comparison with the National Research Council (NRC) Emotion Lexicon was performed. Frequency lists of sarcasm and irony corpora were obtained primarily and a comparison with the EmoLex list performed. This helps to provide accurate emotion to corresponding words and thus is better in distinguishing sarcasm and irony. Observations inferred that sarcastic texts are more emotional than ironic texts. The method also assigns polarity to each word in a tweet and obtained the combined polarity of the tweet. Mostly, the polarity of the sentence will be the polarity of the majority of words. In such a case, sarcastic texts are slightly more positive than ironic texts. In all cases it won't be true because sarcastic tweets can have a change in polarity in different parts. Polarity changes when one moves from the first part to next part of the text. A sarcastic sentence often starts with a positive sentiment and ends with negative sentiment or vice versa.

Joshi et al. (2015) made use of the better future of sentiment category. Their method used the Vader tool in the NLTK natural language processing toolkit and obtained the sentiment polarity of subparts of the sarcastic sentence. Thus, it was observed that sarcastic sentences are more negative than non-sarcastic sentences. Explicit incongruity and implicit incongruity can be used for better detection of sarcasm. Explicit incongruity makes use of sentiment words of both polarities and thwarted expectations. Implicit incongruity is expressed through phrases of implied sentiment rather than using polar words. The main advantage of this method is that it addresses the problem of assigning polarity to a sarcastic sentence due to change in its sentiment in parts.

Barbieri and Saggion (2014) proposed a method for automatic detection of irony and humor. The method used the SentiWordNet sentiment lexicon for obtaining the sentiment of each word. Here, synsets of words were obtained in the sentence and then assigned a sentiment score of positive or negative to the words. There are six features in the sentiment group such as positive sum, negative sum, positive-negative mean, positive-negative gap, positive single gap and negative single gap. Positive sum and negative sum are the sum of all positive and negative scores assigned to words in the sentence respectively. Positive-negative mean is the average of scores obtained from positive sum and negative sum. Positive-negative gap is the difference between the positive and negative sum. The positive single gap is the difference between the most positive word and the mean of all the sentiment scores of words in the sentence. Likewise, for the negative single gap, this considers the most negative word. This also provides a better distinguishing sentiment feature in case of figurative language detection.

Qadir et al. (2015) discuss the effect of polarity in similes. Similes often express positive or negative sentiment toward an entity. This method makes use of sentiment features, such as component sentiment, which assign polarity scores to the components of a simile. A simile consists of different parts, such as subject of comparison, object of comparison, event and a comparator. Another feature is simile connotation polarity which indicates the overall connotation of the simile. The method used by Qadir et al. (2015) makes use of the AFINN sentiment lexicon and the MPQA subjectivity lexicon in order to assign polarity to the words.

2.5 *Datasets for figurative language detection*

Use of figurative language over micro-blogging services is increasing day by day. Identifying and extracting the relevant dataset is necessary for improving the performance of the system.

Similes in social media do not have a built-in dataset. They are extracted from Twitter by looking for tweets that have words such as *as, than* and *#simile*. Then particular rules are applied to filter out the similes. Tweets having certain syntactic patterns were selected by Qadir et al. (2015, 2016).

Irony TwBarbieri 2014 and Irony TwReyes 2013 are two datasets for irony, and Sarcasm Ptacek 2014 and Sarcasm TwRiloff 2013 are two datasets for sarcasm by Khokhlova et al.

Table 1. Overview of the compared features used for detection of several figurative languages.

| Type of feature | Proposed approaches | | Figurative language |
	Model	Highlight of features	
Lexical features	Thu & New (2017)	Unigram, bigram and trigrams are considered	Simile, sarcasm, irony, humor
	Joshi et al. (2015)	Unigrams	Sarcasm
	Barbieri & Saggion (2014)	Presence of frequent and rare words	Irony and humor
	Khokhlova et al. (2016)	N-grams	Irony and sarcasm
Syntactic and semantic features	Barbieri & Saggion (2014)	Synonyms and ambiguity	Irony and humor
	Qadir et al. (2016)	Syntactic rules	Simile
	Khokhlova et al. (2016)	Parts of speech and hashtags	Irony and sarcasm
Pragmatic features	Barbieri & Saggion (2014)	Emoticons, laughs and punctuation marks	Irony and humor
	Joshi et al. (2015)	Count of capital letters	Sarcasm
	Khokhlova et al. (2016)	Presence of interjections	Irony and sarcasm
	González-Ibánez et al. (2011)	Positive and negative emoticons, ToUser	Sarcasm
Emotion-based and sentimental features	Thu & New (2017)	Eight basic emotions from EmoLex and three sentimental features from Vader	Simile, sarcasm, irony, humor
	Khokhlova et al. (2016)	Eight basic emotions from EmoLex	Irony and sarcasm
	Qadir et al. (2015)	Simile component polarity and simile connotation polarity	Simile
	Joshi et al. (2015)	Polarity of each word, explicit and implicit incongruity	Sarcasm
	Barbieri & Saggion (2014)	Polarity of synsets of words, positive sum, negative sum, positive-negative gap, positive-negative mean, positive single gap and negative single gap	Irony and humor

(2016). Irony TwBarbieri 2014 includes 10,000 ironic and 40,000 non-ironic instances. Irony TwReyes 2013 includes 10,000 ironic and 30,000 non-ironic instances. Sarcasm Ptacek 2014 includes 25,000 sarcastic and 75,000 non-sarcastic instances, and TwRiloff 2013 includes 10,000 sarcastic and 40,000 non-sarcastic instances.

The dataset for the combination of all figurative language was created by Thu and New (2017). It includes balanced and class imbalanced datasets. Balanced datasets include tweets having hashtags #simile, #sarcasm, #irony and #humor, collected with no restriction on time or person. The class imbalanced dataset was created by streaming tweets for a particular day. Because the datasets and evaluation criteria used in the methodologies for intention detection discussed above are different, a comparative study based on the results is not possible. Table 1 shows the different features and contributions of different methods according to different features. It also highlights features that better distinguish different figurative language.

3 FURTHER SUGGESTIONS

Identification of occurrence of one figurative language over another can be identified for better understanding of the text. For example, identification of sarcastic similes, ironic similes and humorous similes. Automatically inferring implicit properties in the sentence can be done to improve the efficiency of the system. Figurative language detection can be processed from different angles that make use of cognitive and psycholinguistic information, gestural information, tone and paralinguistic cues. Another suggestion is to build a system that works efficiently on large texts of the documents. All these are possible directions for future research in this area.

4 CONCLUSION

Figurative language provides better analysis and understanding of text than its literal interpretation. There are different types of figurative language terms. But in the case of social media, simile, sarcasm, irony and humor have an impact. Supervised classification methods are present for automatic detection of figurative language. Identification of features that better distinguish figurative language helps in improving the efficiency of the system. Each figurative language has its own features. Through this survey, we have tried to highlight the distinguishing features of each figurative language. This survey is conducted with the hope of shedding some light on the different features of the various figures of speech and how they can be incorporated for the automatic detection of those languages simultaneously.

REFERENCES

Bamman, D. & Smith, N.A. (2015). Contextualized sarcasm detection on Twitter. In *Ninth International AAAI Conference on Web and Social Media* (pp. 574–577). New York, NY: AAAI Press.
Barbieri, F. & Saggion, H. (2014). Automatic detection of irony and humour in Twitter. In S. Colton, D. Ventura, N. Lavrac & M. Cook (Eds.), *Proceedings of the Fifth International Conference on Computational Creativity, Ljubljana, Slovenia, 10–13 June 2014* (pp. 155–162).
Crossley, S.A., Kyle, K. & McNamara, D.S. (2017). Sentiment Analysis and Social Cognition Engine (SEANCE): An automatic tool for sentiment, social cognition, and social-order analysis. *Behavior Research Methods, 49*(3), 803–821.
Davidov, D., Tsur, O. & Rappoport, A. (2010). Semi-supervised recognition of sarcastic sentences in Twitter and Amazon. In *Proceedings of the Fourteenth Conference on Computational Natural Language Learning* (pp. 107–116). Stroudsburg, PA: Association for Computational Linguistics.
Fellbaum, C. (1998). *WordNet: The encyclopedia of applied linguistics.* New York, NY: John Wiley & Sons.
Fersini, E., Pozzi, F.A. & Messina, E. (2015). Detecting irony and sarcasm in microblogs: The role of expressive signals and ensemble classifiers. In *Proceedings of IEEE International Conference on Data Science and Advanced Analytics (DSAA) 2015* (pp. 1–8). New York, NY: IEEE.

González-Ibáñez, R., Muresan, S. & Wacholder, N. (2011). Identifying sarcasm in Twitter: A closer look. In *Proceedings of the 49th Annual Meeting of the Association for Computational Linguistics: Human Language Technologies: Short Papers* (Vol. 2, pp. 581–586). Stroudsburg, PA: Association for Computational Linguistics.

Joshi, A., Sharma, V. & Bhattacharyya, P. (2015). Harnessing context incongruity for sarcasm detection. In *Proceedings of the 53rd Annual Meeting of the Association for Computational Linguistics and the 7th International Joint Conference on Natural Language Processing, Beijing, China, 26–31 July 2015 (Short Papers)* (pp. 757–762). Stroudsburg, PA: Association for Computational Linguistics.

Khokhlova, M., Patti, V. & Rosso, P. (2016). Distinguishing between irony and sarcasm in social media texts: Linguistic observations. In *2016 International FRUCT Conference on Intelligence, Social Media and Web (ISMW FRUCT), St. Petersburg* (pp. 1–6). New York, NY: IEEE.

Liebrecht, C.C., Kunneman, F.A. & van Den Bosch, A.P.J. (2013). The perfect solution for detecting sarcasm in tweets# not. In *Proceedings of the 4th Workshop on Computational Approaches to Subjectivity, Sentiment and Social Media Analysis, Atlanta, Georgia, 14 June 2013* (pp. 29–37). Stroudsburg, PA: Association for Computational Linguistics.

Mihalcea, R. & Pulman, S. (2007). Characterizing humour: An exploration of features in humorous texts. In A. Gelbukh (Ed.), *Computational Linguistics and Intelligent Text Processing. CICLing 2007. Lecture Notes in Computer Science* (Vol. 4394, pp. 337–347). Berlin, Germany: Springer.

Paul, A.M. (1970). Figurative language. *Philosophy & Rhetoric, 3*(4), 225–248.

Qadir, A., Riloff, E. & Walker, M.A. (2015). Learning to recognize affective polarity in similes. In *Conference Proceedings—EMNLP 2015: Conference on Empirical Methods in Natural Language Processing* (pp. 190–200). Stroudsburg, PA: Association for Computational Linguistics.

Qadir, A., Riloff, E. & Walker, M.A. (2016). Automatically inferring implicit properties in similes. In *Proceedings of the 2016 Conference of the North American Chapter of the Association for Computational Linguistics: Human Language Technologies* (pp. 1223–1232). Stroudsburg, PA: Association for Computational Linguistics.

Ramteke, A., Malu, A., Bhattacharyya, P. & Nath, J.S. (2013). Detecting turnarounds in sentiment analysis: Thwarting. In *Proceedings of the 51st Annual Meeting of the Association for Computational Linguistics, Sofia, Bulgaria, 4–9 August 2013* (pp. 860–865). Stroudsburg, PA: Association for Computational Linguistics.

Reyes, A., Rosso, P. & Buscaldi, D. (2012). From humor recognition to irony detection: The figurative language of social media. *Data & Knowledge Engineering, 74*, 1–12.

Reyes, A., Rosso, P. & Veale, T. (2013). A multidimensional approach for detecting irony in Twitter. *Language Resources and Evaluation, 47*(1), 239–268.

Riloff, E., Qadir, A., Surve, P., De Silva, L., Gilbert, N. & Huang, R. (2013). Sarcasm as contrast between a positive sentiment and negative situation. In *Proceedings of the 2013 Conference on Empirical Methods in Natural Language Processing* (pp. 704–714). Stroudsburg, PA: Association for Computational Linguistics.

Taylor, J.M. & Mazlack, L.J. (2005). Toward computational recognition of humorous intent. *Proceedings of the Annual Meeting of the Cognitive Science Society, 27*, 2166–2171.

Thu, P.P. & New, N. (2017). Impact analysis of emotion in figurative language. In *Proceedings of 2017 IEEE/ACIS 16th International Conference on Computer and Information Science (ICIS), Wuhan* (pp. 209–214). New York, NY: IEEE.

Veale, T. & Hao, Y. (2010). Detecting ironic intent in creative comparisons. In *Proceedings of the 19th European Conference on Artificial Intelligence, ECAI 2010* (pp. 765–770). Amsterdam, The Netherlands: IOS Press.

Emerging Trends in Engineering, Science and Technology for Society, Energy and Environment – Vanchipura & Jiji (Eds)
© 2018 Taylor & Francis Group, London, ISBN 978-0-8153-5760-5

A survey of various approaches for interpersonal relationship extraction

M. Nijila & M.T. Kala
Computer Science and Engineering Department, Government Engineering College, Palakkad, Kerala, India

ABSTRACT: Recent efforts to digitize literary works emphasize the importance of a computational route to analyzing large collections of text. To better understand the literary narrative, most research focuses on modeling it from the viewpoint of events or characters. Recently, character-centric approaches have been used to extract relationships between people, which enables better understanding of the narrative. Different methodologies consider distinct types of relationships. The purpose of this paper is to survey the various methodologies that try to infer interpersonal relationships of characters in narratives. The advantages and limitations of each method are compared and the various datasets used in each methodology are discussed in this paper.

1 INTRODUCTION

Interpersonal relationships have long been studied by analysts in several domains to get a better understanding of the narratives. These relationships exhibit a variety of phenomena like family, friendship, hostility, and romantic love. We can see that the narratives are a rich reflection of these relationships and this provides a better medium for the analysis. Two approaches used for natural language understanding are an event-centric approach and a character-centric approach.

An event-centric approach tries to understand the narrative based on the events described within it. These methods aim to demonstrate the given text using sequences of events, their participants, and the relationships between them. Such a representation is called a 'script'. Another representation of the narrative includes frames, plot units, and schemas.

On the other hand character-centric approaches consider the characters involved in the narrative. Identifying the relationship between the characters is a better approach to narrative understanding. Recent works are focused on creating a structure called social networks, sometimes called signed networks, to model relationships between characters. These social networks are constructed based on co-occurrence of characters in conversations, social events, and so on.

In general, characterizing the nature of relationships between individuals can assist automatic understanding of the text by explaining the actions of people mentioned in the text and building expectations of their behavior toward others. Modeling relationships has many real-world applications, such as predicting possible relationships between people using their posts or messages in social media, personalizing newsfeeds, predicting virality, and suggesting friends or topics of interest, for a particular user.

Relationship extraction is commonly achieved through supervised or unsupervised methods. Some may use hybrid approaches for this. These approaches make use of different machine-learning algorithms for classification, for example, Naive Bayes and Support Vector Machines (SVMs). Relationship extraction has the potential to use deep learning models for a better performance.

The purpose of this survey is to discuss different methodologies used for relationship extraction between characters and make a comparison between them. This paper will give an overall idea of recent advances in this area.

The layout of this paper is as follows: Section 2 presents a formal definition of interpersonal relationship extraction and describes various approaches used for extracting relationship between characters; Section 3 presents avenues for future work in this area; Section 4 gives concluding remarks of this work.

2 RELATIONSHIP EXTRACTION BETWEEN CHARACTERS

Many applications in the field of natural language understanding, information extraction, and information retrieval require understanding of the relationship between the entities mentioned in a particular document. Here, entities include a wide grouping which may contain people, locations, organizations, quantities, and so on. This survey focuses only on the methodologies that extract the relationship between the characters. We can formulate the problem of relationship extraction between characters as: given a narrative X, the goal of a relationship extraction system is to identify all the characters mentioned in it, and infer the relationship between these characters. This will provide a better understanding of the narrative. It will help to process a large collection of text in, for example, digital humanity, political sciences, and so on. Figure 1 shows the hierarchy of classification of various methodologies. Figure 2 shows the basic idea of the interpersonal relationship extraction system.

2.1 Supervised approaches

Supervised approaches are the most widely used machine-learning methods. They rely on a training set where domain-specific examples have been tagged. Supervised approaches require a suitable tagged corpus.

Culotta et al. (2006) proposed an approach for relationship extraction between characters using a supervised approach. They proposed a methodology that uses an integrated supervised machine-learning method combining both contextual and relational patterns to extract

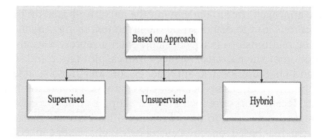

Figure 1.   Classification of methodologies for relationship extraction based on the approach used.

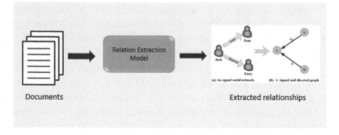

Figure 2.   Basic idea of interpersonal relationship extraction system.

relationships. Traditional algorithms are able to learn relations such as 'son' and 'sibling' from local language clues. But to extract relations like 'cousin' we need additional information.

The methodology used by Culotta et al. (2006) creates a Conditional Random Field (CRF) chain to extract relations while simultaneously uncovering important relational patterns that improve extraction performance. It is limited to biographical text; for example, encyclopedia articles. It extracts relations with a model using CRFs. The advantage of this methodology is that it incorporates relational paths as features to the model. Thus it can learn interesting relational patterns that may have low precision, without reducing extraction performance. In this system each learned path has a weighting associated with it. Low-precision patterns may have a lower weighting than high-precision ones. This methodology will produce better results only when the database contains accurate relational information and the sentence contains limited contextual clues. This is a major drawback of this methodology.

Chaturvedi et al. (2016) proposed a method for relation extraction that considers the evolving nature of the relationship between the characters. Most previous work had made an assumption that relationships between characters are static. But, in actual fact, the relationship evolves as we proceed through the narrative. Here, the relationship between people belongs to two classes, either cooperative or non-cooperative. The approach of Chaturvedi et al. (2016) made use of a collection of fully labeled and partially labeled sentences from a narrative to train a semi-supervised segmentation framework. Inputs to this model are a narrative summary and a pair of characters appearing in it. The model segments these sequences of sentences in the narrative into meaningful and non-overlapping segments, which represent the flow of relationships in the narrative. The proposed methodology assumes each sentence in a segment associates with a particular relationship state.

Two types of features are used in this methodology: (1) content features; and (2) transition features. Content features are used to characterize the textual content of the sentences. It includes action-based features, adverb-based features, lexical-based features and semantic-parse-based features. Action-based features model the actions affecting two characters by identifying all the verbs in the sentence. Adverb-based features are used to model the narrator's bias toward describing the characters. For that, the model extracts all the adverbs that modify verbs extracted in action-based features. Lexical-based features are used to analyze the connotation of all words occurring between a pair of characters. Semantic-parse-based features incorporate a frame-net style parse of the sentence. Transition features are used to remember the long history of relationship states.

The advantage of this model is that it considers the flow of information between sentences to get the evolving relationship. The assumption about the type of relationship is a limitation of this approach. Rather than considering cooperative and non-cooperative relationships, it is more valuable to consider diverse sorts of relationships.

Srivastava et al. (2016) proposed an approach for inferring polarity of the relationship between characters. Their methodology was to formulate the problem as a joint structured problem for each narrative and present a general model that uses linguistic and semantic features and features based on the structure of the community. The model jointly infers all relation labels for a pair of characters in a document. Like the previously discussed approach by Chaturvedi et al. (2016), the methodology of Srivastava et al. (2016) also considers two types of relationships (cooperative and adversarial). For training and inference, this methodology uses a structured perceptron algorithm. It uses a narrative-specific model that allows differential weightings for features depending on the narrative type.

The features used in this model are text-based and structural features. Text-based features are the same as the content features used in the model proposed by Chaturvedi et al. (2016). These text-based features are used to find relationships between a pair of characters in isolation. The approach proposed by Srivastava et al. (2016) is also motivated by the fact that the relationship between the characters can also be inferred from the relationship between others in a narrative. Clique, love triangle, Mexican standoff, and common enemy are the triadic structural features considered in this approach.

This approach considers the type of narrative and this can be considered as a merit of this approach. It also shows the importance of understanding the relationship between two characters within the context of their relationship with others in the narrative. The approach makes an assumption that the relationships between characters are static, which is a major shortcoming.

2.2 *Unsupervised approaches*

Unsupervised approaches try to find hidden structures from unlabeled data. They do not need any training phase and can, therefore, be directly applied to any kind of data.

Krishnan and Eisenstein (2014) proposed an unsupervised approach for relationship extraction. This model produces a signed social network from the content exchanged across the network edges. It incorporates both network structure and linguistic content and connects these signed social networks with address terms. The usage of the address term indicates the relationship between two parties. Names, cities, and placeholder names such as 'dude', are considered in this approach. Identification of address terms contain several subtasks: (1) distinguishing addresses from any mention of other individuals; (2) identifying a lexicon of titles, which either precede name addresses or can be used in isolation; (3) identifying a lexicon of placeholder names.

The approach proposed by Krishnan and Eisenstein (2014) made use of a semi-automated method to construct an address term lexicon, by using bootstrapping from the address term tagger to build candidate lists. Features used in this methodology are address terms, mutual friends (using the Adamic–Adar metric) and triads. This approach is only useful to induce asymmetric signed networks and is a limitation of this method.

Chaturvedi et al. (2017) proposed an approach using unsupervised learning of evolving relationships between literary characters. The methodology proposed by Chaturvedi et al. (2016) and Srivastava et al. (2016) assumes that relationships are only of two types: cooperative or non-cooperative. The proposed methodology in Chaturvedi et al. (2017) relaxes this condition and assumes that relationships can be of multiple types. This model makes a Markovian assumption to capture the 'flow of information' or the historical context between individual sentences.

The model proposed in Chaturvedi et al. (2017) is a Globally Aware General Hidden Markov Model (GHMM), which incorporates a short historical context via Markovian assumptions as well as a longer global context using two distinct local and global components. This is the main advantage of this model.

2.3 *Hybrid approaches*

Makazhanov et al. (2014) proposed an approach for extracting family relationships from novels. This methodology combines word-level techniques with an utterance attribution approach. This model considers the verbal interactions between characters as candidate relations, and the speakers themselves as their potential arguments. It also incorporates detection of vocatives – the explicit forms of address used by the characters in a novel.

The methodology used by Makazhanov et al. (2014) contains four stages for relation extraction: (1) utterance attribution; (2) vocatives detection; (3) relation extraction; (4) relation propagation. For each candidate–utterance pair, a feature vector is created. Features include various quantitative and qualitative characteristics. This methodology is a mixture of heuristic and supervised approaches.

Once all the utterances are recognized the next step is to detect vocatives. Once a basic family relation list is created, we can then proceed to select candidate utterances for the vocative detection task. This task tried both supervised and unsupervised methods.

The advantage of this methodology is that, at each stage, it uses a combination of different methods. So it produces better results at each level and that contributes to the final result. A limitation of this method is inaccurate utterance attribution, a major problem with some vocative utterances.

Devisree and Raj (2016) proposed a hybrid approach to relationship extraction from stories. This methodology combines the features of unsupervised and supervised learning methods and also uses some rules to extract relationships.

The methodology proposed by them is designed to work similarly to how the human reader processes a story to find character relationships. Two phases are included in this method. The first phase is the sentence-classification stage. Sentences which contain mentions of character pairs are passed to this phase and then it classifies these sentences into corresponding relationship classes. If phase one fails to classify these sentences into one of the classes then it is passed to phase two, which is a text semantic-similarity checking stage.

The advantage of this system is that it uses supervised and rule-based approaches to get fairly accurate output when compared to existing systems.

2.4 *Discussion of datasets for relationship extraction*

Articles from the online encyclopedia Wikipedia are used to model the relationship between characters in various approaches. Culotta et al. (2006) sampled 1,127 paragraphs from 271 articles and labeled a total of 4,701 relation instances. The structure of the Wikipedia articles somewhat simplifies the extraction task, because important entities are hyperlinked within the text.

SparkNotes is another dataset ('Plot Overviews') of 300 English novels. This dataset contained 50% of fully annotated sequences (402 sentences) and 50% partially annotated sequences (containing 390 sentences, of which 201 were annotated). Of all annotated sentences, 472 were labeled with a cooperative state. AMT (Amazon Mechanical Turk) is another dataset used for this task. This dataset was used for evaluation only after training on SparkNotes data. The work done by Chaturvedi et al. (2016) uses the SparkNotes dataset for training and the AMT dataset for evaluation.

Srivastava et al. (2016) used the CMU (Carnegie Mellon University) Movie Summary Corpus, a collection of movie plot summaries from Wikipedia, along with aligned metadata, and then set up an online annotation task using the brat tool, resulting in a dataset of 153 movie summaries, consisting of 1,044 character relationship annotations.

Some approaches used a dataset of movie dialogs, including roughly 300,000 conversational turns between 10,000 pairs of characters in 617 movies. The advantage of choosing this dataset is, it not only provides the script of each movie, but also indicates uses this dataset in their work. The methodology used by Devisree and Raj (2016) was tested on Jane Austen's *Pride and Prejudice*. This novel contains a fairly high number of characters and a rich set of family relationships, making this a plausible choice for relationship extraction.

Table 1. Overview of the compared approaches and the datasets used.

| Type of approach | Proposed approaches | | | |
	Model	Dataset/Domain	Extracted relation	Results
Supervised	Culotta et al. (2006)	Articles from wikipedia	Mother, cousin, friend, education, boss, member of and rival	F1 = 0.6363 P = 0.7343 R = 0.5614
	Chaturvedi et al. (2016)	SparkNotes, AMT	Cooperative and non-cooperative	F1 = 0.76 P = 0.76 R = 0.76
	Srivastava et al. (2016)	Dataset of 300 English novel summaries	Cooperative and non-cooperative	F1 = 0.805 P = 0.806 R = 0.804
Unsupervised	Krishnan and Eisenstein (2014)	Dataset of movie scripts	Inducing the social function of address terms	F1 = 0.83
	Chaturvedi et al. (2017)	Dataset of 300 English novel summaries	Familial, Desire, Active, Communicative and Hostile	F1 = 0.55
Hybrid	Devisree and Raj (2016)	Collection of kids' stories	Parent-Child, Friendship, No-Relation	P = 0.87 R = 0.79
	Makazhanov et al. (2014)	Novels	Familial relations	A = 0.78

3 FUTURE RESEARCH

Characterizing the relationships between people can be useful for understanding the text. Most of the methodologies presented so far were a domain-specific approach. So we need to extend this approach to different domains. The current approaches made several assumptions about the types of relationships that can be relaxed in future work. Future work could also focus on studying asymmetric relationships. Other directions of study could include the usefulness of varying text modes (genre, number of characters, time period of novels, etc.) or mining 'relationship patterns' from such texts. Recent advances in deep learning can be used to get a better result in relation extraction between characters. Error analysis shows that mismatched co-reference labeling is the most common source of errors in the existing models. In future, these models could be customized to study the various stages of certain types of relationships. In addition, including more contextual information can improve the characterization of relationships.

4 CONCLUSION

There are different methodologies for extracting relationships between characters. Commonly used approaches for modeling relationships can be classified into two: supervised and unsupervised. Some methodologies use a mixture of approaches and are called hybrid approaches. They take advantage of different methodologies to get better results. Some of the recent approaches for extracting relationships that belong to these three classes are discussed in this paper. Research in this area still has room for improvement.

REFERENCES

Agarwal, A., Kotalwar, A., Zheng, J. & Rambow, O. (2013). SINNET: Social interaction network extractor from text. In *The Companion Volume of the Proceedings of IJCNLP 2013: System Demonstrations, 14–18 October 2013, Nagoya, Japan* (pp. 33–36).

Augenstein, I., Das, M., Riedel, S., Vikraman, L. & McCallum, A. (2017). SemEval 2017 Task 10: Science IE - Extracting keyphrases and relations from scientific publications. *arXiv:1704.02853*.

Bost, X., Labatut, V., Gueye, S. & Linares, G. (2017). Extraction and analysis of dynamic conversational networks from TV Series.

Brennan, J.R., Stabler, E.P., Van Wagenen, S.E., Luh, W.-M. & Hale, J.T. (2016). Abstract linguistic structure correlates with temporal activity during naturalistic comprehension. *Brain and Language, 157,* 81–94.

Chaturvedi, S., Iyyer, M. & Daumé, H., III. (2017). Unsupervised learning of evolving relationships between literary characters. In *Proceedings of the Thirty-First AAAI Conference on Artificial Intelligence (AAAI-17)* (pp. 3159–3165). Palo Alto, CA: AAAI Press.

Chaturvedi, S., Srivastava, S., Daumé, H., III & Dyer, C. (2016). Modeling evolving relationships between characters in literary novels. In *Proceedings of the Thirtieth AAAI Conference on Artificial Intelligence (AAAI-16), 12–17 February 2016, Phoenix, Arizona* (pp. 2704–2710). Palo Alto, CA: AAAI Press.

Culotta, A., McCallum, A., & Betz, J. (2006). Integrating probabilistic extraction models and data mining to discover relations and patterns in text. In *Proceedings of the Human Language Technology Conference of the North American Chapter of the Association of Computational Linguistics* (pp. 296–303). Stroudsburg, PA: Association for Computational Linguistics.

Devisree, V. & Raj, P.R. (2016). A hybrid approach to relationship extraction from stories. *Procedia Technology, 24,* 1499–1506.

Frunza, O., Inkpen, D. & Tran, T. (2011). A machine learning approach for identifying disease-treatment relations in short texts. *IEEE Transactions on Knowledge and Data Engineering, 23*(6), 801–814.

He, H., Barbosa, D. & Kondrak, G. (2013). Identification of speakers in novels. In *Proceedings of the 51st Annual Meeting of the Association for Computational Linguistics* (Vol. 1, pp. 1312–1320).

Hoffmann, R., Zhang, C., Ling, X., Zettlemoyer, L. & Weld, D.S. (2011). Knowledge-based weak supervision for information extraction of overlapping relations. In *Human Language Technologies:*

Proceedings of the 49th Annual Meeting of the Association for Computational Linguistics (pp. 541–550). Association for Computational Linguistics.

Iyyer, M., Guha, A., Chaturvedi, S., Boyd-Graber, J. & Daumé, H., III. (2016). Feuding families and former friends: Unsupervised learning for dynamic fictional relationships. In *Human Language Technologies: Proceedings of the 2016 Conference of the North American Chapter of the Association for Computational Linguistics* (pp. 1534–1544). Stroudsburg, PA: Association for Computational Linguistics.

Krishnan, V. & Eisenstein, J. (2014). You're Mr. Lebowski, I'm the dude: Inducing address term formality in signed social networks. *arXiv:1411.4351.*

Kwon, H., Kwon, H.T. & Yoon, W.C. (2016). An information-theoretic evaluation of narrative complexity for interactive writing support. *Expert Systems with Applications, 53*, 219–230.

Ling, Y. (2017). *Methods and Techniques for Clinical Text Modeling and Analytics* (Doctoral dissertation, Drexel University, Philadelphia, PA).

Makazhanov, A., Barbosa, D. & Kondrak, G. (2014). Extracting family relationship networks from novels. *arXiv:1405.0603.*

Pichotta, K. & Mooney, R.J. (2016). Learning statistical scripts with LSTM recurrent neural networks. In *Proceedings of the Thirtieth AAAI Conference on Artificial Intelligence, 12–17 February 2016, Phoenix, Arizona* (pp. 2800–2806). Palo Alto, CA: AAAI Press.

Ru, C., Li, S., Tang, J., Gao, Y. & Wang, T. (2017). Open relation extraction based on core dependency phrase clustering. In *2017 IEEE Second International Conference on Data Science in Cyberspace (DSC)* (pp. 398–404). New York, NY: IEEE.

Shaalan, K. & Oudah, M. (2014). A hybrid approach to Arabic named entity recognition. *Journal of Information Science, 40*(1), 67–87.

Srivastava, S., Chaturvedi, S. & Mitchell, T.M. (2016). Inferring interpersonal relations in narrative summaries. In *Proceedings of the Thirtieth AAAI Conference on Artificial Intelligence, 12–17 February 2016, Phoenix, Arizona* (pp. 2807–2813). Palo Alto, CA: AAAI Press.

Emerging Trends in Engineering, Science and Technology for Society,
Energy and Environment – Vanchipura & Jiji (Eds)
© *2018 Taylor & Francis Group, London, ISBN 978-0-8153-5760-5*

A survey of morphosyntactic lexicon generation

M. Rahul & S. Shine
Government Engineering College, Palakkad, Kerala, India

ABSTRACT: Morphosyntactic lexicons are the vocabulary of a language and they have a vital role in the field of Natural Language Processing (NLP). They provide information about the morphological and syntactic roles of words in a language. There are various methodologies proposed for the generation of highly accurate and large volumes of morphosyntactic lexicons, which are mainly focused on Machine Learning (ML) approaches. The aim of this survey is to explore various methodologies for morphosyntactic lexicon generation and discuss their advantages and disadvantages.

1 INTRODUCTION

A lexicon can be referred to as a large collection of information about the vocabulary of a language. Information in the sense of the lexical category to which each lexeme belongs. Such a lexical resource often imposes an internal structure called lexical entries such that it can be easily accessed by machines for various applications. Lexicons are the primary resource a Natural Language Processing (NLP) application requires at the critical stages of processing. More precisely, a lexicon is a component of an NLP system (Guthrie et al., 1996) that bears both semantic and grammatical information of individual words in a language. Such a lexical resource is used for NLP applications like machine translation, morphological tagging, part-of-speech tagging, language modeling and dependency parsing (Allen, 1995).

Machine-readable dictionaries can be considered for dealing with NLP applications which require large amounts of lexical information. Dictionaries such as Collins' COBUILD Dictionary (Moon, 2007) and Oxford Advanced Learner's Dictionary (OALD) (Hornby et al., 1995), among others, are developed for the above-mentioned applications. But a morphological and syntactic lexicon, or a morphosyntactic lexicon, can do more in a semantic and grammatical perspective.

The traditional approach to construct such a lexical resource involves large amounts of human intervention. Deep lexicographic and morphological analysis is required for the construction of an error-free, high-quality, vast lexicon. Such a job is not very easy because the vocabulary of a language can be very large. The annotation of words, with corresponding morphosyntactic properties, demands a great deal of time and requires experienced human annotators. The difficulties involved in such a job again increase with the complexity of the language we are dealing with. So, whether a language is an isolating, agglutinating or inflecting one will determine how much effort is necessary to accomplish such a task. Such a lexical resource is not available for every language; the coverage is limited. Because there is a lot of manual work in the background, such resources will be very expensive to obtain. In addition, quality and accuracy are the important factors which highly influence the performance of the NLP applications that require lexical resource at the critical stages. In order to overcome these difficulties, we require a language-independent approach to address the problem of morphosyntactic lexicon generation.

Machine Learning (ML) is a powerful tool for addressing many NLP problems. Here also, we can make use of various techniques in ML to address the problem of morphosyntactic lexicon generation. Beginning with the supervised approach, it requires a large volume of annotated data at the training phase. But with the help of freely available lexical resources like Wiktionary (Zesch et al., 2008) we can construct models that are capable of classifying the instances in a correct manner. Again, issues such as overfitting and large computation time will affect the model. Unlabeled resources are available in large scale but we possess few ways to exploit them. Semi-supervised learning is an efficient approach. It makes use of the concepts of both supervised and unsupervised learning strategies while, at the same time, demanding less human effort and giving higher accuracy (Zhu, 2005). Semi-supervised learning begins with a seed set of annotated instances at the training phase. It uses the various parameters estimated and labels the rest of the instances (which is purely an unsupervised approach). Constructing a seed lexicon, and obtaining unlabeled instances of the same, is not a tedious job for any particular language.

The proposed approaches make use of ML techniques to generate morphosyntactic lexicons in a language-independent manner. Thus they can be applied for any language, no matter how complex it is.

The aim of this survey is to identify the different methodologies for constructing morphosyntactic lexicons and discuss their merits and demerits. This study has high importance in the current scenario because, as already mentioned, many NLP applications require such a lexical resource at different stages of processing.

This paper is organized as follows: Section 2 gives a formal definition of the problem and discusses various classification (machine learning) approaches; Section 3 discusses future research; Section 4 makes brief concluding comments.

2 MORPHOSYNTACTIC LEXICON

2.1 *Definition*

A typical morphosyntactic lexicon contains the base forms and inflected forms of words along with their grammatical and semantic information such as grammatical category and sub-categorization features. Table 1 shows a subset of a morphosyntactic lexicon for the English language.

The given lexicon consists of the base form and all the inflected forms for the word 'cry'. Each lexical entry is associated with certain attributes such as part of speech, number information and gender information. If we can determine the lexical level and the morphological, syntactic, and semantic relation between these words with the help of unlabeled corpora, then the generation of large volumes of morphologically and syntactically annotated corpora are possible. In this survey, the methods proposed are all language-independent so that they can be adapted independently to any language.

2.2 *Semi-supervised methods*

Semi-supervised methods best exploit the unlabeled corpora that are comparatively easy to obtain. Less human effort is required while, at the same time, producing higher accuracy. It also helps to build good classifiers.

Faruqui et al. (2015) proposed an approach in which they used graph-based semi-supervised learning. Graph-based semi-supervised learning has a vital importance in the field of

Table 1. Attributes for base form and its inflected forms.

Word types	Attributes
Cry	POS: VERB ...
Cried	POS: VERB, VFORM: FINITE, TENSE: PAST ...
Crying	POS: VERB, VFORM: GERUND, TENSE: PRESENT ...
Cries	POS: VERB, NUMBER: SINGULAR, DEGREE: THIRD PERSON ...

machine learning. The underlying concept is that, using a small set of annotated data points, or labeled data points, large sets of unannotated or unlabeled data points can be labeled. Because it is a graph-based approach it begins with a graph which is weighted. The assumption is that the edge connected with a large weighting tends to have the same label. In this way, the attribute of a labeled node can be propagated through the higher weighted edge to the unlabeled node.

The method proposed makes use of graph-based semi-supervised learning. While dealing with unlabeled instances, a transductive approach called 'label propagation' is used. It begins with a lexical graph (Polguere, 2009) that contains both labeled and unlabeled nodes. Nodes refer to the word types in a language. The lexical graph here covers the vocabulary of English. Nodes are connected with edges which are labeled with some features shared among them. The features are prefix–suffix information (the English language only contains these two inflections), morphological transformation, and the cluster number to which both words belong. Using the seed set that contains only labeled nodes, the system can automatically generate lexicons which are both syntactically and morphologically annotated.

The model estimation is purely a supervised approach in which only labeled nodes are used for the empirical estimation of various parameters such as attribute vector and weight vector. Using the parameters learned, the unlabeled nodes are labeled in the label propagation phase iteratively. The paradigm projection phase finally determines the most appropriate attribute labels among multiple possible attributes (multi-label classification problem).

The advantage of this method is that, because it is a graph-based approach, the methodology is said to be language-independent. No language-specific assumptions are made at any stage of processing. Semi-supervised learning requires few words to be annotated. It produces a highly accurate, large volume of lexicons. The disadvantage is that the construction of a lexical graph is a somewhat difficult task because a lexical graph almost covers the target language.

Ahlberg et al. (2014) address the problem of paradigm induction with the help of inflection tables. An inflection table contains all the inflected forms of words in a language (Halle & Marantz, 1993). With inflection tables as input, the system produces a generalization of inflection paradigms and this helps in assigning attributes correctly to the unlabeled lexical entries based on the generalization information. Also, with the help of an unannotated corpora, the inflection tables can be expanded and produce more generalizations. The notation for paradigm consists of ordered string subsequences interspersed with variables. The variables can be instantiated over other words.

The paradigm generalization algorithm finds the common subsequences among various words using the Multiple Longest Common Subsequence (MLCS) problem (Pietrzak, 2003; Wang et al., 2011). The MLCS problem is actually an NP-hard problem but this can be solved heuristically. At the prediction phase the candidate word is matched against patterns. In order to enhance the matching, additional information is given; for example, part of speech information. When multiple matching occurs then a confidence scoring method helps to select the most appropriate paradigm.

The advantage of this methodology is that it produces human-readable generalizations. In addition, because it uses abstract paradigm representation, this only considers how the segments vary within the inflection table entries. The approach is language-independent. The disadvantage is that it uses suffix matching to build the tables from base form. It may well work with inflections close to the end of words, as in English, but not for inflections such as infix and circumfix.

Banea et al. (2008) proposed a bootstrapping method to construct subjectivity lexicons (Mihalcea et al., 2007). Because it is a semi-supervised approach, subjective words are used as the seed set. Banea et al. used an online dictionary and a raw corpus to generate a lexicon. At each step of the bootstrapping iteration, the seed set was expanded with related words occurring in the online dictionary, in response to a well-formulated query. The noise is filtered out in the filtering step by calculating the similarity between the extracted candidates and the seed set. For a similarity calculation, Pointwise Mutual Information and Latent Semantic Analysis (LSA) (Bouma, 2009) are used. The advantage is that this method can be applied to languages with scarce resources. The disadvantage is that this methodology requires an online dictionary and only focuses on subjectivity lexicons.

2.3 *Supervised methods*

Durrett and DeNero (2013) proposed a supervised approach to predict the inflected forms of a lexical item. The method consists of acquiring orthographic transformation rules of morphological paradigms automatically from labeled examples. For training they used Wiktionary data. Wiktionary is a crowdsourced lexical resource that contains the root word and all its inflected forms. The first step in the learning procedure consists of identifying interpretable transformation rules, which are generated by aligning each base form to all of its inflected forms. The learning procedure involves three main steps. These are alignment, span merging, and rule extraction. In the alignment step each inflected form is aligned with the base form using an iterative edit-distance algorithm (Marzal & Vidal, 1993). The iterative edit-distance algorithm determines the lowest cost transformation of the base form into its inflected form using single-character operations such as insertion, deletion and substitution. In the prediction phase, transformation rules are applied for an unseen base form, which will generate more candidate inflectional tables. Semi-Markov Conditional Random Fields (CRF) (Sarawagi & Cohen, 2005) are then trained to apply these rules correctly to unseen base forms (prediction). The advantage of this method is that it is a data-driven approach that best exploits the Wiktionary lexical resource.

Ahlberg et al. (2015) proposed paradigm classification in morphology by using a supervised learning method. This method also makes use of inflectional tables and Longest Common Subsequences (LCS) at intermediate steps to address the problem of paradigm classification. The method makes use of paradigm generalization (Ahlberg et al., 2014) for generating generalization paradigms, and a Support Vector Machine (SVM) (Hearst et al., 1998) for classification.

2.4 *Unsupervised methods*

Unsupervised methods require a large volume of unannotated data sets which can easily be obtained for almost every language. Soricut and Och (2015) proposed an unsupervised methodology for morphology induction using concept word embeddings. Here, the algorithm induces morphological transformations between words by learning the rules of those transformations. This method only considers two kinds of morphological transformations, one through adding a prefix and the other through adding a suffix. This also provides a mechanism to apply these rules to known words, unknown words and rare words. Prefix–suffix rules for all possible pairs of words are extracted from the vocabulary and train an embedding space for all words. The skip-gram (Cheng et al., 2006; Guthrie et al., 2006) model is used for training. The advantage of this methodology is that training requires only a monolingual corpus. No language-specific information is required.

2.5 *Discussion on datasets*

ML approaches require a large volume of resources for training and evaluation. These may be annotated or unannotated.

Faruqui et al. (2015) used one year of news articles scraped from different sources for the construction of a lexical graph and clustered these words with the help of an exchange algorithm because the cluster number is one of the edge features.

Durrett and DeNero (2013) and Ahlberg et al. (2015) used inflectional tables scraped from the Wiktionary lexical resource, which contains lexical information such as meaning, definition and description.

The online Romanian dictionary was used by Banea et al. (2008) for expanding the subjectivity lexicon by fetching related words in response to well-formulated queries.

Table 2. Overview of the compared approaches and the datasets used.

Type of approach	Model	Dataset
Supervised	Durrett & DeNero (2013)	Wiktionary
	Ahlberg et al. (2015)	Wiktionary
Unsupervised	Soricut & Och (2015)	Wikipedia
		WMT-2013 Shared Task
		Arabic Gigaword
Semi-supervised	Faruqui et al. (2015)	News articles scraped from different sources
	Banea et al. (2008)	Online Romanian dictionary
	Ahlberg et al. (2014)	Wiktionary

Soricut and Och (2015) used the Wikipedia resource for the English language. For German, French and Spanish, they used WMT-2013 Shared Task, and the Arabic Gigaword corpus was used for the Arabic language. Various other manually created datasets are used for training and evaluation for individual languages.

3 FUTURE RESEARCH

Semi-supervised learning approaches best fit the problem of morphologically and syntactically annotated lexicon generation if corpora are available for individual languages. Most of the approaches discussed so far considered only prefix and suffix transformations. Other morphological transformations should also be considered while constructing the models because the model should be capable of handling any complex language. If the methodology is language-independent, then it can be applied to NLP applications in any language.

4 CONCLUSION

ML approaches can obtain higher accuracy and exhibit better performance. The availability and correctness of the lexical resource during training and evaluation largely influences the NLP applications.

This survey has mainly focused on various approaches to addressing the problem of lexicon generation with syntactic and morphological information using ML techniques such as supervised, semi-supervised and unsupervised methods in which most of the methodologies are language-independent. This makes the adaptation of the various methodologies into any language possible, which, in turn, accelerates the development of various natural language applications in individual languages. Semi-supervised learning methodologies are efficient in this area and can generate more accurate and large volumes of morphosyntactic lexicons.

REFERENCES

Ahlberg, M., Forsberg, M. & Hulden, M. (2014). Semi-supervised learning of morphological paradigms and lexicons. In *Proceedings of the 14th Conference of the European Chapter of the Association for Computational Linguistics* (pp. 569–578). Gothenburg, Sweden: Association for Computational Linguistics.

Ahlberg, M., Forsberg, M. & Hulden, M. (2015). Paradigm classification in supervised learning of morphology. In *Human Language Technologies: The 2015 Annual Conference of the North American Chapter of the Association for Computational Linguistics* (pp. 1024–1029). Stroudsburg, PA: Association for Computational Linguistics.

Allen, J. (1995). *Natural language understanding*. London, UK: Pearson.

Banea, C., Mihalcea, R. & Wiebe, J. (2008). A bootstrapping method for building subjectivity lexicons for languages with scarce resources. In *Proceedings of the 6th International Conference on Language Resources and Evaluation, LREC-08, 26 May – 1 June 2008, Marrakech, Morocco* (pp. 2764–2767). Paris, France: European Language Resources Association.

Bouma, G. (2009). Normalized (pointwise) mutual information in collocation extraction. In *Proceedings of the Biennial GSCL Conference 2009* (pp. 31–40).

Cheng, W., Greaves, C. & Warren, M. (2006). From n-gram to skipgram to concgram. *International Journal of Corpus Linguistics*, *11*(4), 411–433.

Durrett, G. & DeNero, J. (2013). Supervised learning of complete morphological paradigms. In *Human Language Technologies: Proceedings of the 2013 Annual Conference of the North American Chapter of the Association for Computational Linguistics* (pp. 1185–1195). Stroudsburg, PA: Association for Computational Linguistics.

Faruqui, M., McDonald, R. & Soricut, R. (2015). Morphosyntactic lexicon generation using graph-based semi-supervised learning. *arXiv:1512.05030*.

Guthrie, D., Allison, B., Liu, W., Guthrie, L. & Wilks, Y. (2006). A closer look at skip-gram modelling. In *Proceedings of the 5th International Conference on Language Resources and Evaluation (LREC-06)* (pp. 1–4). Paris, France: European Language Resources Association.

Guthrie, L., Pustejovsky, J., Wilks, Y. & Slator, B.M. (1996). The role of lexicons in natural language processing. *Communications of the ACM*, *39*(1), 63–72.

Halle, M. & Marantz, A. (1993). Distributed morphology and the pieces of inflection. In K. Hale & S.J. Keyser (Eds.), *The view from building 20* (pp. 111–176). Cambridge, MA: The MIT Press.

Hearst, M.A., Dumais, S.T., Osuna, E., Platt, J. & Scholkopf, B. (1998). Support vector machines. *IEEE Intelligent Systems and their Applications*, *13*(4), 18–28.

Hornby, A.S., Wehmeier, S. & Ashby, M. (1995). *Oxford advanced learner's dictionary* (5th ed.). Oxford, UK: Oxford University Press.

Marzal, A. & Vidal, E. (1993). Computation of normalized edit distance and applications. *IEEE Transactions on Pattern Analysis and Machine Intelligence*, *15*(9), 926–932.

Mihalcea, R., Banea, C. & Wiebe, J. (2007). Learning multilingual subjective language via cross-lingual projections. *Annual Meeting Association for Computational Linguistics*, *45*(1), 976–983.

Moon, R. (2007). Sinclair, lexicography, and the Cobuild Project: The application of theory. *International Journal of Corpus Linguistics*, *12*(2), 159–181.

Pietrzak, K. (2003). On the parameterized complexity of the fixed alphabet shortest common supersequence and longest common subsequence problems, *Journal of Computer and System Sciences*, *67*(4), 757–771.

Polguere, A. (2009). Lexical systems: Graph models of natural language lexicons. *Language Resources and Evaluation*, *43*(1), 41–55.

Sarawagi, S. & Cohen, W.W. (2005). Semi-Markov conditional random fields for information extraction. In *Proceedings of the 17th International Conference on Neural Information Processing Systems* (pp. 1185–1192). Cambridge, MA: MIT Press.

Soricut, R. & Och, F.J. (2015). Unsupervised morphology induction using word embeddings. In *Human Language Technologies: Proceedings of the 2015 Annual Conference of the North American Chapter of the Association for Computational Linguistics* (pp. 1627–1637). Stroudsburg, PA: Association for Computational Linguistics.

Wang, Q., Korkin, D. & Shang, Y. (2011). A fast multiple longest common subsequence (MLCS) algorithm. *IEEE Transactions on Knowledge and Data Engineering*, *23*(3), 321–334.

Zesch, T., Müller, C. & Gurevych, I. (2008). Extracting lexical semantic knowledge from Wikipedia and Wiktionary. In *Proceedings of the 6th International Conference on Language Resources and Evaluation, LREC-08, 26 May – 1 June 2008, Marrakech, Morocco* (pp. 1646–1652). Paris, France: European Language Resources Association.

Zhu, X. (2005). *Semi-supervised learning literature survey*. Technical Report 1530, Computer Science, University of Wisconsin-Madison. Retrieved from https://minds.wisconsin.edu/bitstream/handle/1793/60444/TR1530.pdf.

Emerging Trends in Engineering, Science and Technology for Society,
Energy and Environment – Vanchipura & Jiji (Eds)
© 2018 Taylor & Francis Group, London, ISBN 978-0-8153-5760-5

Classification of question answering systems: A survey

S. Sandhini & R. Binu
Computer Science and Engineering Department, Government Engineering College, Palakkad,
Kerala, India

ABSTRACT: Question Answering Systems (QAS) are a unique kind of information retrieval. In Question Answering (QA) the system retrieves the precise answer to the questions asked by the user in natural language. QA is multidisciplinary. It involves information retrieval, natural language processing, linguistics, knowledge representation, databases, software engineering, artificial intelligence and so on. This paper classifies QAS based on different techniques used in answer ranking and answer extraction. The survey also provides the main contributions, experimental results, and limitations of various approaches. Finally, we discuss our perspective on the future direction of QAS.

1 INTRODUCTION

Question Answering (QA) is a particular type of Information Retrieval (IR). The main aim of Question Answering Systems (QAS) is to retrieve an exact answer to a question in natural language. The correlation among the QA framework and information retrieval technology, in IR, the input query containing keywords, and the output consists of a set of documents that are important to the query asked by a user. QA is not the same as IR in that the user can ask a question specifically of the system in natural language. The system at that point answers the question as a concise answer extracted from a source document.

QASs developed in different domains, such as information sources, sorts of questions, arrangements of answers, and so on; the quantity of such QASs is too huge. The main type of questions presented by users in natural language are the factoid questions, for example, "When did the Egyptian revolution occur?". Also, QA frameworks are arranged into two main classes, in particular open domain QA systems and closed domain QA systems.

This survey discusses the different methodologies used for answer extraction and the merits and demerits of each. In this paper, Section 2 describes the overview of QAS and different criteria in support of classifying the large number of QASs available. Section 3 discusses various future research directions. Section 4 is the conclusion.

2 QUESTION ANSWERING SYSTEMS (QASs)

Figure 1 shows the architecture of a QA framework. The system comprises of three different modules, for example, question processing, document processing, and answer processing. The "query processing module" deals with question classification, information retrieval is done in the "document processing module", and the answer extraction process is dealt with in the "answer processing module".

The question processing module performs question analysis, question classification, and answer type classification. Question analysis is referred to as the question focus. The question focus is defined as the word or series of words which shows what information is requested in the question (Harabagiu et al., 2000). Keeping in mind that the end goal is to answer a question correctly, we need to know the type of question requested. Answer type classification

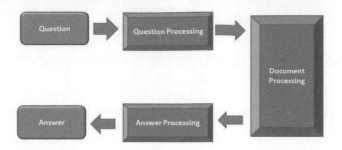

Figure 1.   QA system architecture.

is related to question classification. In document processing, it retrieves a set of rank documents that are relevant to the question. It is referred as a paragraph indexing module. The last phase in the QA design, the answer processing module, is in charge of determining, retrieving and validating the answers from the list of documents retrieved during the document processing module (Allam & Haggag, 2012).

The present search engine or information retrieval system only does the retrieval of the document, i.e. given a few keywords and to answer some relevant information associated with the given keyword. Information retrieval does not retrieve the answer. The user wants an exact response to a question (Hirschman & Gaizauskas, 2001).

The criteria for classifying QAS are the types of techniques used in answer ranking and also in retrieving answers.

2.1   Classification of QAS based on answer ranking

Here the classification is based on the techniques used in answer ranking such as occurrence frequency of candidate answers, relevance between information source and question and similarity between question and answer. Mendes et al. (2014) proposed a method depending on candidate answers occurrence frequency attention about recognition of logical relationship which includes comparability relationship, consideration relationship and incomplete connection between candidate answers based on the fact that logical relationship decide the results of answer arranging. The approach recommended by Toba et al. (2014) and Figueroa and Neumann (2014) depends on the similarity among question and answer gives more comparable is question to answer, greater dependable is the proper response of question. Liu et al. (2014) propose a methodology depends on relevance among data source and question gives extra applicable is question to data source, extra valid is the right response retrieved from the data source.

Mendes et al. (2014) proposes a method to select the answer based on semantic relations. The fundamental answer selection approach depends on the frequency. It comprises three steps. First, the candidate answers are normalized to a canonical format, for example date answers are set to the form D01 M01 Y2015. Second, each two answers are compared. If they are equivalent, then they are accepted to be the same entity and the score of each answer is expanded by one. Here, the score of each answer is its frequency of occurrence. Finally, the relations that exist between each pair of candidate answers are considered. Once more, each pair of candidate answers is analyzed to distinguish if a relation exists. The answer's scores are refreshed, depending upon their relations with different candidates of the event of every candidate answer: the most frequent candidate answer is picked as the relevant answer.

Toba et al. (2014), propose a method to attempt to show every individual question answer combine in an unexpected way will move toward to any other maximum which isn't feasible. In this, they propose a system which initially considers the question contents to manage the correct model selection to be utilized for characterization of answer quality, before accomplishing the appropriate evaluation of the answer quality.

Liu et al. (2014) propose a novel approach recommending essential and complementary Q&A documents to knowledge groups of QA web sites. The original ideas of this approach are as follows: it produces community topic profiles by considering QA gathering factors, such as 'community member', the push ranking of QAs and the gathering time of QAs from the sincerely collected QA facts on particular topics. And it also finds the ranking of QAs based on question similarity and answer relationship. It proposes a QA-based integral approach and topic based correlative way to propose complementary QA documents.

2.2 *Classification of QAS based on answer extraction*

This classifies QAS based on techniques used for retrieving answers. The strategy of answer extraction depends on text patterns, named entity, and similarity computing between sentences. Wu et al. (2015) propose a technique to find the appropriate answer to the question which is most extreme similar to question in Q&A dataset is returned in view of sentence similarity. Ravichandran and Hovy (2002) propose a model for discovering answers by misusing surface text data utilizing manually developed surface patterns. Xu et al. (2003) developed a method to improve the poor recall of the manual hand-making patterns. Liu et al. (2015) propose a method based on similarity computing between sentences. Peng et al. (2005) introduced a way to deal with capturing long-distance dependencies through utilizing linguistic systems to upgrade patterns. Rather than exploiting surface text data utilizing patterns, Lee et al. (2005) utilized the named-substance way to deal with finding an answer.

Ravichandran and Hovy (2002) use patterns to find the answer to a question based on the following steps. First, determine the type of the question. Then the question term is recognized. The third step is to create a query from the question term and perform information retrieval. Then segment the documents into different sentences. After that, replace the question term in each sentence by the question tag. And a pattern table is created for the particular question type. Finally, sort these answers based on the pattern's score and return the top answers.

Xu et al. (2003) introduce definitional QA. This approach combines complementary technologies, including information retrieval and various linguistic and extraction for analyzing text. First, the question type is identified, i.e. whether a question is a *who* or a *what* question. Second, information retrieval pulled documents about the question focus from the TREC corpus. Third, heuristics were applied to the sentences in the retrieved documents to decide whether they say the question target. Fourth, kernel facts that specify the question target were extracted from sentences by a variety of semantic preparing and information extraction tools. Fifth, all kernel facts were positioned by their type and their comparability to the profile of the question. Finally, heuristics were applied to identify repetitive kernel facts.

Lee et al. (2005) developed a hierarchical classifier that classifies questions into fine-grained classes. For classifying Chinese questions, they presented a machine learning approach (SVM) and a knowledge-based approach.

2.3 *Contributions, experiments and limitations*

Mendes et al. (2014)
In this research, the most frequent candidate answer is selected without any other processing. The selection is mainly based on the relation between the answers. If all candidate answers have the same frequency, the system probably choose randomly among all answers. Results show an accuracy of 33.89% for the 438 questions.

Toba et al. (2014)
The framework provides predicting quality of answers in the community QA. The primary objective of this method is to identify the best machine learning algorithm in terms of feature selection. The experiment shows that the best machine learning model that fits this approach is logistic regression.

Liu et al. (2014)

This method is to solve the issue of information overloading in community QA web sites. The experimental results show that the performance of the recommendation approaches. In this, it considers community topic profiles with QA collection elements, and complementary scores of QAs, in step with forms better than traditional group primarily based recommendation methods.

Ravichandran and Hovy (2002)

In this, the technique takes in online data patterns utilizing a few seed questions and answer anchors, without requiring human comment. A TREC10 question data set is used for the two sets of experiments. In the first, the TREC corpus is applied as the data source utilizing an IR factor in their QA framework. Within the second test, the internet is applied as the data source utilizing the AltaVista web engine to carry out IR. The demerits of this method are the system performs only certain types of questions. So, in large definitional questions it performs badly, as the patterns did not deal with lengthy-distance dependencies.

Xu et al. (2003)

This uses various approaches, consisting of data retrieval and distinct linguistic and extraction tools like parsing, name finding, co-reference resolution, extraction of relation, proposition and established patterns, so it adopts a hybrid approach. Here they performed three runs utilizing the F-metric for evaluation. In the main run, BBN2003A, the web was not utilized as part of the answer finding. In the second run, BBN2003B, answers for factoid questions. Finally, BBN2003C was the same as BBN2003B except that if the answer for a factoid question was discovered various times in the corpus, its score was supported. The performance for BBN2003A, BBN2003B, BBN2003C runs was 52.1%, 52.0% and 55.5% respectively. The limitation of this approach, is that the experiments only tested for "what" and "when" questions. It didn't consider other factoid questions such as "when" and "where" questions.

Lee et al. (2005)

Their research proposed hybrid architecture for the NTCIR5 and CLQA to answer Chinese factoid questions. To arrange Chinese questions they displayed a machine learning approach

Table 1. Classification of QA systems.

	Classification based on answer ranking		
Approach	Model	Key problem	Type of information source
Based on occurrence frequency	Mendes et al. (2014)	Logical relationship recognition between candidate answers	Open
Based on similarity between questions	Toba et al. (2014)	Similarity computing	Closed
Relevance between information source and question	Liu et al. (2014)	Relevance computing	Open

	Classification based on answer extraction		
Approach	Model	Information source	Evaluation
Text patterns	Ravichandran and Hovy (2002)	TREC corpus	High recall rate Low precision rate
Similarity computing between sentences	Xu et al. (2003)	Yahoo! Answer	High precision rate Limited answering
Named entity	Lee et al. (2005)	TREC10 corpus	High accuracy

(SVM) and a knowledge-based approach (InfoMap). Various approaches are adopted and integrated including question focus, coarse-fine taxonomy and SVM machine learning. The method accomplished overall precision, for correct answers and for correct unsupported answers of 37.5% and 44.5% respectively. Likewise, utilizing InfoMap gave a precision of 88% for the question classification module and for utilizing support vector machine of 73.5%.

3 CONCLUSION AND FUTURE SCOPE

In this survey, we classified QASs on the basis of criteria such as answer ranking and answer extraction. The occurrence frequency of candidate answers, similarity between question and answer, relevance between information source and question functions are taken from the answer ranking method. And, for answer extraction text patterns, named entity and similarity computing between sentences features are taken. In future, valid answer extraction strategies are needed. For open domain QA systems, it returns fake candidate answers, noisy data, and imprecise candidate answers which influence the final answer.

REFERENCES

Allam, A.M.N. & Haggag, M.H. (2012). The question answering systems: A survey. *International Journal of Research and Reviews in Information Sciences (IJRRIS)*, 2(3).
Figueroa, A. & Neumann, G. (2014). Category-specific models for ranking effective paraphrases in community question answering. *Expert Systems with Applications*, 41(10), 4730–4742.
Gupta, V. & Lehal, G.S. (2009). A survey of text mining techniques and applications. *Journal of Emerging Technologies in Web Intelligence*, 1(1), 60–76.
Hao, T. & Agichtein, E. (2012). Finding similar questions in collaborative question answering archives: toward bootstrapping-based equivalent pattern learning. *Information Retrieval*, 15(3–4), 332–353.
Harabagiu, S.M., Moldovan, D.I., Pasca, M., Mihalcea, R., Surdeanu, M., Bunescu, R.C. & Morarescu, P. (2000). FALCON: Boosting knowledge for answer engines. In *TREC*. 9 (pp. 479–488).
Hirschman, L. & Gaizauskas, R. (2001). Natural language question answering: the view from here. *Natural Language Engineering*, 7(4), 275–300.
Kolomiyets, O. & Moens, M.F. (2011). A survey on question answering technology from an information retrieval perspective. *Information Sciences*, 181(24), 5412–5434.
Lampert, A. (2004). A quick introduction to question answering. *CSIRO ICT Centre*.
Lee, C., Shih, C., Day, M., Tsai, T., Jiang, T., Wu, C., Sung, C. & Hsu, W. (2005). ASQA: Academia Sinica question answering system for NTCIR-5 CLQA. In *Proceedings of NTCIR-5 Workshop Meeting*.
Liu, D.R., Chen, Y.H. & Huang, C.K. (2014). QA document recommendations for communities of question–answering websites. *Knowledge-Based Systems*, 57, 146–160.
Liu, D.R., Chen, Y.H., Shen, M. & Lu, P.J. (2015). Complementary QA network analysis for QA retrieval in social question-answering websites. *Journal of the Association for Information Science and Technology*, 66(1), 99–116.
Liu, Y., Yi, X., Chen, R. & Song, Y. (2016). A survey on frameworks and methods of question answering. In *Information Science and Control Engineering (ICISCE), 3rd International Conference* (pp. 115–119). New York, NY: IEEE.
Lopez, V., Uren, V., Sabou, M. & Motta, E. (2011). Is question answering fit for the semantic web? A survey. *Semantic Web*, 2(2), 125–155.
Mendes, A.C. & Coheur, L. (2011). An approach to answer selection in question-answering based on semantic relations. In *International Joint Conference on Artificial Intelligence (IJCAI)* (pp. 1852–1857).
Mendes, A.C., & Coheur, L. (2013). When the answer comes into question in question-answering: Survey and open issues. *Natural Language Engineering*, 19(1), 1–32.
Peng, F., Weischedel, R., Licuanan, A. & Xu, J. (2005). Combining deep linguistics analysis and surface pattern learning: A hybrid approach to Chinese definitional question answering. In *Proceedings of the Conference on Human Language Technology and Empirical Methods in Natural Language Processing* (pp. 307–314). Association for Computational Linguistics.

Ravichandran, D. & Hovy, E. (2002). Learning surface text patterns for a question answering system. In *Proceedings of the 40th Annual Meeting of the Association for Computational Linguistics* (pp. 41–47). Association for Computational Linguistics.

Stoyanchev, S., Song, Y.C., & Lahti, W. (2008). Exact phrases in information retrieval for question answering. In *Coling 2008: Proceedings of the 2nd workshop on Information Retrieval for Question Answering* (pp. 9–16). Association for Computational Linguistics.

Toba, H., Ming, Z.Y., Adriani, M. & Chua, T.S. (2014). Discovering high quality answers in community question answering archives using a hierarchy of classifiers. *Information Sciences 261*, 101–115.

Voorhees, E.M., & Buckland, L. (2003). Overview of the TREC 2003 Question Answering Track. In *TREC* (pp. 54–68).

Wu, Y., Hori, C., Kashioka, H. & Kawai, H. (2015). Leveraging social Q&A collections for improving complex question answering. *Computer Speech and Language, 29*, 1–19.

Xu, J., Licuanan, A. & Weischedel, R.M. (2003). TREC 2003 QA at BBN: Answering definitional questions. In *TREC* (pp. 98–106).

Zhang, D. & Lee, W.S. (2003). A web-based question answering system. Massachusetts Institute of Technology (DSpace@MIT).

Zheng, Z. (2002). AnswerBus question answering system. In *Proceedings of the Second International Conference on Human Language Technology Research* (pp. 399–404). Morgan Kaufmann Publishers Inc.

Emerging Trends in Engineering, Science and Technology for Society,
Energy and Environment – Vanchipura & Jiji (Eds)
© 2018 Taylor & Francis Group, London, ISBN 978-0-8153-5760-5

A survey of lexical simplification

K.S. Silpa & M. Irshad
Computer Science and Engineering Department, Government Engineering College,
Palakkad, Kerala, India

ABSTRACT: Lexical simplification is the process of identifying complex words in a sentence and replacing them with simpler substitutes to make the text more accessible to people. The possible candidate substitutions are generated in the second stage of a lexical simplification system, known as the substitution generation stage. In this survey, different methodologies used for generating such substitutions are discussed. The merits and drawbacks associated with each method are also described.

1 INTRODUCTION

While reading a document, article or any other text resources, we might struggle with unfamiliar words. Unfamiliar words can be regarded as complex words and these kinds of words make a text difficult to understand. Jargon, technical terminology, and so forth, are more difficult to understand for the general population. If everyone wrote texts in the simplest form, every reader could understand them easily. But these kinds of documents are very rare. Complex words are always a barrier to comprehending a text. If there was a system to identify these complex words and replace them with simpler alternatives, these kind of understandability problems could be rectified easily and the reader could get enough information from the text; otherwise, it may be left unread. So, lexical simplification systems have been introduced, which increase the readability of a sentence by identifying complicated terms and replacing them with a simpler substitute for better understanding.

A typical Lexical Simplification (LS) system consists of four stages. The first stage of any lexical simplification system is Complex Word Identification (CWI) and this is the most important stage of all stages in the LS pipeline. In this stage, the system must identify the complex word; the candidate substitutions for the identified complex word are generated later. But the task is very difficult to implement because there is no correct definition for a complex word. We don't know how to define a complex word. The complexity of a word differs from person to person. Different people have different vocabularies depending, for example, on the newspapers they read, their interaction manner, and so forth.

Either explicitly or implicitly, every lexical simplification system will identify a complex word. Among the different complex word identification methodologies, simplifying everything is the simplest method but it is not the most effective because it will consider all the words as complex.

The second stage of LS is known as Substitution Generation (SG). After correctly identifying a complex word, the next step is to generate a suitable substitute for that word. The substitutes have a similar meaning to the complex word but should be simpler. This can be done with or without considering the context. A method based on context is the extraction of a substitute word from the sentence-aligned parallel corpora of English Wikipedia and Simple English Wikipedia. The context and word alignment in both corpora are considered in order to generate the substitution. WordNet, using SG methods, is purely context-independent. All thesauri-based approaches are like this.

Not all of the generated substitutions are correct replacements of a complex word. So, filtering is required to identify the simplest word. The methods of Substitution Selection (SS) and Substitution Ranking (SR) are used, respectively, to select and rank the candidate substitutions, which are obtained from the SG stage. Most SS strategies have used Word Sense Disambiguation (WSD) techniques to avoid ambiguity. The same word may have different meanings according to the context. So, by using WSD techniques we can select the correct substitute. SR is for ranking the remaining candidate substitutions based on their simplicity. The first-ranked substitute is the simplest substitute and, finally, the complex word is replaced by this one.

Figure 1 describes the lexical simplification pipeline. Here, the complex sentence 'The cat perched on the mat' is taken as the input and, finally, the complex word perched is replaced with the simpler term, sat.

Lexical simplification systems were first developed for aphasic readers as an assistive technology. Aphasia is a disability related to understanding language, often caused by a stroke or head injury. Those who find understanding a text difficult could consider lexical simplification as an assistive technology. In the same way, LS is used by people who suffer from cognitive disabilities such as dyslexia. There are several assistive technology projects in LS, each of them different from one another, based on the needs of the group they are aiming to assist. Many people often find it difficult to understand technical terms such as medical terms, scientific terms, and so on. LS is also a solution for these kinds of problem. Another group of people who have benefited from LS are second-language learners, people with low literacy skills, children, non-native speakers, and so forth. So, LS is relevant to a wide variety of different applications.

There are four stages in the lexical simplification process which have been briefly described. The simplification rules are generated in the second stage, the SG stage. There are several

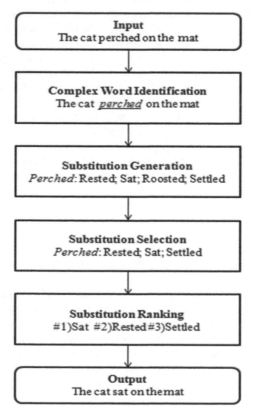

Figure 1. Stages in the lexical simplification process.

approaches for generating simplification rules. This survey aims to discuss the different methodologies for SG and make comparison between them.

The layout of this paper is as follows: Section 2 describes the various approaches for SG; Section 3 discusses future work in this area; Section 4 makes concluding remarks in relation to this work.

2 LEXICAL SIMPLIFICATION

Lexical simplification is the task of making difficult words easier to understand. In the process of lexical simplification, SG is the generation of all possible candidate substitution words for a complex word. Figure 2 shows the different approaches for generating the candidate substitutions.

2.1 *Using word-embedding models*

Word-embedding models are used to convert a word into vector format. This mapping is based on the context of the words. In LS, we are trying to find a simple word bearing a similar meaning to a complex term. Words with similar meanings will be adjacent to each other when converted to a vector and can thereby be extracted. It results in some complex and simple word pairs.

Kim et al. (2016) proposed a LS method for replacing complex scientific terms with simpler alternatives. In this method, parallel corpora of scientific articles and Wikipedia are used to extract word pairs. In order to generate the rule, word-embedding models are used, such as Word2vec and GloVe. Here, Word2vec is used for learning word vectors and it is used with a skip-gram architecture. The word-embedding dimension is set to 300, and the context window size is set to 10. After obtaining the word pairs, some filtering methods are applied to obtain the most appropriate substitutions, which are regarded as a kind of SS and SR approach. This includes calculating the cosine similarity between the words in a rule, elimination of word pairs with different parts of speech, and so forth.

Word2vec is a very effective tool for vector formation and fast training is also possible with it. The methodology of Kim et al. (2016) also provides a higher precision and is more efficient than constructing a co-occurrence matrix as proposed in Biran et al. (2011). However, the limitation of Word2vec is that it will map to antonyms instead of synonyms; in other words, instead of complex word to simpler synonym mapping, it results in complex word to simpler antonym mapping. This is because both antonyms and synonyms of a particular word may have the same context and so they appear adjacent in the vector representation. So, from the simplification rules, the antonyms must be filtered out.

Paetzold and Specia (2016) proposed a context-aware, word-embedding model for SG that uses a sense label assigned to each of the candidate words in the training corpus. Each sense of a particular word has a different numerical vector. The sense labels are similar to the Part-Of-Speech (POS) tag. But this kind of model will identify the inflected form of nouns and verbs as a different tag. To rectify this problem, all tags related to nouns, verbs, adjectives and adverbs are marked as N, V, J and R, respectively. Then the POS-tag annotated corpus is trained with models like GloVe. The Paetzold and Specia (2016) methodology uses a candidate generation

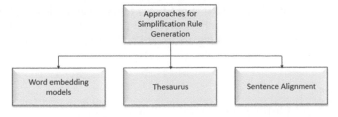

Figure 2. Classification of methodologies used for generating simplification rules.

algorithm, which helps to filter out the false and ungrammatical candidates. Thus, problems associated with ambiguity can be reduced. Traditional word-embedding models cannot accommodate the different senses of a particular word; in other words, for every sense of a word, it has only one numerical vector. So, this model is helpful in eliminating the limitations of traditional word-embedding models. The training corpus is huge, so giving a sense label to each of them is tedious work. This difficulty is a major problem associated with this methodology.

Paetzold and Specia (2017) proposed a methodology for SG, in which the context-aware word embeddings are enhanced with lexicon retrofitting. Retrofitting is always applied as a post-processing task to improve the quality of the vector. In this methodology, the retrofitting is performed with the help of WordNet over lexicon relations such as synonyms. The final result is dictionaries that contain a word-tag pair and its corresponding synonym. Faruqui et al. (2014) introduced a graph-based retrofitting algorithm, which is also used in this methodology. Sometimes, especially with more dimensions, word vectors are much more helpful for extracting semantic information.

2.2 *Using sentence alignment*

The methodology proposed by Horn et al. (2014) uses the sentence-aligned corpora of English Wikipedia (EW) and Simple English Wikipedia (SEW) for rule extraction. The words in EW are taken as normal words and the words in SEW are taken as simple words. There is a sentence-aligned simple sentence corresponding to each normal sentence in the corpus. Due to the same alignment of words in a normal and simple sentence, the candidate pairs can be easily extracted. Using the tool GIZA++, an alignment is automatically induced between the words, which helps to identify word pairs.

It is possible to have errors in word alignment and sentence alignment. So, we cannot guarantee that all alignments are correct and all extracted word pairs are real. Some filtering tasks need to be carried out. One of these is to remove the word pairs in which the normal words already exist in the stoplist. The stoplist is a list which contains only the simplest terms. If the normal word is in the stoplist, no further simplification is required. Similarly, if the normal word is a proper noun, then the corresponding word pairs can also be removed. The POS tags of these two words in a word pair are checked to ensure that they are same.

The main advantage of using the methodology proposed by Horn et al. (2014) is that it is helpful for generating a large number of simplification rules; in other words, the coverage is high. Large numbers of simplification rules lead to a simplification where the preservation of meaning is high. It is necessary to ensure that meaning and grammar is preserved in every LS. The method of Horn et al. performs well in the case of preservation of meaning. For any kind of sentence-aligned approach, there is a problem of one to two mappings of words between the SEW and EW: this will generate false word pairs that need to be filtered out.

Paetzold and Specia (2017) proposed two different methodologies for SG. One methodology is based on sentence alignment. It uses the Newsela corpus for extracting candidate substitutions. The Newsela corpus contains professionally simplified articles. Sentence alignment is then generated on all the pairs of versions of the Newsela articles. The paragraph and sentence-alignment algorithms proposed by Paetzold and Specia (2016) are used in this approach. Term Frequency–Inverse Document Frequency (TF-IDF) similarity is used to identify the correct paragraph and sentence alignment. After producing such alignments, the methodology proposed by Horn et al. (2014) is used to extract the word pairs.

The SG in Paetzold and Specia (2017) produces fewer grammatical and meaning errors and achieves the highest precision and F1 scores. But the disadvantage is that this methodology may cause a ranking error. Ranking errors are replacements which do not simplify the selected sentence.

2.3 *Using thesauri*

De Belder and Moens (2010) proposed a text-simplification system for children, in which both syntactic and lexical simplifications are employed and simplified news articles and

Wikipedia articles are used. In the lexical simplification task, WordNet is used, along with the Latent Words Language Model (LWLM) for extracting the synonyms. This methodology selects words as candidate substitutions which are both identified as a synonym in WordNet and selected by LWLM. Kucera–Francis frequency is used to score the candidates to obtain the simplest word. The word with highest Kucera–Francis frequency is considered to be the simplest synonym among the candidate synonyms.

The main disadvantage of using the methodology of De Belder and Moens (2010) is that it doesn't replace the most complex terms. This is because it is very difficult to identify a synonym for these kinds of complex terms. An alternative method is to replace the complex term with a brief explanation of that term, but this is not considered to be lexical simplification. As mentioned before, when using the methodology of Kim et al. (2016), antonyms can also be wrongly identified as simpler synonyms. These kinds of mistakes are not made in this approach. This is because WordNet contains a correct distinction between synonyms and antonyms.

There are several simplification systems for the English language. Bott et al. (2012) proposed a lexical simplification system for the Spanish language. In this system, on-line dictionaries are used for extracting the synonyms. There are no parallel corpora for the Spanish language, so parallel corpus-based methodologies are not relevant. Like WordNet in English, the Spanish language has Euro WordNet and the Spanish OpenThesaurus. To find a suitable synonym for a word, it checks all the possible alternatives of the given word in the Spanish OpenThesaurus. The selected words are then converted into word vectors. Word frequency and word length-based approaches are used to obtain simpler synonyms.

The main advantage of this methodology is that it notably preserves the meaning of text and it can be applied in any language with the help of resources available in that language. However, problems arise with words that are not in the dictionary. Many words may not appear in the dictionary; in these cases, the number of alternatives is minimized. So, there is a need to apply a word sense disambiguation technique for better performance.

Table 1. Overview of the compared approaches and datasets used for lexical simplification.

| Type of approach | Proposed approaches | | Result |
	Model	Dataset	
Word-embedding model	Kim et al. (2016)	Scientific corpus of nearly 500k publications in Public Library of Science (PLOS) and PubMed Central (PMC), paired with a general corpus from Wikipedia	f-measure – 0.285 Precision – 0.389
	Paetzold and Specia (2017)	Newsela (version 2016-01-29.1) contains 1,911 news articles in their original form, as well as up to five simplified versions	Precision – 0.337 f1 score – 0.256
	Paetzold and Specia (2016)	Corpus of movie subtitles	Precision – 0.118 Recall – 0.161 f1 score – 0.136
Sentence alignment	Horn et al. (2014)	A sentence-aligned data set of English Wikipedia sentences and Simple English Wikipedia sentences, containing 137,000 aligned sentence pairs	Precision – 0.761 Accuracy – 0.663
	Paetzold and Specia (2017)	Newsela (version 2016-01-29.1) contains 1,911 news articles in their original form, as well as up to five simplified versions	Precision – 0.453
Thesaurus	De Belder and Moens (2010)	Simplified news articles and encyclopedia articles	Accuracy – 0.41
	Bott et al. (2012)	On-line dictionary and the Web	Precision – 0.121

3 FURTHER SUGGESTIONS

The degree of simplicity and the preservation of meaning and grammar are the main focuses of any lexical simplification system. If more substitutions are produced, then the degree of meaning preservation will be reduced and vice-versa. This problem is difficult to solve. The performance of SG can be improved by increasing the number of thesauri used. A suitable method for complex word identification is still a problem in LS, because there is no correct definition for a complex word and the complexity varies according to each individual's vocabulary. There are some domain-specific LSs available; it can also be extended to other domains.

4 CONCLUSION

LS systems are used to make a text more accessible. One of the main tasks in LS is substitution generation, which is used to generate the possible complex and simple word pairs. Several methods are used for generating these pairs. The methods include word-embedding models, thesauri and sentence alignment. Among these methods, word embedding is the most widely used. The positives and negatives of the different methods are discussed in this paper. SG methods have an important role in the preservation of grammar and meaning of the generated sentence.

REFERENCES

Adel, H. & Schütze, H. (2014). Using mined coreference chains as a resource for a semantic task. In *Proceedings of the 2014 Conference on Empirical Methods in Natural Language Processing (EMNLP), 25–29 October 2014, Doha, Qatar* (pp. 1447–1452). Stroudsburg, PA: Association for Computational Linguistics.

Biran, O., Brody, S. & Elhadad, N. (2011). Putting it simply: A context-aware approach to lexical simplification. In *Proceedings of the 49th Annual Meeting of the Association for Computational Linguistics: Human Language Technologies: Short papers* (Vol. 2, pp. 496–501). Stroudsburg, PA: Association for Computational Linguistics.

Bott, S., Rello, L., Drndarevic, B. & Saggion, H. (2012). Can Spanish be simpler? LexSiS: Lexical Simplification for Spanish. In *Proceedings of COLING 2012: 24th International Conference on Computational Linguistics: Technical Papers* (pp. 357–374).

Carroll, J., Minnen, G., Canning, Y., Devlin, S. & Tait, J. (1998). Practical simplification of English newspaper text to assist aphasic readers. In *Proceedings of the AAAI-98 Workshop on Integrating Artificial Intelligence and Assistive Technology* (pp. 7–10).

Coster, W. & Kauchak, D. (2011). Learning to simplify sentences using Wikipedia. In *Proceedings of the 49th Annual Meeting of the Association for Computational Linguistics, 24 June 2011, Portland, Oregon* (pp. 1–9). Stroudsburg, PA: Association for Computational Linguistics.

Daelemans, W., Höthker, A. & Erik Tjong Kim Sang. (2004). Automatic sentence simplification for subtitling in Dutch and English. In *Proceedings of the 4th International Conference on Language Resources and Evaluation, Lisbon, Portugal* (pp. 1045–1048).

De Belder, J. & Moens, M.F. (2010). Text simplification for children. In *Proceedings of the SIGIR Workshop on Accessible Search Systems* (pp. 19–26).

Deléger, L. & Zweigenbaum, P. (2009). Extracting lay paraphrases of specialized expressions from monolingual comparable medical corpora. In *Proceedings of the 2nd Workshop on Building and Using Comparable Corpora: From Parallel to Non-Parallel Corpora* (pp. 2–10). Stroudsburg, PA: Association for Computational Linguistics.

Devlin, S. & Unthank, G. (2006). Helping aphasic people process online information. In *Proceedings of the 8th International ACM SIGACCESS Conference on Computers and Accessibility* (pp. 225–226). New York, NY: Association for Computing Machinery.

Elhadad, N. & Sutaria, K. (2007). Mining a lexicon of technical terms and lay equivalents. In *Proceedings of the Workshop on BioNLP 2007: Biological, Translational, and Clinical Language Processing, 29 June 2007, Prague, Czech Republic* (pp. 49–56). Stroudsburg, PA: Association for Computational Linguistics.

Faruqui, M., Dodge, J., Jauhar, S.K., Dyer, C., Hovy, E. & Smith, N.A. (2014). Retrofitting word vectors to semantic lexicons. *arXiv:1411.4166*.

Horn, C., Manduca, C. & Kauchak, D. (2014). Learning a lexical simplifier using Wikipedia. In *Proceedings of the 52nd Annual Meeting of the Association for Computational Linguistics* (Vol. 2, pp. 458–463). Stroudsburg, PA: Association for Computational Linguistics.

Huenerfauth, M., Feng, L. & Elhadad, N. (2009). Comparing evaluation techniques for text readability software for adults with intellectual disabilities. In *Proceedings of the 11th International ACM SIGACCESS Conference on Computers and Accessibility* (pp. 3–10). New York, NY: Association for Computing Machinery.

Jonnalagadda, S., Tari, L., Hakenberg, J., Baral, C. & Gonzalez, G. (2009). Towards effective sentence simplification for automatic processing of biomedical text. In *Proceedings of Human Language Technologies: The 2009 Annual Conference of the North American Chapter of the Association for Computational Linguistics, Companion Volume: Short Papers* (pp. 177–180). Stroudsburg, PA: Association for Computational Linguistics.

Kajiwara, T., Matsumoto, H. & Yamamoto, K. (2013). Selecting proper lexical paraphrase for children. In *Proceedings of the Twenty-Fifth Conference on Computational Linguistics and Speech Processing (ROCLING 2013)* (pp. 59–73).

Kim, Y.S., Hullman, J. & Adar, E. (2015). DeScipher: A text simplification tool for science journalism. *Ann Arbor, 1001*, 48109.

Kim, Y.S., Hullman, J., Burgess, M. & Adar, E. (2016). SimpleScience: Lexical simplification of scientific terminology. In *Proceedings of the 2016 Conference on Empirical Methods in Natural Language Processing, 1–5 November 2016, Austin, Texas* (pp. 1066–1071). Stroudsburg, PA: Association for Computational Linguistics.

Och, F.J. & Ney, H. (2000). Improved statistical alignment models. In *Proceedings of the 38th Annual Meeting of Association for Computational Linguistics* (pp. 440–447). Stroudsburg, PA: Association for Computational Linguistics.

Paetzold, G. & Specia, L. (2015). LEXenstein: A framework for lexical simplification. In *Proceedings of ACL-IJCNLP 2015 System Demonstrations, 26–31 July 2015, Beijing, China* (pp. 85–90). Stroudsburg, PA: Association for Computational Linguistics.

Paetzold, G.H. & Specia, L. (2016). Unsupervised lexical simplification for non-native speakers. In *Proceedings of the Thirtieth AAAI Conference on Artificial Intelligence, 12–17 February 2016, Phoenix, Arizona* (pp. 3761–3767). Palo Alto, CA: AAAI Press.

Paetzold, G.H. & Specia, L. (2017). Lexical simplification with neural ranking. In *Proceedings of the 15th Conference of the European Chapter of the Association for Computational Linguistics, 3–7 April 2017, Valencia, Spain: Short Papers* (Vol. 2, pp. 34–40). Gothenburg, Sweden: Association for Computational Linguistics.

Shardlow, M. (2013). A comparison of techniques to automatically identify complex words. In *Proceedings of the ACL Student Research Workshop, 4–9 August 2013, Sofia, Bulgaria* (pp. 103–109). Stroudsburg, PA: Association for Computational Linguistics.

Thomas, S.R. & Anderson, S. (2012). WordNet-based lexical simplification of a document. In *Proceedings of the 11th Conference on Natural Language Processing (KONVENS 2012), 19 September 2012, Vienna, Austria* (pp. 80–88).

Emerging Trends in Engineering, Science and Technology for Society,
Energy and Environment – Vanchipura & Jiji (Eds)
© *2018 Taylor & Francis Group, London, ISBN 978-0-8153-5760-5*

Fast and efficient kernel machines using random kitchen sink and ensemble methods

P. Melitt Akhil & K. Rahamathulla
Department of Computer Science and Engineering, Government Engineering College, Thrissur,
Kerala, India

ABSTRACT: The introduction of the information era has affected all areas of computer science. This has also affected the approach taken by machine learning researchers. Compared to the early days, researchers now use a generic approach to selecting a machine learning model. As the generic model doesn't contain much domain knowledge, it has been compensated for by a huge dataset. This makes the model optimization more complex and time consuming. We can replace this overhead with randomization instead of optimization. There are many existing methods which apply randomization in machine learning models. But all of these methods compromise on accuracy. This paper shows an efficient way to use randomization along with ensemble methods without a decrease in efficiency.

1 INTRODUCTION

As the entire world is tending toward the Internet of Things and Software Defined Anything, the volume and velocity of data is exploding. This has led to certain changes in a researcher's approach in machine learning area too. In the early days of machine learning, researchers used complex models. Most of the domain knowledge will be embedded in these models. And the models will be application dependent. There won't have been huge datasets available for optimization purposes at that time. But now, with the help of world wide web, the Internet of Things and other information era facilities, there is a possibility of getting huge sets of data. This huge dataset will contain most of the domain knowledge for a particular application. This has made researchers follow a generic approach to model selection.

Because of the availability of huge datasets, there is no need to embed the complete domain knowledge into the model. This reduces the development time of models. The generic models can be reused for similar applications which, in turn, improve a particular type of model as it has been used for many projects. As the training data contains more accurate features, the models optimized using this data will be more accurate.

But the huge size of datasets has increased the training time exponentially. If the input dataset contains N data points, then the training time of a kernel machine will be N^2. Also, it is difficult for storing kernel matrices because of its huge size, which is $N \times N$.

To overcome this overhead, there are many approaches, such as dimension reduction, converting a gram matrix from dense to sparse, decomposition of a gram matrix, random projection and sampling. But all of these methods lack accuracy because when the quantity of training data decreases, features obtained from the dataset also decrease.

In order to have accurate and fast training models, methods which decrease training time and increase accuracy, need to be integrated. This paper proposes a method which uses a sub-sampling method and an ensemble method together to obtain a fast and efficient kernel machine.

Section 2 introduces the Random Kitchen Sink (RKS) and ensemble methods. In Section 3 the new methodology is explained. Section 4 analyses the evaluation criteria for the new model.

2 RELATED WORKS

Random Kitchen Sink (RKS) (see 2.2) and other random approaches (see 2.1) are explained in detail in this section. Different types of ensemble methods (see 2.3) are also studied.

2.1 *Randomization methods*

SVM uses Quadratic Programming (QP) methods to solve the model. But the QP which comes from SVM cannot be solved easily. So, methods like chunking (Vapnik) and decomposition are used widely (Platt, 1999). Chunking which reduced the dimension of the matrix breaks down the large QP into smaller ones. But even chunking cannot handle large data set problems. This motivated the decomposition method which uses a fixed-size matrix for every QP sub-problem. Sequential Minimal Optimization (SMO) (Platt, 1999) also follows the fixed matrix size approach to decomposition. But it chooses a greedy approach in solving optimization problem. SMO solves the smallest optimization problem at each iteration. This avoids numerical QP optimization. Thus, kernel evaluation is the only major factor that contributes to computation time in SMO. Kernel evaluation time can be reduced by increasing the sparseness of the input data.

In order to get a sparse gram matrix many approximation methods were used. One method is Singular Value Decomposition (SVD) which approximates gram matrices. But, due to the large size of gram matrices, finding SVD is a time complex task. This can be avoided by randomly sampling the gram matrix. Rows and then columns are chosen independently according to normal distribution (Frieze et al., 2004). After that it is easy to find the SVD of the resultant matrix as the resultant matrix is of a smaller size.

Also there are methods which formulate the relationship between SVM and a Minimum Enclosing Ball (MEB) problem (Tsang et al., 2005). This leads to a more simple and efficient way of approximating large data sets.

Willims and Seeger (2001) recommended the Nyström method. The Nyström method contains a matrix C of size $n \times c$, which contains randomly chosen columns c. It also has a matrix W, which contains the intersections of randomly chosen columns and their corresponding rows. This method is modified by allowing the column sample to be formed using arbitrary sampling probabilities (Drineas & Mahoney, 2005).

One of the sampling techniques uses a three-tier approach for speeding Kernel Principal Component Analysis (Achlioptas et al., 2001). At the first stage, the gram matrix is sampled with probability factor s. This matrix is then quantized with the maximum value in the matrix. After this, a randomized rounding is conducted for the kernel coefficients. This helps for a fast batch approximation of the kernel function.

Random projection is more efficient than normal sampling techniques. It utilizes mapping elements to a d-dimensional subspace where d is much smaller than the actual dimension (Blum, 2006). The Anchor method is a random projection method (Wang et al., 2017), where randomization is embedded in the machine learning methods. The features are extracted, not for each data point, but for a group of points called anchors. Anchor granularity is decided randomly using geometric distribution. This approach removes over-fitting and noise in the model.

2.2 *Random Kitchen Sink (RKS)*

By mapping the data points from the original low dimension space to a higher infinite dimensional space we can easily separate the data points according to various clusters. This is the main objective of the kernel trick. But this is time consuming and costly. In RKS, the complexity is reduced by mapping to a low dimensional Euclidean inner product space. This uses a randomized feature map z: $R^d -> R^D$, where z is low dimensional.

Some of the randomized maps are sinusoids randomly drawn from the Fourier transform of the kernel function, and randomly shifted grids which partitioned the input space (Rahimi & Recht, 2007).

RKS is a generalized randomization method for any type of machine learning model. Every optimization problem can be viewed as a function fitting problem. Each function fitting problem tries to reduce the empirical risk equation. The generalized empirical risk function contains parameters, ω, and weights, α, over K features. It uses a minimization process for both weights and parameters.

But RKS randomizes the parameters, ω, and minimizes the weights, α, using batch optimization (Rahimi & Recht, 2008b). Randomization instead of optimization decreases the computational complexity of training (Rahimi & Recht, 2008a).

There are also methods which use Hadamard and diagonal matrices instead of Gaussian matrices in RKS (Quoc et al., 2013).

Some of the applications of RKS are retrieval of bio-geo-physical parameters (Laparra et al., 2015), hyper spectral image classification (Nikhila, 2015; Nikhila et al., 2015), Tamil document classification (Sanjanasri & Anand, 2015), offline power disturbance signal classification (Aneesh et al., 2015) and detection of mitotic nuclei in breast histopathology images (Beevi et al., 2016).

2.3 *Ensemble method*

Ensemble classifiers can create a strong learner with the help of weak learners. It will have a base learner. The results of these base learner are combined using different rules. The combining rule can weight the base learners according to the error rate (Xiao & Wang, 2017). Also, top performing base learners can be found out. At last in order to get a single output voting system can be implemented where majority voting output of each base learner will be counted as final output.

It is hard to predict multi-regime time series datasets using a single model. But using ensemble methods we can reduce the error rate. The random forest ensemble method uses bagging and random subspace techniques for achieving diversity (Lin et al., 2017).

Liver tumor detection (Huang et al., 2014) and predicting real exhaust gas temperatures for turbofan engines (Lin et al., 2017) are some of the examples where different ensemble methods are used.

3 PROPOSED METHOD

3.1 *Mathematical model*

Every randomization method comes with an expected error factor. And, in every case, this error factor decreases the accuracy of the model. To overcome this compromise in accuracy, we have integrated the ensemble method with randomization. The randomization method selected for the classification is RKS as this method has less of an error rate.

In order to create the voting system for the entire model first, the voting share of each learner needs to be calculated. It depends on the error rate of each learner (Equation 1).

$$E_j = \frac{1}{m} \sum_{i=1}^{m} I(x_i) - y_i \tag{1}$$

Then the voting share can be calculated with proportions of error rate distributed between the individual learner and entire learners using Equation 2.

$$V_j = \frac{\sum_{i=1}^{n} E_i}{E_j} \tag{2}$$

The training of base learners is done using the kernel which was built by RKS. The steps of the proposed method are described in algorithm 3.1.

Algorithm 1 Proposed algorithm

Input: Data points $\{(x_i, y_i)|i = 1..M\}$, probability distribution $p(w)$
Output: $F(x) = \max_{i=1..n} V_i \sum_{j=1}^{m} f(x_i; e_j) w_j$

1: Split data points into n overlapping samples according to $p(\omega)$.
2: Use RKS algorithm to find out features, e, and weight, w, for each kernel.
3: Train each base learner with its corresponding kernel.
4: Find the voting share of each learner with Equation 1.
5: Use majority voting system to find out final output.

3.2 *System design*

First, the entire data set is partitioned into multiple overlapping small datasets using probability distribution. Then using RKS, different kernel spaces are created in parallel. Using those kernel spaces, base learners are trained. The output of each base learners is fed into the voting system (Figure 1). The voting system gives a weighted voting share to each base learner. It then calculates the final output using majority voting share.

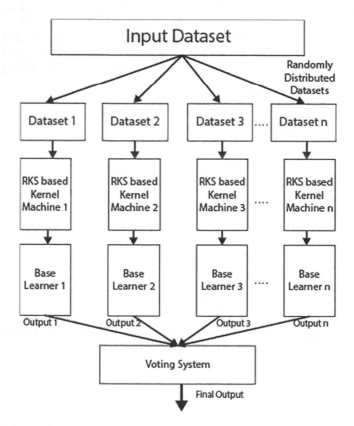

Figure 1.   Model representation.

4 EVALUATION CRITERIA

The error rate of the proposed method depends only on the error rate of each base learner (Equation 3). The randomization process done during the RKS affects this error. But because a weighted voting share system is used, the error rate of each learner will be constant. So, the ultimate error rate of the entire model will be the smallest error rated learner.

$$E = \max_{j=1...n} E_j \times V_j \qquad (3)$$

This error is less than normal randomization methods which use only one learner. Also, in the case of time complexity, our model is faster than the classical kernel machine. It takes only O (dlogd) time, where d is the size of the dataset for each learner which is obtained by partitioning the entire input space. Here, d << n. As the size of data set for each kernel machine is smaller, it takes up much less space to store the gram matrix (d × d). Thus, spatial complexity also decreases.

5 CONCLUSIONS

We proposed a method which combines advantages from both randomization and ensemble methods. Without losing accuracy, the kernel is trained faster than a normal kernel. The algorithm splits the entire dataset and, for each piece, a base learner is trained using RKS. The weighted voting share method is integrated with the system in order to decrease the error rate. The time and space complexity of the problem also decreases due to the randomization process.

REFERENCES

Achlioptas, D., McSherry, F. & Scholkopf, B. (2001). Sampling techniques for kernel methods. In *Advances in Neural Information Processing Systems* (335–342), *14*.

Aneesh, C., Hisham, P.M., Sachin, K.S., Maya, P. & Soman, K.P. (2015). Variance based offline power disturbance signal classification using support vector machine and random kitchen sink. *Procedia Technology, 21*(21), 163–170.

Beevi, K.S., Madhu, S.N. & Bindu, G.R. (2016). Detection of mitotic nuclei in breast histopathology images using localized acm and random kitchen sink based classifier. *38th Annual International Conference of the IEEE Engineering in Medicine and Biology Society (EMBC)*, 2435–2439.

Blum, A. (2006). Random projection, margins, kernels, and feature-selection. *SLSFS '05 Proceedings of the 2005 International Conference on Subspace, Latent Structure and Feature Selection*, 52–68.

Drineas, P. & Mahoney, M.W. (2005). On the Nyström method for approximating a gram matrix for improved kernel-based learning. *Journal of Machine Learning Research*, 2153–2175.

Frieze, A., Kannan, R. & Vempala, S. (2004). Fast Monte-Carlo algorithms for finding low-rank approximations. *Journal of the ACM (JACM) 51*, 1025–1041.

Huang, W., Yang, Y., Lin, Z., Huang, G.-B., Zhou, J., Duan, Y. & Xion, W. (2014). Random feature subspace ensemble based extreme learning machine for liver tumor detection and segmentation. *36th Annual International Conference of the IEEE Engineering in Medicine and Biology Society*, 4675–4678.

Laparra, V., Gonzalez, D.M., Tuia, D. & Camps-Valls, G. (2015). Large-scale random features for kernel regression. *IEEE International Geoscience and Remote Sensing Symposium (IGARSS)*, 17–20.

Lin, L., Wang, F., Xie, X. & Zhong, S. (2017). Random forests-based extreme learning machine ensemble for multi-regime time series prediction. *Expert Systems with Applications, 83*, 164–176.

Nikhila, H., Sowmya, V. & Soman, K.P. (2015). Comparative analysis of scattering and random features in hyperspectral image classification. *Second International Symposium on Computer Vision and the Internet (VisionNet15) 58*, 307–314.

Nikhila, H. (2015). Hyperspectral image classification using random kitchen sink and regularized least squares. *IEEE International Conference on Communication and Signal Processing (ICCSP)*.

Platt, J.C. (1999). Using analytic QP and sparseness to speed training of support vector machines. In *Proceedings of the 1998 Conference on Advances in Neural Information* (pp. 557–563).

Quoc, L., Sarlos, T. & Smola, A. (2013). Fastfood approximating kernel expansions in loglinear time. *30th International Conference on Machine Learning (ICML)*.

Rahimi, A. & Recht, B. (2007). Random features for large-scale kernel machines. In *Advances in Neural Information Processing Systems, 20 (NIPS)* (pp. 1177–1184), Vancouver, British Columbia, Canada.

Rahimi, A. & Recht, B. (2008a). Uniform approximation of functions with random bases. *46th Annual Allerton Conference on Communication, Control, and Computing*, 555–561.

Rahimi, A. & Recht, B. (2008b). Weighted sums of random kitchen sinks: Replacing minimization with randomization in learning. In *Advances in Neural Information Processing Systems, 21 (NIPS)* (pp. 1313–1320), Vancouver, British Columbia, Canada.

Sanjanasri, J.P. & Anand, K.M. (2015). A computational framework for Tamil document classification using random kitchen sink. *International Conference on Advances in Computing, Communications and Informatics (ICACCI)*, 1571–1577.

Tsang, I.W., Kwok, J.T. & Cheung, P.-M. (2005). Core vector machines: fast SVM training on very large data sets. *Journal of Machine Learning Research*, 363–392.

Wang, S., Aggarwal, C. & Liu, H. (2017). Randomized feature engineering as a fast and accurate alternative to kernel methods. *KDD '17 Proceedings of the 23rd ACM SIGKDD International Conference on Knowledge Discovery and Data Mining*, 485–494.

Xiao, Q. & Wang, Z. (2017). Ensemble classification based on random linear base classifiers. *IEEE International Conference on Acoustics, Speech and Signal Processing (ICASSP)*, 2706–2710.

Emerging Trends in Engineering, Science and Technology for Society,
Energy and Environment – Vanchipura & Jiji (Eds)
© 2018 Taylor & Francis Group, London, ISBN 978-0-8153-5760-5

HOG feature-based recognition for Malayalam handwritten characters

E.P. Anjali & Ajay James
Department of Computer Science and Engineering, Government Engineering College, Thrissur, Kerala, India

Saravanan Chandran
Department of Computer Science and Engineering, National Institute of Technology, Durgapur, West Bengal, India

ABSTRACT: Optical Character Recognition (OCR) converts images containing handwritten or printed characters into an editable format. OCR systems have wide applications such as processing bank cheques, or document conversion of legal papers. One major concern in developing a character recognition system is the selection of efficient features. Major challenges in Malayalam handwritten character recognition are varying writing styles, the presence of compound characters and of similarly shaped characters. Mixing up old and new styles of writing adds additional complexity in HCR systems. In this paper, an alternative that allows the use of histogram of oriented gradients is presented. The proposal consists of using HOG for feature extraction, then using SVM as the classifier, Malayalam character recognition can be carried out with more quality than that obtained in the existing methods.

1 INTRODUCTION

The entire world is now moving towards digitization and is growing fast where printed documents and manuscripts need preserved in a digital format. Optical Character Recognition (OCR) is a system by which any document, whether it is handwritten or printed, can be converted into an editable text format. Thus, it helps when searching in documents. Nowadays, OCR is used by a large number of institutions such as banks, insurance companies, post offices and book publishers in order to verify the authenticity of the documents, either handwritten or printed such as cheques, and envelopes. The research and development of OCR is based on progress in fields such as image processing, pattern recognition and machine learning. There are two types of OCR systems: offline and online. In an offline character recognition system, the input image is captured by a scanner and passes through three stages; preprocessing, feature extraction and classification. Online character recognition systems work in real-time.

The first step in every traditional recognition system is the removal of noise. This is followed by other preprocessing techniques such as binarization, thinning, resizing. The preprocessed image is then segmented into lines, words and then characters. Unique features are extracted from each character and the chosen classifier is trained using these vectors. Compared to printed documents, handwritten character recognition is more complex since different people write in different ways and styles. Other challenges in handwritten character recognition are the variation in the fonts/thickness of letters and the difference in the gaps between the letters. Moreover, the skewness of handwritten matter will be very different from person to person (Ryu et al., 2014). To overcome these challenges, researchers are working hard to improve this system.

This paper has other parts such as insight on previous works in OCR using HOG (Sec. 2), proposed method (Sec. 3) and conclusion (Sec. 4).

Elleuch et al. (2017) have extracted features using a HOG descriptor. The image of each character is initially divided into small cells. The HOG of each pixel is then computed using a one-dimensional mask. They have represented a rectangular HOG using three parameters namely, "number of cells per block, number of pixels per cell, and number of channels per cell". The orientation is taken in-between 0 and 180 degrees. They have chosen nine bins to represent orientation. After computing the histogram, normalization is done by L2-norm. They have compared the performance of HOG features with Gabor features. Gabor features are directly extracted from gray scale images. Their system was tested using an IFN/ENIT dataset. The error classification rate of the HOG descriptor with SVM was 1.51% and that of Gabor features was 7.16%.

Elleuch et al. (2015) examined Arabic handwritten script recognition using a multi-class SVM. In the recognition phase, structural features are input to a supervised learning algorithm. SVM is used as the classifier and handwritten Arabic characters are used for testing. Their experimental study has efficiently showed that they have excellent outcomes if compared to the existing Arabic OCR systems.

Ebrahimzadeh and Jampour (2014) have developed an efficient handwritten digit recognition system using HOG and SVM. The input image is partitioned into 9×9 cells and a histogram is computed. Eighty-one features are used to represent each digit. Digit classification is performed using linear multi-class SVM. To validate the model, a MNIST handwritten digit dataset is used. They have achieved 97.25% accuracy. Linear SVM yields better accuracy compared to polynomial, RBF and sigmoid functions.

Kamble and Hegadi (2015) use Rectangular HOG (RHOG) features to recognize handwritten Marathi characters. Normalization will bring all characters to the same size. After extracting RHOG features, the feature vector length is 576. A Sobel mask is used to measure the gradient values. After calculating the gradient and orientation, bins of histograms are computed. SVM is used as the classifier. The performance is compared with a Feed forward artificial neural network. The dataset included 8,000 samples and the neural network-based classification performed better than SVM.

The handwritten character recognition in Kulkarni's (2017) model uses the HOG features for character recognition followed by the center of the mass of the image with SVM algorithm. Otsu's method is used for segmentation. Once regions of interest are obtained, the mean of the weighted mean of white pixels is calculated. The center of image is assigned this value. 9-bit integer values are extracted using the HOG descriptor. For character classification, the HASY dataset is used which contains handwritten alpha numeric symbols. The proposed model is evaluated using SVM and KNN.

Qinyunlong (2008) combines HOG in multiple resolutions with canonical correlation analysis. A Gaussian pyramid is used to get a multi-resolution HOG. Once we have the gradient map, HOG features can be extracted from these maps for each resolution. In preprocessing, Box-Cox transformation is applied. The system is tested with three handwritten databases.

Iamsa-at and Horata (2013) The images are converted into gray scale and after preprocessing, the resized images are at 32×32 pixels. HOG was computed by applying a one-dimensional mask. Based on the intensity of the characters, the gradient is calculated. Each pixel casts a vote for the cell that lies closest to its orientation. L2 norm is used to normalize the histograms of overlapping blocks. The dataset contains Thai and Bangla handwritten characters. The performance of the feedforward-backpropagation neural network and the Extreme Learning Machine (ELM) are compared. Eighty hidden layers were used in the backpropagation neural network and the activation function is logistic. A sigmoid function is used to train the ELM. Their experimental study shows that DFBNN outperforms ELM.

Tikader and Puhan (2014) have modified the traditional HOG feature for recognizing English-Bengali scripts. The input image is not divided into cells. After computing gradients, instead of splitting the cells, binning operation is applied to the whole image. For classification, linear SVM is used. The system performance depends on the number of bins chosen since they are proportional to each other.

Newell and Griffin (2011) proposed multiscale HOG descriptors for robust character recognition. Their work describes two variations of using HOG features. Their first scheme simply extends the histograms across different scale space. Orientation is calculated using Derivative-of-Gaussian filters. In the second method, pairs of oriented gradients across different scales are integrated to generate feature vectors. This is referred to as oriented gradient columns.

3 PROPOSED METHOD

Due to the high variety and complexity of Malayalam handwritten strokes, shapes and concavities, we have selected a HOG feature descriptor which is robust to local displacements, yet still supply discriminating feature vectors as figuration of the handwritten characters. Bhowmik et al. (2014) mention "The main idea behind the HOG descriptors is that the local object appearance and shape within an image can be described by the distribution of intensity or edge directions." Then, obtained feature vectors are fed into an SVM classifier to create the classification. The system works in three steps:

- Preprocessing
- HOG feature extraction
- Classification

In the preprocessing step, we have some basic image processing to separate characters from real samples or preparing data from dataset and then in the second part, we extract HOG features which is very distinguishable descriptor for character recognition where we divide an input image into 99 cells and compute then the histogram of gradient orientations thereby we represent each character with a vector of 81 features. The overall view of the proposed approach has been illustrated in Figure 1. HOG is a fast and reliable descriptor which can generate distinguishable features. Also, SVM is a fast and powerful classifier which is useful to classify HOG features. The subsequent sections explain the steps in detail.

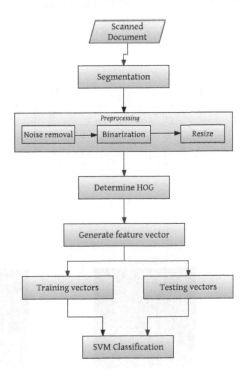

Figure 1.   Steps in OCR.

3.1 Segmentation

The first step in processing is the segmentation of characters from the image. The image is horizontally scanned to obtain the lines from image. Horizontal projection can be used to separate lines since there will be ups and downs in the projection. Line segmentation is followed by word and character segmentation. To segment lines into words and then to characters, vertical projection can be used. Vertical projection will have gaps thereby separating the words and individual characters.

3.2 Preprocessing

3.2.1 Noise removal

When a document is scanned, some noise is unavoidable. Noise may occur due to the quality of the document, scanner etc. Before processing the image, the noise should be removed. A low pass filter can be used to remove the noise thereby smoothing the image.

3.2.2 Binarization

Grayscale images are converted to two tone black and white images by the process called binarization. In order to convert a grayscale image into a binary image, a value is selected as the threshold and all pixel values above the threshold are converted to 1 and those below the threshold are converted to 0.

3.3 HOG

HOG was first suggested (Dalal & Triggs, 2005) for detecting the presence of humans in images. It also has wide applications in computer vision and image processing areas because of its characteristics such as invariance to illumination and local geometric transformations. The input image is partitioned into small square cells and then computes the histogram of gradient directions or edge directions. The orientation of each pixel in a cell is quantized into bins of histograms. Each bin represents an angle range between 0° and 360° or 0° and 180°. A feature vector is formed by combining normalized histograms of each cell.

H

1	2	1
0	0	0
-1	-2	-1

H^T

-1	0	1
-2	0	2
-1	0	1

Figure 2.   Sobel mask.

Figure 3.   HOG of Malayalam character with varying cell size.

HOG is robust to illumination and local geometric changes as the local histograms have been normalized based on the contrast. It is also independent of image size and captures localized information. Compared to the well-known SURF and SIFT operators, a HOG feature is computed across the entire image. Thus, for character recognition systems, HOG features are often better than other structural features since it is invariant to local geometric changes.

HOG features on some Malayalam characters have been illustrated in Figure 3 where a Sobel operator is used to find out the horizontal and vertical component (Kamble & Hegadi, 2015). The Sobel mask is shown in Figure 2.

The gradient along the x and y axes can be estimated as follows:

$$G_x(u,v) = H * I(u,v) \tag{1}$$

and

$$G_y(u,v) = H^T * I(u,v) \tag{2}$$

where $I(u, v)$ is the cropped character. Gradient $G(u, v)$ is calculated as:

$$G(u,v) = \sqrt{G_x^2(u,v) + G_y^2(u,v)} \tag{3}$$

Orientation is measured as:

$$\theta(u,v) = \tan^{-1}\frac{G_y(u,v)}{G_x(u,v)} \tag{4}$$

3.4 Classification

In the third stage, a linear multi-class support vector machine has been employed to classify characters. The optimization criteria of support vector machines are the marginal width between two classes. If the width is higher, so the separability of patterns. The patterns that lie on the soft margin are called support vectors and they characterize the classification function.

3.5 Post processing

Once the characters are recognized, semantic rules should be applied to correct errors caused during classification, i.e. in the case of Malayalam, some characters such as dependent vowels and signs should not appear at the beginning of a word. Also, dependent vowels will appear only with consonants and independent vowels should not appear in-between a word.

4 CONCLUSION

This paper proposes a model for handwritten Malayalam character recognition using histogram of oriented gradients as feature descriptor. HOG is invariant to local geometrical changes and illumination. SVM is used as the classifier since it helps to achieve robust performance.

REFERENCES

Bhowmik, S., Roushan, M.G., Sarkar, R., Nasipuri, M., Polley, S. & Malakar, S. (2014). Handwritten Bangla word recognition using HOG descriptor. In *Fourth International Conference of Emerging Applications of Information Technology (EAIT)*, (pp. 193–197). New York, NY: IEEE.

Dalal, N. & Triggs, B. (2005). Histograms of oriented gradients for human detection. In *IEEE Computer Society Conference on Computer Vision and Pattern Recognition.*, 1, (pp. 886–893). New York, NY: IEEE.

Ebrahimzadeh, R. & Jampour, M. (2014). Efficient handwritten digit recognition based on histogram of oriented gradients and SVM. *International Journal of Computer Applications, 104*(9).

Elleuch, M., Hani, A., & Kherallah, M. (2017). Arabic handwritten script recognition system based on HOG and Gabor features. *International Arab Journal of Information Technology (IAJIT), 14.*

Elleuch, M., Lahiani, H., & M. Kherallah (2015). Recognizing Arabic handwritten script using support vector machine classifier. In *15th International Conference on Intelligent Systems Design and Applications (ISDA)* (pp. 551–556). New York, NY: IEEE.

Iamsa-at, S. & Horata, P. (2013). Handwritten character recognition using histograms of oriented gradient features in deep learning of artificial neural network. In *International Conference on IT Convergence and Security (ICITCS)*, (pp. 1–5). New York, NY: IEEE.

Kamble, P.M. & Hegadi, R.S. (2015). Handwritten Marathi character recognition using R-HOG feature. *Procedia Computer Science, 45*, 266–274.

Kulkarni, R.L. (2017). Handwritten character recognition using HOG, COM by OpenCV and Python. *International Journal, 5*(4).

Newell, A.J. & Griffin, L.D. (2011). Multiscale histogram of oriented gradient descriptors for robust character recognition. In *International Conference on Document Analysis and Recognition (ICDAR)*, (pp. 1085–1089). New York, NY: IEEE.

Qinyunlong, S. (2008). Handwritten character recognition using multi-resolution histograms of oriented gradients. In *11th IEEE International Conference on Communication Technology.* (pp. 715–717). New York, NY: IEEE.

Ryu, J., Koo, H.I. & Cho, N.I. (2014). Language-independent text-line extraction algorithm for handwritten documents. *IEEE Signal Processing Letters, 21*(9), 1115–1119.

Tikader, A. & Puhan, N. (2014). Histogram of oriented gradients for English-Bengali script recognition. In *International Conference for Convergence of Technology (I2CT)* (pp. 1–5). New York, NY: IEEE.

*Emerging Trends in Engineering, Science and Technology for Society,
Energy and Environment – Vanchipura & Jiji (Eds)*
© *2018 Taylor & Francis Group, London, ISBN 978-0-8153-5760-5*

A novel approach for the veracity and impact prediction of rumors

Anju Rose G. Punneliparambil & N.D. Bisna
*Department of Computer Science and Engineering, Government Engineering College,
Thrissur, Kerala, India*

ABSTRACT: Malignant or unplanned falsehood can be spread on social media and can hazardously affect people and society. Models for automated verification of rumors are already developed. But the impacts of rumors are neither analyzed nor predicted with these rumor verification models. Impact prediction of rumor can be used for determining whether the rumor to be responded to or not. This is very relevant for sudden situations in which a rumor with large negative impact has to be addressed. As an extension to the veracity prediction model, impact prediction is proposed.

1 INTRODUCTION

Social media services have enhanced the way individuals get data and news about current occasions. Traditional news media just offers information to individuals. But online social media such as Twitter and Facebook are platforms for people to impart information and their insights about the news occasions. Users on Twitter create their profile and share their status or opinions. These posts are known as tweets. Despite the fact that an extensive volume of content is posted on Twitter, the majority of data is not valid or valuable in giving information about the occasion. There can be noise, spam, ads and personal emotions in tweets. This makes the quality of content on Twitter faulty.

1.1 *What is a rumor?*

Social psychology literature characterizes a rumor as a story or statement whose truth value is unconfirmed or intentionally false (Allport & Postman, 1965). Rumors are spread on the web. Fake stories and false claims affect people's life in harmful ways. False bits of gossip are harming as they cause public panic and social distress. For example, on 25 August 2015, a rumor about "shootouts and kidnappings by drug gangs happening close to school in Veracruz" spread through Twitter and Facebook. This caused chaos in the city including car collisions, since individuals left their cars in the middle of road and hurried to get their children from school. (Ma et al., 2016).

1.2 *Motivation*

Debunking rumors at an early stage of diffusion is especially significant to limiting their unsafe impacts. To recognize bits of rumor from truthful occasions, people and organizations have frequently depended on common sense and investigative journalism. Rumor revealing sites like snopes.com and factcheck.org are such cooperative endeavors. Be that as it may, on the grounds that manual confirmation steps are associated with such endeavors, these sites are not complete in their topical scope and furthermore can have long debunking delay.

In the area of Natural Language Processing (NLP), there is some recent research analyzing and deciding the truth value of social media content. There are several works are already done to predict the veracity of rumor in Twitter. By predicting the impact of a false rumor along with its veracity, the rumor can be stopped immediately before it makes further chaos.

Section 2 presents some previous works on rumor prediction. Section 3 discusses the proposed methodology. The method for impact analysis along with veracity prediction is handled. Section 4 gives brief idea about the data set and the evaluation criteria. The paper concludes in Section 5.

2 RELATED WORK

There are many works about predicting and analyzing the veracity of rumors. Some of the major works are discussed here.

Ma et al. (2016) propose a technique for detecting rumors from microblogs with Recurrent Neural Networks (RNN). It shows a strategy that learns continuous representations of microblog events for detecting rumors. The model depends on RNN for learning the hidden representations that catch the variety of contextual information of important posts after some time. Using RNN, social context information of an event is demonstrated as a variable-length time series. At the point when individuals are exposed to a rumor claim, they will forward the claim or comment on it, in this way making a continuous stream of posts. In this method, both the temporal and textual representations from rumor posts are learned under supervision. The model is also effective for early detection of rumors, where satisfactory precision could be accomplished.

RumourEval is a SemEval shared task and is for identifying and handling rumors and reactions to them, in text (Derczynski et al., 2017). Derczynski et al. propose a shared task where members investigate rumors as cases made in content. In the same task, users react to each other inside discussions attempting to determine the veracity of the rumor. If one needs to evaluate the evidence of a rumor, different sources can be used to make a final decision about the rumor's veracity. SemEval consists of two sub tasks: (a) classifying rumor tweets into support, deny or comment, and (b) veracity classification. Sub task A corresponds to talk around cases to confirm or refute them using crowd response analysis. Sub task B relates to the AI-hard problem of evaluating a claim. Overall, it is a tedious philosophy.

Real time rumor debunking is the proposal of Liu et al. (2015). It is an efficient procedure for mining language features like individuals' opinions, discovering witness accounts and getting the underlying belief from messages. Use of sourcing, network propagation, credibility and other user and meta features help to expose rumors. Their contributions include: (1) approach to automatically debunk rumors on social media; (2) an authentic rumor database constructed on real data, and the process of its creation; and (3) an algorithm for predicting veracity in real time is potentially faster than human verification.

In Vosoughi et al. (2017) a rumor prediction approach named Rumor Gauge is discussed. Identification of salient features of rumors on Twitter is done by looking at a few aspects of diffusion: linguistics and the users involved. Comparison of each aspect with respect to spreading of true and false rumors is made with Rumor Gauge.

A rumor signature can be formed by extracting the time series from these features. Then using Hidden Markov Models (HMMs), the rumor signature can be classified as true or false. This paper suggests an approach to predict the veracity of rumors in reasonable time and with sufficient accuracy. So, an extension to this approach is possible to predict the impact of false rumor. This is to resist the spreading of that rumor.

3 METHODOLOGY

This section discusses both veracity prediction and impact prediction. Veracity prediction is the same as that in the method of Vosoughi et al. (2017).

3.1 *Veracity prediction*

A temporal communication network can be interpreted as a rumor in this case, where nodes correspond to users, communication between nodes can be represented as edges in the network. Time related features capture the propagation of messages through the network. It notes that rumor features would be related to either the nodes or the edges in the network. Features related to users are known as user identities and those related to messages as linguistic.

3.1.1 *Linguistic*

Features from the text of the rumor are analyzed and those are collectively known as linguistic features. Some linguistic features were identified. Features that significantly contribute to the system are mentioned here. Significance can be found using a chi-square test.

Ratio of negated tweets: This is the ratio of tweets having negation over the total number of tweets in a rumor. Negations are detected using the Stanford NLP parser (Chen & Manning, 2014).

Average maturity of tweets: Politeness and elegance considered as maturity of a tweet. There are five indicators of the maturity of a tweet:

1. Smileys.
2. Abbreviations: Number of abbreviations (such as gn for goodnight, sry for sorry, gbu for god bless you) present in the tweet.
3. Vulgarity: Number of vulgar words present in a tweet.
4. Word complexity: Length of words in the tweet are considered for checking the maturity of the tweet.
5. Sentence complexity: Complexity of sentence contributes to the maturity of the tweet.

Ratio of tweets containing opinion and insight: Linguistic Inquiry and Word Count (LIWC) gives a list of insight and opinion words. Words from the tweet are compared to words in the category of LIWC dictionary (Pennebaker et al., 2003).

Ratio of uncertain and guessing tweets: Guessing and uncertain words come under another category of LIWC. This includes words such as perhaps, like, guess, and so on. Each tweet is checked against the guessing and uncertain words from LIWC.

3.1.2 *User identities*

User identity features consist of features of the user who is involved in spreading a rumor. A total of six user features were identified which contribute to the output of model.

Controversiality: This is calculated by counting reactions to the user's tweet. These reactions are then run through a Twitter sentiment classifier (Vosoughi et al., 2015). It classifies them as either positive, negative, or neutral. The count of all positive and negative reactions is taken and based on that a controversiality score for the user can be calculated.

$$\text{Controversiality} = (p + n)^{\min(p/n, n/p)} \tag{1}$$

In Equation 1, p denotes the number of positive reactions and n denotes the number of negative reactions.

Originality: This is the ratio of the number of original tweets a user has posted to the number of times the user posted retweets of someone else's tweet.

Credibility: This checks that the user's account has been officially verified by Twitter.

Influence: Influence is found by the number of followers of a user.

$$\text{Role} = \frac{\text{No of Followers}}{\text{No of Followees}} \tag{2}$$

$$\text{Engagement} = \frac{T + Rt + Rp + F}{\text{AccountAge}} \tag{3}$$

In Equation 4, T denotes the number of tweets, Rt denotes the number of retweets, Rp denotes the number of replies and F denotes the number of favorites.

3.2 *Impact prediction*

Impact prediction is based on three characteristics. Response rate, profile feature and diffusion rate are taken into account for predicting the impact of a false rumor.

3.2.1 *Response rate*
This can be further classified into positive and negative response rates. Positive Response Rate (PPR) means what fraction of people respond to a false rumor in a way they think that is true. Negative Response Rate (NPR) means the fraction of users who considered the rumor as fake itself. PPR is proportional to the impact score, i.e. the higher the PPR, the larger will be the impact of the rumor.

$$PPR = \frac{\text{No of positive reactions}}{\text{Total no of reactions}} \tag{4}$$

$$NPR = \frac{\text{No of negative reactions}}{\text{Total no of reactions}} \tag{5}$$

3.2.2 *Profile analysis*
This feature analyses the profile of the user and his/her followers. It is useful to check similar kinds activities done in the past. If a previous action is found and it did not create any chaos, then the rumor tweet may not be harmful. Like that history information (reactions of user to similar rumors) is also helpful to predict the impact.

3.2.3 *Diffusion rate*
The diffusion rate is the number of reactions (i.e replies, likes, shares, etc.) per an hour. If the diffusion rate is larger, the impact score will be larger too. That means a false rumor with a high spreading tendency will have a very high negative impact on society.

3.3 *Model*

User identity features and linguistic features determine the signature of a rumor. Figure 1 depicts the overview of the proposed model. Some rumor tweets are manually annotated for training of the model. Hidden Markov Model (HMM) is trained using annotated rumors. If a new tweet arrives, the model compares the tweet with the stored collection. Then it predicts the veracity. If the rumor veracity found to be less than 0.2, then it will be considered as fake. Then the corresponding tweet is taken for impact analysis. A separate HMM is needed to

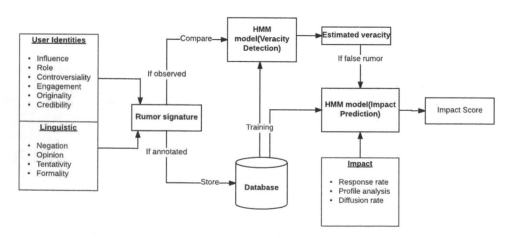

Figure 1.   Overview of proposed methodology.

predict impact. False rumors from the database are used to train the HMM. Then an impact score will be returned. The impact score is based on response rate, profile analysis and diffusion rate. If the impact score is close to 1, then the rumor is considered to be harmful.

4 DATASET AND EVALUATION CRITERIA

4.1 *Dataset*

The dataset is available as a single JSON file. It includes 300,002 tweets about the health and death of Tamilnadu chief minister, Jayalalitha. Every tweet that is associated with the situation is included in the dataset. For predicting the veracity of rumors, the model can also be trained with PHEME dataset of rumors. It contains tweets about eight events. For example, Ferguson unrest in US and the Germanwings plane crash in the French Alps. This dataset is publicly available (Zubiaga et al., 2016).

4.2 *Evaluation criteria*

Wikipedia, Snopes.com, and FactCheck.org (sites that check reliability) are utilized to recover the trusted confirmation sources. These facts considered for manual annotation. There were three main criteria for the evaluation of the proposed model.

1. Precision value of the model in anticipating the rumor veracity before verification by trusted authorities.
2. Contribution of features in predicting veracity of the rumor.
3. Accuracy of the model as a function of elapsed time since the start of the rumor.

5 CONCLUSION

Verification of rumors is a critical task and influential to populations so they can make decisions based on the truth. This paper described a model for the verification of rumors and prediction of the impacts. As a part of the impact analysis, impact features such as response rate, profile feature and diffusion rate are identified. The ability to predict the impact of the rumor along with its impact might be applied in the emergency services. News consumers and journalists can use the proposed model in their field of work.

REFERENCES

Allport, G.W. & Postman, L.J. (1965). *The psychology of rumor*. New York, NY: Russell, Russell.
Chen, D. & Manning, C.D. (2014). A fast and accurate dependency parser using neural networks. (pp. 740–750).
Derczynski, L., Bontcheva, K., M. Liakata, M., R. Procter, R., Hoi, G.W.S & Zubiaga, A. (2017). Semeval-2017 task 8: Rumoureval: Determining rumour veracity and support for rumours. In SemEval@ACL.
Liu, Xiaomo, Nourbakhsh, Armineh, Li, Quanzhi, Fang, Rui, Shah, & Sameena (2015). *Real-time rumor debunking on twitter*. Melbourne, Australia.
Ma, Jing, Gao, Wei, Mitra, Prasenjit, Kwon, Sejeong, Jansen, Jim, Wong, Kam-Fai, Cha, & Meeyoung. (2016). Detecting rumors from microblogs with recurrent neural networks. In *The 25th International Joint Conference on Artificial Intelligence (IJCAI 2016)*.
Pennebaker, J.W., Mehl, M.R. & Niederhoffer, K.G. (2003). Psychological aspects of natural language use: Our words, our selves. *Annual Review of Psychology, 1*, 547–577.
Vosoughi, S., Mohsenvand, M. & Roy, D. (2017). Rumor gauge: Predicting the veracity of rumors on twitter. *ACM Transactions on Knowledge Discovery from Data 4*.
Vosoughi, S., Zhou, H. & Roy, D. (2015). Enhanced twitter sentiment classification using contextual information. pp. 16–24.
Zubiaga, A., Liakata, M. & Tolmie, P. (2016). Pheme rumour scheme dataset: Journalism usecase.

Emerging Trends in Engineering, Science and Technology for Society,
Energy and Environment – Vanchipura & Jiji (Eds)
© 2018 Taylor & Francis Group, London, ISBN 978-0-8153-5760-5

A framework for efficient object classification for images having noise and haze using deep learning technique

A. Bhavyalakshmi & M. Jayasree
Department of Computer Science and Engineering, Government Engineering College, Thrissur, Kerala, India
APJ Abdul Kalam Technological University, Kerala, India

ABSTRACT: Object category classification is one of the most difficult tasks in computer vision because of the large variation in shape, size and other attributes within the same object class. Also, we need to consider other challenges such as the presence of noise and haze, occlusion, low illumination conditions, blur and cluttered backgrounds. Due to these facts, object category classification has gained attention in recent years. Many researchers have proposed various methods to address object category classification. The main issue lies in the fact that we need to address the presence of noise and haze which degrades the classification performance. This work proposes a framework for multiclass object classification for images containing noise and haze using a deep learning technique. The proposed approach uses an AlexNet Convolutional Neural Network structure, which requires no feature design stage for classification since AlexNet extracts robust features automatically. We compare the performance of our system with object category classification without noise and haze using standard datasets, Caltech 101 and Caltech 256.

1 INTRODUCTION

Object category classification is the task of classifying an object into its respective class within a group of object classes. The most important step in this task is the choice of feature extraction method. Obtaining key features from an image-containing object is a burdensome task due to variation in attributes like shape, size and color within the same object class. There exist many conservative feature extraction methods like Scale-Invariant Feature Transform (SIFT) (Lowe, 1999) and Histogram Oriented Gradient (HOG) (Dalal & Triggs, 2005) features for object recognition. To overcome the limitations of low-level features (Chan et al., 2015), dictionary learning and deep learning were introduced in which features were learned from the data itself instead of manually designing features.

In the above specified feature extraction techniques, most of the time is spent in deciding ideal features for all classes of objects and in the selection of a suitable classification method. Although these methods provide hopeful results, due to the inability of these methods to capture most compact and flawless features. Most advanced methods, like deep learning neural networks, replace them with feature extraction for object classification. The main advantage of using deep learning architecture is that it learns compact and flawless features automatically. The main disadvantage of these methods is that they require a huge amount of data and computation power (Hieu Minh et al., 2016).

At the end of 20th century and in first decade of the 21st century, neural networks have shown satisfying results for the object classification problem (Cireşan et al., 2011; Jia et al., 2009). But in recent years the old neural network structures have been replaced by new deep and complex neural network architectures called deep learning neural networks (LeCun et al., 2015). As specified earlier, the main disadvantage of these deep learning structures is that they require extensive training with a huge amount of data to capture flawless and compact features.

During recent years, different deep learning architectures have been developed. AlexNet (Zeiler et al., 2014) and ZFNet (Zeiler et al., 2014) are two examples of such deep learning networks. The AlexNet came top in the ImageNet challenge in 2012 but was outperformed by other deep networks in the following years. Despite this defeat, AlexNet is still used for many image classification problems.

Images are mostly corrupted by Gaussian noise during the acquisition of the image in the camera. The presence of haze can also make the image unclear. So, object categorization in the presence of noise and haze becomes an important issue. This paper deals with the classification of noisy and hazy images by a deep trained network model AlexNet.

This work is mainly divided into two experiments, object classification of noisy and hazy images and object classification of images without noise and haze.

The rest of the paper is arranged as follows: Section 2 gives the literature survey; Section 3 gives the overall system architecture of the proposed system; and finally, Section 4 gives the conclusion.

2 RELATED WORK

Hieu Minh et al. (2016) suggested a technique for feature extraction which combines AlexNet with a Recursive Neural Network (RNN) known as AlexNet-RNN. The AlexNet is used to extract optimal features and which is then fed to the RNN. The RNN has a structure similar to that of an ordinary neural network, the only difference is that it retains the weight learned for a while.

Jian et al. (2016) used an average and weighted feature extraction method. First the behavior of the features is investigated. In the next step, the most powerful features are chosen and after that the average of these best features is used for classification. In weighted average combination some of the features are omitted by assigning a value zero. Finally, all the features are integrated into a k-NN framework for classification purpose.

Shengye et al. (2015) used a two-level feature extraction method for image classification. In the first level, Bag of words (BOW) along with a spatial image pyramid is used to extract first level features. During the second level, the features from the first level are extracted based on dense sampling and spatial area sampling. The second technique adapts a multiple kernel learning that can be used to fuse different image feature to obtain compact features. They used the Caltech 256 and 15 scene datasets, obtaining an accuracy of 54.7% and 89.32% for Caltech 256 and 15 scene respectively.

Transfer learning based on a deep learning approach was employed by Ling et al., (2015). In recent years, deep neural networks have been used to extract the high level, most compact and selective features from images that can be used for different tasks in computer vision. As the image features are passed through different layers, the feature learning takes place at different levels and feature learning in different levels represents different abstractions (Ling et al., 2015). It is very difficult to apply deep learning techniques to vision problems as it requires huge labeled data for training.

Foroughi et al. (2017) proposed an object categorization method which has significant intra-class variation. It uses a joint projection and low-rank dictionary learning method using dual graph constraints (JP-LRDL). It simultaneously learns a robust projection and a discriminative dictionary in the low dimensional space. These can handle different types of variation within the same object class, that raises due to occlusion, changes in viewpoint and poses, size changes and various shape alterations.

Demir and Guzelis (2016) proposed a method for object recognition using two variations of CNN. The first CNN included ten layers and the second one was similar to AlexNet which consisted of nine layers. For feature extraction, the entire image is divided into nine patches and features were extracted from each patch. The feature extraction method uses the above specified CNN variants and BOW. The BOW uses SURF as the feature detector and HOG as the feature descriptor. Finally, the features are supplied to an SVM for classification. Existing methods do not consider the presence of various artifacts such as noise and haze during classification. We are trying to implement a system that classifies objects even in the presence of noise and haze using the deep convolutional neural network architecture, AlexNet. And finally, compare the work with the object classification system with images without any artifacts.

3 PROPOSED FRAMEWORK

In this section, we briefly summarize the essential steps for multiclass image category classification in the presence of noise and haze using a deep learning technique.

3.1 *Object categorization*

We conducted experiments using Caltech 101 and Caltech 256 images. Caltech 101 contains 101 different image classes (examples: faces, bags, watches, etc.). Each class has 40 to 800 images with 300×200 pixel size. Also, it contains background cluttered images. Caltech 256 is a successor of the Caltech 101 dataset. It overcomes certain limitations of Caltech 101 such as the limited number of categories. It has a total of 30,607 images with 256 categories. The minimum number of images per category is 80. This dataset is harder than the former one as it has more categories and no left right aligned images. It also contains new and larger clutter categories.

3.2 *Training and testing sets*

In the first step, the AlexNet framework extracted features from images in Caltech 101 and Caltech 256 with no alterations. For this, we divided the full dataset into training and testing sets. In the second phase different noises like Gaussian noise, salt and pepper noise and haze were added to images. The images containing noises and haze were then randomly divided into testing and training images. The training and testing set was used to extract training and testing features from the AlexNet framework respectively. Finally, we compared the performance of these two methods.

3.3 *Feature extraction using AlexNet*

Convolutional Neural Networks (CNN) are a type of artificial neural networks that are mainly used for object detection (Wanli et al., 2016), object recognition (Ming & Xiaolin, 2015) and image retrieval (Hailiang et al., 2017). Recently many researchers have used CNN for solving other problems. The main layers in CNN include the convolution layer, the pooling layer, the normalization layer and the fully connected layer. The main advantage of CNN is that it extracts robust features automatically and we can reduce the dimension of features learned by increasing the size of the hidden layers. Due to the weight sharing mechanism, we require the minimum amount of parameters and neurons for training the CNN.

AlexNet is a variant of CNN which provides high performance and learns compact and robust features. The basic AlexNet framework proposed by Alex et al. (2012) consists of five convolution layers and three max pooling layers. As in any CNN, the learning takes place in

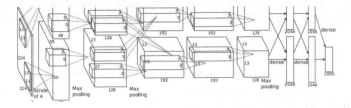

Figure 1.   Structure of AlexNet framework.

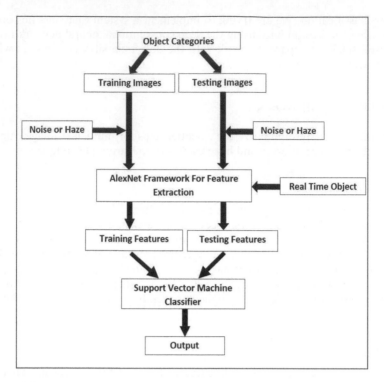

Figure 2. Architecture of the proposed framework.

the convolution layer. The pooling in a CNN can be either max pooling or average pooling. For most of the image classification problems, max pooling is found to be more effective than average pooling. Hence in this paper, we adopt max pooling.

The AlexNet used in this paper consists of one input layer, five convolution layers, three pooling layers, three fully connected layers, along with two normalization layers, seven Rectified Linear Unit (ReLU) layers, two dropout layers and a softmax and output layer. The intermediate layers: convolution, pooling, fully connected and ReLU layers form the bulk of the CNN. The convolution layer performs the convolution operation which is the same as convolution in image processing. What it actually does is remembers the pattern learned during the training process and when a new pattern occurs it tries to label it to the closest pattern. The pooling layer is used to reduce the feature size. It down samples the feature map by dividing it into a rectangular pooling region and selecting the maximum value from each region. When using CNN for image classification, the choice of activation function is very important. Usually, we can use the hyperbolic tangent function (tanh) or ReLu. In our model, we use ReLu as the activation function. In this framework, we use a max pooling layer with size 3×3 and stride 2. The first convolution layer uses 96 kernels with size $11 \times 11@3$, the second uses 256 kernels with size $5 \times 5@48$, the third and fourth uses 384 kernels with size $3 \times 3@256$ and the final one uses 256 kernels with size $3 \times 3@192$. The first two convolution layers are followed by a normalization, ReLu and max pooling layer. The next two are only followed by a ReLu layer. The final convolution layer is followed by one ReLu and pooling layer. After the sixth and seventh ReLu layers we employed two dropout layers of 50% to avoid overfitting.

3.4 Classification

We can use different classifiers such as k-Nearest Neighbor (k-NN) (Yin & Bo, 2009), Naive Bayes (Shih-Chung et al., 2015), Hidden Markov Model (HMM) (Jia et al., 2000) and Support Vector Machine (SVM) (Cortes & Vapnik, 1995).

In our framework, we use SVM for classification. SVM is a supervised machine learning algorithm which is generally used for both classification and regression challenges. SVM uses a hyperplane or set of hyperplanes for classification and regression. Since it tries to maximize the margin size, it is also called maximum margin classifier. The main advantage of using SVM for classification is that it performs classification effectively even if the dimension is high. The responses from the AlexNet are reshaped into feature vectors and passed to the classifier for training and testing purposes. The main reason for using SVM for classification is due to the overfitting issue in the fully connected layers and SVM also only requires a few parameters compared to the fully connected layer for classification purposes.

4 CONCLUSION

In this paper, we have proposed an efficient method to classify a large number of object categories in the presence of noise and haze. Here we use a variant of CNN, AlexNet, for extracting flawless and compact features.

This approach requires no traditional feature extraction during the training stage as CNN learns the features automatically, and classification can be performed very efficiently even in the presence of noise and haze. The output of AlexNet is compact and highly relevant features. Finally, we can use simple multiclass classifiers like SVM for effective classification.

Training and testing was conducted on two benchmark datasets Caltech 256 and Caltech 101, with noise and haze. We assume that there will be no considerable changes in the performance even if artifacts like noise and haze are present.

REFERENCES

Alex, K., Ilya, S. & Geoffrey, E.H. (2012). ImageNet classification with deep convolutional neural networks. In Pereira, F., Burgoo, C., Bottou, L., Weinberger, K., (Eds.), *Advances in Neural Information Processing Systems 25* (pp. 1097–1105). Curran Associates, Inc.

Cireşan, D.C., Meier, U., Masci, J., Gambardella, L.M. & Schmidhuber, J. (2011). Flexible, high performance convolutional neural networks for image classification. pp. 1237–1242.

Cortes, C. & V. Vapnik (1995, Sep). Support-vector networks. *Machine Learning 20*(3), 273–297.

Dalal, N. & Triggs, B. (2005). Histograms of oriented gradients for human detection. *IEEE Computer Society Conference on Computer Vision and Pattern Recognition 1*, 886893.

Demir, Y. & Guzelis, C (2016). Moving towards in object recognition with deep learning for autonomous driving applications. IEEE Conference Publications.

Foroughi, H., Ray, N. & Zhang, H. (2017). Object classification with joint projection and low-rank dictionary learning. *IEEE Transactions on Image Processing 99*, 1–1.

Hailiang, L., Yongqian, H. & Zhijun, Z. (2017). An improved faster r-cnn for same object retrieval. *IEEE Access 5*, 13665–13676.

Hieu Minh, B., L. Margaret, L., N. Eva, Cheng. Katrina, & B. Ian S. (2016). Object recognition using deep convolutional features transformed by a recursive network structure. *IEEE Access 4*, 10059–10066.

Jia, D., Wei, D., Richard, S., Li-Jia, L., Kai, L. & Li, F.-F. (2009). Image Net: A large-scale hierarchical image database. *IEEE Conference on Computer Vision and Pattern Recognition*, 248–255.

Jia, L., N. A, & G.R. M (2000). Image classification by a two-dimensional hidden Markov model. *IEEE Transactions on Signal Processing, 48*, 517–533.

Jian, H., Huijun, G., Qi, X. & Naiming, Q. (2016). Feature combination and the knn framework in object classification. *IEEE Transactions on Neural Networks and Learning Systems, 27*, 1368–1378.

LeCun, Y., Bengio, Y. & Hinton, G. (2015). Deep learning. *Nature, 521*, 436444.

Ling, S., Fan, Z. & Xuelong, L. (2015). Transfer learning for visual categorization: A survey. *IEEE Transactions on Neural Networks and Learning Systems, 26*, 1019–1034.

Lowe, D.G. (1999). Object recognition from local scale-invariant features. *The Proceedings of the Seventh IEEE International Conference, 2*, 11501157.

Ming, L. & Xiaolin, H. (2015). Recurrent convolutional neural network for object recognition. *IEEE Conference on Computer Vision and Pattern Recognition (CVPR)*, 3367–3375.

Shengye, Y., Xinxing, X., Dong, X., Stephen, L. & Xuelong, L. (2015). Image classification with densely sampled image windows and generalized adaptive multiple kernel learning. *IEEE Transactions on Cybernetics, 45,* 381–390.

Shih-Chung, H., I-Chieh, C. & Chung-Lin, H. (2015). Image classification using pairwise local observations based naive Bayes classifier. *Asia-Pacific Signal and Information Processing Association Annual Summit and Conference (APSIPA),* 444–452.

T.H. Chan, K. Jia, S.G.J.L.Z.Z. & Y. Ma. (2015). Pcanet: A simple deep learning baseline for image classification. *IEEE Transactions on Image Processing, 24,* 50175032.

Wanli, O., Xingyu, Z., Xiaogang, W., Shi, Q., Ping, L., Yonglong, T., ... & T. (2016). DeepID-Net: Object detection with deformable part based convolutional neural networks. *IEEE Transactions on Pattern Analysis and Machine Intelligence, 39,* 1320–1334.

Yin, L. & Bo, C. (2009). An improved k-nearest neighbor algorithm and its application to high resolution remote sensing image classification. *17th International Conference on Geoinformatics,* 1–4.

Zeiler, M.D., e. D. Fergus, R., Pajdla, T. Schiele, B. & Tuytelaars, T. (2014). Visualizing and understanding convolutional networks, book Title = Computer Vision – *ECCV 2014:* 13th *European Conference, Zurich, Switzerland, September 6–12, 2014, Proceedings, Part I.* Cham: Springer International Publishing.

Emerging Trends in Engineering, Science and Technology for Society,
Energy and Environment – Vanchipura & Jiji (Eds)
© 2018 Taylor & Francis Group, London, ISBN 978-0-8153-5760-5

Bilingual handwritten numeral recognition using convolutional neural network

Jettin Joy & M. Jayasree

Department of Computer Science and Engineering, Government Engineering College, Thrissur, Kerala, India
APJ Abdul Kalam Technological University, Kerala, India

ABSTRACT: Optical character recognition (OCR) is the process of identifying characters in any form, with the help of a photo-electric device and computer software. Handwritten numeral recognition can be considered as a special area in OCR, where the characters are handwritten numerals. Unlike printed numeral recognition, handwritten numeral recognition is very difficult as different persons have different writing styles. Usually, Conventional OCR includes steps like pre-processing, segmentation, feature extraction and finally classification and recognition. Conventional OCR uses handcrafted features for feature extraction, whereas Convolutional Neural Networks (CNN) extracts features automatically. Hence by using CNN, we can eliminate the need for manual feature engineering. Recently many researchers have used deep learning architectures like CNN, Stacked or Deep Autoencoders and Deep Belief Networks for Handwritten Numeral Recognition and they have shown promising results compared to other methods. No one has yet implemented a system for recognizing Malayalam and Kannada handwritten numerals using CNN. So this is the first project to address Malayalam and Kannada handwritten numeral recognition using deep learning approach. Also, we compare the performance with another deep learning architecture, stacked autoencoder. The sole aim of this project is to automatically recognize and classify Malayalam and Kannada numerals using Convolutional Neural Network.

1 INTRODUCTION

Optical character recognition (OCR) is one of the old and most popular research area. Researchers are trying to improve the performance of existing methods by introducing new methods in segmentation, feature extraction and classification. Handwritten character and/ or numeral recognition is a part of OCR where the characters are handwritten in nature. Unlike printed numeral recognition, recognition of handwritten numerals is very difficult. This is mainly due to the variation in the appearance of numerals, as different persons have different writing styles.

Furthermore, handwritten character and/or numeral recognition can be either online or offline. The online method involves recognition of characters in real-time i.e, the characters are recognized as soon as they are written, usually the input is coordinate values of the characters. On the other hand offline character recognition system takes input in the form of the image, which is obtained from scanning, pre-processing and segmenting the document containing the characters. Offline method is more challenging than the online as it involves noise and other artifacts like document quality, variation in the shape and style of characters etc.

Recently deep learning methods have shown promising performance in areas like object recognition (Ayegl et al. 2016), object classification (Wanli et al. 2016), automatic number plate recognition (Menotti, Chiachia, Falco, & Oliveira Neto 2014, Syed Zain, Guang, Afshin, & Enrique 2017), sentiment analysis in texts (Abdalraouf & Ausif 2017), stock market prediction (Vargas et al. 2017), automatic speech recognition (Palaz et al. 2015) and character recognition (Mehrotra et al. 2013). From these we can conclude that nowadays

Figure 1.   Malayalam and Kannada numerals.

deep learning methods are used in almost all research areas such as Data analytics, natural language processing (NLP), Object and Image classification and character recognition etc. Convolutional Neural Network or CNN or ConvNet is a type of deep learning neural network architecture, that is most suitable for image structure representation and visual imagery analysis. The main advantage of CNN is that it requires minimal amount of pre-processing (Lecun et al. 1998).

In recent years several variants of Convnet has emerged in the ImageNet challenge, some of them include Alex Net (Alex et al. 2012), ZF Net (Zeiler, Fergus, Pajdla, Schiele, & Tuytelaars 2014), Google Net (Christian, Wei, Yangqing, Pierre, Scott, Dragomir, Dumitru, Vincent, & Andrew 2015) and Res Net (He et al. 2015). The main advantage of using deep learning structure like ConvNet is that we can eliminate the need for manual feature extraction. Deep neural networks can outperform the conventional methods (P Nair, Ajay, & Chandran 2017).

Malayalam and Kannada are two of the most commonly used languages in South India and are the official languages of the Kerala and Karnataka state respectively. These two languages belong to Dravidian language family. Malayalam and Kannada language have their own numerals and are more complex compared to modern Hindu-Arabic numerals due to the complex curved nature. Figure 1 shows Malayalam and Kannada Numerals.

To my extent of knowledge, this is the first paper to address Malayalam and Kannada handwritten numeral recognition using Convolutional Neural Network. We are attempting to use CNN to achieve better performance in recognizing Malayalam and Kannada handwritten numerals.

2   LITERATURE SURVEY

Akm and Abdul (2017) used CNN for Handwritten Arabic numeral recognition. They conducted experiment with two models. In the first model, they used they used a variant of multi layer perceptrons (MLP) with notable changes. To avoid over-fitting, they employed dropout mechanism. The second is a ConvNet model with two convolution and two max pooling layer. The first convolution layer includes 30 feature maps each with kernel size 55. The second convolution layer include 15 feature maps each with kernel size 3×3 pixel. Both the max pooling layer had a size of 22. In order to avoid over-fitting a dropout of 25% was used. The accuracy for the first model was 93.8% and that of second model (CNN Model) was 97.4%.

Akhand et al. (2015) used a CNN based architecture for recognizing Bangla handwritten numerals. The CNN included 2 convolution layer with 5×5 kernel and 2 subsampling layer with size of 2×2, which uses average pooling. The first level feature feature maps (6 feature maps) that obtained after first convolution had a size of 24×24. Using the first subsampling layer the feature maps were downsampled to 12×12 and the second convolution produced 12 feature maps of 8×8. Finally, using a second level subsampling it is further downsampled to 4×4. No dropout mechanism were used in this method. The overall accuracy for the system was 97.93%.

Ramadhan et al. (2016) used a similar CNN structure as above for handwritten mathematical symbol recognition, they used 3 feature map in the first convolution and subsampling and 5 feature maps in the second convolution and subsampling and unlike average subsampling used above they employed max pooling. The overall accuracy for training and testing was 93.27% and 87.72%. respectively.

Another deep learning architecture like deep sparse belief network and stacked denoising autoencoders are also used for handwritten digit recognition (Ragheb and Ali 2014). The spare deep belief network consists of three RBMs with 600 hidden units. The training was done in a greedy manner using contrastive divergent method. They also proposed a stacked denoising autoencoder architecture with a corruption rate of 25% and 600 units each hidden layer.

Md et al. (2016) proposed a structure with an autoencoder with 3 convolution layer and 3 max pooling layers, a decoder with 3 deconvolutional layer and 3 upsampling layers. The output of this is passed through a Convolutional layer and finally the output is obtained. They used 3 models for training. The first one is called SCM: Simple convolutional model, the second one is called SCMA: simple convolutional model with augmented images. The final one is called ACMA: Autoencoder with Convolutional Model with Augmented images. And they found that the third model has high efficiency than the former two.

Pal et al. (2007) proposed a system for recognizing handwritten numerals of six Indian scripts. They used two type of feature vectors, first one is of length 64 and is used for fast recognition. The second one is of length 400 and is used for higher accuracy recognition. They used contour points to extract the first features and gray-scale local-orientation histogram of the component to extract second features. The system showed satisfactory performance for all languages.

V Rajashekararadhya and Vanaja Ranjan (2009) proposed a zoning and distance based feature for effectively recognizing Kannada numerals with support vector machine (SVM) classifier. They divided entire image into n zones after calculating the centroid. After that average distance from each pixel to the centroid is calculated and repeated for each zone. and a total of n features are extracted for classification and recognition. Finally, a Support vector machine is used as classifier. The overall performance listed was satisfactory for all class of numerals.

The above two methods use handcrafted feature extraction method.

We are attempting to implement a efficient system for recognizing Malayalam and Kannada numerals using CNN that uses automatic feature extraction methods to improve the existing performance.

3 PROPOSED SYSTEM

The Convolutional Neural Networks need to trained with large image dataset to obtain good performance. This is the main challenge in any deep learning project. i.e, it requires a large dataset. However, there are techniques to increase the size of images in the dataset. There is no open source large dataset available for Malayalam and Kannada numerals. Figure 2 shows the overall architecture of the proposed system.

The proposed system includes the following steps:

3.1 *Data acquisition and dataset creation*

There is no open source large dataset for numeral recognition in Malayalam and Kannada. Hence we require to build it from the scratch. This is the most time consuming and tedious task. In this step, we collect the data from a large population. As CNN requires large dataset, it is important to collect as many numeral writings from different persons. After collecting data the document containing numeral images are scanned and individual numerals are segmented from the scanned image using appropriate segmentation mechanism. The Figures 3 and 4 shows three sets of samples of Malayalam and Kannada handwritten numerals respectively.

3.2 *Pre-processing*

The pre-processing step usually involves resizing, normalization and removing unwanted entities from numeral images, like noise, haze etc. Pre-processing steps helps to ease the numeral recognition. Firstly, the images are resized to an appropriate size, then it is converted to gray scale image. After that, the pixels in images are inverted to obtain the negative of the image.

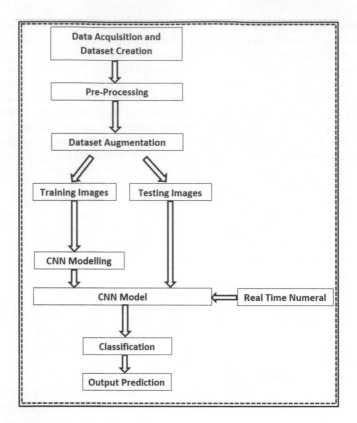

Figure 2. System architecture.

Figure 3. Malayalam handwritten numerals.

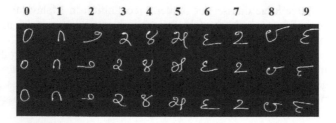

Figure 4. Kannada handwritten numerals.

This is usually done to reduce the storage size. The image size has great influence in the training stage. If it is too large we will require increased training time as computation involved is high. If the size is too small fitting of images into the network becomes difficult, hence it is

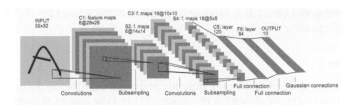

Figure 5.   LeNet-5 architecture.

always good to choose an appropriate size. We can use methods like padding for choosing a standard size.

3.3 *Dataset augmentation*

As mentioned earlier, CNN requires large image dataset for training. To obtain large dataset we use dataset augmentation. For augmenting the images affine transformations like translation, scaling, rotation and shearing can be used. The affine transforms mentioned above generally preserves points, straight lines, and planes. Other methods like blurring, vary the contrast and brightness of the images can be applied to obtain a large dataset from the existing dataset. After the augmentation step, the entire dataset is randomly divided into testing images and training images, e.g 70% for training and remaining 30% for testing.

3.4 *CNN modeling*

This is the most vital step in this project. Here we define various layers in CNN and appropriate network parameters for training the dataset. The main layers in a CNN are described below.

3.4.1 *Convolution layer*
This is the first layer that receives input, the convolution operation tries to label a new image class by referring what it has learned in the past. The main property of convolution operation is that it is translation invariant (Thomas & Helmut 2015).

3.4.2 *Subsampling layer*
The subsampling layer takes input from the convolution layer. It is used to smooth or reduce the feature map size or feature size. pooling can be done in two ways. First one is called max pooling and the later one is called average pooling, for most of the problems max pooling is found to be more effective than average pooling, Hence we adopt max pooling mechanism. Figure 5 shows a simple LeNet (Lecun et al. 1998) CNN architecture.

3.4.3 *ReLu layer*
This layer mainly determines which activations are propagated to next layer. Usually, output signals that are more close to previous references are propagated. There are many activation functions that can be used in which ReLu is most effective for fastening the training process.

3.4.4 *Fully connected layer*
This is the final layer in CNN, which means that every neuron of the current layer receives output from the previous layer. The layers involved in feature learning are Convolution layer, Subsampling layer, and ReLu layer. After this phase, a CNN model with different layer is obtained.

3.5 *Training and testing*

The CNN model is used for training and testing the dataset. In order to avoid overfitting, dropout mechanism can be used i.e, dropping out a certain number of neuron output in each layer (Srivastava, Hinton, Krizhevsky, Sutskever, & Salakhutdinov 2014). The testing images are used to test the model created. A randomization technique is used for dividing the entire dataset into training and testing set.

3.6 Classification

This is the final stage, in which a softmax function is used to classify the output of the CNN. The softmax function output falls within the range [0,1] and the sum of the output of all class is equal to 1. The softmax function classifies the input numeral to a class that has highest output value. The above system can be evaluated by varying the layers in CNN and also varying different parameters like learning rate. Backpropagation with gradient descent is most effective learning rule for image classification problem.

4 CONCLUSION

Handwritten numeral recognition has a large number of applications like ZIP code recognition (LeCun et al. 1989), recognizing numerals in old documents etc. Traditional methods use handcrafted features, which requires a great deal of effort and time. This can be eliminated by introducing new automatic feature learning methods like Convolutional Neural Network, Deep Belief Network, AutoEncoders etc. These deep learning method has shown outstanding performance in recognizing numerals as well as handwritten characters (Shailesh et al. 2015).

Here we have proposed a bilingual handwritten numeral recognition system for Malayalam and Kannada numerals. The dataset creation and CNN modeling is very time consuming. We would provide first ever large open source dataset for Malayalam and Kannada numerals. To reduce the time required for the training stage graphics processing unit (GPU) support is used. Also to avoid overfitting in training phase of the CNN dropout mechanism is applied. CNN have shown better results in recognizing numerals in various scripts, hence has a high probability that it will provide the same for Malayalam and Kannada numerals.

REFERENCES

Abdalraouf, H. & M. Ausif (2017). Deep learning approach for sentiment analysis of short texts. *3rd International Conference on Control, Automation and Robotics (ICCAR)*, 705–710.

Akhand, M.A.H., M. Mahbubar Rahman, P.C. Shill, I. Shahidul, & M.M. Hafizur Rahman (2015). Bangla handwritten numeral recognition using convolutional neural network. *International Conference on Electrical Engineering and Information & Communication Technology (ICEEiCT2015)*, 1–5.

Akm, A. & K.T. Abdul (2017). Handwritten arabic numeral recognition using deep learning neural networks. *CoRR abs/1702.04663*.

Alex, K., S. Ilya, & E.H. Geoffrey (2012). Imagenet classification with deep convolutional neural networks. *In Pereira, F., Burges, C., Bottou, L., Weinberger, K., eds.: Advances in Neural Information Processing Systems 25. Curran Associates, Inc.*, 10971105.

Ayegl, U., D. Yakup, & G. Cneyt (2016). Moving towards in object recognition with deep learning for autonomous driving applications. *International Symposium on Innovations in Intelligent SysTems and Applications (INISTA)*, 1–5.

Christian, S., L. Wei, J. Yangqing, S. Pierre, R. Scott, A. Dragomir, E. Dumitru, V. Vincent, & R. Andrew (2015). Going deeper with convolutions. *IEEE Conference on Computer Vision and Pattern Recognition (CVPR)*.

He, K., X. Zhang, S. Ren, & J. Sun (2015). Deep residual learning for image recognition. *CoRR 7*.

LeCun, Y., B. Boser, J.S. Denker, D. Henderson, R.E. Howard, W. Hubbard, & L.D. Jackel (1989). Backpropagation applied to handwritten zip code recognition. *Neural Comput. 1*(4), 541–551.

Lecun, Y., L. Bottou, Y. Bengio, & P. Haffner (1998). Gradient-based learning applied to document recognition. *86*, 2278–2324.

Md, S., M. Nabeel, & A. Md Anowarul (2016). Bangla handwritten digit recognition using autoencoder and deep convolutional neural network. *International Workshop on Computational Intelligence (IWCI)*, 64–68.

Mehrotra, K., S. Jetley, A. Deshmukh, & S. Belhe (2013). *Unconstrained handwritten Devanagari character recognition using convolutional neural networks*.

Menotti, D., G. Chiachia, A.X. Falco, & V.J. Oliveira Neto (2014). Vehicle license plate recognition with random convolutional networks. *27th SIBGRAPI Conference on Graphics, Patterns and Images 249*, 1530–1834.

Nair P., P., J. Ajay, & S. Chandran (2017). Malayalam handwritten character recognition using convolutional neural networks. *International Conference on Inventive Communication and Computational Technologies (ICICCT)*, 278–281.

Pal, U., N. Sharma, T. Wakabayashi, & F. Kimura (2007). Handwritten numeral recognition of six popular indian scripts. *Ninth International Conference on Document Analysis and Recognition, 2007, ICDAR 2007 2*, 749–753.

Palaz, D., M. Magimai-Doss, & R. Collobert (2015). Convolutional neural networks-based continuous speech recognition using raw speech signal. *IEEE International Conference on Acoustics, Speech and Signal Processing (ICASSP)*, 4295–4299.

Ragheb, W. & L. Ali (2014). Handwritten digit recognition using sparse deep architectures. *9th International Conference on Intelligent Systems: Theories and Applications (SITA-14)*, 1–6.

Rajashekararadhya, V., S. & P. Vanaja Ranjan (2009). Support vector machine based handwritten numeral recognition of kannada script. *IEEE International Advance Computing Conference, 2009. IACC 2009.*, 381–386.

Ramadhan, I., B. Purnama, & S. Al Faraby (2016). Convolutional neural networks applied to handwritten mathematical symbols classification. *4th International Conference on Information and Communication Technology (ICoICT)*, 1–4.

Shailesh, A., P. Ashok Kumar, & G. Prashnna Kumar (Software, Knowledge, Information Management and Applications (SKIMA), 2015). Deep learning based large scale handwritten devanagari character recognition. *9th International Conference on Software, Knowledge, Information Management and Applications (SKIMA), 2015*, 1–6.

Srivastava, N., G. Hinton, A. Krizhevsky, I. Sutskever, & R. Salakhutdinov (2014). Dropout: A simple way to prevent neural networks from overfitting. *J. Mach. Learn. Res. 15*, 1929–1958.

Syed Zain, M., S. Guang, D. Afshin, & G.O. Enrique (2017). License plate detection and recognition using deeply learned convolutional neural networks. *CoRR abs/1703.07330*, 1703.07330.

Thomas, W. & B. Helmut (2015). A mathematical theory of deep convolutional neural networks for feature extraction. *CoRR abs/1512.06293*.

Vargas, M., B. Lima, & A.G. Evsukoff (2017). Deep learning for stock market prediction from financial news articles. *IEEE International Conference on Computational Intelligence and Virtual Environments for Measurement Systems and Applications (CIVEMSA)*, 60–65.

Wanli, O., Z. Xingyu, W. Xiaogang, Q., Shi, L. Ping, T. Yonglong, L. Hongsheng, Y. Shuo, W. Zhe, L. Hongyang, Kun, W., Y. Junjie, L. Chen-Change, & T. (2016). Deepid-net: Object detection with deformable part based convolutional neural networks. *IEEE Transactions on Pattern Analysis and Machine Intelligence 39*, 1320–1334.

Zeiler, M.D., e. D. Fergus, Rob., T. Pajdla, B. Schiele, & T. Tuytelaars (2014). *Visualizing and Understanding Convolutional Networks, book Title=Computer Vision – ECCV 2014: 13th European Conference, Zurich, Switzerland, September 6–12, 2014, Proceedings, Part I*. Cham: Springer International Publishing.

Emerging Trends in Engineering, Science and Technology for Society,
Energy and Environment – Vanchipura & Jiji (Eds)
© 2018 Taylor & Francis Group, London, ISBN 978-0-8153-5760-5

A security framework to safeguard the hybrid model of SDN from malicious applications

C.M. Mansoor & K.V. Manoj Kumar
Department of Computer Science and Engineering, Government Engineering College, Thrissur, Kerala, India

ABSTRACT: The hybrid control model of SDN is designed to improve network productivity by reducing the controller load. When the network is under heavy load, flow rules are installed on a network device by other network devices on behalf of the controller, and in the case of normal load the control is centralized. Thus the controller does not have to program flows to each network equipment one by one, instead it can ask the equipment to spread this flow to other equipment on behalf of the controller. This model is not secure from malicious applications as all the applications are treated in the same way and there is no way to distinguish between a genuine and a malicious application. This paper proposes a permission system to which the applications must subscribe on initialization with the controller and before approving the application commands a permissions check is performed. The priority of the application is also considered while granting the permission in order to deal with policy conflict. This will effectively monitor the working of every application and thus will prevent any unauthorized operations.

1 INTRODUCTION

The main feature of the SDN architecture is the separation of the control plane from the data plane. The data plane devices forward packets on the basis of instructions obtained from a logically centralized controller which also maintains the network state. In traditional networks a change in the device configuration or the routing strategy would mean the modification of the firmware of all the involved data plane devices, which would incur high cost. Since SDN implements the control plane in software, the changes in the routing strategy can be made from a single point i.e. there is centralization of policies. The forwarding devices no longer needs to make decisions and are thus less complex allowing the creation of low cost devices.

Open flow is the most widely used protocol for communication between the control and data plane. In an attempt to efficiently balance the controller load when using open flow, Othman & Okamura (2013a) proposed a hybrid control model of SDN; that allows the regular centralized control model to be used in the normal situations, but at the same time introduces a distributed control model in order to ensure proper working of the network in situations where the controller is under substantial loads and is required to install large number of flows in to the forwarding devices. In such cases, the hybrid control model can be used to relieve the controller from doing any further processing to relocate the flows, and enabling it to install those flows as they are; and relying on the distributed control of the network equipment to solve any issues of network equipment overloading. And thus, the hybrid control model enables the smooth working of the network even in cases of overloading.

The SDN architecture provides both advantages and disadvantages to the security of the network. On one hand, once an attack is detected the global view of the network allows to take countermeasures much quickly and on the other hand it brings with it additional security challenges. One of the issues is the attack by malicious applications. Vulnerability created in the network by granting complete control and visibility of the network to the applications was discussed in X. Wen & Wang. (2013). The interaction of the applications with the network can be monitored by the use of a permissions set which works in the same way as the system used in android play store as shown in S. Scott-Hayward & Sezer (2014). In the Android smart phone scenario, various applications will only be granted minimal set of permissions required for its working. In the same way SDN application designed to read the network topology should not be granted permission to obtain various real time notifications which would make the network vulnerable by exposing sensitive information to the controller.

This paper proposes a security application on top of the hybrid control model of SDN, to safeguard the network from malicious applications by using a permission system. Applications can subscribe to various permissions and can only execute only those operations for which permission has been granted. This allows to prevent the malicious operations and to monitor the working of every application.

2   PROBLEM DESCRIPTION

In the hybrid model of SDN, the applications interact with the network via the controller which provides an abstraction of the data plane elements. The applications can get information such as the current state of the network or can change the routing strategy by asking the hybrid controller to write the flows to various devices. The controller will either spread the flow by itself or in case of heavy load ask other equipment to spread the flows. Every application is treated in the same way and there is no distinction between a genuine application and a malicious application.

Although the effect of some attacks such as dos will be less severe in this model compared to the normal SDN, an attacker can control the entire network by writing the intended flow rules to the network devices via a malicious application. Also there is no control over the type of operations that genuine applications can perform.

In addition to securing the hybrid model of SDN by preventing malicious applications from taking control over the network, this work also enforces priority to different applications to ensure that their operations do not interfere with each other.

3   RELATED WORK

Various security issues and corresponding solutions were discussed in Sandra Scott-Hayward & Sakir Sezer (2016) The problem of treating every application with the same privilege was first identified in X. Wen & Wang. (2013). The authors propose PermOF with a set of permissions which were enforced at the Application Programming Interface (API) entry using an isolation mechanism. It was successful in solving the app privilege problem and thus securing the network. The concept of the permissions system is extended in S. Scott-Hayward & Sezer (2014). It is implemented on top of the Floodlight controller. The authors define the set of permissions to which the application must subscribe on initialization with the controller and introduce an Operation Check point, in which before approving the application commands a permissions check is performed. They also used an unauthorized operations log to examine the malicious activity in order to build a profile for SDN application—layer attacks. Although it discussed about the problem of application priority enforcement, a solution for that specific issue was not presented.

The issue of policy conflict was discussed in P. Porras & G. Gu (2012). The system uses the FortNOX enforcement engine, which handles possible conflicts by using authors security authorization to decide on flow insertion. It checks whether the new flow clashes with the existing flow rule. If the new flow rule is conflicting with the existing one, it will be installed only if it is issued by a higher priority author. The need to resolve the proper authorization level of the flow rule author is a drawback of this method.

Stanford research institute extended the floodlight Controller to develop the Security enhanced floodlight (SEK). An administrator authorizes applications java class, which is digitally verified by the SEK at run time. The application has full control over the network once it is signed and approved.

In an attempt to improve network productivity by reducing the controller load, a hybrid control model was discussed in Othman & Okamura (2013a). In order to ease the controller pressure when the network is under heavy load, flow rules are installed on a network device by another network device on behalf of the controller, and in the case of normal load, the control is centralized. The same authors presented the method to secure the distributed control model in Othman & Okamura (2013b). The central control element is secured by TLS and the transmission of flow installation requests between the network devices is protected using a signature algorithm. The system also uses a centralized trust manager. The solution in the present paper will be implemented on top of Othman & Okamura (2013b).

4 DESIGN

A complete set of permissions is defined to which the applications must subscribe on initialization with the controller. The permission set is similar to that in X. Wen & Wang. (2013) and it includes all types of permissions an application may need to execute. The permissions are stored securely with the application IDs linked to the group of permissions allowed to an application similar to S. Scott-Hayward & Sezer (2014) and the network administrator can add or remove permissions of an application via the user interface of the security application.

The applications are also given priority in order to ensure that the flow rule installation of different applications do not interfere with one another. The priorities are stored together with the permissions and are taken in to account while allowing applications to perform various operations. A sample of the permission system is given in the Table 1.

Before granting the subscribed permission for an app, the security application checks if it interferes with the operation of another application. The permission will only be granted if the operation of a higher priority application is not affected by it. The applications are allowed to ask the security app for a time slot for which to secure their operation, which will be stored by the security app along with the permission set data. If there is a new request for a permission before the expiration of the timeslot, the permission will be granted only if it was requested by a higher priority application. The working of the security application is given in Figure 1.

The security application also allows other applications to request for various permissions and to know the permissions that are currently granted to them.

Table 1. Sample permissions set.

Application ID	Allotted permission	Priority
A	read_topology	X
	read_all_flow	
	flow_mod_route	
	flow_mod_drop	
	flow_mod_modify_hdr	
	modify_all_flows	
B	pkt_in_event	Y
	flow_removed_event	
	error_event	
	topology_event	
C	flow_mod_modify_hdr	Z
	modify_all_flows	
	send_pkt_out	
	set_device_config	
	set_flow_priority	

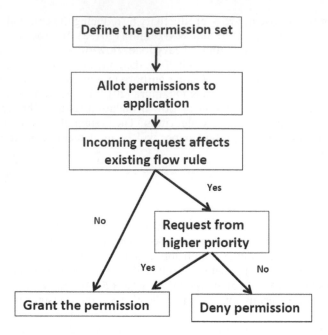

Figure 1. Security application.

5 CONCLUSION

The SDN architecture encourages the deployment of third party applications, but it also introduces the additional task of ensuring which of these are genuine. This paper proposes a security framework to safeguard the hybrid model of SDN by preventing malicious applications from taking control over the network. The hybrid model is an interesting attempt to balance the controller load and once the various security issues are solved, it can be successfully used practically.

REFERENCES

Othman, O.M. & K. Okamura (2013a). Hybrid control model for flow-based networks. *in the international conference COMPSAC 2013 – The First IEEE International Workshop on Future Internet Technologies, Kyoto, Japan.*

Othman, O.M. & K. Okamura (2013b). Securing distributed control of software defined networks,. *Int. J. Comput. Sci. Netw. Security 13,* 5–14.

Porras, P., S. Shin, V.Y.M.F.M.T. &. G. Gu (2012). A security enforcement kernel for open flow networks. *in Proceedings of the 1st workshop on Hot topics in SDN. ACM,* 121–126.

Sandra Scott-Hayward, Member, I.S.N. & I. Sakir Sezer, Member (2016). A survey of security in software defined networks. *IEEE COMMUNICATION SURVEYS & TUTORIALS. 18,* 623–654.

Scott-Hayward, S., C.K. & S. Sezer (2014). Operation check point: Sdn application control. *22nd IEEE International Conference on Network Protocols (ICNP). IEEE,* 618–623.

Wen, X., Y. Chen, C.H.C.S. &. Y. Wang. (2013). Towards a secure controller platform for open flow applications. *in Proceedings of the second ACM SIGCOMM workshop on Hot topics in software defined networking. ACM,* 171–172.

Emerging Trends in Engineering, Science and Technology for Society, Energy and Environment – Vanchipura & Jiji (Eds)
© 2018 Taylor & Francis Group, London, ISBN 978-0-8153-5760-5

A human-intervened CAPTCHA (HI-CAPTCHA) for wireless LAN-based classroom applications

Pradeep Giri, Paras & Rajeev Singh
G. B. Pant University of Agriculture and Technology, Uttarakhand, India

ABSTRACT: CAPTCHA stands for Completely Automatic Public Turing test to tell Computers and Humans Apart. CAPTCHAs are meant to distinguish between humans and software bots and are used to prevent unauthorized access to Internet resources by the bots. A CAPTCHA is a type of challenge-response test, in which a test is generated by the computer program that most humans can pass but computer programs cannot. In this paper, a new CAPTCHA approach is proposed: Human-Intervened CAPTCHA (HI-CAPTCHA). This HI-CAPTCHA strengthens security in Wireless LAN (WLAN)-based mobile applications and systems. For example, in WLAN-based classroom applications, it is used for the identification of the bots as well as for that of genuine users. A genuine user is one who is authorized to use the WLAN system and is responding from inside the classroom.

Keywords: CAPTCHA, HI-CAPTCHA, Turing test, Information security, Mobile-based system

1 INTRODUCTION

CAPTCHA stands for Completely Automatic Public Turing test to tell Computers and Humans Apart. It is a mechanism used to distinguish between human users and computer programs (von Ahn et al., 2004). The CAPTCHA term was coined in 2000 by Luis von Ahn, Manuel Blum, Nicholas J. Hopper (all of Carnegie Mellon University), and John Langford (then of IBM) (von Ahn et al., 2003). It is encountered in one form or another while using different services, such as Gmail and PayPal, or online banking accounts (Baird & Popat, 2002; Datta et al., 2009). CAPTCHA is basically a challenge-response test that most humans can pass but computer programs cannot. The rule of thumb of CAPTCHA is that it should be solved easily by humans but not by software bots. CAPTCHA uses a reverse Turing test mechanism (Coates et al., 2001). It has the following specifications:

- The judge is a machine rather than a human;
- The goal is that all human users will be recognized and can pass the test whereas no computer program will be able to pass the test.

Thus, CAPTCHA helps in preventing automated and Artificial Intelligence (AI) software programs known as bots from conducting unlawful activities on web pages, or in stopping spam attacks on mail accounts. The bots try to automatically register for a large number of free accounts and use these accounts to send junk email messages or to slow down services by repeatedly signing in to accounts and causing denial of service. CAPTCHA stops such autonomous entries and activities by bots in websites or in password-protected accounts. Therefore, many websites utilize CAPTCHA against web bots. Some of the applications (Carnegie Mellon University, 2010) of the CAPTCHA are as follows:

- Preventing comment spam in blogs
- Protecting website registration

- Protecting email addresses from scrapers and search engine bots
- Strengthening online polls
- Preventing dictionary attacks and worms.

CAPTCHA utilizes the fact that it is easy for humans to extract the text from the distorted image whereas it is relatively difficult for the bots to understand the text hidden inside the distorted image. As AI is evolving CAPTCHAs are becoming smarter and more secure. They not only identify bots, but also help in the machine-learning process. For example, Google's reCAPTCHA mechanism not only helps in bot identification but also helps in some other tasks such as Google's book digitalization program, where a reCAPTCHA is used for the learning of unidentified words which are not identified by Optical Character Recognition (OCR) (Azad & Jain, 2013). Thus, the human effort required to solve the CAPTCHA is utilized for some creative work, that is, for machine-learning processes.

Using CAPTCHA is always a win-win situation. If the CAPTCHA is solved by the bot, a hard AI problem is solved; if not, it can be used for stopping bots [3]. A modified CAPTCHA, HI-CAPTCHA, is proposed in this paper, which can be used for the identification of bots as well as for the identification of genuine users. The proposed HI-CAPTCHA system is targeted toward enhancing security in mobile-based classroom systems such as attendance systems and examination systems. These WLAN-based mobile classroom systems face one major challenge: that of identifying an authentic user in the classroom (termed a genuine user) from one that lies outside the classroom (termed an invalid user). The proposed method tries to distinguish these users based upon their responses.

2 RELATED WORK

Several Internet-based CAPTCHA mechanisms that use text, graphics, image, or sound exist and are used by researchers to prevent bots and enhance security. These may be categorized broadly into: first, plain CAPTCHA that is used only for the purpose of detecting and distinguishing humans and bots; second, other CAPTCHA that meets the above purpose and utilizes the human effort in solving some problems. This paper aims to utilize the human efforts used in solving the CAPTCHA for authentication purposes.

reCAPTCHA (Figure 1) was developed by Luis von Ahn, Ben Maurer, Colin McMillen, David Abraham and Manuel Blum at Carnegie Mellon University (CMU). It was acquired by Google in September 2009 (von Ahn et al., 2009). They studied harnessing the time and energy of the human being, which is required for solving CAPTCHA. Every time the CAPTCHA is solved, the effort used is utilized in digitalizing books, annotating images, and building machine-learning data sets. It performs multiple tasks, that is, it identifies a bot, tries to solve AI problems, and helps in machine learning. For example, it helps in utilizing the human effort for digitalizing books in the following manner. The user is provided with two texts: the first text is recognized by OCR while the other one is not. The user has to enter both texts and thus helps in the machine-learning process by entering the non-identified word to the database.

The proposed HI-CAPTCHA system is also designed to utilize human effort to authenticate users in the WLAN environment and to differentiate between a bot and human. It is also designed to differentiate between genuine and invalid users/humans. The latter category of users is not permitted to use/work with the system.

Figure 1. Google's reCAPTCHA (von Ahn et al., 2009).

3 PROPOSED HUMAN-INTERVENED CAPTCHA (HI-CAPTCHA)

3.1 *Proposed HI-CAPTCHA system*

Today, CAPTCHAs are very much more secure. Many websites use CAPTCHA for identifying bots as well as genuine users. Several users interact with these websites and try to solve the CAPTCHA logic. Thus, a lot of human effort is involved to solve the CAPTCHA. This human effort may be utilized in an effective way, as done by Google reCAPTCHA.

The proposed HI-CAPTCHA not only differentiates between a bot and a human but also differentiates genuine and invalid users/humans. A genuine user is one who is authorized, stays in the classroom, and is entitled to use the system. For example, in a mobile-based classroom exam, a user with a mobile device sitting in the class is a genuine user, whereas one who is answering from outside the class is an invalid user. Hence the user's answer to the proposed HI-CAPTCHA determines whether he is genuine or invalid. Thus, the user's effort to answer HI-CAPTCHA is utilized in an effective manner for authentication purposes. The entire HI-CAPTCHA system is shown in Figure 2. The administrator and clients have their own mobile application interfaces to interact with the system. They connect via the WLAN Access Point (AP) to the Spring Tool Suite (STS) webserver and database.

In the proposed HI-CAPTCHA system, questions are generated by the administrator. The generated questions are sent to the user/client. The answer to the question varies as per the administrator's state and other things that can be identified by a person inside the room. For example, in the question, 'What is the color of the administrator's shirt?', the answer to the question is not fixed as the administrator can wear any color shirt on any given day. The answer depends on the administrator's shirt color on a given day. Similarly, in the question, 'Is the administrator moving or standing?', the answer changes according to the administrator's direction of movement. The system works with a local server, where the numbers of users connected to the local server are fixed. The HI-CAPTCHA questions will not only detect the bot but will also help in authenticating the genuine user in selective mobile-based systems. For example, in e-polling, attendance-based systems, online exams, and so on.

3.2 *HI-CAPTCHA question*

A human-intervened CAPTCHA system has customized questions. The questions are generated by the administrator. The questions are designed in such a way that the answer always varies. The answer always depends on the current time, situation and location. The variable answers to the questions provide additional security against bots and invalid users. This system can be used for identifying bots and invalid users in small areas and for performing multiple tasks.

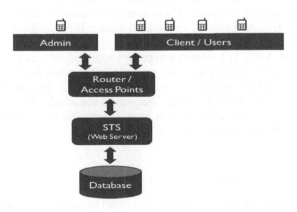

Figure 2.   Block diagram of human-intervened CAPTCHA.

Some of the questions* and probable answers of the HI-CAPTCHA are:

- Is the administrator moving or standing?
  Probable answer is either moving or standing.
  The answer to the question varies according to the movement of the administrator.
- What is the color of the shirt the administrator is wearing?
  Probable answer is red, green, blue, black etc. The answer to the question depends on the color of the shirt the administrator is wearing.
- In which direction is the administrator standing?
  Probable answer is east or west or south or north**. The answer to the question varies according to the location, time and situation of the administrator.

4 HI-CAPTCHA RESULTS

The proposed HI-CAPTCHA system is developed on the Android platform. The main software used for the development of the proposed system are Pivotal tc Server, Spring Tool Suite and Android Studio 2.0. It is assumed that the user outside the room/class is not able to see/perceive things inside the room/class. In HI-CAPTCHA, the administrator (teacher) selects one question from the list of questions. The questions are designed in such a way that the answer to the questions vary according to the administrator's requirement; for example, 'What is the color of the administrator's shirt?' The answer changes according to the color of administrator's shirt and the administrator can wear any color shirt. Hence, the answer is not fixed and bots and outsiders are not able to answer the questions. Thus, the system remains secure against bot attack and users positioned outside the room.

The proposed HI-CAPTCHA system is designed to differentiate between the genuine user and the invalid user. For example, in mobile-based examination systems all users outside the class are invalid users. A genuine user is one who is inside the classroom and authorized to use the system.

In the prototype of the system implemented (Figure 3a), the administrator has three options to select: automated CAPTCHA, HI-CAPTCHA or an administrator task. If only bots are to be restricted, plain automated CAPTCHA is selected. If invalid users are to be restricted, HI-CAPTCHA is selected. The 'administrator task' option is used for changing passwords, registering users, and so on. Once HI-CAPTCHA is selected, other screens (Figures 3b and 3c) are flashed up that ask for the setting up of questions and their answers by the administrator in the current session. The duration in which a client's response is required is also selected (Figure 3d) depending upon network characteristics such as delay or bandwidth. After finalizing the questions, answers and duration, the administrator sends the question to the client to answer (Figure 3e). The client view of the HI-CAPTCHA is shown in Figure 3f. It is used by the client to answer the administrator's question correctly.

During the system run, all users who were in the class are able to see the administrator and hence are able to answer the questions. Users outside the classroom were not able to see the administrator and therefore they were not able to answer the questions and, therefore, were identified as invalid users.

The proposed HI-CAPTCHA system works very well and detects bots and invalid users. During the test, users outside the classroom were identified as invalid users. In Figure 3f, the question selected by the administrator is: 'In which direction is the administrator standing?' The answer to the question depends on the direction of the administrator. All the valid users in the class were able to see the administrator so they were able to answer correctly, while the invalid user was not able to answer correctly.

---

*The HI-CAPTCHA system permits the administrator to design new and different questions as per needs.
**The direction coordinates for this question may be decided by the instructor in the class and the students briefed in advance.

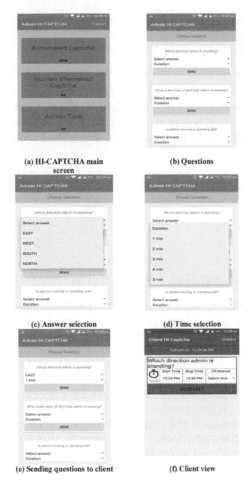

(a) HI-CAPTCHA main screen   (b) Questions

(c) Answer selection   (d) Time selection

(e) Sending questions to client   (f) Client view

Figure 3.   HI-CAPTCHA test case.

HI-CAPTCHA is used for the identification of genuine users. A genuine user in the system is one who is authorized to use the system. For example, in the mobile-based online examination system, all the users inside the class are genuine users, while the users outside the class are not genuine.

The following test case is conducted to prove the result of the HI-CAPTCHA:

- The administrator selects the HI-CAPTCHA to authenticate the genuine user.
- A question selected by administrator is 'Is the administrator moving or standing?'
- Answer to the question is based on the current time, situation or location of the administrator—administrator is standing.
- The time selected by administrator is 1 minute. This is the maximum time the user has to answer the question.

All users inside the class answered correctly, as they can see the administrator. But a user outside the classroom was not able to answer correctly because he was not able to see the movement of the administrator. Hence, the administrator was able to identify the genuine users. Figure 3 shows the diagrammatical view of the HI-CAPTCHA test case, for identifying a genuine user.

All the users inside the class were genuine users as they were authorized to use the proposed system. However, a case may arise in which a user inside the class is not connected to the server. In such a case, the user is not able to utilize the services of the system and will be considered as an unauthorized user. There are a few situations in which this may occur:

- The application may be not installed on the device. The user has to install the HI-Client CAPTCHA application.
- The device is not configured properly to use the system.

If the user is not able to connect to the server, the following condition may be checked:

- The user has to sign up to the HI-Client CAPTCHA application.
- The user inside the class has to log into the client HI-CAPTCHA application.
- If he/she logs in successfully, he/she may continue. But if the user fails to log in, the user has to check the settings of the mobile, that is, check Wi-Fi settings, IP settings, and so on.

In this case, when the user is in the classroom he must be treated as a genuine user. He may ask the administrator to resend the question following proper connection with the server after having been treated as not genuine/unauthorized.

5   CONCLUSION

Consistency of style is very important. Note the spacing, punctuation and capitals in all of the examples above.

CAPTCHA is a very effective way for stopping bots and reducing spam. CAPTCHA keeps web data secure from intruders. Almost every website contains CAPTCHA in one form or other. Every 'Sign in and sign up' or form submission over the Internet contains CAPTCHA. As AI is evolving, the need to develop new and advanced forms of CAPTCHA has arrived. Google's reCAPTCHA is an example of an advanced CAPTCHA. It not only detects software bots, but also helps in the machine-learning process. In the same way, the proposed CAPTCHA system is also designed for the multi-tasking environment. The proposed HI-CAPTCHA not only differentiates between a bot and a human but also differentiates between genuine and invalid users/humans. The proposed HI-CAPTCHA system can be designed for WLAN-based mobile attendance systems, e-polling systems or for generating the details of users. It works very well in a local server environment. In future, it may also be tested along with new CAPTCHA variants such as NO CAPTCHA and reCAPTCHA (Google, 2017).

REFERENCES

Azad, S. & Jain, K. (2013). CAPTCHA: Attacks and weaknesses against OCR technology. *Global Journal of Computer Science and Technology, 13*(3).

Baird, H.S. & Popat, K. (2002). Human interactive proofs and document image analysis. In D. Lopresti, J. Hu & R. Kashi (Eds.), *Document analysis systems V. DAS 2002. Lecture notes in computer science* (Vol. 2423, pp. 507–518). Berlin, Germany: Springer.

Carnegie Mellon University. (2010). CAPTCHA: Telling humans and computers apart automatically. Retrieved from http://www.captcha.net.

Coates, A.L., Baird, H.S. & Faternan, R.J. (2001). Pessimal print: A reverse Turing test. In *Proceedings of 6th International Conference on Document Analysis and Recognition, Seattle, WA* (pp. 1154–1158). New York, NY: IEEE.

Datta, R., Jia, L. & Wang, J.Z. (2009). Exploiting the human-machine gap in image recognition for designing CAPTCHAs. *IEEE Transactions on Information Forensics and Security, 4*(3), 504–518.

Google. (2017). Introducing the new reCaptcha! Retrieved from https://www.google.com/recaptcha/intro/index.html.

von Ahn, L., Blum, M., Hopper, N.J. & Langford, J. (2003). CAPTCHA: Using hard AI problems for security. In E. Biham (Ed.), *Advances in cryptology—EUROCRYPT 2003. Lecture notes in computer science* (Vol. 2656, pp. 294–311). Berlin, Germany: Springer.

von Ahn, L., Blum, M. & Langford, J. (2004). Telling humans and computers apart automatically. *Communications of the ACM, 47*(2), 57–60.

von Ahn, L., Maurer, B., McMillen, C., Abraham, D. & Blum, M. (2009). reCAPTCHA: Human-based character recognition via web security measures. *Science, 321*(5895), 1465–1468.

Emerging Trends in Engineering, Science and Technology for Society,
Energy and Environment – Vanchipura & Jiji (Eds)
© 2018 Taylor & Francis Group, London, ISBN 978-0-8153-5760-5

Semantic identification and representation of Malayalam sentence using SVM

Shabina Bhaskar, T.M. Thasleema & R. Rajesh
Department of Computer Science, Central University of Kerala, Kasaragod, India

ABSTRACT: This paper introduces Semantic Role Labeling (SRL) method for Semantic Identification (SI) and Conceptual Graph techniques (CG) for semantic representation. In Semantic Role Labeling, the semantic roles have been identified as agent, patient etc using the Karaka theory. The performance of SRL is calculated using Yet Another Chunk Annotator, a Support Vector Machine based algorithm which has shown significant improvement over earlier methods. The semantic representation introduced here is a directed graph which shows the relation between concepts and semantic roles.

1 INTRODUCTION

Semantics is the study of linguistic utterance. It refers to the sentence level meaning which is context independent and purely linguistic. The semantic approaches introduced has manifold applications in various NLP areas such as question answering system, machine translation system, text summarization system etc. Semantic identification and representation is a very important issue in natural language processing.

A sentence is represented by predicates and its corresponding arguments. Predicate represents an event and abstract roles are semantic roles that the arguments of a predicate can take in an event. For meaning identification and representation, we must identify the predicates and the abstract roles in a sentence. For high level understanding many question types need to be dealt with such as who did, what, to whom. Semantic Role Labeling, a shallow semantic parsing approach tries to find answers to these questions (Jurasfky & Martin 2002). Thus it takes the preliminary steps in extracting meaning from a sentence by giving generic labels or roles to the tokens of the text (Guilda. & Jurasky 2002).

The Semantic Role labeling is implemented (Kadri et al. 2003) using SVM classifier. The results evaluated using both hand-corrected TreeBank syntactic parses, and actual parses from the Charniak parser shows a precision and recall rate of 75.8% and 71.4% for this work. The semantic role labeling based on syntactic chunks is presented in (Kadri & Wayn 2003) shows some inspirational results with a precision and recall rate of 76.8% and 73.2% respectively.

Semantic role labeling methods in Malayalam is presented in (Dhanya, P.M 2010, Jisha & Satheesh 2016) and roles identified are used for general concept understanding and concept representation (Radhika & Reghuraj 2009). Plagiarism detection in Malayalam document based on extracting the Semantic roles and computing their similarity (Sindhu & Suman mary 2015) is another work related to Semantic role labeling.

The conceptual graph representation proposed by Sowa (Sowa 2008) express the meaning of a sentence in logically precise, humanly readable and computationally tractable form. In CG, we have to contemplate about the concepts and their corresponding semantic relations. This technique has been applied to many real life objects including text (Paola et al. 2008). The CG is used in relation extraction, information extraction and many other concept extraction techniques (Montes et al. 2001). Fact Extraction is a part of more general problem in knowledge extraction from text (Yi Wan et al. 2014). The fact extraction from natural language texts with conceptual graph is explained in (Bogatyrev 2017). For fact extraction, the input data is in the form a text and the text is transformed to a set of conceptual graph and the maximal number of conceptual graph is equal to the number of processed sentence. The concepts are identified from this CG and then presented in the form of a concept lattice. Concept lattice represents a matrix with relation on two sets of objects and their attributes. The facts can be extracted by processing the input textual queries in concept lattice. Concept and relationship extraction from unstructured text data plays a key role in meaning aware computing paradigms (Anoop & Ashraf 2017).

Malayalam, a morphologically rich and agglutinative language contribute to the arduousness of semantic identification. In Malayalam, the root or the stem word by itself gets inflected to change its meaning, or to combine the word with other words. Malayalam being a relatively free word order language, semantic role identification becomes a strenuous activity. The feature of inflection and agglutination nature of Malayalam be partly responsible for computer-based language processing a challenging task. For semantic identification in Malayalam, we have to associate different roles connected with various verbs, which is known as thematic role analysis or case role (Karaka) analysis. Thus karaka analysis plays a pivotal role in finding the relations between verbs and nouns in a sentence and we distinguish abstract semantic roles or thematic roles such as agent, patient etc. According to Paninian perspective there are four levels in the understanding process of a sentence (Bharati et al.). They are surface level (uttered sentence), vibhakthi level, karaka level and semantic level. The karaka level has relationship to semantics on one side and to syntax on the other side. Karaka relation can be identified from post position markers after noun or surface case ending of noun. These markers and case endings are called vibhakthi (Manu & Reghu 2012). In Malayalam karaka relations are analyzed from vibhakthi and post position markers (Archana et al. 2015). But for a sentence level meaning identification we have to consider the semantic roles associated with adjectives and adverbs also. The semantic roles of adjectives, adverbs and postposition markers are also identified in this work.

After Semantic role identification, it will be pursued with the meaning representation. The intermediate representation presented here is a basic level meaning representation of a sentence. The intermediate representation gives us information about semantic roles of nouns, adjectives, adverbs and the verbs in a sentence. And through the introduction of the conceptual graph for meaning representation, this graph manifests how the verb is related to noun, adjectives and adverbs in a sentence.

In section 2 we discussed about the methodologies used in this work. And section 3 and 4 discussed about the results and conclusion.

2 METHODOLOGY

Overall system architecture is given in Figure 1. For syntactic level processing the main steps are tokenization, POS tagging and Morphological analysis. In semantic side, the main steps are semantic role labeling and intermediate representation.

2.1 *Tokenization*

Tokenization is the task of chopping a character sequence into pieces, called tokens, perhaps at the same time throwing certain characters, such as punctuation. For tokenization we developed an algorithm and for compound word splitting SVM machine learning techniques used. It classifies the tokens into two groups compound words and simple words.

2.2 *POS tagging*

The process of assigning part of speech for every word in a given sentence according to the context is called as part of speech tagging. SVM classifier is used for POS tagging and the system is trained with annotated corpora. For annotation we used BIS tagset.

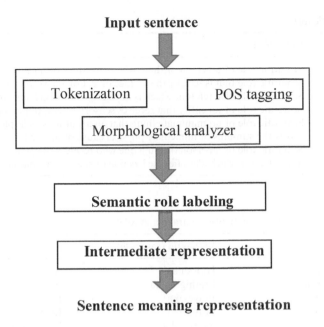

Figure 1.   System architecture.

2.3   *Morphological analysis*

Morphological analysis is the process of recognizing the root stem, and the categorical information of the items that may accompany the root or the stem. Here for Morphological analysis, Paradigm based approach is used and we developed Paradigms for nouns and verbs. This Paradigms includes high level information such as vibhakthi and verb forms other than categorical information.

2.4   *Semantic role labeling*

Semantic Role Labeling is the task of assignment of the semantic roles to the constituents of the sentence. The semantic roles such as agent, patient, beneficiary, experiencer, instruments, recipients and locative are identified in this work. The Table 1 gives you information about the semantic roles used in this work.

2.4.1   *Description about semantic role labeling method*
In Malayalam Karaka relation express the syntactic and semantic relations exists between words in a sentence. This karaka relation and vibhakthi helps us to identify the semantic roles in a sentence. See Table 2.

But in some situation we cannot identify the semantic roles with the help of karaka relations alone. In such cases we have to add some additional semantic information about noun and verb. In this work we have created a lexicon of noun and verb in which we grouped the nouns and verbs which have same semantic property. For example, the words like dance, sing have the same semantic property and it is grouped as artistic performance verbs. The noun class and verb class used for this work is given below.

Verb class

1. Artistic Performance verb (പാടുക, നൃത്തം ചയ്യയുക)
2. Attach Verb (ചരേക്കുക, കട്ടുക, ആകർഷിക്കുക)
3. Beneficiary Verb (നല്കുക, കിട്ടുക)
4. Causative Verb (കരയിപ്പിക്കുക, ചിരിപ്പിക്കുക, സന്തോഷിപ്പിക്കുക)
5. Emotional verb (കരഞ്ഞു, ചിരിച്ചു)

Table 1.   Semantic roles.

Semantic roles	Definition
Agent	Semantic role of a person or thing who is the doer of an event.
Patient	Semantic role that is usually the surface object of the verb in a sentence.
Beneficiary	Semantic role of a referent which is advantaged or disadvantaged by an event.
Experiencer	Semantic role of an entity that receives sensory or emotional input
Instrument	Semantic role of an inanimate thing that an agent uses to implement an event. It is the stimulus or immediate physical cause of an event.
Recipients	Semantic role of a person or thing one who receives the goal of an action.
Locative	Semantic role which identifies the location or spatial orientation of a state or action

Table 2.   Relation between karaka and roles.

Karakas	Vibhakthi (case role)	Semantic role
Karthu	Nideshika	Agent
Karma	Prathigrahika	Patient
Karana	Prayogika	Instrument
Kaarana	Prayogika	Instrument
Sakshi	Samyogika	Experiencer
Swami	Udheshika	Beneficiary
Adhikarana	Adharika	Locative

6. Destroy Verb (കൈ·ാല്ലുക, നശിപ്പിക്കുക)
7. Vehicle Motion Verb (ഓടിക്കുക, സവാരിചെയ്യുക)
8. Weather Verb (മഴ പെയ്യുക, മഞ്ഞുപെയ്യുക)

   Noun class
1. Animals (പട്ടി,പൂച്ച ,പശു)
2. Audible only (കുര, കൈയടി)
3. Birds (കാക്ക,കൊ·ോഴി ,താറാവ്)
4. Buildings (കെട്ടിടം,വീട്,സ്കൂൾ)
5. Collective humans (സമൂഹം, ആൾക്കൂട്ടം)
6. Concepts (വിശ്വാസം, സംഭവം, സന്ദർഭം)
7. Positions (ഡൊ·ോക്ടർ,അദ്ധ്യാപകൻ, വിദ്യാർത്ഥി)
8. Person names (രാമൻ,സീത)
9. Non-living (മണ്ണ്,വെള്ളം,മശേ,കസരേ)

Next we discussed some examples about how we can identify semantic roles with the help of karaka, vibhakthi and semantic property.

1. Nirdeshika എൽസി കരഞ്ഞു Nirdeshika + verb (emotional verb) The semantic role is experiencer. രാമു രാജുവിനെ കരയിച്ചു
   Here the sentence is subject + object + verb form and if the verb is causative verb then the noun with nirdeshika vibakthi take the role as Agent. രാജുവിനെ shows prathigrahika vibakthi and the role is patient.
2. Udeshika അമ്മുവിന് നൽകി Udeshika + verb Here the verb is beneficiary verb then role is beneficiary or recipient.
3. Nirdeshika + nirdeshika + verb രാമു മാങ്ങ തിന്നു
   Here the first noun has the semantic property person name and second is fruit name, in that case person takes the role of agent and the other takes patient.
   മണ്ണ് വെള്ളം കുടിച്ചു

838

Here we check the semantic property of noun both are non-human. So we take the order the first become agent and second become patient.

4. Nirdeshika + prathigrahika രാം രാജുവിനെ കണ്ടു Here the role is agent + patient
5. Nirdeshika + udeshika രാമൻ സീതയ്ക് നൽകി
   Verb is beneficiary so semantic role is beneficiary.
6. Nirdeshika + udeshika + nirdeshika ഉണ്ണി മീനുവിന് പണം നൽകി Role is agent + beneficiary + patient.

2.4.2 *YamCha for semantic role labeling*

YamCha is Yet Another Multipurpose Chunk Annotator which is based on the SVM machine learning algorithm, first introduced by Vapnik in 1995. It Use **PKE/PKI**, which make the classification (chunking) speed faster than the original SVMs. Here the features selected are words in a sentence, their POS tag and semantic roles. In this work we created corpora in which each words and its semantic roles are annotated. Other than the semantic roles of noun we have also identified the semantic roles for adjectives, adverbs and verbs.

2.5 *Intermediate representation*

Intermediate representation is which gives us information about semantic roles, verbs, adjectives and adverbs in a sentence. In this representation we grouped nouns, verbs, adjectives, adverbs and postposition ma rkers separately.

Nouns

- AGENT 1
- AGENT2
- PATIENT1
- PATIENT2
- BENEFICIARY
- EXPERIENCER

Adjectives

- APPEARANCE
- COLOR
- MEASUREMENT

Adverbs

- FREQUENCY
- MANNER

Postpositions

- DESTINATION
- LOCATION

Verbs

- MAIN VERB
- SUB VERB

In this work the maximum number of verbs in a sentence is two, one is main verb and other is sub verb. AGENT1, PATIENT1 are the roles for main verb and AGENT2, PATIENT2 are the role for sub verb. The intermediate representation for a sentence with two verbs is given below.

Input sentence: അമ്മയുടെ പാട്ട് കേട്ട് കുട്ടി ഉറങ്ങി

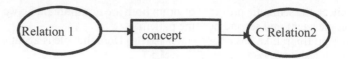

Figure 2.   Conceptual graph.

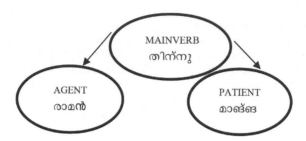

Figure 3.   Semantic representation in the form of a directed graph.

Noun
  AGENT1 കുട്ടി
  AGENT2 അമ്മ
  PATIENT1 NIL
  PATIENT2 പാട്ട്
Adjectives
  NIL
Adverbs
  NIL
Postpositions
  NIL
Verbs
  MAIN VERB ഉറങ്ങി
  SUB VERB കടേട്

2.6   *Conceptual graph representation*

Conceptual Graphs (CG) serve as an intermediate language for translating computer-oriented formalisms to and from natural languages. They serve as a readable, but formal design and specification language. CGs have been implemented in a variety of projects for information retrieval, database design, expert systems, and natural language processing [19]. In diagrammatic representation of CG rectangular box represents the concept node and oval represents the semantic relation (semantic roles). CG are based upon the following general form:

In this work we adopt the techniques of Conceptual Graph representation for meaning representation and here the concepts-semantic relation is expressed in the form of a directed graph. The concepts are represented by vertices and edges represents their corresponding semantic relations. The main verb is considered as the root node and the child nodes are nouns, adjectives and adverbs. But in a complex sentence, the main verb is root node and sub verb is its right child node. Semantic representation of a sentence രാമൻ മാങ്ങ തിന്നു is given in Figure 3.

An algorithm is introduced for semantic representation of sentence in the form a graph.

1. Create a list for nouns, adverbs, adjectives, postpositions and verbs.

2. Set count as the number of items corresponding to each list.

#simple sentence with word length two

3. If noun count=1 and verb count=1, then

> MAIN VERB set as root node and AGENT1 set as the left child

#simple sentence with word length three

Else

4. if noun count=2 and verb count=1

> MAIN VERB is set as root node, AGENT1 set as left child and PATIENT1 set as right child.

4.1 else if noun count=1, verb count=1 and adjective count=1   MAIN VERB is set as root node, AGENT1

> set as left child and ADJECTIVE set as left child of AGENT1.

4.2 else if noun count=1 and verb count=1 and adverb count=1

> MAIN VERB is set as root node, AGENT1 set as left child and ADVERB set as left child of AGENT1.

4.3 else if noun count=2 and postposition count=1 and verb count=1

> MAIN VERB set and root node and AGENT1 set as the left child and LOCATION/INSTRUMENT is set as right child.

#simple sentence with word length four

Else

5. If noun count=3 and verb count=1

> MAIN VERB set as root node, AGENT1 Is set as left child, PATIENT1 is set as right child and BENEFI-CIARY/RECEIPENT set as right child of PATIENT1

5.1 else noun count=3and verb count=1 and postposition count=1

> MAIN VERB set as root node, AGENT1 Is set as left child, PATIENT1 is set as right child and LOCA-TION/INSTRUMENT set as right child of PA-TIENT1

5.2 else if two for noun count=2 and verb count=1 and adjective count=1 or adverb count=1

> ` MAIN VERB set as root node, AGENT1 set as left child and ADVERB set as its left child and PA-TIENT1 is set as right child of MAIN VERB and adjective set as right child of PATIENT1

#complex sentence with word length five

Else

6. noun count=3 and verb count=2

> MAIN VERB set as root node and AGENT1 set as left child of root node. SUB VERB set as right child of root node, AGENT2 is its left child and PATIENT2 is its right child.

7.Stop.

A total of 300 and 50 sentences are used for training and testing respectively. In testing, out of sentences for SRL we took 40 simple sentence with word length from two to four and 10 complex sentence with word length five. The semantic roles corresponding to each word is identified and it is transformed in to a directed graph representation.

3   RESULTS

In this experiment the sentence word length is minimum two and maximum five. We have taken the complex sentence with maximum two verbs. In our study the evaluation is based on the standard metrics such precision and recall.

The precision for a class is the number of true positives divided by the total number of elements labelled as belonging to the positive class (i.e. the sum of true positives and false positives, which are items incorrectly labelled as belonging to the class). Recall in this context is defined as the number of true positives divided by the total number of elements that actually belong to the positive class (i.e. the sum of true positives and false negatives). The precision, recall rate corresponding to each role is given in Table 3.

The result of semantic role labeling and semantic representation in the form of graph is given below

Input sentence: രാമു മനോഹരമായ ചിത്രം വരച്ചു
Semantic roles:

രാമു	AGENT1
മനോഹരമായ	APPEARANCE
ചിത്രം	PATIENT1
വരച്ചു	MAIN VERB

Graph

MAIN VERB വരച്ചു

AGENT രാമു   PATIENT ചിത്രം

APPEARANCE മനോഹരമായ

Table 3.   Precision and recall for semantic roles.

Semantic roles	Precision	Recall
Agent1	81.2%	78.2%
Agent 2	69.0%	65.4%
Patient 1	80.0%	76.4%
Patient 2	75.0%	68.3%
Beneficiary	80%	76.8%
Experiencer	75.5%	60.6%
Instrument	70.6%	64.0%
Appearance	73.0%	68.1%
Colour	67.2%	60.0%
Measurements	60.6%	58.8%
Frequency	60.9%	56.8%
Manner	62.2%	58.0%
Location	66.4%	59.9%
Main verb	80.9%	76.9%
Sub verb	78.8%	73.3%

4 CONCLUSION

The SVM based classifier introduced here considered the Semantic Role Labeling (SRL) as a semantic grouping problem of different words. The SRL identified the semantic roles associated with each words in a sentence. The Conceptual Graph presented here shows the relation between predicates and arguments in a sentence. The CG representation depends on the semantic role labeling. If the roles are identified correctly, then it will lead to more accurate system. By introducing more semantic roles and by increasing the sentence word length we can add improvements in the present study.

REFERENCES

Anoop, V.S. & Ashraf. 2017. Extracting Conceptual Relationships and Inducing Concept Lattices from Unstructured Text. *Journals of intelligent systems.*

Archana, S.M. & Vahad, Naima. & Rekha, Thankappan. & Raseek. C. 2015. A Rule Based Question Answering System in Malayalam corpus Using Vibhakthi and POS Tag Analysis. *International Conference on Emerging Trends in Engineering, Science and Technology (ICETEST – 2015).*

Bharati, Akshar. & Vineeth, Chaithanya. & Sangal, Rajeev. Natural Language Processing: A Paninian Perspective. *Prentice-Hall of India, New Delhi.*

Bogatyrev, Mikhail. 2017. Fact extraction from natural language with conceptual modelling, *Springer international Publishing AG.*

Dhanya, P.M. 2010. Semantic Role labeling Methods: A Comparative Study. In: proceedings of National Conference on Human Computer Interaction and Image Processing. *Vidya Academy of Science & Technology.*

Guilda & Jurasky, Daniel. 2002. Automatic Labelling of semantic roles. 245–288.

Gildea. & Hockenmaier. 2003. Identifying Semantic Roles Using Combinatory Categorical Grammar. 2003. In Proceedings of the 2003 conference on Empirical methods in natural language processing. Association for Computational Linguistics, Philadelphia, USA, pp. 57–64, 2002.

Kadri, Hacioglu. & Pradhan, Sameer. & Wayne, Ward. & Martin, James. & Jurasfky, Daniel. 2003. Shallow Semantic Parsing Using Support Vector Machines. *CSLR Tech. Report.*

Kadri, Hacioglu. & Wayne, Ward. 2003. Target word detection and semantic role chunking using support vector machines. *In Proceedings of HLT-NAACL 2003.*

Jish, Jayan. P. & Kumar, Satheesh. 2016. Semantic role labeling for Malayalam. IJCTA. *pp. 4725–4731* © *International Science Press.*

Jurasfky, Daniel. & Martin, James. 2002. An introduction to Natural Language Processing: Computational Linguistics and Speech Recognition. *Pearson Education.*

Manu, Madhavan. & Reghu, Raj. P.C. 2012. Application of Karaka Relations in Natural Language Generation. *In Proc. of National Conference on Indian Language Computing, CUSAT, Kerala.*

Montes, Y. Gomez. M. & Gelbukh, A. & Lopez-Lopez & Baeza-Yates. Text mining with conceptual graphs. 2001. *IEEE International Conference on Systems, Man and Cybernetics. e-Systems and e-Man for Cybernetics in Cyberspace (Cat. No. 01CH37236) Year: 2001, Volume: 2.*

Paola, Velardi. & Maria, Teresa. Pazienza. & Mario De' Giovanetti. 1998. Conceptual graphs for the analysis and generation of sentence. IBM journal of Research and Development.

Radhika, K.T. & Reghuraj, P.C. 2009. Semantic role extraction and General concept understanding in Malayalam using Paninian Grammar. *International Journal of Engineering Research and Development, vol 9, issue 3.*

Sindhu, L. & Suman Mary, Idicula. 2015. SRL based Plagiarism Detection system for Malayalam Documents. *International Journal of Computer Sciences issues, vol 12, November 2015.*

Sowa. John. 2008. Conceptual Graph. Chapter 5 of the Handbook of Knowledge Representation, ed. by F. van Harmelen & V. Lifschitz, & Porter. B, *Elsevier, 2008, pp. 213–237.*

Stephen, chu. & Branko, Cesnik. 2001. Knowledge representation and retrieval using conceptual graphs and free text document self-organisation techniques, *International Journal of Medical Informatics 62 (2001) 121–133.*

Yi, Wan. & Tingting, He & Xinhui, Tu. 2014. Conceptual graph based text classification. *IEEE International Conference on Progress in Informatics and Computing 2014.*

Emerging Trends in Engineering, Science and Technology for Society,
Energy and Environment – Vanchipura & Jiji (Eds)
© 2018 Taylor & Francis Group, London, ISBN 978-0-8153-5760-5

A survey on relation extraction methodologies from unstructured text

S.K. Bhaskaran & P.C. Rafeeque
Department of Computer Science and Engineering, Government Engineering College, Palakkad, India

ABSTRACT: The web contains very large amount of unstructured text. The process of converting unstructured text to structured with semantic information annotation can provide useful summaries for both humans and machines. The semantic relation is one of the most important parts of semantic information. Hence, extracting semantic relations held between entities in a text is important in many natural language understanding applications like question answering, conversational agents, summarization. There are several proposed methods in this area. Hand built patterns, bootstrapping methods, supervised methods, unsupervised methods, and Distant Supervision (DS) are examples. The purpose of this work is to review various methods used for Relation Extraction (RE). For each approach, respective motivation is discussed and the merits and demerits compared. Discussion about various datasets is also included in this paper.

1 INTRODUCTION

1.1 *Information extraction*

The main goal of an Information Extraction (IE) task is to provide machine-readable summaries (ie, acquiring structured information from unstructured text). Normally, documents on the web are unstructured and are not understandable by a machine. Thus, IE aims to provide machine-readable summaries. Usually, IE means extracting information from text; sometimes called text analytics commercially. Now, what type of information can be extracted? It can extract entities, the relationship between entities and it can also figure out larger events that are taking place. Normally, interested entity types are the names of people, organizations, locations, times, dates, prices or, sometimes, genes, proteins, diseases and medicines. Some of the relations between entities are located_in, employed_by, part_of, married_to etc.

1.2 *Relation extraction*

The amount of unstructured electronic text on the web is very large. So, understanding this text is a complicated task for humans. A popular idea is to transform this unstructured text into structured by annotating semantic information. But human annotation is impossible because of the large volume and heterogeneity of data. It is possible for an agent (or computer) to do all these annotations. So, the agent (or computer) needs to understand how to recognize whether a piece of text holds some semantic property or not. Hence, extraction of the semantic relations between entities is an important task. The Relation Extraction (RE) task aims to recognize and classify relations between pairs of entities found in a text. For instance, the RE system will extract *CEO-of (Jobs, Apple)* from the sentence *"Jobs is the CEO of Apple"*.

Traditional methods for RE handle the task as a pipelined approach. That is, first extract the entities and then recognize the relationship between those entities. This is particularly easy to handle and will assure the flexibility in each subtask. But it has some drawbacks; specifically the relevance of each subtask is neglected and we have to model both subtasks separately. Also, it results in error propagation and redundant information. To tackle the above problems, it is useful to think about a joint learning framework. Here, a single model is used to extract entities together with their relations. It provides an effective integration of both subtasks. However, existing joint extraction methods are feature-based which may require complicated feature engineering and have great dependency on other NLP toolkits. It also produces redundant information.

A neural network-based approach can reduce the manual work in feature extraction. Introduction of a deep learning framework to RE has made a significant performance improvement to traditional methods. In this review, we begin with problem formulation, and we discuss various methods proposed for RE task. We will highlight the merits and demerits of each method and if it is useful to choose a better methodology. It will also help us to make new proposals for existing works.

Organization: The rest of this paper is organized as follows. Section 2 formulates the problem and explain about challenges of the RE problem. It also addresses some applications in NLP. Section 3 discusses various pipelined methods, jointly extracting methods and the end-to-end tagging models. Section 4 discusses future enhancements to existing RE methods. The conclusion is given in section 5.

2 FORMULATION OF THE PROBLEM

The inputs to an RE system are a POS-tagged corpus, C, a Knowledge Base (KB), K, and a predefined relation type set, R. An entity mention is a token span, denoted as e and $r(e1,e2)$ is the relation (binary) between entities $e1$ and $e2$. The manual labeling of training corpora is expensive in terms of time. The method of Distant Supervision (DS) can overcome this problem and it is the reason why DS has achieved significant importance in RE tasks.

DS generates training data by automatic alignment of text with a KB. It can jointly extract entities together with their relations with minimal or no human supervision. Hence, KB is also an input to the RE system.

PROBLEM DEFINITION: *Given a POS-tagged corpus, C, a KB, K, and predefined relation type set, R, the RE task aims to 1) detect entities from C, 2) generate training data D with KB, K, and 3) estimate a relation type r which belongs to R U {None}.*

Current RE tasks with DS face the following limitations when handling a joint extraction task:

- **Domain Restriction**: Most methods (Mintz et al., 2009; Takamatsu et al., 2012) heavily rely on pre-trained named entity recognizers which are typically designed for general types such as person or organization. We need further manual work to deal with domain specific names.
- **Error propagation**: Error can be propagated from upper components to lower components and dependencies among tasks are ignored in most existing methods.
- **Domain-independent systems**: A major challenge is to design a domain-independent system. Most existing methods are domain dependent.
- **Label Noise**: Mapping of relations in the text with KB relations may produce false labels in training corpora. It may cause uncertainty in DS and thereby results in inaccurate models.

RE is the extraction of semantic relations from unstructured text. Once extracted, such structured information is used in many ways. For example, as primitives in IE, building extending KBs and ontologies, question answering systems, semantic search, machine reading, knowledge harvesting, paraphrasing, and building thesauri.

3 RELATION EXTRACTION METHODS

In this section, we briefly explain three methods employed for the problem of RE. The two main frameworks used for this are the pipelined framework and the joint learning framework. Also, some neural network-based methods for extracting entities and relations are explained for understanding the tagging approaches used in this area.

3.1 Pipelined methods

Traditional methods for RE handle the task in a pipelined manner. That is, they perform Named Entity Recognition (NER) first and then classify the relations. The NER task is similar to POS tagging and sequence classifiers such as Hidden Markov Models (HMM) and Conditional Random Fields (CRF) which are used for this task. Recently, Long Short Term Memory (LSTM) for NER has been achieved significant performance improvement over sequence classifiers. Now, we move on to the second subtask, namely relation classification. Existing methods for relation classification use feature-based methods as well as neural network-based methods. Let us discuss some of the pipelined methods.

Mintz et al. (2009) proposed a domain-independent model for RE which uses DS as an alternative paradigm for existing supervised, unsupervised and semi-supervised or bootstrapping methods. DS performs automatic labeling of relations in text and relations in KB. It uses a multi-class logistic classifier for extraction. The method uses relations and relation instances from Freebase, which is an online database of structured semantic data. For a Freebase relation with a pair of entities, all sentences containing those entities are retrieved from the text. It combines advantages of both supervised IE and unsupervised IE.

In this method, textual features like lexical features, syntactic features, and name entity tag features are extracted and trained by the classifier using these features. Unlike a corpus-specific method, the main advantage of this method is that it allows us to extract evidence for a relation from many documents and thereby avoids domain dependence. Another advantage of this approach is that it does not require a labeled corpus. Since it is a feature-based approach, the main limitation is that it needs feature extraction which requires manual work. Moreover, uncertainty in DS is another problem since it exploits only one specific kind of indirect supervision knowledge and ignores very useful supervision knowledge. To overcome this limitation, we can think about global DS which will effectively reduce uncertainty by including additional knowledge.

In the method by Takamatsu et al., (2012), when a sentence refers to an entity pair in a KB, then that sentence is heuristically labeled by DS with the corresponding relation in the KB. However, this labeling will result in wrong labels and thereby poor extraction performance. The model effectively reduces incorrect labeling by predicting whether assigned labels are correct or not. This is done by the model's hidden variables. The advantage is that it reduces wrong labels but it does not address inter-dependencies.

Surdeanu et al. (2012), introduced a challenging learning scenario where the relation held by a pair of entities is unknown for each sentence. The method uses a multi-instance multi-label learning (MIML) framework, which can jointly model all the entities and labels by a graphical model. The model assumes that a relationship between two entities has exactly one label, but the entities can have multiple labels. Two separate feature sets are used for the entity classifier and the relation classifier respectively. One of the merits of this work is it also reduces incorrect labels. The limitation of this approach is that it needs feature extraction for both subtasks.

3.2 Jointly extracting methods

In a joint extraction framework, entities are extracted together with relations between pair of entities.

The method by Li and Ji (2014), is a joint extraction method with DS. It exploits global features in the joint search space. The joint framework is an incremental one and uses a structured perceptron with the advantage of efficient beam-search. A segment-based decoding is another specialty of this model which is based on the idea of a semi-Markov chain and is employed as an opposition to traditional token-based tagging schemes. This method has developed a number of new and effective global features to capture the inter-dependency among entities and relations. In this method, segment-based features are considered along with the global features. Segment-based features are based on the entire mention instead of individual tokens. They are gazetteer features, word case features (case information about all tokens contained), contextual features (neighbor unigrams and bigrams), and parsing features such as the phrase label of common ancestor (NP), the depth of the common ancestor, whether the segment matches a base phrase (true) or is a suffix of a base phrase and the head word of the segment. Global features are dynamically created during the search, and they capture long distance dependencies. One merit of this work is that it captures the inter-dependency among entities and relations. Uncertainty in DS is again the drawback of this method.

Hoffmann et al. (2011), proposed a knowledge-based weak supervision approach. The framework is known as MultiR. It uses heuristic labeling and multi-instance learning algorithms for coping with noisy data. MultiR is a probabilistic, graphical model which employs multi-instance learning technique and handles overlapping relations. It also results in accurate sentence-level predictions, decoded individual sentences, and corpus-level extractions. Here, RE is modeled as a MIML problem, but it learns to use a perceptron algorithm. Also, it introduced at least one deterministic decision rather than using a relation classifier.

CoType is an embedding based framework proposed by Ren et al. (2017). It is a novel distant supervision framework that can extract entity mentions and relations jointly with minimum linguistic supervision. For detecting entities using DS, this work developed a context-agnostic algorithm. An overview of framework is as follows: The algorithm is first run over a POS-tagged corpus which is the input to the CoType. It generates entity relation mentions and extracts text features for each relation. Thus, DS is used for generating training corpora. A joint embedding objective is also formulated for modeling type association, mention feature-co-occurrence, and entity-relation cross-constraints. Experiments on three public datasets have demonstrated that CoType improves extraction problem performance significantly. Advantages: 1) Domain-independent; 2) Modeling of type association and inter-dependencies. The drawback of this work is that it does nothing for reducing false labels in the training data.

3.3 *End-to-end tagging models*

End-to-end tagging models use several deep learning architectures for the extraction problem into a tagging task.

Lample et al. (2016) proposed a neural network-based approach for state-of-art NER. There are two neural networks: one is LSTMs and CRFs based and a transition-based approach is used by the second neural network for constructing and labeling segments. The tagging scheme used here is IOBES (Inside, Outside, Beginning, End, Single). An advantage of this method is that the features extracted are not language specific.

Another method by Zheng et al. (2017) is a joint extraction method based on a tagging scheme. It uses an IOBES tagging scheme and transforms the extraction problem into a tagging task. An LSTM based framework is employed and results show that tagging based methods are more efficient than other pipelined and jointly extracting methods. NER results are extracted using CoType which uses DS. The main advantage is that it ignores complicated feature engineering. But there is a shortcoming for the identification overlapping relations to settle the triplet overlapping problem.

3.4 *Discussion on datasets*

The above discussed methods use publicly available datasets such as New York Times articles (NYT) and Wikipedia articles (Wiki-KBP). Bioinfer is a commonly used dataset from the medical field. But these are not annotated. An effective and efficient way for annotation is DS which is the most accepted approach.

Approaches in Hoffmann et al. (2011), Mintz et al. (2009) and Surdeanu et al. (2012) use Wikipedia datasets and Freebase as the KB. The dataset from Wikipedia is an easy scenario for DS with Freebase because the facts in Freebase are partly derived from Wikipedia. Experiments on a CoType framework by Ren et al. (2017) used three public datasets, NYT, Wiki-KBP, and Bioinfer which contain biomedical paper abstracts. Bioinfer contains several overlapping relations. Out of these three sets, NYT is used with Freebase as the KB in the method by Zheng et al. (2016).

Table 1. Overview of compared methods and datasets used.

| Approach | Proposed methods | | |
	Model	Contributions	Domain/Dataset
Pipelined	Mintz et al., 2009 (9)	DS-logistic, allow corpora of any size	Wikipedia with Freebase
	Takamatsu et al., 2012 (14)	DS, reducing wrong labels	Wikipedia with Freebase
	Surdeanu et al., 2012 (13)	DS, multi-instance multi-learning algorithms	NYT, KBP and Freebase
Joint	Li and Ji, 2014 (7)	DS, exploits global features	ACE 2005
	Hoffmann et al., 2011 (5)	MIML, sentence-level predictions, reducing false labels	Wikipedia with Freebase
	Ren et al., 2017 (10)	DS, CoType, modeling type association, mention feature-co-occurrence, and entity-relation cross-constraints	NYT, Wiki-KBP, Bioinfer
Tagging Based	Lample et al., 2016 (6)	LSTM-CRF, transition-based approach, IOBES tagging	CoNIL 2002, CoNIL 2003
	Zheng et al., 2017 (17)	DS, a novel tagging scheme, Bi-LSTM-LSTM	NYT and Freebase

In the method by Li and Ji (2014) their experiments were done on an ACE 2005 dataset. It contains data from six different domains such as Newswire, Broadcast Conversation, Broadcast News, Telephone Speech, Usenet Newsgroups and Weblogs. Different datasets were used by Lample et al. (2016); CoNIL 2002 and CoNIL 2003 which contain independent named entity labels for English, Spanish, German and Dutch. Since different methodologies used different datasets, a general comparison is not possible. Contributions of the different methodologies discussed and domains or datasets used by them are summarized in Table 1.

4 FUTURE DIRECTIONS

Many natural language understanding applications need competitively structured information. So, RE is always important. Most of the proposed methods employ a joint extraction task with DS. Enhancements in DS will fill the gaps in existing methods. One of the problems with DS is uncertainty which may lead to lack of supervision. Introduction of a pseudo feedback idea to the existing framework may overcome this. Feature enrichment of the current embedding framework is an interesting enhancement. Unlike the method used in Hoffmann et al. (2011), it is possible to develop an end-to-multiple-end model for addressing triplet overlapping problem. Modeling type correlation and performing type inference for test entity and relation jointly will be other interesting future work in this area.

5 CONCLUSIONS

Semantic relations in text are the key meaning component for natural language applications. There are several proposed approaches for the RE task and here we have given a brief summary of some of them. So far, we have reviewed some important aspects of the entity RE problem starting with the problem formulation, discussing the different challenges and applications and finally culminating with a discussion of some important approaches. Various methods of pipelined approaches, joint extraction framework and neural network approaches, which use end-to-end tagging schemes have been discussed, and we can clearly say that end-to-end models can outperform other models. Moreover, deep learning frameworks have replaced most of the existing frameworks because of their popularity.

REFERENCES

Bollacker, K., Evans, C., Paritosh, P., Sturge, T. & Taylor, J. (2008). Freebase: a collaboratively created graph database for structuring human knowledge. In *Proceedings of the 2008 ACM SIGMOD International Conference on Management of Data* (pp. 1247–1250). Association for Computing Machinery.

Gormley, M.R., Yu, M. & Dredze, M. (2015). Improved relation extraction with feature-rich compositional embedding models. *arXiv preprint arXiv:1505.02419*.

Gupta, R. & Sarawagi, S. (2011). Joint training for open-domain extraction on the web: exploiting overlap when supervision is limited. In *Proceedings of the fourth ACM International Conference on Web Search and Data Mining* (pp. 217–226). Association for Computing Machinery.

Hochreiter, S. & Schmidhuber, J. (1997). Long short-term memory. *Neural Computation*, 9(8), 1735–1780.

Hoffmann, R., Zhang, C., Ling, X., Zettlemoyer, L. & Weld, D.S. (2011). Knowledge-based weak supervision for information extraction of overlapping relations. In *Proceedings of the 49th Annual Meeting of the Association for Computational Linguistics: Human Language Technologies-Vol 1* (pp. 541–550). Association for Computational Linguistics.

Lample, G., Ballesteros, M., Subramanian, S., Kawakami, K. & Dyer, C. (2016). Neural architectures for named entity recognition. *arXiv preprint arXiv:1603.01360*.

Li, Q. & Ji, H. (2014). Incremental joint extraction of entity mentions and relations. In *ACL (1)* (pp. 402–412).

Min, B., Grishman, R., Wan, L., Wang, C. & Gondek, D. (2013). Distant supervision for relation extraction with an incomplete knowledge base. In HLT-NAACL (pp. 777–782).

Mintz, M., Bills, S., Snow, R. & Jurafsky, D. (2009). Distant supervision for relation extraction without labeled data. In *Proceedings of the Joint Conference of the 47th Annual Meeting of the ACL and the 4th International Joint Conference on Natural Language Processing of the AFNLP, Vol 2* (pp. 1003–1011). Association for Computational Linguistics.

Ren, X., Wu, Z., He, W., Qu, M., Voss, C.R., Ji, H., ... & Han, J. (2017, April). CoType: Joint extraction of typed entities and relations with knowledge bases. In *Proceedings of the 26th International Conference on World Wide Web* (pp. 1015–1024). International World Wide Web Conferences Steering Committee.

Riedel, S., Yao, L. & McCallum, A. (2010). Modeling relations and their mentions without labeled text. *Machine Learning and Knowledge Discovery in Databases*, 148–163.

Ritter, A., Zettlemoyer, L. & Etzioni, O. (2013). Modeling missing data in distant supervision for information extraction. *Transactions of the Association for Computational Linguistics*, 1, 367–378.

Surdeanu, M., Tibshirani, J., Nallapati, R. & Manning, C.D. (2012). Multi-instance multilabel learning for relation extraction. In *Proceedings of the 2012 Joint Conference on Empirical Methods in Natural Language Processing and Computational Natural Language Learning* (pp. 455–465). Association for Computational Linguistics.

Takamatsu, S., Sato, I. & Nakagawa, H. (2012). Reducing wrong labels in distant supervision for relation extraction. In *Proceedings of the 50th Annual Meeting of the Association for Computational Linguistics: Long Papers-Vol 1* (pp. 721–729). Association for Computational Linguistics.

Wang, C., Fan, J., Kalyanpur, A. & Gondek, D. (2011,). Relation extraction with relation topics. In *Proceedings of the Conference on Empirical Methods in Natural Language Processing* (pp. 1426–1436). Association for Computational Linguistics.

Xu, W., Hoffmann, R., Zhao, L. & Grishman, R. (2013,). Filling knowledge base gaps for distant supervision of relation extraction. In *ACL (2)* (pp. 665–670).

Zheng, S., Wang, F., Bao, H., Hao, Y., Zhou, P. & Xu, B. (2017). Joint extraction of entities and relations based on a novel tagging scheme. *arXiv preprint arXiv:1706.05075*.

Zheng, S., Xu, J., Zhou, P., Bao, H., Qi, Z. & Xu, B. (2016). A neural network framework correlation extraction: Learning entity semantic and relation pattern. *Knowledge-Based Systems*, 114, 12–23.

Emerging Trends in Engineering, Science and Technology for Society,
Energy and Environment – Vanchipura & Jiji (Eds)
© 2018 Taylor & Francis Group, London, ISBN 978-0-8153-5760-5

Decoupling control of TRMS based on a Relative Gain Array (RGA) and Kharitonov theorem

Sumit Kumar Pandey, Jayati Dey & Subrata Banerjee
Department of Electrical Engineering, National Institute of Technology, Durgapur, India

ABSTRACT: This work addresses decoupling control of TRMS. The modeling of TRMS is done through identification algorithm using real-time input-output data. A decoupling technique is proposed on basis of Relative Gain Array (RGA) to eliminate the cross coupling effect. The stabilization of TRMS is achieved with PID controllers, the parameter range of which is obtained with Kharitonov stability criteria. PSO method is further tested for parameters tuning within range obtained earlier. The performance of the controller is tested in simulation and responses are found satisfactory.

1   INTRODUCTION

The TRMS has two control inputs with two outputs, respectively, as pitch (ψ) and yaw (φ) with significant cross coupling between them [1]. The designing of the controller is always a challenging task due to strong coupling effect and highly nonlinear characteristics of the TRMS.

In [2] authors explain method of identification for TRMS. The non-linear least squares identification method is applied for calibration. In [3] identification is performed by neural network approaches for TRMS. The different methods of identification and control technique which are frequently applied in MIMO system is explained in [5, 6]. The relative gain analysis is used for the pairing analysis of the MIMO system [7] and the decoupling technique is applied to eliminate the cross coupling effect of MIMO system.

The aim of the present work is to design PID controller to control the decoupled TRMS plant. First the identification of the TRMS is done and the transfer function is obtained. The decoupling of the TRMS is performed to eliminate the cross coupling effect of the TRMS and this decoupling is validated by the RGA analysis and the simulation results. The PID controller is designed on the basis of Kharitonov theorem through which a robust PID controller range is obtained. Further PSO is employed for the fine tuning of the controller parameters. The remainder is as next section describes the identification process of TRMS followed by the decoupling technique in section 3. In section 4 range of PID parameters are determined using Kharitonov theorem while in section 5 PSO method is implemented to obtain the optimum value of PID parameter. Section 6 depicts the simulation results and at last conclusion.

2   IDENTIFICATION OF TRMS MODEL

The objective of identification process to find the transfer function of TRMS. As there exist cross-coupling in two rotors of TRMS, it is considered as two linear rotor models with two linear couplings in-between. Therefore four linear models have to be identified as u_1 to y_1 and u_2 to y_2 u_2 to y_1 and u_1 to y_2 as shown in Figure 1. An experiment for model identification is carried out with the help of the MATLAB Toolbox using chosen identification models for four transfer functions [4–6]. The TRMS model and the experimental setup are excited with

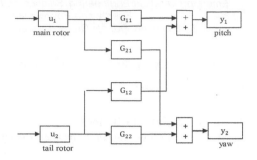

Figure 1.   Transfer function model for TRMS.

same input excitation and their responses are recorded. The excitation signal contains different sinusoids. To estimate accurate model error is minimized between the chosen model and the actual plant output. The optimal model parameters, for which the square of the error is minimal ("least mean square (LMS)" method) is considered as "identified model" [7]. The identified model of TRMS is obtained as,

$$G(s) = \begin{bmatrix} G_{11}(s) & G_{12}(s) \\ G_{21}(s) & G_{22}(s) \end{bmatrix} \tag{1}$$

$$G(s) = \begin{bmatrix} \dfrac{0.01657s^2 + 0.4194s + 2.454}{s^3 + 1.487s^2 + 4.403s + 5.449} & \dfrac{0.04986s + 0.0962}{s^2 + 0.2377s + 4.902} \\ \dfrac{0.02248s + 0.4527}{s^2 + 0.4099s + 0.2181} & \dfrac{0.0009881s^2 + 0.03361s + 0.4065}{s^3 + 1.345s^2 + 0.4568s + 0.3826} \end{bmatrix} \tag{2}$$

3   DECOUPLING OF TRMS

Here a decoupler is designed for TRMS based on generalized decoupling technique to rectify the coupling effect associated with the plant. In the generalized decoupling technique the design of decoupler for any square plant $G(s)$ is based on the formula as described by equation 3. The RGA method is applied to investigate the pairing analysis of the plant and it is verified that the decoupled plant $G_N(s)$ shown by Figure 2 is perfectly decoupled as described by equation (7).

$$G_D(s) = G_I(0) * G_R(s) \tag{3}$$

where, $G(0)$ = steady state gain matrix of $G(s)$, $G_D(s)$ = Decoupling matrix of $G(s)$, $G_I(0)$ = inverse matrix of $G(0)$ & $G_R(s)$ = Diagonal matrix of $G(s)$. Considering G_{11}, G_{22} only, the diagonal form of the plant.

$$G_R(s) = \begin{bmatrix} \dfrac{0.01657s^2 + 0.4194s + 2.454}{s^3 + 1.487s^2 + 4.403s + 5.449} & 0 \\ 0 & \dfrac{0.0009881s^2 + 0.03361s + 0.4065}{s^3 + 1.395s^2 + 0.456s + 0.3826} \end{bmatrix} \tag{4}$$

The decoupler is designed by following equation (3) as,

$$D(s) = G_I(0) * G_R(s) = \begin{bmatrix} \dfrac{0.04s^2 + 1.01s + 5.93}{s^3 + 1.487s^2 + 4.403s + 5.449} & \dfrac{-0.000039524s^2 - 0.00134s - 0.01626}{s^3 + 1.395s^2 + 0.456s + 0.3826} \\ \dfrac{-0.07s^2 - 1.97s - 11.582}{s^3 + 1.487s^2 + 4.403s + 5.449} & \dfrac{0.001007862s^2 + 0.0034s + 0.41}{s^3 + 1.395s^2 + 0.456s + 0.3826} \end{bmatrix} \tag{5}$$

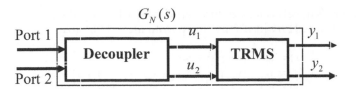

$$G_N(s)$$

Port 1 Decoupler u_1 TRMS y_1

Port 2 u_2 y_2

Figure 2.   Decoupled plant of TRMS.

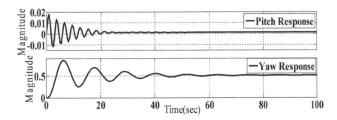

Figure 3.   Output response of decoupled plant when step input is applied to yaw rotor while zero input is applied to pitch rotor.

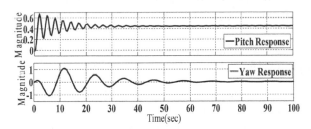

Figure 4.   Output response of decoupled plant when step input is applied to pitch rotor while zero input is applied to yaw rotor.

The decoupled plant is determined as below.

$$G_N(s) = G(s) * D(s) \tag{6}$$

Now, the relative gain array of the above matrix is calculated as described by equation (7). It verifies that upon decoupling the each output of the plant solely depends on the individual single reference input.

$$RGA(G_N(0)) = RGA(G(0) * D(0)) = \left(\begin{bmatrix} 0.4457 & 0.0018 \\ -0.0116 & 1.0196 \end{bmatrix} \right) = \begin{bmatrix} 1 & 0 \\ 0 & 1 \end{bmatrix} \tag{7}$$

It is also verified by simulation results as shown in Fig. 3 when step is tested to port 2 of the decoupler (Fig. 2) and zero input is tested to port 1. The corresponding yaw output shows step response as shown in Fig. 3 while pitch shows zero response signifying that decoupler completely nullifies the coupling effect on output. Similar situation arises when a step is tested to port 1 of the decoupler and zero input to port 2 as result of which corresponding pitch output shows step response while yaw shows zero response shown in Fig. 4.

4   APPLICATION OF KHARITONOV THEOREM TO FIND THE RANGE OF PARAMETERS OF PID CONTROLLER

Let assume the set $\delta(s)$ of real polynomials of degree n of the form as

$$\delta(s) = \delta_0 + \delta_1 s + \delta_2 s^2 + \delta_3 s^3 + \delta_4 s^4 + \ldots\ldots\ldots \delta_n s^n \tag{8}$$

The coefficient is in between as $\delta_0 \in [x_0, y_0], \delta_1 \in [x_1, y_1], \ldots\ldots\delta_n \in [x_n, y_n]$
Set $\delta(s)$ is said to be Hurwitz if following polynomial is Hurwitz.

$$
\begin{aligned}
K_1(s) &= x_0 + x_1 s + y_2 s^2 + y_3 s^3 + x_4 s^4 + x_5 s^5 + y_6 s^6 + \ldots\ldots, \\
K_2(s) &= x_0 + y_1 s + y_2 s^2 + x_3 s^3 + x_4 s^4 + y_5 s^5 + y_6 s^6 + \ldots\ldots, \\
K_3(s) &= y_0 + x_1 s + x_2 s^2 + y_3 s^3 + y_4 s^4 + x_5 s^5 + x_6 s^6 + \ldots\ldots, \\
K_4(s) &= y_0 + y_1 s + x_2 s^2 + x_3 s^3 + y_4 s^4 + y_5 s^5 + x_6 s^6 + \ldots\ldots,
\end{aligned}
\tag{9}
$$

PID controller transfer function is written as below,

$$
H(s) = K_p + \frac{K_i}{s} + K_d s \tag{10}
$$

On the basis of above equation the characteristic equation $1 + G_{11}(s)H(s)$ is written as below.

$$
\begin{aligned}
&2.454 K_i + s(2.454 K_p + 0.4194 K_i + 5.449) \\
&+ s^2 (0.4194 K_p + 0.01657 K_i + 2.454 K_d + 4.40) \\
&+ s^3 (1.487 + 0.01657 K_p + 0.4194 K_d) + s^4 (1 + 0.01657 K_d)
\end{aligned}
\tag{11}
$$

Now the four interval polynomial has been obtained from the above equation where the range of K_p, K_i & K_d are $(K_p^- K_p^+), (K_i^- K_i^+)$ & $(K_d^- K_d^+)$ respectively.

$$
\begin{aligned}
K_1(s) &= 2.454 K_i^- + s(2.454 K_p^- + 0.4194 K_i^- + 5.449) \\
&+ s^2 (0.4194 K_p^+ + 0.1657 K_i^+ + 2.454 K_d^+ + 4.40) \\
&+ s^3 (1.487 + 0.01657 K_p^+ + 0.4194 K_d^+) + s^4 (1 + 0.01657 K_d^-)
\end{aligned}
$$

$$
\begin{aligned}
K_2(s) &= 2.454 K_i^+ + s(2.454 K_p^+ + 0.4194 K_i^+ + 5.449) \\
&+ s^2 (0.4194 K_p^- + 0.1657 K_i^- + 2.454 K_d^- + 4.40) \\
&+ s^3 (1.487 + 0.01657 K_p^- + 0.4194 K_d^-) + s^4 (1 + 0.01657 K_d^+)
\end{aligned}
$$

$$
\begin{aligned}
K_3(s) &= 2.454 K_i^+ + s(2.454 K_p^- + 0.4194 K_i^- + 5.449) \\
&+ s^2 (0.4194 K_p^- + 0.1657 K_i^- + 2.454 K_d^- + 4.40) \\
&+ s^3 (1.487 + 0.01657 K_p^+ + 0.4194 K_d^+) + s^4 (1 + 0.01657 K_d^+)
\end{aligned}
$$

$$
\begin{aligned}
K_4(s) &= 2.454 K_i^- + s(2.454 K_p^+ + 0.4194 K_i^+ + 5.449) \\
&+ s^2 (0.4194 K_p^+ + 0.1657 K_i^+ + 2.454 K_d^+ + 4.40) \\
&+ s^3 (1.487 + 0.01657 K_p^- + 0.4194 K_d^-) + s^4 (1 + 0.01657 K_d^-)
\end{aligned}
$$

$$
\tag{12}
$$

In order to satisfy the condition of stability for the above four intervals polynomial the range of PID controller parameter is obtained as in case of the main rotor are $K_p = [0.1-1]$, $K_i = [0.1-1]$ & $K_d = [0.5-2]$. Similarly, by adopting the same procedure the range of PID controller paramet. obtained for tail rotor is as $K_p = [0.1-2]$, $K_i = [0.1-0.5]$ & $K_d = [2-5]$.

5 APPLICATION OF PSO TECHNIQUE TO TUNE THE PID CONTROLLER PARAMETERS

The PSO algorithm is described by flow chart in Figure 5. The velocity and position formula is calculated here is as below.

$$
V_{id}^{n+1} = V_{id}^n + c_1 rand().(P_{id}^n - X_{id}^n) + c_2 rand().(P_{gd}^n - X_{id}^n) \tag{13}
$$

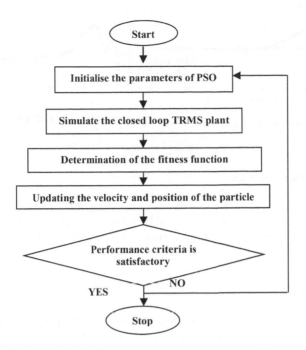

Figure 5. PSO algorithm flow chart.

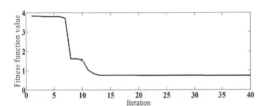

Figure 6. Convergence characteristics of main rotor.

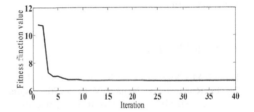

Figure 7. Convergence characteristics of tail rotor.

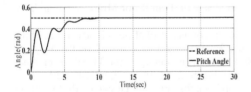

Figure 8a. Step response of the main rotor.

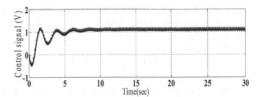

Figure 8b. Control signal of main rotor.

$$X_{id}^{n+1} = X_{id}^n + V_{id}^{n+1} \tag{14}$$

The objective function in the time domain is calculated as W [8]

$$W = (1 - e^{-\beta}).(M_p + E_{ss}) + e^{-\beta}(t_s - t_r) \tag{15}$$

The convergence characteristic of main and tail rotor is shown in Figures 6 and 7 respectively.

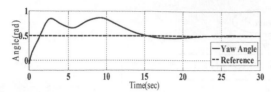 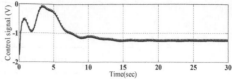

Figure 9a.   Step response of the tail rotor.

Figure 9b.   Control signal of tail rotor.

6 SIMULATION RESULTS

The parameter value of the PID controller is individually determined for both rotors using PSO technique within the ranges as obtained by Kharitonov criteria. The step response and the control signal are exhibited by Figure 8(a) and 8(b) and 9(a) and 9(b), respectively, for main and tail rotors.

7 CONCLUSION

This paper designs a decoupler for TRMS based on RGA technique in order to nullify the cross-coupling effect of the MIMO system. Then two PID controllers derived based on the Kharitonov stability theorem have been tuned using PSO technique. The accuracy of this method is established by results. To further validate the simulation results, the proposed technique will be implemented in the real time system of TRMS at NIT Durgapur Advanced Control laboratory.

REFERENCES

[1] TRMS 33–949S User Manual. Feedback Instruments Ltd., East Sussex, U.K., 2006.
[2] D. Rotondo, F. Nejjari, V. Puig (2013) Quasi-LPV modeling, identification and control of a twin rotor MIMO system, Control Engineering Practice, vol. 21, iss. 6, pp. 829–846.
[3] B. Subudhi & D. Jena (2009) Nonlinear system identification of a twin rotor MIMO system, TENCON 2009 IEEE Region 10 Conference.
[4] M. A. Hossain, A. A. M. Madkour, K. P. Dahal & H. Yu, (2004) Intelligent active vibration control for a flexible beam system, Proceedings of the IEEE SMC UK-RI Chapter Conference, Londonderry, U.K.
[5] I. Z. Mat Darus & Z. A. Lokaman (2010) Dynamic modeling of Twin Rotor Multi System in horizontal motion, Journal Mekanikal, no. 31, pp. 17–29.
[6] I. Z. Mat Darus (2004) Soft computing active adaptive vibration control of flexible structures, Ph.D. Thesis, Department of Automatic Control and System Engineering, University of Sheffield.
[7] A. Rahideh, M.H. Saheed & H.J.C. Huijberts (2008) Dynamic modeling of the TRMS using Analytical and Empirical approaches", Control Engineering Practice, vol. 16, no. 3, pp. 241–259.
[8] Z. L.Gaing (2004) A Particle Swarm Optimization Approach for optimum Design of PID Controller in AVR System, IEEE transcations on Energy Conversion, vol. 19, no. 2, pp. 384–391.

Emerging Trends in Engineering, Science and Technology for Society,
Energy and Environment – Vanchipura & Jiji (Eds)
© 2018 Taylor & Francis Group, London, ISBN 978-0-8153-5760-5

Root ORB—an improved algorithm for face recognition

A. Vinay, Aprameya Bharadwaj, Arvind Srinivasan, K.N. Balasubramanya Murthy
& S. Natarajan
Center for Pattern Recognition and Machine Intelligence, PES University, Bangalore, India

ABSTRACT: The biggest challenge faced by Face Recognition is to extract important facial features efficiently. Existing methods like SIFT, Root SIFT perform quite accurately, but fail in real world applications that require real-time processing. The ORB feature detector-descriptor performs quite well in this regard. We have modified the ORB feature detector to Root ORB and used it as a pre-processing step for a study of classification algorithms for face recognition. Our results show that the Root SIFT feature detector-descriptor performs only 6.25% to 12.5% as fast as the Root ORB detector-descriptor. We have chosen to work with detector-descriptors like ORB rather than using Neural Networks to extract facial features. Compute-intensive methods like Neural Networks require a lot of training data and computational time to obtain these features.

1 INTRODUCTION

Face recognition is the method or technique of detecting and recognizing a person from an image. A popular approach is to extract features from an image and use these to match with other images.

It has many applications in security, identification systems, and surveillance. For example, the FBI has a program to include face recognition along with other biometrics to retrieve records from its database. We also find everyday uses of face recognition like authentication mechanisms in mobile devices.

Face recognition can be traced back to the 1960s. The objective then, was to select a small set of images from a large database that could possibly contain an image to be matched. Since then, we've come a long way. Now, features extracted from a single image can give us a good idea about the image. We've been able to overcome hurdles like change in illumination, change in facial expressions, and motion of the head. Also, it is relatively easier to obtain datasets today.

We have modified the existing ORB detector-descriptor to Root ORB in the hope of improving accuracy and decreasing computational time. We have used the Bag of Words model before classification to create image histograms. We have then used machine learning classifiers to classify the images in the dataset. We have conducted a comparative study of Root SIFT versus Root ORB. ORB was used as it is very fast and open source.

2 TECHNICAL BACKGROUND

In face recognition, faces in the image are identified and represented by features and descriptors extracted from the image. This involves two steps: Feature detection and Description.

In detection, points of interest are determined. In description, attributes of these points are ascertained and stored in a vector. This vector is then used for applications like image classification. The computational efficiency of this task depends on the feature detector-descriptor algorithms we use, and the learning algorithms we apply.

Popular detector-descriptor methods include: Scale Invariant Feature Transform (SIFT) (Lowe (2004)), Speeded-Up Robust Features (SURF) (Bay et al 2008)), Oriented FAST and Rotated BRIEF (ORB) (Rublee et al. (2011)). Another important feature detector-descriptor is a modification of SIFT called Root SIFT (Arandjelović et al. (2012)). The main difference between these are the size of the feature vector obtained and the computational time.

SIFT returns a 128-dimensional vector of features. SURF takes lesser time as compared to SIFT as it returns a 64-dimensional vector of features. ORB is the fastest as it uses Features from Accelerated Segment Test (FAST) (Rosten et al. (2006)) detector and Binary Robust Independent Elementary Features (BRIEF) (Calonder et al. (2010)) descriptor and returns a 32-dimensional vector.

Root SIFT improves the accuracy of SIFT but still requires a similar computational time as it also returns a 128-dimensional vector.

Here, we use Root ORB detector-descriptor and some machine learning classifiers on the images of two well-known datasets. As Root ORB returns a 32-dimensional feature vector, we hope to achieve accuracy in the neighbourhood of Root SIFT when used along with these classifiers and optimize on the computational time.

The machine learning classification algorithms we use are:

a. Support Vector Machine Algorithm using Additive Chi Squared Kernel
b. K-Nearest-Neighbors Algorithm
c. Naïve Bayes Algorithm.

2.1 *SIFT*

The scale-invariant feature transform (SIFT) is an algorithm to extract and characterize local features in images. SIFT is scale and rotation invariant, and is therefore capable of delivering promising accuracies. It extracts the key-points within the image. These key points are then used for comparison, classification and matching. It performs the following functions to extract the key points. First, it identifies the interest points on the image by using Gaussian Difference. Then, the location and scale for these interest points are determined. Next, an orientation is assigned to each of these interest points. Finally, gradients are measured around these points and the image is transformed to minimize distortion.

2.2 *Root SIFT*

For certain areas in computer vision, using Euclidean distance to measure similarity between descriptors often yields inferior performance. In these areas, using the Hellinger kernel yields a superior performance when compared to using Euclidean distances. Root SIFT is a descriptor which is an element wise square root of the L1 normalized SIFT vectors.

2.3 *ORB*

The ORB algorithm makes use of the FAST detector and BRIEF descriptor. First, FAST is applied to obtain all key points. Then, the Harris corner measure is applied and the best key points among them are found. ORB makes use of the BRIEF descriptor. Although BRIEF performs poorly on rotation, ORB stabilizes it by using the generated key points.

2.4 *FAST*

FAST detector uses circles to classify key points as corner points. To check whether a pixel is a corner point, we compare the intensity of the pixel with the intensities of the circle of points around it. If they are considerably brighter or darker, we can flag this pixel as a corner point. FAST is renowned for its high computational efficiency. It also performs efficiently when used with machine learning algorithms.

2.5 *BRIEF*

SIFT uses 128-dimensional float vector for descriptors. Similarly, SURF uses 64-dimensional float vector. Working with these vectors for a large number of features requires a lot of space. It also requires a long time to match these features. Furthermore, all of these dimensions are not required for matching. BRIEF is a feature descriptor which is used along with feature detectors like FAST. It is generally used in applications which require real time processing.

3 PROPOSED METHOD

In subsequent paragraphs, we discuss in detail the methods used in our proposed system.

3.1 *Root ORB*

We applied the existing root algorithm from Root SIFT to the ORB algorithm. Here, as mentioned earlier, using the Hellinger kernel yields a superior performance when compared to using Euclidean distances. Hellinger kernel for L1 normalized histograms x and y is given by:

$$H(x,y) = \sum_{i=1}^{n} \sqrt{x_i y_i} \qquad (1)$$

After L1 normalization is done to all the ORB vectors, element wise square root is taken for all these vectors.

In the feature map space, calculating the Euclidean distance is analogous to calculating the Hellinger distance in the original space as:

$$x'^T y' = H(x,y) \qquad (2)$$

Where x' is the ORB vector after L1 normalization and computing element wise square root.

Key points are obtained for the images after Root ORB is applied. Root ORB uses a 32-dimensional vector to store the descriptors obtained and also uses the FAST detector. Due to this, the time required for classification is much lesser than compared to Root SIFT.

3.2 *Bag of words model*

In face recognition, we can represent a face by a bag of words. These visual words are a discriminative representation of the key points extracted from the face. These key points are

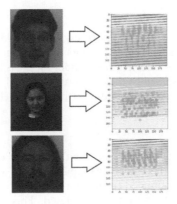

Figure 1.   When Root ORB is applied to images of the datasets.

encoded by a visual vocabulary. Next, all these features are pooled into an image representation. We use this model by building a large vocabulary of visual words. We analyze each image by observing the frequency of the words appearing in the image.

3.3 Support vector machine

Support vector machines (SVM) (Hearst et al. (1997)) are supervised learning models which are generally used for classification. When you provide data that consists of data points belonging to different, known categories, the SVM algorithm builds a model which can assign categories to new data points. When you represent the SVM model in space, you see that the algorithm works in such a manner as to maximise the gap between the different classes. A new data point added to this space can be classified by determining the category which is closest to this data point.

SVMs, by default, use linear classification methods. Using different kernels, we can also perform non-linear classification.

The additive chi squared kernel (Maji et al. (2008)) is often used in computer vision. Since the kernel is additive, it is possible to treat all components separately. This makes it possible to sample the Fourier transform in regular intervals. It is given by:

$$k(x,y) = \sum_i \frac{2x_i y_i}{x_i + y_i} \tag{3}$$

3.4 Naïve bayes

Naive Bayes (Lewis (1998)) is a set of algorithms used for classifying features which uses the Bayes' theorem of probability. The features being classified are assumed to be independent of each other.

Naïve Bayes is scalable and can be extended to a large number of features.

Here we make use of the Gaussian Naïve Bayes where the likelihood of the features is assumed to be Gaussian in nature. It is given by:

$$P(x_i \mid y) = \frac{1}{\sqrt{2\pi\sigma_y^2}} \exp\left(-\frac{\left(x_i - \mu_y\right)^2}{2\sigma_y^2} \right) \tag{4}$$

The parameters σ_y and μ_y are estimated using maximum likelihood.

3.5 K-nearest neighbors

In machine learning, K-Nearest Neighbors (Guo et al. (2003)) algorithm is used for classification. The output indicates whether it belongs to a certain class or not. For a given data point, its K-nearest neighbors are determined. Then, the class that occurred most frequently amongst these neighbors is assigned to the data point. If $k = 1$, then the object is assigned to that particular class. If k = 0 then it indicates that it doesn't belong to that class.

3.6 Pipeline

After Root ORB was applied to the dataset, the descriptors obtained were clustered using the bag of words model. Then, the model was separated into a training set and a testing set. Finally, we studied the results obtained from each of the three classifiers we used – K-NN, Naïve Bayes, SVM. These results were then tabulated. This was done for the Faces95 and Grimace datasets.

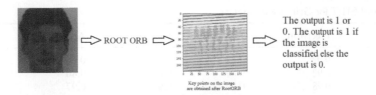

The output is 1 or 0. The output is 1 if the image is classified else the output is 0.

Key points on the image are obtained after RootORB

Figure 2. An overview of the entire process.

Figure 3. Sample of Faces95 dataset [11].

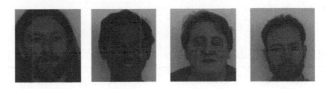

Figure 4. Sample of Grimace dataset [12].

4 DATASETS

The datasets used are:

1. Faces95 dataset – This dataset contains of 72 different people with 20 pictures each (both male and female). The size of each image is 180×200. The background is red. There is large head scale variation and change in lighting between images. There is slight variation of expression and motion of face in the images.
2. Grimace dataset – This dataset contains of 18 different people with 20 pictures each (both male and female). The size of each image is 180×200. The background is plain. There is little head scale variation and change in lighting between images. There is a large variation of expression and motion of face in the images.

5 RESULTS

All our programs were executed on Intel i7 core, using Windows10. The packages which were used are Opencv3 (3.2.0), scikit-learn (0.18.1), numpy (1.11.0), joblib (0.11) and inbuilt packages in python which include OS, time, random and math. Python3.5 was used.

First, we present the results for Root SIFT when applied with K-NN, Naïve Bayes, and SVM classification algorithm:

1. When Root SIFT is used on Grimace database with the following machine learning algorithms applied independently:
2. When Root SIFT is used on Faces95 database with the following machine learning algorithms applied independently:
3. When Root ORB is used on Grimace database with the following machine learning algorithms applied independently:
4. When Root ORB is used on Faces95 database with the following machine learning algorithms applied independently:

Table 1. Root SIFT for Grimace dataset.

Method	Precision	Recall	F1-score	Accuracy	Time taken
K-NN	0.92	0.89	0.88	0.88	16.11
Naïve Bayes	0.92	0.87	0.86	0.86	15.80
SVM	0.98	0.97	0.97	0.97	15.78

Inference: The SVM Classifier was the most accurate.

Table 2. Root SIFT for Faces95 dataset.

Method	Precision	Recall	F1-score	Accuracy	Time taken
K-NN	0.94	0.91	0.91	0.90	140.08
Naïve Bayes	0.83	0.76	0.76	0.76	53.93
SVM	0.99	0.99	0.99	0.98	61.09

Inference: The SVM Classifier was the most accurate.
The K-NN classifier was the slowest.

Table 3. Root ORB for Grimace dataset.

Method	Precision	Recall	F1-score	Accuracy	Time taken
K-NN	0.89	0.87	0.85	0.86	1.52
Naïve Bayes	0.88	0.87	0.86	0.86	1.49
SVM	0.99	0.99	0.99	0.98	1.59

Inference: The SVM Classifier was the most accurate.

Table 4. Root ORB for Faces95 dataset.

Method	Precision	Recall	F1-score	Accuracy	Time taken
K-NN	0.89	0.87	0.86	0.87	8.69
Naïve Baycs	0.94	0.91	0.91	0.90	7.71
SVM	0.97	0.97	0.97	0.97	8.12

Inference: The SVM Classifier was the most accurate.

6 CONCLUSIONS

The Faces95 dataset poses the challenges of large head scale variation and illumination changes, and the Grimace dataset focusses on large variation in expression and the motion of the face. From our results, we see that the accuracy of our method is high in all our test cases. Hence, we can say that our method overcomes the above hurdles.

From Table 2, we notice that the K-NN Classifier has performed much slower as compared to the other classifiers. Upon closer observation, we see that it performs the slowest in all the cases. This could be attribute to the fact that finding the nearest neighbor in a high dimensionality space requires a lot of time.

From the results, we notice that the SVM Classifier is the most accurate classifier. This could be due to the Additive Chi Squared Kernel which applies Fourier Transforms at regular intervals to the dataset.

When we compare Tables 1 and 2 with Tables 3 and 4 respectively, we observe that the accuracy of the Root ORB method is about the same, or sometimes greater than the accuracy of the Root SIFT method. However, when the computational times are compared, the Root ORB method is significantly faster than the Root SIFT method.

Also, the key point vector size in the Root ORB method is 75% smaller than the vector size in the Root SIFT method.

Hence, we see that the Root ORB method is both space and time efficient, in obtaining key points for classifying images as compared to the Root SIFT method.

REFERENCES

Arandjelović, Relja, and Andrew Zisserman. "Three things everyone should know to improve object retrieval." In *Computer Vision and Pattern Recognition (CVPR), 2012 IEEE Conference on*, pp. 2911–2918. IEEE, 2012.

Bay, Herbert, Tinne Tuytelaars, and Luc Van Gool. "Surf: Speeded up robust features." *Computer vision–ECCV 2006*(2006): 404–417.

Calonder, Michael, Vincent Lepetit, Christoph Strecha, and Pascal Fua. "Brief: Binary robust independent elementary features." *Computer Vision–ECCV 2010* (2010): 778–792.

Faces95 Dataset—http://cswww.essex.ac.uk/mv/allfaces/faces95.html.

Grimace dataset—http://cswww.essex.ac.uk/mv/allfaces/grimace.html.

Guo, Gongde, Hui Wang, David Bell, Yaxin Bi, and Kieran Greer. "KNN model-based approach in classification." In *CoopIS/DOA/ODBASE*, vol. 2003, pp. 986–996. 2003.

Hearst, Marti A., Susan T. Dumais, Edgar Osuna, John Platt, and Bernhard Scholkopf. "Support vector machines." *IEEE Intelligent Systems and their applications* 13, no. 4 (1998): 18–28.

Lewis, David D. "Naive (Bayes) at forty: The independence assumption in information retrieval." In *European conference on machine learning*, pp. 4–15. Springer, Berlin, Heidelberg, 1998.

Lowe, David G. "Distinctive image features from scale-invariant keypoints." *International journal of computer vision*60, no. 2 (2004): 91–110.

Maji, Subhransu, Alexander C. Berg, and Jitendra Malik. "Classification using intersection kernel support vector machines is efficient." In *Computer Vision and Pattern Recognition, 2008. CVPR 2008. IEEE Conference on*, pp. 1–8. IEEE, 2008.

Rosten, Edward, and Tom Drummond. "Machine learning for high-speed corner detection." *Computer Vision–ECCV 2006*(2006): 430–443.

Rublee, Ethan, Vincent Rabaud, Kurt Konolige, and Gary Bradski. "ORB: An efficient alternative to SIFT or SURF." In *Computer Vision (ICCV), 2011 IEEE international conference on*, pp. 2564–2571. IEEE, 2011.

Emerging Trends in Engineering, Science and Technology for Society,
Energy and Environment – Vanchipura & Jiji (Eds)
© *2018 Taylor & Francis Group, London, ISBN 978-0-8153-5760-5*

Face recognition using SURF and delaunay triangulation

A. Vinay, Abhijay Gupta, Harsh Garg, Shreyas Bhat, K.N. Balasubramanya Murthy &
S. Natarajan
Center for Pattern Recognition and Machine Intelligence, PES University, Bangalore, India

ABSTRACT: Face recognition is transforming the way people are interacting with machines. Earlier it was used in specific domains like law enforcement but with extensive research being done in this field, it is being extended to various applications like automatic face tagging in social media, surveillance systems in airports, theaters and so on. Local feature detection and description is gaining a lot of significance in the face recognition community. Extensive research on SURF and SIFT descriptors have found widespread application. Key points matched by SURF and triangulated using Delaunay Triangulation boosts the interest points detected. Other modern techniques like machine learning and deep learning require huge amount of training data and computational capabilities which sometime becomes a limitation of its usage. In contrast, hand-crafted models like SURF, SIFT overcomes the requirement of training data and computing power. The pipeline proposed in the paper reduces the average computational power and increases accuracy.

1 INTRODUCTION

One of the most challenging problem faced by the computer vision community is Face Recognition. Decades of work in this field has made face recognition system almost as smart as human beings. Face Recognition has been increasingly applied in surveillance systems and authentication services in order to prevent loss of sensitive information and curb security breaches leading to loss of money. Apart from these applications, Face Recognition is also extensively being used by social networking sites such as Facebook to tag friends. Tech giants such as Google, Microsoft use face recognition for image based search and to provide authentication services respectively.

Although face recognition has been effectively used in a number of applications, its performance tend to decline when the image shows significant variations in pose, expression, scale, illumination and translation. Most of the real world problems that require face recognition possesses these variations. Computer Vision enthusiasts are trying to make this system invariant to all these challenges. To overcome the problems associated with pose, expression, scale and upto some extent illumination, we propose a robust model which performs well when experimented with datasets which contain images variant to above mentioned constraints.

In any face recognition system, the most crucial step is to locate interest points in an image. A vast variety of keypoint detectors and feature descriptors have been extensively used by researchers in literature (eg. Bay et al. (2008), Lindeberg (1998), Lowe (2004)). In recent years, considerable amount of work has been done on Speeded-Up Robust Features (SURF) which is built upon previous works (eg. SIFT) to speed up and incorporate invariance in scale and in-plane rotation. The proposed model combines several algorithms and mathematical functions to boost the robustness and veracity of the system.

2 RELATED WORK

Delaunay Triangulation find its use in finger print identification, logo recognition and other object recognition applications (e.g Miri and Shiri (2012)). In Bebis et al. (1999) the Delaunay Triangulation is computed using ridge endings and ridge bifurcation minutiae represented in form of its coordinates. This index-based new approach achieves average accuracies of 86.56%, 93.16%, 94.12% for 3, 5, 7 imprints per person on testing set of size 210, 150 and 90 respectively. It characterizes better index selectivity, low storage requirements, minimal indexing requirements and fast identification for fingerprint recognition. Kalantidis et al. (2011) uses a novel discriminative triangle representation using multi-scale Delaunay Triangulation. Indexing is done using inverted file structure for robust logo recognition. On adding 4k distractor classes to the Flickr dataset the performance of the proposed multi-scale Delaunay Triangulation approach drops by 5.5% as compared to the base line Bag of Words model.

SIFT (Geng and Jiang (2009a), Geng and Jiang (2009b)) introduced enhancements like Partial-Descriptor-SIFT, Key-Points-Preserving-SIFT and Volume-SIFT (VSIFT) by keeping all the initial keypoints, preserving the interest points on a large scale or near face boundaries and by removing unreliable keypoints based on their volume respectively. They reduce the error rate by 4.6% and 3.1% in KPSIFT and PDSIFT on the AR dataset. SIFT in combination with a bag of words model has been deployed in (Sampath et al. (2016)) to detect household objects. Modern techniques such as Convolutional Neural Network is outperformed by a hybrid model (Al-Shabi et al. (2016))) using SIFT.

The effectiveness of combining Delaunay Triangulation and SIFT has been demonstrated in Dou and Li (2012). With the increase in number of viewpoints, the accuracy of SIFT algorithm decreases. To overcome this, Delaunay Triangulation (DelTri) exploits the overlapped regions in different images. Due to uniqueness of the DelTri, the overlapped region of an image pair overcomes the changes which were generated by different viewpoints at certain degree, thus increases the correct ratio of the result matches from 58.8% (SIFT+RANSAC) to 88.2% (SIFT+DelTri). This combination has also been used in Liu et al. (2017) for face alignment along with Convolutional Neural Networks.

Recent descriptors such as SURF is improvised upon SIFT by restricting the total number of reproducible orientation through the use of information from circular area around the interest points. After this step, construction of square region is performed. Further, the descriptors are extracted from these regions. SURF is used in various object tracking algorithms (Shuo et al. (2012)), where interest points in the defined object are matched between consecutive frames by obtaining the Euclidean distance between their descriptor.

3 METHOD PROPOSED

A robust approach is proposed with better results by combining Bilateral Filters for image smoothing, SURF to detect facial keypoints, PCA to minimize the number of key points, FLANN and Delaunay Triangulation for face matching in the dataset. The main steps of our model is depicted in Fig. 1.

3.1 Bilateral filter

Filtering is mostly used for reducing noise in the nearby pixel values which are mutually less correlated than the signal values. Tomasi and Manduchi (1998) smoothen the image by utilizing

Figure 1. Proposed approach – BF + SURF + PCA + FLANN.

a non-linear combination derived by averaging the smooth regions of the nearby image values, preserving edges. Other filtering techniques such as Gaussian blurring (Mostaghim et al. (2014)) which performs linear operation, neglecting edges is determined by sigma and non-linear filters like median filter (Dreuw et al. (2009)) which replaces the pixel values with the median value available in the local neighborhood are outperformed by Bilateral filters. In Bilateral Filtering weight of the intensity values of the surrounding pixels replace the intensity value of a pixel that is based on Gaussian distribution. This system loops over each pixel and simultaneously adjust weights of the neighboring pixels, retaining sharp edges.

Bilateral filtering combines the range and spatial domain filters

$$h(x) = \frac{1}{n(x)} \int_{-\infty}^{\infty} \int_{-\infty}^{\infty} g(\xi)c(\xi,x)s(g(\xi),g(x))d \tag{1}$$

where is $n(x)$ the normalization factor, calculated as

$$n(x) = \frac{1}{n(x)} \int_{-\infty}^{\infty} \int_{-\infty}^{\infty} c(\xi,x)s(g(\xi),g(x))d\xi \tag{2}$$

In which $c(\xi, x)$ evaluates the geometric closeness between the neighboring center x and a nearby point ξ. $s(g(\xi), g(x))$. evaluates the photometric similarity between the pixel at the neighborhood center x and that of a nearby point ξ.

3.2 *SURF*

Speeded-up robust features is invariant to in-plane rotation, contrast, scale and brightness. It comprises of keypoint point detector which interpolates the highly discriminative facial points. Further the descriptor extracts the features of the keypoint by constructing feature vectors. To minimize computation time SURF employs fast Hessian Matrix approximation and the scale space is examined by up scaling the integral image based filter sizes to detect interest points.

Given a point $a = (x, y)$ in an image I, the Hessian matrix H(a; σ) in at scale is defined as follows:

$$H(a,\sigma) = \begin{bmatrix} G_{xx}(a,\sigma) & G_{xy}(a,\sigma) \\ G_{xy}(a,\sigma) & G_{yy}(a,\sigma) \end{bmatrix} \tag{3}$$

where $G_{xx}(a, \sigma)$, $G_{xy}(a, \sigma)$ and $G_{yy}(a, \sigma)$ are the convolutions of the Gaussian second order partial derivatives with the image I in point a respectively.

To minimize computation time, Gaussian is approximated as a set of box filters which denote the lower scale to compute the blob response maps which are represented by $D_{xx}(a, \sigma)$, $D_{xy}(a, \sigma)$ and $D_{yy}(a, \sigma)$. The Hessian Matrix is estimated as:

$$det(H_{approx}) = D_{xx}D_{yy} - (\omega D_{xy})^2 \tag{4}$$

where ω represents the weight for the energy conservation between the actual and the approximated Gaussian kernels.

Interest points in an image are found at varied scales, where implementation of scale space is done through an image pyramid. Gaussian smoothing and sub-sampling are used to generate the pyramid labels.

The SURF descriptor uses the following methodology to find features in an image:

Step I: Setting a reproducible orientation through the use of information from the circular area around the derived keypoint.

Step II: A square region is constructed, according to the chosen orientation.

867

Step III: Extraction of the SURF descriptor.

For instance, the rotated neighborhood is divided into 16 sub-squares which is further divided into four squares. The descriptor for sub-square A_i is:

$$A_i = (\sum d_x, \sum d\_x, \sum d_y, \sum d\_y) \tag{5}$$

where d_x and d_y are HaarWavelet responses in horizontal and vertical directions respectively. The descriptor emphasizes on the spatial distribution of gradient information inside the interest point neighborhood.

3.3 *Reducing features using PCA*

Principal Component Analysis, used for dimensionality reduction, helps to linearly project high dimensional samples into low dimensional feature space, which help in creating a low dimensional structure of facial pattern. This methodology drops useless information and decomposes the facial key points into uncorrelated components.

The major objective of this paper is to reduce the cost, especially for SURF descriptor. A coarse matching between the key points is obtained by applying PCA on two sets of SURF features of images. Next, we calculate Kullback-Leibler (KL) divergence similarity score to improve matching accuracy. Experimental results of our proposed technique concludes that

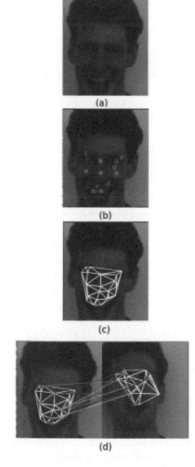

Figure 2.   (a) applying bilateral filtering to the input image-, (b) finding key-points using SURF-PCA, (c) applying Delaunay triangulation in (b), (d) matching faces with the dataset.

it can reduce the dimension of SURF features and the related matching cost in contrary to the conventional approach by approximately the same precision.

3.4 FLANN

Once facial key points are found, a matching algorithm is used to search in the dataset. There exist various approximating methods for computing nearest neighbors like brute force search, Support Vector Machines, FLANN (Muja and Lowe (2009)). They are typically used for computing the top k-nearest neighbors efficiently ignoring the rank of dataset. To compute short list of nearest neighbors we use FLANN implementation of k-d tree algorithm.

Rather than considering all neighbors in a Rank-Order clustering as given below:

$$d(a,b) = \sum_{i=1}^{O_a(b)} O_b(f_a(i)) \tag{6}$$

where $f_a(i)$ is the ith face in the neighbor list of a, and $O_b(f_a(i))$ gives the rank of face $f_a(i)$ in face bs neighbor list.

We only use sum of the top k-neighbors where their presence or absence on the short list is considered more significant than the numerical rank. A distance measure, by summing the presence or absence of shared nearest neighbors rather than ranks is given by the following distance function:

$$d_m(a,b) = \sum_{i=1}^{\min(O_a(b),k)} I_b(O_b(f_a(i)),k) \tag{7}$$

where indicator function,

$$I_b(x,k) = \begin{cases} 0, & \text{if x in } b\text{'s top nearest neighbor.} \\ 1, & \text{otherwise.} \end{cases} \tag{8}$$

3.5 Delaunay triangulation

Let P be the set of keypoints of an image which is sufficiently dense, then a good approximation is contained in Del(P). This means that an ample subset of the triangular surface of the Delaunay Triangulation are selected to get an accurate representation of the image. Since a triangle has only three vertices only 6 key points are matched for a pair of images in a viewpoint. Through this, we can achieve robust matching.

We assume projective transformation between the two input images, as given below:

$$\begin{bmatrix} xh \\ yh \\ k \end{bmatrix} = H \begin{bmatrix} x \\ y \\ 1 \end{bmatrix} \tag{9}$$

$$H = \begin{bmatrix} h_{11} & h_{12} & h_{13} \\ h_{21} & h_{22} & h_{23} \\ h_{31} & h_{32} & h_{33} \end{bmatrix} \tag{10}$$

Computing x', y' that is non-homogeneous coordinates as:

$$\begin{aligned} x' &= \frac{xh}{k} = \frac{h_{11}x + h_{12}y + h_{13}}{h_{31}x + h_{32}y + h_{33}} \\ y' &= \frac{yh}{k} = \frac{h_{21}x + h_{22}y + h_{23}}{h_{31}x + h_{32}y + h_{33}} \end{aligned} \tag{11}$$

Table 1. Accuracies of proposed model on different datasets.

Dataset	PCA	min-Hessian	Accuracy
FACE95	0.8	565	75.2
FACE96	0.8	565	76.8
GRIMACE	0.8	565	79.1
ORL	0.8	565	84.4

where $(x', y') \Leftrightarrow (x, y)$ are pixel-point correspondences and H is Homography transformation matrix.

The symmetric transfer error is calculated using Euclidean distance and transformation matrix, $d(x, H^{-1}x')^2 + d(x', Hx)^2$, for matching keypoints in a pair of image. The inlier's having values lesser than threshold are counted.

4 DATASET

To test the robustness and effectiveness of the proposed model we use the FACES95, FACES96, the ORL Database of Faces and GRIMACE which show variance in pose, expression, rotation and illumination altogether.

- FACES95 contain 1440 images of 72 individuals; 20 images each. The data set was constructed by taking snapshots of 72 individual with a delay of 0.5 seconds between successive frames in the sequence. Significant head movement variations was introduced between images of the same individual. Similar to the above methodology FACES96 was constructed for 3040 images.
- GRIMACE consist 360 images of 18 individuals. The images taken are variant to scale, lighting and position of face in the image. In addition the subject made grimaces after moving his/her head which gets extreme towards the end of the sequence.
- The ORL Database contain images of 40 individuals with total number of images equal to 400. The pictures were taken at different time, light conditions, facial expression and facial details. The database was utilized in a face recognition project carried out in an association with the Speech, Vision and Robotics Group of the Cambridge University Engineering Department.

5 RESULTS AND CONCLUSION

We executed the proposed model over every group of images present in the four benchmark databases, namely, FACES95, FACES96, GRIMACE and ORL. The results obtained using our technique on the four datasets are tabulated in Table 1. The table compares accuracy of our model over different datasets. The retained variance for PCA is set to 0.8 which implies 20% of the total variance is retained. Threshold for min-Hessian value is set to 565 which accepts the most salient keypoints. The ORL database which is variant to pose, expression and scale performed well using our method. The other three data sets which in addition to pose, expression and scale are also variant to illumination which perform moderately low on our proposed approach as compared to ORL. So the proposed model fails for illumination and hence should not be used if they are in varied lightning conditions. The usage of FLANN in our model also increases the speed of matching the images from the database.

REFERENCES

Al-Shabi, M., W.P. Cheah, & T. Connie (2016). Facial expression recognition using a hybrid cnn-sift aggregator. *arXiv preprint arXiv:1608.02833*.

Bay, H., A. Ess, T. Tuytelaars, & L. Van Gool (2008). Speeded-up robust features (surf). *Computer vision and image understanding 110*(3), 346–359.

Bebis, G., T. Deaconu, & M. Georgiopoulos (1999). Fingerprint identification using delaunay triangulation. In *Information Intelligence and Systems, 1999. Proceedings. 1999 International Conference on*, pp. 452–459. IEEE.

Dou, J. & J. Li (2012). Robust image matching based on sift and delaunay triangulation. *Chinese Optics Letters 10*(2102), S11001.

Dreuw, P., P. Steingrube, H. Hanselmann, H. Ney, & G. Aachen (2009). Surf-face: Face recognition under viewpoint consistency constraints. In *BMVC*, pp. 1–11.

Geng, C. & X. Jiang (2009a). Face recognition using sift features. In *Image Processing (ICIP), 2009 16th IEEE International Conference on*, pp. 3313–3316. IEEE.

Geng, C. & X. Jiang (2009b). Sift features for face recognition. In *Computer Science and Information Technology, 2009. ICCSIT 2009. 2nd IEEE International Conference on*, pp. 598–602. IEEE.

Kalantidis, Y., L.G. Pueyo, M. Trevisiol, R. van Zwol, & Y. Avrithis (2011). Scalable triangulationbased logo recognition. In *Proceedings of the 1st ACM International Conference on Multimedia Retrieval*, pp. 20. ACM.

Lindeberg, T. (1998). Feature detection with automatic scale selection. *International journal of computer vision 30*(2), 79–116.

Liu, Y., A. Jourabloo, W. Ren, & X. Liu (2017). Dense face alignment. *arXiv preprint arXiv:1709.01442*.

Lowe, D.G. (2004). Distinctive image features from scale-invariant keypoints. *International journal of computer vision 60*(2), 91–110.

Miri, S.S. & M.E. Shiri (2012). Star identification using delaunay triangulation and distributed neural networks. *International Journal of Modeling and Optimization 2*(3), 234.

Mostaghim, M., E. Ghodousi, & F. Tajeripoor (2014). Image smoothing using non-linear filters a comparative study. In *Intelligent Systems (ICIS), 2014 Iranian Conference on*, pp. 1–6. IEEE.

Muja, M. & D.G. Lowe (2009). Fast approximate nearest neighbors with automatic algorithm configuration. *VISAPP (1) 2*(331–340), 2.

Sampath, A., A. Sivaramakrishnan, K. Narayan, & R. Aarthi (2016). A study of household object recognition using sift-based bag-of-words dictionary and svms. In *Proceedings of the International Conference on Soft Computing Systems*, pp. 573–580. Springer.

Shuo, H.,W. Na, & S. Huajun (2012). Object tracking method based on surf. *AASRI Procedia 3*, 351–356.

Tomasi, C. & R. Manduchi (1998). Bilateral filtering for gray and color images. In *Computer Vision, 1998. Sixth International Conference on*, pp. 839–846. IEEE.

Emerging Trends in Engineering, Science and Technology for Society,
Energy and Environment – Vanchipura & Jiji (Eds)
© *2018 Taylor & Francis Group, London, ISBN 978-0-8153-5760-5*

Opportunities and challenges in software defined networking and network function virtualization

Aiswarya Roy, V.K. Asna, N. Nimisha & C.N. Sminesh
Department of Computer Science and Engineering, Government Engineering College, Thrissur, Kerala, India

ABSTRACT: Software Defined Networks (SDN) and Network Function Virtualization (NFV) are the two emerging paradigms in networking. The control plane is physically separated from the forwarding plane and logically centralized in the SDN architecture. SDN overcomes many limitations in traditional network infrastructures by separating the network's control plane from the routers and switches. With the decoupling of control plane and data plane, the entire network is controlled by a centralized controller and network switches become simple forwarding devices. NFV is the initiation to give a virtualized platform to the network which is presently carried out by proprietary hardware. The NFV concept was introduced to increase the feasibility and scalability of networks. This paper mainly focuses on the research opportunities and challenges in the control plane and data plane in SDN and NFV.

1 INTRODUCTION

Software Defined Networking (SDN) is an emerging network architecture. One distinct change in SDN compared with traditional networking is that the control plane is removed to a set of controllers (Akyildiz et al., 2014). For small networks, a single centralized controller is used. But when the size of the network increases a single controller is not enough (Vissicchio et al., 2015). Multiple controllers are used to solve scalability and reliability issues. One of the key problems in SDN is the placement of controllers as this could have an impact on network performance and cost (Braun & Menth, 2014). Another problem in the control plane is the load balancing among controllers. The data plane uses forwarding devices for the processing and delivery of packets. The data plane enables data transfer to and from clients.

The NFV concept was introduced to increase the feasibility and scalability of network. In the traditional networking model each network function is implemented by different physical nodes. By virtualizing a single physical node can be used for different functions. These virtual function nodes are known as Virtualized Network Function (VNF), which may run on a single or a set of Virtual Machines (VMs). Increased elasticity, increased service agility, improved operational simplicity and faster innovation are the major advantages of NFV. SDN and NFV technologies are synergistic and they could offer improved programmability and faster service enablement.

This paper is structured as follows: Section 2 discusses the research and challenges in the control plane; Section 3 discusses the data plane in SDN; Section 4 discusses NFV and the promising combination of SDN and NFV; and Section 5 concludes the survey.

2 CONTROL PLANE IN SDN

In SDN, the control plane is decoupled from the data plane. It removes the control plane from the hardware and moved to a set of controllers. The control plane takes the decision about where and how to forward packets. The small networks need only a centralized controller to work properly. But when the size of the network increases, a single controller is not enough. So multi-controller deployment is used to solve the scalability and reliability issues raised by the centralized architecture of SDN.

When using multiple controllers, challenges like the controller placement need to be addressed. Ksentini et al. (2016) proposed a methodology to find the optimal placement of SDN controllers. In this method the main focus is on the following performance metrics:

The communication overhead and latency between controllers and switches
The communication overhead and latency between controllers
The load balancing between controllers

Objectives of this approach are optimizing the performance of the control plane and to reduce the cost. This method helps to minimize latency and communication overheads among controllers and ensure load balancing.

Galich et al. (2017) proposed ways to reduce control traffic latency in an OpenFlow-based SDN. They derived an algorithm for processing ARP packets using an OpenDaylight controller. The algorithm computes the number of OpenFlow control traffic packets. The total control traffic latency is dependent on latencies occurring at switches and the controller and the number of service traffic packets in the network. The main idea proposed is that if the total control traffic latency exceeds the standard values per approved regulations, it can be reduced by reducing either switch or controller latency. Finding the location of controllers and switches is the main challenge in this methodology.

Han et al. (2016) proposed an algorithm which optimizes minimum-control-latency based on greedy controlling pattern design. The latency measurement will not be precise enough when measuring only based on ping results. In this paper, a methodology is proposed based on active latency measurement to obtain more accurate control latency. The main objective of this methodology is to obtain a partition with the optimal number of controllers while minimizing the control latency. A minimum-control-latency optimized algorithm and a control latency bounded algorithm are used in this methodology. Minimum-control-latency optimizing can improve the imbalance when partitioning SDN domains. However, to consider hundreds of topologies in the internet will be a challenge to this methodology.

Load balancing among controllers is another challenge that arises due to multi-controller deployment. ZHou et al. (2017) propose an effective load balancing mechanism based on a switches group. The mechanism not only balances the load among controllers, but also solves the load oscillation and improves time efficiency. A switches selection algorithm and a target controllers selection algorithm are used for this purpose. The load measurement component runs on every local controller to measure the load of the controller and periodically sends load information to the super controller. The current load press of a controller can be calculated as:

$$p_i = \frac{c_i}{c_{imax}} \tag{1}$$

Table 1. Opportunities and challenges in SDN control planes.

Literature on control plane	Opportunities	Challenges
(Ksentini et al. 2016)	Optimizing the performance of control plane	Controller placement problem
(Galich et al., 2017)	The total control traffic latency can be reduced by reducing either switch or controller latency	Finding the location of controllers and switches
(Han et al., 2016)	Minimize the control latency with optimal number of controllers	Considering hundreds of topologies
(Zhou et al., 2017)	Balancing among controllers	Proper arrangement of controllers and switches

where c_i is the current load of the controller i and c_{imax} is the maximum load capability of the controller i.

From the above literature it is observed as given in Table 1.

One of the key problems in SDN is the placement of controllers as this could have an impact on network performance and cost. Therefore, they should be placed in a way to minimize control traffic latency among the switches and the controllers and load balance among controllers. These observations are shown in Table 1.

3   DATA PLANE IN SDN

The data plane is the part of the network that carries user traffic. The data plane enables data transfer to and from clients, handling multiple conversations through multiple protocols, and manages conversations with remote peers. Data plane traffic travels through routers. It uses forwarding devices for the processing and delivery of packets.

Bozakov and Rizk (2013) propose that in SDN applications switches with various capacities for control message processing cause unpredictable delays due to their concurrent operation. This methodology uses a queuing model to characterize the service of a switch's control interface to concentrate on this issue. To improve predictability in terms of expected delays, it implements the control connection to the switch and also enables applications to easily adapt to the control message processing rates. To transmit controller messages to a specific switch over an established control connection controller interface typically implemented as function, send to switch. Therefore, the developer can make sure that a control message has been accepted by the interface or not. And if a switch is operated over its processing limits, instead of allowing the messages to queue up at the switch, the socket will get blocked at the application level. As a result, the application may adjust its sending rate much quicker. The current SDN abstraction model is maintained which is the main benefit of this methodology.

Pontarelli et al. (2017) present a method to implement complex tasks into stateful SDN programmable data planes. The presented method proposes to use the internal microcontroller, typically used to configure the programmable data plane, to also perform some complex operations that do not require to be executed on each packet. These operations can be executed on a set of data gathered by the data plane and processed in a time scale that is much higher than the time window of a packet, but is much less than time scale needed for an external SDN controller. Moreover, the use of the configuration microcontroller instead of an external SDN controller to avoid the exchange of data on the control links permits a fine grain tuning of the operations to perform from the timing point of view. Extending the original stateless SDN data plane paradigm enables the execution of simple control functions directly into the fast path. In this paper, a different approach is to extend stateful SDN data plane with the support of a set of lazy operations. The term lazy refers to a set of operations that are triggered by the reception of a packet inside the pipeline, but return its result after a certain amount of time without blocking the forwarding of the packet itself inside the switch.

Zhang et al. (2014) proposed Big Switch abstraction as a specification mechanism for high-level network behavior. This specification allows the operator to define end-to-end flow policies. This can be used for placing rules on individual switches by the network operating system. This is forced to do so by the limited capacity of the Ternary Content Addressable Memories (TCAMs) used for rules in each switch. Using a centralized rule management system, it is compiled down to individual rules on different switches. Permitted (PERMIT rule) or dropped (DROP rule) are the rules related to each packet. For placing rules on switches the authors proposed a solution based on Integer Linear Programming (ILP). While optimizing the number of rules and maintaining the switch capacity constraints, this can be applied on a given firewall policy. Switch priority, capacity, and policy constraints are satisfied by ILP. And this also optimizes certain objective functions such as minimizing the total number of rules. But complex rule placement constraints such as monitor certain packets, that do not want to let firewall rules to block packets before they reach the monitoring rules is not supported by this concept.

Table 2. Opportunities and challenges in SDN data plane.

Literature on data plane	Opportunities	Challenges
(Pontarelli et al., 2017)	Because almost all pro grammable Data planes already haves an internal micro-controller used to configure the memory tables, method proposed in the paper could be widely applied.	To implement complex tasks into a stateful SDN programmable data plane.
(Zhang et al., 2014)	Applicable to real-sized net works.	The meaning of the original policies have to maintained by the results obtained from this method.
(Bozakov & Rizk, 2013)	Extension of this method can manage more complex control net work topologies and distributed controllers.	Device hetero—geneity as an inherent property of SDN which must be considered.

From the above literature it is observed that the key challenges in SDN data plane are implementation of complex tasks into a SDN programmable data plan and sharing of rules across different paths. Proper handling of these problems leads to better opportunities. These observations are shown in Table 2.

4 NETWORK FUNCTION VIRTUALIZATION

The NFV concept was introduced to make the network more flexible and simpler than the traditional network concept which is dependent on hardware constraints (Lopcz, 2014). NFV is about providing network functions by software rather than using a number of hardware. Most importantly, the benefit of NFV is the flexibility to easily, rapidly, and dynamically provision and instantiate new services in various locations.

SDN is another network paradigm. The combination of both SDN and NFV started a new era of networks. The combination of SDN principles and NFV infrastructure can change the way of building, deploying and controlling network applications, which is described by King et al. (2015). To provide flexibility, scale, distribution and bandwidth to match the user demands, flexible and dynamic optical resources are used in the current transport networks. Due to the requirement of significant engineering resources, these are non-real-time capabilities and often lack the flexibility for dynamic scenarios. To overcome these challenges, Daniel et al. propose a new architecture with the combination of SDN, NFV, flexi-grid, and ABNO (Application-Based Network Operations). This new architecture has the capability to deploy a vCDN (virtualized Content Distribution Network) which is capable of scaling the user bandwidth demand and control resources programmatically.

Another combination of SDN and NFV with IoT is proposed by Ojo et al. (2016). A typical Io-T architecture has a 3-layered structure, which are the perception layer, the network layer and the application layer. By adopting the SDN concept in this Io-T architecture the network layer is divided into two planes; control plane and data plane. The data layer comprises SDN routers and switches to forward packets. These switches and routers in the data layer are programmatically controlled in the control layer. By enabling NFV on this new SDN adopted IoT architecture, the network agility and network efficiency of IoT applications increases. The IoT—gateway becomes dynamic, scalable and elastic by virtualization. Virtualization makes the infrastructure more flexible and sustainable by decoupling the network control and management function from the hardware.

The future internet scenario named virtual presence, by leveraging on the joint SDN/NFV paradigm, together with a fog computing approach is proposed by Faraci and Lombardo (2017). Virtual presence is achieved by exporting a real hardware or software resource, that

Table 3. Opportunities and challenges of NFV.

Literature on NFV	Opportunities	Challenges
(King et al., 2015)	Capable to respond to high bandwidth real time and predicted video stream demands	NFV environment is more dynamic than the traditional one
(Ojo et al., 2016)	SDN and NFV concepts help to solve new challenges of IoT	Security issues of NFV
(Faraci & Lombardo, 2017)	Smart device sharing service could increase the resource utilization	Having toco exist in a cloud in tegrated environment

is physically or virtually located in a personal network of a given user such as a smart TV or a virtual set-top box, to another personal network. The proposed architecture allows a virtual presence to be instantiated in a personal network, in such a way that it allows using it as present locally in the visited personal network. In this proposed architecture, SDN and NFV concepts help to run both network and application functions in a virtualized form on the nodes at the edge of the network.

A comparison of SDN and NFV for redesigning the LTE (Long Term Evolution) packet core is done by Jain et al. (2016). The experiment result shows that an NFV based implementation is better suited for networks with high signaling traffic, because handling the communication with the SDN controller quickly becomes the bottleneck at the switches in the SDN-based EPC (Evolved Packet Core). On the other hand, an SDN-based EPC is a better choice when handling large amounts of data plane traffic, because SDN switches are often more optimized for packet forwarding than virtualized software appliances.

Along with the opportunities and benefits, NFV poses significant security problems. The security issues in Virtualized Network Functions (VNFs), which is an important part of NFV's architecture is discussed by Aljuhani and Alharbi (2017). These security issues are classified into three: insider attack, outsider attack, and between VNFs attack.

The combination of SDN and NFV provides vast opportunities in networking. It could make the transport network more dynamic and feasible as described by King et al. (2015). Adopting SDN and NFV concepts in IoT could provide solutions to address the challenges in IoT (Ojo et al., 2016). The virtual presence concept could provide a lot of opportunities (Faraci & Lombardo, 2017). Along with a lot of opportunities, NFV also faces challenges. These opportunities and challenges are shown in Table 3. The major problems in dealing with NFV are the security issues.

5 CONCLUSION

The paper mainly focuses on the research opportunities and challenges in SDN and NFV. From conducting the literature survey on the control plane and the data plane in SDN, identified research opportunities and challenges are summarized. NFV provides how network functions can be done using software rather than hardware. It is observed that a combination of NFV and SDN is really a promising technology to solve many of the complex networking issues in the existing network architectures.

REFERENCES

Akyildiz, I. F., Lee, A., Wang, P., Luo, M. & W. Chou, W. (2014). A roadmap for traffic engineering in SDN-Openflow networks. *Computer Networks, 71*, 1–30.

Aljuhani, A. & Alharbi, T (2017). Virtualized network functions security attacks and vulnerabilities. In *IEEE 7th Annual Computing and Communication Workshop and Conference (CCWC)* (pp. 1–4). New York, NY: IEEE.

Bozakov, Z. & Rizk, A. (2013). Taming SDN controllers in heterogeneous hardware environments. In *Second European Workshop on Software Defined Networks (EWSDN)* (pp. 50–55). New York, NY: IEEE.

Braun, W. & Menth, M. (2014). Software-defined networking using Openflow: Protocols, applications and architectural design choices. *Future Internet, 6*(2), 302–336.

Faraci, G. & Lombardo, A. (2017). An NFV approach to share home multimedia devices. In *IEEE Conference on Network Softwarization (Net-Soft)* (pp. 1–6). New York, NY: IEEE.

Galich, S., Deogenov, M. & Semenov, E. (2017). Control traffic parameters analysis in various software-defined networking topologies. In *International Conference on Industrial Engineering, Applications and Manufacturing (ICIEAM)* (pp. 1–6). New York, NY: IEEE.

Han, L., Li, Z., Liu, W., Dai, K. & Qu, W. (2016). Minimum control latency of SDN controller placement. In *IEEE Trustcom/BigDataSE/ISPA* (pp. 2175–2180). New York, NY: IEEE.

Jain, A., Sadagopan, N., Lohani, S.K. & Vutukuru, M. (2016). A comparison of SDN and NFV for redesigning the LTE packet core. In *IEEE Conference on Network Function Virtualization and Software Defined Networks (NFV-SDN)* (pp. 74–80). New York, NY: IEEE.

King, D., Farrel, A. & Georgalas, N. (2015). The role of SDN and NFV for flexible optical networks: Current status, challenges and opportunities. In *17th International Conference on Transparent Optical Networks (ICTON)* (pp. 1–6). New York, NY: IEEE.

Ksentini, A., Bagaa, M., Taleb, T. & Balasingham, I. (2016). On using bargaining game for optimal placement of SDN controllers. In *IEEE International Conference on Communications (ICC)* (pp. 1–6). New York, NY: IEEE.

Lopez, D.R. (2014). Network functions virtualization: Beyond carrier-grade clouds. In *Optical Fiber Communications Conference and Exhibition (OFC)* (pp. 1–18). New York, NY: IEEE.

Ojo, M., Adami, D. & Giordano, S. (2016). A SDN-IoT architecture with NFV implementation. In *IEEE Globecom Workshops (GC Wkshps)* (pp. 1–6). New York, NY: IEEE.

Pontarelli, S., Bruschi, V., Bonola, M. & G. Bianchi, G. (2017). On offloading programmable SDN controller tasks to the embedded microcontroller of stateful SDN data planes. In *IEEE Conference on Network Softwarization (NetSoft)* (pp. 1–4). New York, NY: IEEE.

Vissicchio, S., Tilmans, O., Vanbever, L. & Rexford, J. (2015). Central control over distributed routing. *ACM SIGCOMM Computer Communication Review 45*(4), 43–56.

Zhang, S., Ivancic, F., Lumezanu, C., Yuan, Y., Gupta, A. & Malik, S. (2014). An adaptable rule placement for software-defined networks. In *44th Annual IEEE/IFIP International Conference on Dependable Systems and Networks (DSN)* (pp. 88–99). New York, NY: IEEE.

Zhou, Y., Wang, Y., Yu, J., Ba, J. & Zhang, S. (2017). Load balancing for multiple controllers in SDN based on switches group. In *19th Asia-Pacific Network Operations and Management Symposium (AP-NOMS)* (pp. 227–230). New York, NY: IEEE.

Emerging Trends in Engineering, Science and Technology for Society,
Energy and Environment – Vanchipura & Jiji (Eds)
© 2018 Taylor & Francis Group, London, ISBN 978-0-8153-5760-5

A parallel framework for maximal clique enumeration

Thilak Anusree & K. Rahamathulla
Department of Computer Science and Engineering, Government Engineering College, Thrissur,
Kerala, India

ABSTRACT: Graphs play a major role in modeling many real-world problems. Due to the availability of huge data, the graph processing in serial environment become more complex. Thus, fast and efficient algorithms which work effectively utilizing the modern technologies are required. A maximal clique problem is one of the graph processing methods which is used in many applications. The Bron-Kerbosch (BK) algorithm is the most widely used and accepted algorithm for listing each and every maximal clique in a graph. Here an idea of a parallel version of BK algorithm is proposed which will reduce the computation time to a large extent than its serial implementation. It utilizes the cluster computing strategy.

1   INTRODUCTION

1.1   *Maximal clique enumeration problem*

The listing of all maximal cliques which are present in a graph is called the maximal clique enumeration problem. Let G be a graph, where V and E are the set of vertices and edges of G respectively. A clique in graph G is a subset of vertices of G where each and every vertex belonging to the subset is connected to all other vertices within the subset.

1.2   *BK algorithm*

Of all the algorithms for maximal clique enumeration, the Bron-Kerbosch (BK) algorithm is the most widely used method. Designed by two Dutch scientists, Joep Kerbosch and Coen-raad Bron, this algorithm was published in 1973. However, although many different algorithms, which are theoretically better than BK on the inputs, have been proposed, BK still remains better in practical applications for listing all the cliques.

1.3   *Parallel approach*

In real-world scenarios, the sizes of graphs are so large that the number of nodes range from hundreds to thousands. The processing of such large graphs using a serial algorithm may consume a lot of time.

Because of this higher time complexity of serial algorithms, a parallel approach can be used to overcome this difficulty and to give the result in much less time. Here we try to implement a parallel version of a BK algorithm which will efficiently enumerate all cliques in any given graph that are maximal.

1.4   *Graph based applications*

Structure alignment of 3-D protein, gene expression analysis, social hierarchy detection, and genome mapping are some of the applications which use graph processing. Maximal clique enumeration is the main method which has been used in these applications.

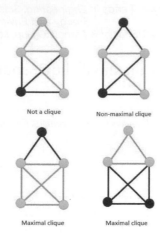

Figure 1. Maximal cliques in a graph.

1.5 *Organization of paper*

This paper is being organized as follows: the background works are discussed in Section 2; Section 3 gives the literature survey where some approaches are discussed; Section 4 outlines the proposed method; And Section 5 details the conclusion.

2 BACKGROUND

A maximal clique is a completely connected subgraph which is not the subset of any other bigger clique in that graph, i.e. we can never expand an already existing clique (which is maximal in nature) by adding one more neighboring vertex to it. Every graph having n vertices can have a maximum of 3_3 maximal cliques (Tomita et al., 2006).

The biggest clique in a graph is the maximum clique. The maximum common subgraph problem can be reduced to maximal clique enumeration problem (Jayaraj et al., 2016) where the former is an NP-complete problem. Figure 1 shows maximal cliques in an example graph G.

The maximal clique enumeration problem originally arose in different areas of research as a set of related problems. The algorithms for solving those issues can be regarded as the initial algorithms for clique detection. The algorithm of Harary and Ross, (1957) was the first broadly acknowledged endeavor for listing all of the cliques which are maximal in a graph. which displays a technique for discovering connections between individuals utilizing the sociometric information between them which will form a clique.

3 LITERATURE SURVEY

With the introduction of the backtracking search strategy, the maximal clique enumeration problem gained momentum to a great extent. The efficiency of combinatorial optimization algorithms is improved by using backtracking which limits the search space size of the algorithm. A set of viability criteria is established that is used to help the backtracking algorithm from explorating the non-promising paths (Schmidt et al., 2009).

3.1 *Maximal clique enumeration methods*

The BK algorithm lists all the cliques present in a graph exactly once (Jayaraj et al., 2016). It uses three sets for finding maximal cliques using recursion. It uses an approach in which only the nearest neighbors of the node are considered. A node is taken and its neighbors are explored recursively to find whether they form a clique or not. The BK algorithm is given in Algorithm 1 (Bron & Kerbosch, 1973).

Algorithm 1 Serial Bron-Kerbosch algorithm
LISTCLIQUE(Q, R, P)
P, Q, R and N are set of vertices
which N:q's neighbors in graph G
P:belong to current clique Q:can
be used to complete P R:cannot
be used to complete P

```
1:  Let Q be the set of n nodes {q1, q2,...qn}
2:  if set Q and R is empty
3:      CLIQUE-FOUND;
4:  else
5:      for i ← 1 to n do
6:          Q ← Q \ {qi }
7:          S ← Q
8:          T ← R
9:          N ← { v εV | {qi, v} ε E}
10:         LIST-CLIQUE(S ∩ N, T ∩ N, P U {qi})
11:         R← R U { qi}
12:     end for
13: end if
```

Wen et al. (2017) propose a Multi Objective Evolutionary Algorithm (MOEA) based on maximal cliques to detect the overlapping community. In comparison with existing systems, the MOEA has low computational cost and high partition accuracy for detecting the structure of overlapping communities. Both the efficiency and effectiveness of MOEA was validated using real-world and synthetic networks.

3.2 *Parallel methods for solving clique problems*

In a large graph to enumerate all the maximal bi-cliques, Mukherjee and Tirthapura (2017) proposed a technique using MapReduce. The enumeration is parallelized by using the basic clustering framework. This is followed by optimizations one each for load balancing improvement and redundant work reduction. The algorithm was found be effective in handling large graphs in the experimental results. This is the first successful work to list all the bi-cliques using a parallel implementation in large graphs. The previous works were on smaller graphs which used mainly sequential methods for the enumeration of bi-cliques.

Xu et al. (2016) first proposed a distributed algorithm for the computation of cliques which are maximal. As the vertices in a graph can have high degrees the problem of skewed workloads can arise. This problem has been efficiently addressed here. Thus, in the case of real-world graphs, the time complexity in the worst case for computing the cliques which are maximal has been reduced effectively. Next, to analyze efficiently the maximal cliques set, they proposed some fundamental algorithms to process the fundamental query operations. In a case were the graph is being updated, algorithms were devised to efficiently update the maximal cliques set.

The efficiency of the algorithms was verified using real-world graphs under different domains for finding, querying, and updating the maximal cliques set.

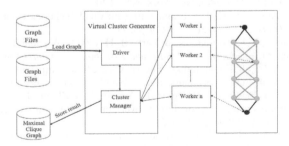

Figure 2.   System design.

4 PROPOSED METHOD

Here we would like to implement a new parallel version of the already existing **BK** algorithm, which will be efficient in maximal clique enumeration. Here we take each node in the graph and explore its neighbors in a parallel manner, i.e. each node will be explored by an independent worker node thus reducing the overall computation time. Here we have a cluster manager who divides the work between the worker nodes. The worker nodes, which are many in number, work simultaneously to produce a faster result.

In the proposed system, the graph is stored in a file as an edge list, i.e., each edge will be represented individually by a set of the two vertices it connects. The graph will be processed using branch and bound criteria and a tree will be created. Each worker node will traverse the tree in a bfs manner to find the cliques in it. Later, the result from all the worker nodes will be combined to form the list of all maximal cliques.

Here we can implement this by using any parallel framework such as graph X in Apache Spark, or by using CUDA. Graph X is built on top of Apache Spark as an embedded framework and is used for processing graphs. It's a distributed dataflow system which is widely used. The sufficient graph abstraction that is required to express existing graph APIs is being provided by graph X (Gonzalez et al., 2014).

CUDA is the parallel computing architecture of **NVIDIA** that by harnessing the power of the graphics processing unit enables the users to have an increased computing performance. This can also be used to implement an efficient parallel version of **BK** algorithm.

5 CONCLUSIONS

Many real-world problems can be modeled using graphs. And, maximal clique enumeration problem has a wide range of applications in the fields of drug discovery and its analysis, social hierarchy detection and many more. The serial computation of maximal cliques is a time-consuming task when the sizes of graphs are large and we know that the real-world graphs are mostly large in size. Thus, here we propose a parallel algorithm for the already existing **BK** algorithm for the processing of large graphs. The **BK** algorithm is one of the most widely accepted graph algorithms for listing all the maximal cliques present in a graph.

REFERENCES

Bron, C. & Kerbosch, J. (1973). Algorithm 457: finding all cliques of an undirected graph. *Communications of the ACM, 16*(9), 575–577.

Gonzalez, J.E., Xin, R.S., Dave, A., Crankshaw, D., Franklin, M.J. & Stoica, I. (2014). Graphx: Graph processing in a distributed dataflow framework. In *OSDI, Volume 14*, (pp. 599–613).

Harary, F. & Ross, I.C. (1957). A procedure for clique detection using the group matrix. *Sociometry 20*(3), 205–215.

Jayaraj, P., Rahamathulla, K. & Gopakumar, G. (2016). A GPU based maximum common subgraph algorithm for drug discovery applications. In *IEEE International Parallel and Distributed Processing Symposium Workshops* (pp. 580–588). New York, NY: IEEE.

Mukherjee, A. & Tirthapura, S. (2017). Enumerating maximal bicliques from a large graph using MapReduce. *IEEE Transactions on Services Computing.*

Schmidt, M.C., Samatova, N.F. Thomas, K. & Park, B.-H. (2009). A scalable, parallel algorithm for maximal clique enumeration. *Journal of Parallel and Distributed Computing, 69*(4), 417–428.

Tomita, E., Tanaka, A. & Takahashi, H. (2006). The worst-case time complexity for generating all maximal cliques and computational experiments. *Theoretical Computer Science, 363*(1), 28–42.

Wen, X., Chen, W.-N., Lin, Y., Gu, T., Zhang, H. Li, Y., Yin, Y. & Zhang, J. (2017). A maximal clique based multi objective evolutionary algorithm for overlapping community detection. *IEEE Transactions on Evolutionary Computation, 21*(3), 363–377.

Xu, Y., Cheng, J. & Fu, A.W.-C. (2016). Distributed maximal clique computation and management. *IEEE Transactions on Services Computing, 9*(1), 110–122.

Emerging Trends in Engineering, Science and Technology for Society,
Energy and Environment – Vanchipura & Jiji (Eds)
© 2018 Taylor & Francis Group, London, ISBN 978-0-8153-5760-5

Retrieving and ranking similar questions and data-driven answer selection in community question answering systems

V. Dhrisya & K.S. Vipin Kumar

Department of Computer Science and Engineering, Government Engineering College, Thrissur, Kerala, India

ABSTRACT: Different techniques have been utilized to study how to find similar questions from recorded archives. These similar questions will have multiple answers associated with them so that end-users have to carefully browse to find a relevant one. To tackle both problems, a novel method of retrieving and ranking similar questions, combined with a data-driven approach of selecting answers in Community Question Answering (CQA) systems, is proposed. The presented approach for similar question retrieval combines a regression procedure that maps topics determined from questions to those found from question-answer pairs. Applying this can avoid issues due to distinctions in vocabulary used within question-answer sets and the inclination of queries to be shorter than their answers. To alleviate answer-ranking problems, a scheme via pairwise comparisons is presented. In the offline learning component, the scheme sets up positive, negative, and neutral training samples represented as preference pairs by means of data-driven perceptions. The model incorporates these three sorts of training samples together. At that point, utilizing the offline prepared model, the answer candidates are sorted to judge their order of preference.

1 INTRODUCTION

One of the fastest developing customer-generated content portals, the Community Question Answering (CQA) system has emerged as a huge market that satisfies complex information needs. CQA provides a platform for customers to ask questions on any topic and also answer others as they wish. It also enables a search through the recorded past Question-Answer (QA) set. Conventional factual QA can be answered by simply retrieving named entities or content from available documents, while CQA extends its significance to answer complicated questions such as reasoning, open-ended, and advice-seeking questions. CQA places few restrictions, if any, on who can post and who can answer a query and is thus quite open. Both the general CQA sites such as *Yahoo! Answers* and *Quora*, and the specialized ones like *Stack Overflow* and *HealthTap*, have had a significant influence on society in the last decade.

Even though there is active user participation, certain phenomena result in question deprivation in CQA portals. For example, users have to wait a long time before getting responses to their queries and a considerable number of questions never get any answer, and the askers are left unsatisfied. The situation is probably caused by the following: (1) the posted queries may be ambiguous, ineffectively stated or may not invoke curiosity; (2) the CQA systems may not effectively direct recent questions to the appropriate answerers; (3) the potential answerers, having the required knowledge, may not be available or are overwhelmed by the number of incoming questions. This third situation often arises in specialized CQA portals, where answering is restricted to authorized specialists only. With reference to the first case, question quality modeling can check the question quality and can assist in requesting that askers restructure their queries. For the other two cases, the situation can be addressed by means of question routing. Question routing is performed by expertise matching and consideration of the likelihood of potential answerers. It works by exploring the human resources currently associated with the system. Besides that, solved past queries can be reused to answer newly presented ones. As time goes on, an enormous set of QA pairs are recorded in databases. New questions have an increased chance of directly obtaining a relevant answer by browsing from the archives, in lieu of waiting for a while to get a fresh response. This proposal has attracted the attention of many researchers who work to make the automatic extraction of recorded, relevant information from the CQA database possible.

Taking into account the QA structure and its meaning, questions or answers with similar meaning may not be lexically similar, and this is defined as a 'lexical chasm'. Consider an example: the questions such as "Where can I learn python on the internet for free?" and "Are there any sites for learning python?" mean the same thing, that is, they are semantically similar but lexically different. The reverse case can also arise where questions have words in common but imply varied meanings. Beyond the requirement for precisely determining a question's meaning, the solution should also handle spell checking, uncertainty avoidance, and short questions. Similar questions can be obtained by correlating the newly presented question to the meaning of recorded questions, because many previous works show that it is inefficient to find similar questions based only on their answers. Recently, this problem has been related to the topic of modeling because it scales down the dimensionality of textual information by comparison with conventional techniques such as 'bag-of-words' (BOW), and it handles uncertainty and similarity comprehensively. The contribution of the present work as a whole can be said to be twofold: first, topics of the questions and answers in the database are modeled by applying Latent Dirichlet Allocation (LDA), and second, the QA topic distribution of a new question is estimated using a regression step.

2 LITERATURE REVIEW

State-of-the-art techniques such as BOW, with its specific properties, term frequency–inverse document frequency (tf-idf) and BM25 (Robertson & Walker, 1997), can calculate lexical similarity between two documents, but they do not consider their semantic and contextual information. In the past decade, topic modeling has been used as an important technique in the field of text analysis. The topics that characterize documents can be treated as its semantic representation. Therefore, to find the semantic similarity between documents, we can use topic distributions obtained using LDA. Further, various approaches for applying topic modeling to historical QA have been proposed. For finding similar questions to the problem, topic modeling along with topic distribution regression (Chahuara et al., 2016) is used.

Consider a corpora C of size L with C consisting of many question-answer pairs:

$$C = \{(q_1, a_1), (q_2, a_2),....,(q_L, a_L)\}$$

where $Q = \{q_1, q_2,..., q_L\}$ and $A = \{a_1, a_2,..., a_L\} \ \forall \ (q_i, a_i) \in C: q_i \in Q, a_i \in A$ are, respectively, question and answer sets.

Answers compared with questions in such portals are likely to be longer and the questions may have only limited relevant words. This can restrict a model's capacity to detect hidden trends. A way of overcoming this is by inferring each question q_i such that it has its text along with a title and description. Moreover, every q_i may be associated with multiple answers and these answers are concatenated to obtain each term a_i. This is done so as to best determine the question's relevance based on the contextual details provided by them. Figure 1 illustrates the proposed framework. The job of retrieving similar questions can be optimized to the task of ranking the QA pairs contained in the created set C, assigning a similarity to question q, and generating a result having its top-ranked element as that with the highest similarity found.

In the learning phase of the task of extracting similar questions, the set C already created is used in training two topic models: first, LDA on the question set Q; second, LDA on the question-answer pair set QA. The learning phase provides topic distributions associated with the sets Q and QA, θ_i^Q and θ_i^{QA} as the result. Using these topic distribution samples, a regression

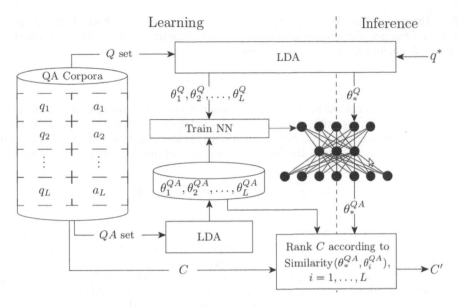

Figure 1. An analysis of the system's learning and deduction.

model is trained to learn the translation function between the Q and QA distributions. During deduction, using the Q set LDA model, we determine the topic distribution of a question (θ_*^Q), which is mapped to its probable QA topic distribution (θ_*^{QA}) using the trained regression model. Last, according to the similarity between each pair's topic distribution and the new question's QA topic distribution, a similarity value is calculated and the value is used to rank the QA corpora. The questions, respective to the QA distribution, that were found similar to the presented question can be considered as the output of the primary batch of processing.

The answer-selection problem in CQA is analogous to the conventional ranking task, where the given question and its set of answers are comparable to a query and a set of relevant entities. The objective is thus optimized to find an ideal ranking system of the answer candidates according to their pertinence, exactness and quality with respect to the given query. A ranking function which uses relevance intuition can be designed in the following three ways: (1) pointwise – in this type of method (Dalip et al., 2013; Shah & Pomerantz, 2010) the relevance measure of each individual QA pair is estimated by a standard classification or regression model; (2) pairwise – in these kinds of methods (Bian et al., 2008; Hieber & Riezler, 2011; Cao et al., 2006; Li et al., 2015), the preference of two answer candidates is predicted using a 0–1 classifier; (3) listwise – in this type, the integrated ranking of all candidate answers to the same question is performed at the same time (Xu & Li, 2007).

The answer selection in the problem is addressed using the novel PLANE model proposed by Nie et al. (2017). Given a question, it requires a set of top k relevant questions $Q = q_1, \ldots, q_k$ from the QA repositories and, according to Nie et al. (2017), this is done using a question-matching algorithm k-NN. According to this, question q_i is assumed to have a set of $m_i \geq 1$ answers, represented by $A_i = a^0, a^1, \ldots, a^{mi}$ whereby a_i^0 is the answer of q_i selected as best by community users. From the identified relevant questions, a learning-to-rank design is developed to sort all the answers associated with them. Two training sets X and U are built from the set of QA pairs. x^1 and x^2 denote the N-dimensional feature vectors of the two QA pairs that are compared in a single comparison. Also, y is denoted as the preference relationship of x, whose value is found as below:

$$y = +1, \text{ if } x^1 > x^2$$
$$y = -1, \text{ if } x^2 > x^1$$

With these variables, $X = \{(x_i, y_i)\}^N_{i=1}$ with preferable labels are built. It can be inferred that this preference relationship does not work with the unimportant answers of a particular question. To include those answers, it is essential to have:

$$(q_i, a^j_i) \cong (q_i, a^k_i)$$

where $\cong$ represents a neutral preference relationship between the two QA pairs under consideration. $u^{(1)}$ and $u^{(2)}$ represent, respectively, the N-dimensional feature vectors of the two QA pairs. Taking into account all pairs of comparisons in this pattern, $U = \{(u_j, 0)\}^M_{j=1}$ is created. Jointly incorporating X and U, the following pairwise learning-to-rank model is proposed:

$$\min_w \sum_{i=1}^{N} [1 - y_i w^T x_i] + \lambda \|w\|_1 + \mu \sum_{j=1}^{M} |w^T u_j|,$$

where $x_i = x_i^1 - x_i^2 \in R^N$ and $u_j = u_j^1 u_j^2 \in R^D$ denote the two training instances from X and U, respectively; symbols N and M denote, respectively, the number of preference pairs in X and U, and the desired coefficient vector is represented by $w \in R^N$. The first term in the equation indicates a hinge loss function, which helps in the binary preference judgment job, and it gives a relatively rigid and convex upper limit on the binary indicator function. The conventional formulation for the Support Vector Machine (SVM) can be considered equivalent with empirical risk minimization of this loss. Pertaining to the support vectors, points lying outside the margin boundaries that are properly classified will not be penalized, while points on the wrong side of the hyperplane or within the margin boundaries will be penalized in a linear mode, proportional to their distance from the proper boundary. The second term represents a *l1* norm. It helps in feature selection and in regularizing the coefficient value's summation that helps in penalizing the preference distance between unimportant answers of the same question.

3 PROPOSED FRAMEWORK

The system tries to solve the problem of finding similar questions and selecting an answer from them using the proven efficient methods of the relevant area. For the purpose of training, we require a large amount of data. The historical archive data from online general CQA websites such as *Quora* and *Yahoo! Answers*, and specialized ones such as *Stack Overflow* and *HealthTap*, are suitable as sources of data for the system. Retrieving this data and processing it into the form required comprises the first phase of the framework.

The data is then fed to the topic modeling phase using the LDA method. LDA works well for topic modeling with data from a variety of topics. As the data is from the CQA, where questions from different fields appear, the LDA should be efficient. After finding the translation function between questions and question-answer pairs, as described in Section 2, using the regression model, the topic model of the expected answer of the new question is determined. Those question-answer pairs that are similar to this found topic are considered to be "similar questions".

Now, using these question-answer pairs, the PLANE model is trained in the form of positive, negative and neutral preference pairs as described in Section 2. The PLANE model is trained offline using the constructed pairs. On providing the new input question, it returns the ranked answers. The best answer will be the one that solves the similar question and that was chosen by users. Finally, the system is evaluated for its performance by comparison with systems with other techniques. Accuracy, precision and recall are the metrics that can measure the performance of such learning systems. Practically, user feedback can be collected from user satisfaction with the related answer shown. On new questions arriving at the site under the same topic, a trace-back mechanism can increase the efficiency of prediction.

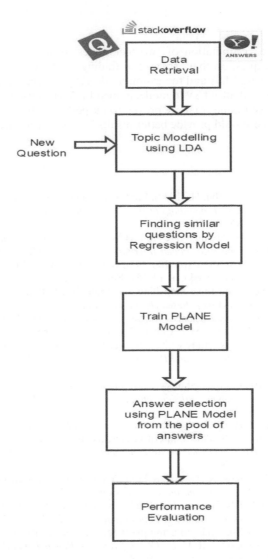

Figure 2. Proposed framework flowchart.

4 EVALUATION OF PERFORMANCE

The data required for the system is from the online community question-answering websites. As it tries to solve a real-time problem of finding a suitable answer at the time a new question arrives, the efficiency of the system can be measured primarily from user feedback. The objectives for which the model was designed can be evaluated to check whether they were met:

- to return the best existing relevant information from the historical archive data;
- to reduce the waiting time to get answers to the question.

Because the two phases in the system use novel techniques that are said to outperform other methods for the same problem, the combination of both should give better efficiency.

5 CONCLUSION

This paper proposes a combination of two techniques to address answer retrieval from archive data for a new question that arrives. For finding similar questions from the existing data, topic modeling using LDA and regression models is preferred. In the successive phase, for ranking answers, an enhanced SVM model named PLANE is used. As mentioned above, the system uses novel techniques that are proven to outperform other methods, and the proposed system as a whole should provide better results in terms of accuracy.

REFERENCES

Bian, J., Liu, Y., Agichtein, E. & Zha, H. (2008). Finding the right facts in the crowd: Factoid question answering over social media. In *Proceedings of the 17th International Conference on World Wide Web* (pp. 467–476). New York, NY: ACM.

Cao, Y., Xu, J., Liu, T.Y., Li, H., Huang, Y. & Hon, H.W. (2006). Adapting ranking SVM to document retrieval. In *Proceedings of the 29th Annual International ACM SIGIR Conference on Research and Development in Information Retrieval* (pp. 186–193). New York, NY: ACM.

Chahuara, P., Lampert, T. & Gancarski, P. (2016). Retrieving and ranking similar questions from question-answer archives using topic modelling and topic distribution regression. In N. Fuhr, L. Kovács, T. Risse & W. Nejdl (Eds.), *Research and advanced technology for digital libraries*. TPDL 2016. Lecture Notes in Computer Science (Vol. 9819, pp. 41–53). Cham, Switzerland: Springer.

Dalip, D.H., Goncalves, M.A., Cristo, M. & Calado, P. (2013). Exploiting user feedback to learn to rank answers in Q&A forums: A case study with stack overflow. In *Proceedings of the 36th International ACM SIGIR Conference on Research and Development in Information Retrieval* (pp. 543–552). New York, NY: ACM.

Hieber, F. & Riezler, S. (2011). Improved answer ranking in social question-answering portals. In *Proceedings of the 3rd International Workshop on Search and Mining User-Generated Contents* (pp. 19–26). New York, NY: ACM.

Li, X., Cong, G., Li, X.L., Pham, T.A.N. & Krishnaswamy, S. (2015). Rank-geoFM: A ranking based geographical factorization method for point of interest recommendation. In *Proceedings of the 38th International ACM SIGIR Conference on Research and Development in Information Retrieval* (pp. 433–442). New York, NY: ACM.

Nie, L., Wei, X., Zhang, D., Wang, X., Gao, Z. & Yang, Y. (2017). Data-driven answer selection in community QA systems. *IEEE Transactions on Knowledge and Data Engineering, 29*(6), 1186–1198.

Robertson, S.E. & Walker, S. (1997). Some simple effective approximations to the 2-Poisson model for probabilistic weighted retrieval. *Readings in Information Retrieval, 345*, 232–241.

Shah, C. & Pomerantz, J. (2010). Evaluating and predicting answer quality in community QA. In *Proceedings of the 33rd International ACM SIGIR Conference on Research and Development in Information Retrieval* (pp. 411–418). New York, NY: ACM.

Xu, J. & Li, H. (2007). AdaRank: A boosting algorithm for information retrieval. In *Proceedings of the 30th Annual International ACM SIGIR Conference on Research and Development in Information Retrieval* (pp. 391–398). New York, NY: ACM.

Emerging Trends in Engineering, Science and Technology for Society,
Energy and Environment – Vanchipura & Jiji (Eds)
© 2018 Taylor & Francis Group, London, ISBN 978-0-8153-5760-5

Convolutional Neural Network (CNN) framework proposed for Malayalam handwritten character recognition system using AlexNet

J. Manjusha
Department of Computer Science and Engineering, Government Engineering College, Thrissur, India
APJ Abdul Kalam Technological University, Kerala, India

A. James
Department of Computer Science and Engineering, Government Engineering College, Thrissur, India

Saravanan Chandran
National Institute of Technology, Durgapur, West Bengal, India

ABSTRACT: Handwritten character recognition has wide application in today's scenarios such as bank cheque k processing and tax returns. The phases in character recognition are image scanning, noise removal, segmentation, extracting features, and classification. Different methods are used for recognizing characters in different languages. Most of them use handcrafted feature extraction methods which are time-consuming. Here we use automatic feature extraction using a Convolutional Neural Network (CNN) which take less time for training and testing of data and produces more error-free results compared to other methods. Very few works on character recognition using CNN have been reported to have a high accuracy of prediction. The model used here is AlexNet architecture which considerably reduces time and errors in predicting the output. Compound Malayalam character recognition is also focused on in this work using CNN.

1 INTRODUCTION

Optical Character Recognition (OCR) is one of the challenging areas of computer vision and pattern recognition. It is the process of converting text or handwritten documents into a scanned and digitized form for easy recognition. It is categorized as offline and online based on the method by which image acquisition is carried out. Malayalam handwritten character recognition is very important because of its use by a large population of people but the variability in writing style and representation make it difficult for recognition. Some of the applications of OCR is data entry for business applications like cheques processing and passport verification. Handwritten character recognition includes like image acquisition, preprocessing, segmentation of characters, feature extraction, and recognition. Compared to all other stages, feature extraction plays an important role in determining the accuracy of recognition. Some of the challenges faced by handwritten Malayalam character recognition systems includes the un-availability of a standard data set and the unlimited variations in human handwriting. The traditional methods usually require artificial feature design and manually tuning of the classifier. Particularly, the performance of the traditional method can be determined to a large extent by using empirical features. The traditional method has reached its limit through decades of research while the emergence of deep learning provides a new way to break this limit. In this paper, a CNN-based (Convolutional Neural Network-based) handwritten character recognition framework is proposed. In this framework, proper sample generation, training scheme and CNN network structure are employed according to the properties of handwritten characters. Here we discuss handwritten character recognition using a CNN model called AlexNet.

2 RELATED WORKS

In the area of Malayalam character classification, previous works have mainly focused on traditional feature extraction methods, which take the most time since it contains the larger number of character classes. Different proposed methods having high recognition rates were reported for the handwritten recognition of Chinese, Japanese, Tamil, Bangla, Devanagari and Telugu using CNN.

El-Sawy (2017) proposed handwritten Arabic character recognition using CNN. The CNN model was trained and tested using a database of 16,800 handwritten Arabic character images with an average misclassification error of 5.1% for the test data.

Tsai's (2016) model is a Deep Convolutional Neural Network (D-CNN) for recognizing handwritten Japanese characters. A VGG-16 network with 11 different convolutional neural network architectures were explored. The general architecture consists of a relatively small convolutional layer followed by an activation layer and a max pooling layer with a final FC (Fully Connected) layer having the same number of channels as the number of classes. It achieved an accuracy rate of 99.53% for overall classification.

Roy et al. (2017) introduced a layer-wise deep learning approach for isolated Bangla handwritten compound characters. Supervised layer-wise trained D-CNNs are found to outperform standard shallow learning models such as Support Vector Machines (SVM) as well as regular D-CNNs of similar architecture by achieving an error rate of 9.67%, and thereby setting a new benchmark on the CMATERdb 3.1.3.3 with a recognition accuracy of 90.33%, representing an improvement of nearly 10%.

For Chinese (Xuefeng Xiao et al., 2017) to incur the high computational cost of deeper networks a Global Supervised Low-Rank Expansion (GSLRE) method and an Adaptive Drop-Weight (ADW) technique was used. For HCCR with 3,755 classes a CNN network with nine layers was adopted, that can reduce the networks computational cost by nine times and compress the network to 1/18 of the original size of the baseline model, with only a 0.21% drop in accuracy.

Md. Mahbubar Rahman et al. (2015) considered a CNN-based Bangla handwritten character recognition system. The normalized data was used and employed CNN to classify isolated character recognition. 20,000 handwritten characters with different shapes and variations were used in this study. The proposed BHCR-CNN misclassified 351 cases out of 2,500 test cases and achieved an accuracy of 85.96%. On the other hand, the method misclassified 954 characters out of 17,500 training characters giving an accuracy rate of 94.55%.

An integrated two classifier – CNN and SVM – model for Arabic character recognition was introduced by Mohamed Elleuch and Kherallah (2016) with a dropout technique thereby reducing the over fitting. The performance of the model was compared with the character recognition accuracies gained from state-of-the-art Arabic Optical Character Recognition. The error rate without dropout layer was recorded as 14.71% and a considerable reduction error rate of 5.83% using dropout was reported.

An unsupervised CNN model for feature extraction and classification of multi script recognition was proposed by Durjoy Sen Maitra and Parui (2015). For a larger character class problem, they performed a certain amount of training for a five-layer CNN. SVM was used as the classifier for six different character databases all of which have achieved an error rate less than 5%.

Another approach for handwritten digit recognition using CNN and SVM for feature extraction and recognition respectively (Chunpeng Wu et al., 2014) achieved a recognition rate of 99.81% without rejection, and a recognition rate of 94.40% with 5.60% rejection.

Jinfeng Bai et al. (2014) proposed a Shared Hidden Layer deep Convolutional Neural Network (SHL-CNN), which recognizes both English and Chinese image characters, produced a reduced recognition error of 16–30%, compared with models trained by characters of only one language using conventional CNN, and by 35.7% compared with state-of-the-art methods.

Zhong et al., (2015) presented a deeper CNN architecture for handwritten Chinese character recognition (denoted as HCCR-GoogLeNet) which uses 19 layers in total.

The work also uses certain directional features like HoG, Gabor filters and gradient features to improve the performance of HCCR-GoogLeNet. It involves 7.26 million parameters, and has shown that with the proper incorporation with traditional directional feature maps, the proposed model achieved new state-of-the-art recognition accuracy of 96.35% and 96.74%, respectively, outperforming the previous best result by a significant gap.

An offline English word recognition method using neural networks has been proposed by Gupta et al. (2011). It divides the word recognition process into two classes, segmentation based and holistic based. A second approach is used in identification of fixed size vocabulary where global features extracted from the entire word image are considered. The segmentation based (Bhanushali et al., 2013) strategies, on the other hand, employ bottom-up approaches, starting from the stroke or the character level and going toward producing a meaningful word.

Another framework for handwritten English character recognition presented by Yuan et al. (2012) use a modified LeNet-5 CNN model. It uses a different architecture of neurons for each layer. An error-samples-based reinforcement learning strategy is developed for training of the CNN. UNIPEN lowercase and uppercase datasets were used for evaluating the experiments, with recognition rates of 93.7% for uppercase and 90.2% for lowercase letters.

The majority of the published neural networks employed only fully connected layers, while most of the efforts were focused on feature extraction.

3 PROPOSED METHOD

Handwritten Malayalam character recognition faces challenges such as similarity in structure of characters, unlimited variations in human handwriting, un avail ability of standard datasets and much more. Even though many handcrafted feature extraction methods were used for Malayalam character recognition none has achieved 100% accuracy. Here we are proposing an automated feature extraction technique using CNN that efficiently extracts features from character images and produces higher recognition rates. This work also focuses on compound Malayalam character recognition using CNN. The overall flow of data is shown in Figure 2.
The proposed method has the following stages:

3.1 *Data collection*

Malayalam character recognition has been a recent topic of research and it poses the limitation of unavailability of standard datasets for classification. Our initial work is to collect the required amount of handwritten character data from different people. Collecting data from different people will help the system to learn variations and to predict exactly while testing. A larger dataset results in more accurate prediction of the test images.

3.2 *Data augmentation*

Unavailability of standard datasets and limitations for collecting a large amount of handwritten dataset causes problems when using CNN since it needs a huge amount of data for training and testing. To transform and modify the available data is one method to increase the size of dataset. Here we use affine transformation which is a linear transformation method to augment the data. To preserve points, straight lines, and planes, a linear mapping method is used called affine transformation. Translation, scaling, sheering and rotation are the four major affine transformations.

3.3 *Preprocessing*

Performance of any character recognition system is directly dependent upon the quality of the input documents. To remove the noise from the character images a data preprocessing method is used. Salt and pepper noise are the most common noise elements present in an image and these can be removed by some contrast enhancement processes. The input image also requires some resizing. The image size should fit the network and also be recognizable.

3.4 *Model training and optimization*

The model used for training the data is the AlexNet model which is a basic framework of CNN. The AlexNet architecture (shown in Figure 1) consists of eight learned layers—five convolutional and three fully connected (Krizhevsky et al., 2012). Some of the most important layers are ReLU that uses max(0,x) function which train several times faster than their equivalents with tanh units. Also, they do not require input normalization to prevent them from saturating. The next layer is max-pool layers, which summarize the outputs of neighboring groups of neurons in the same kernel map. To reduce over fitting, we use dropout layers. To reduce the model training error, we use a weight decay mechanism. To improve the training and testing accuracy after each process, the parameters set has to be optimized with different values to check the point which produces the maximum accurate predictions. Some of the parameters used here are batch size, momentum, weight decay, learning rate, iteration index and epoch.

3.5 *Testing*

Testing module deals with the test images. Test images are obtained by splitting the augmented dataset randomly. It will first preprocess the input image and it will classify the unlabeled test

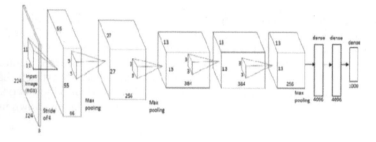

Figure 1.   AlexNet architecture.

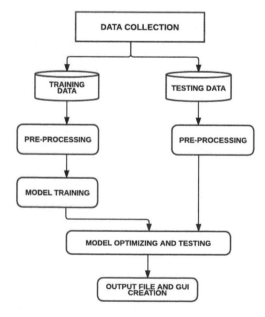

Figure 2.   Sketch of flow configuration.

data. Test data is not labeled in the sense it should be recognized by the machine. Labels are assigned to each of the test images by the network and then the accuracy is measured. In the post processing stage of the proposed system, the classifier output is mapped to the character Unicode. The output of the classifier will be some integer labels. This integer label should be converted into the corresponding character Unicode. The Unicode is written in a text file.

3.6 Classification

The final layer of the CNN is a softmax layer (Pranav P Nair et al., 2017) and this softmax layer is used for classifying the given input image. This softmax layer is used to classify the character. The softmax function has a value between 0 and 1. The sum of the output of all the classes is also 1. The class with the maximum value will be selected as the class for a particular input image.

4 CONCLUSIONS AND FUTURE WORK

Character recognition has a wide range of applications in postal automation, tax returns, bank cheque processing and many more. Even though works on Malayalam character recognition have been reported none has achieved 100% accuracy and most of them take a large processing time.

Our proposed method uses automatic feature extraction using CNN that considerably reduces the training and testing time and produces a more accurate result. Here compound Malayalam character recognition is also considered, which makes the system more useful for real time application processing. The model needs a system with Core i7 PC of 2.6 GHz with 64GB memory with CUDA enabled GPU for parallel processing.

REFERENCES

Bhanushali, S., Tadse, V. & A. Badhe, A. (2013). Offline handwritten character recognition using neural network. In *Proceedings of National Conference on New Horizons in IT-NCNHIT* (pp. 155).

Chunpeng Wu, Wei Fan, Y.H.J.S.S.N. (2014). Handwritten character recognition by alternately trained relaxation convolutional neural network. In *14th International Conference on Frontiers in Handwriting Recognition* (pp. 291–296).

Durjoy Sen Maitra, U.B. & Parui, S.K. (2015). CNN based common approach to handwritten character recognition of multiple scripts. *13th International Conference on Document Analysis and Recognition (ICDAR)* (pp.1021–1025).

El-Sawy, A. (2017). Arabic handwritten characters recognition using convolutional neural network. *WSEAS Transactions on Computer Research 5*, 2415–1521.

Gupta, A., Srivastava, M. & Mahanta, C. (2011). Offline handwritten character recognition using neural network. In *IEEE International Conference on Computer Applications and Industrial Electronics (ICCAIE)* (pp. 102–107). New York, NY: IEEE.

Jinfeng Bai, Zhineng Chen, B.F.B.X. (2014). Image character recognition using deep convolutional neural network learned from different languages. In *IEEE International Conference on Image Processing (ICIP)* (pp. 2560–2564).

Krizhevsky, A., Sutskever, I. & Hinton, G.E. (2012). Imagenet classification with deep convolutional neural networks. In Pereira, F., Burges, C., Bottou, L. & Weinberger, K. (Eds). *Advances in Neural Information Processing Systems, 25*, pp. 1097–1105. Curran Associates, Inc.

Md. Mahbubar Rahman, M.A.H. Akhand, S.I.P.C.S. (2015). Bangla handwritten character recognition using convolutional neural network. *International Journal of Image, Graphics and Signal Processing 8*, 42–49.

Mohamed Elleuch, R.M. & Kherallah, M. (2016). A new design based-SVM of the CNN classifier architecture with dropout for offline Arabic handwritten recognition. In *International Conference on Computational Science (ICCS) 80*, (pp. 1712–1723).

Pranav P Nair, Ajay James, C.S. (2017). Malayalam handwritten character recognition using convolutional neural network. *International Conference on Inventive Communication and Computational Technologies (ICICCT 2017)* (pp. 278–281).

Saikat Roy, Nibaran Das, M.K.M.N. (2017). Handwritten isolated Bangla compound character recognition: A new bench-mark using a novel deep learning approach. *Elsevier Pattern Recognition Letters, 90*, 15–21.

Tsai, C. (2016). Recognizing handwritten Japanese characters using deep convolutional neural networks. http://cs231n.stanford.edu/reports/2016/pdfs/262_Report.pdf.

Xuefeng Xiao, Lianwen Jina, Y.Y.W.Y.J.S.T.C. (2017). Building fast and compact convolutional neural networks for offline handwritten Chinese character recognition. *Elsevier Pattern Recognition 72*, 72–81.

Yuan, A., Bai, G., Jiao, L. & Liu, Y. (2012). Offline handwritten English character recognition based on convolutional neural network. In *10th IAPR International Workshop on Document Analysis Systems (DAS)* (pp. 125–129). New York, NY: IEEE.

Zhong, Z., Jin, L. & Xie, Z. (2015). High performance offline handwritten Chinese character recognition using GoogLeNet and directional feature maps. In *13th International Conference on Document Analysis and Recognition (ICDAR)* (pp. 846–850). IEEE.

Emerging Trends in Engineering, Science and Technology for Society,
Energy and Environment – Vanchipura & Jiji (Eds)
© *2018 Taylor & Francis Group, London, ISBN 978-0-8153-5760-5*

A novel technique for script identification in trilingual optical character recognition

A. Rone Maria & K.J. Helen
Department of Computer Science and Engineering, Government Engineering College,
Thrissur, Kerala, India

ABSTRACT: In multilingual environment searching, editing and storing of documents is made easier by script identification. It also an aid for selecting script specific Optical Character Recognition (OCR) for multilingual documents. India is a multilingual country so documents may contain more than one script. In Kerala, a state in India, the documents may contain text in three languages: Malayalam, the—official language of the state; Hindi, the national language; and English, the global language. For processing such multiscript documents, it is necessary to identify the script before feeding the text line to specific OCRs. This paper presents a novel and efficient technique for script identification in English, Hindi, Malayalam trilingual documents. Features for classification are extracted from horizontal projection of text images. Training and testing are done on our own data set developed from documents containing these three languages.

1 INTRODUCTION

Optical Character Recognition (OCR) is the technology of translating text in an image to machine encoded form, which has gained huge importance in the development of document analysis systems for applications such as digitization, machine translation, and cross-lingual information. The idea of an OCR system is to identify alphabets, numbers, punctuation marks, or special characters, present in digital images, without any human involvement. This is achieved through a matching process between the extracted features of a given character's image and the library of image models. Previously, most of the research work of OCR has dealt with only monolingual documents, but actual documents may contain more than one script. In multilingual imaged document analysis and processing there are many fields where good research has been carried out. Some of the applications to recognize multilingual text are for postal addresses, data entry forms, railway reservation forms, bank cheques, application forms, question papers, government forms, receipts, magazines, and newspapers which contain text written in two or more languages. In India every government office uses at least three languages: English' Hindi, the official language of the nation; and the official language of the corresponding state (Padma et al., 2009). In Kerala, a state in India, the documents may contain text in three languages; Malayalam, the official language of the state; Hindi, the national language; and the English global language. Figure 1 shows some examples of trilingual documents in Kerala. There are two distinct methods to build Multilingual Optical Character Recognitions (MOCRs): i) a single OCR is trained for all the scripts, or ii) train an isolated OCR for each script (Mathew et al., 2016). In the first method, the classifier would need a huge collection of data and the output space is much larger. The training of the classifier becomes time-consuming and computationally expensive with larger output space. But, in the second method, the data required for training would be much lesser and output space would be smaller. Most of the work done in recognizing multilingual scripts has been restricted to using a script identification stage before recognition. This is due to the fact that either the segmentation or the recognition relied on script-dependent features. Other groups

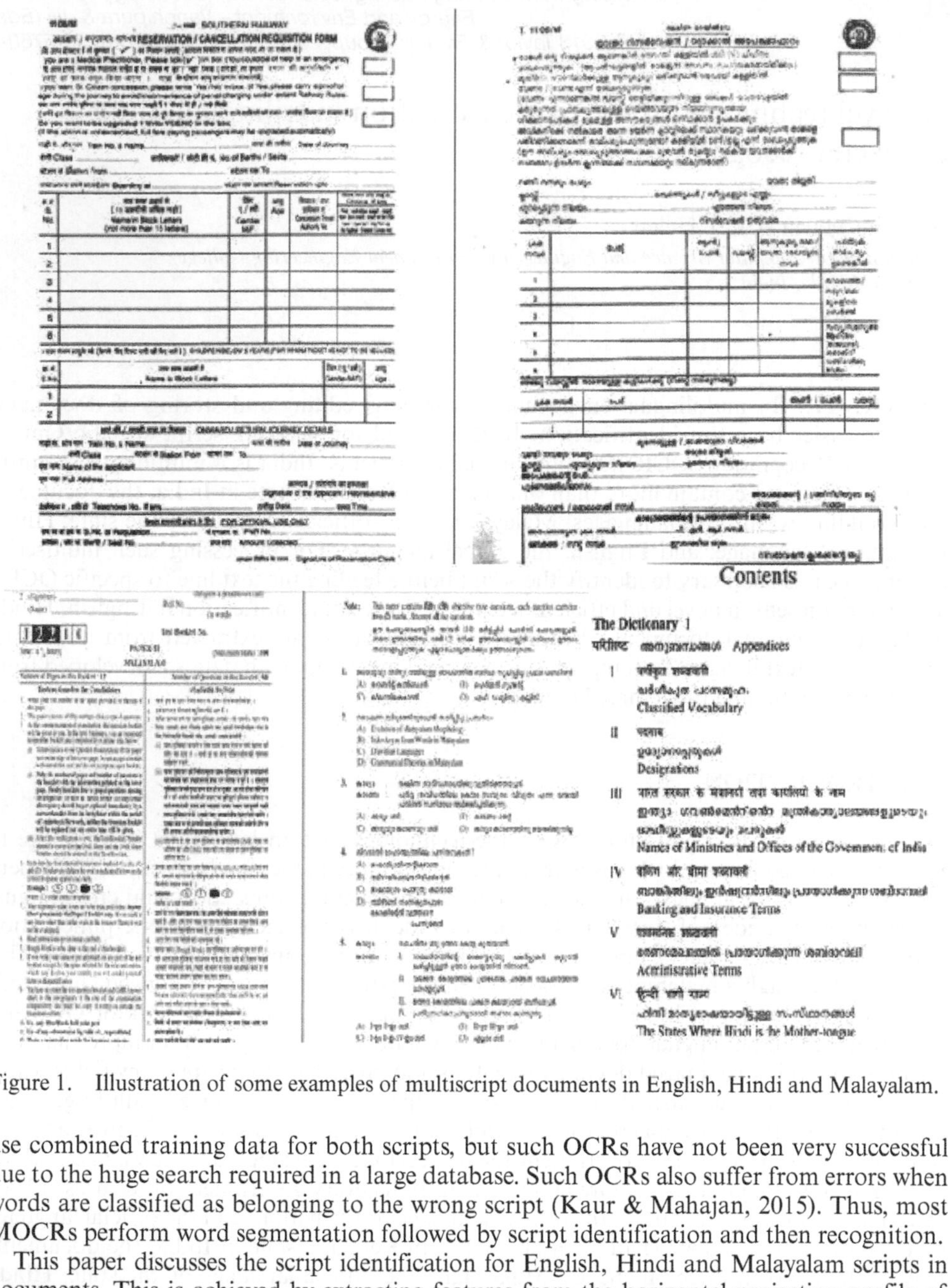

Figure 1. Illustration of some examples of multiscript documents in English, Hindi and Malayalam.

use combined training data for both scripts, but such OCRs have not been very successful due to the huge search required in a large database. Such OCRs also suffer from errors when words are classified as belonging to the wrong script (Kaur & Mahajan, 2015). Thus, most MOCRs perform word segmentation followed by script identification and then recognition.

This paper discusses the script identification for English, Hindi and Malayalam scripts in documents. This is achieved by extracting features from the horizontal projection profile of text lines.

The rest of the paper deals with some of the previous works in section 2 and proposed work in section 3. This paper concludes in section 4.

2 PREVIOUS WORKS

Philip and Samuel (2009) used dominant singular values and Gabor features for classification of printed English and Malayalam script. To identify the script the text is segmented to word level and Gabor features are extracted. Dominant singular values are used for character recognition.

Rahiman et al. (2011) proposed another method for an English-Malayalam bilingual OCR. Here the image is scanned from top left to bottom right and segmented into line level, word and character level. This segmented character is resized to a 16 X 16 bitmap. A database is created for characters in different fonts for both languages. The resized image is compared with characters in the database. A pixel match algorithm is used for this comparison and the matched character is displayed.

Aithal et al. (2010) also proposed a simple and efficient method for script identification from a trilingual document in English, Hindi and Kannada languages. The proposed system uses a horizontal projection profile for script identification.

Mohanty et al. (2009) have developed a bilingual OCR for recognizing English and Oriya texts. Structural features such as the presence of loops, concavities, lines present or absent on the left and right part of characters, upper portion open are selected as features in this work.

A method for identifying languages from documents printed in three languages English, Hindi, Kannada has been proposed by Padma et al. (2009). The proposed approach is based on the characteristic features of the top and bottom profiles of input text lines.

Kaur and Mahajan (2015) have proposed an approach for the identification of English and Punjabi scripts at line level. Character density features and headlines are used for script identification.

Dhandra et al. (2007) proposed a method for script identification at word level. They deal with scripts such as Kannada, Tamil, Devnagari and English numerals using visual appearance. Seven features are extracted in real values and form a vector using these for each training word. K nearest neighbor classifier is used for classifying test words.

3 PROPOSED WORK

This paper presents a novel script identification technique for identifying English, Hindi and Malayalam scripts from multiscript documents. The steps of the proposed system are shown in Figure 2.

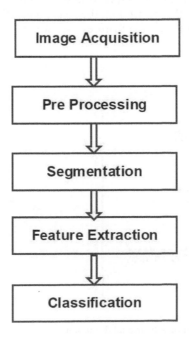

Figure 2. System architecture.

3.1 *Image acquisition*

The input is the image of the trilingual document. Converting the paper document into an image by scanning is the first step. A scanned imaged can be in any format, jpg, jpeg or png. The scanned image is transformed to a gray scale image.

3.2 *Preprocessing*

Image preprocessing is the process of improving image quality for better understanding using predefined methods. Commonly used preprocessing methods are noise removal, binarization, and skew correction. The noise may occur during scanning or transferring of document images. Smoothing operations are used for noise removal. Converting gray scale image from 256 gray levels to two levels is called binarization. The binarization is generally done by taking a threshold value for an image and set intensity values to one for pixels which have larger value than threshold value. Set intensities to zero if it is less than the threshold value. Skew is a deformation that is introduced while scanning a document. It is necessary for aligning the text lines to the coordinate axes.

3.3 *Segmentation*

White gap between text lines are used for segmentation. Horizontal projection of the scanned image is used for line segmentation. Horizontal projection of a trilingual document is shown in Figure 3. The zero value points in the figure represents the spaces between text lines. Line segmentation is performed at these points.

3.4 *Feature extraction*

Feature extraction is an important phase of any character recognition system. The objective of feature extraction is to ensure pattern identification with a minimum number of features. In this proposed system, the features are extracted from the horizontal projection profile. The horizontal projection profile is obtained by counting the number of black pixels in each row of the image (Puri and Singh, 2016). Figure 4 shows feature extraction from the horizontal projection of text lines. From this projection, the peak points P1 and P2 are determined. The values at peak points P1 and P2 are the features used for classifying the English, Hindi and Malayalam scripts.

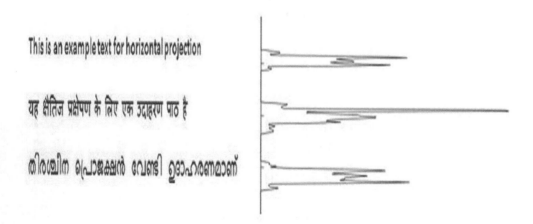

Figure 3.   Horizontal projection of a trilingual document.

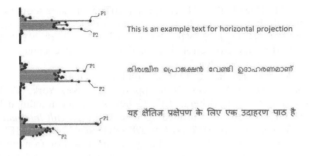

Figure 4. Feature extraction from horizontal projection.

Table 1. Range of P value for different scripts.

Script	Range of P value
English	1.10 to 1.30
Malayalam	0.85 to 0.95
Hindi	2.00 to 2.50

3.5 *Classification*

Classification is based on the features extracted in the previous step. For this proposed system, a rule-based classifier is used.

STEP 1: For each text line determine the first and second peak points P1 and P2 respectively.
STEP 2: Find the values at P1, value (P1), and P2, value (P2).
STEP 3: Calculate the decision parameter Pvalue from these values as

$$P \text{ value} = \frac{\text{value (P1)}}{\text{value (P2)}} \tag{1}$$

STEP 4:
If P value falls in the range of 0.85 to 0.95 then the text is identified as Malayalam.
If P value falls in the range of 1.10 to 1.30 then the text is identified as English.
If P value falls in the range of 2.00 to 2.50 then the text is identified as Hindi.

The range of Pvalues for English, Hindi and Malayalam scripts are shown in Table 1. Based on these observed values, rules are made and fed into the rule-based classifier.

4 CONCLUSIONS

In this paper, a novel technique for script identification in English, Hindi and Malayalam multilingual documents has been proposed. Features for classification are extracted from the horizontal projection of scripts. A rule-based classifier is built from the knowledge obtained from the sample data.

REFERENCES

Aithal, P.K., Rajesh, G., Acharya, D.U. & Subbareddy, N.K.M. (2010). Text line script identification for a tri-lingual document. In *International Conference on Computing Communication and Networking Technologies (ICCCNT)* (pp. 1–3). New York, NY: IEEE.

Dhandra, B., Hangarge, M., Hegadi, R. & Malemath, V. (2007). Word level script identification in bilingual documents through discriminating features. In *International Conference on Signal Processing, Communications and Networking (ICSCN'07)* (pp. 630–635). New York, NY: IEEE.

Kaur, I. & Mahajan, S. (2015). Bilingual script identification of printed text image. *International Journal of Engineering and Technology 2*(3), 768–773.

Mathew, M., Singh, A.K. & C. Jawahar, C. (2016). Multilingual OCR for Indic scripts. In *12th IAPR Workshop on Document Analysis Systems (DAS)* (pp. 186–191). New York, NY: IEEE.

Mohanty, S., Dasbebartta, H.N. & Behera, T.K. (2009). An efficient bilingual optical character recognition (English-Oriya) system for printed documents. In *Seventh International Conference on Advances in Pattern Recognition (ICAPR'09)* (pp. 398–401). New York, NY: IEEE.

Padma, M., Vijaya, P. & Nagabhushan, P. (2009). Language identification from an Indian multilingual document using profile features. In *International Conference on Computer and Automation Engineering (ICCAE'09)*. (pp. 332–335). New York, NY: IEEE.

Philip, B. & Samuel, R.S. (2009). A novel bilingual OCR for printed Malayalam-English text based on Gabor features and dominant singular values. In *International Conference on Digital Image Processing* (pp. 361–365). New York, NY: IEEE.

Puri, S. & Singh, S.P. (2016). Text recognition in bilingual machine printed image documents—challenges and survey: A review on principal and crucial concerns of text extraction in bilingual printed images. In *10th International Conference on Intelligent Systems and Control (ISCO)* (pp. 1–8). New York, NY: IEEE.

Rahiman, M.A., Adheena, C., Anitha, R., Deepa, N., Kumar, G.M. & Rajasree, M. (2011). Bilingual OCR system for printed documents in Malayalam and English. In *3rd International Conference on Electronics Computer Technology (ICECT), 3* (pp. 40–45). New York, NY: IEEE.

Emerging Trends in Engineering, Science and Technology for Society,
Energy and Environment – Vanchipura & Jiji (Eds)
© 2018 Taylor & Francis Group, London, ISBN 978-0-8153-5760-5

An optimal controller placement strategy using exemplar-based clustering approach for software defined networking

A.G. Sreejish & C.N. Sminesh
Department of Computer Science and Engineering, Government Engineering College, Thrissur, Kerala

ABSTRACT: Software Defined Networking (SDN) is a next-generation paradigm that emphasizes the decoupling of networking control plane and the data plane. The decoupling allows managing the network using software programs in a flexible and efficient manner. The feasibility, scalability, and performance of large-scale SDN are confronted with critical challenges, need multiple controllers to impact performance outlook of SDN. The controller placement and load balancing in SDN are critical problems in providing a reliable network with high resource utilization. To resolve these problems, an effective controller placement strategy is needed. This paper focuses on minimizing latency between controllers and switches and maximizing load balance ability of the controllers. In order to obtain the optimal number of controllers and its corresponding locations, a modified affinity propagation algorithm is proposed which takes into consideration the computation complexity of the actual network state. The proposed mechanism provides a solution by considering the trade-offs between latency and load-balance.

1 INTRODUCTION

The emerging technology Software Defined Networking (SDN) emphasizes the decoupling of control plane and the data plane. This separation helps to provide only an abstract view of network resources and its state to the external applications. In large-scale SDN, the control plane is composed of multiple controllers that provide the global view of the entire network. The controllers are programmable, which acts as an intermediary between the network administrator and data plane. In control plane, control intelligence is decoupled. Therefore, The controllers install the rules in the flow table that are used to forward the traffic flows entering the network and switching is done by OpenFlow (OF) switches in the forwarding plane. However, the separation of control-data plane introduces performance limitations and reliability issues (Yeganeh et al. 2013, Jarschel et al. 2012). They are:

a. The nodes in the network must be continuously monitored and controlled by using a proactive or reactive method. The nodes communicate with the corresponding controllers to obtain the new forwarding rules to be installed. Based on these rules, the nodes process various new flows arriving at it. The response time of the overall system will increase when the communication overhead between controllers and switches is high. This is because the controller has a limited processing power (Yeganeh et al. 2013), with respect to the number of nodes assigned to it or the number of flow queries is too high.

b. In large-scale SDN, the density of network elements and traffic flows are really high. If we are using a single physically centralized controller, there is a chance for Single Point of Failure (SPoF). So multiple controllers have to be placed to ensure that SPoF will be eliminated from the network. In SDN, control plane creates the logically centralized view of the entire network. Inorder to create this view, the controllers need to communicate with each other and update/synchronize their databases (Tootoonchian and Ganjali 2010). To reduce the inter-communication among the controllers, we can create an overlay network linking the controllers (Shi-duan 2012).

c. The failure of controller results in disconnection of data plane and control plane easily may lead to packet loss and performance degradation (Yeganeh et al. 2013, Heller et al. 2010). Consider the scenario that some of the nodes are still alive, if they are unable to communicate with the controllers, they may fail to process the newly arrived flows.

The controllers can be placed optimally by finding the solution for (a), (b), and (c), which will be an NP-hard problem (Heller et al. 2010). As a result, various solutions have been suggested focusing on performance, (problem (a), (b)), and performance plus reliability (problem (c)).

In this paper, we analyze some of the solutions and then proposes a controller placement strategy that improves availability of the network by considering both performance and reliability. The controller placement strategy affects all aspects of performance of SDN like the delay between a controller and a node, network availability, and other performance metrics. So the aim is to develop an algorithm that computes the number of controllers needed and the location of controllers for a large-scale network where the placement results in significant improvement of both performance and reliability of the network. To reduce the computation cost of finding the optimal placement of controllers an exemplar-based clustering technique called affinity propagation is used.

In this paper, the remaining sections are organized as follows. Section 2 presents the related works. Section 3 provides an overview of the proposed system and several metrics that are used to formulate the final algorithm. Section 4 comprises of conclusion and future work.

2 RELATED WORKS

This section gives an overview of some related works on controller placement in large-scale SDN. The main problem is to find the number of SDN controllers that are required for a given network topology and where to place them so that it resolutely maintains the performance even in cases of failure

In WANs, if the placement of controller introduces a significant increase in path delays, then it will postpone the time taken for control plane to achieve the steady state. It has been explained in (Heller et al. 2010) which is theoretically not new. If only propagation latency is considered, the issue is akin to a warehouse or facility location problem, which can be solved by the use of Mixed Integer Linear Program (MILP) tools.

The Heller et al. prior work (Heller et al. 2010) encourages to consider the problem of controller placement and measures the impact of controller placement on existing topologies like Internet2 (Yeganeh et al. 2013) and on various cases available in the Internet Topology Zoo (Knight et al. 2011). In reality, the main purpose was not to locate the ideal positions for the controllers but to provide an evaluation of a major design problem which requires further examination. It has been demonstrated that optimal solutions can be discovered for practical network instances in failure-free cases by figuring out the complete solution with offline calculations. This work also affirms that in most topologies the existing response time requirements cannot be satisfied using a single controller, although they didn't consider the resiliency aspects.

(Zhang et al. 2016) address Multi-objective Optimization Controller Placement (MOCP) problem and focus on three objectives; maximize controller load balance capacity, maximize network reliability and minimize control path latency. It provides an optimal controller placement strategy in such a way that the routing requests are optimally distributed among multiple controllers. The work converted the MOCP into a mathematical model as the optimization objective function and developed Adaptive Bacterial Foraging Optimization (ABFO) algorithm to resolve it, claiming those above objectives are optimized efficiently and effectively. But in this method, the optimal number of controllers needed not is identified dynamically. For a large-scale network, this identification becomes exhaustive and will reduce the reliability of the solution.

The work (Borcoci et al. 2015) presented an analytical view on using multi-criteria decision algorithms (MCDA) to choose an optimal solution from several controller placements solutions in large-scale SDN with the help of some weighted variables. The advantage of MCDA is that it can produce an optimal result at the same time as considering numerous criteria. This method claims that it can be applied to different scenarios including reliability-aware or failure-free assumption ones, primarily based on more than one metrics supported by using the reference model MCDA. This work deals with several metrics such as average-case latency or worst-case latency in failure-free situations, worst-case latency when a controller fails, nodes or links failures, inter-controller latency, controller load imbalance and multi-path connectivity metric, but some of these metrics are considered only when the service provider policy demands it. By using this method different network/service providers gain the ability to maintain the quality of service (QoS) provided by them. This is achieved by assigning weights to criteria when selecting the optimal solution. The limitation of this work is that it demands to pre-compute all possible solutions for the given network topology, thereby increasing the computational cost.

In order to reduce the computational complexity, cluster-based approaches can be used. For dividing a large-scale SDN into small network domains, an efficient Spectral Clustering placement algorithm is used (Xiao et al. 2014). With the aid of dividing WAN into various SDN domains, the controller with fewer nodes can gain maximum performance. Also, there will be a sharp reduction in the inter-controller communications. For examining metrics of controller placement problem for SDN, they have considered the propagation latency and controller performance and used Internet2 topology to probe the performance of the system. The number of controllers as input for partitioning the network was the limitation of this work.

In (Zhao et al. 2016), a multi-controller placement problem that considers minimization of the latency between controllers and switches is discussed. A modified version of affinity propagation (AP) is used to resolve the controller placement problem. Affinity propagation is an exemplar-based clustering method. The advantage of this method is that it can adaptively compute the most favorable number and placement of controllers based on the network topology. When comparing latency minimization with k-median and k-center methods, it has size-able simulation results. This implies that this technique can offer more steady and precise outcomes.

But, the propagation latency considered in above work is calculated only by considering the shortest path between nodes. The latency can be affected by both the distance between nodes and the bandwidth of the corresponding link. The objective of the proposed system is to formulate a new model that takes both distance and bandwidth into consideration for computation of similarity values used in affinity propagation.

3 PROPOSED SYSTEM

The main goal of the proposed system is to design an approach that automatically computes the optimal number of controllers that are needed to manage the given SDN network and its corresponding locations.

3.1 *Proposed system design*

This section outlines the overview of the architecture and some useful metrics that can be used to evaluate the proposed system. The ultimate challenge is to maximize the Control Plane performance. Among several metrics considered in (Borcoci et al. 2015), latency and load of the controller are the major factors that affect the overall performance of SDN. For the reason that the controller is responsible for producing forwarding rules, and the mismatched packets are buffered or discarded until the corresponding flow entries are installed.

The minimization of average-case latency can be considered as a k-median problem and minimization of worst-case latency as k-center problem. However, to solve both k-median and k-center problem, two input parameters; the number of controllers which is denoted by k and the set of initial cluster centers which is denoted by C are needed. Another challenge is controllers load balancing. For identifying the maximum possible load a controller can

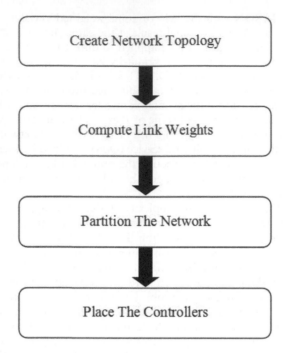

Figure 1.   Overview of proposed system.

process, the bandwidth of links connected to it can be considered. The overall design of the proposed method is given in Figure 1.

The proposed system is advanced in four submodules; network topology creation, link weight calculation, network partition and controller placement. In the first module, network topologies are created using GARR and GEANT (Knight et al. 2011). The bandwidth of the links present in these topologies and latency are considered to derive link weight. The third module partition the network using modified affinity propagation. Finally, the controllers are placed in the locations of exemplars.

For the modified-Affinity Propagation method, there is no need to initialize the number of controllers and its locations. Affinity Propagation (AP) is an exemplar-based clustering approach that has numerous benefits which include performance, no need to initialize the value of k, and ability to obtain exemplars with high accuracy. The AP is modified in such a way that it adapts to the problem of controller placement in SDN. Especially, the similarity measurement between two nodes adopts both latency and bandwidth of the links.

In a real network topology, there may be no direct links between two nodes. But they are reachable via other links present in the network. So the reachable shortest path distance $L(u, v)$ is considered as latency (Zhao et al. 2016), which can be found by using the Floyd-Warshall all-pairs shortest paths algorithm (Cormen et al. 2009). The bandwidth of the links is already available in the internet topologies GARR and GEANT.

The exemplar-based clustering problem is formulated by using the link weight which is assigned to each edge, considering the fact that it minimizes the latency and equalizes the load of the controller. Link weight is composite metric, which is computed using Eq. 1. An assumption is made that the locations of controllers are the same as some of the nodes.

$$Link\,Weight\,(u,\,v) = I(u,\,v)(BW(u,\,v) + L(u,\,v)) \qquad (1)$$

where $I(u, v)$ is used to indicate the presence of a link between u and v. $BW(u, v)$ is the bandwidth of the link present in between u and v and $L(u, v)$ is the latency.

3.2 Performance metrics

Measuring the performance of the proposed technique is vital to evaluating how well it solves the controller placement problem. The metrics that can be used to evaluate the proposed method are average and worst case latency, controller imbalance factor and inter-controller latency which is discussed in the following section.

3.2.1 Average and worst case latency

The network can be represented by using an undirected graph $G(V, E)$ where V is the set of nodes and E is the set of links present in it. n denotes the number of nodes and it is defined as $n = |V|$.

Latency is a measure of how much time it takes for a packet of data to reach from one node to some other node, which can also be computed in terms of distance. Here, the latency is considered as the shortest path distance between two nodes u and v. There are two variations of latencies that can be defined for a particular controller placement; average case latency and worst case latency (Borcoci et al. 2015).

Average Case Latency:

$$L_{avg} = \frac{1}{n} \sum_{v \in V} \min_{c \in C} L(v, c) \tag{2}$$

Worst Case Latency:

$$L_{worst} = \max_{v \in V} \min_{c \in C} L(v, c) \tag{3}$$

Any of above mentioned latencies need to be minimum for the optimization algorithm that should obtain a placement of controller.

3.2.2 Controller imbalance factor

A properly designed system would require more or less same load on all controllers, i.e., consistency of the node to controller distribution. A metric can be introduced to measure the imbalance between controllers for a given controller placement. It is defined as the difference between the most and minimal number of nodes assigned to a controller (Zhao et al. 2016).

$$C_{imbalance} = \max_{c \in C} n_c - \min_{c \in C} n_c \tag{4}$$

where n_c denotes the number of nodes under the controller c. For Eq. 4, need to take consideration that during the case of failures the control of nodes may be moved from primary controller to other controllers. This reassignment can increase the load of respective controllers. An optimization algorithm should minimize Eq. 4 in order to find a controller placement that provides good performance.

3.2.3 Inter-controller latency

The inter-controller latency (Borcoci et al. 2015) will affect the response time of the inter-controller communications, which are performed to update the global view of the network status. For a given placement, it can be computed as the maximum latency between two controllers:

$$L_{inter-controller} = \max L(c_u, c_v) \tag{5}$$

where $L(c_u, c_v)$ is the distance between two controllers c_u and c_v. For minimizing Eq. 5, we need to place the controllers close to each other. But the problem is that, this may increase the node-controller latencies given by Eq. 2 and Eq. 3.

The modified-AP is implemented using python and above performance metrics are used to evaluate the proposed system using Mininet on two topologies GARR and GEANT. GARR topology is composed of 52 nodes and 66 edges, and there are 40 nodes and 61 edges present in GEANT topology. We can compare the results of the modified-AP algorithm with k-median and k-center to verify the stability and accuracy of the proposed system.

4 CONCLUSION AND FUTURE WORK

This paper has shown a work (in progress) study on the application of several weighted criteria for placement of controller in large-scale SDN. The proposed method dynamically compute the number of controllers needed and its location for a given network using modified affinity propagation (modified-AP) clustering. This approach additionally specifies the most suitable controller for each switch. This method can be applied to a scenario that considers failure-free assumptions, given that it attains an overall optimization.

As future work, the proposed method can be extended to include fault-tolerance and reliability aspects to improve the efficiency of the controller placement. The simulations can be conducted on large-scale SDN by considering additional metric like the capacity of the controllers.

REFERENCES

Borcoci, E., R. Badea, S.G. Obreja, & M. Vochin (2015). On multi-controller placement optimization in software defined networking-based wans. *ICN 2015: The Fourteenth International Conference on Networks*, 261–266.

Cormen, T., C. Leiserson, & R. Rivest (2009). *Introduction to algorithms*. Massachusetts, USA: The MIT Press.

Heller, B., R. Sherwood, & N. McKeown (2010). The controller placement problem. In *Proc. HotSDN*, pp. 7–12.

Jarschel, M., F. Lehrieder, Z. Magyari, & R. Pries (October 2012). A flexible openflow-controller benchmark. In *Proc. European Workshop on Software Defined Networks (EWSDN)*, Darmstadt, Germany, pp. 48–53.

Knight, S., H.X. Nguyen, N. Falkner, R. Bowden, & M. Roughan (2011). The internet topology zoo. *IEEE JSAC 29*, 1765–1775.

Shi-duan (October 2012). On the placement of controllers in software defined networks. *ELSEVIER, Science Direct 19*, 92–97.

Tootoonchian, A. & Y. Ganjali (2010). Hyperflow: a distributed control plane for openflow. In *Proc. INM/WREN*.

Xiao, P.,W. Qu, H. Qi, Z. Li, & Y. Xu (2014). The sdn controller placement problem for wan. In *Communications in China (ICCC), 2014 IEEE/CIC International Conference on*, pp. 220–224. IEEE.

Yeganeh, S.H., A. Tootoonchian, & Y. Ganjali (February 2013). On scalability of software-defined networking. *IEEE Comm. Magazine 51*, 16–141.

Zhang, B., X. Wang, L. Ma, & M. Huang (2016). Optimal controller placement problem in internet-oriented software defined network. In *Proc. International Conference on Cyber-Enabled Distributed Computing and Knowledge Discovery (CyberC)*, pp. 481–488.

Zhao, J., H. Qu, J. Zhao, Z. Luan, & Y. Guo. (2016). Towards controller placement problem for software-defined network using affinity propagation. *Electronics Letters 53*, 928–929.

Emerging Trends in Engineering, Science and Technology for Society,
Energy and Environment – Vanchipura & Jiji (Eds)
© 2018 Taylor & Francis Group, London, ISBN 978-0-8153-5760-5

Identifying peer groups in a locality based on Twitter analysis

K. Sreeshma & K.P. Swaraj
Department of Computer Science Government Engineering College, Thrissur, Kerala, India

ABSTRACT: Twitter is one platform where people express their thoughts on any trending topics they are interested in. The exploration of this data can help us to find peer groups or group of users with similar interests. As in any other social network, this is also subjected to various spam attacks. So before identifying peer groups, the accounts that are ingenuine or regularly involved in spamming activities has to be filtered out. The main idea is to make use of the URLs the accounts share and their frequency to identify the account type. Here instead of focusing on one account, a group of accounts or a campaign is identified based on the similarity of the accounts. The similarity measure is calculated by applying Shannon's Information theory to estimate the amount of information in a URL and then using the value to find out information shared by each account. Once similar accounts are identified a graph is plotted connecting those accounts who have a similarity measure above a threshold. The potential campaigns are identified from this graph. Then they are classified to spammers and normal users using ML algorithms. The normal users we thus identify are members who have similar interests. To further improve the efficiency these members are grouped together based on their location, so peer groups in a locality are identified. This peer group identification can help in connecting those people with similar interests in a locality.

1 INTRODUCTION

The exploding growth of data has always opened new opportunities for those who indulge. The role of social networks in the life of a person is unfathomable these days. In fact, the scenario is like ones virtual friend knows about them more than their own friends and family. So the behavioral data of an individual is all available in his social network accounts. The right assessment of his social media activity can be helpful to identify his character and interests.

Twitter is a microblogging social network where most of the users express their opinion on some trending topics. The tweets they post or share will help us find the interest of the users. The idea here is to identify a group a people with similar interest. The tweets generally allow text and URLs only. So the URLs play a significant role in the characteristics of a tweet (Zhang et al. 2016). To be able to identify the common URLs shared between the users will be an important measure in establishing a relation between similar users.

Being a very popular social media network it has its disadvantages as well. It is also subjected to spam attacks. The spamming in social networks can cause much more harm than traditional spamming like email spamming (Grier et al. 2010). There will be a lot of ingenuine users with fake profiles who like to promote illegitimate contents. So when we are to peer groups based on their social network activity we should be able to genuine accounts from spammers efficiently. After filtering out the spammers we need to rightly group the authentic users with similar interests and then we can apply geotagging techniques to group them based on their geographic coordinates.

In this article, we are here to identify peer groups in a locality based on their Twitter usage. For this purpose, we are making use of the URLs that are part of their tweets. So as the first step we need to connect the users which share similar interests based on an account similarity measure derived on the basis of the number of common URLs they share and the information contained in each URL. Once we obtain an account similarity graph, next step is to classify potential campaigns to normal and abnormal and finally we group accounts in each campaign according to their locations so as to obtain peer groups in one locality. The rest of the paper carries the details for each of the aforementioned stages.

2 RELATED WORK

Much related works have been done in identifying spammers in a social network. Since most of the efforts lie in, correctly classifying normal and abnormal groups based on their activities, here importance is given to rightly classify spammers and non spammers. Most traditional methods try to detect the spammers based on their messages sent or the activities of their account. The message level detection (Benevenuto et al. 2009) checks each tweet posted for any discrepancies or spam contents in any URLs mentioned. But this may require a real-time processing as umpteen tweets are getting posted every hour.

The account level detection methods (Lee et al. 2010) examine the activity of the user accounts, such as whether they have promoted spam contents, to find the authenticity of an account and thus identify if it's a spammer or not. But both of these methods have left many spams unidentified at the end of the evaluation. Instead of classifying individual messages and accounts, some papers have proposed identifying spam campaigns. A campaign refers to a group of accounts who purposefully work towards the same goal. Spam campaigns often contain accounts which post harmful information such as malware, virus etc.

(Hatanaka & Hisamatsu 2010) proposed a method to group users into distinct blacklist groups based on the degree of similarity in their bookmarks and they reduced the rank of the bookmarks promoted by this blacklists. A detection framework for spam campaigns based on the similarity of the URLs shared by the accounts is proposed by (Gao et al. 2010). This framework quantifies the similarity measure between accounts based on the URLs they share, to draw a similarity graph and also put forward some characters of spam campaigns. (Lee et al. 2010), (Lee et al. 2013) has proposed a content-driven approach for identifying spam campaigns and categorizing it. This method employs a strategy to group users based on text similarity but classification is done by manual inspection.

(Zhang et al. 2016), (Zhang et al. 2012) proposes a multilevel classification method for identifying spammers and non-spammers where the first level includes classification into normal and abnormal campaigns and further classification to identify spam and promotion campaigns. This method makes use of the similarity measure between the users to plot a similarity graph. The similarity measure was calculated based on the common URLs shared by the accounts and later they extended the work to consider the timestamp as well. (Jiang et al. 2016) identify the importance of finding peer groups in a social network for a friend recommendation system.

3 PROPOSED METHOD

In this section, a new method is proposed for identifying peer groups in a locality based on the Twitter activity of the users. First, The account similarity is estimated based on the common URLs shared by them. A similarity graph is plotted based on these measurements and then we have to extract potential campaigns from it. Secondly, for classification of this campaigns into spammers and non spammers, we have to apply machine learning techniques and classify them into normal and abnormal campaigns. In the final step, now that we have obtained normal campaigns and filtered the spam campaigns, we have to make use of the geotags to group the users in a normal campaign according to their locality. Thus we will obtain peer groups in a locality.

The process flow is depicted in 1.

3.1 Estimation of account similarity

For estimating the account similarity, we make use of the URLs shared in the tweets. Those accounts who share same URLs tends to be a part of the same campaign. For quantifying the information contained in a URL, we have to make use of Shannon's Information theory. The information contained in a URL u, is defined as:

$$I(u) = -logP(u) \quad (1)$$

and P(u) is the probability of u and dened as

$$P(u) = \frac{\#u}{N} \quad (2)$$

where # u is the number of tweets containing URL u in the corpus and N is the number of all tweets containing URL(s).

In order to calculate the amount of information contained in all URLs shared posted by account a_i, we make use of the below formula,

$$I_a(i) = \sum_{u \in U_i} Num_i(u) * I(u) \quad (3)$$

where $Num_i(u)$ is the number of tweets containing URL u posted by account a_i. The amount of information shared by accounts a_i and a_j through the sharing of common URLs is collectively summated as

$$I_a(ij) = \sum_{u \in U_i \cap U_j} (Num_i(u) + Num_j(u)) * I(u) \quad (4)$$

Finally, to estimate the similarities between accounts, we use:

$$S_{ij} = \frac{I_a(ij)}{I_a(i) + I_a(j)} \quad (5)$$

where $0 \le S_{ij} \le 1$

Now that we have obtained the measure of the similarity between various accounts, we need to plot a graph combining the accounts which has a similarity measure above a particular threshold. The obtained graph is used to identify potential campaigns. The campaigns are those areas in the graph which are very dense. We identify them using the concept of maximal

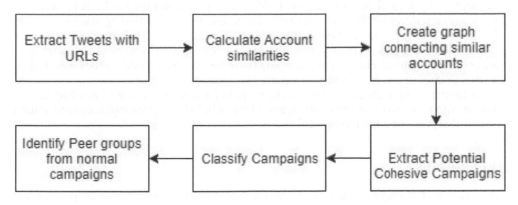

Figure 1.   Process flow diagram.

coclique. Maximal coclique is defined as (v_i, v_j) for all neighbors, which represents the size of the largest clique containing both vertexes v_i and v_j, and v_j is a neighbor vertex of v_i.

3.2 Classification of potential campaigns

The potential campaigns have been identified from the similarity graph. Next is another important step which classifies each of these into normal or spam campaigns. Here we are listing some of the features which will help us to classify the campaigns. These features are used to train some classifier based on ML algorithms and used for prediction of the type of campaign. Those features include:

Average Posting Interval (API): It is the average time difference between posting tweets with URLs. The spammers tends to have a low time difference between posting tweets.

Average Number of tweets containing URLs (TUNum): This will give the count of the no. of tweets posted by the account that is having URLs in them. It tends to be higher for spam accounts compared to normal accounts.

Posting frequency of tweets with URLs (TUFrequency): This measure is similar to the previous one. But this gives a weightage instead of giving an exact number. This value is also higher for spam accounts.

Average of distinct URLs posted by each account (URLNum): Spammers will post URLs from various domains and hence will have higher URL-Num value.

VURatio: This is the ratio of valid URLs out of the total URLs shared by the account. The number of invalid or broken URLs shared by spammers tends to be higher.

Blacklisted URLs: The spam accounts might share URLs which are blacklisted in black-lists such as Google Sage Browsing, PhishingTank etc. Any account with intentions of mis-guiding a user will make use of these URLs.

3.3 Identifying peer groups based on the locality

Now we have classified the potential campaigns and identified the normal campaigns. This is a group of people with similar interests which is evident from the number of the common URLs posted by them. This group is subjected to further refining by grouping them based on their locality. The locality of the account is available in the tweet data. Based on this we can finally identify peer groups of a particular locality. This can be made use of by various recommendation systems which would like to recommend products or plans on a location specific basis.

4 CONCLUSION

The peer group identification based on Twitter is proposed and the methods involved for each stages are detailed. We have used URL based estimations for finding similar accounts and created a graph. The cohesive campaigns are then extracted. The extracted campaigns are classified using some very important features and categorized as normal and spam campaigns. The accounts in each normal campaigns are further categorized on the basis of their location and we obtain peer groups in a particular location.

The twitter analysis throws insight to the nature of the twitter user and can help him to connect with people having similar thoughts. This can be helpful to various recommendation systems that work on a location specific basis.

REFERENCES

Benevenuto, F., T. Rodrigues, V. Almeida, J. Almeida, & M. Goncalves (2009). Detecting spammers and content promoters in online video social networks. *In Proceedings of the 32nd International ACM Conference on Research and Development in Information Retrieval(SIGIR09) ACM*, 620–627.

Gao, H., J. Hu, C. Wilson, Z. Li, Y. Chen, & B. Zhao (2010). Detecting and characterizing social spam campaigns. *In Proceedings of the 10th Annual Conference on Internet Measurement(IMC10). ACM,* 35–47.

Grier, C., K. Thomas, V. Paxson, & M. Zhang (2010). Spam: The underground on 140 characters or less. *In Proceedings of the 17th ACM Conference on Computer and Communications Security (CCS10) ACM,* 27–37.

Hatanaka, T. & H. Hisamatsu (2010). Method for countering social bookmarking pollution using user similarities. *In Proceedings of the 2nd Conference on Networked Digital Technologies (NDT) Springer,* 523–528.

Jiang, F., C.K. Leung, & A.G. M Pazdor (2016). Big data mining of social networks for friend recommendation. *IEEE/ACM International Conference on Advances in Social Networks Analysis and Mining (ASONAM) IEEE,* 3 pages.

Lee, K., J. Caverlee, Z. Cheng, & D. Z Sui (2013). Campaign extraction from social media. *ACM Transactions on Intelligent Systems and Technology 5.*

Lee, K., J. Caverlee, & S. Webb (2010). Uncovering social spammers: Social honeypots + machine learning. *In Proceedings of the 33rd International ACM Conference on Research and Development in Information Retrieval ACM,* 435–442.

Zhang, X., Z. Li, S. Zhu, & W. Liang (2012). Detecting spam and promoting campaigns in the twitter social network. *In Proceedings of the IEEE International Conference on Data Mining (ICDM12). IEEE,* 1194–1199.

Zhang, X., Z. Li, S. Zhu, & W. Liang (2016). Detecting spam and promoting campaigns in twitter. *ACM Transaction. Web 10,* 28 pages.

Emerging Trends in Engineering, Science and Technology for Society,
Energy and Environment – Vanchipura & Jiji (Eds)
© 2018 Taylor & Francis Group, London, ISBN 978-0-8153-5760-5

Route choice analysis of metros using smart card data

Sreemol Sujix
Computer Science and Engineering, Thejus, India Engineering College, Thrissur, Kerala, India

ABSTRACT: Route choice behavior is an interesting topic in the transportation area. Metro systems have become a great solution for meeting the emerging traffic needs of large cities. The arrival of metros indeed indicates the development strategy of a country. Given the 'origin to destination' pairing of a metro trip, there are often multiple routes between such pairings. Different routes are chosen by passengers with different preferences, and such preferences have always been a point of interest in the transportation area. The data in this study has been collected from passenger smart cards. This analysis will help operators to improve their passenger service, and will also help passengers to plan their journeys.

Keywords: Route choice analysis, Smart card data, Data mining, Big data

1 INTRODUCTION

Metros have become the most demanded mode of transport for passengers due to their speed, efficiency, time management, comfort, capacity to accommodate more passengers, and so forth. It has become a necessary piece of infrastructure for a growing metropolitan city. The use of metros has not only helped in decreasing road traffic but has also paved the way to pollution-free transport when compared to cars and other vehicles that pollute the air by emitting harmful carbon monoxide, which can create holes in the ozone layer. Therefore, using metros is more advantageous, safe and eco-friendly.

The pattern of traffic in a metro is usually very complex because the trains and routes chosen by a passenger are unknown. Route choice analysis is a study that is related to the distribution of passengers in the different routes and the trains chosen by them. Dealing with such abstract and diverse data to infer the required information and modeling of route choice behavior are two major challenges faced in public transport management.

The emergence of big data analytics has helped to store, process and manage this complex data, whereas traditional data processing applications are inadequate to handle it. Conducting route choice analysis is of primary importance to both passengers and metro operators. For train operators, this analysis will help them to understand how passenger flow takes place in the metro network and hence improve service reliability. For metro passengers it will be of great use in trip planning. Indeed, this study can help urban administrators in route suggestions and managing emergency situations.

A metro generally provides its passengers with a smart card facility. A smart card is a pocket-sized card with an embedded circuit. Every time a smart card is swiped at the station gate, details of the trip being made with that card are recorded and the monetary value is stored and debited from the card. This smart card data is used for data collection processes and hence contributes to the analysis of travel behavior.

In this paper, the probability of passengers choosing a particular route for an Origin to Destination (OD) pairing with multiple routes is shown in Figure 1. Here, big data analytics have been employed to deal with such vast and complex data. The Hadoop framework has been preferred for this implementation because it supports batch processing on enormous amounts of information. The Hadoop framework consists of a distributed file system and a MapReduce

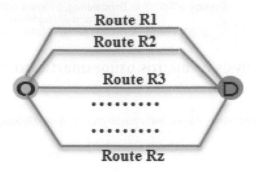

Figure 1.　Routes in an OD pair.

function. These two elements of Hadoop can be used to enable the storage of huge amounts of data and for performing parallel processing to save time and improve efficiency.

2　BACKGROUND

Traditional approaches are no longer scalable. The old method of route choice study was to collect the information from surveys conducted by asking passengers about their routes and trains. This method was a tedious process and could not yield the best possible results as it was limited to persons, places and times.

Automated Fare Collection (AFC) systems were then used for the analysis, which gave broader information regarding the travel pattern of passengers. A drawback related to these AFC systems was that they did not give any information regarding the train and route chosen by passengers but just provided the details of the origin and destination stations traveled by the passengers. In earlier studies, the walking time between the swiping gate and the platform, and transfer time between platforms, were ignored. These have been considered in this study.

Here we consider three cases and make a comparative study. In one case, route choice analysis is done using smart card data only; another case study is done using smart card and timetable data, and the final study uses smart card data, timetable data and MCL data, which is obtained by means of conductor checks.

3　LITERATURE SURVEY

3.1　*Big data*

With advancements in technology, enormous amounts of data are generated every second of every day, as shown in Figure 2. Previously, landline phones were widely used, but advances in technology have led to them being replaced by smart phones and Android phones, which have, arguably, made peoples' lives smarter as well as their phones smarter. There were also bulky desktops, which were in great demand and were used for processing megabytes of data. Floppy disks were also in use, and then hard disks evolved and now cloud technology has emerged for storing huge amounts of data. Enormous amounts of data are generated daily from the use of smart phones alone. Our every action with a smart phone, such as sending a video through WhatsApp or any other messenger application, or a post on Facebook chat, is also generating lots of data. The data generated is not in a format that a relational database can handle. The volume of data has also increased exponentially and it has become difficult to deal with such enormous amounts of data. Consider, for example, the case of self-driving cars that have sensors that constantly record details such as the size of an obstacle and the distance from it before, finally, considering all these factors and deciding how to react: a huge amount of data is generated for each kilometer that the car has driven.

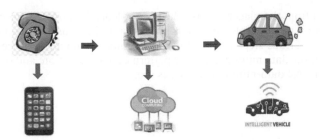

Figure 2. Advances in technology.

Huge quantities of passive data streams are collected by smart cards, GPS, Bluetooth and mobile phone systems all over the world. This data happens to be very useful to transport planners, because of the valuable spatial and temporal information it contains. Big data plays an important role in storing and processing huge amounts of data in a way that existing systems cannot. The classification of data as big data is based on volume, veracity, velocity and value. Volume refers to the exponentially rising amount of data. Variety refers to the multiple sources from which the data is emerging. The data so obtained are of different kinds and can be classified as structured data, which includes tables, Excel spreadsheets and so on, semi-structured data, consisting of cstv, emails, XML files and so on, and unstructured data, which includes video, images and so forth. Data is being generated at an alarming rate. Velocity refers to the speed at which the data is produced. Value refers to the mechanisms that derives correct meanings from the data that is extracted. Veracity considers the difficulty of extracting such value from the data and helps in handling this.

3.2 *Hadoop*

Hadoop is one of the frameworks used to handle big data. It consists of a distributed file system for storing large amounts of data and a MapReduce function to perform parallel processing on the collected data by assigning work to each processor connected in the network. Hadoop is an open source framework and provides distributed storage and computation across clusters of computers.

3.3 *Smart card data*

AFC systems adopt smart card platforms for payment processes in metro systems and trams all over the world. Only a passenger with a card with sufficient credit can travel on the metro. The smart card information helps to track the travelers' behavior in an efficient manner. It records information regarding the origin station from which the passenger has boarded, the travel time, the destination station from which the passengers exited, their credit balance, and the smart card ID. All this information is recorded as and when the passenger swipes the card in the origin and destination stations. This mechanism also helps to calculate the fare of the journey and automatically deducts this from the card balance.

There are various benefits to using smart cards for travel purposes. The waiting time spent at ticket counters can be avoided as the card can be recharged over the internet from anywhere. In addition, it is possible for a passenger to terminate the journey at any time and the fare will be taken only for the traveled distance. Smart cards are transferrable as they can be used by friends or family members. All these benefits of smart cards make it more advantageous for travelers to use metros, buses and tram systems.

3.4 *Data mining*

Data mining is a powerful tool for extracting valuable information from raw data. Data mining, which is also known as knowledge discovery, resembles the gold-mining process in which gold is extracted from the placer deposit; here knowledge is extracted from data. In today's scenario, the data is rich but information is poor, so raw data has to be converted into useful information. The various steps involved in data mining are data pre-processing, data selection, transformation, data mining, building patterns, data evaluation, and data visualizations. Some of the principal data-mining techniques are:

- association;
- clustering;
- classification.

Association is a data-mining process that is often used to analyze sales transactions. Items that are purchased together, for example bread and jam, are associated items and placing them together in a market yields more profit. Clustering is an unsupervised learning technique: similar data is grouped into clusters. Classification is a supervised learning technique performed on the assumption that some knowledge of the data is available. Decision trees, neural networks, and naive Bayes algorithms are some of the methods involved in this area.

4 METHODOLOGIES

4.1 *Analysis using smart card and train timetable data*

In this route choice analysis, only the OD pair is taken into account. Several routes exist between the origin and destination stations. This analysis will help in obtaining the probability that a passenger will choose a route in the OD pair as shown in Figure 3.

Two possible journeys can be made by the passengers; either a direct journey from origin to destination or by making a transfer journey, as shown in Figures 4a and 4b. Consider a day, I, that is divided into half-hour intervals as $\{I1, I2, I3, I4, \ldots I48\}$. The routes chosen in a particular time interval Ij are considered for the route study.

4.1.1 *Datasets*

The route choice analysis conducted here utilizes smart card data as well as timetable data. Timetable data is maintained by the stations to inform passengers of train numbers, train routes, arrival times, departure times and the metro line name or number. Clubbing both the dataset the route choice study will be enhanced.

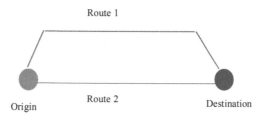

Figure 3. Route graph.

Figure 4a.   Direct journey.

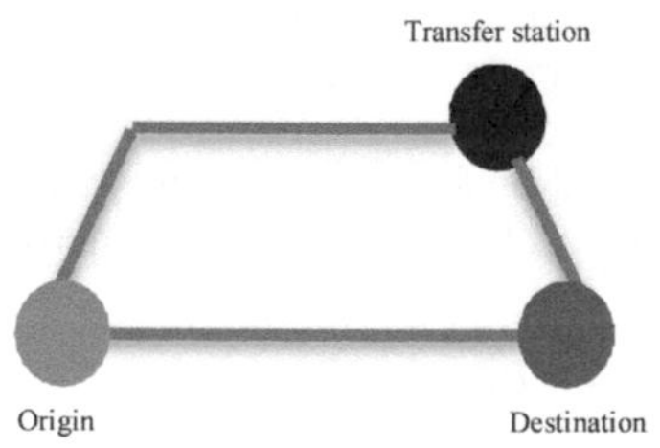

Figure 4b.   Transfer journey.

4.1.2 Notations and assumptions

In this system, the smart card data of a passenger, along with the train timetable data, is used to determine the route information. A probabilistic model has been developed to estimate the routes and trains chosen by the passenger. Two polynomial distributions are considered here, that is, the number of trains a passenger waits at his station and at the transfer station. Consider the trip, x, of a passenger; the beginning time of the trip is considered as $x.b$, the origin is $x.o$, the end time is $x.e$, and the destination is $x.d$. This information is acquired by joining together the tap-in and tap-out data from the smart card, as shown in Figure 5. If the trip has i parts, this means that the trip will need $(i-1)$ transfers. The set of effective routes of an OD pair is $R = \{R1, R2, \ldots Rz\}$. It is assumed that the routes being chosen in a particular time slot Ij obey the polynomial distribution with parameter $\alpha j = \{\alpha j1, \alpha j2, \ldots \alpha jz\}$ where $\sum \alpha jz = 1$. Given the train timetable and the trip of the passengers $X = \{x1, x2, \ldots xq\}$ at time slot Ij, it is possible to calculate αj, which is the maximum likelihood function.

Two assumptions which are usually ignored are considered here: first, the time taken for the departure of two adjacent trains will be greater than that of the time taken by passengers to walk between the platform and the charge gates. Second, the majority of passengers leave the station when they reach their destination.

To calculate αj of an OD pair, all effective routes between the OD pair are needed. Then the journeys in a particular time slot Ij and the train timetable are used to calculate the maximum likelihood estimation.

4.1.3 Maximum likelihood estimation

Maximum Likelihood Estimation (MLE) is a technique used for estimating the parameters of a given distribution, using some observed data. Maximum likelihood estimation is the log of the likelihood function. The likelihood function is the probability of getting the data when the probability distribution model for that data is given.

Passengers entering the metro gate at a time slot j in station s and choosing metro line l are represented by (s, l, j), which is known as the tap-in information, through the charge gate.

Passengers who transfer from metro line l to l' in transit station s at time slot j are represented by (s, l, l', j), which is known as the transfer information.

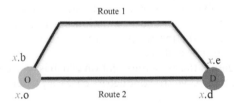

Figure 5.   Route notations.

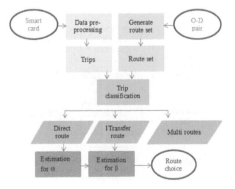

Figure 6.   Processing chart.

To calculate the maximum likelihood *αj* of an OD pair firstly, all effective routes between the OD pair should be calculated. Secondly, the trips made between the OD pair on a particular day *Ij* are considered. By knowing the trips and the routes, all the possible trains and train combinations chosen by the passenger can be analyzed by matching the tap-in and transfer information. Thus, it is possible to calculate the maximum likelihood of routes chosen by calculating the probabilities retrieved from θ and β values as shown in Figure 6. Finally, considering both θ and β values, the maximum estimation value *αjz* can be calculated using Equation 1:

Maximum likelihood function = Log of likelihood

$$L(X, \text{Tab}, \alpha j) = \log \sum_{Rz \varepsilon R} (\alpha jz * \Pr(\text{xq.e} \mid \text{Tab}, \text{xq.b}, Rz)) \tag{1}$$

where Pr (*x*q.e|Tab, *x*q.b, *R*z) represents the possibility that a passenger *x*q passes through exit gate at time *x*q.e on condition of Tab, *x*q.b and the route chosen, *R*z. So, Pr(*x*q.e|Tab, *x*q.b, *R*z) can be calculated by summing up the probabilities of all plans.

The smart card data which acts as the input for the route study includes several kinds of errant data such as missing data, duplicated data and data with logical errors. In the data pre-processing step this erroneous data will be filtered out and the trips will be extracted from it. In the next step in generating the route, a dataset from another input, that is, the train timetable, and certain shortest path algorithms are used. The routes which are not used in the OD pair are also filtered. Finally, the trips are classified as follows:

- direct route;
- one-transfer route;
- multi-transfer route.

The trips such as the direct route and the one-transfer route help to estimate the value of θ and β, respectively. With the known θ and β value, the probability of choosing each route in an OD pair is calculated.

4.2 *Analysis using smart card, train timetable and MCL data*

The route choice study of passengers provides operators with the opportunity to improve their passenger service. Analysis using smart card data and train timetable data only gives information regarding a trip's origin and destination but not the train information. So, in order to make an accurate prediction we prefer to consider the smart card data, the train timetable, and the conductor check data.

The conductor check data is the extra information which is collected from a passenger during his or her trip on the metro. Metro conductors check a passenger's ticket using a mobile

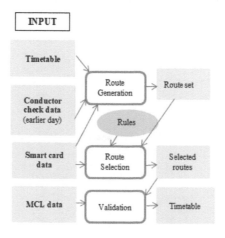

Figure 7. Processing chart.

chip card reader. When the passenger's card is swiped in the reader, information such as the smart card number, the current time, and the number of the current train will be recorded.

The steps involved in this analysis are route generation, route selection and validation. The inputs used for this analysis are smart card data, the train timetable, and conductor check data from earlier days and a specific day d. Routes are generated by using various route-generation methods such as the Bellman–Ford algorithm or Dijkstra algorithm. In the route-selection process, the generated route sets are further shortlisted by using certain rules. Then the validation process is done by checking the selected route sets with the conductor check data of the specific day d, as shown in Figure 7.

4.2.1 Route-generation methods
Initially, the train timetable data is converted into an event activity network which consists of the basic network, BN, and the extended network, EN, as shown in Figure 8. The BN consists of information regarding the trains and the station, and the EN consists of more detailed information such as the wait time, and transfer stations, along with the train and the station information. Route-generation methods work by considering factors such as minimum cost, extended route search, and link learning.

4.2.2 Route-selection methods
After the route-generation mechanism, the multiple routes generated are shortlisted by applying certain rules to mine the required routes. Some of the selection rules used are:

- First Departure (FD)
- Last Arrival (LA)
- Least Transfers (LT)
- Selected Least Transfers Last Arrival (STA)

Based upon these rules, the routes are filtered and selected from the route set. The FD rule selects the route whose departure time is closest to the time of check-in. The LA rule selects the routes that are arriving last and are closest to the time of check-out. The LT rule selects the route with minimum transfers. STA is a combination between the maximum time duration route and the LT rule.

4.2.3 Validation
Due to the availability of conductor checks on the specific day d, it is possible to evaluate the performance of the analysis by comparing the selected route with MCL data and filtering the required route.

4.2.4 Shortest path algorithms
Routes between an OD pair can be retrieved by referring to the train timetable information of the stations. The use of shortest path algorithms such as Dijkstra or Bellman–Ford help to find the minimum-cost routes quickly when applied to a BN or an EN formed by converting the

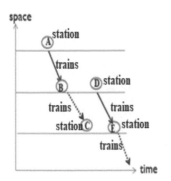

Figure 8a.   Basic network.

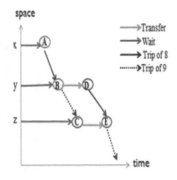

Figure 8b.   Extended network.

train timetable. The running time of the Bellman–Ford algorithm is known to be O(|V| × |E|). A graph with negative weights cannot be executed using the Dijkstra algorithm, so the Bellman–Ford algorithm is used for this purpose. The Bellman–Ford algorithm is also much simpler than the Dijkstra algorithm. However, the time complexity of the Bellman–Ford algorithm is O(VE), which is greater than the Dijkstra algorithm.

4.3 *Analysis using smart card data alone*

The metro users' mobility patterns can be studied from different angles using smart card information. The passengers' spatio-temporal travel patterns inside a metro can be inferred from this. This study will also be helpful in anomaly detection. The same smart card can be used by a passenger to travel in a metro as well as on a bus or tram. So, two types of smart card information can be collected from a single passenger's smart card: metro travel and bus travel. The additional bus travel information will help in producing clustering with higher precision.

4.3.1 *Data-mining tools*
K-means clustering is the data-mining technique that is employed for the route study to find similar groups of data. Clustering is the process of grouping data points into similar clusters. Data points are assigned to clusters based on the calculation of the square of Euclidean distance, which is the minimum distance from the data points to the cluster. The data points are assigned to clusters with low inter-group similarity and high intra-group similarity. This is preferred because of its simplicity and efficiency.

4.3.2 *Temporal analysis*
An individual passenger is concentrated upon here and his or her mobility patterns on the metro and bus are considered. First, we find all the trips that belong to a specific passenger from the transaction record. This record contains the boarding and exiting information taken from the smart card used for the bus and metro journey. The second step is pre-processing, where the passengers who rarely travel are ignored or filtered as they contribute little information on the temporal or spatial characteristics. So, only passengers with a card containing more than six active days are considered. Temporal analysis selects features that have small values and convey more temporal information. Such features will help improve the scalability of analysis. The k-means clustering algorithm is used to cluster all passengers based on their four-dimensional temporal features. As a result, four groups will be formed, such as TGrp1, TGrp2, TGrp3 and TGrp4, as shown in Figure 9.

The proportion p of active days is categorized as Very Frequent (VF), Relatively frequent (RF), Frequent (GF), and Least Frequent (LF) on the basis of the study conducted. These proportions help to classify the TGrp groups into any of these categories based on the rules listed in Table 1.

Based upon the proportions of Table 1, the temporal groups fall into the following categories as shown in Table 2.

By also incorporating the bus data, the analysis can be more accurately completed. With bus data and TGrp as centers, all passengers are re-clustered to form four groups BTGrp1,

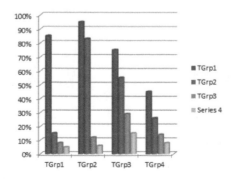

Figure 9.   Temporal analysis chart.

BTGrp2, BTGrp3 and BTGrp4. The passengers are then compared and matched with TGrp and BTGrp. Thus, an inference can be made from this study that some passengers prefer to travel by metro alone, whereas others prefer to complete their journey by using both metro and bus, and there are also travelers who use the bus alone to make their journey. A comparison of the joint probability distribution of passengers is done and information is derived about their travel pattern, which is shown in Table 3.

4.3.3 *Spatio-temporal analysis*

Metro passengers are spatially clustered into four groups using a k-means algorithm on the four-dimensional spatial features of SGrp1, SGrp2, SGrp3 and SGrp4, as shown in Figure 10.

The conditional probability of SGrp given TGrp is calculated, which shows how SGrp and TGrp are connected with each other. From this inference we discover that group 1 and group 2 passengers are regular in their spatio-temporal characteristics. Group 3 passengers are relatively regular and group 4 passengers are irregular in their temporal and spatial characteristics. There are three group labels for every metro user, which are denoted, respectively, by TGrp, BTGrp, and SGrp.

Table 1. Proportions.

Group	VF	RF	GF	LF
p	$p \geq 80\%$	$70\% \leq p < 80\%$	$50\% \leq p < 70\%$	$p < 50\%$

Table 2. Group characteristics.

Group	1 st	2 nd	3 rd	4 th
TGrp1	VF	LF	LF	LF
TGrp2	VF	VF	LF	LF
TGrp3	RF	GF	LF	LF
TGrp4	LF	LF	LF	LF

Table 3. Joint probability distribution (%).

Group	BTGrp1	BTGrp2	BTGrp3	BTGrp4	BTGrp5
TGrp1	5.48	7.78	2.54	0.04	15.83
TGrp2	0.00	31.70	0.94	0.00	32.64
TGrp3	0.03	3.55	23.52	0.02	27.11
TGrp4	0.58	5.7	15.17	2.94	24.41
Total	6.08	48.75	42.17	3.00	100.00

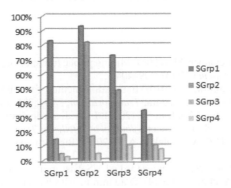

Figure 10. Centers of SGrp.

Table 4. Class groups.

Class	TGrp	BTGrp	SGrp
Class-1	1	2	1
Class-2	2	2	2

Table 5. Comparative study.

Maximum likelihood estimation	K-means clustering	Bellman–Ford algorithm
Estimates the likelihood of route selection.	Partitions data into clusters and uses the joint and conditional probabilities for estimation.	Finds effective paths chosen based on minimum path, cost, etc.
Uses smart card and train timetable data.	Uses smartcard data.	Uses smartcard data, train timetable and MCL data.
Applicable to different types of data.	Applicable to all data but outliers don't work well.	Applicable to network models.
Precise estimation.	Simple and efficient method.	Maximizes performance.

Class-1 represents those who take the metro on one trip and take a bus in another trip; Class-2 represents those who take the metro in round trips, as shown in Table 4. The comparisons of average cost and travel time across five weekdays are shown in Figure. It is clear that Class-1 is lower than Class-2 in terms of cost and travel time. It was learned that, in Shenzen, the average cost of taking the metro is higher than taking the bus and, for economic reasons and if time permits, some passengers will choose the bus.

5   ANALYSIS OF METHODS

A comparison of route choice analysis methods using smart card data is presented in Table 5.

6   CONCLUSIONS

With the development of computer technology there has been a tremendous increase in the growth of data. Big data analytics is an emerging field in intelligent transportation systems nowadays. The objective of this paper was to make a study on route choice behavior using smart card information, which helps in analyzing the travel patterns of the passenger, their route, and their train selection. This inference will, indeed, be useful for operators in terms of improving the services given to passengers and will also be helpful to passengers in trip planning. A comparative study has been conducted here to understand the route choice behavior of passengers.

REFERENCES

Agard, B., Morency, C. & Trépanier, M. (2006). Mining public transport user behavior from smart card data. *IFAC Proceedings Volumes, 39*(3), 399–404.

Kusakabe, T., Iryo, T. & Asakura, Y. (2010). Estimation method for railway passengers' train choice behavior with smart card transaction data. *Transportation, 37*(5), 731–749.

Trépanier, M., Tranchant, N. and Chapleau, R. (2007). Individual trip destination estimation in a transit smart card automated fare collection system. *Journal of Intelligent Transportation Systems, 11*(1), 1–14.

Tsai, C.W., Lai, C.F., Chao, H.C. & Vasilakos, A.V. (2015). Big data analytics: A survey. *Journal of Big Data, 2*, 21.

White, T. (2012). *Hadoop: The definitive guide* (3rd ed.). Sebastopol, CA: O'Reilly Media.

Wu, X., Zhu, X., Wu, G.Q. & Ding, W. (2014). Data mining with big data. *IEEE Transactions on Knowledge and Data Engineering, 26*(1), 97–107.

International Conference on Changing Cities-Architecture and Energy Management (ICCC)

Architectural applications

Emerging Trends in Engineering, Science and Technology for Society,
Energy and Environment – Vanchipura & Jiji (Eds)
© 2018 Taylor & Francis Group, London, ISBN 978-0-8153-5760-5

Ecological conservation of sites through responsive design: Case of the Muziris heritage interpretation center

Malavika Gopalakrishnan & S. Surya
School of Architecture, Government Engineering College, Thrissur, Kerala, India

ABSTRACT: This paper outlines the process by which the ecological conservation of the site of the Muziris Interpretation Center and the maritime museum at Pattanam, North Paravur was proposed, through site responsive design. The proposed site was unique, with a major part of the site being an unrealized pisciculture plot (traditional '*chemmeenkettu*'). The ecological conservation of the '*chemmeenkettu*' was done after conducting buildability studies of the site and formulating a proposal for the future realization of the site to its full potential. The main museum was designed as a floating structure, with minimum damage to the site and utilizing only a percentage of the waterlogged area. The museum was designed as a climatically responsive building by using renewable energy systems in floating structures.

1 INTRODUCTION

The Muziris Heritage Project aims at reinstating the historical and cultural significance of the legendary port of Muziris. The region is dotted with numerous monuments of a bygone era that conjure up a vast and vivid past, and the Muziris Heritage Project is one of the biggest conservation projects in India, aiming to preserve a rich culture that is around 3,000 years old. The material evidence unearthed at the excavation site, located about 25 km north of Kochi, points to the possibility that Pattanam may have been an integral part of the legendary Port of Muziris, and thus the interpretation center at Pattanam helps to promote awareness and understanding of the cultural distinctiveness and diversity of Muziris.

The Muziris, located along the west coast of Kerala, has the most productive waters in the world. It paved the way for the supreme diversity and abundance of both fishes and fisher folk. The indigenous and conventional methods of fishing are rapidly declining and are in need of conservation. Brackish water fish farming and cage fish farming are unique methods of culturing fishes and these eco-friendly methods of fishing provide livelihood security for the fisher folk in the region, as well as enhancing its biodiversity by supporting varied ecosystems. The proposed site for the interpretation center is one such, traditionally known as a '*chemmeenkettu*'.

2 ARCHITECTURAL CONCERNS OF THE PROJECT

2.1 *Understanding the tangible and intangible elements of Muziris heritage*

The Muziris Heritage Site is an outstanding example of the buildings, archaeological sites and landscapes that illustrate a significant stage in the history of Kerala. The area bears an exceptional testimony to a culture that is fast disappearing. The built heritage of Muziris is extensive and spreads across the site. While the physical remains of Muziris are outstanding and constitute a unique ensemble, there are also intangible associations and traditions that form an important part of Muziris. The practice of fishing is thousands of years old. One such practice is the unique Chinese nets and brackish water shrimp farming.

2.2 Recreating their lost identity

The Kottapuram and Paravur regions were thriving commercial centers, whose fabric had a variety of Portuguese, Dutch and traditional Kerala influences. The market streets portray the pressures and problems that it faced in earlier times. The culture and the traditions are the significant factors that make it a unique destination. The development that took place, in terms of scale, design and typology, was unsympathetic to the original character of the region, affecting the identity of the place. Reviving the lost traditions, and at the same time creating future opportunities, are the key aspects of recreating the glory of this once prestigious port city.

3 LOCATION AND CHARACTERISTICS OF THE SITE

The site is located at Pattanam, North Paravur, in Kerala, with an area of 49 acres and 92.84 cents. It is located about 1 km from the Kodungallur–Paravur route. The site is along the banks of the Kollam–Kottapuram waterway (National waterway III), encompassing the west and south side of the site.

3.1 Features of the proposed site

The majority of the site (68%) is an unrealized brackish water shrimp farming plot in a total area of 50 acres. The site is also surrounded by Chinese fishing nets and local fishing centers. The site is quiet and serene, away from the city. Even though the location is remote, it can be easily accessed by the upcoming Muziris heritage circuit. The site is also near to the Pattanam excavation site.

In Kerala, there is a long history of the traditional system of salt water aquaculture being practiced in the seasonal Pokkali fields and the perennial fields lying adjacent to the coastal and backwater areas. In these fields, seed shrimps and fishes are allowed to enter through tidal water and are then trapped. After a short duration of growth, they were periodically harvested during full and new moon periods, from November–April.

The site has an average depth of 4–5 ft. (1.2 m–1.5 m) and a maximum depth of 2 m. It is also known locally as 'chemmeenkettu'. The fishing season starts from November and lasts until April, and the harvest is done at 'thakkam' (7 pm-4 am). The water in the chemmeenkettu is recycled from the river during the day.

3.2 Architectural character of the site

The architectural pattern around the site is a traditional Kerala style with inspiring forms of traditional architecture and pristine backwaters. There is also a significant colonial influence, with elements of Dutch and Portuguese architecture. The material typology has sunburned and laterite bricks with non-imposing sloping roofs done in Mangalore tiles and sheet roof-

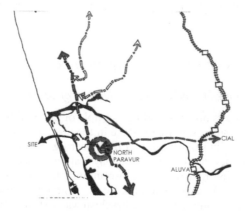

Figure 1. North Paravur site map.

Figure 2. Google earth image showing site.

Figure 3. Site image.

Figure 4. Site image.

ing. Some private residences around the site are mostly **MIG** (Middle Income group) and **LIG** (Low income Group) occupancy with single storied flat roofs.

3.3 *Neighborhood context*

Pattanam means 'town' in the local language of Malayalam and it is located 8 km south of Kodungallur and 25 km north of Kochi in the Ernakulum District in the southern Indian state of Kerala. The site is located about 1 km from the Pattanam excavation site. Paravur market and synagogue are also in close proximity. The Sahodaran Ayyapan museum is located opposite the site. Recent archaeological excavations have unearthed signs of early Roman trade, which was part of Muziris that flourished during the reign of the first Chera dynasty of South India.

National Waterway No 3 is a 168 km stretch of this inland navigational route located in Kerala, India and runs from Kollam to Kottapuram. It was declared a National Waterway in 1993, and connects the industrial centers of Kochi to the Kochi port. The majority of those inhabiting the region were engaged in traditional industries, such as coir, cashew, brick-making and fishing.

4 ECOLOGICAL CONSERVATION OF THE SITE AND RESPONSIVE DESIGN

4.1 *Buildability studies of the site*

The unique nature of the site required the conducting of buildability studies in order to determine a suitable location for the project with minimum intervention. The key aspect of

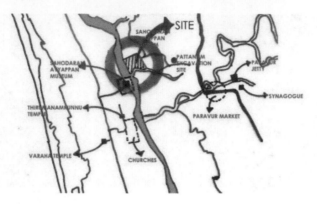

Figure 5. Pattanam and nearby landmarks.

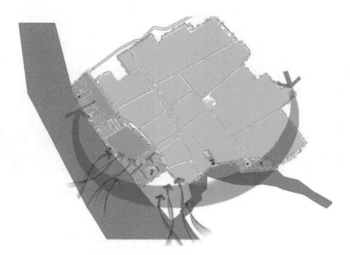

Figure 6. Site showing land area and waterlogged area.

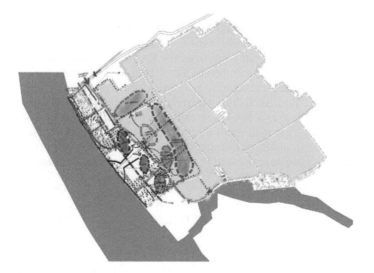

Figure 7. Site zoning.

conservation is centered on not undertaking any landfilling or affecting the shrimp farming practice. The entire site is demarcated via a bund system for pisciculture, but due to a lack of resources and poor maintenance, only 50% of the site is utilized. There is no commercial fish farming and, during the off season, the site remains as a waterlogged area, not realizing its full potential.

To understand whether alternative construction techniques, such as stilt systems, pier and floating systems, were possible, a detailed 'depth analysis' was undertaken for each bund. Figure 7 shows the various depths in each part of the site. As a floating system is the most

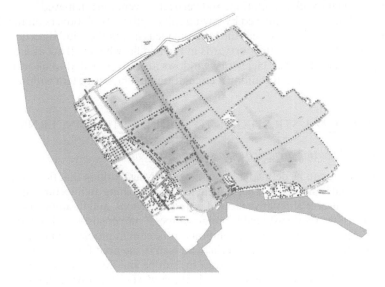

Figure 8. Depth analysis of the site.

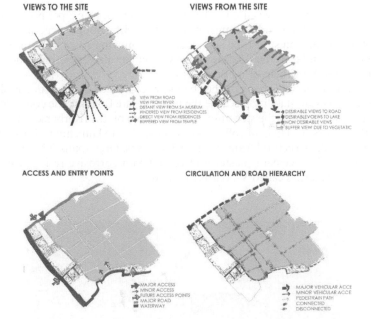

Figure 9.   Site analysis (a) views to the site (b) views from the site (c) access and entry to the site (d) circulation and road hierarchy within the site.

sustainable method to use for minimum disruption to the natural setting, the minimum depth required was 1 m. As seen from the studies, the area toward the left had an average depth of 3 m and, thus, a floating system was proposed for the museum.

Further, the views, access, entry, circulation and road hierarchy were analyzed to understand the suitable orientation and placement of the buildings.

4.2 *Architectural language achieved: Site responsive design considering local climatic factors*

Considering the tropical climate, natural ventilation, a sloping roof, long overhangs, passive and active cooling through orientation, and vertical louvers were achieved.

The natural flow of air is achieved by having full length vertical louvers on the river-facing side. The building is oriented so that the longest side is facing the wind direction.

The building is L-shaped with courtyards in the middle, which will act as a pressure point for catching the wind. Further, clerestory windows are also provided to allow the hot air to rise and flow outside, thus maintaining a circular air circulation system within the building.

5 ARCHITECTURAL STRATEGIES TO ADDRESS THE CONCERNS

5.1 *Relevance of floating architecture in sustainable design*

Floating architecture can be defined as a building for living/working that floats on water with a flotation system and is moored in a permanent location. Floating architecture on water has been emerging as a sustainable alternative around waterside regions, and floating architecture can be regarded as one of the most sustainable building types if proper sustainable factors are applied. Floating architecture could be a future solution for the current problems in many districts, cities and landscapes. Such problems can especially be seen in the need for additional housing areas and construction grounds in some countries in Europe and Asia, as a result of the growing population and/or the slowly rising sea levels in the context of worldwide climate change.

5.2 *Construction and design of floating structures*

Steel pontoon float technology is used to keep the museum in its proper position while floating on the river. While the buildings must be designed with enough structure for lateral and gravitational loads, they must also be stabilized in the moving river. The museum floats on a pontoon system, similar to the floating runways used for airplanes, and is held to the river bed by anchoring with the help of a pile system.

Additionally, the angle and shape of the chains are controlled by Dynamic Positioning System technology to ensure stability. Because the buildings are floating, the construction process requires the prefabricating of the structure of the museum on the riverbanks, and then launching them in the river using a field of rollers as casters, similar to launching a boat into the water.

Construction of the structural frame is in steel to enable large spans. They are then covered in mesh wrap and given cement plaster and coated with corrosive resistant paint. For the walls, lightweight materials called V-boards are used.

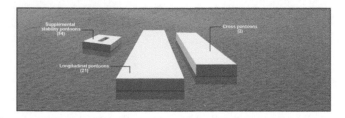

Figure 10.   Types of steel pontoons.

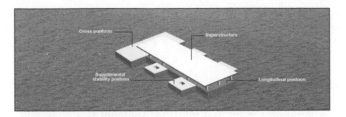

Figure 11.   Steel pontoons assembled together.

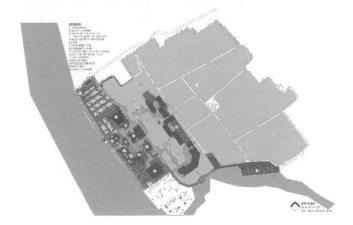

Figure 12.   Site layout.

5.3   The floating mechanism

Airbags are placed, each measuring 2 meters in diameter and 12 meters in length, under the giant buoy. The airbags are made of a special rubber that can support a 2,000 ton buoy, and will enable it to keep afloat.

The buoy is then embedded to the river bed using anchors (coconut piles) with metal chains to prevent it from moving or washing away. The chains are designed so as to automatically tighten when the water level increases.

The key principle is to balance the weight of the building to that of the water displaced. If the weight of the building is less than that of the water displaced, it will float.

5.4   Use of renewable energy in floating structures

5.4.1   Solar energy

Solar photovoltaic power generation has long been seen as a clean energy technology that draws upon the planet's most plentiful and widely distributed renewable energy source—the sun. Cells require protection from the environment and are usually packaged tightly in solar panels. The main museum receives its energy from the solar farm, proposed at the site, making it more sustainable.

5.4.2   Hydrothermal cooling

Hydrothermal temperature control involves pumping the river water from the Kollam–Kottapuram waterway through a vein-like system in the wall and floor, and this cools the building structure down by 15 degrees Celsius (from 45 degrees to 30 degrees), reducing the cost of cooling by around 40 percent. Humidity control is usually considered to be the most expensive aspect of environmental control. The optimum temperature and humidity for a museum is 18–20 degrees Celsius and 45–50% respectively.

Temperature and humidity control has been achieved to a great extent with the help of climatically responsive design. In addition to that, dehumidifiers can be placed in every gallery.

6 CONCLUSION

The Muziris Interpretation Center and maritime museum is a conscious effort to achieve a site responsive design, stressing the importance of sustainable and eco-friendly architecture. The proposal is aimed at inducing a deep appreciation for the lost heritage of Muziris, through understanding the context of the site and celebrating the essence of its uniqueness. The aspect of conservation leads the project by challenging the conventional construction practices.

One of the most challenging aspects of the design was the distinctiveness of the site. As the site is a pisciculture plot, the focus was on nurturing and protecting the natural setting. The site buildability studies and analysis revealed that a larger part of the site was better left as it is, and this can be developed further in the future for aqua tourism. On the remaining plot, the museum was proposed as a floating structure with minimum intervention to the site. Each gallery was visualized as an island floating amidst the tranquil setting, connected through bridges. This not only makes the museum stand out, but elevates the visitor experience.

An ecological and climatically responsive design practice ensures maximum sustainability and retains the natural setting of the site. Further, renewable energy sources, such as solar and hydrothermal energy, are also incorporated in the design, in order to uplift the design.

REFERENCES

Adler, A.D. (1998). *Metric handbook planning and design data* (3rd ed.), Routledge; 5 edition (5 March 2015).

Cherian, P.J., Selvakumar, V., and Shajan K.P., 2007a "The Muziris Heritage Project: Excavations at Pattanam- 2007", in Journal of Indian Ocean Archaeology, New Delhi, Vol 4, pp. 1–10.

Cherian, P.J. (2014). *Unearthing Pattanam, catalogue for 2014 exhibition*. National Museum, New Delhi.

Habibi S (2015) Floating Building Opportunities for Future Sustainable Development and Energy Efficiency Gains. J Archit Eng Tech 4:142. doi:10.4172/2168-9717.1000142.

Kuriakose, B. (2009). *Conservation development plan for Muziris heritage sites* (Consulation draft), Chennai.

Emerging Trends in Engineering, Science and Technology for Society,
Energy and Environment – Vanchipura & Jiji (Eds)
© 2018 Taylor & Francis Group, London, ISBN 978-0-8153-5760-5

Design strategies for daylighting in tropical high rises

Nishan Nabeel, Anju John & A.K. Raseena
DG College of Architecture, Chelambra, Kerala, India

ABSTRACT: Daylighting is one of the measures adopted to take advantage of the climate and environment in designing buildings. Here emphasis is given to the strategies that can be adopted to bring daylight effectively to the tropical high rises thereby reducing the energy consumption due to artificial lighting. The tropical climate differs substantially from conditions of other climates. These differences should be taken into account while designing the facades of high rises in this region. This paper mentions some of the terms regarding the daylighting components, daylighting systems, principles and their analysis in the context of tropical high rise. It also analyses some of the case studies and derives design strategies that can be adopted for daylighting in tropical high rises.

1 INTRODUCTION

Daylighting is the controlled admission of natural light, direct sunlight and diffused skylight into a building to reduce electric lighting which in turn saves energy of the building. If used efficiently, it helps to create productive environment for the building occupants as well as reduce one third of the total building costs. Daylighting design focuses on how to provide enough daylight to an occupied space without undesirable side effects like heat gain or glare. Therefore it balances heat gain or heat loss, variation in daylight availability and glare control.

The main aims in daylighting a building are to (1) get significant quantities of daylight as deep into the building as possible, (2) to maintain a uniform distribution of daylight from one area to another, and (3) to avoid visual discomfort and glare[1].

2 DAYLIGHT MEASUREMENT

2.1 *Daylighting data*

The quantity and quality of the light generated by all the sources at a particular point is difficult to calculate. Luminance is the first factor to be considered. The intensity of illumination from direct sunlight on a clear day varies with the thickness of the air mass it passes through—a function of the angle of the sun with respect to the surface of the earth[2]. Light is intense at noon than at sunrise and sunset, and intense at lower latitudes than at higher ones.

It is difficult to codify the variations in the sky luminance caused by weather, season and time of the day. Several standard sky models have been produced to meet this difficulty like Uniform Luminance Sky Distribution, CIE (International Commission for Illumination) Standard Overcast Sky Distribution and Clear Blue Sky Distribution. For apertures facing away from the sun, reflected light is an important source of indoor daylight. External reflected light includes the light reflected from surrounding surfaces like ground, water, vegetation, other buildings.

1. Lee Jin You, Roger, Lee Ji Hao, Theophilus, Sun and architecture, heavenly mathematics.
2. The European Commission Directorate-General for Energy, Daylighting for buildings.

2.2 Daylight factor

(DF) is the ratio of the light level inside a structure to the light level outside the structure. It is defined as:

$DF = (Ei/Eo) \times 100\%$ where, Ei = illuminance due to daylight at a point on the indoors working plane, Eo = simultaneous outdoor illuminance on a horizontal plane from an unobstructed hemisphere of overcast sky. To calculate Ei, requires knowing the amount of outside light received inside of a building. Light can reach a room via through a glazed window, roof light, or other aperture via three paths:

- Direct light from a patch of sky visible at the point considered, known as the sky component (SC),
- Light reflected from an exterior surface and then reaching the point considered, known as the externally reflected component (ERC),
- Light entering through the window but reaching the point only after reflection from an internal surface, known as the internally reflected component (IRC). The sum of the three components gives the illuminance level (typically measured in lux) at the point considered:

$$Illuminance = SC + ERC + IRC[3]$$

2.3 Sky component

Sky component depends directly on the position of the sun in the sky and the angle in which the sun's rays enter the building. The sun's path varies with location (local latitude), rising and setting position (based on the time of the year), duration of the day and night.

To measure the angle of the sun in its motion across the sky, we need to take its altitude and azimuth reading. Altitude is the angular distance above the horizon measured perpendicularly the horizon. It has a maximum value of 90^0 at the zenith, which is the point overhead. Azimuth the angular distance measured along the horizon in a clockwise direction[4].

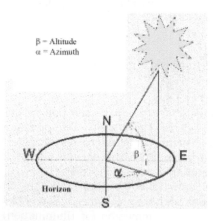

Figure 1.   Azimuth and altitude.
Source: Lee Jin You, Roger, Lee Ji Hao, Theophilus, Sun and architecture, heavenly mathematics.

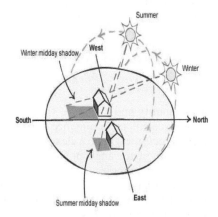

Figure 2.   Angle of the sun in summer and winter.
Source: http://www.build.com.au/window orientation-and-placement.

---

3. https://en.wikipedia.org/wiki/Daylight_factor.
4. Lee Jin You, Roger,Lee Ji Hao, Theophilus, Sun and architecture, heavenly mathematics.

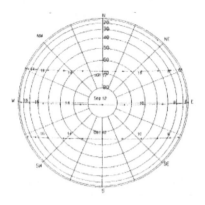

Figure 3.   Singapore sun path.
Source: H. Schepers, Mc Clintock, J. Perry, Daylighting design for tropical facades.

3   POSITION OF SUN IN THE TROPICS—SINGAPORE AS AN EXAMPLE

Singapore located on the equator has on average 6 hours of sunshine in a day and high proportion of clear and partly clear blue skies. The sun path in the figure below shows that the sun is typically high overhead and traces a path that is almost directly East to West[5]. Hence to block the direct solar penetration on North and South facades, horizontal shading may be effective. The East and West facades will be exposed to low angle sun for few hours in the morning and evening which need not be treated.

A clear sky indicates high luminous intensity in the zone directly adjacent to the sun. Hence skylight from a clear sky often penetrate deeper into a building which may also cause glare to the occupants that have direct view to the sky. So in such regions, high light transmitting glazing should be used in combination with the blinds.

There are many new design opportunities which respond to climate and solar path of the tropics. A good understanding of sun and sky luminance is necessary to evaluate these opportunities. The statistical information on the sky distribution of tropics is very little and more research is needed to be carried out in this area.

4   DESIGN STRATEGIES USING DAY LIGHT

Design strategies for daylighting includes daylight optimized footprint, efficient window design, high performance glazing, passive or active skylights, tubular daylight devices, daylight redirection devices, solar shading devices, daylight responsive electric lighting controls, daylight optimized interior design which includes furniture design, space planning and room surface finishes.

4.1   *Building footprint*

- The building should be oriented along the East West direction so that the longer façade is well exposed to the north light.
- Deviation from the south should not exceed 15 degree in either direction for best solar access and ease of control.
- The floor depth should not be more than 60 ft (18.2 m)

---

5. H. Schepers, Mc Clintock, J. Perry, Daylighting design for tropical facades.

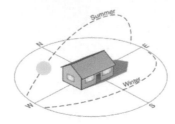

Figure 4.   Building footprint.

Table 1.   Minimum glazed areas for view when windows are restricted to one wall.

Depth of room from outside wall (max)	Percentage of window wall as seen from inside
< 8 m	20%
8–11 m	25
11–14 m	30
> 14 m	35

4.2   *Efficient window design*

One of the principal challenges of daylight design is achieving good daylight levels away from the external walls as daylight factors decrease as distance from windows increases. The daylight penetrates significantly about 4 to 6 m from the external walls in a conventional building. Generally a room will be adequately lit to a depth 2 to 2.5 times the height of the window from the floor, so taller rooms can be daylit to greater depth. Simply increasing overall window size will not be productive if it raises light levels close to the window more than it raises light levels deeper in the room. So the reflectance values of internal materials should be as high as possible.

To design windows intended to view outdoors, the factors like nature of exterior landscape, size and proportion of the interior space, and the positions and mobility of the people who occupy it should be taken into consideration. Minimum areas of glazing for rooms which are lit from one side only as shown in Table 1. This total area should be distributed so as to provide some view from all occupied parts of the room[6].

Daylight optimized fenestration where the window serves the both functions of daylight delivery and view to the occupants is another option. This type of window has two components—daylight window and view window. Daylight window starts at a height of 2.2 m above the finished floor with 50% to 70% transmittance and view window below that with a 40% transmittance. Efficient window design will be different with increasing levels as the angle of the sun differs at different heights.

4.3   *High performance glazing*

High performance glazing admit more light and less heat when compared to the typical window which allows daylighting without negative impact on building cooling load in summer. It is achieved through spectrally selective films and usually consists of a double pane insulated glazing unit.

4.4   *Passive and active skylights*

Passive skylight will be covered with clear diffusing medium which allows light to penetrate. It can be double layered to increase the insulation. Active skylights consist of a mirror system that tracks the sun and is designed to increase the performance by channeling sunlight down into skylight well.

---

6. The European Commission Directorate-General for Energy, Daylighting for buildings.

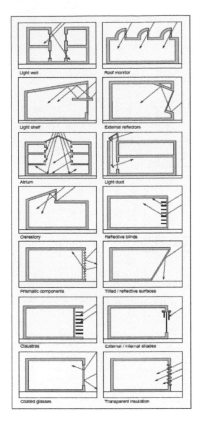

Figure 5. Daylighting devices.
Source: The European Commission Directorate-General for Energy, Daylighting for buildings.

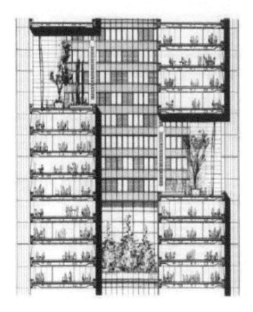

Figure 6. Atria in high rises. Source: The European Commission Directorate-General for Energy, Daylighting for buildings.

4.5 *Other daylighting devices*

The Fig 5 shows some of the daylighting devices used in typical buildings. Well, in high rises only a few of the devices can be used. For example, top lighting is mostly preferred for low rises but they can be provided on top of atrias in high rises.

But a continuous atria might give a tunneling effect for the occupants. Hence instead of continuous atrias, splitted atrias can be provide which enables both side lighting and top lighting as shown in Fig 6.

All the types of sidelighting can be provided on the facades of high rises keeping in mind the building orientation. Each façade may be treated differently according to the intensity of daylight and glare factor.

4.6 *Shading devices*

Shading devices are very much required in high rises to shade the large amount of façade exposed to the sun. They play a major role in creating relaxed inner surroundings that is cool in summer and warm in winter. Discount of 5% to 15% in annual cooling electricity intake has been reported when shading devices are used. They also help to differentiate one façade from another. It helps in maintaining comfortable indoor environment, reduces heat gain and improves the daylighting quality of building interiors.

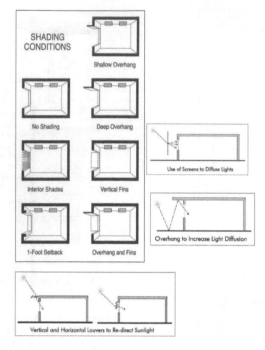

Figure 7.   Shading devices.
Source: Canada-Daylighting-Guide-shading-5.jpg.

Trees Planted serve as shading devices and beautifies landscape and provides oxygen to the occupants. Internal shading devices also help to create a sense of privacy. One of the disadvantage of using shading devices is that it obstructs outdoor views for occupants in some cases. The solar geometry explains that the publicity of each facade to the sun is specific, and varies through orientation.[7] Each façade should be treated differently. For example, facades facing north in the northern hemisphere would not need shading devices as solar penetration is restricted to only few months of summer. Whereas in south elevation, solar penetration should be controlled. Horizontal shading devices above windows are best suited here. The length of the projection depends on the height of the window and the attitude of elevation of the solar at sun noon. It should be designed in such a way that it absolutely get rid of solar penetration in summer time and allows complete solar penetration in winter.

Steps to consider while designing shading device involves 1 Understand the sun path of the environment 2 Select the shading type – Horizontal, -vertical, -egg rate 3 Identify category- Fixed shading devices, – Adjustable shading device, – Movable shading, device-Dynamic shading, device-Automatic shading device 4 Calculate the design dimensions – To understand horizontal and vertical shadow angles.[8]

4.7 Daylight responsive lighting controls

It consists of continuous dimming controls or stepped controls by sensing the availability of natural daylight and the occupant. The photocells in the light fixtures sense he available light and dim or turn off the electric light in response which enhance the energy reduction.

---

7. Mustapha Adamu Kaita, Dr. Halil Alibaba, research paper on Shading Devices in High Rise Buildings in the Tropics.
8. Mustapha Adamu Kaita, Dr. Halil Alibaba, research paper on Shading Devices in High Rise Buildings in the Tropics.

Table 2. Shading strategy by window location.

Window orientation	Shading strategy
North	Usually not needed
South	Overhang, horizontal louvers, trellis over window
East/West	Vertical louvers, horizontal slats, deciduous trees

Source: Canada-Daylighting-Guide-shading-5.jpg.

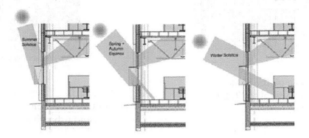

Figure 8. Seasonal performance of shading, redirection devices.
Source: Whole building design guide by Institute of building science, California.

4.8 *Daylight optimized interior design*

Furniture layout in an interior space can regulate the incoming of daylight. For example in case of offices, arrangement of the cubicle partitions play important role in achieving continuous daylight in all seasons. Heights of the partitions running parallel to south façade will be limited and positioned at a distance from the façade, which enhances the use of smaller shading devices. A minimum of enclosed offices are provided to allow penetration of daylight in combination with highly reflective surfaces to bounce and distribute the redirected daylight.

5 EVALUATION OF CASE STUDIES

After the literature study, how these strategies have been adopted practically has to be evaluated. For the purpose, 5 different tropical high rises have been selected and analysed in terms of daylighting strategies adopted.

5.1 *The Hansar, Bangkok, Thailand (Residential)*

Building footprint: Sky gardens are introduced at every 5th floor across east and west elevations allows ample penetration of low density light. Daylighting for all apartments has been ensured by organising the units around a central core and courtyards coupled with vertical slots throughout the building heights.

Shading devices: Balconies acts as shading devices in most apartments across their frontage. This works in combination with delicate sunshades to keep the interior cool at all times. A double layer system was devised to provide shading. As high rise buildings in the tropics gain almost their entire solar load from vertical surfaces rather than its roof, shading of walls is much effective than of roofs.

Façade treatment: A porous façade in the form of a metal mesh screen has been designed to hang off projected ledges at every floor serving as the buildings outer skin. This façade helps in regulating light and air throughout its fully vertical surfaces.

Window design: These external sunscreens overlap with the inner window layer and also function as privacy screens creating a foreground that frames the city views from within the units.

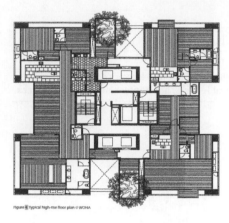

Figure 8 Typical high-rise floor plan © WOHA

Figure 9.   Typical plan of the Hansar.
Source: International journal for Tall buildings and urban habitat.

Figure 5. External sun screens © WOHA

Figure 10.   External sunscreens at Hansar.
Source: International journal for Tall buildings and urban habitat.

Its staggered placement creates visual interest from exterior while neatly concealing a/c condenser units and services in the background. The sun screens are specially coated with metallic bronze colour.

5.2   *Marina bay sands, Singapore, Malaysia (Hotel)*

There are three concrete towers. Each concrete tower hotel is designed at a height of 55 stories. Spanning across the top of the three towers is a 3 acre sky park—a space framing large urban windows between the towers.

Building footprint: The structure is oriented in North South direction with longer facades facing East and West. Here importance was given to the views to the sea.

The three void spaces are connected by one continuous and conditioned glazed atrium which allows large penetration of light to the restaurants, retail spaces and a public thoroughfare.

High performance glazing: Largest amount of heat gain occurs in the west façade. The design solution implemented was custom double glazed unitized curtain wall.

Shading devices: Perpendicular to the façade, glass fins were installed to provide shading. The east façade utilises deep planted terraces which follows the sloping radial geometry of the buildings profile. The planter's help to create microclimate cooling and the deep overhangs of the balconies shade the hotel rooms from direct sun.

5.3   *Kohinoor square, Mumbai, India (Mixed use)*

Daylighting devices: The main building houses spacious lobby and double height landscaped Sky Gardens and a double height terraces with floor to ceiling glazing on every alternate floor to act as tranquil and refreshing breakout zones that allows entry of daylight to the interiors. This follows the concept of splitted atrias discussed earlier.

High performance glazing: The façade consists of faceted unitized aluminum curtain walls with provisions for high performance double glass façades on the tower.

5.4   *Kanchanjunga apartments, Mumbai, India*

Building footpint: Kanchanjunga apartments are oriented in East West direction to face the sea.

Window design: Maximum apertures are provided in these facades. In the South and North facades, apertures in the form of balconies are only provided, to allow adequate amount of daylight along with shading from the hot sun

Figure 11. View of Marina sands bay.
Source: International journal for Tall buildings and urban habitat.

Figure 12. East façade.
Source: International journal for Tall buildings and urban habitat.

Figure 13. Double height terraces.
Source: Kohinoor square, a multifaceted development—research paper by Sandeep shikre.

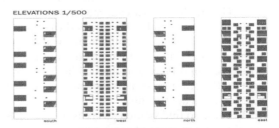

Figure 15. South, West, North and East Elevations.
Source: https://identityhousing.wordpress.com.

Figure 14. Daylight distribution.
Source: https://identityhousing.wordpress.com.

Figure 16. Air wells between towers.
Source: 2013 On Site Review Report on Met tower by Zainab Faruqui Ali.

Figure 17. South façade detail.
Source: 2013 On Site Review Report on Met tower by Zainab Faruqui Ali.

East and West facades are enhanced with lot of apertures both in the form of windows and balconies to enable the entry of low intensity morning and evening daylight.

5.5 Met tower, Bangkok (Residential)

Building footprint: The staggered configuration of the plans coupled with the separation of three towers by air wells provides ample natural ventilation and daylight indoors.

Daylight devices: Balconies allows daylight penetration along with shading in the south façade. The apartments in the Met are naturally ventilated, with access to light and air on all four sides because the tower effectively has no circulation core, in the traditional sense.

Shading devices: Expanded metal mesh was selected for the sun-shading canopies of the units.

6 CONCLUSION

From the above study, we can understand that, in the case of high rises, design strategies for day lighting has to be adopted mainly in the building footprint and the façade treatment. Although we can find a lot of day lighting devices, the ones that we can use in high rises are limited. For example, courtyards with skylights are one of the major day lighting devices used in low rises which is not possible in high rises due to its tunneling effect.

The day lighting devices that can be adopted in high rises of tropical region mainly include the use of balconies, sky gardens, air wells, garden pockets splitted atrias etc... These strategies have many functions like daylight penetration, sun shading, maximize outdoor views, acts as relaxation space and provides natural ventilation too. More than efficiency of window design, balcony design is emphasized in high rises.

Another point to be considered while designing the shading devices are that it should be designed keeping in mind the angle of the sun that changes with increasing height in high rises. The shading devices adopted in 4th floor will be different from that on the 60th floor.

From the case studies we can find that another important strategy used in high rises are the high performance glazing mainly in south and west facades to allow daylight penetration without solar gain. North south orientation which is desirable for daylight penetration is given less importance and the buildings are oriented in such a way that outdoor views like view to the sea, rain penetration etc … are given importance like in Marina bay sands and Kachenjunga apartments.

We can also find splitted atrias as in case of Kohinoor apartments which can be used efficiently as a common entertainment space as well as a day lighting device.

REFERENCES

Adamu Kaita, Mustapha,. Alibaba Halil, (2016, October) *Shading Devices in High Rise Buildings in the Tropics, International Journal of Recent Research in Civil and Mechanical Engineering*, Vol. 3, Issue 2, pp: (37–46).
Ander, Gregg, (2016) *Daylighting, Whole building design guide*, National Institute of Building sciences.
Daylight_factor (2013, July) retrieved from https://en.wikipedia.org/wiki/Daylight_factor.
H. Schepers, Mc. Clintock, J. Perry; *Daylighting design for tropical facades*.
Lee Jin You, Roger; Lee Ji Hao, Theophilus; *Sun and architecture, heavenly mathematics.*
Retrievedfrom https://identityhousing.wordpress.com.
Wong, Mun Sum (2012), Case study—*The Hansar, Bangkok*, CTBUH journal.
Shikre, Sandeep,Kohinoor square, a multifaceted development.
Sun orientation-and-placement (2013) retrieved from http://www.build.com.au/window Sun orientation-and-placement.
The European Commission Directorate-General for Energy, Daylighting for buildings.

Emerging Trends in Engineering, Science and Technology for Society,
Energy and Environment – Vanchipura & Jiji (Eds)
© 2018 Taylor & Francis Group, London, ISBN 978-0-8153-5760-5

Internalizing the concept of sustainability—redefining the curriculum of building construction in architecture schools

Om Prakash Bawane
R V College of Architecture, Bangalore, India

ABSTRACT: Environmental consequences of contemporary construction practices have been scientifically established. Construction activities constitute about 40% of global CO_2 emissions. The seriousness of the issue has given rise to the promotion of sustainable construction technologies worldwide. It is imperative that curriculum content in the subject 'Construction Technologies and Materials' is critically examined to ensure that the issues of sustainability are adequately addressed. The coursework in building construction must help internalize the concept of sustainability in the architecture profession. Sensitizing the future generation of architects toward environmentally friendly construction practices would go a long way in mitigating the adverse impacts of construction on the environment. This paper is an attempt to examine the course content in the subject 'Building Construction and Materials' typically adopted at undergraduate programs in architecture schools in India, in the context of sustainability.

Keywords: Sustainability, Construction, Building Materials, Environmental Impact, Course Contents

1 INTRODUCTION

The subject of 'Building Construction and Materials' occupies an important place in the under graduate architecture curriculum. In terms of the number of teaching hours and credits, the subject is placed second in order next to the core subject of architectural design. A sequential exposure to the process of construction through seven to eight semesters of structured syllabi helps students appreciate various aspects of the constructability of a design idea. Construction is the process and means of transforming the architectural ideas into a built product. In architectural projects, methods and materials of construction could be vital elements in shaping the overall design concept. Architects are trained to be conscious of this fact and exercise their prerogative on the selection of construction systems and materials that would enhance the architectural quality of the build environment. Ironically, any act of construction has a bearing on the natural environment and ecology. The design idea and visualization of the overall build environment are what finally influence the architects' decisions on methods and materials of construction.

1.1 *Weightage to the subject*

Architecture curriculum at undergraduate level in India is largely guided by the 'Minimum Standards of Architectural Education' laid down by the Council of Architecture [1]. In the recommended allocation of teaching hours, 'Building Construction and Materials' has been assigned about 504 hours, which makes a significant share of approximately 12% of the 4,176 total teaching hours.

A closer look at the architecture curriculum of 'Building Construction and Materials' in architecture schools across India would establish the importance of this subject in the training of prospective architects. The curriculum introduces the subject right from the first semester with few exceptions in some schools where the subject is taught from the second semester onwards. The contents of the course are typically spanned over five to seven semesters. In the scheme of teaching, the subject on average is assigned six hours per week. The scheme of teaching of all sample institutions indicates that in terms of weekly teaching hours, 'Building Construction and Materials' is placed second after the core subject of architectural design. In the absolute marking system, the subject enjoys second highest weightage, which is very explicit.

The weightage assigned to the subject in terms of the number of teaching hours and marks highlights its overall importance in the architecture curriculum. There exists a close connectivity between the proceedings of design studios across the semesters and the sequence in which the various concepts of the construction systems are introduced to the students. It is imperative for prospective professionals to appreciate the implications of the construction system on architecture and also the intricacies of integrating the construction and design ideas.

2 UNDERSTANDING SUSTAINABILITY

The word sustainability finds it origin in the Latin word *sustenere* meaning to hold, bear, endure, etc. [5]. The term sustainable has a much philosophical connotation meaning any concept, product or process that would not undermine the capacity of the earthE's ecosystem to maintain its essential functions. In context of the definition of sustainable development advocated by the World Commission on Environment and Development, sustainability is perceived as a holistic concept encompassing social, environmental and economic aspects also referred to as the three pillars of sustainability. The training of architects needs to recognize the most sensitive aspect of sustainability, namely the environment, since construction is one of the major consumers of natural resources and at the same time a major pollutant of the natural environment.

2.1 *Construction and environment*

The construction industry accounts for 40–50% of global energy usage, nearly 50% of world water usage and around 60% of the total usage of raw materials activities (Wilmot Dixon, 2010). Consequently, at a global level construction activities constitute 32% of global CO_2 emissions, 40% of drinking water pollution, 30–40% of greenhouse gases and 40–50% of solid waste generation.

2.2 *Sustainable construction*

Sustainability is still a new concept in construction and yet to be accepted as an integral part of decision making processes. Considering the enormity of adverse impacts of construction on the environment, it is imperative that construction in its totality embraces sustainability. There are little options except sustainable construction practices. Sustainable construction is a holistic approach that considers the reproduction capacity of nature with regard to the materials drawn from the natural stock, energy spent on making a finished product available on the construction sites, impacts of the construction process on the environment and finally the life cycle cost.

3 COURSEWORK IN CONSTRUCTION

The contents and structure of coursework in the subject of 'Building Construction and Materials' are found to be long-established in most institutions. In certain universities the sequence of exposure aligns with the sequence of onsite construction processes, for example, they begin with study of the different types of foundations (substructure), walls and frames (superstructure), including other elements, such as stairs, floors and finally the roofing systems. The other approach is material-based formulation, which begins with construction with basic materials, namely clay, stone, timber at lower semesters, followed by construction in concrete, steel and aluminum. The approach toward delivery of contents appears rather mechanically sequenced and devoid of any stance on issues of sustainability.

3.1 *Orienting the course toward sustainability*

The natural resources are finite and bound to exhaust if their rate of consumption exceeds the reproduction capacity of the earth. Sources of building materials extracted form earth's strata are depleting at a rapid pace and consumption of green stock like timber has exceeded the reproduction capacity of our forests. It is therefore crucial to sensitize the students about sustainable consumption of natural resources. Energy-related air pollution is linked to the overdependence on energy derived from the non-renewable sources, the fossil fuels. The coal-based thermal power plants 54% of commercial energy produced in India. The environmental consequences of using coal are serious. The estimates show that the total CO_2 emissions from thermal power plants in the year 2010 is 498,655.78 Gg, the SO_2 emissions is estimated to be about 3,840.44 Gg [3]. The significant portion of commercial energy is spent on the production of building materials in usable form and their transportation to the construction sites. The concept of embodied energy must be introduced at the early stage of the construction and materials syllabi. Table 1 indicates embodied energy contained in some of the conventional building materials.

The environmental consequences of employing the materials must be included as topics of debate and short research assignments. Every process or system of construction must be presented with environmentally friendly construction systems. Owing to unsustainable consumption patterns and practices, water is becoming an increasingly scarce resource. *In-situ* construction system consumes large amounts of water. The water scenario demands a more conservative approach in its use in construction processes. The introduction of dry construction systems at various levels is imperative. Construction and demolition waste is posing a new challenge in urban centers across India. Estimated waste generation during construction is 40–60 kg per m². Similarly, waste generation during renovation and repair work is estimated to be 40–50 kg per m². The highest contribution to waste generation comes from the demolition of buildings. Demolition of pucca (permanent) and semi-pucca buildings, on average generates between 300 kg per m² and 500 kg per m² of waste, respectively [4]. The dimension of the C&D waste problem is a compelling factor to bring in chapters construction waste management built around the concept of the three R's. Table 2 Summarizes salient aspects and topics relevant to sustainable construction.

Table 1. Embodied energy in building materials.

Materials	Embodied energy in MJ
Burnt brick	3.75–4.5 per unit
Cement	5.85/kg
Lime	5.63/kg
Steel	42.0/kg
Aluminum	236.8/kg
Glass	25.83/kg
Sand (Bangalore)	206/Cu.m.
Stabilized mud block $230 \times 190 \times 100 \times 2.60$	2.6/block
Hollow concrete block $400 \times 200 \times 200$	12.30/block

Table 2.   Summary of salient aspects and topics relevant to sustainable construction.

Aspects of sustainable construction	Topics relevant to sustainable construction
Source of materials	Finite vs. infinite sources.
Embodied energy	Low energy vs. high energy building materials.
Environmental impacts	Environmentally friendly construction techniques.
Water conservation	Dry construction systems.
The three R concepts	Reduce, reuse and recycle concept in construction.

Table 3.   Sustainable construction materials and techniques.

Material/construction technique	Application
Natural clay, natural clay blended with sand and lime, quarry dust with lime/cement	Substitute with conventional mortar.
Natural soil blocks made of laterite, stabilised mud blocks, fly-ash walling blocks	Masonry work.
Bamboo and agricultural waste	Roofing trusses, roofing, walling and flooring construction.
Dry construction	Construction of walls, roofs and floors using precast and prefabricated components.
Alternative materials	Hollow clay roofing and walling, soil-cement compressed walling and roofing techniques, ferro-cement concrete, etc.
Recycling of construction and demolition waste	Reuse of walling blocks, timber components, metallic fixtures and concrete waste, etc.
	Recycling of waste timber, glass, plastics, concrete etc to produce new recycled walling, flooring and roofing materials.
Vernacular Techniques	Adobe construction, Dhajji wall construction.

4   FRAMEWORK FOR CURRICULUM IN SUSTAINABLE CONSTRUCTION

The course content in sustainable construction can be carefully structured by identifying the materials and methods for traditional and contemporary practices. Relevance of indigenous technologies needs to be re-established and the curriculum should serve as an instrument to transfer the environmental construction technologies to the field. Table 3 identifies materials and methods that are deemed sustainable; however the list is only a representative one.

5   NEED TO MOVE BEYOND TOKENISM

The issue of sustainability in construction syllabi needs to be pursued more aggressively. The course curriculum in sample institutions offer subjects titled as, energy efficient buildings, sustainable architecture/planning, Green architecture/buildings. However, only one such course of semester duration is made available, in many cases as an elective at higher semesters. These courses provide only an overview of the subject. The concept and content of sustainability needs to be entwined across the width and the depth of the entire course curriculum.

6   CONCLUSION

Climate change, rise in sea levels, rise in the earth's average temperature and ozone layer depletion are some definite indicators of the extent of the harm that has been caused by human activities. These damages are irreversible and man can only be wise not to worsen the scenario any further. Construction activities are a major factor in causing adverse environmental consequences. The construction sector needs to be supplied with adequately trained professionals to implement sustainable construction practice on sites. The obvious responsibility of capacity building in the fields of sustainable construction lies with universities and part of that responsibility must be shared with architecture institutions, since in most projects the architects will have the final decision with regard to construction techniques and materials. Construction is fundamentally an act against nature, hence the onus lies with the architecture and construction community to make this act less damaging. Sustainable architecture and green buildings are going to be the norms in the construction sector. The university education in architecture has a crucial role in capacity building, spreading awareness, promoting research and development and the transfer of sustainable construction technologies.

REFERENCES

[1] Council of Architecture. (2013). *Minimum standards of architectural education–2008. Handbook of Professional Documents*.
[2] Jagadish, K.S, et al. (2009). *Alternative building materials and technologies*. New Age International Pvt Ltd. Publishers.
[3] Mittal, Moti L., et al. (2012). *Estimates of emissions from coal fired thermal power plants in India*. Retrieved from http://www.epa.gov/ttnchie1/conference/ei20/ session5/ mmittal.pdf.
[4] Waste Management World. (2011). *Rebuilding C&D waste recycling efforts in India, 12*(5). Retrieved from http://www.waste-management-world.com. http://www.etymonline.com.

Policy and assessment scenario

Emerging Trends in Engineering, Science and Technology for Society,
Energy and Environment – Vanchipura & Jiji (Eds)
© 2018 Taylor & Francis Group, London, ISBN 978-0-8153-5760-5

Effectiveness of housing schemes in rural India: A case study of Vellanad, Kerala

N. Vijaya, Priyanjali Prabhakaran & A. Lakshmi
Department of Architecture, College of Engineering, Thiruvananthapuram, India

ABSTRACT: The government of Kerala announced a new housing mission in 2016 called "LIFE", i.e. Livelihood Inclusion and Financial Empowerment, in accordance with the national policy of Housing for All 2022. As a preliminary step to the scheme, all incomplete houses started under previous housing schemes had to be completed. This provided an opportunity for a detailed study of the incomplete houses. Studying the reasons for these houses not being completed can prevent such failures in the upcoming housing schemes. A survey and assessment of these incomplete houses was conducted in Vellanad block panchayat in the Thiruvananthapuram District. The paper outlines the study and the reasons for the non-completion of houses under various rural housing schemes in Vellanad block panchayat.

Keywords: housing schemes, effectiveness

1 INTRODUCTION

Rural housing is an indispensable part of community development and village planning. Rural housing programs are intended to fulfill the housing needs in rural areas, according to the resources available to the community development agency at the level of the block and the grama panchayat in the form of finance, technical advice, demonstration, provision of improved designs and layouts and local building materials and technology. Ideally, such housing developments should not just provide a roof over their head, but should assist the rural community in sustaining their livelihoods. Rural housing development usually incorporates community participation, both physically and financially, in the construction process. This is evident in the Indian housing policies, which underwent constructive transformations from the early phases until the recent ones. It played the role of provider in the early phases of housing development schemes, while it shifted to a facilitator role later, after the decentralization that was caused by the 73rd Constitutional Amendment Act. Fiscal policies have also played a role in the shift from free housing to the introduction of schemes, such as the credit cum subsidy schemes, which involve the participation of the beneficiary. The decentralization also gave local urban bodies powers and responsibilities, and one such item is housing for the poor and homeless.

India faced acute housing shortages immediately after independence and, since then, housing has been a major area of focus of the government as an instrument for poverty alleviation and for the nation's development agenda. The early phase of the Five Year Plans, i.e. from independence until the 1960s, looked into housing migrants and industrial housing as its main agenda. The phase from the 1960s–1990s saw a change in approach in the plans, with the establishment of various Housing Boards that had the objective of taking up housing activities for all sections of society, with a special focus on Lower Income Groups (LIG). Later, the ninth Five Year Plan declared housing to be a subject for the state, following the decentralization caused by the 73rd Constitutional Amendment. The rural housing program, as an independent program, started with Indira Awaas Yojana (IAY) in January 1996. IAY made a good impact on housing the homeless through its three tier system of releasing installments in three stages, as well as completing the house within a time span of three years. Although IAY addressed the housing needs in rural areas, certain gaps were identified during the concurrent evaluations and the performance audit by the Comptroller and Auditor General (CAG) of India in 2014. Indira Awaas Yojana was followed by Pradhan Mantri Awaas Yojana, which has now undertaken a rural version known as PMAY (G) with effect from April 1, 2016.

The homeless statistics in India vary between states, the highest being in Bihar and the lowest in Daman and Diu (Census of India, 2001). It is of serious concern that the real cause of homelessness has not really been addressed by our housing experts. Our national policies have addressed the issues of homelessness, but only on a broad perspective. Planners and housing experts are not participating in the stage where suitable land parcels are identified. Land unsuitable for construction, in terms of access to transportation, topography and infrastructure, is being selected for housing. Construction on such land escalates the cost of building the house due to high transportation prices. The fund then becomes insufficient to complete each stage. Beneficiaries of these housing schemes are always from the economically weaker sections and low income groups. They are forced to contribute not only physically but also financially to complete the house construction, but fail in most of the cases. This is due to the mismanagement of factors such as the elderly age of the beneficiary, lack of motivation and lack of support to start a livelihood to generate an additional income.

2 HOUSING SCHEMES IN KERALA AND THEIR STATUS

State intervention in the housing sector began in Kerala during the 1950s. Many innovative housing programs were developed in congruence with the national policies, raising hope among the homeless poor of becoming house owners. A housing boom began in the state in the mid-seventies. Public housing schemes have also had an impressive record during the past two decades in terms of investment and physical achievements (Gopikuttan, 2002). As the gap between need and supply decreased, the inequality in housing conditions widened in Kerala. The poor have become progressively incapable of self-help and mutual help for solving their housing problems due to various reasons. Thus they have become dependent on a supporting agency for the execution of their housing projects. The absence of professional agencies to take up such roles has created unfortunate situations, such as the mismanagement of funds, poor access to infrastructure, etc.

In India, 32% of rural households live in kutcha structures, but the figure reduces to 19% in Kerala (Panchayat Level Statistics, 2011). Indira Awaas Yojana (IAY), with its inception in 2007, has been the most successful housing scheme in Kerala, decreasing the number of homeless and also establishing the policy of gender mainstreaming. The LIFE mission is the newest housing scheme for the economically disadvantaged and homeless population in Kerala as envisioned by the state government. The LIFE mission survey enlists two lakh homeless populations in Kerala to benefit in the next five years in order to fulfill the national policy of Housing for All 2022. The research objective of the paper is to explore the reasons for the incompleteness of houses in Vellanad and summarize strategies to solve it.

3 METHODOLOGY

As part of the LIFE housing mission of the government of Kerala, it was decided, as a first step, to complete all of the incomplete houses sanctioned under the earlier public housing schemes. A socio-economic study was done by 18 postgraduate students of Planning (Housing) at the College of Engineering Trivandrum, guided by two faculties of the department. The survey was conducted at the block panchayat area to assess the condition of incomplete houses and to examine the reasons behind them. Initially, a pilot assessment of five houses was conducted and a detailed questionnaire was developed. After this, the questionnaire was modified to include all of the necessary data. The research team conducted a two day complete survey of the incomplete houses in Vellanad block on August 10, 2017. The researchers were divided into eight groups to cover the eight gram panchayats of the Vellanad block under the guidance of the village extension officers of the respective gram panchayats. Tribal houses were excluded from the study due to their distinctive housing and socio-economic character and they thus require a different approach.

Vellanad is one of the 12 block panchayats of the Thiruvananthapuram District, Kerala. It is 17 kilometers distant eastwards from the district headquarters. Vellanad is surrounded by Nedumangad and Thiruvananthapuram Taluks to the west and, Nemom and Perumkadavila Taluk to the South, and toward the South. Vellanad was chosen as the model cluster by the state government of Kerala after the preparation of the Integrated Cluster Action Plan for socio-economic and infrastructure planning and the initiation of spatial planning. Vellanad block was formed in the year 1962 (Figure 1) and has had housing schemes for the poor under various government missions. Currently it has several unfinished houses that were begun under various housing schemes, such as the EMS Bhavana Padhati (state government scheme) and Indira Awaas Yojana (IAY) from 2007 to 2015.

The construction of houses had started after identifying eligible beneficiaries under various schemes, such as the EMS housing scheme and Pradhan Mantri Awaas Yojana (PMAY) housing. A primary survey was done in the eight grama panchayats in Vellanad block, namely Aryanad, Kattakkada, Kuttichal, Poovachal, Tholikkode, Uzhamalakkal, Vellanad and Vithura, using the questionnaire prepared. Later, a socio-economic analysis of Vellanad block was done to understand the reasons for the delay. Also, a comparison of various factors linked with the delay was identified. Various analytical methods were used and inferences arrived upon.

From secondary data analysis, it is evident that rural housing in Vellanad had been showing a backward trend from 2006 until 2010, but took a forward leap during 2010–11. The recorded number of dwelling houses constructed and completed in Vellanad rural sector was

Figure 1.   Regional setting of Vellanad block panchayat.

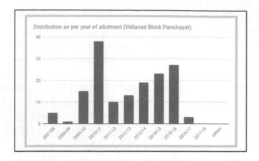
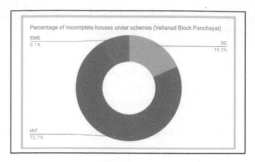

Figure 2. Distribution of incomplete houses under various schemes.

Figure 3. Percentage of incomplete houses under various schemes.

highest in the year 2010–11. The ownership shows a changing pattern, with male ownership decreasing from 70% to 61% and female ownership increasing from 28% to 38% of the total number of houses completed during the years from 2006 to 2011. The number of pucca houses in Vellanad shows an increasing rate from 66% in the year 2006–07 to 82% in the year 2010–11. Kutcha houses increased to 26% during the year 2007–08, but have decreased to 18% during 2010–11. The primary survey summarizes that the national housing mission, Indira Awaas Yojana (IAY), has been actively executed in the Vellanad block during 2010–11, thus affirming that IAY has been able to effectively improve the rural housing scenario in Vellanad by completing a high number of pucca houses with female ownership. The year 2016 saw a change in the strategy of rural housing when the government of Kerala launched its exclusive housing mission named LIFE (livelihood inclusion financial empowerment), which incorporates livelihood inclusivity and financial empowerment as its strategies. The LIFE mission also includes the landless population and dilapidated houses. This mission will empower the beneficiaries to manage to have a pucca house and increase their living conditions by the year 2022, coinciding with the national mission of Housing for All 2022.

4 POST SURVEY ANALYSIS AND INFERENCES: QUANTITY OF INCOMPLETE HOUSES IN VELLANAD, KERALA

Post survey, the team met and had a group discussion to exchange the observations they had made regarding their grama panchayat.

Certain common characteristics could be identified regarding the issues and, thus, the analysis criteria were decided based upon them. Under the IAY scheme, 73% of the incomplete houses were allotted. Various analytical graphs were generated to analyze the physical infrastructure, as well as the socio-economic conditions (Figures 2, 3).

5 ISSUES IDENTIFIED IN HOUSING SCHEMES IN VELLANAD, LEADING TO INCOMPLETE HOUSES

5.1 *Choice of unsuitable land*

Plots ranging in size from 3 to 6 cents were purchased, depending upon the financial efficiency of the beneficiary. Those belonging to the Scheduled Caste received financial aid up to Rs. 75, 000 from the Scheduled Caste Welfare Board. Thus, it is evident from this data that the site selection happened in remote areas, mostly without vehicular access, because of the lack of sufficient funds for this. One unique case has been noted in Poovachal grama panchayat, where 95% of the housing scheme fund disbursed has been used up toward the construction of a retaining wall, the site being on an extremely sloping terrain.

954

5.2 Lack of professional advice and involvement

The survey noticed that the beneficiaries are identified by the village extension officers and the plans approved by the block extension officers.

None of these stages involved getting expert advice from a professional, such as an urban planner or a housing expert. The site selection and preparation of the house plans was also not being undertaken by experienced professionals, such as architects. This led to plans with excessive floor areas that did not match the funds available from the mission and meant that the beneficiaries were not able to complete their houses. In certain cases the plan had not considered the fact that there was only a single occupant in the house, and the floor area was higher than that required.

Poovachal has the highest number of incomplete houses with a floor area exceeding that specified by the government housing schemes (Figure 4). Apparently, IAY does not specify a maximum limit for the floor area for the houses built under it. One specific case was noted in Vellanad grama panchayat where the beneficiary was a woman aged 65 years and the housing plot was located about 700 m inside a rubber plantation, without vehicular access. The location of the house itself causes issues of safety and security.

5.3 Age and health of the beneficiary

A large number of the beneficiaries are senior citizens. Single women who are senior citizens have been allotted single occupancy houses, which are located in plots without proper access (Figure 5).

Families with young children may also find the location inconvenient, where mothers cannot get any assistance for babysitting or for participation in economic activities. About 63% of the beneficiaries in Vellanad block are middle-aged.

It was evident from the analysis that health issues have played a vital role in creating incomplete houses in Vellanad (Figure 6). Two beneficiaries in Vellanad grama panchayat

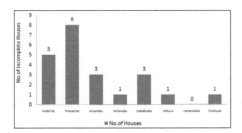

Figure 4. Distribution of incomplete houses based on built area exceeding 66 sqm.

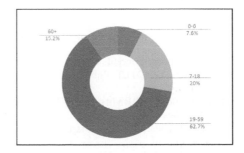

Figure 5. Age category of the beneficiaries.

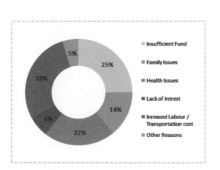

Figure 6. Summary of various reasons for incompleteness of houses.

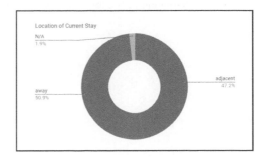

Figure 7. Distance of current stay.

were facing serious health issues. They needed external support to proceed with the management of their house construction.

5.4 *Reduced work participation*

Beneficiary participation in house construction is affected in many ways. The majority of households (50.9%) are staying away from the construction site (Figure 7). This has prevented them from participating in the construction work. There has been a low rate of financial contribution toward the house construction from the beneficiaries, especially the women. Most of the women are either not employed or physically unfit to join the construction due to health reasons. Informal sector financing institutions exist in the block panchayat and most of the beneficiaries have taken loans from such institutions. This has led to a diversion of earnings toward the interest payments of such loans rather than toward house construction.

6 CONCLUSION

Vellanad represents a rural area in Kerala. The majority of the population is engaged in agriculture or related activities. It exhibits a variety of study options along with an active housing mission status. Four reasons have been identified that affect housing schemes for the poor in Kerala: (1) inappropriate plot allocation, (2) lack of expert professional advice, (3) the age and health of the beneficiary, and (4) reduced beneficiary participation. The identified issues can be solved through two strategies. A rework in the team leading the housing scheme is one solution, by including urban planners or housing experts to guide the site selection as well as in preparing appropriate floor plans. Importance is not given to associated home-based income generating activities, which are usually undertaken by the poor. The research leaves scope for further interrogation into aspects such as the scope of livelihood inclusion and methods to incorporate alternative building materials and technology into similar rural housing development. It is hoped that such studies are undertaken before formulating housing schemes.

REFERENCES

Census of India, Concepts and Definitions. (2001). Government of India.
Gopikuttan, G. (2002). Public housing schemes for rural poor in Kerala, A critical study of their suitability.
Housing and Land Rights Network. (2016). The Human Rights to Adequate Housing and Land in India. New Delhi.
Integrated Cluster Action Plan, Shyama Prasad Mukherji Rurban Mission. Ministry of Rural Development.
Kannan, K.P. & Imran. Khan. (2016). *Housing condition in Kerala with special focus on rural area and socially disadvantaged sections*. Thiruvananthapuram: Laurie Baker Centre for Habitat Studies.
Panchayat Level Statistics. (2011). Thiruvananthapuram: Government of Kerala.
Thiruvananthapuram: Centre for Development Studies.

Emerging Trends in Engineering, Science and Technology for Society,
Energy and Environment – Vanchipura & Jiji (Eds)
© 2018 Taylor & Francis Group, London, ISBN 978-0-8153-5760-5

Planning for solid waste management in Kuttanad wetland region, Kerala State

A. Sanil Kumar & V. Devadas
Department of Architecture and Planning, Indian Institute of Technology, Roorkee, Uttarakhand, India

ABSTRACT: Solid waste is one of the biggest issues in wetland regions worldwide due to their high-water content and scarcity of land. The inefficiency in the waste management often leads to pollution of the environment and subsequent degradation of such regions. The study examines the problems prevalent in the Kuttanad Wetland Region, which is known as the 'Rice Bowl of Kerala State'. It is observed that solid waste and its improper handling have emerged as one of the biggest challenges in this region which is affected by severe deterioration in its water sources and pollution of its natural elements. The study is based on primary and secondary data collected from the region and presents a profile of successful solid waste management strategies through the case discussion of The Netherlands. The paper recommends strategies for solid waste management in wetland regions which could lead to a plausible sustainable development of the region.

1 INTRODUCTION

Municipal Solid Waste (MSW) is one of the biggest challenges in the development of urban pockets worldwide. The composition of municipal solid waste varies greatly from municipality to municipality, and it changes significantly with time (Kumar, et al., 2016). The process of waste management includes an array of tasks which are focused towards generation, prevention, characterisation, monitoring, treatment, handling, reuse and residual disposition of solid wastes. As compared to the natural ecosystem and the ecological cycle, waste management focuses on the waste generated from the man-made activities and processes. This process faces a lot of challenges in situations like India with increasing population, squalor and dwindling economic conditions. The system of governance is also a big hurdle consisting of corrupt practices, rapid unplanned urbanisation and lack of resource management. The development trend has exerted more pressure on the existing system which requires better management and technological interventions. The management of waste in a wetland region brings a set of uncertain factors for effective execution. The scope of the research includes solid waste disposal strategies pertaining to a wetland region. The study is based on primary and secondary data collected pertaining to the Kuttanad Wetland Region (KWR) which surrounds the Vembanad Lake.

1.1 *Characteristics of solid waste in Kerala State*

In the case of Kerala State, the quantum of MSW generation varies between 0.21–0.35 kg/capita/day in the urban centres and it goes up to 0.5 kg/capita/day in large cities (NEERI, 1996). The average daily per capita generation of solid waste comes to 0.178 kg with a very high variation from 0.034 kg (Koothuparamba) to 0.707 kg (Thalassery) (CESS, 2001; Padmalal & Maya, 2002; Varma & Dileepkumar, 2004). In the municipalities located within the Kuttanad Wetland Region, the total waste generation varies between 5 to 43 tonnes per day while the per capita waste generation amounts to an average of 240 to 250 grams per day (NEERI, 1996). The latest report (generated in the year 2012) by the Central Pollution

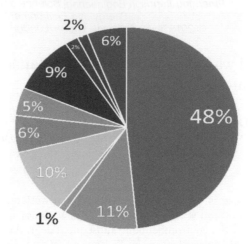

Figure 1.   Quantum of municipal solid waste generation in Kerala State. The majority of the waste generated (48%) is through domestic sources, followed by commercial establishments (11%), hotels & restaurants (10%), street sweepings (9%), markets, construction, (6%), institutions (5%), among other sources.

Table 1.   Municipal solid waste generation as per the standard norms (based on reports from NEERI, 1996; Varma & Dileep, 2004; SEUF, 2006).

S. No	Name of municipality	Population (2001)	MSW generation tonnes/day	Per capita generation (kg/per person/day)
1	Alappuzha	177079	43	0.24
2	Kottayam	60725	15	0.25
3	Chenganassery	51960	13	0.25
4	Thiruvalla	56828	14	0.25
5	Cherthala	45102	11	0.24
6	Vaikkom	22637	5	0.22
7	Kayamkulam	65299	16	0.25
Total		479630	117	0.24

Control Board puts the State average figure at around 249 grams per day per person (CPCB, 2012). It is observed that there is no drastic change in the total MSW generation in the region. The waste sourced from domestic sources forms the majority (48 per cent) of the MSW, followed by commercial establishments, hotels and restaurants, street sweepings, with waste coming from markets, construction and demolition activities, institutions/schools, proving to be the other important components. In the major cities of the State, around 80 per cent of the waste is compostable organics enabling high level of recycling in the form of manure or fuel (KSUDP, 2006 & Varma, A., 2009).

2   KUTTANAD WETLAND REGION

The Kuttanad Wetland Region (KWR) is spread over 1471 square kilometres, and is a unique site where below sea level paddy farming is conducted in India. It is also recognised as a Globally Important Agricultural Heritage System (GIAHS) by FAO (Food and Agricultural Organization) of United Nations. The KWR is spread over three districts (Alappuzha, Kottayam and Pathanamthitta) with the river waters reaching the Vembanad Lake from six districts. The region consists 78 villages, 25 census towns and seven municipalities.

2.1   Existing facilities

The district of Alappuzha has a windrow composting facility with an input of 50 tonnes/day which caters to the existing population of Alappuzha Municipality. Kayamkulam Municipality has a landfill site at Murikkummoodu which is spread over an area of 0.82 acres catering to about 12 tonnes/day. There are four vermi-composting plants in the Alappuzha Municipality. They have a combined capacity of 10 tonnes/day. Other landfill facilities include Chengannur Municipality (Perunkulam Paadom—capacity of 8 tonnes/day), Mavelikkara Municipality (capacity of 9 tonnes/day) (TCPO, 2011). The Alappuzha municipality has been recognized as a zero-waste town owing to its successful implementation of solid waste management strategies. The municipality set up biogas plants, pipe compost units in households and aerobic composting units in public places. It also set up surveillance cameras across the city, linked to the police control room, to catch those littering public places (Philip, 2014). The success of this model has resulted in the village panchayats surrounding Alappuzha taking up aerobic composting units at various locations along the National and State highways.

2.2   Case discussion—The Netherlands

The selection of The Netherlands as a case discussion for solid waste management strategies was based on the fact that the region has a running coastline of 451 kilometres which is similar to the Kerala State with coastline of 580 kilometers. In a similar manner, most of its landmass is coastal lowland and reclaimed land (polders) and it is prone to flooding. Despite the threat from the surrounding waters, they are one of the leading nations in their waste management strategy. Their planning was shaped by scarcity of land, proximity to water and their concerns for the environment—features common to a wetland region. The Dutch Waste Management Association (DWMA) interacts with about 50 companies which are involved in collecting, recycling, processing, composting, incinerating and land filling waste. The waste disposal model considered landfilling waste as the last objective relying on recycling strategies for various waste categories and taxing each tonne of landfilled waste (Feller, 2014). The Dutch waste management strategy is backed by a strong legislation brought forth through the efforts of Ad Lansink (known as Lansink's Ladder strategy) which imbibes stringent waste management norms in all aspects of daily life.

Some of the waste treatment innovations such as roads made of recycled plastic are presently being adopted in Kerala State. The identification of waste into different categories will lead to its effective utilisation in the wetland region. The segregation and recycling of these waste categories can be executed through the linking of these items with existing taskforces such as Kudumbasree units, Green Army, self-help groups in individual villages and townships with opportunities for income generation, among others.

Some of the unique waste management innovations brought forth by The Netherlands waste treatment agencies which can be adopted in the near future include:

- Underground refuse containers with e-access—General use (records each user and their waste dumping pattern). This system can be adopted in the newly planned housing townships and large-scale technology parks.
- Styrofoam recycling units—Package industry (recycles 100 per cent of the waste). This can be adopted with a complete recording of the input and output data pertaining to the seven municipalities (first phase).
- Waste sorting machines—construction industry (recycles 95–98 per cent of the waste). There can be an amendment to the existing rules and regulations related to the construction industry to adopt materials with high recyclability.
- Online reporting of waste accumulation—response time linked to online data. This system can be adopted for major junctions and intersections in the region in the initial phase.
- Methane extraction—from various landfill sites. This was one of the biggest problems which led to the change in collection and disposal patterns within the Alappuzha municipality. However, it can be adopted to the various landfill sites with added infrastructure.

- Nerada technology—converts waste water reusable for agriculture (1/3rd size of a conventional plant). This would be critical to the regions such as Kuttoor panchayat (near Thiruvalla) and locations surrounding Alappuzha municipality where waste water is directly dumped into the existing canals.

2.3 Data collection and analysis

2.3.1 Primary survey

A socio-economic survey was conducted in Kuttanad Wetland region with responses sought from various households distributed within the location. The information collected was part of a larger survey format conducted as part of a study on development for an integrated planning framework for Kuttanad Region. The survey locations included Ramankary, Kainakary, Nedumudi, Champakulam, Thakazhy, Edathua, Thalavady, Vechoor, Thalayazham, Mannancherry, Thaneermukkom, Karuvatta, Thiruvarppu, Kottayam and Nedumpuram. The sites selected were based on the premise of having optimum representation from various locations along with a consideration for most afflicted regions in the wetland based on primary observation and secondary data. A total of 441 households (1891 respondents) were covered based on random sampling method in the duration from June to November 2017 and their responses collated. They were inquired about questions related to waste disposal methods, agency through which waste was disposed, frequency of collection or disposal, among others. The respondents were also inquired about their perception regarding the quality of air, water and soil conditions within the region. They were also requested to record their remarks regarding the key problems and their root causes within the study region. The following observations are suggested as per the primary survey and reconnaissance studies within the region:

- The majority of the residents (75 per cent) of the KWR are burning their waste on site, which is followed by people who bury their waste on site in plastic bags (18 per cent), followed by households who throw their waste at random sites (4 per cent) and families with waste collection facilities (3 per cent).
- Waste collection, wherever applicable, was being conducted daily or on a weekly basis. Most of the residents (93 per cent) disposed off their waste on a regular basis using burning or burial methods. The frequency of burning or burial varied anywhere between twice a week or once a week in most cases.
- There are no charges or expenditure related to waste disposal incurred by the residents.
- Wastes such as bottles or glasses are discarded in an improper manner but burning plastics and organic waste is prevalent.
- About 54 per cent of the respondents have rated the air quality of the region as Good, while 43 per cent have rated it as satisfactory while 3 per cent have termed it as poor.
- Regarding the soil quality, 55 per cent have termed it as satisfactory while 41 per cent rate it good, while the remaining 4 per cent term it as poor.
- In questions regarding quality of life within the region, a majority (89 per cent) have termed it as good (34 per cent) and satisfactory (55 per cent) while 11 per cent of the sample population rates it as poor.
- The respondents were asked about the imminent problems within the region. Based on the ranking method, the improper handling of waste was rated to the biggest problem, followed by the salinity in water (2), pollution of water sources (3), scarcity of water for domestic use (4), and an increase in pests (5), among other issues. The possible causes for these problems were rated in the following descending order—lack of waste management (1), increase in illegal waste dumping (2), improper agricultural activities (3), lack of integrated planning (4) and obstructed water channels (5), among other causes.
- The commercial establishments (hotels, restaurants, shops) dump their liquid waste directly into the natural water channels without any treatment.

2.3.2 Field discussions

The author conducted discussions with block and panchayat members from the selected regions before commencing on the survey. The field study helped in gathering information from the following region related to persistent problems within the region and the upcoming projects proposed by the members which are to be implemented in the coming five to ten years. Further, the details regarding the region were discussed with various experts from fields of urban planning, agriculture and allied sciences, scientific research, along with government body officials including the municipality, Town and Country Planning department, among others to ascertain the issues related to developmental hurdles and solid waste strategies within the region. These discussions were also conducted with local NGOs and the findings documented in this section.

It was observed that solid waste disposal strategies were restricted to the municipalities with a majority of the villages and panchayats lacking these methods of proper waste disposal. The quality of life of the residents was severely affected owing to the poor quality of water and excessive waste being dumped directly into the water sources. In the ecological system, the region is moving towards a great imbalance with a spike in the number of debris, weeds and floating plants thriving in the environment leading to severe damage to the water quality. Since, the water channels are connected to the main river streams flowing through the region, the existing water channels, which have been converted into drains, in turn, pollute the larger water bodies within the region. The presence of plastic (in all its forms) within the region was resulting in many of the paddy fields turning into dumping grounds for these waste items. The empty plots within the regions were also being used as dump sites by residents. It was suggested by one of the experts that the timely cleaning of canals on a regular/annual basis could be a major step to the solving of other problems related to the region (Kumar & Devadas, 2015).

2.4 *Critical issues*

Based on expert discussions and personal interviews, the following issues were projected as critical regarding efficient waste management:

- Sensitive nature (ecology) of the site
- Human errors in handling of solid waste
- Existing waste at various locations
- Budget constraints owing to unique geographical conditions
- Scarcity of land
- Excess presence of water
- Lack of recycling practices
- Imminent threat of flooding.
- Toxic nature of some wastes.
- Will of the governing authorities.
- Lack of effective or stringent regulations (Kumar & Devadas, 2015).

3 METHODS OF SOLID WASTE DISPOSAL

A comparison was conducted between the various solid waste disposal methods and their advantages and disadvantages based on their utility in the study area. The various methods considered included Landfills, Incineration, Pyrolysis, Deep-well injection and Deep-sea waste disposal. Based on the analysis and expert inputs, it was observed that landfilling and incineration were more feasible options for large scale solid waste disposal strategies for the wetland regions.

4 STRATEGIES FOR WASTE MANAGEMENT

Following the analysis of the existing situation, the following recommendations are forwarded as sustainable strategies for solid waste management.

- Integrated Planning for Waste Management—Solid waste management can be taken up as a priority in planning realizing its potential as a key component to environmental protection and income generation.
- Waste management through policy planning—A step in this direction is implemented through the 'Swachh Bharat Abhiyan' (Central scheme) and 'Suchitwa Mission', 'Green Protocol' or 'Haritha Keralam' (State schemes). The various policies are facilitating liquid and solid waste management through setting up of stable infrastructure for waste disposal. This has to be extended further considering the harmful nature of certain waste components which would result in higher investment in disposal than manufacturing costs.
- Minimisation of waste generation through organised processes—Waste minimisation is defined as the reduction, to the extent feasible, of pollutant waste that is generated or subsequently treated, stored or disposed of. (Polprasert, 1996).
- Waste segregation at source—The segregation of waste helps in planning of the separate treatment methods for each of the waste categories. The cost of transporting mixed garbage is high and segregation at source can reduce the total cost by 30 per cent. (The Hindu, 2013). Many countries which have planned for waste segregation by marking bins with different colours as per the waste category.
- Waste management as a community initiative—Waste collection and disposal involves the contribution of all members of the society. Thus, the planning agencies have to project waste management as a community initiative to consider its best utility.
- Usage of waste for power generation—In the post-segregation scenario, the 'segregated' waste items can be processed and utilised as fuel for powering turbines to generate electricity.
- Composting strategies—One of the primary strategies for a wetland region is the concept of composting. Considering wetland regions, these facilities needs to be organised in a planned manner at locations which are less prone to flooding but are generally accessible.
- Decentralised waste treatment strategies—In wetland regions, it is easier to focus on micro-waste processing units which can convert the waste into a usable form and make it less harmful for nature.
- Technological Intervention—Waste segregation and its management can be further reinforced through the use of technology enabling users to know various inputs regarding waste disposal strategies.

5 CONCLUSION

The concept of solid waste management in wetland regions has many facets for consideration. The success of the Alappuzha Municipality in adopting the Zero-Waste strategy reflects that a series of different techniques can be combined together for effective waste management. The future research could focus on sustainable technologies which encourage user participation in waste management strategies. This research suggests that solid waste management strategies have to combine strong legislative procedures, modern technology and participatory approach to achieve plausible solutions.

REFERENCES

CESS. 2001. *Carrying capacity based development planning of Greater Kochi Region (GKR)*, Rep. Centre for Earth Science Studies, Thiruvananthapuram. p. 269.
CPCB. 2012. *Status Report on Municipal Solid Waste Management*, Rep. Central Pollution Control Board. New Delhi. p. 9.
Feller, G. 2014. Dutch Successes [Journal]. - *Northbrook: Waste Management World*, 2014. - 1: Vol. 11.
KSUDP. 2006. *Solid waste management of Kollam, Kochi, Thrissur and Kozhikkode Corporations of Kerala*. Dft. Detailed Project Report. Local Self Government Department, Government of Kerala & Asian Development Bank.

Kumar, S. & Devadas V. 2015. Waste Management in Kuttanad Wetland Region. In: Sustainable & Smart Cities Conference 2015. Surat: SVNIT, Surat.

Kumar, S; Dhar, H.; Nair, V.V.; Bhattacharyya, J.K.; Vaidya, A.N.; Akolkar, A.B. 2016. Characterization of municipal solid waste in high-altitude sub-tropical regions. *Environmental Technology*. 37: 2627–2637.

NEERI. 1996. *Municipal solid waste management in Indian Urban Centres*. Rep. National Environmental Engineering Research Institute. Nagpur.

Padmalal, D.; Narendra B.; Maya K.; K., R. Reghunath; Mini, S.R.; Sreeja, R. & Saji, S. 2002. *Municipal solid waste generation and management of Changanasseri, Kottayam and Kannur Municipalities, Kerala*. Rep. Centre for Earth Science Studies, Thiruvananthapuram. CESS PR-02-2002. p. 47.

Philip, S. 2014. 'Clean Home, Clean City': Alappuzha municipality shows the way. *The Indian Express*. [online] pages. Available at: http://indianexpress.com/article/india/india-others/clean-home-clean-city-alappuzha-municipality-shows-the-way/ [Accessed October 10, 2017].

Polprasert, C. 1996. *Organic Waste Recycling—Technology and Management* [Book]. - England: John Wiley and Sons, Vol. 2nd edition.

Special Correspondent. 2013. Waste segregation at source cheaper, effective: expert. *The Hindu*.[online]. Available at: http://www.thehindu.com/news/cities/Mangalore/Waste-segregation-at-source-cheaper-effective-expert/article11613715.ece [Accessed October 10, 2017].

TCPO. 2011. *IDDP Alappuzha Draft Plan* - Alappuzha: yet to be published.

Varma, A. & Kumar, D. 2004. *A handbook on solid waste management*. Clean Kerala Mission, Govt. of Kerala. p. 78.

Varma, A. 2009. *Status of Municipal Solid Waste Generation in Kerala and their Characteristics*. [online]. Available at: http://www.seas.columbia.edu/earth/wtert/sofos/Ajaykumar_Status%20of%20 MSW%20Generation%20in%20Kerala%20and%20its%20Characteristics.pdf [Accessed October 1, 2017].

Emerging Trends in Engineering, Science and Technology for Society,
Energy and Environment – Vanchipura & Jiji (Eds)
© 2018 Taylor & Francis Group, London, ISBN 978-0-8153-5760-5

Green infrastructure as a tool for urban flood management

Reshma Suresh & C.A. Biju
Government Engineering College, Thrissur, Kerala, India

ABSTRACT: Floods have been a frequent occurrence in cities, be they planned or unplanned. This can be attributed to the increase in constructed spaces, which causes a decrease in the permeable land. For a long time, governments implemented gray solutions, such as drains and levees, to mitigate flood risk. However, with the adverse impacts of floods growing, a need has arisen for a more integrated approach to urban flood risk management. Green Infrastructure (GI) solutions have emerged as a key component of this integrated approach.

Keywords: Urban flood, Greay solutions, Green Infrastructure

1 INTRODUCTION

With the increase in impervious surfaces in urban areas, the storm water runoff is overwhelming the existing infrastructure, which causes flooding and sewer overflows. Under these conditions, cities are under pressure to find cost-effective, sustainable and socially responsible solutions to urban flood management. This can be done through Green Infrastructure (GI), which can complement or augment the present solutions for urban flood management. The term "green infrastructure," when used for storm water management, denotes techniques, such as rain gardens, green roofs, permeable pavements, street trees, and rain barrels, that infiltrate, evapotranspirate, capture, and reuse storm water onsite. GI allows for both "a reduction in the amount of water flowing into conventional storm water systems (and thus a reduction in the need to build or expand these systems) and a reuse of storm water at the source."

2 GREEN INFRASTRUCTURE

2.1 *Definition*

Green infrastructure refers to natural or semi-natural ecosystems that provide water utility services that complement, augment or replace those provided by gray infrastructure. A GI framework can be developed on any scale, including multinational, national, regional, local community or on an individual plot. Have suggested three scales: individual, community and statewide scales. The framework is applied depending upon the relevant goals of the community and the benefits to the environment.

2.2 *Elements of GI*

A wide variety of green infrastructure elements are present and, depending upon the location, preferences, living standards and the goals which communities require, these elements may differ. All these GI elements are intended to provide a safe environment and may increase the economic status of the community.

The GI elements are as follows:

2.2.1 Green roofs

These are vegetated roofs that are partially or completely covered with plants. They grow in 3–15 inches of soil, sand, or gravel planted over a waterproof membrane. It can be intensive (175–220 kg. per sq. ft.) or extensive (33–110 kg. per sq. ft.). It reduces the annual storm water runoff by 50–60% on average, offers the retention of 90% of volume for storms less than one inch and at least 30% for larger storms, and increases the lifespan of the roof by 2–3 times. Additional benefits include protection from wind damage and UV rays, and regulating temperature impacts by as much as 21°C.

2.2.2 Rain gardens

Rain gardens are also known as bio-retention basins. These are designed to absorb rain water that drains from impervious areas (roofs, parking areas, streets, and lawn areas). These are shallow depressions that are planted with native flowering plants and grasses. They can reduce runoff because storm water drains into the soil, leading to a reduction in erosion. It can be made effective by disconnecting downspouts from homes and commercial buildings that once directed water into the existing storm water management system. Rain gardens can reduce 40–70% of runoff.

2.2.3 Porous pavements

Porous pavements are permeable pavements that allow rainwater to infiltrate into the soil and these can be used as pavements, parking lots and pedestrian pathways instead of conventional pavements. They can be made of pervious concrete, porous asphalt, and permeable interlocking pavers and can reduce runoff volume by 70 to 90%. They can infiltrate 3 inches of rainwater from a 1-hour storm and have an average life expectancy of 30 to 35 years.

2.2.4 Vegetated swales

These are linear features that have wide and shallow channels present on side slopes, with natural vegetation to promote infiltration and reduce the flow velocity of the runoff. These can be used in parking lots, since storm water picks pollution from vehicles, and these can be treated before entering the watershed. It can reduce the runoff into drains by 50%.

2.2.5 Wetlands

Wetlands receive and treat storm water that is drained from limited impervious areas. These are aesthetically pleasing and suitable for small wildlife habitats. These do not require a large amount of space and can be useful in congested urban areas. It is an effective means of managing the more intense and frequent precipitation events. It helps in reducing peak flows and reducing the intensity of flood events in urban areas.

2.2.6 Planter boxes

Planter boxes are rain gardens that have vertical walls that close at the bottom and they absorb runoff from sidewalks, parking lots and streets. They are designed to store water temporarily and slowly infiltrate it into the soil. They help to discharge water from terraces into the soil, rather than into drains.

Figure 1. Porous pavement.

2.2.7 Green parking

These are used in parking lot designs that incorporate green infrastructure in their functional requirements. This also helps in the management of storm water and reduces the heat island effect. These can be used in parking lots, cycle tracks and in pedestrian walkways. Impermeable land can be minimized and reduce the runoff by 25%.

2.2.8 Green streets

Green streets help in integrating all the green infrastructure elements to store, infiltrate and evapotranspire storm water. Planter boxes, pervious pavements and bio-swales can be used in the design of green streets. They help to reduce the storm water runoff to a maximum of about 70 to 90%.

2.3 Benefits of green infrastructure

GI has a predominant importance in creating sustainable communities and in tackling climate change. It offers a wide variety of benefits to communities and ecosystems. The removal of pollution, functioning of ground water recharge, flood control and the management of storm water are the overall benefits of GI. It provides a much better quality of life, a healthy environment and an efficient transportation system. GI implementation offers multiple benefits in terms of social, health, environmental, economic and climate aspects. A list of summarized benefits are described below to show its importance in each individual aspect.

2.4 Role of GI in flood management

In urban areas, up to half of the area does not allow rainwater to seep into the ground. This is due to impervious surfaces, such as roads, roofs, parking areas, sports facilities, etc.

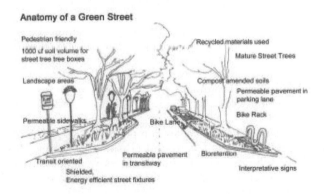

Figure 2. Anatomy of a green street.

Table 1. Benefits of GI.

Type	Improve air quality	Reduce heat island effects	Improve water quality	Improve energy efficiency	Reduce pollution	Flood control
Porous paving		Yes	Yes		Yes	Yes
Bio-swale		Yes	Yes			Yes
Planters	Yes	Yes				Yes
Rain gardens		Yes	Yes			Yes
Wetlands		Yes	Yes	Yes		Yes
Green parking	Yes	Yes			Yes	Yes
Green roof		Yes		Yes		Yes
Green streets	Yes	Yes		Yes	Yes	Yes

Source: The Center for Clean Air Policy, February 2011.

Table 2. Benefits of GI elements of flood management.

Storm water control method	Average peak flow (% removed)
48″ Soil bio-retention	85
30″ Soil bio-retention	82
Constructed wetlands	81
Porous pavement	79
Green roof	65
Bio-swales	50
Green parking	47

Source: The Center for Clean Air Policy, February 2011.

Obviously, urban areas in more developed economies may have a higher share of built-up areas with hard surfaces. Since the ground is unable to absorb rainwater, the water flows directly into rivers, streams, and sewers.

The United States Environmental Protection Agency (EPA) estimates that a typical American city block "generates five times more runoff than a woodland area of the same size, while only about 15 percent (of rainwater) infiltrates into the ground for groundwater recharge". During major rainfall events, this additional runoff overwhelms rivers, streams, and sewers and causes severe flooding. Additional risks include drought (due to reduced groundwater recharge and reduced surface water storage) and negative impacts on water quality.

The combination of increasing flood risk, the potential for major human and economic losses, and the unevenness in the efficacy and costs of gray infrastructure has led to a growing interest in exploring other approaches.

GI solutions focus on managing wet weather impacts by using natural processes. As part of an integrated flood risk management framework, they can also deliver environmental, social, and economic benefits, and can be cost-effective, low in impact, and environmentally friendly (sustainable). In contrast, traditional approaches, such as levees and dams, focus on changing the flow of rivers and streams to protect local communities, and use piped drainage systems in urban areas to quickly move storm water away from the built environment. As an unintended consequence, fast drainage of water may result in drought problems, and drought may reduce the ability of existing green spaces to provide important services, such as reducing heat stress.

Gray infrastructure solutions remain a key component of flood risk management frameworks and are necessary in many situations. But GI solutions can be a valuable part of an integrated approach. GI solutions include, among others, wetlands, bio shields, buffer zones, green roofing, tree pits, street side swales, porous pavements, and the use of green materials (wood, bamboo, coconut nets, etc.). These measures not only help to reduce flood impacts but also produce environmental and health benefits.

3 FLOODING IN INDIA

India is the most flood affected nation in the world after Bangladesh. It accounts for 1/5th of the global deaths by floods every year and on average 30 million people are evacuated every year. The area vulnerable to flood is 40 million hectares and the average area affected by floods is 8 million hectares. Unprecedented floods take place every year in one place or another. The most vulnerable states of India are Uttar Pradesh, Bihar, Assam, West Bengal, Gujarat, Orissa, Andhra Pradesh, Madhya Pradesh, Maharashtra, Punjab, and Jammu and Kashmir. A history of floods can be seen from ancient times. In the independent India, the first major flood occurred in 1953. After this a series of floods happened every year.

3.1 *Cause of urban flooding in India*

About 63% of the area of India is made up of urban areas.

The area under urban settlement in India has increased from 77,370.50 sq. km in 2001 to 102,220.16 sq. km in 2011, showing that 24,850.00 sq. km of additional land area has been brought under urban use. The land use changes show that there has been a 74.84% increase in built-up areas and a 42.8% decrease in open spaces between the years 1990 and 2015, with a substantial increase in urbanization. The vegetation had decreased by 63% by 2015. In the past 43 years, the increase in peak runoff and runoff volume is marginally varied by 3.0% and 4.45% and the total flood hazard area has increased by 22.27% in urban areas. It is also seen that there has been a 35% decrease in wetland areas from 1990 to 2015, which is the main natural basin for storing the rainwater.

Thus, it can be summarized that the major causes of urban flooding in India are:

- land use changes
- surface sealing due to urbanization (which increases runoff)
- occupation of flood plains and obstruction of flood flows
- urban heat island effect (which has increased the rainfall in and around urban areas)
- sudden release of water from dams located upstream of citizen towns and the failure to release water from dams resulting in a backwater effect
- the indiscriminate disposal of solid waste into urban water drains and channels.

4 GI IN INDIA

Green infrastructure has been practiced in India for a long time, but not in a co-ordinated manner. Not all the elements of GI are implemented in a place, which can lead to multiple and long-term benefits. Many of the elements are being successfully implemented, but are not using the benefits of their combined use. For example, green roofs are being practiced effectively in Bangalore, Hyderabad. The rain gardens are implemented in Samshabad air-port. The rainwater harvesting is also an element of GI that is being widely practiced in India. Porous pavements, on the other hand, have not been used in India until now, so their potential for urban flood management has not yet been explored.

5 CASE STUDIES AND ANALYSIS

5.1 *Case studies and analysis*

There are various examples from across the world where green infrastructure has scored over the contemporary gray infrastructure. They throw light on the integration of each element of GI and how successful each project was.

5.2 *Case study themes for consideration*

It can be difficult to compare GI projects—each is very context specific and the process through which each comes into function depends on the existing planning system. However, a number of common themes arise that can be used to make GI projects successful. They are 1) Approach 2) Regulatory framework 3) Implementation and monitoring 4) Results.

5.3 *Case studies and analysis: Methodology*

The case studies of GI across various projects were considered and categorized. The general details of the city where GI was initiated, origin of intervention, approaches, effects and its role in urban flood management were considered for each of the projects.

5.3.1 *Portland, Oregon*
Portland is the biggest city in the northwestern state of Oregon. It has a population of over 600,000, and is highly urbanized. The city has a history of major floods. Portland has 50% impermeable land surfaces (60% attributed to streets, 40% to rooftops). In 2004, Portland

experienced 50 overflow events. The city has failed to deal with it by the construction of dams, flood walls, etc. The city is a prime example of green storm water management.

Approach: The city adopted a comprehensive multifaceted approach, which includes regulations, incentives and the monitoring of results. In 2007 a green streets program was started to incorporate the use of green street facilities in public and private developments.

Regulatory framework: A storm water management manual and code, which outline the requirements. In order to develop or redevelop over 500 ft² of impervious surface, the city requires pollution reduction and runoff control standards.

Implementation and monitoring: To monitor progress and compile good practices, a floor area bonus was provided for green roofs, which incentivized residents to create an eco roof in exchange for an increase in a building's allowable area.

Results: GI of Oregon is a mature and comprehensive GI program. By 2014, 330,000 ft² of eco roof installations came into existence. Green street projects retain and infiltrate about 43 million gallons per year, with the potential to manage 40% of Portland's runoff annually. Portland invested 52 crores in green infrastructure to save 1,621 crores in hard infrastructure.

5.3.2 *Milwaukee, U.S.*

Milwaukee is the largest city in the state of Wisconsin and the fifth largest city in the Midwestern United States. The Milwaukee region has steadily grown since the year 2000. As the region continues to grow, so does its impervious surfaces. The more housing, pavements and concrete increase, the more absorbent wetlands diminish. The region has continuously faced threats of flash flooding, causing damage to buildings and vehicles. A minimum of 10 floods in a year were faced by the city.

Approach: Focused on downspout disconnections and rain barrels. Sixty rain gardens were installed to control runoff, with twenty thousand square foot of green roof on a local housing project that will retain 85% of runoff, with the remaining 15% redirected to rain gardens and retention basins for onsite irrigation.

Regulatory framework: Several programs and initiatives were carried out. "Every Drop Counts"- shows residents how to reduce storm water by installing rain barrels and planting rain gardens. In co-operation with the Regional Planning Commission, the sewerage district also implemented the green streets program, which involved the purchase and protection of undeveloped lands and open spaces adjacent to streams, shorelines, and wetlands. In 2010, the Milwaukee Metropolitan Sewerage District awarded 24 crores in green infrastructure grants to 14 groups. Through the grants, 7,500 square feet of permeable pavements, 4,000 square feet of green roofs, 1,100 square feet of bio-swales and rain gardens, and two 1,000 gallon rain harvesters were implemented.

Results: GI in Milwaukee provides a 31 to 37% reduction in storm water flow, a 5 to 36% reduction in peak flows, and a 14 to 38% reduction in Combined Sewer Overflow volume.

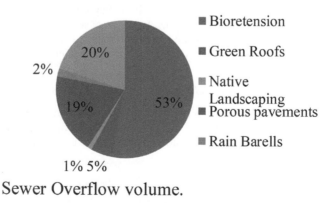

Sewer Overflow volume.

Figure 3.   Existing green infrastructure types.

5.3.3 *Toledo, Spain*

Toledo is a municipality located in central Spain. It is about 232.1 km² in area with a population of 83,459 in 2016. The entire city continuously faces flooding, even during small amounts of rainfall, which affects the traffic movement. In January 2005, some of the major roads were closed for 24 hrs due to flooding. Since 2006, the use of policies and demonstration projects to help promote green infrastructure has drastically reduced Combined Sewer Overflow.

Approach: Institutionalizing green infrastructure through demonstration and restoration projects. New storm water fee systems by using green parking, green roofs and pervious pavements, with the goal of reducing combined sewer overflows while maximizing the net present value of benefits.

Regulatory framework: Stringent storm water regulations for all new construction and redevelopment. A number of ordinances to accelerate green infrastructure investments and water billing fees will be determined by calculating the amount of impervious cover on a given property. Financial incentives to retrofit properties with green infrastructure were also provided.

Results: There are 280 crores in benefits, compared to 778 crores for the gray infrastructure option. A 60% reduction in CSO volume and a 25% reduction in storm water flow.

6 INFERENCE

From the studies it can be inferred that:

- Indian floods are mainly due to the increase in urbanization, with related factors such as land use conversion, decrease in open spaces, decrease in permeable land, etc.
- From the case studies it can be inferred that about 30–40% of the urban floods can be reduced by incorporating green infrastructure practices
- Also, the context in India is comparable with that of the cities where green infrastructure is applied
- With effective recommendations, urban floods in India can be tackled to an extent

7 RECOMMENDATIONS FOR INCORPORATING GI FOR URBAN FLOOD MANAGEMENT

The development pattern in India shows an increasing trend toward urbanization and there has been a major investment in gray infrastructure. Green infrastructure has not been coordinated or integrated at any levels, even though awareness about sustainable development is on the rise. There is no special policy or strategy for Indian cities. The various elements of GI are fragmented and the potential is not fully utilized.

1. A limit to the amount of impervious space must be incorporated in guidelines based on the context.
2. To make it more successful, incentives and subsidies must be given to those who are involved in such practices.
3. The cities can be categorized in the order of flood severity and the elements of GI can be incorporated accordingly, based on the context.
4. Find opportunities in existing regulations: examine whether/how current permits and byelaws can cover the new activities and create awareness of the same.
5. The approach must only be limited to open spaces, roads, apartments and houses. It must also be implemented in government buildings.
6. A town planning scheme can be the method through which GI can be integrated and can act as a platform to improve the resources of GI.

8 CONCLUSION

Over the last two decades, urban planning orthodoxy has promoted a compact urban form and higher densities to reduce energy consumption and the ecological footprint of cities.

However, as outline, densification efforts often pose problems for urban drainage systems, while brownfield sites that are targeted for development may actually serve more important functions in terms of water retention, recreational uses and urban cooling.

This paper focuses on the GI approaches as a means to manage urban flood risk. The paper does not oppose the use of traditional gray infrastructures, as they can be the most appropriate solution in certain circumstances. However, the paper challenges the dominance of gray solutions, such as levees and dam flood walls, as a solution for flood management rather than sustainable green solutions.

ACKNOWLEDGMENT

The authors gratefully acknowledge the contributions of faculty, Government Engineering College, Thrissur, friends and family for successful completion of the study.

REFERENCES

Chatburn, & Craig (2010). Green Infrastructure Specialist, Seattle Public Utilities. Interview by Sarah Hammitt.

Hoang, L. & Fenner, R.A. (2015). System interactions of storm water management using sustainable urban drainage systems and green infrastructure. *Urban Water Journal, 2016,* 1–21.

Lennon, M. & Scott, M. (2014). Urban design and adapting to flood risk: The role of green infrastructure. *Journal of Urban Design, 19*(5), 745–758.

Opperman, J.J. (2014). *A flood of benefits: Using green infrastructure to reduce flood risks.* Arlington, Virginia: The Nature Conservancy.

Soz, S.A. & Kryspin-Watson, J. (2016). The role of green infrastructure solutions in urban flood risk management. *Urban Flood Community of Practice, 25*(3), 12–20.

Valentine, L. (2007). *Managing urban storm water with green infrastructure: Case studies of five U.S. local governments.*

Emerging Trends in Engineering, Science and Technology for Society,
Energy and Environment – Vanchipura & Jiji (Eds)
© 2018 Taylor & Francis Group, London, ISBN 978-0-8153-5760-5

Accessibility and interactions with urban blue spaces: A case of Conolly Canal

Shahana Usman Abdulla & C.A. Bindu
Government Engineering College, Thrissur, Kerala, India

ABSTRACT: Urban waterways represent a potential site for interaction with nature in a busy urban environment, what we here refer to as "blue spaces". Blue spaces need to be viewed as amenities and given importance as green spaces. Few research has been done on the accessibility of and interaction with blue spaces in an urban area. Through this paper, we are trying to study how a blue space in an urban area is perceived by the nearby residents and the reasons behind such a perception, whether positive or negative. A case of Conolly Canal passing through Kozhikode city is studied and analyzed to find whether it is perceived as a positive or a negative amenity and what factors lead to such a social impact.

1 INTRODUCTION

The study of different aspects of urban planning are often more complex as they need to take a number of details into consideration. Normally, attention is given to the physical, social and economic environment. Social environment has to be given major priority as all the interventions are ultimately meant for social welfare. Thus research has to be focused more on how people perceive different elements of physical environment in order to direct planning interventions in the right way.

There is a growing literature showing how proximity to urban green space can produce improved health outcomes like reductions in obesity, diabetes and cardiovascular morbidity (Cutts, Darby, Boone, & Brewis 2009; Ngom, Gosselin, Blais, & Rochette 2016). Urban green spaces are not limited to terrestrial parks and open areas, but also include urban waterways. The benefits provided by water features have been widely acknowledged, both as ecological services (e.g., carbon sequestration, oxygen production, noise reduction, microclimates, etc.) and as places that are used for recreation and social interaction (e.g., exercise, sport, etc.) (Kumar 2010, Kondolf & Pinto 2016). In this paper, we are trying to explore how local residents experience an urban blue space in a sample of neighborhoods in Kozhikode.

2 BLUE SPACES

As blue spaces, we consider hydrographic features that can be waterbodies (e.g., estuaries, ice masses, lakes and ponds, playas, reservoirs, and swamps and marshes) or flowlines that make up a linear surface water drainage network (e.g., canals and ditches, coastlines, streams and rivers) (USGS 2015).

In the sparse blue space literature that does exist, coastal waterways were shown to provide quality of life benefits, and residents most frequently visited waterways closest to where they lived (Cox, Johnstone & Robinson 2006). Another study explored distance to stormwater ponds in Florida, finding that economically stressed census block groups in the inner-city community tended to be located closer to stormwater ponds with less quality, diversity, and size (Wendel, Downs, & Mihelcic 2011). Meanwhile, inland urban waterways such as rivers and canals remain understudied as neighborhood amenities with potential impacts on

Figure 1. Conolly Canal in Kozhikode town.

urban households. Two meta-analyses focusing on the impacts of blue space on mental health (Gascon et al. 2015) or long-term human health (Völker & Kistemann 2011) found inadequate evidence due to the limited amount of empirical research on the topic.

The factors like accessibility of households to the blue space, whether they interact with or use the blue space, purpose of visit, time spent and influence on daily lives are being studied in this paper. A similar study in Northern Utah, United States, showed the blue space they had considered as a positive amenity based on social perception and accessibility.

3 STUDY AREA

Conolly Canal, usually called as Canoly Canal, is the part of the West coast canal network of Kerala and runs through the Kozhikode city. It was constructed in the year 1848 under the orders of the then collector of Malabar, H.V. Conolly.

The canal stretching through Kozhikode town is about 11.4 km long and connects Akalapuzha in the north and Kallai puzha in the south of Kozhikode town. The width of canal varies between 6 to 20 m and the water depth during the monsoon ranges between 1 to 3 m.

4 METHODOLOGY

The survey was conducted by a team of 8 members, out of which 3 were engineers of Kerala State Pollution Control Board, an official from Town Planning Department and 4 post graduate students.

The study comprised of a socio economic survey covering the residents and stakeholders living along the banks of the stretch. The stakeholders involved included not just households, but commercial establishments, hospitals and industries. The different types of interaction with the blue space by the stakeholders and the factors responsible were enquired through the survey. This paper concentrates on the social perception of the urban blue space and the factors responsible for such a perception, positive or negative.

The canal portion to be surveyed was divided into 8 stretches, each stretch being approximately 2 km. The stretches are in between the following points as shown in Figure 2.

1. Eranjikkal
2. Kunduparamba
3. Modappattupalam
4. Ashirwad lawns
5. Sarovaram Biopark
6. Kalluthamkadavu

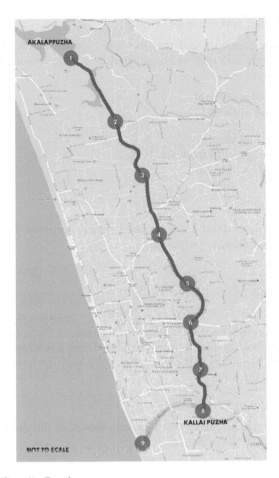

Figure 2. Stretch of Conolly Canal.

7. Mooriyad
8. Kallai
9. Kothi bridge

There are 97 industries on the banks of the canal and river stretch of which 54 were wood based units. Also there are 501 residences and 208 commercial establishments and seven hospitals. 835 stakeholders including residents and businesses located on the banks of the waterway stretching from Elathur to Kothi estuary, were taken for the survey.

The survey characterized households on the basis of having direct access to the canal. Respondents were asked whether the canal affects their lives or not and if yes, how? They were asked if they use the canal so as to analyse the purpose for which it was used. The analysis also explores why the canal is not being used or accessed. Finally, the involvement of people in the revival of the canal is assessed.

5   RESULTS

5.1   *Do you have direct access to canal?*

Majority of the nearby households are not having spatial access to the canal. The households nearby the waterway having direct access are said to experience the impacts of the blue space and themselves have an influence on the water body.

975

5.2 Does the canal affect your life in any way?

The majority of the respondents said that they lives are affected by the canal in one way or the other. While some said they are unaffected by it. The reasons mentioned for being affected by the canal were:

- Foul smell due to wastes dumped
- Flooding during rainy seasons
- Mosquito breeding
- Stagnant water with foul smell
- Disinterest in accessing a polluted water body
- Mixing of canal water in the well

5.3 Do you use the canal?

Majority of the respondents doesn't use the canal for any purposes. The few people who use the canal used it for the following purposes:

- Seasoning of wood
- Dumping domestic waste
- While using the bridge
- Drainage

The majority does not use the canal stating various reasons, the major one being pollution of water creating an unhealthy environment. Conolly Canal has a history of people engaged in fishing, bathing and swimming in the water. But these activities have ceased now due to the unhealthy condition of the water.

5.4 Do you use canal for waste disposal?

Even though majority have claimed not disposing wastes in the canal, few have admitted disposal of the following types of wastes:

- Wastes from wood industries
- Domestic wastes
- Septic discharge
- Sewage

Figure 3.   Access to canal.

Figure 4.   Response to influence of canal on lives.

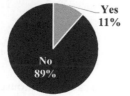

Figure 5.   Response to use of canal.

Figure 6.   Response to waste disposal.

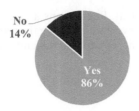

Figure 7. Response to revival.

- Drainage
- Industrial wastes
- Hospital wastes

5.5 *Major causes of pollution*

The major causes of pollution of water in canal are said to be due to dumping of plastic wastes, industrial wastes, hospital wastes, slaughter waste, market waste, domestic waste, stagnant water, septic discharge, sewage disposal, inefficiency of waste management and lack of awareness about importance of canal.

5.6 *Would you cooperate with any revival projects for Conolly Canal?*

There are damages in the lining provided along the canal and this was majorly attributed to the encroachment by properties along the stretch of canal. This kind of an illegal intrusion might create an aversion among the people involved to refrain from cooperating with any revival projects.

The positive response shows the concern and will of the people to retrieve health of the urban ecosystem. The very few who restrain from opting to cooperate with any measures to revive the canal say so as they fear losing their land through land acquisition and due to the fear of getting caught of illegal encroachment of canal area.

6 CONCLUSIONS

The results show that the urban blue space that we have taken is perceived as a negative amenity by the people of the neighborhood. The major reason for such a result is attributed to the increase in water pollution due to improper management. People seem to be disinterested in accessing or spending time at the canal. Though, a blue space has multiple ecological, social and recreational benefits, the canal under study turned out to have a negative influence on the lives of urban residents.

Similar study done in Northern Utah indicated the perception of the urban blue space studied, as a positive amenity. This shows that an ecologically healthy or restored waterway with public access opportunities can contribute to an aesthetically pleasing experience. On the other hand, unmonitored or poorly managed urban waterways can be sites of flooding risk, insect pests, pollution and/or waste disposal. Finally, even ecologically sound wetland systems can be perceived by humans as disamenities, due to the smell of anaerobic decomposition and the insect populations that thrive in them.

A blue space is an important asset for an urban environment. Planning interventions can turn any negative influence a blue space has into positive. Planners need to promote use of and familiarity with urban waterways in order to maximize benefits to local residents and communities. Restoration is increasingly advocated as a strategy facilitating public access and use of urban waterways. Infrastructure development like provision of towpaths, navigational aids, fencing and lighting can improve accessibility and safety. The water quality can be improved through solutions like swinging weed gates, air curtains, physical removal, flushing channels and waste management. Regulations and awareness is required for the proper conduct near the waterway. Any development along the waterway should follow the pillars of sustainability, namely the economic, environmental and social justice.

7 LIMITATIONS

The results of the study may or may not be generalizable to other regions. It will depend on the social structure, built environment and trajectories of urban growth of the region under consideration. The study only covers aspects of social perception of urban blue space. Also, the study is limited to canal in an urban area. The results might change with difference in the type of urban blue space.

8 FUTURE RESEARCH

Further research can be done considering more aspects of social, ecological and recreational aspects. Interventions necessary to revive and elevate the potential of a blue space in an urban area can be studied.

ACKNOWLEDGEMENT

The authors are immensely grateful to Mr. K. V. Abdul Malik, Regional Town Planner, Kozhikode for granting access to data concerning Conolly Canal. We acknowledge the contributions of faculty, Government Engineering College, Thrissur, friends and family for the successful completion of the study.

REFERENCES

Cox, M.E., Johnstone, R., & Robinson, J. 2006. Relationships between perceived coastal waterway condition and social aspects of quality of life. *Ecology and Society* 11(1): art35.

Gascon, M., Triguero-Mas, M., Martínez, D., Dadvand, P., Forns, J., Plasència, A., et al. 2015. Mental health benefits of long-term exposure to residential green and blue spaces: A systematic review. *International Journal of Environmental Research and Public Health* 12(4): 4354.

Historic Alleys, Historic Musings from a Malabar Perspective. 2017. Retrieved from: http://historical-leys.blogspot.in/2017/07/conolly-and-calicut-canal.html.

Kondolf, G.M., & Pinto, P.J. 2016. The social connectivity of urban rivers. *Geomorphology* 277: 182–196.

Kumar, P. 2010. The economics of ecosystems and biodiversity: Ecological and economic foundations.

La Rosa, Daniele. 2014. Accessibility to greenspaces: GIS based indicators for sustainable planning in a dense urban context. *Ecological Indicators* 42: 122–134.

Melissa Haeffnera, Douglas Jackson-Smith, Martin Buchert, Jordan Risley. 2017. Accessing blue spaces: Social and geographic factors structuring familiarity with, use of, and appreciation of urban waterways. *Landscape and Urban Planning* 167: 136–146.

USGS (2015). National hydrography dataset. Retrieved from: http://Nhd.usgs.gov.

Völker, S., & Kistemann, T. 2011. The impact of blue space on human health and wellbeing—Salutogenetic health effects of inland surface waters: A review. *International Journal of Hygiene and Environmental Health* 214(6): 449–460.

Wendel, H.E.W., Downs, J.A., & Mihelcic, J.R. 2011. Assessing equitable access to urban green space: The role of engineered water infrastructure. *Environmental Science & Technology* 45(16): 6728–6734.

*Emerging Trends in Engineering, Science and Technology for Society,
Energy and Environment – Vanchipura & Jiji (Eds)
© 2018 Taylor & Francis Group, London, ISBN 978-0-8153-5760-5*

Energy efficiency governance in an Indian context

V.P. Shalimol
Urban Planning, Government Engineering College, Thrissur, Kerala, India

K.M. Sujith
School of Architecture, Government Engineering College, Thrissur, Kerala, India

ABSTRACT: The sensible consumption of energy plays a major role in providing sustainable development and this responsibility comes through good governance and practice. Governance has been known to India from past millennia through Kautilya. In the present world, accountability, transparency, inclusiveness, equitability, etc. are the key ingredients of good governance. Thus, energy efficiency, sustainability and governance are interconnected. These concepts have brought together carbon emissions, climate change, adaptation and mitigation, as well as employment and poverty reduction. The concept of energy efficiency interlinks these thoughts. It involves legislative frameworks, funding mechanisms and institutional arrangements, which go together to support the implementation of Energy Efficiency (EE) strategies, policies and programs. The government, EE stakeholders and the private sector should work together to achieve this. However, India's population makes up 18% of the world's population, and its energy consumption is 6% of the world's primary energy use, which makes it one-third of the global average. Energy consumption is always on the rise. Energy efficiency and its influence on the governance sector is analyzed through this paper, which includes the laws and decrees, strategies and action plans, funding mechanisms, implementing agencies, internal assistance, etc.

1 INTRODUCTION

The dimensions concerned with governance include the possession of puissance, the competency to make decisions, how the people's voices are perceived aurally and how accounts are rendered. The qualities of good governance were long ago expounded by Kautilya in his treatise Arthashastra as follows: "In the jubilance of his subject lies his jubilance, in their welfare his welfare, whatever please himself he shall not consider good".

The twelfth five year plan (2012–2017) defines good governance as an essential aspect in order for society to be well functioning. It provides legitimacy to the system by providing citizens with a way of effectively using resources and by the deliverance of services. The key ingredients of good governance include accountability, transparency, inclusiveness, equitability, sustainable development, etc. Good governance has always played a critical role in advancing sustainable development. Thus, good governance and sustainability are the two faces of a coin. For the sustainable emancipation of institutions and countries, energy efficiency undertakes major functions. Thus, energy efficiency, sustainability and governance can be viewed in line with carbon reduction, climate change, adaptation and mitigation, as well as employment and poverty reduction. The way by which energy efficiency can be incorporated with sustainable development is by energy efficiency governance.

2 ENERGY EFFICIENCY GOVERNANCE

The International Energy Agency (IEA), with financial support from the European Bank for Reconstruction and Development (EBRD) and the Inter-American Development Bank (IDB), conducted a study on energy efficiency governance. Energy Efficiency (EE) governance includes a combination of legislative frameworks and funding mechanisms, institutional arrangements and co-ordination mechanisms, which will fortify the implementation of energy efficiency strategies, policies and programs by collaboration. From individual households to immensely colossal factories, EE governance cumulates technology development, market mechanisms and regime policies. In order to achieve sustainable economic development, the regime, EE stakeholders and the private sector must collaborate at the required scale and timing of energy efficiency ameliorations. The EE governance policy landscape and its understanding help to develop its efficacy.

Various policies and strategies have been followed throughout the world. They include energy efficiency codes, labels, and incentives. Some examples of this are:

Russia
Code

- Thermal performance of buildings

Labels

- Energy efficiency class of multifamily buildings
- Green standards (2010)

Canada
Code

- Alberta building code (2011)

Labels

- BOMA Best (Building Environmental Standards) Version 2
- ENERGY STAR Portfolio Manager Benchmarking Tool
- LEED Canada (2009)
- LEED Canada (Existing Building: Operations & Maintenance)

Incentives

- EcoENERGY Retrofit (2007)

2.1 *Drivers of the energy efficiency policy*

There are countrywide variations in the context of energy efficiency, and the typical drivers are broadly categorized below:

- Energy security
- Economic development and competitiveness
- Climate change
- Public health

2.2 *Barriers to energy efficiency*

The factors influencing energy efficiency improvement include market, financial, institutional and technical barriers. Every country has these barriers. In order to overcome these barriers each country will have their energy efficiency policies. The barriers to energy efficiency include: market barriers, financial barriers, information and awareness barriers, regulatory and institutional barriers, and technical barriers.

2.3 *Policies used to address these barriers*

Pricing mechanisms
The introduction of variable tariffs where higher consumption levels invoke higher unit prices.

Regulatory and control mechanisms
Energy audits and energy management, Minimum Energy Performance Standards (MEPS), energy consumption reduction targets, and EE investment obligations on private companies.

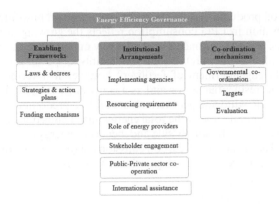

Figure 1. Framework of energy efficiency governance.

Fiscal measures and tax incentives
Grants, subsidies and tax incentives for energy efficiency investments. Direct procurement of EE goods and services.

Promotional and market transformation mechanisms
Public information campaigns and promotions, inclusion of energy efficiency in the school curricula, appliance labeling and building certification.

Technology development
This includes the development and demonstration of EE technologies.

Commercial development and capacity building
This involves the creation of Energy Service Companies (ESCOs), training programs, and the development of the EE industry.

Funding remediation
The introduction of revolving funds for EE investments, project preparation facilities and contingent financing facilities come under this heading.

2.4 *Framework of Energy Efficiency (EE) governance*

Enabling frameworks, institutional arrangements and co-ordination mechanisms are the three main governance areas (Energy efficiency governance, Handbook).

3 ENERGY EFFICIENCY GOVERNANCE IN INDIA

India is home to 18% of the world's population and its primary energy use is 6% of the world's consumption, therefore the energy per capita consumption is only one-third of the global average. India has been responsible for virtually 10% of the incrementation in global energy demand since 2000. Its energy demand in this period has virtually doubled, pushing the country's share of global demand up to 5.7% in 2013 from 4.4% at the commencement of the century. As the country is progressing, with rising incomes and a better quality of life, there will be a greater demand for energy. Coal now accounts for 44% of the primary energy mix. Oil consumption in 2014 stood at 3.8 million barrels per day (mb/d), 40% of which is utilized in the transportation sector. Demand for diesel has been particularly strong, now accounting for some 70% of road transport fuel use. This is due to the high quota of road freight traffic, which tends to be diesel-powered, in the total usage and additionally to regime subsidies that kept the price of diesel relatively low (this diesel subsidy was removed at the cessation of 2014; gasoline prices were deregulated in 2010). On both the supply and demand sides, India is trying to meet its demand. LPG use has increased rapidly since 2000, reaching over 0.5 mb/d in 2013 (LPG is second only to diesel among the oil products, pushing gasoline down into an unwontedly low third place). Nevertheless, the elevation in LPG consumption reflects the growing urbanization, as well as perpetuated subsidies. Natural gas makes up a relatively minute portion of the energy mix (6% in 2013, compared with 21% globally). It is utilized mainly for power generation and as a feedstock and fuel for the engenderment of fertilizers, however, it nevertheless has a diminutive but growing role in the residential sector and as a conveyance fuel. Hydropower, nuclear and modern renewables (solar, wind and geothermal) are utilized predominantly in the potency sector, but play a relatively minute role in the total energy mix. This is done by encouraging investments in renewable energy on the supply side and increasing energy efficiency and conservation on the demand side. The reduction of carbon emissions plays a significant role. There are four basic standard policy measures for reducing carbon emissions:

- Energy efficiency improvement measures
- Command and control measures
- Domestic carbon taxes
- International emission trading regime of the kind envisaged for the Annex B countries.

The most desirable among these are the energy efficiency measures. The energy efficiency of the governance of India is analyzed under the following headings.

3.1   *Energy efficiency laws and decrees in India*

Energy Conservation Act, 2001
This act was enacted in October 2001, to provide for the efficient utilization of energy, its conservation and matters connected therewith. The Bureau of Energy Efficiency (BEE) was established with effect from March 1, 2002. Its mission is to develop policies and strategies with a thrust on self-regulation and market principles, within the overall framework of the act. The prime objective is to reduce the energy intensity of the Indian economy.

The Electricity Act, 2003
This act was enacted to harmonize and rationalize the provisions of existing laws and to reform legislation by the promotion of efficient and environmentally benign policies; the act mandates efficiency in various forms in generation, transmission and distribution. Under the provisions of section 3(1) of the act, the central government brought out the National Electricity Policy for the development of the country's power system, based on the optimal utilization of resources.

3.2   *Strategies and action plans in India*

The main objective of the eleventh plan was to reduce the energy intensity per unit of Greenhouse Gas (GHG) by 20% from the period 2007–08 to 2016–17. When GOI launched the National Action Plan for Climate Change (NAPCC), this formally addressed India's objective for GHG emission reduction. The NAPCC relies on eight missions, of which the National Mission for Enhanced Energy Efficiency (NMEEE) is a critical one.

Eight national missions
The eight national missions that form the core of the National Action Plan are;

- National Solar Mission
- National Mission for Enhanced Energy Efficiency
- National Mission on Sustainable Habitat
- National Water Mission
- National Mission for Sustaining the Himalayan Ecosystem
- National Mission for Green India
- National Mission for Sustainable Agriculture
- National Mission on Strategic Knowledge for Climate Change.

3.3   *Funding mechanisms in India*

A steady and reliable source of funding is essential for energy efficiency institution programs. This can be co-ordinated with the budget allocation, which is undertaken annually.

Venture Capital Fund for Energy Efficiency (VCFEE)

BEE has institutionalized VCFEE in India to encourage equity investment in EE projects. The fund shall provide last mile equity support to specific energy efficiency projects, limited to a maximum of 15% of the total equity required, through special purpose vehicles or Rs. 2 Cr, whichever is less. The support has only been provided to government buildings, private buildings and municipalities.

ICEEP (India Chiller Energy Efficiency Project)

In association with the Ministry of Environment and Forest (MoEF), GOI and IDBI Bank Ltd by World Bank, ICEEP was implemented from August 2009. The phasing out of use of Chlorofluorocarbon (CFC), under the Montreal Protocol, and the achievement of energy efficiency in the refrigeration and air conditioning sector is the objective of ICEEP.

3.4 *Implementing agencies in India for EE*

The Ministry of New and Renewable Energy (MNRE), the Ministry of Environment and Forests (MoEF), the Ministry of Power (MoP) and the Ministry of Urban Development are the implementing agencies that introduced sustainability components with overarching policy objectives to promote energy conservation in buildings.

3.5 *International assistance in India for EE*

Partnership to Advance Clean Energy (PACE)

The Partnership to Advance Clean Energy (PACE) is the lead program on clean vitality between the U.S. and India to mutually chip away at the scope of issues related to energy security, clean energy and environmental change. PACE looks to quicken comprehensive, low-carbon development by supporting exploration and the sending of clean energy innovations and strategies. PACE consolidates the endeavors of a few government and non-government partners on both the U.S. and Indian sides and incorporates three key segments: Research (PACE-R), Deployment (PACE-D), and Off-Grid Energy Access (PEACE). PACE includes a USD 20 million, five year specialized help (TA) program, which is driven by the U.S. Organization for International Development (USAID) and the U.S. Bureau of State and is executed in association with the Ministry of Power (MoP) and the Ministry of New and Renewable Energy (MNRE). The program partners for this include state, public and private sector agencies. Their energy efficiency initiatives include:

- Support for the development of the Smart Grid Regulatory framework was provided
- Assisting BEE in the technical update of the Energy Conservation Building Code
- Supporting the Nalanda University and Uttar Haryana Bijili Vitran Nigam headquarter to became zero energy user
- Supporting the development of a waste heat utilization policy
- Promoting market transformation activities for heating, ventilation and air conditioning.

International Solar Alliance (ISA)

The Alliance, India's brainchild to bring together 121 solar-rich countries on a single platform to give a push to solar energy, was jointly launched by the Prime Minister Narendra Modi and the then French President Francois Hollande in Paris on November 30, 2015.

International Solar Alliance (ISA) was conceived as a coalition of solar resource-rich countries to address their special energy needs and it will provide a platform to collaborate on addressing the identified gaps through a common, agreed approach. It will not duplicate or replicate the efforts that others (such as the International Renewable Energy Agency (IRENA), Renewable Energy and Energy Efficiency Partnership (REEEP), International Energy Agency (IEA), Renewable Energy Policy Network for the 21 st Century (REN21), United Nations bodies, bilateral organizations, etc.) are currently engaged in, but will establish networks and develop synergies with them and supplement their efforts in a sustainable and focused manner.

The vision and mission of the International Solar Alliance is to give a devoted stage for the participation of sun-based asset-rich nations, where the worldwide group, including reciprocal and multilateral associations, corporates, industry, and different partners, can influence a positive commitment to help to accomplish the shared objectives of expanding the utilization of sun-powered vitality in addressing the vitality needs of the forthcoming ISA partner nations in a sheltered, advantageous, reasonable, evenhanded and practical way.

ISA is intended to be a multicountry partnership organization with membership from solar resource-rich countries between the two tropics. ISA's proposed governance structure would consist of an Assembly and a Secretariat.

3.6 *Public–private sector co-operation*

Public–private is a form of co-operation between public authorities and the private sector that aims to modernize the delivery of energy services. This can be a long-term or short-term contractual relationship between a public entity and a private organization where risks are shared and there is increased financing of energy efficiency (EE).

Solarization of CIAL

The use of solar energy at airports has developed gradually. Airports experimented with installations that provided a few hundred kilowatts of peak power at the beginning of this century. Nowadays, two, five or ten megawatt installations are not uncommon and the economics are much improved as grid parity is approached.

Cochin International Airport (CIAL), serving the city of Kochi in the Indian state of Kerala, is the busiest and largest airport in the state and the fourth busiest in the country. The airport serves more than five million people annually.

The CIAL (Cochin International Airport Limited) set up a 100 KW PV predicated solar power plant on the rooftop of the advent block as a pilot project during March 2013. It required 400 panels, each with a capacity of 250 Wp. The installation was designed and executed by Kolkata based Vikram Solar Power. This facility engendered 400 units of power annually, the absence of battery backup reducing the capital cost drastically.

The next logical step was to go for full generation of the power required for the entire operation of CIAL, integrating up to around 48,000 kilowatt hours per day from a PV predicated solar system in its own backyard. To make the airport grid neutral, the capacity of the system was to be about 12 MWp, with a capacity to generate 50,000 kilowatt hours. This was done with permission from KSEB (Kerala State Electricity Board) to bank the electricity.

Bosch Limited was endowed with this esteemed activity through a transparent tender process. They did it with élan and with the immaculate specialized flawlessness that Germans are acclaimed for. The total project cost about `62 Cr. at about `5.17 Cr/MW, which is substantially less than the benchmark set by the regulator. The project payback period is under six years. M/s Bosch handles the system maintenance as well, on a contract, at a cost of `50 Lac per annum.

3.7 *Stake holder engagement in EE initiatives in India*

As the major energy consumers, the corporates no doubt have many significant contributions toward the sustainable development of delivering energy efficiency. The CSR (Corporate Social Responsibility) can be best implemented toward this. Data analyzed by the Ministry of Corporate Affairs for the CSR expenditure of all Indian companies in 2014–2015 showed that 14% (Rs. 1213 Cr) of total CSR spending in India was made on activities focusing on conserving the environment. It was the third highest expenditure on a social impact issue, after education (32%) and health (26%), and was greater than the amount spent on rural development (12%).

3.8 *Governmental co-ordination*

The government's basic leadership on energy at the central level is appropriated between the Ministry of Petroleum and Natural Gas, the Ministry of Coal, the Ministry of Non-Conventional Energy Sources, the Ministry of Environment and Forests, the Ministry of Atomic Energy, and the Ministry of Power. Inside the Ministry of Power, the Central Electricity Authority (CEA), the specialized wing, works intimately with singular state power sheets (SEBs) and utilities in the controlling, transmission, and dissemination of power. At the state level, there exists different divisions, offices and specialists chipping away at different sub-segments of energy. Amusingly, the Ministry of Panchayati Raj, working at the town level, does not have any part to play in rural electrification. The obligation regarding thorough rural electrification (including the quality and collection of power) is scattered between various Ministries in an awkward manner.

Being a concurrent subject under the Constitution, the states share powers with the center only for power, not for the other energy sources. However, they do share powers on environmental regulation. This distribution of powers makes co-ordination between the different energy sources, over the country and between states, difficult.

Environmental clearance is required for all types of power projects, including nuclear, hydro and thermal, under the Environmental Protection Act 1986 (EIA notification in 1994) from MoEF, GoI.

3.9 *Evaluation*

For effective policy-making, compliance evaluation is critical. The importance of evaluation in policy-making has been well established through the examples of Denmark and Sweden. Denmark's energy efficiency program evaluation has made a concerted effort in developing policy and long-term strategy. Compliance evaluation can assist the Indian policy makers with identifying potential issues in the execution of the Energy Conservation Building Code (ECBC) and help them to make necessary changes. Compliance evaluation will likewise enable India to accomplish its proposed energy savings and emission reductions through the ECBC.

The Indian Energy Security Scenarios (IESS)
The Indian Energy Security Scenarios (IESS), 2047, developed by NITI Aayog (erstwhile planning commission) is a tool to assess the government's role in energy use.

This tool is used for tending to requirements, and to give 'all energy demand' and 'all energy domestic supply' points of view under different scenarios/suspicions, on a 5 yearly premise up to 2047 (100th year of India's independence), with 2012 as the base year. It is used to teach the nation about its energy status, and enable them to create pathways that India may embrace for improving its residential energy reliance.

The tool has been used with the assistance of a wide pool of information accomplices from the government, industry, think tanks, non-governmental associations, international research organizations and the scholarly community. Broad stake holder consultations were held with industry, academia, the government, the overall population and different specialists (organizations or people) in various fields of energy to vet the legitimacy of particular directions and suppositions taken. A few between ecclesiastical counsels inside the government were likewise directed to validate the information for particular expansive effort workshops have been led to advance the utilization of this instrument and include more individuals in the activity for agreement fabricating and making mindfulness about energy arrangements.

3.10 *Energy efficiency targets of India*

The first cycle of the government's Perform, Achieve and Trade (PAT) energy efficiency scheme, which ran from 2012–2015, contributed to an emissions reduction of 31 million tons of CO_2, energy savings of 8.67 million tons of oil equivalent, and avoided capacity addition of about 5.6 gigawatts. The scheme also resulted in monetary savings of Rs. 37685 Cr into energy efficiency technologies by the participating industrial units. The PAT scheme uses a market-based mechanism to enhance energy efficiency. Large industrial consumers of energy (called designated consumers or DCs) are given energy efficiency targets. Those who exceed their targets are awarded energy efficiency savings certificates, which they can sell to those who fail to meet theirs. At the end of the first cycle, 3.8 million certificates were issued and one ton of oil equivalent certificates were issued, one ton of oil equivalent being equal to EScert.

4 ANALYSIS AND CONCLUSION

This paper has two sections, with the first one being about detailing the concept of energy efficiency governance in a global aspect. Energy efficiency governance is a new concept that has evolved linking energy efficiency with governance, which will mainly focus on sustainability. For any country to be energy efficient it should implement policies and programs under the umbrella of the government through governance. India also undertook many initiatives in energy efficiency indirectly, under the head of energy efficiency governance. This paper intends to give the energy efficiency framework of the Indian energy efficiency governance. India exhibits good practices in public–private participation, international assistance, government co-ordination, stakeholder engagement, etc., but lacks efficiency in managing the energy sources. India adopts transparent and accountable systems. The overall institutional governance is weak in energy efficiency. Separate local level energy efficiency policies have to be created to allow the actions to begin from the bottom level of governance. There is no constitutional support for beginning the energy planning at the local level, which will impart energy efficiency. The energy resourcing and the evaluation of energy efficiency is found to be inefficient in the Indian context. In order to qualitatively analyze the efficiency, not only at the building level but at the area level also needed so as to confirm to the energy demand and its resource requirements.

ACKNOWLEDGMENT

The authors gratefully acknowledge the contributions of faculty, Government Engineering College, Thrissur, friends and family for successful completion of the study.

REFERENCES

ACEEE (American Council for an Energy Efficient Economy). (2010). America. State Energy Efficiency Scorecard for 2010, October.
BEE (Bureau of Energy Efficiency). (2010). *National mission for enhanced energy efficiency—mission document: Implementation framework*, Ministry of Power, Government of India. agency, I. e. (2015). *India energy outlook.* France: Directorate of global energy economics.
FACT SHEET: The United States and India—Moving Forward Together on Climate Change, Clean Energy, Energy Security, and the Environment (2012, September 22). Retrieved January 12, 5, from https://in.usembassy.gov/fact-sheet-united-states-india-moving-forward-together-climate-change-clean-energy-energy-security-environment/.
International Energy Agency. (2010). *Hand book of energy efficiency governance. Australia.*
International Solar Alliance: Indias brainchild to become a legal entity on Dec 6. (2017, November 14). Retrieved March 04, 2018, from https://economictimes.indiatimes.com/news/environment/developmental-issues/international-solar-alliance-indias-brainchild-to-become-a-legal-entity-on-dec-6/articleshow/61647360.cms?utm_source=contetofinterest&utm_medium=text&utm_campaign=cppst.
ISA mission. (2016, August 2). Retrieved January 5, 2018, from http://isolaralliance.org/.
Ministry of New and Renewable Energy (GOI), Ministry of Power (GOI), USAID. (2016). *Partnership to Advance Clean Energy-Deployment (PACE-D) Technical Assistance Program.*
Mohan, B. & George, F.P. (2016). *Airport solarization: CIAL steals the thunder.*
Niti Aayog to rank states on energy efficiency. (2017, February 22). Retrieved March 04, 2018, from http://economictimes.indiatimes.com/industry/energy/power/niti-aayog-to-rank-states-on-energy-efficiency/articleshow/57301449.cms.
Rao, S.L. (2012). *Coordination in energy sector and its regulation in India.* Institute for social and economic change. 113–120.
Reddy, B.S. (2014). *Measuring and evaluating energy security and sustainability: A case study of India.*
Statistics. (2017, December 5). Retrieved January 5, 2018, from https://www.iea.org/publications/.
Together on Climate Change, Clean Energy, Energy Security, and the Environment. (2012, September 22). Retrieved January 12, 5, from https://in.usembassy.gov/fact-sheet-united-states-india-moving-forward-together-climate-change-clean-energy-energy-security-environment/.
Yu, S., Evans, M. & Delgado, A. (2014). *Building energy efficiency in India: Compliance evaluation of energy conservation building code.* U.S Department of Energy.

Emerging Trends in Engineering, Science and Technology for Society,
Energy and Environment – Vanchipura & Jiji (Eds)
© 2018 Taylor & Francis Group, London, ISBN 978-0-8153-5760-5

Environmental management of a blue-green network: A case of Valapad in Thrissur, Kerala

Nahlah Basheer
Masters of Urban Planning, Government Engineering College, Thrissur, Kerala, India

C.A. Bindu
School of Architecture, Government Engineering College, Thrissur, Kerala, India

ABSTRACT: Natural resources and the extent of their management are important for any region, particularly to meet the demand for resources in these times of change. There should be a ground for effective planning and management of these resources. They have environmental, ecological, socio-cultural and economic roles to play among many others. The need for an integrated development plan for Valapad, a coastal *gramapanchayat* of Thrissur district in Kerala, India highlighted a requirement for a detailed on-site study of the environmental sector of the area. The focus was on developing effective water management and green infrastructure simultaneously. The changes that have taken place in the land use due to the changes in land utilization, levels of encroachment, low public awareness, destruction of flora and fauna, for example have contributed toward diverse effects that are irreversible. This paper focuses on the amount of environmentally sensitive areas present in the region, the existing scenario and the strategies that can be adopted for the management of the same.

1 INTRODUCTION

Long term sustainability of any urban or semi urban area is often related to that of economic growth. But in reality it is crucially dependent on the social, economic and environmental dimensions of the environmental sector. A multidimensional process like planning and development requires an in-depth probe into matters of environment and protection. The case of Valapad, a coastal *gramapanchayat* in the Thrissur district of the south Indian state of Kerala, studied for the formulation of an integrated development plan, is explored in the paper.

2 REGIONAL SETTING

Valapad *gramapanchayat* lies in the Manappuram region of the west coast in Thrissur, which has a separate island-like formation. This seaboard tract extends from Chettuva in the north, to Azhikode (Munambam) in the south. The total length of the belt is approximately 56 km and the total width varies from four to eight km. The water bodies around the belt are Karuvannur River, Canoli Canal and the Arabian Sea. The Karuvannur River flows encircling this formation in the northern side, joined by the Canoli Canal from almost the mid portion, flowing southwards. When it reaches the Munambam region, it is flushed by the adjoining waters of the Periyar River. On the western side of the land is the Arabian Sea. It is this island-like encapsulated land that we call the Manappuram region. Valapad is situated toward the middle of the Manappuram region. The approximate width observed is 5 km.

Valapad followed a ridge and valley terrain in topography where valley areas were drained throughout by natural streams (known locally as *thödu*). Some of these were fit for travel. Other valley areas were wetlands and cultivable paddy fields. The depressions were filled with natural ponds. Today, almost all of the valley areas of yesteryears, such as the wetlands and, paddy fields, have undergone land filling and most of the natural drains have become unfit for regular water passage. This has brought about a considerable change in the topographical pattern of the region. The area now consists of almost flat land. There is no considerable slope observed. The type of soil is sandy, which is acidic in character. So, during the summer, frequent irrigation is a must in the region.

Valapad has a well-established road network comprising of a National Highway, Major District Road and Village roads. It has a transitional character in its location as it is located in between several urban centers. NH 66 lies on the eastern side of the panchayat and hence, the road network forms a grid iron pattern here. This gives rise to easy inter as well as intra connectivity within and outside the region.

3 NATURAL RESOURCES IN VALAPAD

3.1 *Pond*

There are numerous ponds in the *gramapanchayat* area and it is therefore a complete natural ecosystem. In earlier times, almost all households had one or more ponds, of which one source was for bathing and the other for drinking. Most of the ponds of Valapad are mapped in Figure 2, with Muriyathoodukulam, Kothakulam, Ambalakkulam and, Thirunellikkulam being the important ponds in the area. Although there are numerous ponds here, most of these are not in good condition.

Figure 1 details a few ponds by their nature of ownership (for example private, public and temple ponds), their usage, features and pollution rates. It should be noted that the commonly observed threats to ponds here are pollution, land use changes, climate change, inefficient water management techniques, improper managements of pond water, intensive use for irrigation, fish overstocking, and degraded buffers.

3.2 *Wetlands and paddy fields*

The *Nancha* and *Vayal* areas of the panchayat have been showing a dwindling trend for a long time. Agriculture is no longer a source of income here. According to the present statistics, only seven hectares of land in the panchayat is under paddy category. This area is not used for paddy cultivation now. The Kuttippaadam in Ward 15 of the panchayat is one such example.

3.3 *Sacred groves*

Sacred groves are a very ancient and widespread phenomenon, and there are numerous sacred groves in Valapad panchayat, although it should be noted that there are only three to four

Sl. No	Name of the waterbody	Water type Fresh/ oorujalam (lake)/sea	Features	ownership	Uses	Users	Treatment	Pollution			Dependent system	Remarks
								Type	Source	Extent		
1	KOTHAKULAM	Fresh	Very big & old pond. 20 cents extent.	Public	Aarrattupuzha Pooram – Holy dip wash of the idol of Thevar.	Public	Surrounded by compound wall .	Wastes from Commercial stores.	Man	High	Organisms	—
2	SETHUKULAM				Aarrattupuzha Pooram – Holy dip wash of the idol of Thevar.							Sathakalam Aaraattu
3	BLAAHAKULAM											
4	VALLIPPARAMBIL KULAM	Fresh	10 cent	Valliparambil Korunakaran	Irrigation	Man, organisms	Cleaning (every year)	None	—	—	Man, Organisms	—
5	THANDAAPARAMBIL KULAM	Fresh	12 cent	Thandaaparambil Rajeev	Bathing, washing, irrigation	Owners	Cleaning (every year)	Household solid waste	Man	Low	Man, Organisms	Ancient pond
6	THIRUPAZHANCHERI AMBALAKKULAM	Fresh	5 cent	Thirupazhancheri ambalam	Bathing, rituals.	Public	Dug and Cleaning annually	None	—	—	Man, Organisms	Bath space of Thiruvanl before the start of the day to day rituals at the temple.
7	VENNIKKAL TEMPLE POND	Fresh		Vennikkal temple	Bathing, rituals	Public		None	—	—	Man, Organisms	
8	STREAMS(THÓDU) LINKED TO THE ARAPPATHODU	Oorujalam	6 ft long	Panchayat	Washing clothes, fishing	Public	MGNREGA works	Household solid & liquid, commercial waste	Household & commercial	Very high	Public	Width decreasing towards the West. With the input of wastes, water has become unfit for use
9	CHITTEZHATH KULAM	Fresh	Very old. 10 cent	Private-Adimakkanj Valapad	Washing clothes, fishing, Irrigation, rituals	Owners	Dug and Cleaning annually	Household solid waste	Household	Low	Owners	Ancient. In the past, used for watering coconut fields and tapioca plantings. Used for bathing in the pond before rituals related to the dead.
10	PONNIKKULAM	Fresh	5 cent	Private-Adipparambil	Bathing, fish farming, swimming	Public	—	None	Man	Medium	Public	Legend related to the pond.
11	EDAMUTTAM AMBALAKULAM	Fresh	3 cent	Edamuttam ambalam	Bathing, rituals	Public	Compound wall & steps for protection	None	Man	Medium	Public	Useage related to the Aaraattu
12	VEERANKAADUKULAM	Fresh	Very old, 3 cent	Private-Adipparambil	Irrigation, fish farming (Karimeen)	Owners	In private land within boundaries.	None	—	—	Owners	60-70 years old

Figure 1. Ponds: Usage, features and pollution rates.

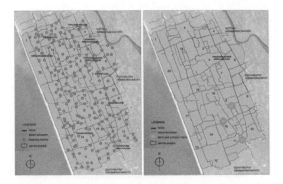

Figure 2.   Location: Existing ponds and wetlands/paddy fields in Valapad *gramapanchayat*.

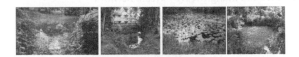

Figure 3.   Used and the majority of unused ponds in Valapad.

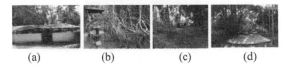

(a)   (b)   (c)   (d)

Figure 4.   Some sacred groves of Valapad *gramapanchayat* (a) Thekkiniyedath Naagakkaavu (b) Arayamparambil Kaavu (c) Paarekkaatt Kaavu (d) Cheeramkaattil Kaavu.

well maintained ones. All the others are cleared of the lush vegetation and reduced to a single platform for worship. The few notable ones are shown in Figure 4. Of these, the Adipparambil Kaavu (*Nagayakhi—Sarpakkaavu*) of the Arayamparambil family is the richest with more than 100 varieties of medicinal plants and other flora in the thirty cent premises. According to the studies conducted by Jincy, T.S. and Subin, M.P. (2015), the area of this sacred grove has been reduced to thirty cents from eighty-four cents in 1998, following land partition.

3.4   *Streams*

The 3,880 m long Pannatthödu, the 5,320 m long Paalamthödu, the 7640 m long beach thödu, the 1,200 m long Netkot thödu, and the 9620 m long *Arappathodu* flowing through wards are the most important streams in Valapad. However, there are other multiple feeder and connector streams to each of these main streams. Streams are numerous but with respect to the flow of water and connectivity, the case exhibits a rather degrading scenario. Neglect and unrestricted waste dumping has resulted in blockage. Restricted water flow during the monsoons lead to flooding of streams and as a result, the people resort to unscientific water clearances. Known as the breaking of *Arappa* (estuary), the rainwater collected and flooding the lands, is allowed to flow directly into the sea, thereby, giving no chance for groundwater retention.

3.5   *Mangroves*

Thrissur district consists of very low numbers of mangrove in the state. Presently, mangroves are confined to the backwaters of Chettuwai, Azhikkodu, Kodungallur and few patches in Venkidang and Pavaratty Panchayats. Valapad is abundant with an ecosystem favorable for the growth of mangroves. Mangroves are present along the Kothakulam *Arappa* area, bordering the Kothakulam beach ward, located approximately 500 m away from the sea. This is

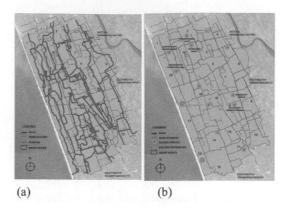

(a)   (b)

Figure 5.   (a) Sacred groves and mangroves (b) stream network.

Figure 6.   Existing condition of streams: Waste accumulation hindering flow.

one of the most picturesque spots in the panchayat. However, there are no buffers against the mangroves and sooner or later, it may suffer from encroachment as elsewhere.

3.6   *Surface water conditions*

The surface water sources in Valapad *gramapanchayat* are the water from the Karuvannur River, Periyar River, and Canoli Canal. The Karuvannur River is the fourth largest river in Thrissur district, formed by the confluence of the Manali and Karumali rivers.

The instruments of natural surface water collection are the natural ponds, *Arappa*, and the various natural drains or the thödu. The Valapad manappuram region comes as a micro watershed of Karuvannur puzha. Hence, it should be noted that the only sources of water to the water bodies is from the surface run off and rainfall.

The *Arappathodu* once filled, leads to stagnation issues and many wards suffer. Hence, connectivity issues arise. As already mentioned (in section 3.4) residents resort to breaking the waters. The collected water is immediately drawn to the sea. There would be no collected water for groundwater recharge. This has led to the lowering of the freshwater table. Water recharging is the only possible way for the regeneration of water to the waterbodies in the panchayat. In Valapad, there are no rainwater collection pits, no rooftop rainwater harvesting systems: neither at household levels nor at panchayat levels. From the primary survey conducted, it was found that only 2% of the people have set a rainwater harvesting system in Valapad. Apart from the existing saltwater intrusion, this may lead to acute water shortage in the near future.

3.7   *Biodiversity*

Being a coastal panchayat, Valapad has the potential to serve a large amount of aquatic and terrestrial biodiversity. The changes in the land use and encroachment over the protected areas can be attributed to the depleting wilderness and ecologically sensitive areas. These lead to endangered plant and animal life. There is also reduction in the amount of medicinal plants that were once easily and largely available in the panchayat.

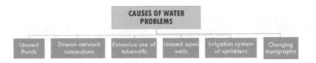

Figure 7. Surface water and shortage.

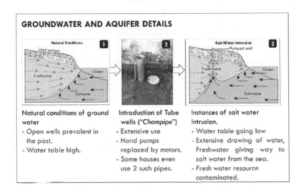

Figure 8. Water problems: Groundwater and aquifer details.

4   LINKAGES WITH OTHER DEVELOPMENT SECTORS

In Valapad *gramapanchayat*, the environment related issues of water shortage, pollution, and waste management are also linked with the shortage of physical infrastructure. The problems of waste management are acute and this has in turn led the people to resort to means of burning plastic wastes and also dumping them into the water bodies. This pollutes the waters and also disrupts the connectivity of flowing water in the streams. Pollution in the form of air and noise along the NH at peak traffic hours is one of the negative effects of traffic and transportation on the environment sector.

Beach and beachfront ambience is the main attraction of Valapad for tourism. Incorporating the principles of the environment and ecology to suit nature tourism, responsible eco-tourism projects can take up momentum in the area. Integrated tourism planning is essential for this and such initiatives can help in the sustainable development of the area.

5   STRATEGIES FOR REVIVAL

The strategies pertaining to the protection, conservation and maintenance of the natural resources that shall be adopted in Valapad, includes those related to ponds, streams, sacred groves, mangroves, flora and fauna, groundwater recharge, waste management and so on, for a better living environment. In the case of Valapad, bringing reasonable participation of the stakeholders, through early and effective consultations by all partners is crucial to frame a partnership. The actions required for this effective management shall be further exercised on three levels, as shown below in Figure 9.

Ponds: Though Valapad is known to have thousands of ponds within the *gramapanchayat*, the area lacks an action plan to maintain these assets. Table 2 below gives relevant strategies that can be adopted for effective maintenance of ponds and the lifeline of water retention in the panchayat.

Streams: Table 3 below gives relevant strategies that can be adopted for effective maintenance of streams here. For any action, the existing conditions of water channels and their connectivity need to be studied and mapped.

Figure 9.   Levels of management to be exercised.

Table 1.   Focus areas at each level of management plan.

	At level	Focus areas shall be
Levels	1 Physical changes and improvements	Infrastructure, planting, and asset management.
	2 Policy and planning	Changes in land use, infrastructure and movement.
	3 Implementation agency plan for sectoral intermix	Identification of agencies for sectoral mix in order to consider the strategies for an execution plan.

Table 2.   Strategies for maintaining ponds.

Levels	Strategies
1	Regular cleaning (especially of the ponds used for the holy dip of *Aaraattpooram*, like Kothakulam, and Sethukulam).
2	Resource mapping of existing unfilled ponds. Curb on unscientific landfilling of waterbody. Introduce functional interfaces along public ponds in a regulated manner.
3	Device a maintenance partnership plan through efforts from beneficiaries (private owners) and panchayat at a reasonable share.

Table 3.   Strategies for maintaining streams.

Levels	Strategies
1	Regular cleaning (especially of the feeder streams to the *Arappathodu*).
2	Network-mapping of existing streams (feeders and connectors included).
	Redefining of edge character with buffers to curb waste disposal into the waters (resulting in blockage).
	Introducing functional interfaces promoting public open spaces and regulated tourism activity along *Arappathodu* parts (for instance, suitable in Kothakulam).
	Integrating streams to be a known and usable part of the panchayat.
3	Devising a management plan incorporating MGNREGA (Mahatma Gandhi National Rural Employment Guarantee Act) workers to solve connectivity issues. Private partnership may not be viable.

Sacred Groves: The present conditions throw light on the fact that sacred groves of Valapad have attained religious recognition but no environmental recognition. As such, due to strong religious beliefs surrounding the same, private owners put efforts into their conservation. Only as it moves forward to attain the level of an environmentally sensitive spot, will the asset become a social responsibility.

Mangroves: Mangroves are one group of natural elements which aid in many ways against environmental and natural hazards. Table 5 gives relevant strategies that can be adopted for their conservation in Valapad gramapachayat.

Medicinal Plants: Table 6 below lists relevant strategies that can be adopted for the revival of medicinal plants that once adorned the panchayat area.

Water Shortage: The integration of projects like Mazhapolima with a cost managed like other sample areas of Guruvayur Municipality, Engandiyoor Panchayat and others. A plan such that 25–30% of the project cost (for an individual household) be provided by the beneficiary contribution shall be proposed. The beneficiary list once submitted to the

Table 4. Strategies for conserving sacred groves.

Levels	Strategies
1	Promote afforestation of medicinal plants and trees inside the groves for better environmental stability.
	Manage a data bank of existing well managed sacred groves in the *gramapanchayat* by laying them out into categories of size, density of vegetation and expanse.
2	General awareness about the social and environmental need for conservation (apart from the religious concepts) and existing assistance schemes.
	Spell out protection and conservation projects in the panchayat by providing funding from the local Panchayat funding. There ought to be protection using buffers (natural/artificial) according to the data gathered.
3	Devise a conservation plan for the principle of partnership with efforts from beneficiaries (private owners), local communities around the nearest vicinity and local authority (panchayat) at a reasonable share.

Table 5. Strategies for conserving mangroves.

Levels	Strategies
1	Boost the existing and initiate planting of new saplings in the Kothakulam *Arappa* area, at the same time, protect the already planted sapling areas in Palappetty.
2	Identify feasible zones for the planting of mangroves
3	Devise a maintenance partnership plan through efforts from beneficiaries (private owners) and panchayat at a reasonable share

Table 6. Strategies for reviving medicinal plants.

Levels	Strategies
1	Through clearing large sites of neglect in each locality, there shall be enough space for planting.
2	Identify suitable areas of potential for a botanical medicinal plant garden.
3	MGNREGA workers and youth clubs can work together to site setting. Devise a management plan involving the women folk of Kudumbashree, environment clubs of schools, Government Ayurveda Hospital Valapad and the many practitioners of naturopathic therapy.

gramapanchayat; a technical team of the project can provide the necessary support for the installation of the open well recharge units.

For initiating the project, the panchayat area can be broken down into clusters and these clusters can be selected based on land use or population density. For example, if on the basis of land use, clusters can be as in Figure 10 below, namely cluster 1 (residential), cluster 2 (residential, commercial/industrial) and cluster 3 (residential, commercial/industrial, public/semi-public use) etc. It shall be seen that in cluster 1 itself, there are three different zones in varying scales. In stage 1, a small set of houses can be selected for initiating the project. In the next three months, the stage 2 zone can be integrated with the already existing trial area. Finally, the final set of the stage 3 zone can be added within the next five months to complete a complete zone. The need to expand the zones one by one is very important in this project because from previous successful projects we know that, the success rates of groundwater recharging can only be achieved from collective participation and involvement from a wide area. Within a span of two years, the whole panchayat area shall be successfully integrated under the project.

Waste Management: To be commenced at the household levels. With most households with ample land parcels being able to manage their own food wastes and replacing plastic with reusable/biodegradable materials wherever possible can be linked to better results. Also, scope for the use of modern technologies such as those in bailing machines, will exist in the near future.

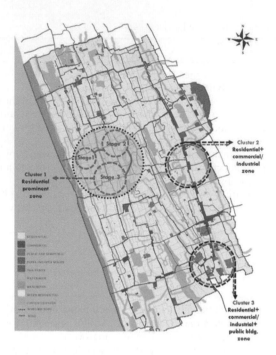

Figure 10.   Proposed clusters for rainwater harvesting.

6   CONCLUSION

The cases of land and water pollution are acute in Valapad, due to the improper and unscientific waste disposal techniques adopted. However, there are instances of certain public ponds being cleaned by schedule under the MGNREGA works. But at the household levels, most ponds are used for waste dumping and left for filling. In terms of disaster susceptibility, the coastal areas along the Kothakulam and Nattika *Arappa* (estuary) portions are prone to storm surges and accretion. In a time of reduced rainfall and drought, this is a very serious problem thet requires higher regard and concern. The existence of a strong blue-green network is a backbone for the development of any area. The management and protection of the same through effective planning strategies and policies will help strengthen not only the life of these networks, but also of the people of Valapad, especially because of the multiple benefits these elements shall serve. This is indeed a vision toward an ultimate living environment for the present as well as generations to come.

REFERENCES

Biodiversity Register. (2010). *Flora and fauna.* -Valapad: Valapad *gramapanchayat.*
CWRDM. (2013). *Impact assessment of Mazhapolima Project on groundwater regime in Thrissur District.*
Interview by Numerous interviewers. (2016). Primary survey on Integrated development plan for Valapad.
Mohandas, M. & Lekshmi, T. (2012). Kerala mangroves—pastures of estuaries—their present status and challenges. *IJSR.*
Subin, M.P. & Jincy, T.S. (2015). A detailed survey on sacred groves of different extends in the Coastal Belt of Thrissur District, Kerala, India. *Global Journal for Research Analyses 163.*
Sudheer, M. (2016). Interview by Nahlah Basheer. Water issues in Valapad *gramapanchayat.*
Valapad K.B. (2016). Interview by N. Basheer-Siji Mohan. Ongoing plan for natural resource maintenance at Valapad.

Emerging Trends in Engineering, Science and Technology for Society,
Energy and Environment – Vanchipura & Jiji (Eds)
© 2018 Taylor & Francis Group, London, ISBN 978-0-8153-5760-5

Urban metabolism and food security: Emergy as a metrics link connecting the food security with the urban spatial aspects for enhanced livability and sustainability

Bijey Narayan & J. Jayakumar
Department of Architecture, College of Architecture Trivandrum, Trivandrum, India

ABSTRACT: Urban food insecurity is a major challenge associated with the phenomenon of global urbanization. Several such multi-scale socioecological challenges have necessitated the re-emergence of the concept of urban metabolism, which essentially deals with the flow of energy and materials into and out of the city. In cities, the production and processing of food, though being away from the consumers, its consumption and disposal are still with them, which shows there is an obvious rift in the urban metabolism in terms of the food flow. The authors, with the help of the emergy model establish a strong link between urban food systems and urban land use pattern, highlighting the importance of sustainable urban planning that is also sustainable from social, economic and ecological perspectives. This model envisages a city that produces its own food and nutrition as well as distributes, consumes and disposes of it by virtue of the efficient systems generated by the urban resources.

Keywords: urban metabolism, emergy model, emergy analysis, urban agriculture, closed-loop food system, urban food security, urban land resources, urban land use plan, urban planning

1   INTRODUCTION

According to UN reports on world urbanization prospects, around 66% of the world's population will live in cities by the year 2050. The majority of this urban growth will take place in low and middle income countries. The projected urbanization shows an alarming rate and scale that can raise slum populations to 2 billion people (United Nations Report, 2014). Urban planning, design and governance are becoming centrally critical for human survival. The statistical evidence is in general pointing to the fact that humans as socialized animals are inhabiting smaller land parcels where the whole of the existing urban areas is only around 1–3% of the land area of the earth. The non-convertible wastes generated here are also solely by humans. All human made systems inadvertently over-depend or exploit nature, considering it as the inexhaustible vessel to meet their material needs.

The urban data also mentions another unpleasant fact that many of the city-dwellers, particularly those living in slums, still suffer from malnutrition (Amerinjan, 2017). Urban dwellers, especially urban poor are less privileged in terms of the accessibility to healthy nutritious food than rural people. This can be attributed to many factors, including the fast pace of urban life and to the skewed socioeconomic bias of the urban land use pattern. It is considered that the economy of a city depends mostly on the income generated by the business, commercial and industrial activities taking place in its built-up land. Hence incorporating or retaining green areas in the urbanized areas has less priority and consideration. As production of food is possible primarily with agricultural land being made available along with labor, skill and infrastructure facilities, urban areas by default become consumers of imported rural food and highly processed agricultural products. In general, the high energy transportation cost of food products from outside, which in many cases is another continent, drastically reduces the required accessibility. Hence much needs to change in today's urban spatial structure and organization in order to keep cities fed with healthy and nutritious food.

In countries like India, where the pattern of development has always been haphazard and equally unpredictable, the process of restructuring urban spatial organization will be quite complex. The majority of the medium cities of India lack a land use plan even today (Niti AAayog, 2017). Considering the unprecedented trends in urbanization, a mechanism or model looks inevitable to relate the needs of the urban dwellers to their consumption pattern (material flow into and out of the cities) that can be converted to a value which in turn forms the key parameter for an urban resource allocation strategy, especially urban land resource planning. The authors are suggesting incorporating a less data intensive tool to assess, estimate and balance the inputs and output in the urban system.

Food security being the focus of this paper, the exploration here is into the flow of food material in the urban areas and its relationship to urban land resource allocation strategies.

2 FOOD SECURITY

Definition of food security has changed over the years. It was defined in the 1974 World Food Summit as the 'availability at all times of adequate world food supplies of basic foodstuffs to sustain a steady expansion of food consumption and to offset fluctuations in production and prices' (FAO, 2003). In 1983, Food and Agricultural Organization of the United Nations (FAO) defined it as 'ensuring that all people at all times have both physical and economic access to the basic food that they need' and thus expanded the concept to include access by vulnerable communities or people to available supplies, implying that the demand and supply side of the food security equation should be balanced. The 1994 United Nations Development Program (UNDP) Human Development Report added a number of component aspects to human security of which food security was only one. This concept, that related food to the human rights, helped start discussions about food security. The 1996 World Food Summit adopted a still more complex definition: 'Food security, at the individual, household, national, regional and global levels [is achieved] when all people, at all times, have physical and economic access to sufficient, safe and nutritious food to meet their dietary needs and food preferences for an active and healthy life' (FAO, 2003).

The State of Food Insecurity refined the definition in 2001 as: 'Food security [is] a situation that exists when all people, at all times, have physical, social and economic access to sufficient, safe and nutritious food that meets their dietary needs and food preferences for an active and healthy life' (FAO, 2003).

Hence the accepted definition of food security can be expressed as: food security exists when all people, at all times, have physical, social and economic access to sufficient, safe and nutritious food that meets their dietary needs and food preferences for an active and healthy life. Household food security is the application of this concept to the family level, with individuals within households as the focus of concern, and 'Food insecurity exists when people do not have adequate physical, social or economic access to food as defined above' (FAO, 2003). In the state of Kerala, food insecurity has become a huge challenge because of the toxic content and questionable safety of the food imported from the neighboring states. Of late, Kerala has witnessed a change in the vegetable production in its urban areas intended for local consumption rather than relying on the food imports from distant places.

3 URBAN INDIA

According to the 2011 census, 377 millions of Indian people live in urban areas compared to 286 millions in 2001. For the first time since independence, the absolute increase in population is more in urban areas that in rural areas (Census India, 2011). A study of Indian cities by McKinsey Global Institute, reveals that the current performance of Indian cities is poor across key indicators of quality of life, such as water supply quantity, public transportation,

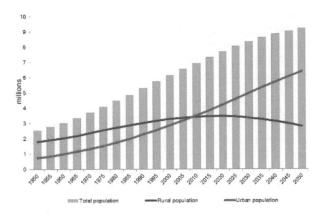

Figure 1.
Source: UN DESA Rome, 2010.

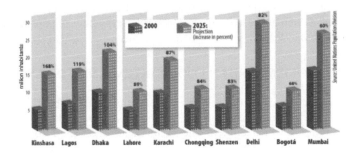

Figure 2.
*United Nations Department of Economic and Social Affairs.
Source: *UN DESA Rome, 2010.

parks and open spaces. On current trends the quality of urban services is expected to fall further by 2030 (McKinsey, 2010) with reasons obvious even today.

The mentioned scale and complexity of urbanization in India demands a comprehensive strategy required in addressing the urban challenges, especially that of food security. Intensive data preparation is necessary to devise strategies for urban resource allocations and the use of accurate metrics is required to ensure the effective execution of urban strategies.

4 LAND USE PLANNING IN INDIA

Town or urban planning in India is still in its infancy. While an urban master plan governs the growth of the urban areas, a Town Planning Scheme or TP Scheme is also widely adopted in various states. Land use planning, the most important feature of all development plans dealing with spatial planning, is very critical for all development purposes since it involves assigning the particular activity to a given parcel of land based on the concept of optimal land rent where the very definition of optimal use is limited to economics. The authors are of the opinion that there has to be another dimension to it, by adding the component of solar energy falling on the urban land surface for the production of food from within, which is important for sustainable development.

Zoning and land use planning help to ensure dedicated land for segregated and planned activities that have positive impacts on the economy and to avoid conflict in activities that affect the quality of life in those areas. Unfortunately many of the fast urbanizing, middle sized towns do not have a comprehensive master plan with a well-developed land use plan (Niti AAayog, 2017). As there is not much credible work done in the urban land use plans or the development of a master plan in the various cities and towns of India (Niti AAayog, 2017), it is imperative that there has to be an alternative system of metrics to link and measure the needs, flow and use of urban resources and materials.

5 URBAN METABOLISM

A city can be better understood if compared with a complex organism with various metabolic processes. Howard T. Odum, an American ecologist, proposed the conceptual model of metabolism (Odum, 1996) where materials from within the city or from outside getting transformed by a series of urban activities, and finally get converted to waste and then released into the environment. Urban metabolism is composed of built-up land, farmland, and unused land (Odum, 1996).

The inner environment of cities by itself cannot support all the metabolic activities; materials, and energy from outside is also needed for it. Additional mechanisms are required to expel the wastes into the environment. Hence urban metabolism is understood and expressed in terms of production, consumption, and processing of urban internal resources completed with waste disposal, as well as the flow of material and energy between the internal and external environment (Odum, 1996).

6 URBAN METABOLISM, URBAN FOOD SYSTEM AND FOOD SECURITY

Urban Food security depends on the design and resilience of the urban food system, which in turn depends on the urban metabolism. Metabolic rifts are caused due to the linear, interrupted food loop systems, which happens due to the exclusion of food from the urban policy and planning agenda for a long time.

The urban food system has a dynamic structure that exists between land, population, food distribution and production processes, resources, technology, economy and employment (Armendáriz et al, 2016).

A resilient urban food system addresses all four levels of the food system: food production, processing, distribution and consumption, (Amerinjan, 2017) which implies a closed-loop food system. In the cities, policies can be established to incentivize the local production of healthier food options and limit unhealthy food imports (Amerinjan, 2017). The Edo period in Japan could be a good model for this. Recycling and living with 'just enough' were made part of public policy when Edo was facing an imminent environmental crisis (Brown, 2010). Almost every material in Edo was made to recycle and none went to waste. The need to recycle was reflected in technologies and practices and the way things were made in the first place. This could arguably be the predecessor of the circular economy that we are trying to engineer today. Similarly, we need a new paradigm within the urban planning that will help ensure food security and thereby sustainability in the cities. Metric-based urban planning is required where the urban metabolic density can be evaluated and allocation of land resources can be done accordingly.

7 'EMERGY' AS METABOLIC METRICS

Emergy is defined as the total amount of available energy (or exergy) of one kind that is used up directly or indirectly in a process to deliver an output, product, flow, or service (Odum, 1996).

Emergy analysis is a method to measure the value or quantity of material energy that is used to transform the different (2 or uniform measurement standard), by the use of specific conversion factors and by combining socio economic with eco-environmental systems, to analyze the flows and transformations of materials and energy quantitatively (Huang et al, 2015).

The formula is given as

$Em = \tau Ex$

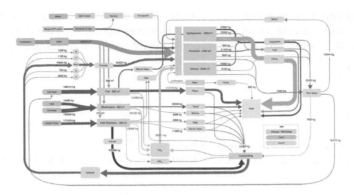

Figure 3.   Scenario journal 15.

where Em is the emergy of one material or energy, Ex is the available joules of one material or energy and τ is the emergy transformity constant of material energy, which is the solar emjoules it needs to transform (Huang et al, 2015).

According to the emergy model, different forms of energy, materials, human labor and economic services are all evaluated on a common basis (the environmental support provided by the biosphere) by converting them into equivalents of only one form of available energy, the solar kind, expressed as solar equivalent Joule (seJ).

The concept of "available energy" allows the analyst to account for all kinds of resources used (minerals, water, organic matter), not only energy carriers (Huang et al, 2015).

A typical material flow study can be seen from the figure. Food materials can be filtered from it and studied separately. Using the emergy database, the solar transformity value can be found for each material, and that can be multiplied with the available amount of the material to find its emergy. Various materials and energy are categorized into different groups for ease and standardization in calculation, such as renewable, non-renewable, industrial and labor.

The solar transformity of some materials are as given below.

1. Sunlight – 1
2. Agricultural production = 1.43×10^5
3. Livestock production = 9.15×10^5
4. Fisheries production = 3.36×10^6

In this way, the emergy-based evaluation of urban land can also be done. By calculating the correlation coefficient of the increments in various urban lands, such as built-up land, its impact on the total metabolic density of the city can be found out. If a city's metabolic activities depend on the activities in the built-up land as per land use, it can be concluded that the changes in the activities on the built-up land will affect the urban development more significantly than the other lands. Considering the environmental load of the built-up land, planners can insist on more farmlands/agricultural land to be included in the urban master plan, to match the emergy value of the food materials coming in and required to be disposed of in the urban area. To address the space shortage in cities, appropriate policies may be designed and implemented to make sure that there are enough sun exposed roof spaces for urban agriculture/farming. This will further necessitate the need for scientific disposal and recycling of wastes that can in turn be used for the production of food items in the urban area itself. Thus, urban food systems can be made a closed-loop system, fostering sustainability and resilience and eventually, a food secure urban community.

8   CONCLUSION

This paper has conceptually explored emergy as a metrics parameter in urban resource allocation to achieve urban food security. Emergy is useful in evaluating and relating the environmental and economic aspects of a triple bottom line development model but has its limitations in evaluating the social component. However, considering the rapid urbanization, urban migration, environmental deterioration and increase in GDP in India, a metrics driven execution of urban strategies definitely has an advantage in the development of cities that are data deficient.

REFERENCES

Aayog, N. (2017). Documents, Government of India.

Armendáriz, V., Armenia S. & Atzori, A.S. Systemic analysis of food supply and distribution, systems in city-region systems—An examination of FAO's policy guidelines towards sustainable agri-food systems (December 7, 2016).

Brown, A. (2010). *Just enough: Lessons in living green from traditional Japan*. Tokyo: Kodansha International Ltd. Retrieved from http://sedac.ciesin.columbia.edu/data/collection/gpw-v4.

Census India (2011). Reports. Retrieved from http://censusindia.gov.in/2011-prov-results/paper2/data_files/india/Rural_Urban_2011.pdf.

Food and Agricultural Organization of the United Nations, Rome (2003). *Food Security: Concept and measurements*. Retrieved from http://www.fao.org/docrep/005/y4671e/y4671e06.htm.

Food and Agricultural Organization of the United Nations. (2010). *Policy Brief 10*. Retrieved from http://www.fao.org/economic/es-policybriefs/multimedia0/presentation-urban-agriculture-and-food-security/en/.

Food and Agricultural Organization of the United Nations. *Urban agriculture* (n.d) Retrieved from http://www.fao.org/urban-agriculture/en/.

https://esa.un.org/unpd/wup/publications/files/wup2014-highlights.pdf.

Huang, Q., Zheng, X. & Hu, Y. (2015). *Analysis of land-use emergy indicators based on urban metabolism: A case study for Beijing. Sustainability*. Retrieved from http://www.mdpi.com/journal/sustainability.

McKinsey Global Institute. (April, 2010). *Report on India's urban awakening: Building inclusive cities, sustaining economic growth*.

Odum, H.T. (1996). *Environmental accounting: Emergy and environmental decision making*. NY, USA: John Wiley and Sons Inc.

Van Ameringen, M. (2017). Global alliance for improved nutrition. [Blog]. Retrieved from https://www.gainhealth.org/knowledge-centre/urban-food-systems-that-promote-good-nutrition/.

Wiskerke J.S.C. (2014). Urban food systems. *World urbanization prospects: United Nations report* (pp. 1–25).

Smart city initiatives

Emerging Trends in Engineering, Science and Technology for Society,
Energy and Environment – Vanchipura & Jiji (Eds)
© 2018 Taylor & Francis Group, London, ISBN 978-0-8153-5760-5

Smart-cities and smart-villages in the Indian context: Some behavioral aspects

P. Varghese

Karpagam University, Coimbatore, Tamil Nadu, India

ABSTRACT: Questions have been asked regarding the feasibility of smart-cities; it has been asked whether the policies could be lopsided when seen alongside normally developing urban processes, which adopt preconceptions or a uniform distribution of resources, or next to traditional lifestyles. Some assumptions adopted by Glaeser (2011), as well as the McKinsey report (2010), regarding urbanization are reviewed, especially within the Indian context. This needs to be seen within a historical/traditional perspective and does not fully require the smart-city development policy as a given. It is argued instead that providing urban amenities in rural areas could also be an alternative or parallel option, for multiple reasons.

1 INTRODUCTION

1.1 *The future of urban India*

The process of urbanization has been stated to be a real and continuing phenomenon the world over; in India it has been recorded at about 31% as of 2011, is projected to be 50% by about 2030, of 590 million, and 70% by 2050 (UNFPA, 2007). As per the last census, 377 million persons (31,2% of the population) live in urban areas (GoI, 2011). It is expected that by 2031, approximately 590 million persons (40%) will be living in urban areas. In 1951, only 6.2% of the population were living in urban areas—this change represents approximately a sixfold increase; meanwhile, towns and urban areas have doubled in number (Ravi et al., 2016).

Urbanization as a naturally occurring phenomenon cannot be negated or overlooked. Current urban perspectives are looked at through the arguments of the economic urbanist Glaeser (2011) in his book *The Triumph of the City*; he believes that compact cities optimize real estate prices and reduce congestion, slums and sprawl. A similar argument for increasing urbanization, the McKinsey report (2010), forms the basis for the development of smart-cities in India.

It is not being argued here that normal rates of urbanization need to be hindered, which is not feasible within democratic set ups; it is argued that urbanization has to be undertaken, especially considering that current urban requirements often lack standard infrastructure or amenities at minimal levels.

What is being questioned, however, is the development of the so-called smart-cities program, which allocates funds to chosen cities as opposed to all cities, which also need similar funds for much needed infrastructure and developmental works.

1.2 *The needs of rural India*

The needs of the rural parts of the country are real; India primarily builds on its agricultural base, where people have been farming for ages. This background is not merely for economic ends, but there is an existing attachment to the land, to nature, to the environment and to the idea of continuity. There could be a cultural bias, but it can be argued that it is an ethos that drives the ordinary man to the land and to nature. Glaeser's (2011) perspective is certainly humanistic, despite him being a rational economist, and he is real in his logical arguments. But it probably does not accept the reality of generations of farmers who have adopted this as a way of life. It is easy to rationalize that all of existence should be viewed from the perspective of *demand and supply*, which would be the base that economists would adopt. This is perceived only by visible, practical aspects that probably miss the deeper links, which need a sociological perspective and analysis to also come up with theories of an *urban bias* by policymakers.

1.3 *Non-economic perspectives of the urban processes*

Glaeser's (2011) point of view is logical in stating that the need for development in cities is separate from rural areas. The McKinsey report (2010) is a pure, rational, economic argument that projects current global trends, and extends the view that the same holds good for India. One cannot fault the views from a logical standpoint, which shows realism, but it is not necessarily the only and definitive perspective.

Instead, alternative socio-cultural possibilities also need to be taken into account, since it almost assumes that urbanization is the answer to rural poverty; possibly even stemming from failed agricultural investments, which depend substantially on the monsoon and on short-term weather predictions. It is being argued instead that providing urban amenities in rural areas would aid rural folk to continue to remain there, and additionally, help preserve traditional knowledge of the world, of nature and of societies that have continued over generations. The use of modern technology to document and analyze traditional wisdom, culture and lifestyles can be prioritized from this viewpoint.

2 URBAN DEVELOPMENT IN INDIA

2.1 *Urbanization in India*

Currently, India has an annual nominal per capita GDP of US\$ 1,408/- (~Rs. 93,000/-) and approximately 363 million persons (30% of the population) live in poverty, about 1.77 million persons are homeless, and 4.9% of the population (aged 15+) are unemployed, while even those who are employed do not have stable incomes. In the rural areas, where wages are even lower, possibly by up to half, and more so for women than for men, about 48% of households survive with little or no socio-economic services (UNFPA, 2007).

Urban growth in India has not developed to the uniform extent that it should have due to several conflicting factors. The regions of the country have been shaped over time by various characteristics: the history, topography and climate, as well as traditional and cultural norms.

With regards to the economy, central and state allocation of funds were the main sources until the recent adoption of the 74th Amendment, when urban areas were expected to build their own reserves. Strangely, even today there are several bodies within a single area that have overlapping jurisdictions and work in unco-ordinated ways. This would inevitably lead to the duplication of administration, unco-ordinated work, a lack of responsibility, inefficiency, and wastage.

A major factor that is lacking is the dearth of qualified officials with a background in areas such as physical planning, transportation, etc.; bureaucracy and a lack of co-ordination seems to plague the official administration. Much of this happens at a pace that is prone to delays, and the adoption of current technological/social media that would help to keep pace with needs is not looked into; training for such things is possibly lacking. The whole official machinery and bureaucratic system needs a rethink and an overhaul to be up-to-date.

In such settings, perhaps the world over, it is no surprise that private enterprise takes up the initiative to create socio-economic change, but with a profit motive; the private sector runs by eliminating redundancy and wastage of resources and time. The judicial processes are also long and drawn out, often taking years for a decision.

Governance in India is fragmented, since most departments work independently without links to others. Many bodies do not have planners with backgrounds in the physical, economic or social sectors, which ultimately dictates the success of policy. The government needs to incorporate such social aspects into the smart-city program. Smart-cities, even in a world context, are not about technological solutions, but ultimately about the social services bettering the quality of life.

2.2 *The rural/agricultural background*

Agriculture and trade in farm produce is universal, but that does not capture the links to other spheres of activities; for example, the traditional and historic development associated with land stems from this base. The land provides not only sustenance in terms of food, or economic support on the sale of the produce, but also knowledge systems of resources, such as water, and of time in terms of the passing of the seasons, for planting, harvesting, etc. Also, many cultural norms in terms of food habits, clothes, shelter, medicine (*ayurveda*), festivals, etc. were derived from such rural foundations. It would thus be a limited viewpoint to assume that, in terms of a pure economic advantage, a wholesale migration would occur to urban areas.

In terms of social indicators, over much of the 20th century (GoI, 2011) birth rates have decreased from 40/1000 in 1971 down to 20.22/1000 in 2013; the infant mortality rates have decreased from 165/1000 live births in 1950–55 to 38/1000 in 2015; and average life spans have increased from 34 years in 1911 to 69 years in 2009. The effective reduction in the number of farmhands does not necessarily indicate that rural folk feel the need to migrate to cities.

Migration or *mobility* can be viewed as being caught in the opposing centrifugal/centripetal forces of the rural–urban divide. It can be highlighted that it is secondary factors that are, when true, enhancing rural to urban mobility—such as opportunity, education or, to a lesser extent, lifestyle. It should be mentioned that technology and mechanization has ensured that fewer hands are required; this would validate some of the changing demographics, but not necessarily all of the migration. Sociological studies to consolidate these assumptions in the context of India need to be undertaken more thoroughly, both in rural and urban areas, in order to comprehend the underlying conditions. Further, it could be necessary to understand whether or not these migrations are temporary—seasonal or otherwise. It is true that education, job skills and similar factors could make people move to urban areas, but it is not fully known whether or not these are permanent.

Conversely, it can be argued that farmers have not been provided with the necessary level of institutional support that is required. For example, an input–output related *minimum support price* that ensures that the farmer gets the benefit of market prices; a forum that provides mechanisms for fair trade, regular payment, or is GST (Goods and service tax) enhanced; institutional mechanisms developed to ensure insurance coverage in case of bad monsoons or seasonal failures; or award benefits of a bumper crop. With technology, a farmer is allowed to control the mechanisms of trade, instead of traders and middlemen making much of the profits. The government needs to develop robust mechanisms for the storage/export of agricultural items to international markets, or to less developed countries in the form of aid.

It even happens that the government (or representatives), in the name of infrastructure or project acquisitions, takes over the land of farmers with no compensation, or compensation that is either minimal or delayed. These kinds of displacements are common, but the government often subsequently turns a deaf ear to such occurrences.

The frequent occurrence of farmer suicides are more often than not because they are unable to pay back the sometimes paltry loans taken for buying seed or fertilizer, which the government itself could subsidize; however, it is often the case that the government is seen as ready to write off the loans of the corporate world. Social justice cannot play favorites.

3 THE ROLE OF THE PRIVATE SECTOR

3.1 *Land, speculation and investment*

Urban growth in India happens without much official involvement. It could seem that the private sector is ahead of the curve in anticipating change or pre-empting moves by the government. On the one hand, the government would have a minimum amount of official information, future plans, and records or processes in place, but much of the expansion seems to happen without administrative curbs/restrictions. Private individuals are able to create change, often on a small scale, but sometimes large, without the seeming knowledge of officials; such changes are often difficult to reverse and are sometimes irreversible.

The expansion of slums in cities is an example that, in apathy to their official responsibilities, as well as finding temporary but real solutions to actual needs, manifests itself in a physical form. The numbers living in slums was estimated as 40.3 million in the census of 2001. In 2007, 32.1% of the urban population of India was recorded as living in slums, down from 54.3% in 1990 (UNFPA, 2007).

As against a current worldwide statistic of 50%, the rate of urbanization in India seems relatively slow, having reached 35%. There could be several reasons for this, not all explained. As mentioned, primary among them could be that the country as a whole stems from that rural/ agricultural base. This has several possible inferences: one is that, with the tradition of farming, there is an attachment to the land and a knowledge base that is associated with the seasons and nature. Such a tradition dictates much of their lives, in daily, seasonal and lifelong associations. Although it could sound basic and naive, it is hardly that; this base dictates their entire life, and it would not be something that they are willing to give up spontaneously for a job in the city. Therefore, this link is quite strong. (Glaeser et al. 2015), however, wonder why urbanization in India seems to be hampered—data indicates that mobility in India is much lower than theoretically expected. Secondly, socio-cultural identity and connections, familial, caste-based, or historical/traditional, are often binding to the village or to the region. A third factor, which acts together with the previous one, is the idea of linguistic connectivity; this is an important part of the decision-making process, especially in relation to interstate mobility (especially for rural to urban). The limitations caused by this are manifold.

In-state migration makes this easier, especially for those who are not sufficiently literate, whose linguistic ability is limited to their native language, and this depends on literacy/educational levels. The average literacy rate in the country is 73%, and this is often a limiting factor on mobility. A person knowing two or more languages would be comfortable to move to other states, not necessarily to urban areas within the region, where s/he is confident that communication will not be an impediment to meeting up with others. Someone who is confident with more than one regional language, as well as another (such as Hindi or English), would be willing to migrate to many of the metropolitan areas within the country, as long as it has necessary perks. Intra-state migration is more common for women, possibly because of family ties or needs, and men are more accepting of interstate mobility.

There are also tertiary factors, such as age, gender, marital status, distance/connectivity from home, local social networks, etc. The current use of telecommunications and social networking has changed this factor substantially. As one advances in age, people are unwilling to move far from primary social ties, such as family, relations, childhood associations, etc. It takes a substantial incentive to stay aloof from such social networks without a definite reason.

India has a housing shortage of 18.78 million units (Jha et al., 2015), which is effectively an annual construction requirement of 2.34 million units per year, as against an ongoing rate of only 1.2 million units per year. People migrating from the rural areas primarily require affordable and basic housing with at least the basic services, which the housing authorities should try to provide. The country looks to the private sector to fill most of this gap. The national/urban bodies are unable to take up the role, but may step in using private sector agencies.

4 THE ECONOMIC CITY

4.1 *Urbanization and economics*

Cities are called the centers for economic progress—in the form of trade, finance or, now to a lesser extent, manufacturing. Preliminary modern models of growth are centered on industry. In the early years of development, the coal and steel industries were part of the basic growth in the eastern parts of the country; mining and heavy industry was centered close to the location of resources, which could be remote from areas where previous urban expansion had occurred, and close to transportation networks or trade routes. Mining and similar industries also meant substantial pollution, and therefore were a lower choice for urbanization.

On the other hand, secondary industries such as spinning mills or goods for daily needs sprang up after agricultural produce, the primary resource; other trading commodities expanded to building materials, etc. Classical economists talk about the basics as *demand and supply,* which gives value to goods and services, and on which economies prosper or decline. Today's valuable products are those of the information age, such as software, networking, etc. Linking these are infrastructure and networks for transportation, power, telecommunications, etc.

4.2 *Urban economics: Feasibility*

One hears the complaints of national or urban bodies concerning the lack of necessary funds for basic infrastructure or its maintenance; it is speculated that the much needed capital in urban areas can be generated by the proper utilization of resources, such as the land held by public institutions belonging to the local, state or national bodies. While it is not necessary to actually sell these assets, other innovative methods, such as renting or leasing to other institutions, public or private, could be a viable method of raising much needed capital (Detter & Fölster, 2017). The government would then have a financial infrastructure in place that caters to official needs and need not go to the private sector to raise funds. Many countries have municipal bonds that are backed by the government to develop such things as infrastructure, but this has not become common in this country.

For instance, the Indian Railways sits on prime real estate in cities, which can be capitalized upon by either building, renting or leasing it to institutions/private enterprises. These locations fit into the needs of travelers for both affordable accommodation and restaurants, proximate to their travel plans. The rail stations at locations at Navi Mumbai, planned by CIDCO, have office spaces that are adjacent, or even directly above, which serve the needs of offices and of commuters. The railways have ventured into the laying of telecommunication lines (Railtel) along their tracks, which could provide an essential infrastructure for the country's needs. Similarly, the almost defunct post and telegraph (P&T) offices have not updated their services to include the current requirements of electronic networks or transportation, instead a new corporate (VSNL/BSNL/MTNL), similar to the private sector, was developed. Private agencies began taking over their roles, even though their earlier assets, such as real estate, remain intact. Such symbiotic situations will further the essential requirements of the country, as well as help to piggyback solutions that are in plain sight but rarely utilized.

5 THE CASE OF INDIA'S SMART-CITIES

5.1 *Technological urbanization*

The idea of smart-cities, a recent worldwide phenomenon, has a problem of definition; various entities and countries have generated their own to fit their needs. As a technological development, smart-cities are expected to incorporate e-governance and 24/7 services, such as superior public transport, water, power and broadband connectivity. The approximated worldwide market for smart-cities has been pegged at US$ 1.7 trillion.

The government initiated a US$ 15 billion, 100 smart-cities program in 2015, due to an election promise made by the new administration; this was a parallel of China's US$ 300 billion, 193 smart-cities investment (Li et al., 2015). The Chinese government is financially capable of expanding such a program from the ground up (or *greenfield*), because of its established, top-down decision-making procedures; doing so in India would be expensive. On an experimental basis, the GIFT-city in Gujarat was begun by the state government, but to get that greenfield project going has been time-consuming and taken much effort, though there could be lasting advantages. Instead, the Smart Cities Mission allocated funds for existing cities as part of their urban renewal process, for the development of these *brownfield* ones. In addition, other cities were also helped to build up over several phase- and state-wise competitive allocations through the AMRUT (Atal mission for urban rejuvenation and transformation) program under India's Ministry of Urban Development (MoUD, 2017). The development is planned for satellite towns, urban renewal and fixes to tier-II/III cities, rather than to build from the ground up. This also applies to the overall inclusion of the UN's Sustainable Development Goals (SDGs).

From the previous term, about 80% of the money allocated for a US$ 20 billion Jawaharlal Nehru National Urban Renewal Mission (JNNURM), introduced in 2006, was used mainly in the transportation sector for widening roads and for flyovers; this included many expensive schemes, such as metros and Bus Rapid Transit systems (BRTs). A review of global cities along ten indices, such as infrastructure, transparency, long-term thinking, and so on, which distilled criteria from about 300 factors that was collated from diverse government, research and academic studies, indicated that none of the cities in the country figured in the top 10/20 listed worldwide, except for a claim of having some of the highest rental values (Feenan et al., 2017).

An underlying issue is that cities in India never urbanized synchronously with advances in technology, the provision of infrastructure, governance, or the conversion of economic assets into on-the-ground development for the public or common good. Theories of spatial equilibrium (Chauvin et al., 2016) could be more relevant to other regional/global models, including that of China, but seem inconsistent in the case of India. An answer to Glaeser (2007), as mentioned earlier, would be differences in language and culture, which would inhibit mobility. Many such factors, it would seem, do not apply uniformly to developing countries, especially in the Indian context considering its population scale and social diversity. In many developing countries, social ties to home communities are often robust, which gives steadiness and support. It was theorized that with progress into market economies with uniform human capital, perhaps spatial equilibrium could evolve in India. The smart-cities mission, or its other arms, must also address issues of infrastructure and services, such as the management of solid waste, use of energy-efficiency and renewables, urban design/planning, including housing and slums, public–private participation, and e-governance in the local bodies, and not merely look for solutions to technical issues.

5.2   *The case of Dholera*

Dholera in Gujarat should be a case study in point regarding smart-city development. Sited on the Delhi–Mumbai Industrial Corridor (DMIC) and initially a collection of 22 villages of 38,000 inhabitants, it was envisaged by the bureaucrats, without the knowledge or involvement of the inhabitants, to become a showcase for a new 'Gujarat model of development', including the use of blanket, quasi-legal procedures called 'Special Purpose Vehicles' (SPVs). It was expected that this would grow as a greenfield smart-city and by 2040 become larger than Delhi or Mumbai, if plans were to materialize (Datta, 2015). However, the answer to the question not addressed at this stage—which was what was to happen to the current and displaced residents, and what was to be their fate—is still unknown. It might follow the dealings with earlier inhabitants, mainly tribal and rural groups displaced by the Sardar Sarovar Dam, who have still not been rehabilitated or received compensation after many decades; many of them were cheated out of their traditional lands by scheming middlemen.

5.3   *Alternatives—smart-villages*

An alternative to such cities, smart, hi-tech or otherwise, is instead the possibility of smart-villages. Home-grown solutions, which have been proposed by technologists who want to see change from the ground up, from the grass-roots, indicate that this development should happen not merely in selected areas or metropolises of the country, but all over. One such case is that of PURA (Providing Urban Amenities to Rural Areas), proposed by A.P.J. Abdul Kalam, an eminent space technologist, and also a past president of the country. He envisaged that such technological developments should happen from the level of the villages themselves;

with the ongoing development in both hardware and software, the benefits of connectivity should proportionately flow to the lowest clusters of society—the villages. Essentially, it was thought that smart-villages need to be developed in order to directly connect the village with towns and cities throughout the country. It is that knowledge that is at the heart of such developments, and from knowledge/learning and the understanding of issues and problems, local or otherwise, technology would only need to be the connector or link within society. This would probably be an inversion of the current understanding, but the logic of it means that the two-thirds of the population that still resides in rural areas would be sufficiently empowered to not need to move to the towns and cities. With the provision of amenities, (wired/wireless) networks and infrastructure similar to that available in the cities, there would be no real incentive or need to move.

The arguments are several; on the one hand, economic necessity can be alleviated by the provision of the forms of income that the technology could bring; technological connectivity could provide partial answers to rural problems, including that of jobs; self-employment and innovation could be a partial answer, rather than seeking employment in urban areas; and solutions to local problems, even relating to society, culture or to epistemological issues, could be worked out at the rural level itself.

The normal process of a distressed farmer moving to a town to seek employment, temporary or otherwise, who ends up living in a slum with others who have migrated, with no promise of the basic services of water, power or sanitation, sometimes with no roof or shelter over their heads, could be minimized.

5.4   *Technological tragedies and urbanization*

Current forecasts regarding technology indicates automation as being part of this progress. While this is happening worldwide, being advanced by the developed world, what it implies has universal implications, even for those countries. Automation is the use of computational methods for design, production and manufacturing; essentially, much of the work that can be done by computers will taken over by them and away from human-centered processes, from accounting to routine assembly or iterative processes that can be programmed. Depending on the complexity, it could also substitute for surgeons or entertainment. This will have economic and social implications. Could the educational system prepare the job-seekers with skills for processes that are still unknown? What are educational systems currently teaching?

Current estimates are that automation could replace most jobs, leaving only 30% of present jobs intact. From this scenario, if the developmental systems forecast that urbanization is happening at an unprecedented rate, then why does the future of urbanization seem so certain? In India, a priority could be to continue maintaining the upgraded knowledge and skills of the population in the towns and villages, but also to provide them with facilities that will give them the advantage of both education-wise, job-wise and future-proofing them. The knowledge base of the world, especially of traditional systems, of indigenous plants, seasonal farming methods, and local culture and its arts, still needs to be documented and propagated. It is difficult to say that urban life will provide the answers to the individual or to a society that needs to adapt its way of life for an uncertain future.

5.5   *A dichotomy of urban intent*

If current forecasts regarding technology are indicators, then the country seems to be banking on the idea that rural–urban migration is this inevitable reality, that the country can build smart-cities to upgrade them to world-class levels. On the other hand, the plight of existing cities is of how to accommodate the growing numbers of people. The country anticipates a future increase in jobs to be taken up by the increasing number of graduates being churned out by the rising number of sanctioned (engineering/other) colleges. In contrast, the modern use of Artificial Intelligence (AI), Machine Intelligence (MI), or automation dictates that repetitive jobs at the manufacturing level, on the shop-floor or in fields such as accounting will be reduced by the introduction of automation and robots. Tasks that humans currently do will be done better by automated systems. Looking at this dilemma, why has the government not looked into its crystal ball to make definite plans for the future? Considering the vastness of the issue, smart-cities are only a partial solution.

6 CONCLUSION

The idea that urbanization is inevitable is only preparing the individual for an uncertain future, especially those from rural backgrounds. Will the move to the city ensure that the individual will be able to fight the odds to make a living? Is farming the land an outdated lifetime concept or pursuit? Are we as a society preparing future generations for an inevitability that no one knows? The answers are not clear.

Glaeser (2011) examines the city from a post-established perspective and rationalizes its existence; the McKinsey (2010) perspective is neither a ground up viewpoint nor multidimensional and proposes the smart-city as a 21st century solution to a socio-economic inevitability. This approach is fallacious, especially within the Indian context. Other models of development also need to be looked at concurrently. In the final analysis, smart-cities could become a reality, but maybe for the wrong reasons—it need not develop for the reasons mentioned. However, the investment in smart-villages could, it is rationalized, have a greater benefit to the population. National policies ought to have a multidirectional perspective, which will drive current thinking and actions for a more definite future.

REFERENCES

Chauvin, J.P., Glaeser, E.L., Ma, Y. & Tobio, K. (2016). *What is different about urbanization in rich and poor countries? Cities in Brazil, China, India and the United States* (NBER Working Paper No. 22002). Cambridge, MA.

Datta, A. (2015). *New urban utopias of postcolonial India: 'Entrepreneurial urbanization' in Dholera smart city, Gujarat.* Leeds, UK: University of Leeds.

Detter, D. & Fölster, S. (2017). *The public wealth of cities: How to unlock hidden assets to boost growth and prosperity.* Brookings Institution Press.

Feenan, R. et al. (2017). *Decoding city performance: The universe of city indices 2017.* Chicago: Jones Lang Lasalle Ip, Inc. and The Business of Cities Ltd.

Glaeser, E.L. (2007). *The economics approach to cities* (Working Paper 13696).

Glaeser, E.L. (2011). *The triumph of the city.* London, UK: Macmillan.

Glaeser, E.L. (2013). *A world of cities: The causes and consequences of urbanization in poorer countries* (NBER Working Paper No. 19745).

GoI. (2011). *Census of India.* [Online] Retrieved from http://www.censusindia.gov.in/2011census/ [Accessed 25 10 2017].

Jha, A., Sharma, S. & JLL-ASSOCHAM. (2015). *Housing for all: Catalyst for development and inclusive growth.* New Delhi: JLL-ASSOCHAM.

Li, Y., Lin, Y. & Geertman, S. (2015). *The development of smart cities in China.* Cambridge, MA: MIT.

McKinsey Global Institute. (2010). *India's urban awakening: Building inclusive cities, sustaining economic growth.* New Delhi: McKinsey Global Instiute.

MoUD. (2017). *http://smartcities.gov.in/content/* [Online] Retrieved from http://smartcities.gov.in/content/ [Accessed 25 October 2017].

Ravi, S., Tomer, A., Bhatia, A. & Kane, J. (2016). *Building smart cities In India: Allahabad, Ajmer, and Visakhapatnam.* Brookings India.

UNFPA. (2007). *State of world population.* [Online] Retrieved from www.unfpa.org [Accessed 25 10 2017].

Emerging Trends in Engineering, Science and Technology for Society,
Energy and Environment – Vanchipura & Jiji (Eds)
© 2018 Taylor & Francis Group, London, ISBN 978-0-8153-5760-5

Environment management through meditation: A sustainable approach

Madhura Yadav

School of Architecture and Design, Manipal University, Jaipur, Rajasthan, India

ABSTRACT: The environment of our planet is degrading at an alarming rate. One of the greatest problems that the world is facing today is that of environmental pollution. The problem of environmental pollution is addressed by many scholars in different ways. But the origin of such pollution lies not in the space but in the human mind thus the mind is polluted first and then the outer environment. Hence until the thought process is cleaned and purified; all measures of environmental protection will not be addressed in a sustainable way. The environment is a very vast field. For the purpose of this study only the physical environment is addressed. Spiritual response is needed to manage the environment. The study will be based on secondary data. The experiment of meditation on water and farming are studied in detail to arrive at suitable guidelines for environment management.

Keywords: Environment, Meditation, Environment Management, Water, Agriculture

1 INTRODUCTION

Environment management is not a new concept; the Vedic vision to live in harmony with the environment was not merely physical but far wider and much more comprehensive. The Vedic message is clear that the environment belongs to all living beings, so it needs protection by all, for the welfare of all. The Mahabharata, Ramayana, Vedas, Upanishads, Bhagavad Gita, Puranas and Smriti comprise the original messages for preservation of the environment and ecological balance. Nature or Earth, has never been considered a hostile element to be conquered or dominated. In fact, man is forbidden from exploiting nature. He is taught to live in harmony with nature and recognize that divinity prevails in all elements, including plants and animals. The relation of human beings with the environment is very natural as he cannot live without it. From the very beginning of creation he wants to know about it for self-protection and benefit. Awareness of the philosophy of karma is fundamental to the working of the universe. The law of karma (and Newton's third law of motion) simply states that for every action, there is an equal and opposite reaction, and that this is true in all parts of the universe. If we understand this law to be universal and immutable, then we would never think of harming the earth or its people.

With the ever increasing development by modern man, human activity has interfered with the fragile and complex interrelationship of our holistic universe and damaged the ecological system and natural processes. Humanity has gradually lost its sensitivity to the world of vibration. We have become more concerned with material development and insensitive toward nature. Nature is made up of five elements, those being space, air, fire, water, and earth, which are the foundation of an interconnected web of life, interconnectedness of the cosmos and the human body. The five great elements (space, air, fire, water, and earth) that constitute the environment are all derived from Prakriti, the primal energy. Environment study deals with the analysis of the processes in water, air, land, soil and organisms, which leads to pollute or degrade the environment. Nature has maintained a status of balance between and among these constituents or elements and living creatures. A disturbance in percentage of any constituent of the environment beyond certain limits disturbs the natural balance and any change in the *natural balance* causes lots of problems to the living creatures in the universe. Different constituents of the environment exist with set relationships with one another. The disequilibrium in nature is due to disequilibrium in man's mind. However, the great danger to the world comes from the fact that mankind does not recognize that mental pollution has gained dangerous proportions. Another fact that is generally unknown or is little realized is that the vibrations from a polluted mind pollute the atmosphere: they make it tense and vicious. Thus it is mental pollution that has spoiled the whole socio-economic atmosphere and climate everywhere. We need to reconsider our priorities and values to transform the way we think. Attitudinal change is essential. Dharma—often translated as 'duty'-can be reinterpreted to include our responsibility to care for the earth.

Meditation is deep and purposeful thought of eternal truths. It is a method to dwell on the landscape of your mind with understanding and a means to access your subconscious. It teaches you to have concentration, single track thinking and instant recognition of deviations from your intended focus. You learn how to distinguish between thought patterns and to select those that are positive, useful and lead you to your chosen goals. Meditation restores spiritual self-awareness and power to the human soul. As we meditate the soul dives in divine power which is rejuvenating and brings back all your list qualities and inner power. It is a spiritual law that the practice of meditation is essential to spread positive vibration in the environment.

2 METHODOLOGY

This paper is based on a conceptual framework. It gives emphasis on theory development, historical research, literature reviews, and critical analyses. It analyses existing theories about how particular interventions like meditation may work. It establishes a linkage between environment and meditation and suggests how environment can be managed in a sustainable way with meditation being a powerful tool.

At the United Nations conference on Environment and Development held in Rio-de-Janeiro, Earth summit, the Kyoto Protocol on sustainable development emphasized the key issues of global environmental concerns. No government at its own level can achieve the goal of environment protection, until the public participates. Public participation is possible only when the public is aware of the ecological and environmental issues. The most important and neglected type of pollution is mental pollution.

2.1 *Mental pollution and trickle-down effects*

Deep thinking would reveal that most of the present global problems are due to mental pollution; this pollution destroys a person's sense of being impartial, considerate and co-operative, and makes him callous, inimical, violent-prone and vicious. It influences economics, politics, commerce, business, and all relationships, and destroys real love, kindness and human values. Stockpiling of nuclear weapons is due to suspicion, fear, rivalry and violent tendencies. Poverty is due to lack of feeling on the part of rich for the poor. It is also due to the greed of the wealthy; all kinds of violence is due to anger, hatred or enmity. Thus, all global problems are in some form of thought pollution or mental pollution.

The remedy to this problem lies in realizing the self and also the relationship with other human beings. Without a spiritual orientation of the relationship between man and man, moral and spiritual values cannot be brought into play or be sustained. It is, therefore essential for man to understand his real and intrinsic nature. If this is not done, it is possible that science without spirituality would lead mankind to a nuclear catastrophe, population explosion or environmental upheaval. It must be borne in mind that mental pollution is the real and most dangerous enemy of mankind. Meditation and spiritual wisdom alone can wash the soul of the thought pollution or what is called 'mental pollution'.

The global environmental crisis is an external signal calling upon us to address an inner spiritual crisis. We need or take personal responsibility for lobbing and thinking according to the natural principal of human values. This is only possible when each of us empowers ourselves through spiritual study such as meditation.

2.2 *Meditation and environment*

Meditation is a deep and purposeful contemplation about eternal truths. In meditation practice human being shift from body consciousness to soul consciousness. Meditation is an effective tool to realize our self. When we meditate, or we remember God, our thoughts are holy, pure and unconditional, so we do not create bad, but only holy thoughts and powerful blessings. So when these powerful blessings and thought power are generated around us and within our environment, then it's raised up. Then other people, who are holy and practicing, also generate the same thought, the same blessing, and the same power. And like attracts like, and those attracting each other become a very powerful force full of positive energy, which defeats all the evil, negative influence in this world. That is how the world will become purified and improved day by day. The positive power generated by spiritual practice can prevent some of the disasters in the world. We still see disasters happen as if nothing has helped. But, if we did not practice spiritually, there would be even more disasters, enough to have destroyed the whole planet long ago.

2.3 *The impact of meditation on the environment*

The universe consists of five basic elements, these being, earth or land, water, fire, air, and ether. Each of the five elements represents a state of matter. Earth is not just soil, but it is everything in nature that is solid. Water is everything that is liquid. Air is everything that is a gas. Fire is that part of nature that transforms one state of matter into another. For example, fire transforms the solid state of water (ice) into liquid water and then into its gaseous state (steam). Withdrawing fire recreates the solid state. Fire is worshipped in many religious rituals because it is the means by which we can purify, empower, and control the other states of matter. Space is the mother of the other elements. The experience of space as luminous emptiness is the basis of higher spiritual experiences. The effect of mediation on water and on farming is summarized below.

2.4 *Experiments with water*

The Japanese researcher Masaru Emoto undertook extensive research of water around the planet. Mr. Emoto took water samples from around the world, slowly froze them and then photographed them with a dark field microscope. The following few pictures are of water crystals from different sources. Generally, clean healthy water creates beautifully formed geometrical crystals, while polluted water is too sick to form any crystals at all.

Water was exposed to different types of music. Distilled water was placed in a bottle between two speakers and specific music was played for one hour. The water was then frozen and photos taken. See the following pictures.

Mr. Emoto decided to see how thoughts and words affected the formation of untreated distilled water crystal. This was done by typing words onto paper, then taping the paper onto glass bottles overnight.

Yusui Moutain   Fountain in Lourdes,   Polluted water -   Mt Cook Glacier,
Spring, Japan   France   Yodo River, Japan   New Zealand

Figure 1. Microstructure of water molecule from different geographic locations of the world.

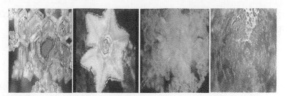
Bach's Air for Kawachi Folk Music for Healing Heavy Meta
String

Figure 2.   Effect of music and dance on water.

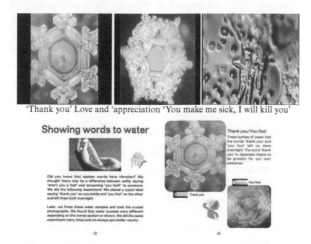
'Thank you' Love and 'appreciation 'You make me sick, I will kill you'

Figure 3.   Effect of thoughts on water crystals.

Polluted water from   Dam water after a Buddhist
the Fujiwara Dam,   prayer is offered

Figure 4.   a-b Effect of prayer and written label on water crystals.

It is observed that the negative words 'You make me sick, I will kill you' resemble an image of polluted water. Now we see the effect of thoughts and intent on the water. Below left is an image of very polluted and toxic water from the Fujiwara Dam. Below right is the same water after a Buddhist monk (Reverend Kato Hoki, chief priest of Jyuhouin Temple) offered a prayer over it for one hour. Prayer, that is sound coupled with intention, seems to have an extraordinary ability to restore the water back to its natural, harmonious, geometric symmetry.

A bowl containing mineral water was placed on the table in front of Dadi Janki, Administrative Head of Brahma Kumaris World Spiritual University.

With regard to the molecular structure of water, our intent (thoughts), words, ideas and music have a profound healing or destructive effect on the water, since the human body is made up of 70% water. Ultimately it means that what we think creates our reality, not just emotionally but physically. Since the planet is also made up of 70% water these experiments have profound implications for the environment too.

These experiments are indeed being done in Japan through Mr. Emoto's work. 300 people, usually led by someone proficient in focus and meditation will gather around a polluted stream and focus healing and wellbeing onto the water for approximately one hour. Mr. Emoto has reported and documented this on film. According to eyewitnesses the polluted water would become visibly clearer approximately 15 minutes after the meditation.

3 EFFECTS OF MEDITATION ON AGRICULTURE

The Rural Development Wing (RDW) of the Rajyoga Education and Research Foundation (RERF) is working in partnership with government institutions, NGOs and research institutes empower thousands of farmers by reconsidering thought-based technology (meditation) combined with organic farming used in rural India for some years.

To face the challenges in agriculture, some farmers decided to undertake experiments on their farms using Rajyoga Meditation practice in their daily activities.

Farmers using this technique reported considerable improvements in resistance to disease, pests and adverse weather conditions. The meditation involves creating the awareness of being the subtle conscious being rather than the physical body and then directing thought energy (peace, love and power) from the Divine Source to the crops.

Now more than 400 farmers all over India who are practicing sustainable yogic and natural farming techniques. In India, a number of Agriculture Universities and Researchers have also taken up the research in order to measure and quantify the actual advantages.

Some preliminary findings have been noted as follows:

- Germination rate of meditated seeds was at 93.33% in contrast with non-meditated seeds which were at a rate of 86.67%. Faster growth rate of meditated seeds, which took six days less to germinate.
- Significant growth in friendly insects' population
- Micronutrient content of meditated crops of wheat showed higher amounts of iron, increased oil content in oil seeds, improved protein and vitamins in vegetables thus increasing the energy value.
- Soil microbial population showed higher population of rhizobium, azotobacter and azospirllum.

Figure 5. Effect of Rajyoga meditation on water.

Figure 6.   Meditation in farming at Mount Abu, India.

Figure 7.   Yogic farming at Mount Abu, India.

Other factors, such as root length and seed weight, were greater in meditated samples. A significant drop in the pest damage was noted. (Results extracted from preliminary report published by SD Agriculture University).

Basic methodology for the application of positive and elevated conscious thought in farming, as recommended by the RERF. The seeds are placed in the BK center for up to a period of one month before sowing. This is then followed by weekly meditations taking place in the fields by groups of BK teachers and students throughout the crop growth cycle.

Over the last two years, Italy, Greece and South Africa are amongst some of countries experimenting with these techniques.

4   OVERVIEW OF RESEARCH IN SOUTH AFRICA

This study reviews by Dr. Ndiritu, South Africa of sound based (acoustic frequency) and thought-based (meditation) technology in improving crop production and assesses their potential. Published experiences of five farmers report substantial improvements in yield and resistance to disease, pests and drought. Published data reports yield improvements of up to 32% and up to 146% in nutritional constituent concentration. The technique is being promoted non-commercially by an NGO and thousands of farmers have been trained to use it. The improvements in crop production from acoustic frequency and meditation techniques are found to be comparable to those from biotechnology. Their potential in mitigating the global food crisis is considered large. Research and community-based initiatives to promote the two techniques are therefore recommended.

5   DISCUSSION AND CONCLUSION

The experiment by Dr. Emmoto on water proved that with regard to the molecular structure of water, our intent (thoughts), words, ideas and music have a profound healing or destructive effect on the water. The human body is made up of 70% water. Ultimately it means that what we think creates our reality, not just emotionally but physically. Since the planet is also made up of 70% water these experiments have profound implications for the environment also.

Brahmakumaris experience of yogic farming substantiates that faster growth rate of meditated seeds, significant growth in friendly insects' population, micronutrient content of meditated crops is higher thus increasing the energy value. Root length and seed weight were greater in meditated samples. A significant drop in the pest damage was noted.

Meditation training is found to be largely non-profit driven and no additional resources are required. It is probably the only technology with an effectively negative carbon footprint. Meditation techniques are adapted for sustainable farming in rural areas to improve crop production.

Human beings have a greater power of thought than any other living creature. Our thoughts create subtle vibrations of positive and negative energy and effect our environment to a great level. Thoughts have power to change the world and therefore it is imperative to understand that *world transformation is through self-transformation.*

The process of transformation will only be possible through meditation wherein we are consciously aware of our true self and our environment. We consciously and subconsciously create thoughts which are generally not channelized and full of positivity. We do not know the actions of these thoughts and the environment. Knowledge of generating creative conscious and pure thoughts full of energy, positivity and radiating them is possible by the means of meditation. Change happens from the inside out, from a seed of awareness to a thought, to an intention and to an action. One of the real benefits of meditation and other reflective practices is the elevated awareness of the internal and external states of the world.

We have discussed the effects of meditation on water and agriculture. If meditation can purify the water and crop production in agriculture, why not the physical environment. From studies it can be concluded that the environment can be managed by meditation.

It is therefore need of an hour to bring it and ourselves to its original state of being the only way is to get back to silence, restore energy and radiate energy to its zero vibration mode to stabilize the nature and help itself heal.

Therefore, a mere physical approach is inadequate to manage the environment, but a holistic approach involving both spiritual and physical aspects is required.

The atmosphere we experience is a reflection of our attitude with the experience of the real power of meditation; we can change deep rooted Sanskaras within the soul and can help others and even the elements to change. Create silence within and create a feeling of love for the elements. Talk to them like service companions. Each element is made of protons neutrons and electrons that are all moving. Their movement is based on our vibrations. Nature serves us tirelessly; we also need to serve nature. By spreading peaceful vibrations of peace we can restore her balance. Let us maintain respect for nature in our hearts and do something to restore her beauty and balance.

We have to serve five elements of nature to manage environment as below:

- **AIR** is everywhere and gives unconditionally to all living things. It does not withhold its love from anyone. Instead of polluting air with negative thoughts, to give it the fragrance of good wishes is to show respect. Ultimately, these reach all living thing through air.
- **EARTH** is stable and nourishing. It gives our feet a place on which to stand. It nourishes all living things. Just like the earth, our stage should be stable, so that we are not shaken by anything. Just like the earth, our thoughts and actions should nourish both people and the environment. On the physical level we have to observe cleanliness.
- **WATER** is flexible and has the power to neutralize. An ocean is a symbol of unlimited virtues, and like water we should be flexible in dealing with others. We can learn from the Great Ocean and neutralize negativity. On a physical level even to let the tap drip is to disregard the importance of water.
- **FIRE** cleanses, purifies and transforms. Fire is a symbol of yoga; our yoga should be intense like fire so that we are transformed by it. On a physical level to abuse fire is to be irreverent. Fire rebels against our burning of trees, for example, by filling out air with smoke.

- **ETHER** is subtle yet fat reaching. Ether is so broad that even the plants were thought to move through it. To become angels our intellect should be like ether—subtle and broad. To sanctify silence is to respect ether.

The human body is composed of and related to these five elements, and connects each of the elements to one of the five senses. This bond between our senses and the elements is the foundation of our human relationship with the natural world. Nature and the environment are not outside us, not alien or hostile to us. They are an inseparable part of our existence, and they constitute our very bodies.

All spiritual paths in the world, understand the law of karma to be the incognito mechanism that explains the unfolding of events in the world. The actions we take today write the script for the circumstances we face tomorrow. If we understand this to be true, then we might see today's world as the inevitable result of the choices we made in the past, and in that instant of understanding, we might see that the only way forward to a peaceful and loving world would be through our elevated awareness and loving behavior toward our earth and our human family.

6 FUTURE SUGGESTED RESEARCH

The broader plan is to use the results of these experiments to

- Generate enough interest for other research groups at the university to try out 'mind-matter' research
- Apply for research funding for field experimentation on yogic farming and analysis of the impact of meditation on other aspects of water management (governance, dealing with water-related disasters, climate change etc.).

REFERENCES

Ball State University sustainability webpage. (n.d.). Retrieved from http://www.bsu.edu/ceres/sustainability.

Brimblecombe, P. (1994). *Environmental encyclopedia. Air pollution.* Detroit: Gale Research International Limited.

Center for energy and environmental sustainability webpage. (n.d.). Retrieved from http://www.cisat.jmu.edu/cees/.

Deshpande, J. (2007). *Meditation Environment through Vipashana Meditation.* Mumbai: Daya Publishing House.

Environment at T. Retrieved from http://www.businessdictionary.com/definition/environment.html#ixzz2PIVbGDJh.

Henepola, G. (1988). *The Jhanas in Theravada Buddhist Meditation* (Wheel No. 351/353). Kandy, Sri Lanka: Buddhist Publication Society. Retrieved from http://www.accesstoinsight.org/lib/authors/gunaratana/wheel351.html.

Kamalashila (1996, 2003). *Meditation: The Buddhist art of tranquility and insight.* Birmingham: Windhorse Publications. Retrieved from http://kamalashila.co.uk/Meditation_Web/index.htm.

Kapleau, P. (1989). *The three pillars of Zen: Teaching, practice and enlightenment.* NY: Anchor Books.

Mosher, H. (2004). *Meditations on the five animals of Tai Chi.* Retrieved from Peaceful Warrior Training.com.

Shashi T. *Origin of Environment Science from Vedas.* Retrieved from http://www.sanskrit.nic.in/svimarsha/v2/c17.pdf.

Singh, R.P. (2008). *Environmental concerns (The Vedas)—A lesson in ancient Indian history.* In Society and environmental ethics 'as part of literature published in the International Conference on Environmental Ethics Education (ICEEE) held on 1617 November, Banmas Hindu University.

Steger, W. & Bowermaster, J. (1990). *Saving the earth.* New York: Bryon Pries.

Tyler Miller Jr. G. (1987) *Living in the Environment.* Belmont: Wadsworth Publishing Company.

Vipassana Research Institute (VRI) (n.d.). Bhikkhuvaggo. In *The Majjhima Nikaya* (second chapter of the second volume). Retrieved from http://www.tipitaka.org/romn/cscd/s0202m.mull.xml.

*Emerging Trends in Engineering, Science and Technology for Society,
Energy and Environment – Vanchipura & Jiji (Eds)*
© 2018 Taylor & Francis Group, London, ISBN 978-0-8153-5760-5

Bicycles for green mobility in urban areas

M. Harisankar
Urban Planning, Government Engineering College, Thrissur, Kerala, India

C.A. Biju
School of Architecture, Government Engineering College, Thrissur, Kerala, India

ABSTRACT: The last few years have seen many countries focusing on sustainability in their developmental agenda. In developed countries the cycling policy has in recent years evolved from a peripheral matter into one to be considered as a priority, in line with the policies for other means of transport. Being a clean, green and healthier mode, bicycling holds the key for sustainable urban mobility and better environmental qualities for our urban areas. This paper concentrates on the need for urban planners to solve the urban mobility problems with the introduction and integration of bicycle oriented transit facilities with the existing transportation network in Indian cities. Planning interventions are very essential in providing traffic calming and infrastructural support coupled with the appropriate policy backup, so as to encourage a modal shift from motorized transport to bicycling, which is discussed in detail in this paper citing some successful examples around the world.

1 INTRODUCTION

The rapid urban growth and increased use of motor vehicles that most countries have experienced in the recent years has created urban sprawl and higher demand for motorized travel, leading to a range of environmental, economic and social consequences. This effect is much pronounced in the case of developing countries. As a sustainable alternative, bicycles can replace or reduce the usage of automobile transport so there is an urgent need to integrate bicycles into the transportation system of our cities. This can enhance the mobility of people in urban areas and can also safeguard the accessibility of our congested cities provided suitable facilities are adopted for a bicycle infrastructure. It provides freedom of movement to rich and poor, young and old alike.

The National Urban Transport Policy (NUTP) has stressed the need for an approach in transport planning that focusses on people and not vehicles. The united Nations Habitat Global Report on human settlements also highlighted urban transport with a focus on reduction in pollution and congestion as a core area for advancing sustainable development in its Five Year Action Agenda 2012–17 (Planning & Design for Sustainable Urban Mobility). Indian cities have a latent demand for cycle and walking trips and a topography that suits it well which can be exploited in the form of strategic planning and designing of suitable infrastructural support for bicycles, which helps enhanced urban mobility to a great extent. This study discusses the suitability of bsicycle as a sustainable mode for urban mobility in urban areas, how far they can be integrated with the existing conditions, their planning and design considerations and programs and techniques for promoting a bicycle infrastructure.

2 BICYCLES IN AN URBAN SCENARIO

Policies for car oriented transport development have resulted in more and more road construction that have clearly failed to cope with the ever increasing demand for rapid motorization that has made our roads more and more congested.

Lewis Mumford in his book *The Roaring Traffic Boom* sarcastically stated this situation, 'adding more lanes in highways to accommodate the increasing number of automobiles are like loosening ones belt as a cure for obesity'. Increasing the lanes in highways and construction of flyovers can be considered only as short term solutions that can ultimately result in more vehicles on the road. Long term solutions in solving the mobility problem in urban centers lies in creating more and more facilities for active modes of transport like bicycles, thereby reducing the demand for motorized travel.

Cycling offers many benefits to the problem of urban mobility. It is one of the ideal modes of transportation for trips up to five km. Figure 1 shows the distance/travel time ratio for different modes of transport. Considering the congestion in our cities, it can be seen that for up to five km the bicycle is in fact faster than motorized modes of transport.

The dimension of social inclusion is also very important in India when large numbers of current bicycle users do not cycle out of choice but remain captive as they have no other viable option. These captive riders are predominantly poor workers and students. Indian cities have mixed land use patterns and low income households living in urban slums, which are in close proximity to the planned residential areas. Captive riders use the bicycle for even longer trips for work (up to 50 km in a day) such is the compulsion of captive riders (Tiwary et al. 2008). Presently the car centric transportation planning in our urban areas has neglected these groups and have made their mobility even more miserable. Thus, improving conditions for cyclists can improve the quality of life for poorer groups, those who are less likely to have cars.

Concern about worsening air quality has increased significantly all over the world in the recent years with the undoubted benefits of cleaner engines and fuels often being undermined by sheer growth in the volumes of traffic. The transportation sector contributes about 20% of CO_2 emission worldwide and 15% in India (Tiwary et al. 2008) and it is a growing source

Figure 1.   Vicious circle of car oriented transport development (Source: Torsten et al 2012).

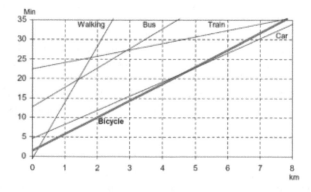

Figure 2.   Distance/ravel time ratio for different transport modes (Source: Torsten et al, 2012).

1020

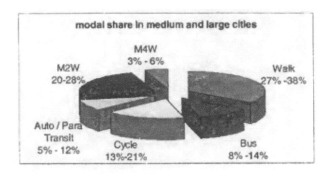

Figure 3. Trends in bicycle modal share (Source: 2).

of greenhouse gas emissions that contribute to climate change. Cycling provides an excellent opportunity for individuals to incorporate physical activity into daily life. The human scale urban environments that support cycling and discourage car use can improve social interactions and increase community attachment, livability and amenity (Litman et al., 2009).

The medium and large cities have a typical bicycle modal share of 13–21% (Figure 3). Cycle trips might be as low as 6–8% in mega cities, however the absolute numbers are about a million bicycles. Most of the medium and large cities in India have about 56–72% trips, which are short trips (less than 5 km in length), offering a huge potential for bicycle use.

3 PLANNING FOR BICYCLES

Planning for cycling involves a combination of infrastructural and non-infrastructural (soft) measures backed by strong policies to support cycling. For the bicycle to be useful for transportation, they need adequate route infrastructure (LTN, 2008). Traffic calming makes people more comfortable on the roads. It can enhance the safety as well as the perceived safety; and it can also reduce the damage caused by accidents when the traffic speeds are low enough. The concept of BFZ (Bicycle Friendly Zone) is where a standard bicycle symbol is used in roads or lanes designed for lower traffic speeds as an important measure in traffic calming. It acts as advice or a warning to vehicle drivers that cyclists are likely to be ahead and this can also narrow the appearance of the available road space while preserving the space for cyclists (McClintock, 2002).

Making the environment outside the front door into an environment safe for cycling is the first step in making the choice of cycling a possible one. Another popular measure that was originally developed in Dutch towns was 'home zones'/'*woonerf*' – the living yard. This is based on the idea that residential streets are planned and reconfigured such that 'residents come before cars'. Most of the streets in Amsterdam, the Dutch capital, also known as the bicycle city of the world, are traffic calmed at 30 km/h, including home zones with a 7 km/h limit. 'Bike streets' are also used here, where bicycles are considered the primary modes of transport and others are treated as guests. Removing the center line and well-designed road humps offer speed reduction benefits whilst being more comfortable to the cyclists. Flat topped road humps can be used as pedestrian/cyclist crossings. A cycle bypass allows the humps to be avoided altogether. Chicanes are usually constructed using two or more build outs alternating between each side of the road.

A defined set of bicycle routes makes it possible to travel around a region by bicycle in a safe and connected manner. In Delhi the BRT (Bus Rapid Transit) corridor was established and has been operating since April 2008. This corridor is designed to be fully supportive to NMT users including bicycle lanes and footpaths. Special accommodations for cyclists at intersections include advance stop lines, bike boxes at intersections, bicycle access lines and green wave signal timing for cyclists along the route. Network management can help deliver a pleasant environment for cyclists and can also reduce the need for a cycle specific

Figure 4. Bicycle Friendly Zones (BFZ) in shared roads. (McClinto).

Figure 5. Bicycle tracks in Gurgaon, Haryana (Source: Car-free Gurgaon).

Figure 6. Provision for bike racks in buses (Source: Pucher & Buhler).

infrastructure. This can be done by vehicle restricted areas, redistribution of carriageway space, self-calming roads where the use of physical features encourages lower speeds and filtered permeability to certain roads. The control of parking through charges or limiting capacity and duration of stay can encourage cycling.

4 NON-INFRASTRUCTURAL/SOFT MEASURES

Most of the countries give as much importance to soft measures as they give to infrastructural measures. Such programs concentrate on the positive sides of cycling. There will be increased opportunities to promote cycling through sustainable transport schemes such as "safe-routes-to-school", "safe-routes-to-leisure-projects" and "bike-to-work-schemes". Also it focusses on the promotion of cycling for other purposes other than leisure and work, like shopping and other utility trips, mainly targeting cycling for women and school children. Many of the cities in the UK have been framing green commuter plans that assist staff in managing journeys to and from work in an environmentally sustainable way. Small cities like Davis in the USA, have been particularly focusing on soft measures. Amsterdam offers bike training for all children in schools and special bike training programs target groups that cycle less. Mega cities like London introduced widespread bicycle training in recent years. The city organizes cycling 'Ciclovias' in which part of road network is closed to motorized traffic and the roads are reserved for bicyclists and other non-motorized modes. The cycling advocacy group—London Cycling Campaign (LCC) organizes the London Skyride with about 500 local rides and events, as well as promoting cycling training, running repair workshops, and providing input to hundreds of city traffic schemes with bike route projects.

5 INTEGRATION WITH PUBLIC TRANSPORT

The integration of cycling with public transportation helps cyclists cover trip distances that are too long to be made by bike alone. Public transportation can also provide convenient alternatives when cyclists encounter bad weather, difficult topography, gaps in the cycle network and mechanical failures. There are four main approaches: provision of bike parking at train stations and bus stops, bike racks on buses, permission and storage space to take bikes on board trains, and the co-ordination of cycle route networks so that paths and lanes lead to public transportation stops. There are many different kinds of bike parking, ranging from simple bike racks on sidewalks near public transportation stops to advanced full service bike stations. There are multi storey bike parking stations being increasingly used in Amsterdam as an answer to the rising demand of bike parking at important transit stations. The serious problem of bike theft in most countries has increased the demand for secure parking. In India such secure bike parking stations need to be set up to increase the cycling levels in our cities. The overcrowded public transportation system and difficulty in design modifications of trains and buses makes taking bikes on board trains and buses increasingly challenging.

A public bicycle sharing system has a major role in solving the last mile problems in cities. It enables the public to pick up the bicycle from any provided self-service docking station, which can then be returned to the same or any other docking station. The concept of bike sharing has also been recently introduced in India and few cities have already experimented, including FreMo in Thane, Green Bike Cycle Rental and Feeder Scheme, Delhi Metro Cycle Feeder Service in Delhi, Namma cycle in Bangaluru and Cycle Chalao in Mumbai.

Bicycles are an important means of transport in all urban areas in India. At present, most of the residents in India depend upon non-motorized transport to meet their transportation needs. The average share of cycles in medium cities of India varies between 3% and 7%. The years after the industrial revolution showed a massive increase in the number of motor vehicles accompanied with a dip in the average share of bicycles in the major cities of India in the 1980s and 1990s. A large amount of utility cycling is still present in Indian cities as it is

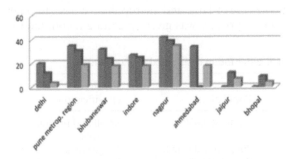

Figure 7.   Trends in bicycle modal share in 1980, 1990 and 2000 (Source: 20).

the most affordable form of transport available to low income households and plays a greater role in social inclusion. Cyclists in India also face a high risk of getting involved in fatal traffic accidents. According to Tiwary et al. (2008), cyclists are involved in 5–10% of the total road related fatalities in medium and large cities. The share of bicycle traffic in many Indian cities continues to be substantial despite the lack of planned facilities for them.

Considering the international trends some of the cities in India have recently adopted bicycle infrastructure facilities with their transportation network. In Delhi, the BRT corridor is designed to be fully supportive to NMT users including bicycle lanes and footpaths. In Lucknow 31 km of road stretch has already been constructed with cycle tracks. Historically known as the "Cycle City" of India the city of Pune has about 90 km of cycle tracks. A bicycle master plan is under construction, which will contain plans for the creation of a city-wide cycle track network and cycle safe streets, a city wide public bicycle scheme, bicycle parking facilities, integration with public transport, cycle promotion institutional mechanisms, capacity building and financial planning for implementing the plan, and awareness and education campaigns.

6 CASE STUDY INFERENCES

Case studies are undertaken based on the method of provision of bicycling facilities in cities of different sizes. The smaller cities have shorter trip distances, decreased levels of pollution and less stressful traffic conditions, while large and mega cities with a larger geographical area and greater trip distances make the provision of such facilities more challenging. Correspondingly, the census classification of Indian cities is used to match the Indian conditions.

7 ANALYSIS

The major findings from the foreign case studies are compared with the existing condition in India to examine how far they can be applied in Indian context. In India due to the higher population and population density compared with the European case study cities, the census classification is used to divide the cities in India based on size. So further analysis is based on this.

7.1 *Traffic calming*

In small cities separate segregated facilities decrease the need for traffic calming while in the large and metropolitan cities strict hierarchy is observed in the provision of bicycling facilities. Most of the street roads are traffic calmed while the coarser network of segregated traffic routes are traffic calmed at intersections.

7.2 *Network management and bicycle infrastructure*

In small cities more importance was given to recreational cycling along the green belt and segregated cycling facilities owing to greater space availability, lower population density and better social bonds. In the case of large and metropolitan cities, although a mix of segregated and shared paths are used, priority was given to getting a continuous network connecting

Table 1. Classification of cities for analysis based on case studies.

European case study cities classification	Population	Census of India classification	Population
Small cities	Up to 300,000	Small and medium towns	Up to 500,000
Large cities and mega cities	300,000–1,000,000 Over 1,000,00	Large cities Metropolitan cities	500,000–1,000,000 Over 1,000,00

major transit points and educational institutions. The network management is carried out in such a way so as to give an added advantage to cyclists in terms of time and distance. Measures like congestion charging and travel demand management, along with making CBD free for motor vehicles can bring positive changes in Indian cities. GIS can also be used in effective decision making process.

7.3 Soft measures

In the case of small cities, the case study cities irrespective of their size focused on family oriented bike training programs and cycling safety training among children. Imparting cycling safety awareness is given the prime consideration in devising such policies and programs. The cities also organize mega events for popularizing bicycling among motorists and bringing about a modal shift. In Indian conditions it is important to identify the major employers in the city, promote bike to work programs and commuter subsidies to encourage employees to use bicycle as a commuting mode. The catchment areas of all schools could be made bike friendly to encourage more children to use bicycles to commute to school.

7.4 Public transportation integration

In small cities there is less scope for integration with public transport owing to the lower geographical area and smaller trip distances, they may compete with each other but in the case of large and mega cities, their larger geographic area and greater trip distances makes it difficult to cycle the whole distance and necessitates integration with public transit. This can be done by making provisions for bicycle mounting in public transit vehicles and parking facilities in transit stations and other areas of commercial activity. In Indian conditions bike racks or bikes inside public transport buses may not be possible due to overcrowding but a small portion of area in motorized vehicle parking stations can be set aside exclusively as bike parking stations to provide cyclists a safe and secure place for keeping their bicycles. Where there are space constraints, multi storey parking stations can be used to increase the efficiency.

7.5 Public bicycle sharing system

The smaller geographical extend and less scattered concentration of transportation terminals makes a public bicycle sharing system a less attractive option in small cities. However, they can be integrated with the existing transportation system in large and mega cities to help improve multi modal connectivity. This is also done by private partnership as their names can also be included with the name of the bicycle sharing system. However, in Indian conditions, theft and vandalism may prove to be a deterrent factor in providing such facilities. Use of information technology and provision of safe and secure parking facilities can help to address such issues. The user fee charged may be free for the first few hours and subsequently charged, but lower than the costs of public transport.

8 STRATEGIES

Newer cities are likely to have more space for infrastructure, given the wider streets that are common in these cities. But a lower density, single use development pattern means that destinations are likely to be more dispersed. In such cities, a bicycle infrastructure may be used more for recreation than transportation, especially if the city builds off-street bicycle paths. For older cities, more compact, mixed-use development patterns keep destinations within the cycleable distance but the city might not have enough space for incorporating a bicycle infrastructure or have greater intensity of vehicular traffic to protect the cyclists. Distances may make utilitarian cycling feasible in these cities, but creating a perception of cycling is challenging in these cities. The strategies are given for small and medium towns, and large and metropolitan cities separately along with some common strategies. Although these classifications may not be strictly followed, as one can see, some of the strategies mix with each other to provide a general framework for the planning of bicycle facilities in various classes of cities.

Common strategies:

- Stress the benefits of cycling to the individual and society with community participation at all stages of planning and development
- Develop a bicycle masterplan for urban areas and periodic updates
- Carry out controversial decisions in stages
- Combine incentives for cycling with distinctive car use
- Nurture coalitions with bicycle groups, city administration, politicians, other bike-friendly groups, national organizations and technical people right at the beginning of implementation

Strategies for small and medium towns:

- Promotion of cycling among school children by expanding bike-to-school programs and cyclist education
- Creating safe cycling conditions by physical segregation from high speed motor vehicles
- Planning for greenways around the outskirts of the city

Strategies for large and metropolitan cities:

- Build a network of integrated bikeways with intersections that facilitate cycling with special consideration of the areas of low income households
- Integration with public transportation and implementation of bike sharing programs
- Physical segregation of fast moving and slow moving vehicles along BRT corridors
- Restrict car use through traffic calming, car free zones and less parking areas and maintenance of bike lanes already provided

9 CONCLUSION

The extent to which the bicycling facilities can be integrated with the existing scenarios of the cities depends on the city size, its inherent qualities and the existing transportation facilities in the cities. Indian cities truly reflect the large potential for bicycle use with a huge latent demand for bicycle facilities and infrastructure. The mixed land use and poly nucleated city structures with a higher percentage of short trips gives an added advantage to providing bicycle facilities while the large percentage of informal housing, high population density and lack of space presents unique challenges. At the same time, there is a great need to provide such facilities in Indian scenarios. So, the planning for bicycle facilities cannot be carried out in isolation but in an integrated approach, addressing the needs of all road users. It is a combination of various infrastructural measures coupled with policies and promotional support programs, backed up with strong political will and community participation. Cycling offers a healthy, cost effective, equitable way to improve the sustainability of urban transportation systems and build more livable cities.

REFERENCES

Belter, T., von Harten, M. & Sorof, S. (2012). Costs and benefits of cycling. *European Union.*
Cycling infrastructure design—how bicycling can save the economy, TSO publishing house.
McClintock, H. (2002). *Planning for cycling.* Cambridge, England: Woodhead Publishing in Environment Management.
Pucher, J. & Buehler, R. (2012), *City cycling.* Cambridge, Massachusetts: The MIT Press.
Rastogi, R. (2011). Promotion of non-motorized modes as a sustainable transportation option: Policy and planning issues. *Current Science*, *100*(9), 1,340–1,347.
Tiwary, G., Arora, A. & Himani Jain, H. (2008). *Bicycling in Asia.* TRIPP, IIT Delhi.

Emerging Trends in Engineering, Science and Technology for Society,
Energy and Environment – Vanchipura & Jiji (Eds)
© 2018 Taylor & Francis Group, London, ISBN 978-0-8153-5760-5

Impact of foreign direct investment on the city: Form and growth of cities

Amal Mathew & Manoj Kumar Kini
Department of Architecture, College of Engineering, Trivandrum, Kerala, India

ABSTRACT: The economic policies introduced by the government of India in 1991 played a great role in shaping the cities of modern India. The flow of revenue from foreign countries resulted in a major change in city structure. Availability of basic infrastructure was the major determinant for the inflow of Foreign Direct Investment (FDI) which influenced the states to improve the facilities. The Multi-National Corporation (MNC) investment region witnessed large-scale development resulting in large demographic changes due to migration. Booming land prices in the area forced growth in a vertical direction, which has impacted the city form. The biggest challenge in FDI-induced urbanization is making the city inclusive. Developments should be made with a vision for the future. A spatio-economic development policy wherein the spatial structure dictates the growth of the economy by attracting the investments can dictate the growth of the city by enhancing the positive and reducing the negative impacts of FDI.

1 INTRODUCTION

1 1 *Foreign direct investment*

Foreign Direct Investment (FDI) or foreign investment refers to the net inflows of investment to acquire a lasting management interest (10% or more) in an enterprise operating in an economy other than that of the investor.

FDI has turned out to be a major economic policy of the world that ultimately had a lot of spatial relations. The growth pattern in the host city varies with the type of investments it procures. Foreign investments bring with them new trends and techniques in construction, market structure and quality of life and lifestyle. FDI has resulted in the creation of polycentric cities. As India is one of the top countries to attract foreign investment, a case study like this could help in understanding and formulating guidelines for the development of a city in receipt of FDI. FDI brings about many impacts in various areas such as social, economic and physical. This study is limited to the impacts in physical form whereby the changes in the city's structure and form is studied.

2 IMPACTS OF FDI

2.1 *Economic impacts*

With the influx of foreign investment, the economy of a region can elevate to a higher level. The companies that need a lot of skilled labor will provide a good salary to their workers whereas unskilled labor will be negatively affected as it will be so easily available at their disposal. This would again result in the widening of the gap between the rich and the poor. The people who best benefit from these developments are the middle class who can enjoy a better buying power as the Multi-National Corporations (MNCs) bring in a lot of offers and discounts in order to capture the local market. The local players in the same industry will be affected and they will have to come up with measures to compete with these companies. The scenario of branding in the organized retail sector happened in this regard.

2.2 Social impacts

The MNC investments in the region cause the formation of a new middle class/high-level professional workforce that occupies white-collar jobs and is characterized by a shared education, lifestyle, consumption patterns and occupations. Social polarization in society becomes stronger as changes in the absolute and relative size of different income groups develops over time—directly inferred through the structure of the high-income lifestyles of those employed in the service sector (e.g. high-income residential and commercial gentrification). Infrastructure and streetscape are developed to match world class standards to attract a safe and standardized environment. To provide safe and secure environment to the MNC, the security of the areas are improved and is the first consideration of the governing authority.

2.3 Physical impacts

Emergence of new retail and wholesale: New buildings and building typologies are becoming common (e.g. malls, hypermarkets). In order to compete with upcoming foreign brands and retail chains the big, native retailers have also stepped into branding and started retail chains, for example, Spencer's, More, Reliance Fresh, and Unlimited Stores.

Huge warehouse-type blocks are introduced in the cities to house multi-brand retail and wholesale stores such as Walmart and Metro. Isolated knowledge campus with various relaxation policies are publicized and supported like Special Economic Zones (SEZs). The development pattern is mainly focused around the area with MNC investments like IT or industrial hubs, with the emergence of new housing typologies (service apartments) and isolated neighbourhoods (fragmented development pattern).

Emergence of cloned built spaces: To attract the new middle class of the city as well as the foreign visitors, cloned spaces are developed in the city, for example, McDonald's, Domino's, KFC, Levi's, Zara, Audi, and BMW.

The physical impacts of FDI can be broadly classified into macro and micro levels, wherein the macro level affects the whole structure and functioning of the city and the micro level impacts predominantly the built form characteristics. The impacts caused in these areas form the crux of this study and formulate the guidelines needed for enhancing the positives and reducing the negatives, which is also a concern.

3 MACRO-LEVEL PHYSICAL IMPACTS

3.1 Emergence of polycentric cities

Formation of polynuclear cities as a result of new developments which require large parcels of land. Growth of cities toward the periphery and around major junctions where the new developments take place. Formation of squatter settlements around the newly developed regions to cater the poor.

   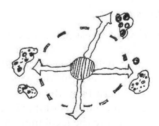

Figure 1.   Polycentric city.   Figure 2.   Formation of squatter settlements.

3.2 *Agglomeration economy*

Companies will be interested in investing in areas where similar types of industry are present. A region within a city often becomes the hub of a particular type of industry.

3.3 *Changes in land use*

As the demand for industrial use increases, the land available will be under pressure and conversion to industrial and commercial land use takes place. High increases in land values pave the way for the creation of luxury gated communities. Unaffordability of housing stock in the city forces the working class to migrate toward the periphery and outskirts.

3.4 *Grain and texture*

The new building typologies which occupy large footprints result in a change from fine grain to coarse grain. Planned areas like townships induce a new texture.

The skyline of the city changes with huge structures now forming the backdrop of the city.

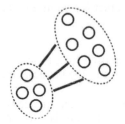　　　　

Figure 3.　Agglomeration economy.　　　Figure 4.　Industrial agglomeration.

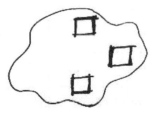　　　　

Figure 5.　Land use changes.　　　Figure 6.　Shift towards the outskirts.

Figure 7.　Change in the grain.

Figure 8.　Change in skyline.

4 MICRO-LEVEL PHYSICAL IMPACTS

4.1 *Cloned typology*

The establishment of international franchises who follow the same design language every-where, irrespective of the location, has resulted in the creation of buildings that are cloned from a foreign setting.

Foreign franchises will attract local customers and the local players will be forced to improve their standards to match the foreign players, which helps in improving the overall quality and competitiveness of the local players. This results in an increase in the cost of the products, which in turn increases the cost of living.

4.2 *Warehouse typology*

Retail and wholesale markets become dominated by the foreign players and they offer a wide variety of products within their showrooms, which are like large warehouses. Large retail and wholesale shops become a hub of activity or a node. Large local players will try to imitate the foreign players, while the smaller players get eliminated.

This will provide the middle class with better buying power and better products. Heavy traffic congestion in areas housing this type of hypermarket is a common phenomenon.

4.3 *Gated communities*

Large plots of lands become gated communities containing all sorts of amenities for the resi-dents. As the number of gated communities increases (due to the increase in housing needs), this has a huge impact on the skyline of the city.

These gated communities will be located at prime locations and hence the prices will be very high, which makes housing unaffordable to the lower middle class forcing them to move to the outskirts.

Figure 9.   Cloned typology.

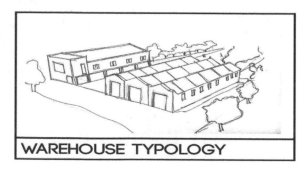

Figure 10.   Warehouse typology.

4.4 *Shopping malls*

Shopping malls emerge as a result of local market agglomeration. A large impact on the city structure occurs as the shopping malls become the anchors and are usually placed at a strategic location.

Because the city lacks parks due to the very high land prices, malls become a place of recreation. All this contributes toward the traffic congestion in areas nearby, also influencing the movement pattern.

4.5 *Office buildings*

Major players such as IT businesses require large office spaces, resulting in huge structures, often with a glass façade. The city's skyline is highly affected by these massive structures, which are situated in dominant positions in the city.

Figure 11. Gated community.

Figure 12. Shopping malls.

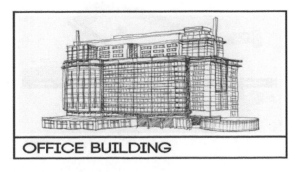

Figure 13. Office building.

Increases in the temperature of a city with many reflective glass façades, along with the high energy consumption of air conditioning within them, are causes for concern with these buildings.

The emergence of new typologies in the built fabric as a result of FDI brings in changes in the image ability of the city. The buildings that spring up reflect global standards, making the city global. New typologies such as cloned buildings, warehouse typology, office buildings, shopping malls and gated communities occur as an impact of FDI.

5 GUIDELINES FOR FUTURE DEVELOPMENT

5.1 Need for regionally global developments

The major impact on the growth and built fabric of the city is the formation of global cities. The Indian city, with its inclusive society whereby the various classes of people live in an inclusive city, is being affected by the very idea of the global city. The city becomes a place for the rich and the well-to-do. This can be understood through the large morphological changes that are happening in the city.

An area where a SEZ is proposed, in order to attract FDI, goes through a sudden change. Speculative investments occur in that region so as to reap the benefits in the future. Once the development starts the change will be drastic. A region with fine grain and texture gets converted into a region of coarse grain with the building of large structures such as office buildings and malls, which are part and parcel of any foreign investment.

The so-called global city structures, which are basically glass towers, change the skyline and character of the region. The traditional and regional style of architecture is completely forgotten and an alien setting is created in the name of the global city. No mandate is given to the investors to come up with glass towers; thus buildings responding to the regional climate and setting need to be proposed in these areas in order to make the area regionally global.

5.2 Open space

When an investment comes into an area, the land value shoots up, which makes people want to build structures wherever possible. This phenomenon leads to the major issue of lack of open space. Cities which grew without a proper master plan witnessed such a growth. Proper measures should be taken so that the open spaces needed for the development are defined in the master plan stage. This lack of open space is clearly visible in Gurgaon, where no consideration for open space was given in the initial stages.

5.3 Envisioned connectivity

In an existing city, new developments might be on the outskirts, the periphery or in the suburbs and these developments in turn become the anchor functions which make the city

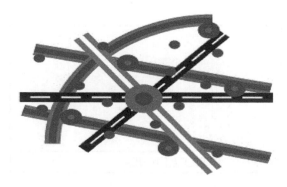

Figure 14.   Connectivity to various city centres.

polynuclear. Envisioning this type of growth at the time of inception of the master plan should be a given so that, once they develop, connectivity with the initial core should not be an issue. Traffic congestion in the newer developments like Whitefield in Bangalore can be accredited to this scenario.

5.4 Squatter settlements

Formation of squatter settlements is a common phenomenon in all emerging cities. When a new master plan is prepared there should be a mandate to allocate affordable housing units for the poor so as to avoid the formation of these types of settlements. All cities have this problem, which needs to be addressed.

5.5 Farmland

Any development should address the ecological sensitivity of the region; productive farmland should not be acquired for the development process. The agricultural production and livelihoods of many have been lost in the past with the setting up of SEZs and other developments. First preference in development should be given to unfertile land so that farming activity is not lost, and the fertile land continues to be farmland. The past experience of Nandigram in West Bengal points toward this aspect.

5.6 Security

Security issues are a major concern of these newly developed cities. In the case of Gurgaon, the city faces lots of security issues as any nightlife is present only within the glass buildings. The segregation and planning made the streets dead at night. This scenario, along with the gated communities that act as another entity completely secluded from the happenings outside their gates, increases security issues. Events outside these compounds are unknown to their residents, making the streets unsafe at night. Introducing various functions at the same place at various point of times of day is an option in tackling this situation. Enhancement of the night life in the cities is an important aspect to be considered.

5.7 Anchor points in a polynuclear city

When a development such as a shopping mall occurs, it becomes an anchor point in the city. This happens because developments like these become a hub of activity and recreation for the city dwellers because the city lacks open spaces and public recreation facilities. This scenario introduces many other impacts to the city because the traffic structure and the movement pattern are influenced as these structures become anchor functions. Such strategic developments should be allowed only in areas where the city's movement will not be hindered.

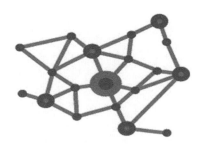

Figure 15.   Poly nuclear city with its anchor points.

6 CONCLUSION

The spatial manifestations of FDI are so evident that it can dictate the way in which a city develops. FDI-induced development can be seen all around the world wherever the policy of FDI is enacted. The growth of Chinese cities is the best example in this regard.

The biggest challenge in FDI-induced urbanization is making the city inclusive. When the Indian city, with its chaos and disorder, is converted into a global one or a new development such as SEZ, the low-income groups are not given much consideration. FDI developments bring a lot of employment opportunities especially the white-collar jobs. The job options for the poor and the ones affected by land acquisition are limited to jobs such as security guards, and gardeners, making the division in society more pronounced. When the city develops over a certain limit, it becomes a place for the rich only. Even the middle class finds it difficult to afford housing in the cities.

The major factor that should be considered is that the developments should be planned. Developments should be made with a vision for the future. Traffic congestion in many cities, when the city expands beyond borders, is due to the lack of vision in the early planning stage.

There is a need for a spatio-economic development policy whereby the spatial structure dictates the growth of economy and the attraction of investment. Western countries have started developing in this manner because the shortcomings of the older developments are becoming more and more visible. Thus, to conclude, FDI has a large impact on the urban form and structure with a number of positives and negatives. It is hoped that a proper spatio-economic policy to guide the same can enhance the positives and reduce the negatives.

REFERENCES

Chadchan, J. & Shankar, R. (2009). Emerging urban development issues in the context of globalization. *Journal of ITPI (Institute of Town Planners India)*, 6(2), 78–85.

Chen, Y. (2009). Agglomeration and location of foreign direct investment: The case of China. *China Economic Review*, 20(3), 549–557. doi:10.1016/j.chieco.2009.03.005.

Eldemery, I.M. (2009). Globalization challenges in architecture. *Journal of Architectural and Planning Research*, 26(4), 343–354.

Mathur, O.P. (2005). Impact of globalization on cities and city-related policies in India. In H.W. Richardson, C-H.C. Bae (Eds.), *Globalization and urban development (Advances in spatial science)*. Berlin, Germany: Springer-Verlag.

Mukherjee, A. (2011). Regional inequality in foreign direct investment flows to India: The problem and the prospects. *Reserve Bank of India Occasional Papers*, 32(2), 99–127.

Shivam, N. & Keskar, Y.M. (2014). Impact of foreign direct investment over the city form, a case of Hyderabad City, India. *International Journal of Innovative Research in Science, Engineering and Technology*, 3(2), 9674–9682.

Wei, W. (2005). China and India: Any difference in their FDI performances? *Journal of Asian Economics*, 16(4), 719–736. doi:10.1016/j.asieco.2005.06.004.

Zhu, J. (2002). Industrial globalisation and its impact on Singapore's industrial landscape. *Habitat International*, 26(2), 177–190. doi:10.1016/S0197-3975(01)00042-X.

Emerging Trends in Engineering, Science and Technology for Society,
Energy and Environment – Vanchipura & Jiji (Eds)
© *2018 Taylor & Francis Group, London, ISBN 978-0-8153-5760-5*

Author index